21st Century Canon of Solar Eclipses

Deluxe Edition

Fred Espenak

Edition 1.0
September 2016

21st Century Canon of Solar Eclipses – Deluxe Color Edition

Astropixels Publishing
P.O. Box 16197
Portal, AZ 85632

www.astropixels.com/pubs

This book may be ordered at: *www.astropixels.com/pubs/21CCSE.html*

More about solar eclipses of the 21st Century can be found at:

www.eclipsewise.com/solar/SEcatalog/21CCSEcat.html

Printed in the United States of America

ISBN 978-1-941983-13-3

First Edition (Version 1.0a)

Front Cover: A time sequence shows the partial phases and totality during the total solar eclipse of 2001 Jun 21 from Chisamba, Zambia. Photo copyright ©2001 by Fred Espenak.

Back Cover: Portrait of Fred Espenak, Copyright © 2016 by Patricia Totten Espenak.

Table of Contents

The Sun's corona is only visible to the naked eye for a few minutes during a total eclipse of the Sun. This is an HDR composite image of the total solar eclipse of August 21, 2017 from Casper, Wyoming.
© 2017 F. Espenak, www.MrEclipse.com

Section 1: Solar Eclipse Fundamentals

1.1 Introduction

The Moon's orbit is tilted about 5.1° to Earth's orbit around the Sun. The points where the two orbits appear to cross are called the nodes. When the New Moon occurs near one of these nodes, the Moon's shadow can sometimes fall on some portion of Earth and a solar eclipse takes place.

The Moon's shadow is composed of three cone-shaped components. The outer or penumbral shadow is a zone where the Sun's rays are partially blocked. Nested within the penumbra is the umbral shadow — a region where direct rays from the Sun are completely blocked. The conical umbra tapers to a point beyond which extends an expanding cone called the antumbra. From within this third shadow, the Moon appears smaller than the Sun and is seen in silhouette against the solar disk.

Figure 1–1: Geometry of an Annular Solar Eclipse

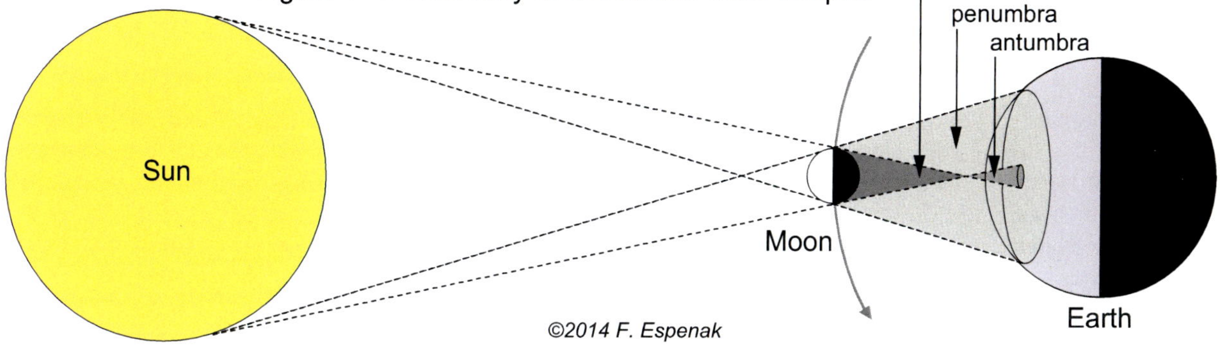

Figure 1–1 illustrates the geometry of an annular solar eclipse. A partial eclipse is visible from within the large penumbral shadow, while the annular eclipse is confined to the much smaller antumbral shadow.

Figure 1–2: Geometry of a Total Solar Eclipse

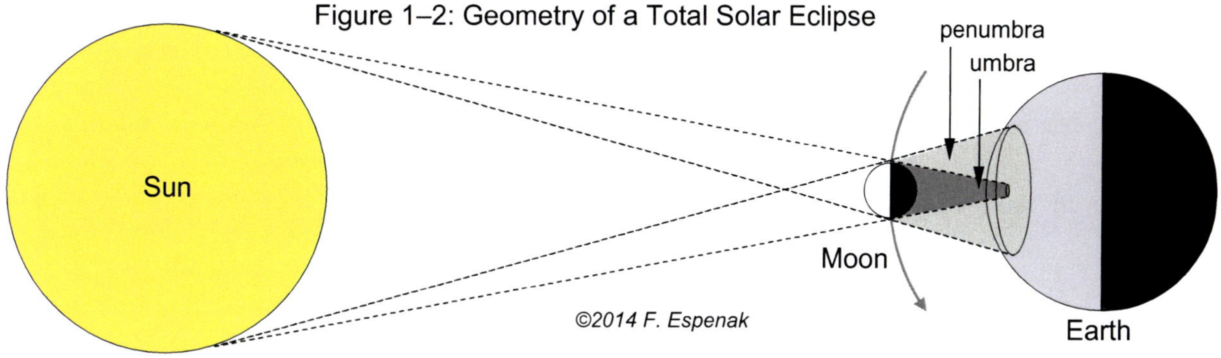

Figure 1–2 illustrates the geometry of a total solar eclipse. A partial eclipse is visible from within the large penumbral shadow, while the total eclipse is only seen from the much smaller umbral shadow.

1.2 Classification of Solar Eclipses

There are four basic types of solar eclipses:

1. **Partial Solar Eclipse** — The Moon's penumbral shadow traverses Earth while the umbral and antumbral shadows miss Earth. A portion of the Sun's disk is obscured from within the penumbra.
2. **Annular Solar Eclipse** — The Moon's penumbral and antumbral shadows traverse Earth. The Moon's umbral shadow completely misses Earth. The Moon's disk appears smaller than the Sun so a bright

ring surrounds the Moon when viewed from within the antumbral shadow. A partial eclipse is seen within the penumbral shadow (Figure 1–1).

3. **Total Solar Eclipse** — The Moon's penumbral and umbral shadows traverse Earth. The Moon's antumbral shadow extends beyond Earth's surface. The Moon's disk appears larger than the Sun and completely covers the solar disk when viewed from within the umbral shadow. A partial eclipse is seen within the penumbral shadow (Figure 1–2).
4. **Hybrid Solar Eclipse** — The Moon's penumbral, umbral and antumbral shadows all traverse parts of Earth. The curvature of Earth's surface brings some regions into the umbra and others into the antumbra. The eclipse is total within the umbra and annular within the antumbra. A partial eclipse is seen within the penumbral shadow. Hybrid eclipses are also known as annular-total eclipses.

Annular, total and hybrid eclipses are sometimes referred to as *central* eclipses[1]. However, on rare occasions it is possible to have an annular or total eclipse that is non-central (Sect. 3.1).

1.3 Visual Appearance of Partial Solar Eclipses

When the Moon's penumbral shadow strikes Earth, a partial eclipse of the Sun is visible from that region. The Moon's apparent motion with respect to the Sun is relatively slow — the partial phases can last up to three hours or more. During this time, the Moon's dark limb slowly creeps across the Sun's disk.

Partial eclipses are dangerous to look at with the unprotected eye because the Sun is still extremely bright. Special techniques are needed to safely view the eclipse (Sect. 1.6). Even when a partial eclipse reaches its maximum phase, the sky and landscape remain bright. But careful inspection of dappled sunlight beneath a leafy tree will reveal multiple images of the eclipse. The gaps between the tree leaves act like pinhole cameras and project images of the eclipse onto the ground below. If the eclipse occurs on a mostly cloudy day, brief views of the partial phases may be possible as the Sun passes though the more translucent clouds.

The Moon's penumbral shadow is 6700 to 7300 kilometers in diameter and can cover a significant fraction of the daytime hemisphere of Earth. Consequently, partial eclipses may be visible from large geographic areas as the penumbra sweeps across Earth's surface.

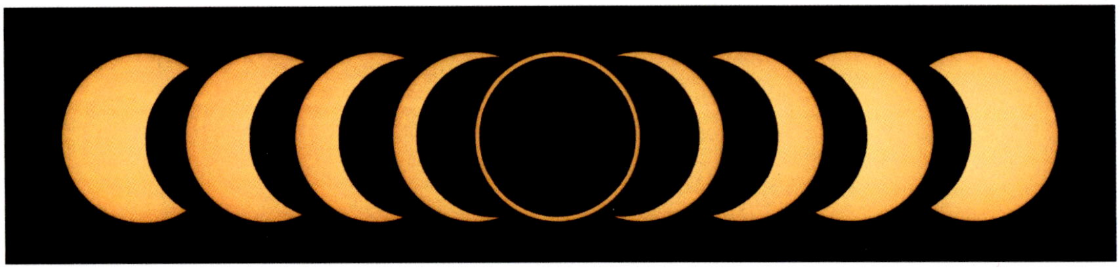

Photo 1–1 shows various phases of the Annular solar eclipse of 2005 Oct 03. ©2005 F. Espenak

1.4 Visual Appearance of Annular Solar Eclipses

During an annular eclipse, the Moon's penumbral and antumbral shadows sweep across Earth. Compared to the penumbra, the antumbra is much smaller and has a maximum diameter of 374 kilometers. Because of this, the antumbra covers a tiny fraction of Earth's surface.

A partial eclipse is visible within the penumbral shadow, but only observers located in the much narrower track of the antumbra will see an annular eclipse. For this reason, the antumbra's trajectory across Earth is called the path of annularity.

---

[1] A central eclipse is one in which the central axis of the Moon's umbral/antumbral shadow intersects with Earth's surface.

All annular eclipses begin with a series of partial phases lasting about an hour. At the peak of the eclipse, the Moon's disk can be seen in complete silhouette against the Sun. The remaining solar photosphere appears as a ring of intensely bright light surrounding the Moon. The annular phase can last a maximum of 12 ½ minutes, but is more typically 3 to 6 minutes in length. After annularity, another series of partial phases occurs as the Moon gradually uncovers the Sun.

Special precautions must be used to watch the eclipse (Sect. 1.6). Even during the annular phase, the Sun is dangerously bright and cannot be viewed without a solar filter. The landscape and sky remain bright throughout the eclipse, giving little indication of the celestial event in progress.

Photo 1-2 shows various phases of the total solar eclipse of 1999 Aug 11. ©1999 F. Espenak

1.5 Visual Appearance of Total Solar Eclipse

During a total eclipse, the Moon's penumbral and umbral shadows fall upon Earth. The umbra has a maximum diameter of 273 kilometers. The narrow track traced out as the umbra sweeps across Earth's surface is called the path of totality. The Sun is completely obscured by the Moon from within this zone. The total phase can last up to 7 ½ minutes, but is more typically 2 to 3 minutes in length.

Total eclipses all begin and end with a series of partial phases lasting about an hour. But this is where the resemblance between partial and annular eclipses ends — the total phase is the most spectacular astronomical event visible to the naked eye. During totality, the Sun's outer atmosphere — the solar corona — appears as a gossamer halo surrounding the Moon, and bright stars and planets are visible.

The eclipse begins to take on a unique character about five minutes before the total phase commences. Sunlight takes on a foreboding quality and casts abnormally sharp shadows. The approaching lunar umbra darkens the western sky and the air temperature is noticeably cooler. A minute before totality, ghostly shadow bands[2] ripple across the ground.

The ambient light grows feeble even though the crescent Sun is still too bright to see. In the final seconds, the Sun's corona emerges from the glare as the solar crescent shrinks to a brilliant jewel. This celestial diamond ring[3] lingers for a moment before the sunlight is extinguished and totality begins.

The Sun's glorious corona is now displayed to full advantage in the darkened sky. Standing within the Moon's umbra affords the rare and unprecedented opportunity to gaze directly at the glowing million-degree plasma surrounding our star. Twisted, tortured, and constrained by the Sun's enormous magnetic fields, the solar corona is revealed to the naked eye only during the brief seconds when the Moon completely blocks the brilliant disk of the Sun. An eerie twilight bathes the landscape and the colors of dusk surround the horizon.

Minutes race by like seconds. Suddenly, a sparkling bead of sunlight reappears along one edge of the Moon and quickly grows to blindingly bright proportions. Daylight returns as the corona fades and the total phase ends. Another hour of partial phases remains before the eclipse ends, it is anticlimactic after totality.

---

[2] Shadow bands are caused by the thin slit-like solar crescent illuminating Earth's atmosphere moments before and after a total solar eclipse. They appear as thin parallel undulating lines of alternating light and dark that race across the ground. This motion is caused by atmospheric winds.

[3] This phenomenon is commonly known as the *diamond ring effect.*

While filters are required for viewing the partial phases, they must be removed for totality. The total phase is the only time it is completely safe to look directly at the Sun without protection. In fact, the total phase is not even visible through solar filters because the Sun's corona is a million times fainter than the photosphere[4].

Figure 1–3: Path of Totality

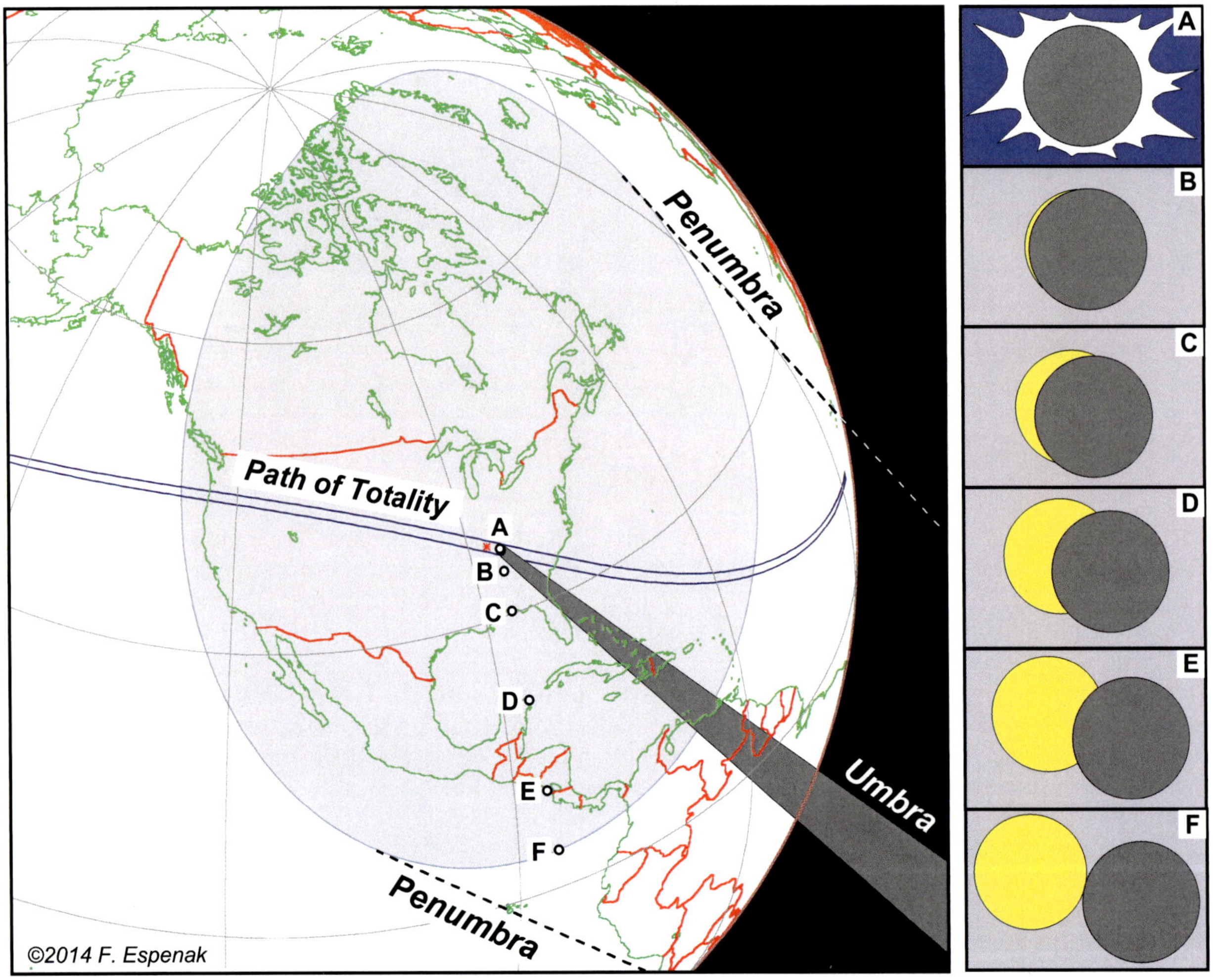

Figure 1–3 illustrates the path of totality for the total solar eclipse of 2017 Aug 21. The appearance of maximum eclipse from six different geographic locations on the map (labeled A through F) is shown in the side bar. The total eclipse is visible from position A because it lies in the path of totality.

1.6 Safely Observing Solar Eclipses

Partial eclipses, annular eclipses, and the partial phases of total eclipses are never safe to watch without special precautions. Even when 99% of the Sun's surface is obscured during the partial phases, the remaining photospheric crescent is intensely bright and cannot be viewed safely without eye protection (Chou, 1981; Marsh, 1982). Do not attempt to observe the partial or annular phases of any solar eclipse with the naked eye. Failure to use appropriate filtration may result in permanent eye damage or blindness. The only time it is safe to view the Sun directly with the naked eye is during the brief period of the total eclipse phase when the Sun's disk is completely covered by the Moon.

[4] The photosphere is the visible surface of the Sun's disk.

The same equipment and techniques used to observe the Sun outside of eclipse can be used for viewing partial phases and annular eclipses (Littmann, Espenak & Willcox, 2008, Reynolds & Sweetsir, 1995; Pasachoff & Covington, 1993). The most inexpensive of these methods is by projection — a pinhole or small opening is used to project the Sun's image onto a screen placed two or more feet beyond the opening. Projected images of the Sun can even be seen on the ground by creating small openings between interlaced fingers, or in the dappled sunlight beneath a leafy tree.

Observing the Sun directly is possible when using filters specifically designed for this purpose. Such filters usually have a thin layer of metal deposited on their surfaces that attenuates both visible and infrared energy. Another widely available filter is a number 14 welder's glass, obtained from welding supply outlets. Black polymer has recently become a popular, inexpensive alternative. This material can be cut with scissors and adapted to any kind of viewing device. No filter is safe to use with an optical device (i.e., telescope, binoculars, etc.) unless it has been specifically designed for that purpose. Sources for filters can be found at:

eclipsewise.com/extra/equipment.html#Solar_Filters

Local science museums, planetariums and amateur astronomy clubs are good sources for additional information. Remember, only the total phase of an eclipse can be safely viewed without a filter.

1.7 Central Line and Duration of Totality

The axis of the Moon's shadow determines the central line of the path of totality (and annularity). For the purposes of this discussion, it is assumed that the central line lies in the middle of the eclipse path.[5]

The duration of totality is longest on the central line, so great effort is often made to get as close to it as possible. However, the duration actually drops off quite slowly with distance from the central line. For example, a location 20% from the center to the edge of the path still has a totality lasting 98% of the central line duration. Even if one travels half way to the path limit, the duration is still 87% of the central line value.

Although figure 1-4 plots data for the 2017 total solar eclipse, the same relationship holds for all total eclipse paths. The solid curve is calculated assuming a smooth lunar limb. In reality, the Moon's limb profile has mountains that extend the duration and valleys that shorten it. These topographic features typically change the duration by a second or two for most of the path — their effects become significantly larger near the umbral path limits. The dotted line in figure 1-4 has been calculated using the Moon's true profile.

The following expression is useful for calculating the duration of totality for any location at a perpendicular distance of δ kilometers from the central line. It assumes a smooth profile for the Moon's limb.

$$d = D \times [\,1 - (\,\delta/Z\,)^2\,]^{1/2} \text{ seconds} \qquad (1\text{-}2)$$

Where: d = duration of totality at point of interest (seconds)
D = duration of totality on the central line (seconds)
δ = perpendicular distance from the central line to position of interest (kilometers)
Z = perpendicular distance from the central line to the path edge (kilometers)

The Moon's limb profile varies with the lunar libration, which changes from eclipse to eclipse. Consequently, the effects of the limb profile are different at each eclipse.

---

[5] The position of the central line is actually offset from the center. The difference is greatest in cases where the Sun's altitude is low. Curvature of Earth's surface is responsible for this shift.

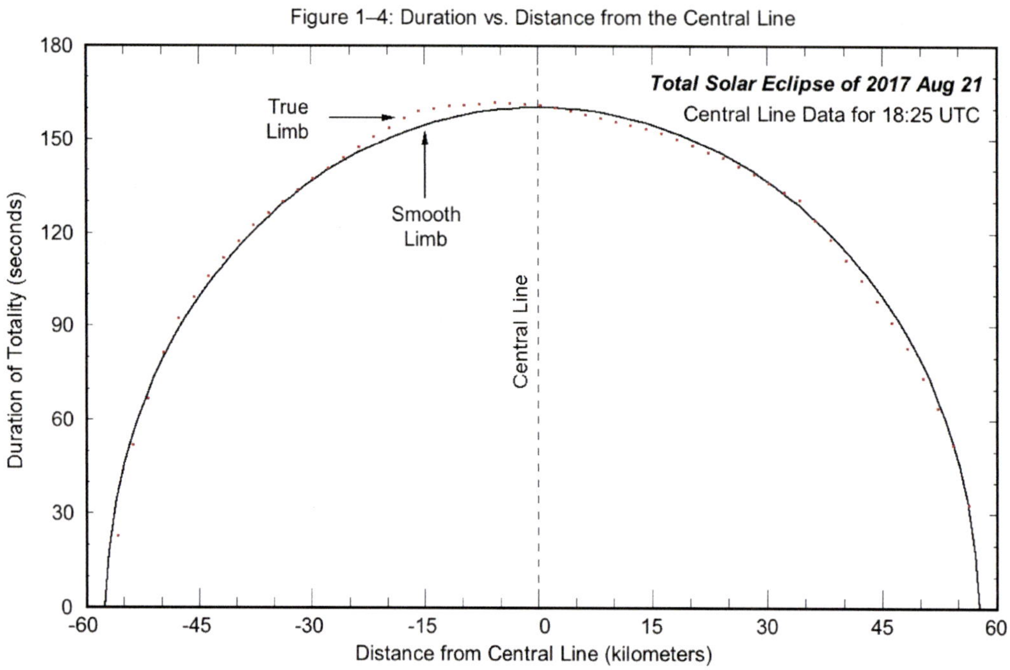

Figure 1–4 shows how the duration of totality changes with distance from the central line. The solid curve represents the duration using the Moon's mean limb. At a distance ~87% from the central line to the path limit, the duration drops to 50%. It is only when the distance from the central line to the path limit reaches ~90% that the duration takes a precipitous drop. But even at a distance of 96%, the duration is still 28% of the central line value. The dotted curve uses the Moon's true limb to calculate the duration using Kaguya/Herald limb data.

Photo 1–5 *Pat and Fred Espenak gaze at the Sun's corona during the total solar eclipse of 2008 August 01 from Jinta, China. ©2008 F. Espenak*

Section 2: Solar Eclipse Predictions

2.1 Solar Eclipse Contacts

During the course of a solar eclipse, the instants when the Moon's disk becomes tangent to the Sun's disk are known as eclipse contacts. They mark various stages or phases of a solar eclipse.

Partial solar eclipses have two primary contacts.

First Contact (C1) — Instant of first exterior tangency of the Moon with the Sun (Partial Eclipse Begins)
Fourth Contact (C4) — Instant of last exterior tangency of the Moon with the Sun (Partial Eclipse Ends)

Central solar eclipses (total, annular or hybrid) have four primary contacts. Contacts C2 and C3 mark the instants when the Moon's disk is first and last internally tangent to the Sun. These are the times when the annular or total phase of the eclipse begins and ends, respectively.

First Contact (C1) — Instant of first exterior tangency of the Moon with the Sun (Partial Eclipse Begins)
Second Contact (C2) — Instant of first interior tangency of the Moon with the Sun (Annular or Total Eclipse Begins)
Third Contact (C3) — Instant of last interior tangency of the Moon with the Sun (Annular or Total Eclipse Ends)
Fourth Contact (C4) — Instant of last exterior tangency of the Moon with the Sun (Partial Eclipse Ends)

Figure 2–1: Solar Eclipse Contacts

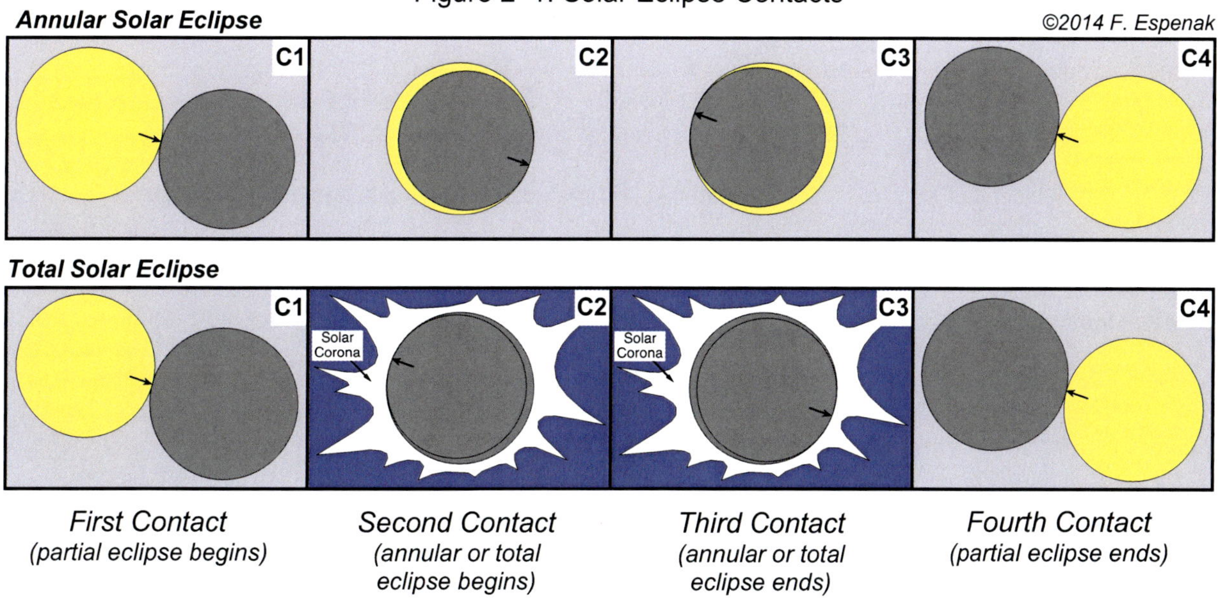

Figure 2-1 illustrates the four contacts for annular and total solar eclipses. The arrows indicate the contact point of the Sun's limb in each diagram.

2.2 Mean Lunar Radius

The International Astronomical Union (IAU) has adopted a value of k=0.2725076 for the mean lunar radius. This value is currently used by the Nautical Almanac Office for all solar eclipse predictions (Fiala and Lukac, 1983) and is believed to be the best mean radius, averaging mountain peaks and low valleys along the Moon's rugged limb. However, this introduces a problem in predicting the character and duration of central eclipses, particularly total eclipses. A total eclipse can be defined as an eclipse in which the Sun's disk is completely occulted by the Moon. This cannot occur so long as any photospheric rays are visible through deep valleys along the Moon's limb (Meeus, Grosjean and Vanderleen, 1966).

This work uses the IAU's accepted value of k (k=0.2725076) for all penumbral (exterior) contacts. In order to avoid eclipse type misidentification and to predict central durations, which are closer to the actual durations observed at total eclipses, we depart from convention by adopting the smaller value for k (k=0.272281) for all central (interior) contacts. This is consistent with predictions published in *Thousand Year Canon of Solar Eclipses* (Espenak 2014). Consequently, the smaller k produces shorter central durations and narrower paths for total eclipses when compared with calculations using the IAU value for k. Similarly, the smaller k predicts longer central durations and wider paths for annular eclipses.

2.3 Solar and Lunar Coordinates

The coordinates of the Sun and the Moon used in the eclipse predictions presented here have been calculated with the JPL DE405 (Jet Propulsion Laboratory Developmental Ephemeris 405). The DE405 is based upon the International Celestial Reference Frame (ICRF), the adopted reference frame of the IAU. The DE405 includes both nutation or libration and has an absolute accuracy of several kilometers for planetary positions,. In most cases this corresponds to a small fraction of an arc-second.

The Moon's center of figure does not coincide with its center of mass. To compensate, an empirical correction is sometimes added to the Moon's center of mass position. Unfortunately, the large variation in lunar libration from one eclipse to the next minimizes the effectiveness of this empirical correction. Because of this, no correction has been made to the Moon's center of mass position in the *21st Century Canon.*

2.4 Measurement of Time

The most natural form of time measurement is the solar day (usually measured from solar noon to solar noon). The length of the solar day varies during the year because of the eccentricity of Earth's orbit around the Sun. Mean solar time resolves this problem by using an average to define the mean solar day.

In 1884, Greenwich Mean Time (GMT) — the mean solar time on the Greenwich Meridian (0° longitude) — was adopted as the standard reference time for clocks around the world. A fundamental basis of GMT is the assumption that Earth's rotation on its axis is constant. It wasn't until the mid-twentieth century that astronomers realized the rotation period is gradually increasing. Earth is slowing down because of tidal friction with the Moon.

For purposes of orbital calculations, time using Earth's rotation was abandoned for a more uniform time scale based on Earth's orbit about the Sun. In 1952, Ephemeris Time was introduced to address the problem. The ephemeris second was defined as a fraction of the tropical year[6] for 1900 Jan 01 as calculated from Newcomb's *Tables of the Sun* (1895). Ephemeris Time was used for Solar System ephemeris calculations until 1979.

Terrestrial Dynamical Time (TD) is the modern replacement for Ephemeris Time and is used in theories of planetary motion in the Solar System. TD is based on International Atomic Time (TAI), which is a high-precision standard using several hundred atomic clocks worldwide. To ensure continuity with Ephemeris Time, TD was defined to match ET for the date 1977 Jan 01. In 1991, the IAU refined the definition of TD to make it more precise. It was also renamed Terrestrial Time (TT) although the author prefers to use the older name Terrestrial Dynamical Time.

Civilian time used throughout the world is still based on mean solar time, although indirectly. While Greenwich Mean Time was determined though observations of the Sun, its modern day replacement, Universal Time (actually UT1) is based on Earth's rotation using observations of distant quasars. UT1 is a nonuniform time

[6] The tropical year is the length of time that the Sun takes to return to the same position in the cycle of seasons, as seen from Earth (e.g., the time from vernal equinox to vernal equinox).

because Earth is gradually slowing down at an irregular rate. At present (2016), the accumulated error in the rotation of Earth in the course of one year is ~0.5 seconds.

Coordinated Universal Time (UTC) is derived from International Atomic Time (TAI). The length of the UTC second is defined in terms of an atomic transition of cesium and is accurate to approximately one nanosecond (billionth of a second) per day. UTC was defined to closely parallel UT1. However, the two time systems are intrinsically incompatible since UTC is uniform while UT1 is based on Earth's rotation, which is gradually slowing. In order to keep the two times within 0.9 seconds of each other, a leap second is added to UTC as needed (currently once every few years).

Today, UTC is the time standard used to define time zones around the world. It is the time reference for GPS satellites and aviation, and is used to synchronize the clocks of computers across the Internet.

2.5 ΔT (Delta T)

The orbital positions of the Sun and the Moon, required by eclipse predictions, are calculated using Terrestrial Dynamical Time (TD) because it is a uniform time scale. However, world time zones and daily life are based on Universal Time[7] (UT1). In order to convert eclipse predictions from TD to UT1, the difference between these two time scales must be known. The parameter ΔT (Delta T) is the arithmetic difference, in seconds, between the two as:

$$\Delta T = TD - UT1 \qquad (2\text{–}1)$$

Past values of ΔT can be deduced from historical records. In spite of their relatively low precision, these data represent the only evidence for the value of ΔT prior to 1600. In the centuries following the introduction of the telescope (circa 1609), thousands of high quality observations have been made of lunar occultations of stars, affording valuable data with increased accuracy in the determination of ΔT.

In modern times, the determination of ΔT is made using atomic clocks and radio observations of quasars. From 1955 to 2010, the average 1-year change in ΔT ranges from 0.18 seconds to 1.06 seconds. Future changes in ΔT are unknown since theoretical models of the physical causes are imprecise. Extrapolations from the table weighted by the long period trend from tidal braking of the Moon offer estimates of +71 seconds in 2024, +85 seconds in 2050, and +127 seconds in 2100.

Polynomial expressions for ΔT based on this data can be found at: *eclipsewise.com/help/deltatpoly2014.html*

2.6 Date Format

There are a number of ways to write the calendar date through variations in the order of day, month, and year. The International Organization for Standardization's (ISO) 8601 advises a numeric date representation, which organizes the elements from the largest to the smallest. The exact format is YYYY-MM-DD where YYYY is the calendar year, MM is the month of the year between 01 (January) and 12 (December), and DD is the day of the month between 01 and 31. For example, the 27th day of April in the year 1943 would then be expressed as 1943-04-27. The ISO convention is adopted here, but the month number has been replaced with the three-letter English abbreviation of the month name for additional clarity. From the previous example, the date then is expressed as 1943 Apr 27.

[7] World time zones are actually based on Coordinated Universal Time (UTC). It is an atomic time synchronized and adjusted to stay within a second of astronomically determined Universal Time (UT1) through the addition of an occasional "leap second" to compensate for the gradual slowing of Earth's rotation.

Section 3: Solar Eclipse Statistics

3.1 Statistical Distribution of Eclipse Types

Eclipses of the Sun can only occur during the New Moon phase. It is then possible for the Moon's penumbral, umbral and/or antumbral shadows to sweep across Earth's surface thereby producing an eclipse. There are four types of solar eclipses:

1. **Partial** — Moon's penumbral shadow traverses Earth (umbral and antumbral shadows miss Earth)
2. **Annular** — Moon's penumbral and antumbral shadows traverse Earth (Moon is too far from Earth to completely cover the Sun)
3. **Total** — Moon's penumbral and umbral shadows traverse Earth (Moon is close enough to Earth to completely cover the Sun)
4. **Hybrid** — Moon's penumbral, umbral and antumbral shadows traverse Earth (eclipse is annular or total along different sections of its path). Hybrid eclipses are also known as annular-total eclipses.

During the 100-year period from 2001 to 2100, Earth experiences 224 eclipses of the Sun. The statistical distribution of the four eclipse types over this interval is shown in Table 3-1.

Table 3-1: Distribution of Basic Eclipse Types

Eclipse Type	Abbreviation	Number	Percent
All Eclipses	—	224	100.0%
Partial	P	77	34.4%
Annular	A	72	32.1%
Total	T	68	30.4%
Hybrid	H	7	3.1%

All partial eclipses are events in which some portion of the Moon's penumbral shadow passes across Earth's surface. In comparison all annular, total and hybrid eclipses can be characterized as events in which some portion of the Moon's umbral and/or antumbral shadow crosses Earth.

In the case of umbral or antumbral eclipses (annular, total or hybrid), they can be further categorized as:

1. **Central (two limits)** — The central axis of the Moon's umbral or antumbral shadow traverses Earth thereby producing a central line in the eclipse track. The umbra or antumbra falls entirely upon Earth producing a ground track with both a northern and southern limit.
2. **Central (one limit)** — The central axis of the Moon's umbral or antumbral shadow traverses Earth. However, a portion of the umbra or antumbra misses Earth throughout the eclipse thereby producing a ground track with just one limit.
3. **Non-Central** — The central axis of the Moon's umbral or antumbral shadow misses Earth. However, one edge of the shadow grazes Earth producing a ground track with one limit and no central line.

Using these categories, the distribution of the 72 annular eclipses appears in Table 3-2.

Table 3-2: Statistics of Annular Eclipses

Annular Eclipses	Number	Percent
All Annular Eclipses	72	100.0%
Central (two limits)	68	94.4%
Central (one limit)	2	2.8%
Non-Central (one limit)	2	2.8%

Central annular eclipses with one limit include: 2003 May 31, and 2044 Feb 28. Non-central annular eclipses include: 2014 Apr 29, and 2043 Oct 03.

Similarly, the distribution of the 68 total eclipses is shown in Table 3–3.

Table 3–3: Statistics of Total Eclipses

Total Eclipses	Number	Percent
All Total Eclipses	68	100.0%
Central (two limits)	67	98.5%
Central (one limit)	0	0.0%
Non-Central (one limit)	1	1.5%

There are no central total eclipses with a single limit during the 21st Century. However, there is one non-central total eclipse (one limit) on 2043 Apr 09.

All 7 hybrid eclipses are central with two limits. Hybrid eclipses with a single limit (both central and non-central) are exceedingly rare. An estimate of the mean frequency of non-central hybrid eclipses is one out of every 600 million eclipses or once every 250 million years (Meeus, 2002).

3.2 Eclipse Frequency and the Calendar Year

There are 2 to 5 solar eclipses in every calendar year. Table 3–4 shows the distribution in the number of eclipses per year for the 100 years covered in the *21st Century Canon.*

Table 3–4: Number of Eclipses per Year

Number of Eclipses per Year	Number of Years	Percent
2	82	73.2%
3	12	16.1%
4	6	10.7%
5	0	0.0%

When two eclipses occur in one calendar year, they can be any combination of P, A, T or H (partial, annular, total or hybrid) with the one exception — they can not both be T.

When three eclipses occur in one calendar year, there are 14 possible combinations of P, A, T or H. In most cases all three eclipses are partial.

Years in which four eclipses take place are: 2011, 2029, 2047, 2065, 2076, and 2094.

The maximum number of five solar eclipses in one calendar year is quite rare. No year during the 21st Century has five solar eclipses. The last time this occurred was in 1935 and the next instance is in 2206.

3.3 Extremes in Greatest Central Duration and Eclipse Magnitude

The longest and shortest central eclipses of the century as well as largest and smallest partial eclipses appear in Table 3-5.

Table 3-05: Extremes in Central Duration and Eclipse Magnitude

Extrema Type	Date (Dynamical Time)	Central Line Duration	Eclipse Magnitude
Longest Annular	2010 Jan 15	11m08s	—
Shortest Annular	2085 Dec 16	00m19s	—
Longest Total	2009 Jul 22	06m39s	—
Shortest Total	2068 May 31	01m06s	—
Longest Hybrid	2013 Nov 03	01m40s	—
Shortest Hybrid	2067 Dec 06	00m08s	—
Largest Partial	2051 Apr 11	—	0.9849
Smallest Partial	2098 Oct 24	—	0.0057

*Central line duration at the instant of greatest eclipse

3.4 Eclipse Seasons

Because of its ~5.1° inclination the Moon's orbit crosses the ecliptic at two points or nodes. If New Moon takes place within approximately 17° of a node[8], a solar eclipse will be visible from some location on Earth.

The Sun makes one complete circuit of the ecliptic in 365.24 days, so its average angular velocity is 0.99° per day. At this rate, it takes 34.5 days for the Sun to cross the 34° wide eclipse zone centered on each node. Because the Moon's orbit with respect to the Sun has a mean duration of 29.53 days, there will always be one and possibly two solar eclipses during each 34.5-day interval when the Sun passes through the nodal eclipse zones. These time periods are called eclipse seasons.

The mid-point of each eclipse season is separated by 173.3 days because this is the mean time for the Sun to travel from one node to the next. The period is a little less that half a calendar year because the lunar nodes slowly regress westward by 19.3° per year.

Figure 3–1: Eclipse Seasons and Orbital Nodes

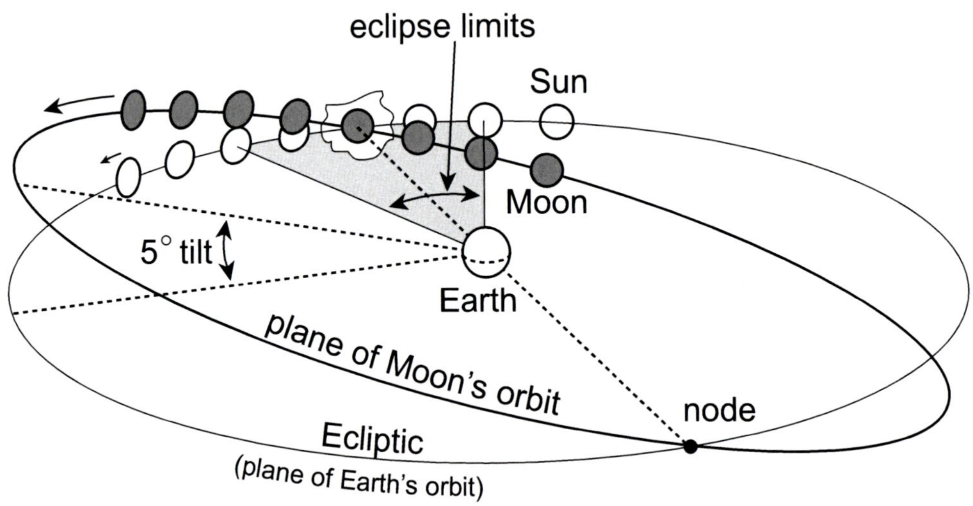

[8] The exact angular distance from the node depends on the distances of the Sun and the Moon from Earth, which determine their angular diameters.

3.5 Quincena

The mean time interval between New Moon and Full Moon is 14.77 days. This is less than half the duration of a 34.5-day eclipse season. As a consequence, the same Sun–node alignment geometry responsible for producing a solar eclipse always results in a complementary lunar eclipse within a fortnight. The lunar eclipse may precede or succeed the solar eclipse. In either case, the pair of eclipses is referred to here as a quincena[9]. The QLE (Quincena Lunar Eclipse parameter) identifies the type of the lunar eclipse and whether it precedes or succeeds a particular solar eclipse. There are three basic types of lunar eclipses:

1. penumbral lunar eclipse (n) — Moon passes partly or completely within Earth's penumbral shadow
2. partial lunar eclipse (p) — Moon passes partly within Earth's umbral shadow
3. total lunar eclipse (t) — Moon passes completely within Earth's umbral shadow

The QLE is a two character string consisting of one or more of the above lunar eclipse types. The first character in the QLE identifies the type of lunar eclipse preceding a solar eclipse. The second character identifies the type of lunar eclipse succeeding a solar eclipse. In most instances, one of the two characters is "–" indicating no lunar eclipse occurs. For example, a QLE of "–p" means that no lunar eclipse precedes a solar eclipse, but a partial lunar eclipse follows the solar eclipse 15 days later.

On rare occasions, a double quincena occurs in which a solar eclipse is both preceded and succeeded by a lunar eclipse.

Photo 4–2 Fine details in the solar corona are captured during the total solar eclipse of 2006 March 29 from Jalu LIBYA.

[9] Quincena is a Spanish word meaning *a period of fifteen days*. This also happens to be the time interval, rounded to the nearest day, between New Moon and Full Moon, or Full Moon and New Moon. So *quincena* is a convenient and appropriate term for describing a pair of eclipses (one solar and one lunar) separated by this period.

Section 4: Explanation of Solar Eclipse Catalog in Appendix A

4.1 Introduction

Earth experiences 224 eclipses of the Sun during the 21st Century. The catalog in *Appendix A* summarizes the principal characteristics of each eclipse and complements the maps in *Appendices B, C* and *D*.

Each line in the catalog corresponds to a single eclipse and provides concise parameters to characterize the eclipse. The calendar date and time of the instant of greatest eclipse are given, along with the value of Delta T . The lunation number and Saros series are listed along with the eclipse type. Gamma is the distance of the shadow axis from Earth's center at greatest eclipse, while the eclipse magnitude is the fraction of the Sun's diameter obscured at that instant. The geographic latitude and longitude of the umbral or antumbral axis are given for greatest eclipse, along with the Sun's altitude and azimuth, the width of the path, and the central line duration of totality or annularity. For both partial and non-central total or annular eclipses, the latitude and longitude correspond to the point closest to the shadow cone axis at greatest eclipse. The Sun's altitude is always 0° at this location. Detailed descriptions of each field in the catalog appear in the following sections.

4.2 Cat Num (Catalog Number)

The catalog number is the sequential number assigned to each eclipse from 1 to 224.

4.3 Calendar Date

Gregorian calendar date at greatest eclipse is given in the ISO order of YEAR-MONTH-DAY.

4.4 TD of Greatest Eclipse (Terrestrial Dynamical Time of Greatest Eclipse)

The instant of greatest eclipse occurs when the distance between the axis of the Moon's shadow cone and the center of Earth reaches a minimum. For partial eclipses, this instant differs slightly from the instant of greatest magnitude due to Earth's flattening. For total eclipses, this instant differs slightly from the instant of greatest duration, although the differences are relatively small.

Greatest eclipse is given in Terrestrial Dynamical Time or TD (Sect. 2.4), which is a time system based on International Atomic Time. As such, TD is the modern equivalent to its predecessor Ephemeris Time and is used in theories of planetary motion in the Solar System. To determine the geographic visibility of an eclipse, TD is converted to Universal Time (UT1) using the parameter Delta T (Sect. 2.5).

4.5 ΔT (Delta T)

ΔT (Delta T) is the arithmetic difference, in seconds, between Terrestrial Dynamical Time (TD) and Universal Time (UT1). For more information on ΔT, see Section 2.5.

4.6 Luna Num (Lunation Number)

The lunation number is the number of synodic months or lunations since New Moon on 2000 Jan 06. It can be converted to the Brown Lunation Number[10] by adding 953.

[10] The *Brown Lunation Number* defines lunation 1 as beginning at the first New Moon of 1923, the year when Ernest W. Brown's lunar theory was introduced in the major national astronomical almanacs.

4.7 Saros Num (Saros Series Number)

Each eclipse belongs to a Saros series using a numbering system first introduced by van den Bergh (1955). The eclipses with an odd Saros number take place at the ascending node of the Moon's orbit; those with an even Saros number take place at the descending node. This relationship is reversed for *lunar* eclipses.

The Saros is a period of 223 synodic months (~ 18 years, 11 days, and 8 hours). Eclipses separated by this period belong to the same Saros series and share similar geometry and characteristics.

4.8 Ecl Type (Solar Eclipse Type)

The first value in this 2-character parameter gives the eclipse type. The four basic types of solar eclipses are:

1. **Partial Solar Eclipse (P)** — The Moon's penumbral shadow traverses Earth; the Moon's umbral and antumbral shadows completely miss Earth
2. **Annular Solar Eclipse (A)** — The Moon's penumbral and antumbral shadows traverse Earth; the Moon is too far from Earth to completely cover the Sun
3. **Total Solar Eclipse (T)** — The Moon's penumbral and umbral shadow traverse Earth; the Moon is close enough to Earth to completely cover the Sun
4. **Hybrid Solar Eclipse (H)** — The Moon's penumbral, umbral and antumbral shadows traverse different parts of Earth; eclipse appears either total or annular along different sections of its path; hybrid eclipses are also known as annular-total eclipses

The second character of the eclipse type is a qualifier defined as follows.

1. m = Middle eclipse of Saros series
2. n = Central eclipse with no northern limit
3. s = Central eclipse with no southern limit
4. + = Non-central eclipse with no northern limit
5. – = Non-central eclipse with no southern limit
6. 2 = Hybrid eclipse path begins total and ends annular
7. 3 = Hybrid eclipse path begins annular and ends total
8. b = Saros series begins (first eclipse in a Saros series)
9. e = Saros series ends (last eclipse in a Saros series)

Qualifiers 1 through 5 are used with annular, total or hybrid eclipses. Qualifiers 6 and 7 apply only to special classes of hybrid eclipses while qualifiers 8 and 9 are used exclusively with partial eclipses.

4.9 QLE (Quincena Lunar Eclipse Parameter)

A lunar eclipse always occurs within 15 days of a solar eclipse. The Quincena Lunar Eclipse parameter (QLE) identifies the type of the lunar eclipse and whether it precedes or succeeds a particular solar eclipse. There are three basic types of lunar eclipses:

1. **Penumbral Lunar Eclipse (n)** — The Moon passes partly or completely within Earth's penumbra
2. **Partial Lunar Eclipse (p)** — The Moon passes partly within Earth's umbra
3. **Total Lunar Eclipse (t)** — The Moon passes completely within Earth's umbra

The QLE consists of a two-character string. The characters identify the type of lunar eclipse preceding and succeeding a solar eclipse, respectively. In most instances, one of the two characters in the QLE is "–" indicating no lunar eclipse occurs. On some occasions, a double quincena occurs in which a solar eclipse is both preceded and succeeded by a lunar eclipse. The QLE then consists of two characters identifying the types of the two lunar eclipses. (Section 3.5).

4.10 Gamma

Gamma is the minimum distance from the lunar shadow axis to the center of Earth, in units of Earth's equatorial radius. This distance is positive or negative, depending on whether the axis of the shadow cone passes north or south of Earth's center. If gamma is between +0.997 and −0.997, the eclipse is a central one (either total, annular, or hybrid). The limiting value 0.997 differs from unity because of the flattening of Earth.

4.11 Ecl Mag (Eclipse Magnitude)

The eclipse magnitude is defined as the fraction of the Sun's diameter occulted by the Moon. For partial eclipses, the eclipse magnitude at the instant of greatest eclipse is given for the geographic position closest to the Moon's shadow axis. For total, annular, and hybrid eclipses, the eclipse magnitude is replaced by the ratio of the topocentric apparent diameters of the Moon and the Sun at greatest eclipse. The eclipse magnitude is always < 1.0 for partial and annular eclipses, but ≥ 1.0 for total and hybrid eclipses.

4.12 Lat Long (Latitude and Longitude)

The latitude and longitude correspond to the position intersected by the lunar shadow axis at greatest eclipse.

4.13 Sun Alt (Altitude of Sun)

The Sun's altitude at the geographic position intersected by the lunar shadow axis is given at the instant of greatest eclipse. For partial eclipses, the Sun's altitude is always 0° because the shadow axis misses Earth. In this case, the geographic position corresponds to the point closest to the shadow axis.

4.14 Sun Azm (Azimuth of Sun)

The Sun's azimuth at the geographic position intersected by the lunar shadow axis is given at the instant of greatest eclipse. The values 0°, 90°, 180°, and 270° correspond to the north, east, south and west, respectively.

4.15 Path Width

For total, annular, and hybrid eclipses, the width of the path of totality or annularity (kilometers) is given at the geographic position intersected by the lunar shadow axis at the instant of greatest eclipse.

4.16 Central Line Dur (Central Line Duration)

For total, annular, or hybrid eclipses, the central line duration of the total or annular phase (minutes and seconds) is given at the geographic position intersected by the lunar shadow axis at greatest eclipse.

For total and hybrid eclipses, this duration is close to the maximum of the total phase along the entire path. For annular eclipses, the duration at greatest eclipse may be near either the minimum or maximum duration of the annular phase along the path. If the annular phase duration exceeds approximately 2.3 min, then it is close to the maximum duration along the central line track. If the annular phase duration is less, however, then it corresponds to a minimum and the annular duration increases towards the ends of the path.

4.17 EclipseWise.com and the 21st Century Canon

Much of the content of the *21st Century Canon of Solar Eclipses* is based on the eclipse predictions website *www.EclipseWise.com.* A plain text file containing the entire solar eclipse catalog in *Appendix A* can be downloaded at: *www.EclipseWise.com/solar/SEpubs/21CCSE.txt.*

A web based version of the catalog is available at: *www.eclipsewise.com/solar/SEcatalog/21CCSEcat.html*

Section 5: Explanation of Solar Eclipse Maps in Appendices B, C and D

5.1 Explanation of Small Global Eclipse Maps in Appendix B

Earth experiences 224 eclipses of the Sun during the 21st Century. An individual global map for each eclipse appears in *Appendix B.*

The geographic visibility of each eclipse is illustrated with an orthographic projection map of Earth showing the path of the Moon's penumbral (partial) and umbral/antumbral (total, hybrid, or annular) shadows with respect to the continental coastlines, political boundaries (circa 2016) and the Equator. North is to the top and the daylight terminator is drawn for the instant of greatest eclipse. An 'x' symbol marks the sub-solar point where the Sun appears directly overhead at that time. The salient features of the eclipse maps are identified in Figure 5–1, which serves as a key.

Figure 5—1: Key to Small Global Solar Eclipse Maps (Appendix B)

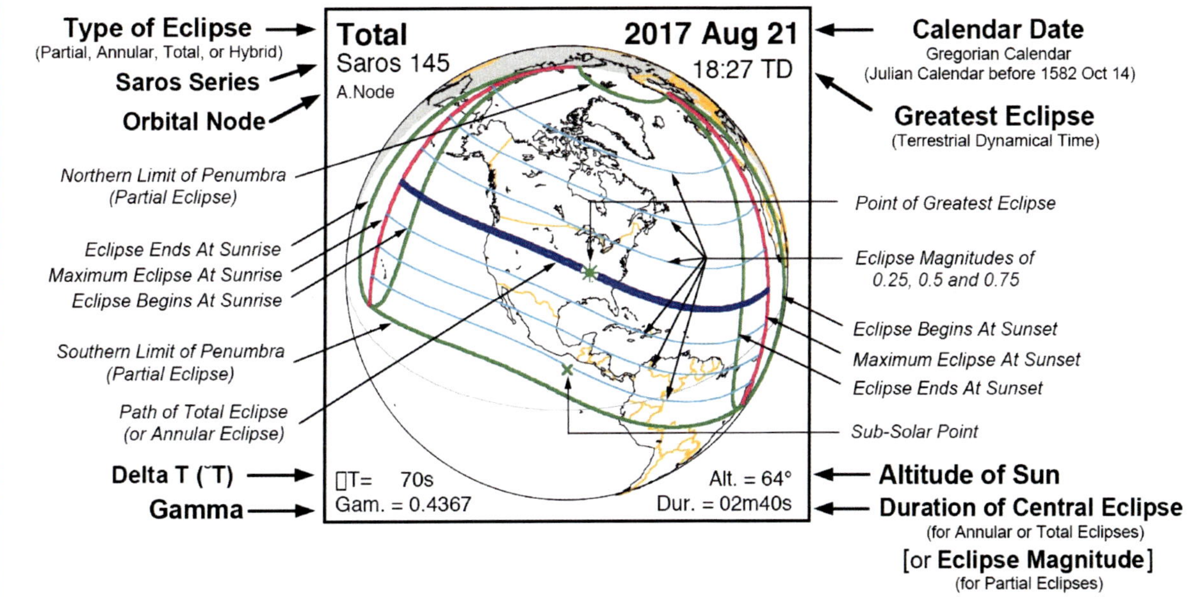

The limits of the Moon's penumbral shadow delineate the region of visibility of a partial solar eclipse. This irregular or saddle shaped region often covers more than half the daylight hemisphere of Earth and consists of several distinct zones or limits. At the northern and/or southern boundaries lie the limits of the penumbra's path. Partial eclipses have only one of these limits, as do central eclipses when the Moon's shadow axis falls no closer than about 0.45 radii from Earth's center. Great loops at the western and eastern extremes of the penumbra's path identify the areas where the eclipse begins/ends at sunrise and sunset, respectively. If the penumbra has both a northern and southern limit, the rising and setting curves form two separate, closed loops (e.g., 2017 Aug 21). Otherwise, the curves are connected in a distorted figure eight (e.g., 2019 Jul 02). Bisecting the *eclipse begins/ends at sunrise and sunset* loops is the curve of maximum eclipse at sunrise (western loop) and sunset (eastern loop).

The eclipse magnitude is defined as the fraction of the Sun's diameter occulted by the Moon. A curve of constant eclipse magnitude delineates the locus of all points where the local magnitude at maximum eclipse is equal to a constant value. The maps include *curves of constant eclipse magnitude* for values of 0.25, 0.5 and 0.75. These curves run exclusively between the curves of maximum eclipse at sunrise and sunset. They are approximately parallel to the northern/southern penumbral limits and the umbral/antumbral paths of central eclipses. The northern and southern limits of the penumbra may be thought of as curves of eclipse magnitude of 0.0. For total eclipses, the northern and southern limits of the umbra are curves of eclipse magnitude of 1.0.

Greatest eclipse is the instant when the axis of the Moon's shadow cone passes closest to Earth's center. Although greatest eclipse differs slightly from the instants of greatest magnitude and greatest duration (for total eclipses), the differences are relatively small. The point on Earth's surface intersected by the axis of the Moon's shadow cone at greatest eclipse is marked by an asterisk symbol '*'. For partial eclipses, the shadow axis misses Earth entirely, so the point of greatest eclipse lies on the day/night terminator and the Sun appears on the horizon.

Data relevant to an eclipse appear in the corners of each map. In the top left corner are the eclipse type (total, hybrid, annular, or partial), the Saros series of the eclipse, and the node of the Moon's orbit where the eclipse occurs. To the top right are the Gregorian calendar date, the time of greatest eclipse (Terrestrial Dynamical Time), and the value of Delta T (ΔT).

The bottom left corner lists gamma (the minimum distance of the Moon's shadow cone axis from Earth's center, in Earth equatorial radii). The Sun's altitude at the geographic position of greatest eclipse is found to the lower right. The content of the final datum in the bottom right corner depends on the type of eclipse. If the eclipse is partial then the eclipse magnitude is given. If the eclipse is total, hybrid or annular, then the duration of the total or annular phase is given at the instant of greatest eclipse.

The list below briefly summarizes the parameters appearing in each global eclipse map.

Solar Eclipse Type – One of four basic types of solar eclipses: Partial, Annular Total or Hybrid (See Sect. 4.1).

Saros Series – The Saros series that the eclipse belongs to. Eclipses with an odd number take place at the Moon's ascending node, while those with an even number are at the descending node (See Sect. 4.7).

Node – The orbital node near which the eclipse takes place. The ascending node (A. Node) is the point where the Moon travels from south to north through Earth's orbital plane. Similarly, the descending node (D. Node) is the point where the Moon travels from north to south.

Calendar Date – Gregorian calendar date at greatest eclipse is given in the ISO order of YEAR-MONTH-DAY.

Greatest Eclipse – The instant (Terrestrial Dynamical Time) when the distance between the axis of the Moon's shadow cone and the center of Earth reaches a minimum (See Sect. 4.4).

ΔT (Delta T) – the arithmetic difference, in seconds, between Terrestrial Dynamical Time (TD) and Universal Time (UT1), (See Sect. 2.5).

Gamma – The minimum distance from the lunar shadow axis to the center of Earth, in units of Earth's equatorial radius (See Sect. 4.10).

Altitude of Sun – The Sun's altitude at the geographic position intersected by the lunar shadow axis is given at the instant of greatest eclipse (See Sect. 4.13).

Duration of Central Eclipse – The central line duration of the total or annular phase (in minutes and seconds) is given at the instant of greatest eclipse (See Sect. 4.16).

Eclipse Magnitude – The fraction of the Sun's diameter occulted by the Moon at the instant of greatest eclipse (See Sect. 4.11).

5.2 Explanation of Large Global Eclipse Maps in Appendix C

The heart of the *21st Century Canon of Solar Eclipses – Deluxe Edition* is the set of 224 full page global maps appearing in Appendix C — one map for every solar eclipse from 2001 through 2100. This feature is similar to the *Fifty Year Canon of Solar Eclipses: 1986 – 2035* (Espenak 1986) and makes *21st Century Canon* its modern successor.

The salient features of the large global maps are identified in Figure 5–2, which serves as a key.

Figure 5—2: Key to Large Global Solar Eclipse Maps (Appendix C)

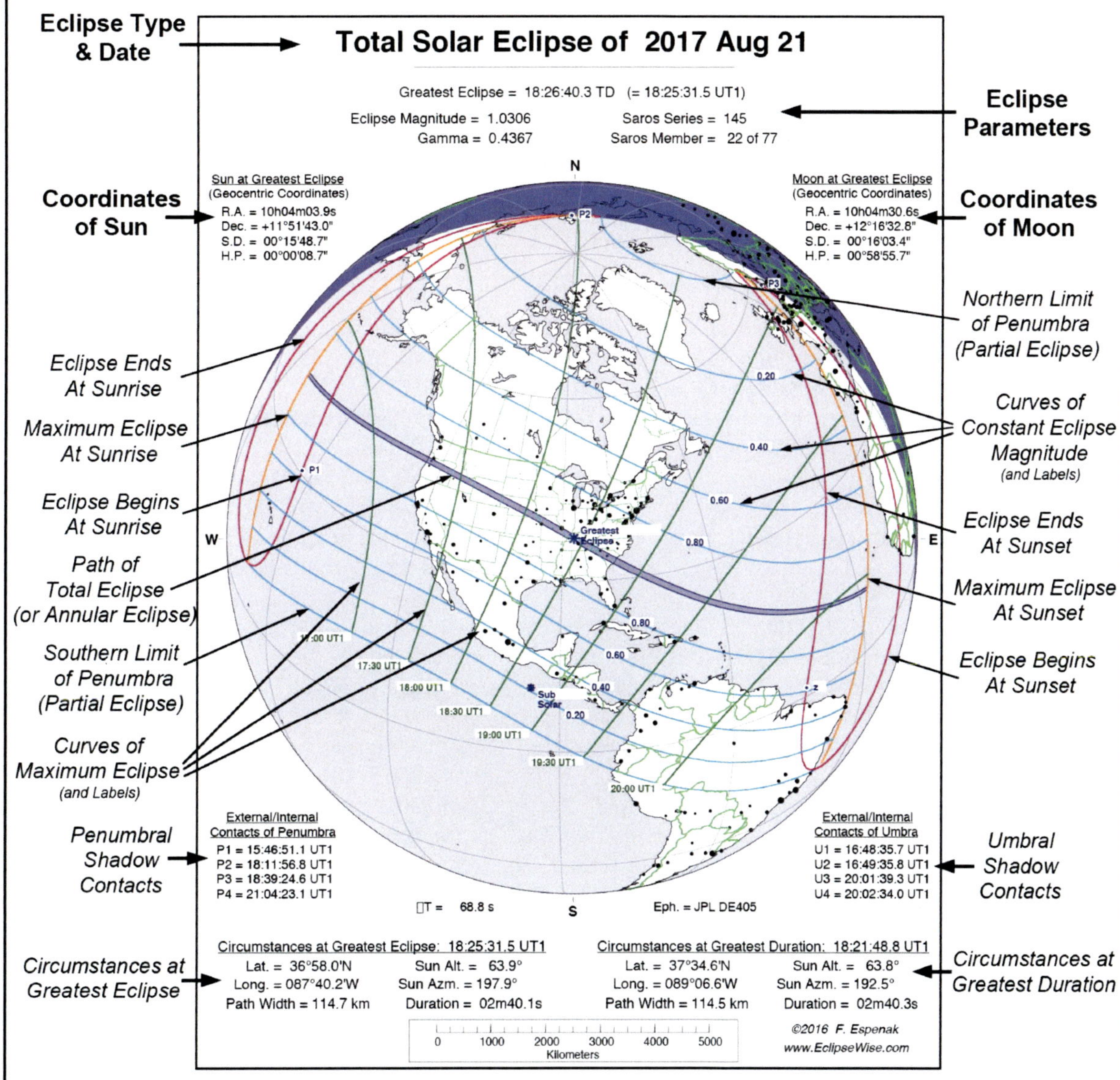

Each orthographic projection map shows the entire path of penumbral (partial) and umbral (total) or antumbral (annular) eclipse with respect to the continental coastlines, political boundaries (circa 2016), and major cities. North is to the top and the daylight terminator is plotted for the instant of greatest eclipse. An asterisk "*" indicates the sub-solar point on Earth (i.e., Sun in the Zenith).

The limits of the Moon's penumbral shadow delineate the region of visibility of the partial solar eclipse. This irregular or saddle shaped region often covers more than half of the daylight hemisphere of Earth and consists of several distinct zones or limits. At the northern and/or southern boundaries lie the limits of the penumbra's path. Partial eclipses have only one of these limits, as do central eclipses when the Moon's shadow axis falls no closer than about 0.45 radii from Earth's center.

Great loops at the western and eastern extremes of the penumbra's path identify the areas where the eclipse begins/ends at sunrise and sunset, respectively. If the penumbra has both a northern and southern limit, the rising and setting curves form two separate, closed loops. Otherwise, the curves are connected in a distorted figure eight. Bisecting the 'eclipse begins/ends at sunrise and sunset' loops is the curve of maximum eclipse at sunrise (western loop) and sunset (eastern loop. The points *P1* and *P4* mark the coordinates where the penumbral shadow first contacts (partial eclipse begins) and last contacts (partial eclipse ends) Earth's surface. If the penumbral path has both a northern and southern limit, then points *P2* and *P3* are also plotted. These correspond to the coordinates where the penumbral shadow cone is internally tangent to Earth's disk.

A curve of maximum eclipse is the locus of all points where the eclipse is at maximum at a given time. Curves of maximum eclipse are plotted at each half-hour Universal Time and are labeled with the time. They generally run between the penumbral limits in the north/south direction, or from the maximum eclipse at sunrise and sunset curves to one of the limits.

The curves of constant eclipse magnitude delineate the locus of all points where the magnitude at maximum eclipse is constant. These curves run between the curves of maximum eclipse at sunrise and sunset . They are roughly parallel to the northern/southern penumbral limits and the umbral paths of central eclipses (total, annular, and hybrid). In fact, the northern and southern limits of the penumbra are curves of constant magnitude of 0.0. The adjacent curves are for magnitudes of 0.2, 0.4, 0.6 and 0.8 and are labeled as such.

For total eclipses, the northern and southern limits of the umbra are curves of constant magnitude of 1.0. Antumbral path limits for annular eclipses are curves of maximum eclipse magnitude. The path of each total eclipse is darkly shaded. Similarly, the path of each annular eclipse is lightly shaded.

Greatest eclipse is defined as the instant when the axis of the Moon's shadow passes closest to Earth's center. Although greatest eclipse differs slightly from the instants of greatest magnitude and greatest duration (for total eclipses), the differences are small. An asterisk '*' marks the point on Earth's surface intersected by the lunar shadow axis at greatest eclipse. For partial eclipses, the shadow axis misses Earth entirely, so the point of greatest eclipse lies on the day/night terminator and the Sun appears on the horizon.

Data pertinent to the eclipse appear with each map. At the top is listed the instant of greatest eclipse, expressed in both Terrestrial Dynamical Time and Universal Time. The eclipse magnitude is defined as the fraction of the Sun's diameter obscured by the Moon at greatest eclipse. For central eclipses, the magnitude listed is actually the geocentric ratio of diameters of the Moon and the Sun. Gamma is the minimum distance (in Earth radii) of the Moon's shadow axis from Earth's center at greatest eclipse. The Saros series of the eclipse is listed, followed by the member position. The first number identifies the sequence position of the eclipse in the Saros, while the second number is the number of eclipses in the entire series.

In the upper left and right corners are the geocentric coordinates of the Sun and the Moon, respectively, at the instant of greatest eclipse. They are:

R.A. – Right Ascension
Dec. – Declination
S.D. – Apparent Semi-Diameter
H.P. – Horizontal Parallax

To the lower left are exterior/interior contact times of the Moon's penumbra with Earth, which are defined:

P1 - Instant of first exterior tangency of Penumbra with Earth's limb. (Partial Eclipse Begins)
P2 - Instant of first interior tangency of Penumbra with Earth's limb.
P3 - Instant of last interior tangency of Penumbra with Earth's limb.
P4 - Instant of last exterior tangency of Penumbra with Earth's limb. (Partial Eclipse Ends)

Not all eclipses have *P2* and *P3* penumbral contacts. They are only present in cases where the penumbral shadow falls completely within Earth's disk. For central eclipses, the lower right corner lists exterior/interior contact times of the Moon's umbral shadow with Earth's limb which are defined as follows:

U1 - Instant of first exterior tangency of Umbra with Earth's limb.
(Umbral [Total/Annular] Eclipse Begins)
U2 - Instant of first interior tangency of Umbra with Earth's limb.
U3 - Instant of last interior tangency of Umbra with Earth's limb.
U4 - Instant of last exterior tangency of Umbra with Earth's limb.
(Umbral [Total/Annular] Eclipse Ends)

At bottom left are the geographic coordinates of the position of greatest eclipse along with the local circumstances at that location (i.e., Sun altitude, Sun azimuth, path width and duration of totality/annularity). For central eclipses, similar data are also given at the bottom right for the position of greatest duration.

5.3 Explanation of Central Solar Eclipse Maps in Appendix D

The final series of 12 maps in Appendix D show the path of every central solar eclipse (total, annular and hybrid) with respect to continental coastlines, political boundaries (circa 2016), and major cities. The maps employ a transverse Mercator projection and are centered on three geographic regions:

1. North and South America
2. Europe and Africa
3. Asia and Australia

Furthermore, each map covers a 25-year period:

1. 2001–2025
2. 2026–2050
3. 2051–2075
4. 2076–2100

Each eclipse is identified by its date at the instant of greatest eclipse. An asterisk symbol marks the location of greatest eclipse.

The paths of total, annular, and hybrid eclipses can be identified as follows.

Total Eclipses – the path is shaded blue in (Color & Deluxe Editions);
the date appears in bold and is preceded by the letter **T**

Annular Eclipses – the path is shaded orange (Color & Deluxe Editions);
the date appears in italics and is preceded by the letter **A**

Hybrid Eclipses – the path is shaded blue along the total sections and
shaded orange along the annular sections;
the date appears in bold-italics and is preceded by the letter **H**

References

Astronomical Almanac for 1986, Washington: US Government Printing Office; London: HM Stationery Office (1985).

Astronomical Almanac for 2011, Washington: US Government Printing Office; London: HM Stationery Office (2010).

Espenak, F., *Fifty Year Canon of Solar Eclipses: 1986–2035,* Sky Publishing Corp., Cambridge, Massuchusetts (1987).

Espenak, F., *Thousand Year Canon of Solar Eclipses: 1501 to 2500,* Astropixels Publishing, Portal, Arizona (2014).

Espenak, F., *Thousand Year Canon of Lunar Eclipses: 1501 to 2500,* Astropixels Publishing, Portal, Arizona (2014).

Espenak, F., and Meeus, J., *Five Millennium Canon of Solar Eclipses: –1999 to +3000 (2000 BCE to 3000 CE),* NASA Tech. Pub. 2006–214141, NASA Goddard Space Flight Center, Greenbelt, Maryland (2006).

Espenak, F., and Meeus, J., *Five Millennium Canon of Lunar Eclipses: –1999 to +3000 (2000 BCE to 3000 CE),* NASA Tech. Pub. 2006–214172, NASA Goddard Space Flight Center, Greenbelt, Maryland (2009a).

Espenak, F., and Meeus, J., *Five Millennium Catalog of Lunar Eclipses: –1999 to +3000 (2000 BCE to 3000 CE),* NASA Tech. Pub. 2006–214173, NASA Goddard Space Flight Center, Greenbelt, Maryland (2009b).

Espenak, F., and Meeus, J., *Five Millennium Catalog of Solar Eclipses: –1999 to +3000 (2000 BCE to 3000 CE),* NASA Tech. Pub. 2006–214174, NASA Goddard Space Flight Center, Greenbelt, Maryland (2009c).

Explanatory Supplement to the Ephemeris, H.M. Almanac Office, London (1974).

Littmann, M., Espenak, F., and Willcox, K., *Totality—Eclipses of the Sun,* 3rd Ed., Oxford University Press, New York (2008).

Meeus, J., *Mathematical Astronomy Morsels,* Willmann-Bell, pp. 56–62 (1997).

——, *More Mathematical Astronomy Morsels,* Willmann-Bell, pp. 70–72 (2002).

——, Grosjean, C.C., and Vanderleen, W., *Canon of Solar Eclipses,* Pergamon Press, Oxford, United Kingdom (1966).

Mucke, H., and Meeus, J., *Canon of Solar Eclipses: -2003 to +2526,* Astronomisches Büro, Vienna (1983).

Newcomb, S., "Tables of the Motion of the Earth on its Axis Around the Sun," *Astron. Papers Amer. Eph.,* Vol. 6, Part I (1895).

van den Bergh, *Periodicity and Variation of Solar (and Lunar) Eclipses,* Tjeenk Willink, and Haarlem, Netherlands (1955).

van der Sluys, M. , *http://hemel.waarnemen.com/Computing/deltat.html* (2010).

von Oppolzer, T.R., *Canon der Finsternisse,* Wien, (1887); Gingerich, O., (Translator) *Canon of Eclipses,* Dover Publications, New York (1962).

Appendix A

Solar Eclipse Catalog: 2001 to 2100

Key to Solar Eclipse Catalog

Cat Num — sequential Catalog Number assigned to each eclipse from 1 to 224

Calendar Date — Gregorian date of Greatest Eclipse

TD of Greatest Eclipse — Terrestrial Dynamical Time of Greatest Eclipse

ΔT — arithmetic difference between Terrestrial Dynamical Time Universal Time (seconds)

Luna Num — number of synodic months, or lunations, since New Moon on 2000 Jan 06

Saros Num — Saros Series Number of eclipse

Ecl Type — Solar Eclipse Type

P = Partial Solar Eclipse
A = Annular Solar Eclipse
T = Total Solar Eclipse
H = Hybrid Solar Eclipse
m = Middle eclipse of Saros series
n = Central eclipse of with no northern limit
s = Central eclipse of with no southern limit
+ = Non-central eclipse of with no northern limit
– = Non-central eclipse of with no southern limit
2 = Hybrid eclipse path begins total and ends annular
3 = Hybrid eclipse path begins annular and ends total
b = Saros series begins (first eclipse in a Saros series)
e = Saros series ends (last eclipse in a Saros series)

QLE — Quincena Lunar Eclipse Parameter

n = Penumbral Lunar Eclipse
p = Partial Lunar Eclipse
t = Total Lunar Eclipse

Gamma — minimum distance from the axis of the lunar shadow to the center of Earth

Ecl Mag — Eclipse Magnitude; fraction of the Sun's diameter obscured by the Moon

Lat & Lng — latitude and longitude where the Sun appears in zenith at greatest eclipse

Sun Alt & **Sun Azm** — altitude and azimuth of the Sun at greatest eclipse

Path Width — width of the central path (km) at greatest eclipse (total, annular & hybrid eclipses)

Central Line Dur — Central Line Duration (minutes. seconds) at greatest eclipse

Cat Num	Calendar Date	TD of Greatest Eclipse	ΔT s	Luna Num	Saros Num	Ecl Type	QLE	Gamma	Ecl Mag	Lat °	Long °	Sun Alt °	Sun Azm °	Path Width km	Central Line Dur
001	2001 Jun 21	12:04:46	64	18	127	T	-p	-0.5701	1.0495	11.3S	2.7E	55	355	200	04m57s
002	2001 Dec 14	20:53:01	64	24	132	A	-n	0.4089	0.9681	0.6N	130.7W	66	188	126	03m53s
003	2002 Jun 10	23:45:22	64	30	137	A	nn	0.1993	0.9962	34.5N	178.6W	78	169	13	00m23s
004	2002 Dec 04	07:32:16	64	36	142	T	n-	-0.3020	1.0244	39.5S	59.6E	72	16	87	02m04s
005	2003 May 31	04:09:23	64	42	147	An	t-	0.9960	0.9384	66.6N	24.5W	3	35	-	03m37s
006	2003 Nov 23	22:50:22	65	48	152	T	t-	-0.9638	1.0379	72.7S	88.4E	15	111	496	01m57s
007	2004 Apr 19	13:35:05	65	53	119	P	-t	-1.1335	0.7367	61.6S	44.3E	0	295		
008	2004 Oct 14	03:00:23	65	59	124	P	-t	1.0348	0.9283	61.2N	153.7W	0	253		
009	2005 Apr 08	20:36:51	65	65	129	H	-n	-0.3473	1.0074	10.6S	119.0W	70	332	27	00m42s
010	2005 Oct 03	10:32:47	65	71	134	A	-p	0.3306	0.9576	12.9N	28.7E	71	209	162	04m32s
011	2006 Mar 29	10:12:23	65	77	139	T	n-	0.3843	1.0515	23.2N	16.7E	67	149	184	04m07s
012	2006 Sep 22	11:41:16	65	83	144	A	p-	-0.4062	0.9352	20.6S	9.1W	66	31	261	07m09s
013	2007 Mar 19	02:32:58	65	89	149	P	t-	1.0728	0.8756	61.0N	55.5E	0	92		
014	2007 Sep 11	12:32:24	65	95	154	P	t-	-1.1255	0.7507	61.0S	90.2W	0	80		
015	2008 Feb 07	03:56:10	65	100	121	A	-t	-0.9570	0.9650	67.6S	150.5W	16	269	444	02m12s
016	2008 Aug 01	10:22:12	66	106	126	T	-p	0.8307	1.0394	65.7N	72.3E	34	235	237	02m27s
017	2009 Jan 26	07:59:45	66	112	131	A	-n	-0.2820	0.9282	34.1S	70.2E	73	337	280	07m54s
018	2009 Jul 22	02:36:25	66	118	136	T	nn	0.0698	1.0799	24.2N	144.1E	86	198	258	06m39s
019	2010 Jan 15	07:07:39	66	124	141	A	p-	0.4002	0.9190	1.6N	69.3E	66	165	333	11m08s
020	2010 Jul 11	19:34:38	66	130	146	T	p-	-0.6788	1.0580	19.7S	121.9W	47	14	259	05m20s
021	2011 Jan 04	08:51:42	66	136	151	P	t-	1.0626	0.8576	64.7N	20.8E	0	155		
022	2011 Jun 01	21:17:18	66	141	118	P	-t	1.2130	0.6011	67.8N	46.8E	0	6		
023	2011 Jul 01	08:39:30	66	142	156	Pb	t-	-1.4917	0.0971	65.2S	28.6E	0	21		
024	2011 Nov 25	06:21:25	67	147	123	P	-t	-1.0536	0.9047	68.6S	82.4W	0	165		
025	2012 May 20	23:53:54	67	153	128	A	-p	0.4828	0.9439	49.1N	176.3E	61	171	237	05m46s
026	2012 Nov 13	22:12:55	67	159	133	T	-n	-0.3719	1.0500	40.0S	161.3W	68	11	179	04m02s
027	2013 May 10	00:26:20	67	165	138	A	pn	-0.2694	0.9544	2.2N	175.5E	74	350	173	06m03s
028	2013 Nov 03	12:47:36	67	171	143	H3	n-	0.3272	1.0159	3.5N	11.7W	71	192	58	01m40s
029	2014 Apr 29	06:04:33	67	177	148	A-	t-	-1.0000	0.9868	70.6S	131.3E	0	319	-	-
030	2014 Oct 23	21:45:39	68	183	153	P	t-	1.0908	0.8114	71.2N	97.2W	0	231		
031	2015 Mar 20	09:46:47	68	188	120	T	-t	0.9454	1.0446	64.4N	6.6W	18	135	463	02m47s
032	2015 Sep 13	06:55:19	68	194	125	P	-t	-1.1004	0.7875	72.1S	2.3W	0	77		
033	2016 Mar 09	01:58:19	68	200	130	T	-n	0.2609	1.0450	10.1N	148.8E	75	162	155	04m09s
034	2016 Sep 01	09:08:02	68	206	135	A	-n	-0.3330	0.9736	10.7S	37.8E	70	16	100	03m06s
035	2017 Feb 26	14:54:33	69	212	140	A	n-	-0.4578	0.9922	34.7S	31.2W	63	340	31	00m44s
036	2017 Aug 21	18:26:40	69	218	145	T	p-	0.4367	1.0306	37.0N	87.7W	64	198	115	02m40s
037	2018 Feb 15	20:52:33	69	224	150	P	t-	-1.2116	0.5991	71.0S	0.6E	0	228		
038	2018 Jul 13	03:02:16	69	229	117	P	-t	-1.3542	0.3365	67.9S	127.4E	0	8		
039	2018 Aug 11	09:47:28	69	230	155	P	t-	1.1476	0.7368	70.4N	174.5E	0	321		
040	2019 Jan 06	01:42:38	69	235	122	P	-t	1.1417	0.7146	67.4N	153.6E	0	178		
041	2019 Jul 02	19:24:07	70	241	127	T	-p	-0.6466	1.0459	17.4S	109.0W	50	359	201	04m33s
042	2019 Dec 26	05:18:53	70	247	132	A	-n	0.4135	0.9701	1.0N	102.2E	66	184	118	03m39s
043	2020 Jun 21	06:41:15	70	253	137	Am	nn	0.1209	0.9940	30.5N	79.7E	83	174	21	00m38s
044	2020 Dec 14	16:14:39	70	259	142	T	n-	-0.2939	1.0254	40.3S	68.0W	73	10	90	02m10s
045	2021 Jun 10	10:43:07	70	265	147	A	t-	0.9152	0.9435	80.8N	66.8W	23	90	527	03m51s
046	2021 Dec 04	07:34:38	71	271	152	T	p-	-0.9526	1.0367	76.8S	46.2W	17	115	419	01m54s
047	2022 Apr 30	20:42:37	71	276	119	P	-t	-1.1901	0.6396	62.1S	71.5W	0	304		
048	2022 Oct 25	11:01:20	71	282	124	P	-t	1.0701	0.8619	61.6N	77.3E	0	244		
049	2023 Apr 20	04:17:56	71	288	129	H	-n	-0.3952	1.0132	9.6S	125.8E	67	334	49	01m16s
050	2023 Oct 14	18:00:41	71	294	134	A	-p	0.3753	0.9520	11.4N	83.1W	68	208	187	05m17s
051	2024 Apr 08	18:18:29	71	300	139	T	n-	0.3431	1.0566	25.3N	104.1W	70	149	198	04m28s
052	2024 Oct 02	18:46:13	72	306	144	A	p-	-0.3509	0.9326	22.0S	114.5W	69	31	266	07m25s
053	2025 Mar 29	10:48:36	72	312	149	P	t-	1.0405	0.9376	61.1N	77.1W	0	83		
054	2025 Sep 21	19:43:04	72	318	154	P	t-	-1.0651	0.8550	60.9S	153.5E	0	89		
055	2026 Feb 17	12:13:06	72	323	121	A	-t	-0.9743	0.9630	64.7S	86.7E	12	268	616	02m20s
056	2026 Aug 12	17:47:06	72	329	126	T	-p	0.8977	1.0386	65.2N	25.2W	26	248	294	02m18s
057	2027 Feb 06	16:00:48	73	335	131	A	-n	-0.2952	0.9281	31.3S	48.5W	73	334	282	07m51s
058	2027 Aug 02	10:07:50	73	341	136	T	nn	0.1421	1.0790	25.5N	33.2E	82	202	258	06m23s
059	2028 Jan 26	15:08:59	73	347	141	A	p-	0.3901	0.9208	3.0N	51.6W	67	161	323	10m27s
060	2028 Jul 22	02:56:40	73	353	146	T	p-	-0.6056	1.0560	15.6S	126.7E	53	17	230	05m10s

Cat Num	Calendar Date	TD of Greatest Eclipse	ΔT s	Luna Num	Saros Num	Ecl Type	QLE	Gamma	Ecl Mag	Lat °	Long °	Sun Alt °	Sun Azm °	Path Width km	Central Line Dur
061	2029 Jan 14	17:13:48	73	359	151	P	t-	1.0553	0.8714	63.7N	114.2W	0	145		
062	2029 Jun 12	04:06:13	74	364	118	P	-t	1.2943	0.4576	66.8N	66.2W	0	355		
063	2029 Jul 11	15:37:19	74	365	156	P	t-	-1.4191	0.2303	64.3S	85.6W	0	30		
064	2029 Dec 05	15:03:58	74	370	123	P	-t	-1.0609	0.8911	67.5S	135.6E	0	177		
065	2030 Jun 01	06:29:13	74	376	128	A	-p	0.5626	0.9443	56.5N	80.1E	55	176	250	05m21s
066	2030 Nov 25	06:51:37	74	382	133	T	-n	-0.3867	1.0468	43.6S	71.2E	67	7	169	03m44s
067	2031 May 21	07:16:04	74	388	138	A	nn	-0.1970	0.9589	8.9N	71.7E	79	354	152	05m26s
068	2031 Nov 14	21:07:31	75	394	143	H	n-	0.3078	1.0106	0.6S	137.6W	72	189	38	01m08s
069	2032 May 09	13:26:42	75	400	148	A	t-	-0.9375	0.9957	51.3S	7.1W	20	345	44	00m22s
070	2032 Nov 03	05:34:13	75	406	153	P	t-	1.0643	0.8554	70.4N	132.6E	0	218		
071	2033 Mar 30	18:02:36	75	411	120	T	-t	0.9778	1.0462	71.3N	155.8W	11	111	781	02m37s
072	2033 Sep 23	13:54:31	76	417	125	P	-t	-1.1583	0.6890	72.2S	121.3W	0	91		
073	2034 Mar 20	10:18:45	76	423	130	T	-n	0.2894	1.0458	16.1N	22.2E	73	162	159	04m09s
074	2034 Sep 12	16:19:28	76	429	135	A	-p	-0.3936	0.9736	18.2S	72.6W	67	18	102	02m58s
075	2035 Mar 09	23:05:54	76	435	140	A	n-	-0.4368	0.9919	29.0S	155.0W	64	340	31	00m48s
076	2035 Sep 02	01:56:46	76	441	145	T	p-	0.3727	1.0320	29.1N	158.0E	68	199	116	02m54s
077	2036 Feb 27	04:46:49	77	447	150	P	t-	-1.1942	0.6286	71.6S	131.5W	0	242		
078	2036 Jul 23	10:32:06	77	452	117	P	-t	-1.4250	0.1992	68.9S	3.5E	0	19		
079	2036 Aug 21	17:25:45	77	453	155	P	t-	1.0825	0.8622	71.1N	47.0E	0	309		
080	2037 Jan 16	09:48:55	77	458	122	P	-t	1.1477	0.7049	68.5N	20.8E	0	166		
081	2037 Jul 13	02:40:36	77	464	127	T	-p	-0.7246	1.0413	24.8S	139.1E	43	3	201	03m58s
082	2038 Jan 05	13:47:11	78	470	132	A	-n	0.4169	0.9728	2.1N	25.5W	65	179	107	03m18s
083	2038 Jul 02	13:32:55	78	476	137	A	nn	0.0398	0.9911	25.4N	21.9W	88	179	31	01m00s
084	2038 Dec 26	01:00:10	78	482	142	T	n-	-0.2881	1.0269	40.3S	163.9E	73	5	95	02m18s
085	2039 Jun 21	17:12:54	78	488	147	A	p-	0.8312	0.9454	78.9N	102.1W	33	153	365	04m05s
086	2039 Dec 15	16:23:46	79	494	152	T	p-	-0.9458	1.0356	80.9S	172.7E	18	123	380	01m51s
087	2040 May 11	03:43:02	79	499	119	P	-t	-1.2529	0.5306	62.8S	174.4E	0	313		
088	2040 Nov 04	19:09:02	79	505	124	P	-t	1.0993	0.8074	62.2N	53.4W	0	234		
089	2041 Apr 30	11:52:21	79	511	129	T	-p	-0.4492	1.0189	9.6S	12.2E	63	337	72	01m51s
090	2041 Oct 25	01:36:22	80	517	134	A	-p	0.4133	0.9467	9.9N	162.8E	66	206	213	06m07s
091	2042 Apr 20	02:17:30	80	523	139	T	n-	0.2956	1.0614	27.0N	137.3E	73	151	210	04m51s
092	2042 Oct 14	02:00:42	80	529	144	A	n-	-0.3030	0.9301	23.7S	137.8E	72	30	273	07m44s
093	2043 Apr 09	18:57:49	80	535	149	T+	t-	1.0031	1.0096	61.3N	151.9E	0	74	-	-
094	2043 Oct 03	03:01:49	81	541	154	A-	t-	-1.0102	0.9497	61.0S	35.2E	0	98	-	-
095	2044 Feb 28	20:24:40	81	546	121	As	-t	-0.9954	0.9600	62.2S	25.6W	4	260	-	02m27s
096	2044 Aug 23	01:17:02	81	552	126	T	-t	0.9613	1.0364	64.3N	120.5W	15	264	453	02m04s
097	2045 Feb 16	23:56:07	81	558	131	A	-n	-0.3125	0.9285	28.3S	166.2W	72	331	281	07m47s
098	2045 Aug 12	17:42:39	82	564	136	T	-n	0.2116	1.0774	25.9N	78.6W	78	206	256	06m06s
099	2046 Feb 05	23:06:26	82	570	141	A	p-	0.3765	0.9232	4.8N	171.4W	68	157	310	09m42s
100	2046 Aug 02	10:21:13	82	576	146	T	p-	-0.5350	1.0531	12.7S	15.1E	58	21	206	04m51s
101	2047 Jan 26	01:33:18	82	582	151	P	t-	1.0450	0.8908	62.9N	111.7E	0	135		
102	2047 Jun 23	10:52:31	83	587	118	P	-t	1.3766	0.3129	65.8N	178.0W	0	346		
103	2047 Jul 22	22:36:17	83	588	156	P	t-	-1.3477	0.3605	63.4S	160.1E	0	40		
104	2047 Dec 16	23:50:12	83	593	123	P	-t	-1.0661	0.8817	66.4S	6.6W	0	188		
105	2048 Jun 11	12:58:53	83	599	128	A	-p	0.6468	0.9441	63.7N	11.5W	49	184	271	04m58s
106	2048 Dec 05	15:35:27	83	605	133	T	-n	-0.3973	1.0440	46.1S	56.4W	66	1	160	03m28s
107	2049 May 31	13:59:59	84	611	138	A	nn	-0.1187	0.9631	15.3N	29.9W	83	358	134	04m45s
108	2049 Nov 25	05:33:48	84	617	143	H	n-	0.2943	1.0057	3.8S	95.2E	73	185	21	00m38s
109	2050 May 20	20:42:50	84	623	148	H	t-	-0.8688	1.0038	40.1S	123.8W	29	352	27	00m21s
110	2050 Nov 14	13:30:53	85	629	153	P	t-	1.0447	0.8874	69.5N	1.0E	0	206		
111	2051 Apr 11	02:10:39	85	634	120	P	-t	1.0169	0.9849	71.6N	32.1E	0	63		
112	2051 Oct 04	21:02:15	85	640	125	P	-t	-1.2094	0.6024	72.0S	117.7E	0	105		
113	2052 Mar 30	18:31:53	85	646	130	T	-n	0.3239	1.0466	22.4N	102.6W	71	161	164	04m08s
114	2052 Sep 22	23:39:10	86	652	135	A	-p	-0.4480	0.9734	25.7S	174.9E	63	20	106	02m51s
115	2053 Mar 20	07:08:19	86	658	140	A	n-	-0.4089	0.9919	23.0S	82.9E	66	341	31	00m50s
116	2053 Sep 12	09:34:09	86	664	145	T	n-	0.3140	1.0328	21.5N	41.7E	72	199	116	03m04s
117	2054 Mar 09	12:33:40	87	670	150	P	t-	-1.1711	0.6678	72.0S	97.9E	0	256		
118	2054 Aug 03	18:04:02	87	675	117	Pe	-t	-1.4941	0.0656	69.8S	121.4W	0	31		
119	2054 Sep 02	01:09:34	87	676	155	P	t-	1.0215	0.9793	71.7N	82.4W	0	296		
120	2055 Jan 27	17:54:05	87	681	122	P	-t	1.1550	0.6932	69.5N	112.3W	0	154		

Cat Num	Calendar Date	TD of Greatest Eclipse	ΔT s	Luna Num	Saros Num	Ecl Type	QLE	Gamma	Ecl Mag	Lat °	Long °	Sun Alt °	Sun Azm °	Path Width km	Central Line Dur
121	2055 Jul 24	09:57:50	88	687	127	T	-p	-0.8012	1.0359	33.3S	25.7E	37	8	202	03m17s
122	2056 Jan 16	22:16:45	88	693	132	A	-n	0.4199	0.9760	3.9N	153.6W	65	175	95	02m52s
123	2056 Jul 12	20:21:59	88	699	137	A	nn	-0.0426	0.9878	19.4N	123.8W	88	3	43	01m26s
124	2057 Jan 05	09:47:52	88	705	142	T	n-	-0.2837	1.0287	39.2S	35.1E	73	359	102	02m29s
125	2057 Jul 01	23:40:15	89	711	147	A	p-	0.7455	0.9464	71.5N	176.3W	41	177	298	04m22s
126	2057 Dec 26	01:14:35	89	717	152	T	p-	-0.9405	1.0348	84.9S	21.7E	19	141	355	01m50s
127	2058 May 22	10:39:26	89	722	119	P	-t	-1.3194	0.4141	63.5S	61.1E	0	322		
128	2058 Jun 21	00:19:35	89	723	157	Pb	t-	1.4869	0.1261	65.9N	9.8E	0	13		
129	2058 Nov 16	03:23:07	90	728	124	P	-t	1.1224	0.7644	62.9N	174.1E	0	225		
130	2059 May 11	19:22:16	90	734	129	T	-p	-0.5080	1.0242	10.7S	100.5W	59	340	95	02m23s
131	2059 Nov 05	09:18:15	90	740	134	A	-p	0.4454	0.9417	8.7N	47.0E	63	203	238	07m00s
132	2060 Apr 30	10:10:00	91	746	139	T	n-	0.2422	1.0660	28.0N	20.8E	76	154	222	05m15s
133	2060 Oct 24	09:24:10	91	752	144	A	nn	-0.2625	0.9277	25.8S	28.0E	75	28	281	08m06s
134	2061 Apr 20	02:56:49	91	758	149	T	t-	0.9578	1.0476	64.5N	59.1E	16	97	559	02m37s
135	2061 Oct 13	10:32:10	92	764	154	A	t-	-0.9639	0.9469	62.1S	54.5W	15	79	743	03m41s
136	2062 Mar 11	04:26:16	92	769	121	P	-t	-1.0238	0.9331	61.0S	147.2W	0	263		
137	2062 Sep 03	08:54:27	92	775	126	P	-t	1.0192	0.9749	61.3N	150.2E	0	286		
138	2063 Feb 28	07:43:30	93	781	131	A	-p	-0.3360	0.9293	25.2S	77.6E	70	329	280	07m41s
139	2063 Aug 24	01:22:11	93	787	136	T	-n	0.2771	1.0750	25.6N	168.3E	74	209	252	05m49s
140	2064 Feb 17	07:00:23	93	793	141	A	p-	0.3597	0.9262	7.0N	69.6E	69	154	295	08m56s
141	2064 Aug 12	17:46:06	94	799	146	T	p-	-0.4652	1.0495	10.9S	96.1W	62	24	184	04m28s
142	2065 Feb 05	09:52:26	94	805	151	P	t-	1.0336	0.9123	62.2N	22.0W	0	125		
143	2065 Jul 03	17:33:52	94	810	118	P	-t	1.4619	0.1639	64.8N	71.7E	0	336		
144	2065 Aug 02	05:34:17	94	811	156	P	t-	-1.2758	0.4903	62.7S	46.4E	0	49		
145	2065 Dec 27	08:39:56	95	816	123	P	-t	-1.0688	0.8769	65.4S	149.3W	0	198		
146	2066 Jun 22	19:25:48	95	822	128	A	-p	0.7330	0.9435	70.1N	96.5W	43	198	309	04m40s
147	2066 Dec 17	00:23:40	95	828	133	T	-n	-0.4043	1.0416	47.4S	175.6E	66	355	152	03m14s
148	2067 Jun 11	20:42:26	96	834	138	A	nn	-0.0387	0.9670	21.0N	130.3W	88	2	119	04m05s
149	2067 Dec 06	14:03:43	96	840	143	H	n-	0.2845	1.0011	6.0S	32.5W	74	181	4	00m08s
150	2068 May 31	03:56:39	96	846	148	T	p-	-0.7970	1.0110	31.0S	123.1E	37	357	63	01m06s
151	2068 Nov 24	21:32:30	97	852	153	P	t-	1.0299	0.9109	68.5N	131.2W	0	194		
152	2069 Apr 21	10:11:09	97	857	120	P	-t	1.0624	0.8992	71.0N	101.4W	0	50		
153	2069 May 20	17:53:18	97	858	158	Pb	t-	-1.4852	0.0879	68.8S	70.1W	0	342		
154	2069 Oct 15	04:19:56	97	863	125	P	-t	-1.2524	0.5298	71.6S	5.6W	0	119		
155	2070 Apr 11	02:36:09	98	869	130	T	-n	0.3652	1.0472	29.1N	134.9E	68	162	168	04m04s
156	2070 Oct 04	07:08:57	98	875	135	A	-p	-0.4950	0.9731	32.8S	60.2E	60	21	110	02m44s
157	2071 Mar 31	15:01:06	98	881	140	A	n-	-0.3739	0.9919	16.7S	37.2W	68	342	31	00m52s
158	2071 Sep 23	17:20:28	99	887	145	T	n-	0.2620	1.0333	14.2N	76.9W	75	198	116	03m11s
159	2072 Mar 19	20:10:31	99	893	150	P	t-	-1.1405	0.7199	72.2S	30.5W	0	270		
160	2072 Sep 12	08:59:20	100	899	155	T	t-	0.9655	1.0558	69.8N	101.8E	14	240	732	03m13s
161	2073 Feb 07	01:55:59	100	904	122	P	-t	1.1651	0.6768	70.5N	114.7E	0	141		
162	2073 Aug 03	17:15:23	100	910	127	T	-t	-0.8763	1.0294	43.2S	89.6W	28	14	206	02m29s
163	2074 Jan 27	06:44:15	101	916	132	A	-n	0.4251	0.9798	6.6N	78.6E	65	171	79	02m21s
164	2074 Jul 24	03:10:32	101	922	137	A	nn	-0.1242	0.9838	12.8N	133.5E	83	7	58	01m57s
165	2075 Jan 16	18:36:04	101	928	142	T	n-	-0.2799	1.0311	37.2S	94.3W	74	354	110	02m42s
166	2075 Jul 13	06:05:44	102	934	147	A	p-	0.6583	0.9467	63.1N	95.0E	49	186	262	04m45s
167	2076 Jan 06	10:07:28	102	940	152	T	p-	-0.9373	1.0342	87.2S	173.9W	20	203	340	01m49s
168	2076 Jun 01	17:31:22	102	945	119	P	-t	-1.3897	0.2897	64.4S	51.4W	0	331		
169	2076 Jul 01	06:50:43	103	946	157	P	t-	1.4005	0.2746	67.0N	98.3W	0	3		
170	2076 Nov 26	11:43:01	103	951	124	P	-t	1.1401	0.7315	63.7N	39.9E	0	215		
171	2077 May 22	02:46:05	103	957	129	T	-p	-0.5725	1.0290	13.1S	148.1E	55	343	119	02m54s
172	2077 Nov 15	17:07:56	104	963	134	A	-p	0.4705	0.9371	7.8N	71.0W	62	199	262	07m54s
173	2078 May 11	17:56:55	104	969	139	T	n-	0.1838	1.0701	28.1N	93.9W	79	158	232	05m40s
174	2078 Nov 04	16:55:44	104	975	144	A	nn	-0.2285	0.9255	27.8S	83.5W	77	25	287	08m29s
175	2079 May 01	10:50:13	105	981	149	T	p-	0.9081	1.0512	66.2N	46.5W	24	108	406	02m55s
176	2079 Oct 24	18:11:21	105	987	154	A	t-	-0.9243	0.9484	63.4S	160.8W	22	72	495	03m39s
177	2080 Mar 21	12:20:15	106	992	121	P	-t	-1.0578	0.8734	60.9S	85.7E	0	271		
178	2080 Sep 13	16:38:09	106	998	126	P	-t	1.0724	0.8743	61.1N	25.6E	0	277		
179	2081 Mar 10	15:23:31	106	1004	131	A	-p	-0.3653	0.9304	22.4S	36.9W	68	329	277	07m36s
180	2081 Sep 03	09:07:31	107	1010	136	T	-n	0.3379	1.0720	24.6N	53.4E	70	211	247	05m33s

Cat Num	Calendar Date	TD of Greatest Eclipse	ΔT s	Luna Num	Saros Num	Ecl Type	QLE	Gamma	Ecl Mag	Lat °	Long °	Sun Alt °	Sun Azm °	Path Width km	Central Line Dur
181	2082 Feb 27	14:47:00	107	1016	141	A	p-	0.3361	0.9298	9.4N	47.3W	70	152	277	08m12s
182	2082 Aug 24	01:16:21	108	1022	146	T	n-	-0.4004	1.0452	10.3S	151.5E	66	26	163	04m01s
183	2083 Feb 16	18:06:36	108	1028	151	P	t-	1.0170	0.9433	61.6N	154.3W	0	116		
184	2083 Jul 15	00:14:23	108	1033	118	Pe	-t	1.5464	0.0169	64.0N	37.9W	0	327		
185	2083 Aug 13	12:34:41	108	1034	156	P	t-	-1.2064	0.6146	62.1S	67.7W	0	58		
186	2084 Jan 07	17:30:24	109	1039	123	P	-t	-1.0715	0.8723	64.4S	68.3E	0	209		
187	2084 Jul 03	01:50:26	109	1045	128	A	-p	0.8208	0.9421	75.0N	169.3W	35	222	377	04m25s
188	2084 Dec 27	09:13:48	110	1051	133	T	-n	-0.4094	1.0396	47.3S	47.4E	66	349	146	03m04s
189	2085 Jun 22	03:21:16	110	1057	138	A	nn	0.0453	0.9704	26.2N	131.0E	87	186	106	03m29s
190	2085 Dec 16	22:37:48	111	1063	143	A	n-	0.2786	0.9971	7.3S	161.1W	74	176	10	00m19s
191	2086 Jun 11	11:07:14	111	1069	148	T	p-	-0.7215	1.0174	23.2S	12.2E	44	2	86	01m48s
192	2086 Dec 06	05:38:55	111	1075	153	P	p-	1.0194	0.9271	67.4N	96.0E	0	182		
193	2087 May 02	18:04:42	112	1080	120	P	-t	1.1139	0.8011	70.3N	127.3E	0	37		
194	2087 Jun 01	01:27:14	112	1081	158	P	t-	-1.4186	0.2146	67.8S	165.1E	0	354		
195	2087 Oct 26	11:46:57	112	1086	125	P	-t	-1.2882	0.4696	71.0S	130.8W	0	132		
196	2088 Apr 21	10:31:49	113	1092	130	T	-p	0.4135	1.0474	36.0N	14.8E	65	163	173	03m58s
197	2088 Oct 14	14:48:05	113	1098	135	A	-p	-0.5349	0.9727	39.7S	56.3W	57	21	115	02m38s
198	2089 Apr 10	22:44:42	113	1104	140	A	n-	-0.3319	0.9919	10.2S	155.0W	71	344	30	00m53s
199	2089 Oct 04	01:15:23	114	1110	145	T	n-	0.2167	1.0333	7.4N	162.5E	77	197	115	03m14s
200	2090 Mar 31	03:38:08	114	1116	150	P	t-	-1.1028	0.7843	72.1S	156.6W	0	284		
201	2090 Sep 23	16:56:36	115	1122	155	T	t-	0.9157	1.0562	60.7N	40.8W	23	218	463	03m36s
202	2091 Feb 18	09:54:40	115	1127	122	P	-t	1.1779	0.6558	71.2N	18.0W	0	128		
203	2091 Aug 15	00:34:43	116	1133	127	T	-t	-0.9490	1.0216	55.6S	150.2E	18	23	237	01m38s
204	2092 Feb 07	15:10:20	116	1139	132	A	-n	0.4322	0.9840	9.9N	49.0W	64	168	62	01m48s
205	2092 Aug 03	09:59:33	116	1145	137	A	nn	-0.2044	0.9794	5.6N	30.0E	78	10	75	02m31s
206	2093 Jan 27	03:22:16	117	1151	142	T	n-	-0.2737	1.0340	34.1S	136.1E	74	350	119	02m58s
207	2093 Jul 23	12:32:04	117	1157	147	A	p-	0.5717	0.9463	54.6N	1.0E	55	191	241	05m11s
208	2094 Jan 16	18:59:03	118	1163	152	T	p-	-0.9333	1.0342	84.8S	10.9W	21	267	329	01m52s
209	2094 Jun 13	00:22:11	118	1168	119	P	-t	-1.4613	0.1618	65.3S	163.9W	0	341		
210	2094 Jul 12	13:24:35	118	1169	157	P	t-	1.3149	0.4225	68.0N	152.5E	0	352		
211	2094 Dec 07	20:05:56	119	1174	124	P	-t	1.1547	0.7046	64.7N	95.3W	0	205		
212	2095 Jun 02	10:07:40	119	1180	129	T	-p	-0.6396	1.0332	16.7S	36.9E	50	347	145	03m18s
213	2095 Nov 27	01:02:57	120	1186	134	A	-p	0.4903	0.9330	7.2N	169.5E	61	195	285	08m47s
214	2096 May 22	01:37:14	120	1192	139	T	nn	0.1196	1.0737	27.3N	153.1E	83	162	241	06m06s
215	2096 Nov 15	00:36:15	121	1198	144	A	nn	-0.2018	0.9237	29.7S	163.0E	78	22	294	08m53s
216	2097 May 11	18:34:31	121	1204	149	T	p-	0.8516	1.0538	67.4N	149.8W	31	121	339	03m10s
217	2097 Nov 04	02:01:25	121	1210	154	A	t-	-0.8926	0.9494	65.8S	86.5E	26	68	411	03m36s
218	2098 Apr 01	20:02:31	122	1215	121	P	-t	-1.1005	0.7984	61.0S	38.4W	0	280		
219	2098 Sep 25	00:31:16	122	1221	126	P	-t	1.1184	0.7871	61.1N	101.3W	0	268		
220	2098 Oct 24	10:36:11	122	1222	164	Pb	t-	-1.5407	0.0057	61.8S	95.8W	0	116		
221	2099 Mar 21	22:54:32	123	1227	131	A	-p	-0.4016	0.9318	20.0S	149.4W	66	329	275	07m32s
222	2099 Sep 14	16:57:53	123	1233	136	T	-n	0.3942	1.0684	23.4N	63.1W	67	211	241	05m18s
223	2100 Mar 10	22:28:11	124	1239	141	A	n-	0.3077	0.9338	12.0N	162.8W	72	151	257	07m29s
224	2100 Sep 04	08:49:20	124	1245	146	T	n-	-0.3384	1.0402	10.5S	38.7E	70	28	142	03m32s

Appendix B

Small Global Solar Eclipse Maps: 2001 to 2100

Key to Small Global Solar Eclipse Maps

Figure 5—1: Key to Small Global Solar Eclipse Maps (Appendix B)

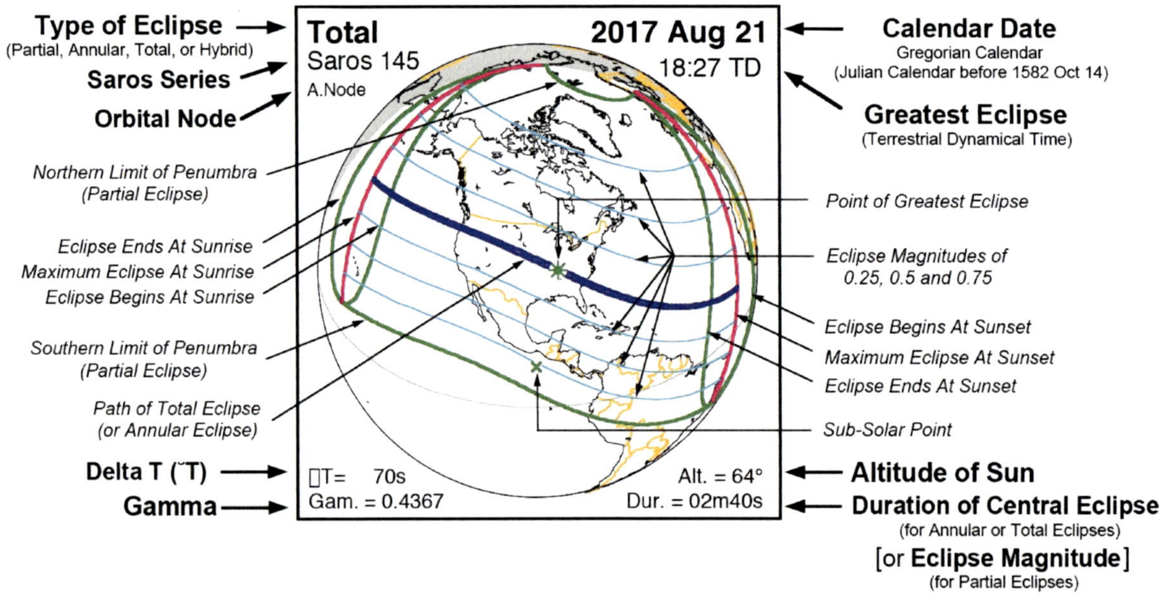

Solar Eclipse Type – One of four basic types of solar eclipses: Partial, Annular Total or Hybrid (See Sect. 4.1).

Saros Series – The Saros series that the eclipse belongs to. Eclipses with an odd number take place at the Moon's ascending node, while those with an even number are at the descending node (See Sect. 4.7).

Node – The orbital node near which the eclipse takes place. The ascending node (A. Node) is the point where the Moon travels from south to north through Earth's orbital plane. Similarly, the descending node (D. Node) is the point where the Moon travels from north to south.

Calendar Date – Gregorian calendar date at greatest eclipse is given in the ISO order of YEAR-MONTH-DAY.

Greatest Eclipse – The instant (Terrestrial Dynamical Time) when the distance between the axis of the Moon's shadow cone and the center of Earth reaches a minimum (See Sect. 4.4).

ΔT (Delta T) – the arithmetic difference, in seconds, between Terrestrial Dynamical Time (TD) and Universal Time (UT1), (See Sect. 2.5).

Gamma – The minimum distance from the lunar shadow axis to the center of Earth, in units of Earth's equatorial radius (See Sect. 4.10).

Altitude of Sun – The Sun's altitude at the geographic position intersected by the lunar shadow axis is given at the instant of greatest eclipse (See Sect. 4.13).

Duration of Central Eclipse – The central line duration of the total or annular phase (in minutes and seconds) is given at the instant of greatest eclipse (See Sect. 4.16).

Eclipse Magnitude – The fraction of the Sun's diameter occulted by the Moon at the instant of greatest eclipse (See Sect. 4.11).

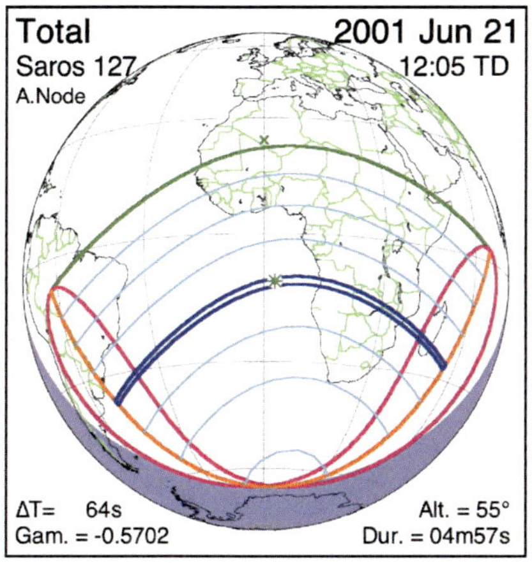
Total
2001 Jun 21
Saros 127
12:05 TD
A.Node
ΔT= 64s
Alt. = 55°
Gam. = -0.5702
Dur. = 04m57s

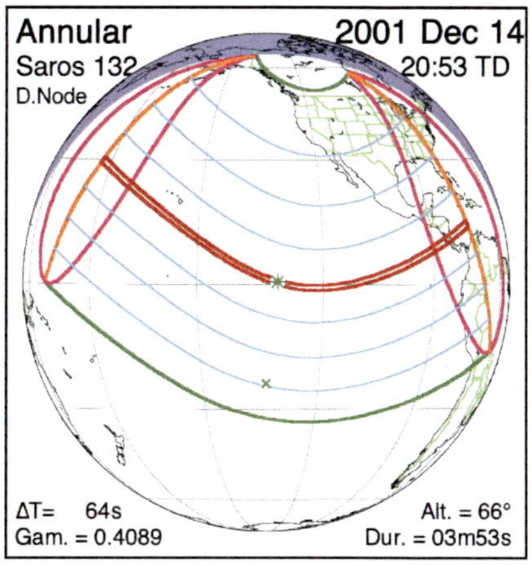
Annular
2001 Dec 14
Saros 132
20:53 TD
D.Node
ΔT= 64s
Alt. = 66°
Gam. = 0.4089
Dur. = 03m53s

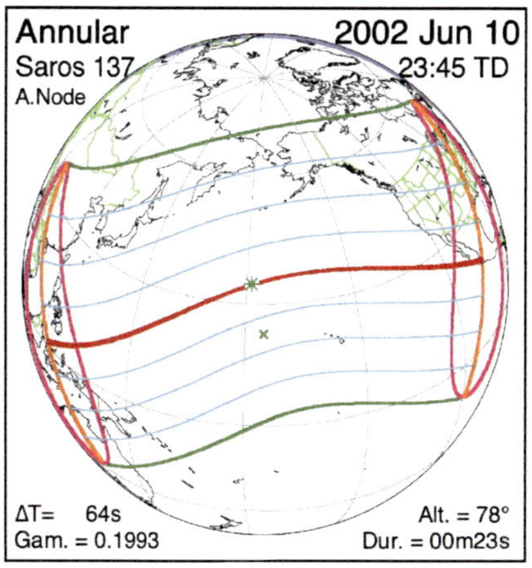
Annular
2002 Jun 10
Saros 137
23:45 TD
A.Node
ΔT= 64s
Alt. = 78°
Gam. = 0.1993
Dur. = 00m23s

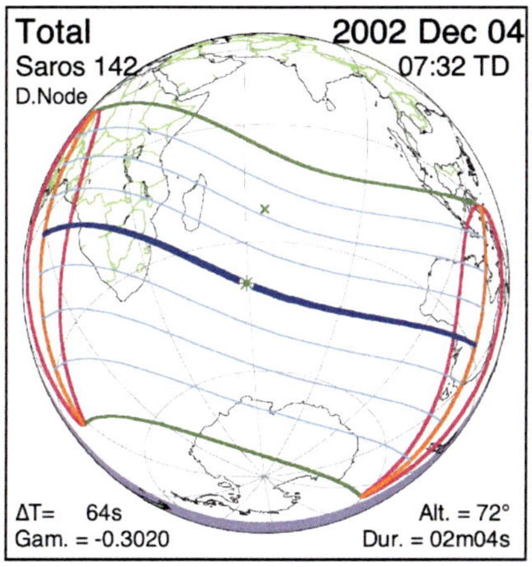
Total
2002 Dec 04
Saros 142
07:32 TD
D.Node
ΔT= 64s
Alt. = 72°
Gam. = -0.3020
Dur. = 02m04s

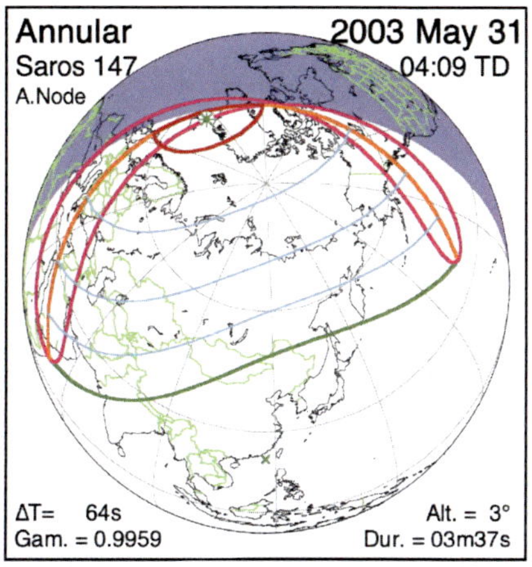
Annular
2003 May 31
Saros 147
04:09 TD
A.Node
ΔT= 64s
Alt. = 3°
Gam. = 0.9959
Dur. = 03m37s

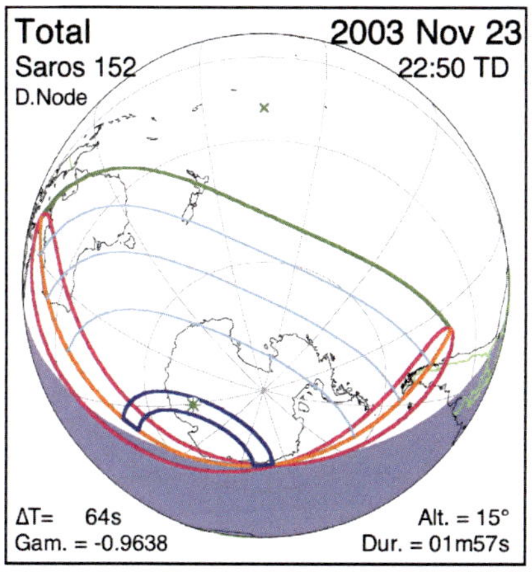
Total
2003 Nov 23
Saros 152
22:50 TD
D.Node
ΔT= 64s
Alt. = 15°
Gam. = -0.9638
Dur. = 01m57s

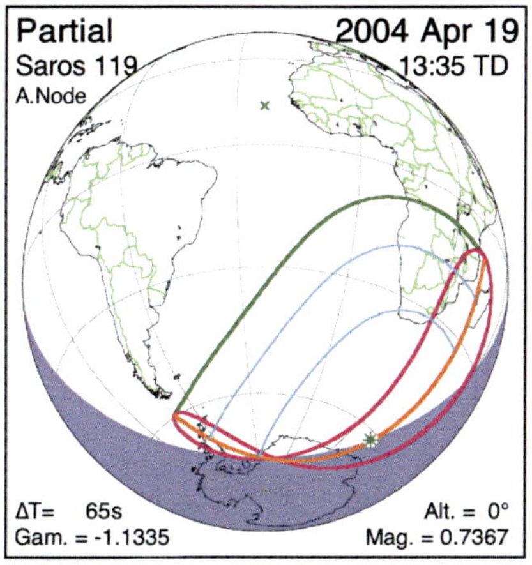
Partial
2004 Apr 19
Saros 119
13:35 TD
A.Node
ΔT= 65s
Alt. = 0°
Gam. = -1.1335
Mag. = 0.7367

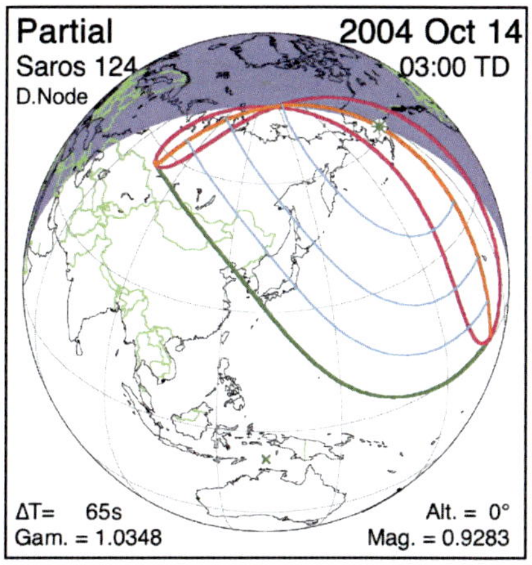
Partial
2004 Oct 14
Saros 124
03:00 TD
D.Node
ΔT= 65s
Alt. = 0°
Gam. = 1.0348
Mag. = 0.9283

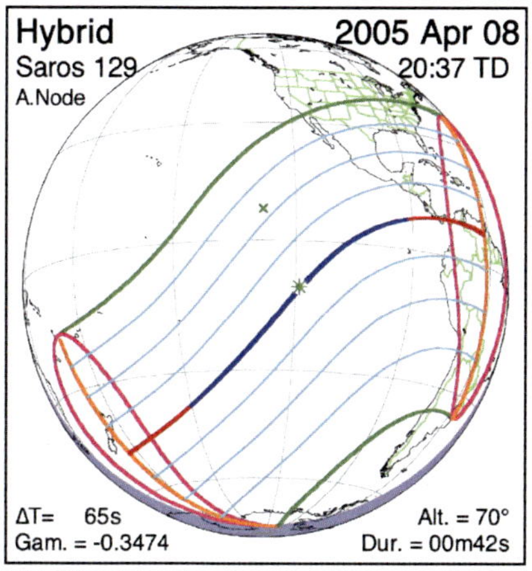
Hybrid
2005 Apr 08
Saros 129
20:37 TD
A.Node
ΔT= 65s
Alt. = 70°
Gam. = -0.3474
Dur. = 00m42s

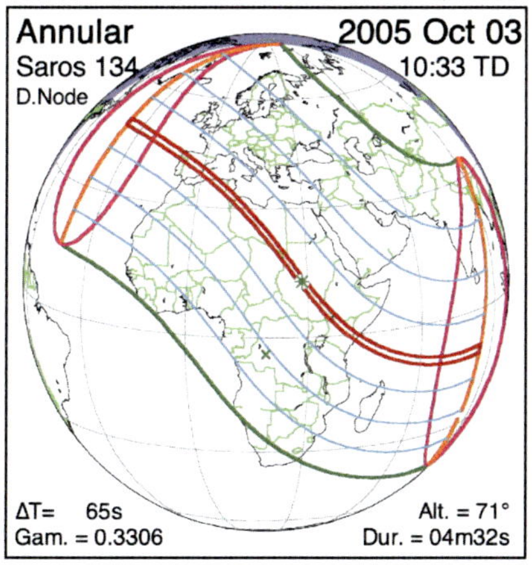
Annular
2005 Oct 03
Saros 134
10:33 TD
D.Node
ΔT= 65s
Alt. = 71°
Gam. = 0.3306
Dur. = 04m32s

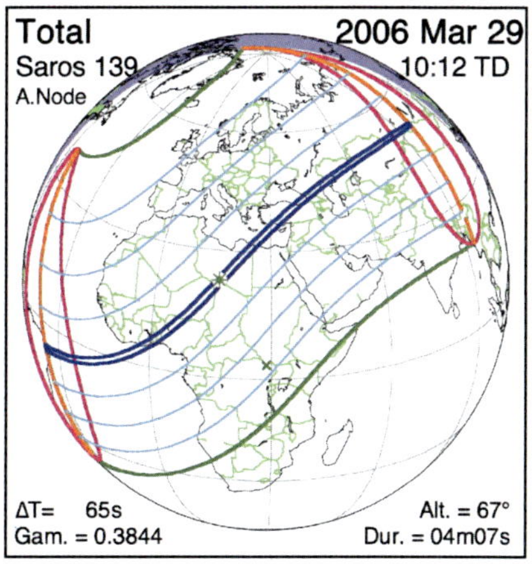
Total
2006 Mar 29
Saros 139
10:12 TD
A.Node
ΔT= 65s
Alt. = 67°
Gam. = 0.3844
Dur. = 04m07s

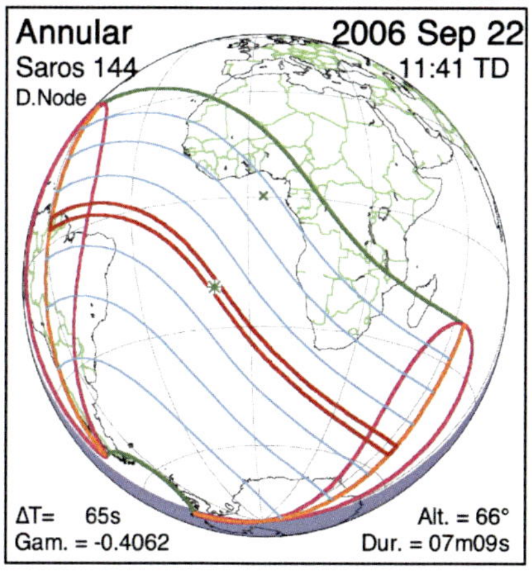
Annular
2006 Sep 22
Saros 144
11:41 TD
D.Node
ΔT= 65s
Alt. = 66°
Gam. = -0.4062
Dur. = 07m09s

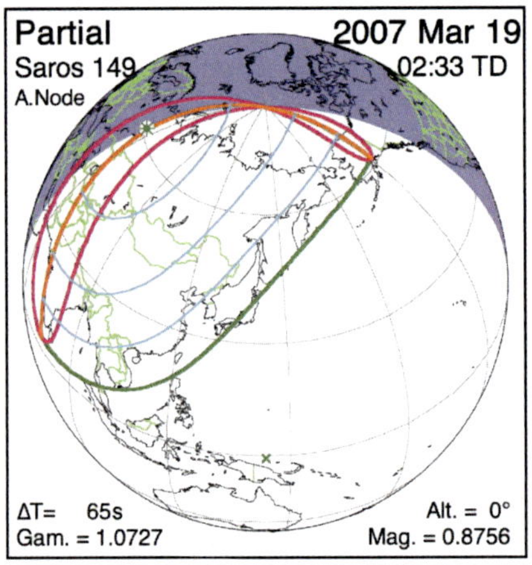
Partial
2007 Mar 19
Saros 149
02:33 TD
A.Node
ΔT= 65s
Alt. = 0°
Gam. = 1.0727
Mag. = 0.8756

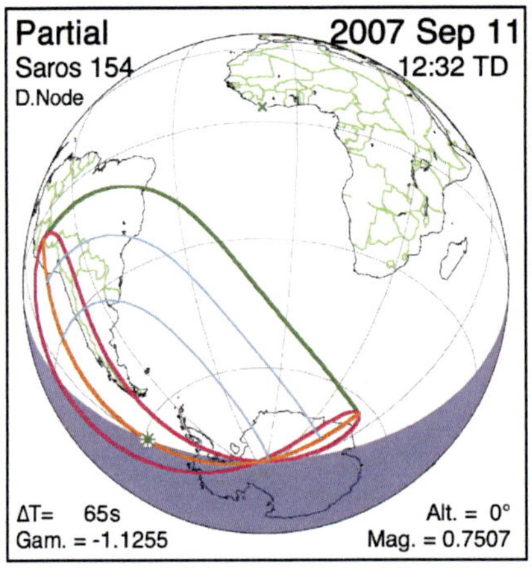
Partial
2007 Sep 11
Saros 154
12:32 TD
D.Node
ΔT= 65s
Alt. = 0°
Gam. = -1.1255
Mag. = 0.7507

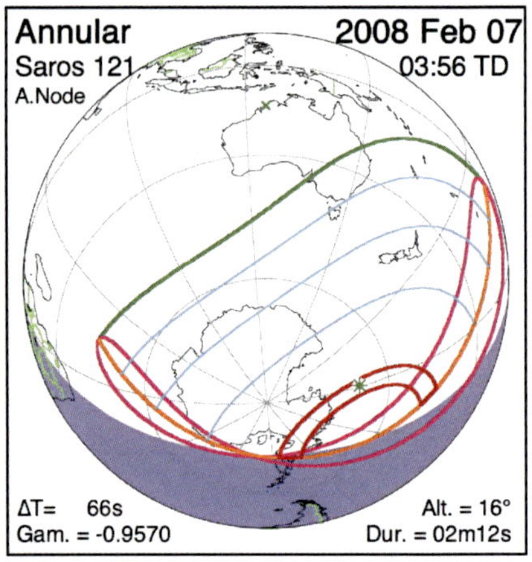
Annular
2008 Feb 07
Saros 121
03:56 TD
A.Node
ΔT= 66s
Alt. = 16°
Gam. = -0.9570
Dur. = 02m12s

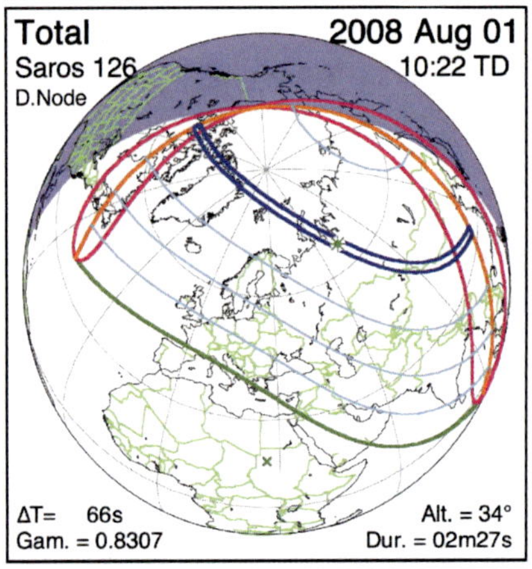
Total
2008 Aug 01
Saros 126
10:22 TD
D.Node
ΔT= 66s
Alt. = 34°
Gam. = 0.8307
Dur. = 02m27s

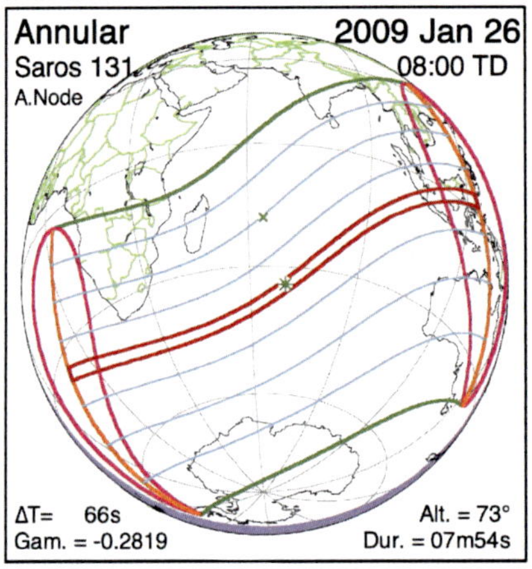
Annular
2009 Jan 26
Saros 131
08:00 TD
A.Node
ΔT= 66s
Alt. = 73°
Gam. = -0.2819
Dur. = 07m54s

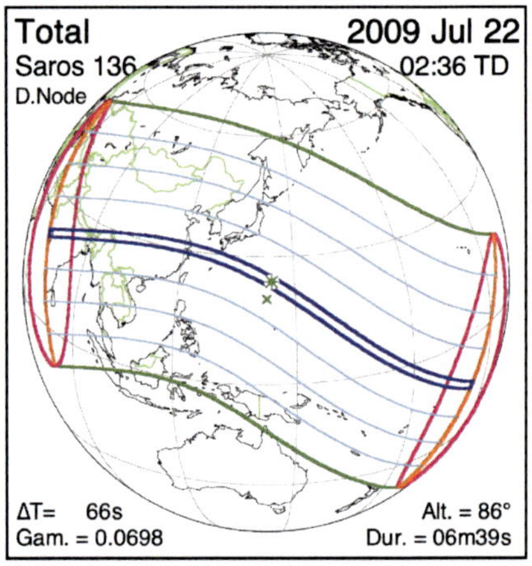
Total
2009 Jul 22
Saros 136
02:36 TD
D.Node
ΔT= 66s
Alt. = 86°
Gam. = 0.0698
Dur. = 06m39s

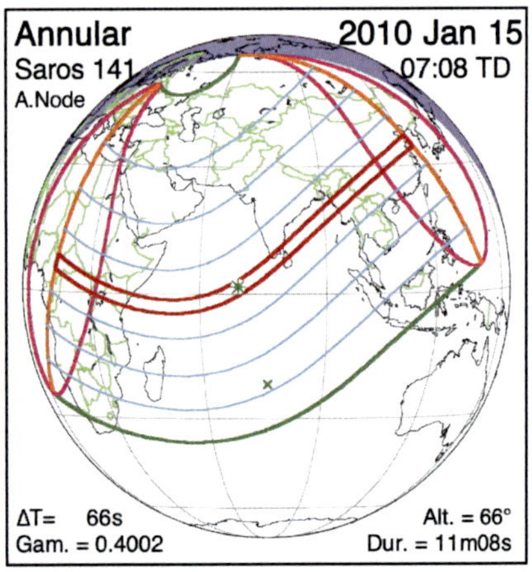
Annular
2010 Jan 15
Saros 141
07:08 TD
A.Node
ΔT= 66s
Alt. = 66°
Gam. = 0.4002
Dur. = 11m08s

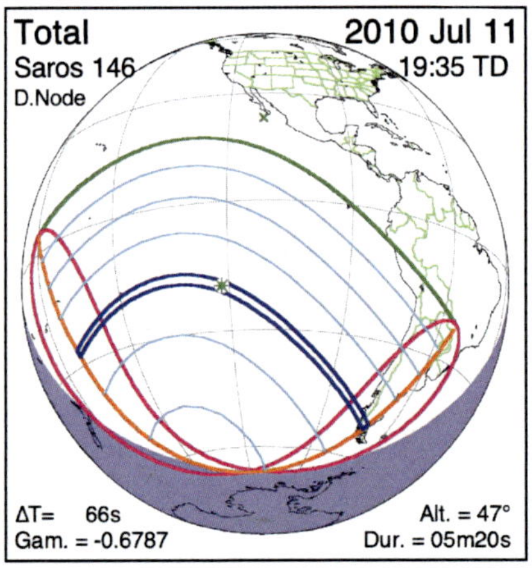
Total
2010 Jul 11
Saros 146
19:35 TD
D.Node
ΔT= 66s
Alt. = 47°
Gam. = -0.6787
Dur. = 05m20s

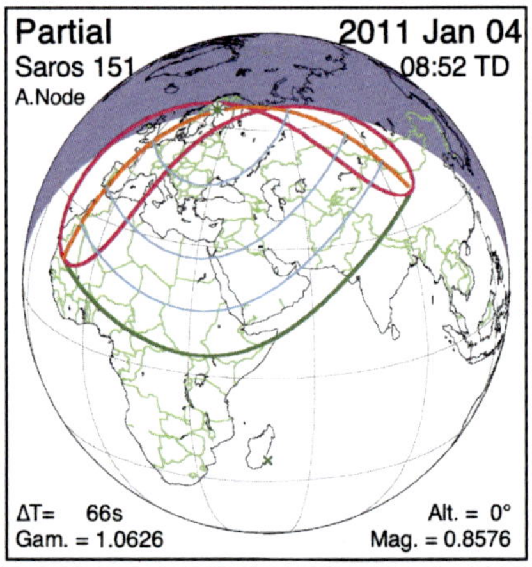
Partial
2011 Jan 04
Saros 151
08:52 TD
A.Node
ΔT= 66s
Alt. = 0°
Gam. = 1.0626
Mag. = 0.8576

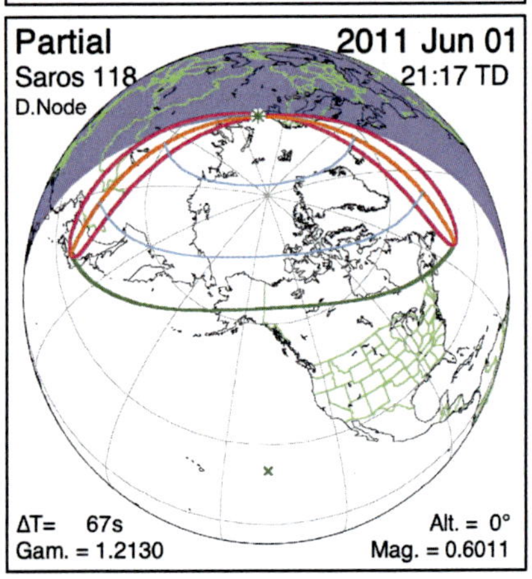
Partial
2011 Jun 01
Saros 118
21:17 TD
D.Node
ΔT= 67s
Alt. = 0°
Gam. = 1.2130
Mag. = 0.6011

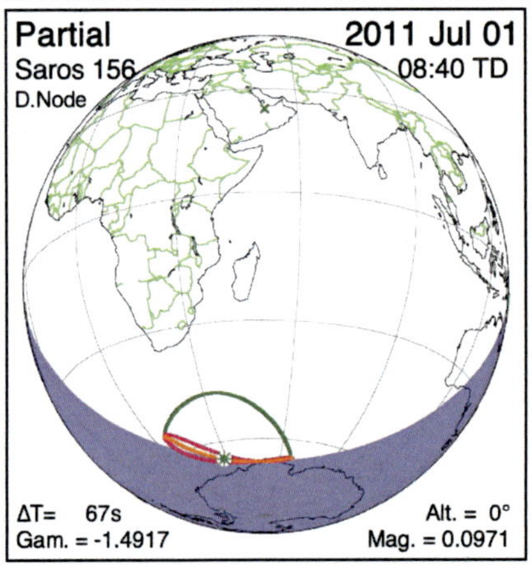
Partial
2011 Jul 01
Saros 156
08:40 TD
D.Node
ΔT= 67s
Alt. = 0°
Gam. = -1.4917
Mag. = 0.0971

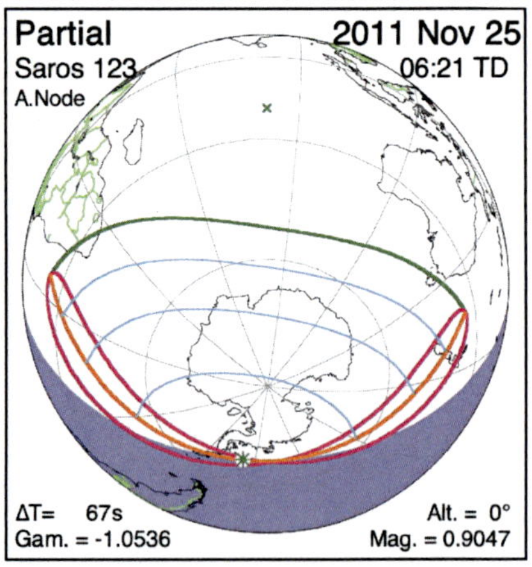
Partial
2011 Nov 25
Saros 123
06:21 TD
A.Node
ΔT= 67s
Alt. = 0°
Gam. = -1.0536
Mag. = 0.9047

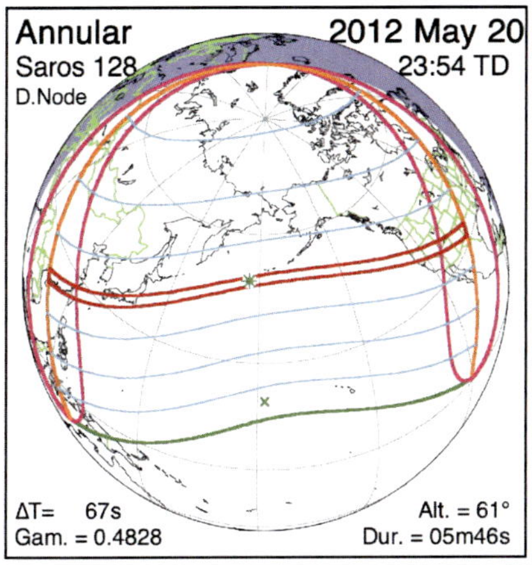
Annular
2012 May 20
Saros 128
23:54 TD
D.Node
ΔT= 67s
Gam. = 0.4828
Alt. = 61°
Dur. = 05m46s

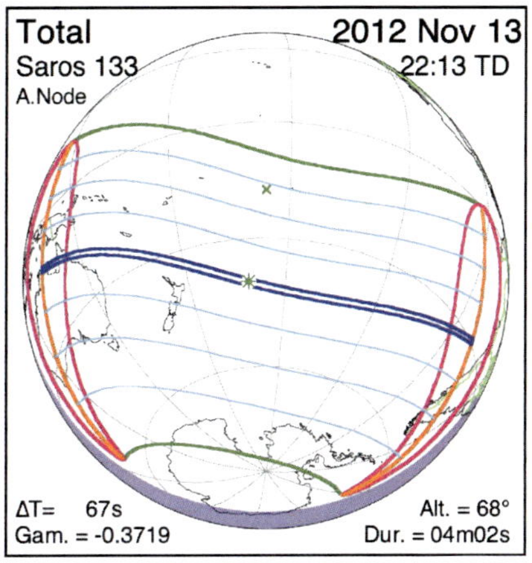
Total
2012 Nov 13
Saros 133
22:13 TD
A.Node
ΔT= 67s
Gam. = -0.3719
Alt. = 68°
Dur. = 04m02s

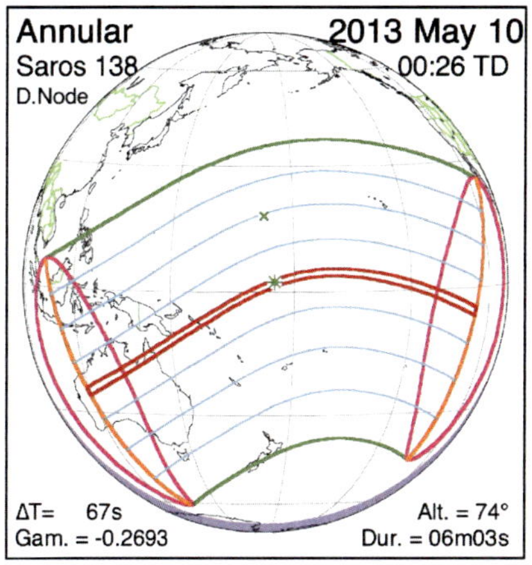
Annular
2013 May 10
Saros 138
00:26 TD
D.Node
ΔT= 67s
Gam. = -0.2693
Alt. = 74°
Dur. = 06m03s

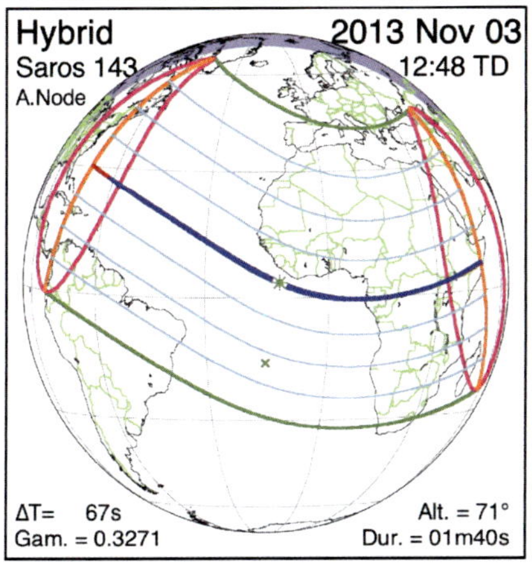
Hybrid
2013 Nov 03
Saros 143
12:48 TD
A.Node
ΔT= 67s
Gam. = 0.3271
Alt. = 71°
Dur. = 01m40s

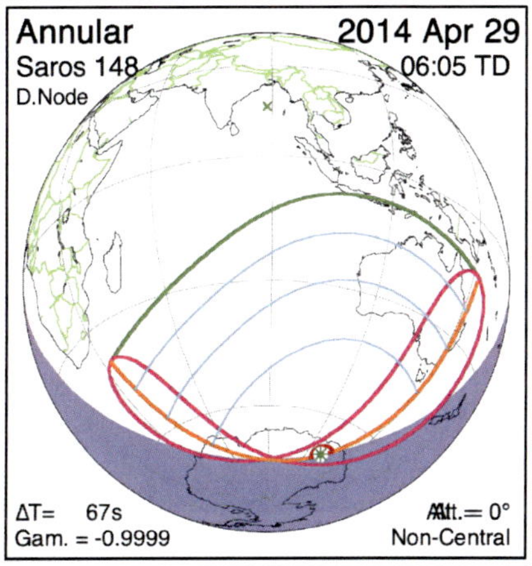
Annular
2014 Apr 29
Saros 148
06:05 TD
D.Node
ΔT= 67s
Gam. = -0.9999
Alt. = 0°
Non-Central

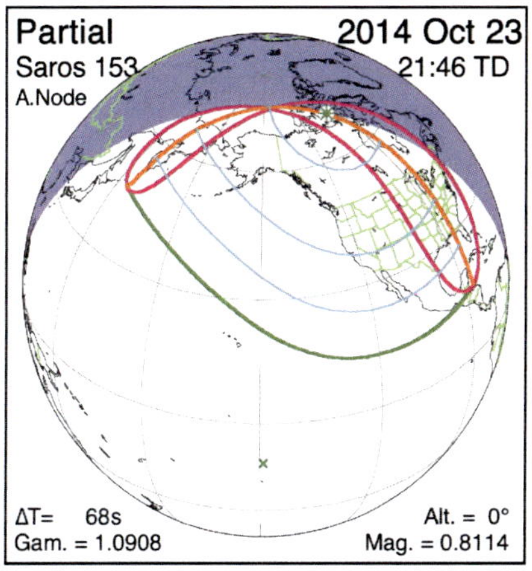
Partial
2014 Oct 23
Saros 153
21:46 TD
A.Node
ΔT= 68s
Gam. = 1.0908
Alt. = 0°
Mag. = 0.8114

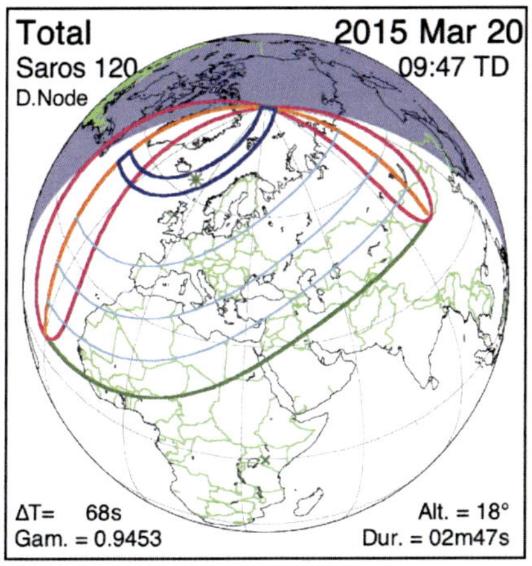
Total
2015 Mar 20
Saros 120
09:47 TD
D.Node
ΔT= 68s
Gam. = 0.9453
Alt. = 18°
Dur. = 02m47s

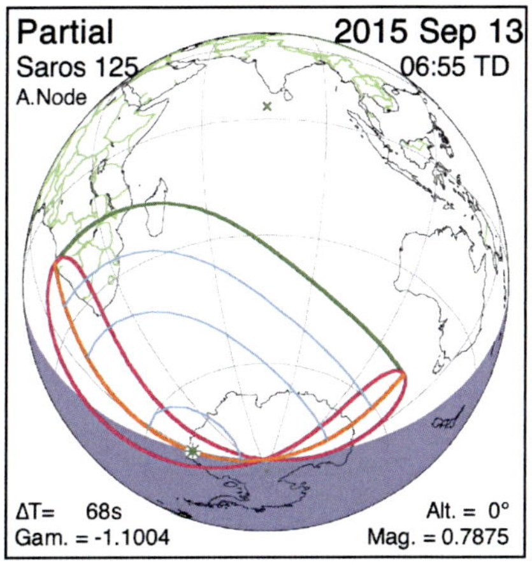
Partial
2015 Sep 13
Saros 125
06:55 TD
A.Node
ΔT= 68s
Gam. = -1.1004
Alt. = 0°
Mag. = 0.7875

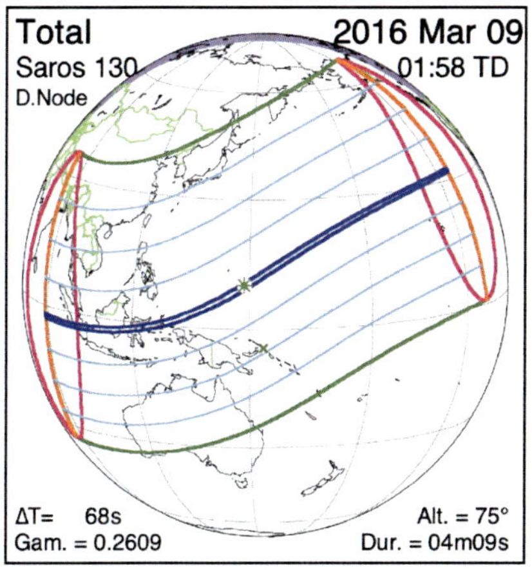
Total
2016 Mar 09
Saros 130
01:58 TD
D.Node
ΔT= 68s
Gam. = 0.2609
Alt. = 75°
Dur. = 04m09s

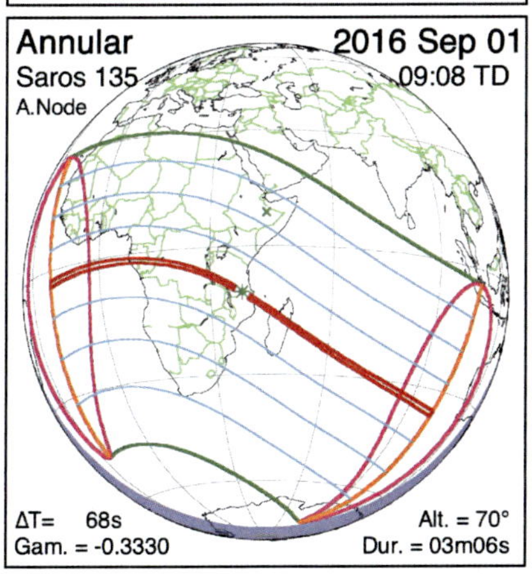
Annular
2016 Sep 01
Saros 135
09:08 TD
A.Node
ΔT= 68s
Gam. = -0.3330
Alt. = 70°
Dur. = 03m06s

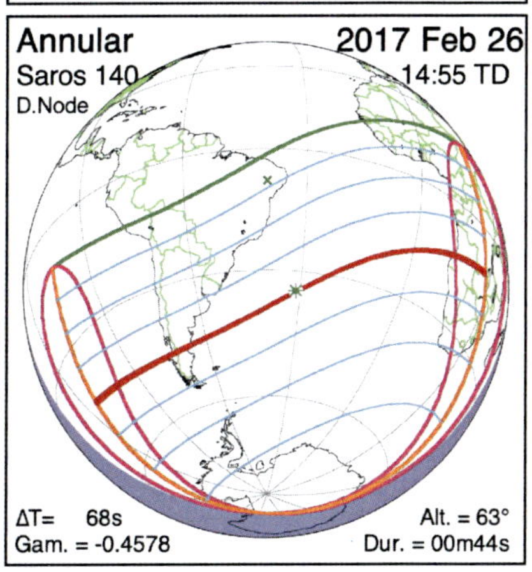
Annular
2017 Feb 26
Saros 140
14:55 TD
D.Node
ΔT= 68s
Gam. = -0.4578
Alt. = 63°
Dur. = 00m44s

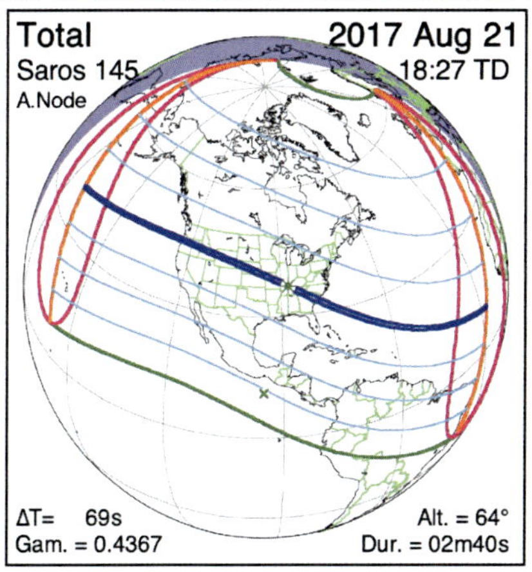
Total
2017 Aug 21
Saros 145
18:27 TD
A.Node
ΔT= 69s
Gam. = 0.4367
Alt. = 64°
Dur. = 02m40s

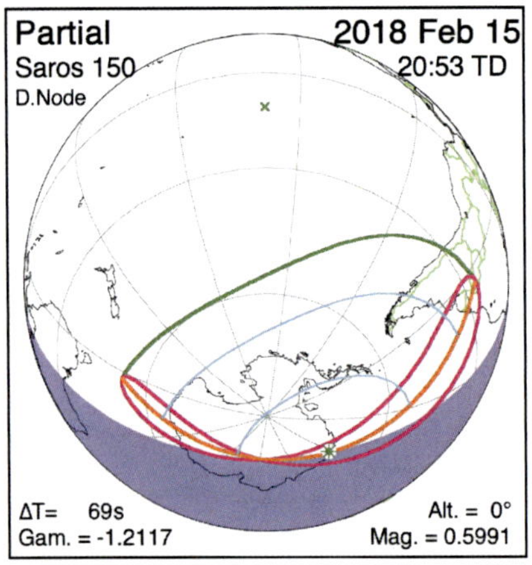
Partial
2018 Feb 15
Saros 150
20:53 TD
D.Node
ΔT= 69s
Alt. = 0°
Gam. = -1.2117
Mag. = 0.5991

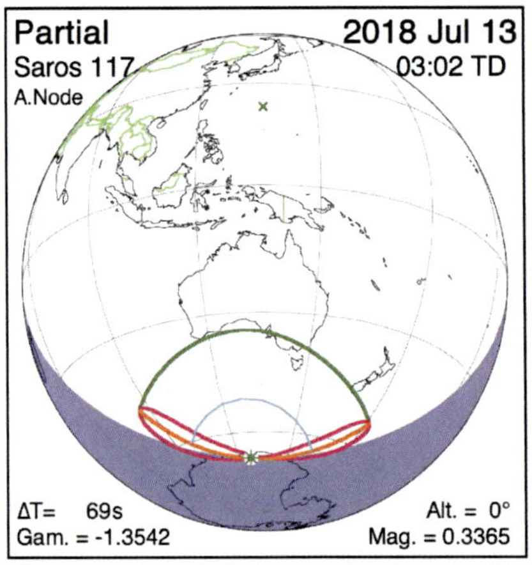
Partial
2018 Jul 13
Saros 117
03:02 TD
A.Node
ΔT= 69s
Alt. = 0°
Gam. = -1.3542
Mag. = 0.3365

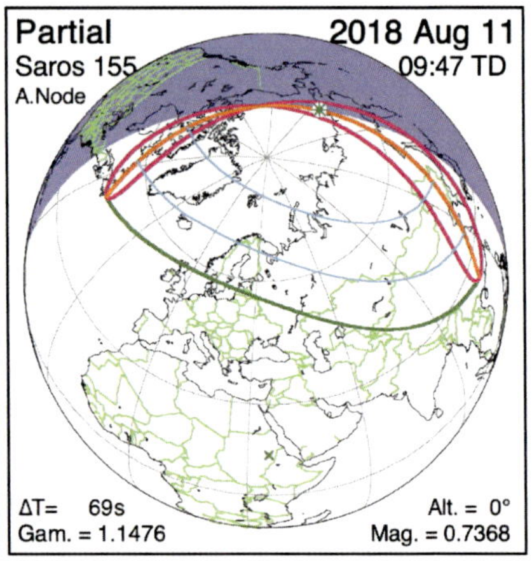
Partial
2018 Aug 11
Saros 155
09:47 TD
A.Node
ΔT= 69s
Alt. = 0°
Gam. = 1.1476
Mag. = 0.7368

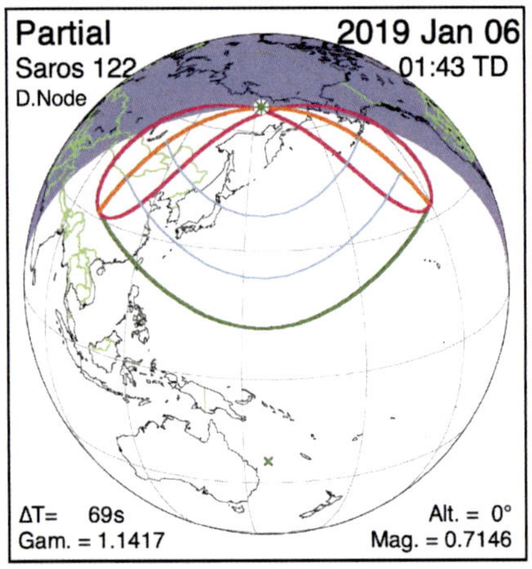
Partial
2019 Jan 06
Saros 122
01:43 TD
D.Node
ΔT= 69s
Alt. = 0°
Gam. = 1.1417
Mag. = 0.7146

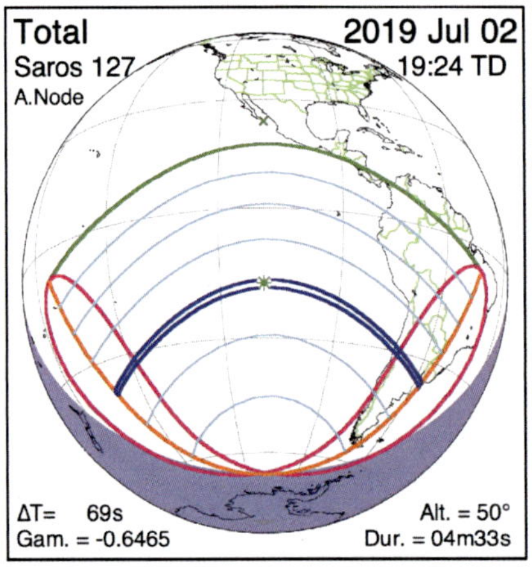
Total
2019 Jul 02
Saros 127
19:24 TD
A.Node
ΔT= 69s
Alt. = 50°
Gam. = -0.6465
Dur. = 04m33s

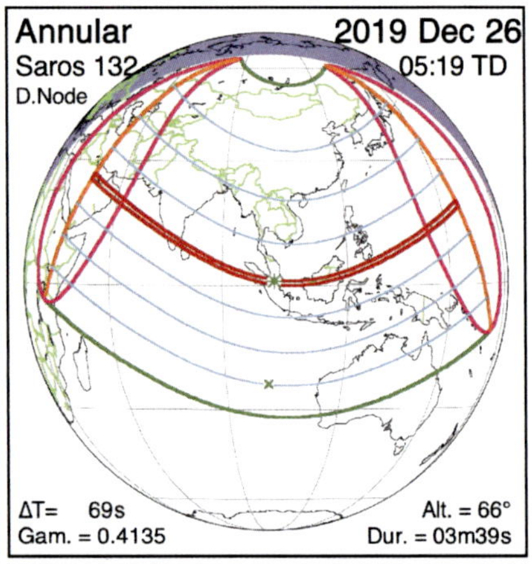
Annular
2019 Dec 26
Saros 132
05:19 TD
D.Node
ΔT= 69s
Alt. = 66°
Gam. = 0.4135
Dur. = 03m39s

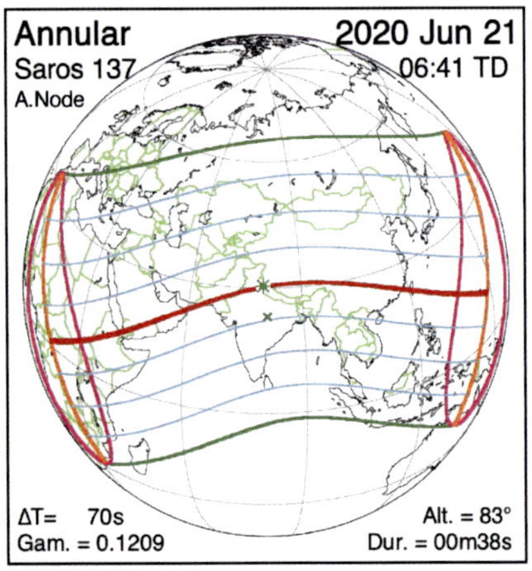
Annular
2020 Jun 21
Saros 137
06:41 TD
A.Node
ΔT= 70s
Alt. = 83°
Gam. = 0.1209
Dur. = 00m38s

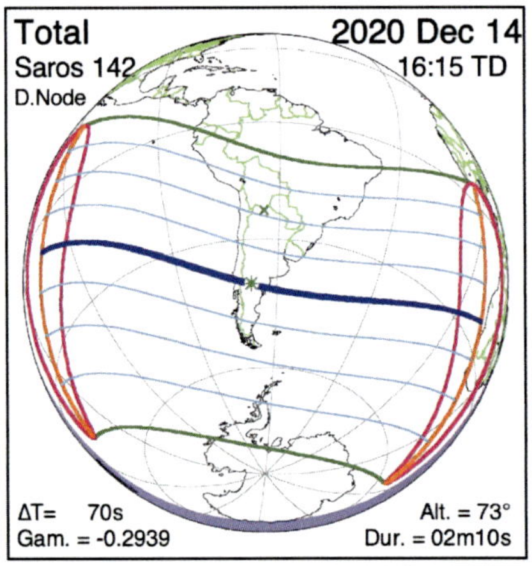
Total
2020 Dec 14
Saros 142
16:15 TD
D.Node
ΔT= 70s
Alt. = 73°
Gam. = -0.2939
Dur. = 02m10s

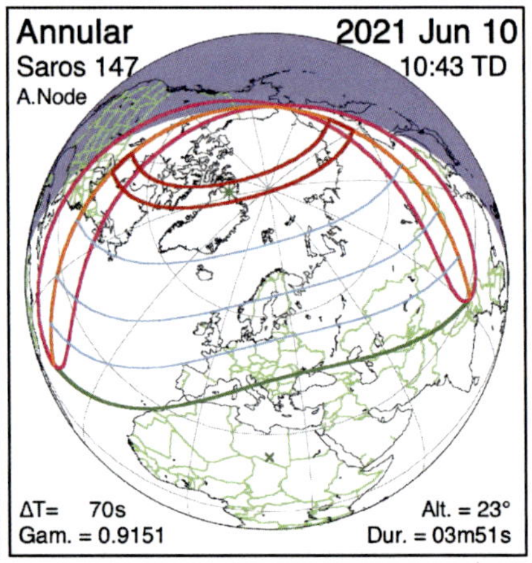
Annular
2021 Jun 10
Saros 147
10:43 TD
A.Node
ΔT= 70s
Alt. = 23°
Gam. = 0.9151
Dur. = 03m51s

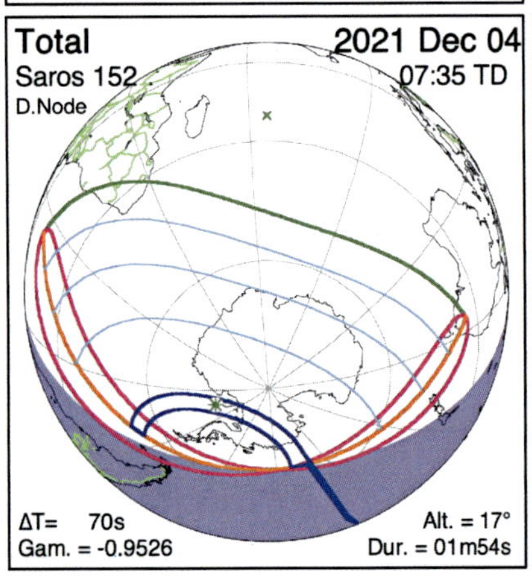
Total
2021 Dec 04
Saros 152
07:35 TD
D.Node
ΔT= 70s
Alt. = 17°
Gam. = -0.9526
Dur. = 01m54s

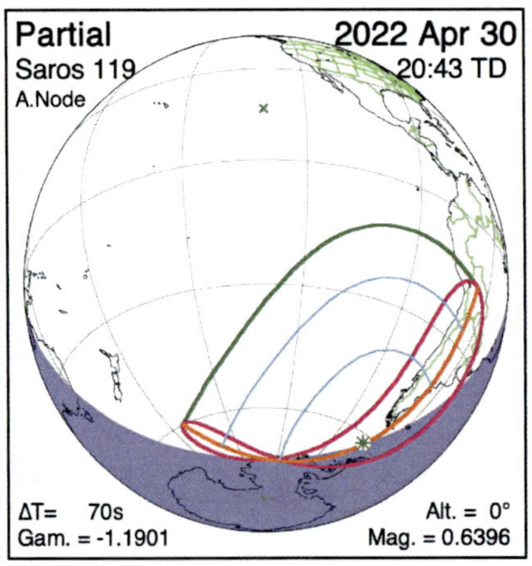
Partial
2022 Apr 30
Saros 119
20:43 TD
A.Node
ΔT= 70s
Alt. = 0°
Gam. = -1.1901
Mag. = 0.6396

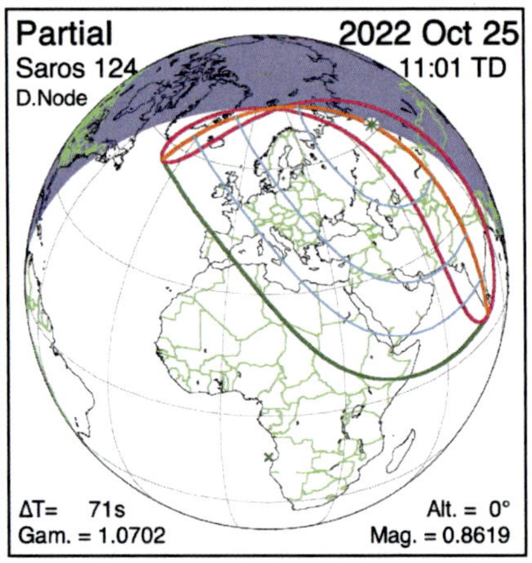
Partial
2022 Oct 25
Saros 124
11:01 TD
D.Node
ΔT= 71s
Alt. = 0°
Gam. = 1.0702
Mag. = 0.8619

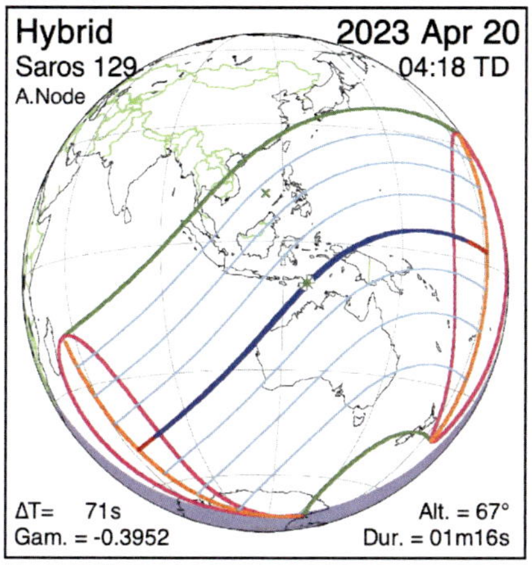
Hybrid
2023 Apr 20
Saros 129
04:18 TD
A.Node
ΔT= 71s
Alt. = 67°
Gam. = -0.3952
Dur. = 01m16s

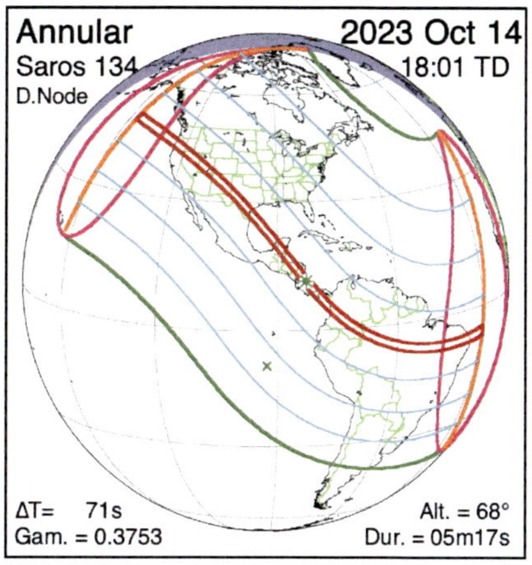
Annular
2023 Oct 14
Saros 134
18:01 TD
D.Node
ΔT= 71s
Alt. = 68°
Gam. = 0.3753
Dur. = 05m17s

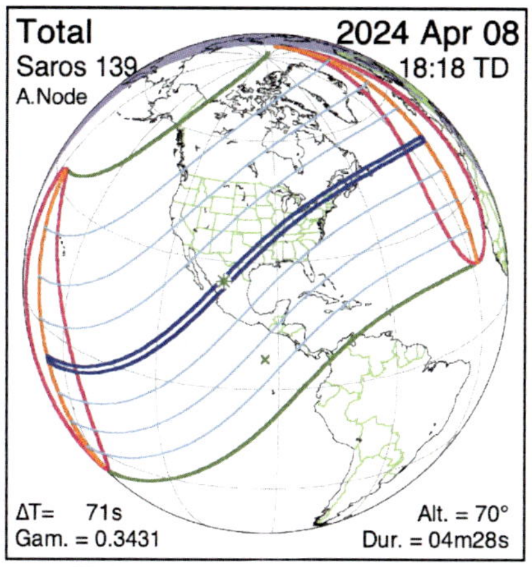
Total
2024 Apr 08
Saros 139
18:18 TD
A.Node
ΔT= 71s
Alt. = 70°
Gam. = 0.3431
Dur. = 04m28s

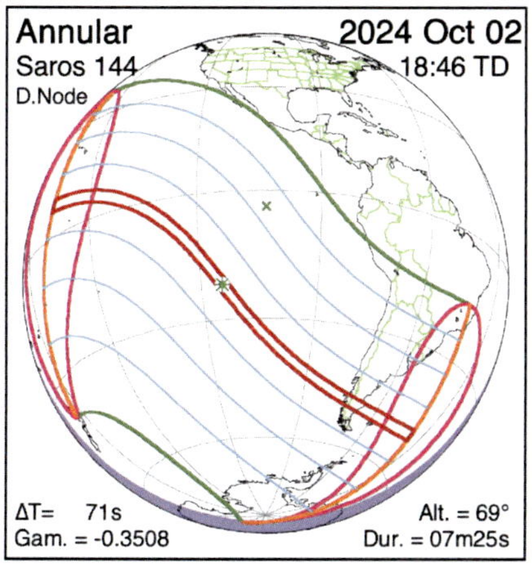
Annular
2024 Oct 02
Saros 144
18:46 TD
D.Node
ΔT= 71s
Alt. = 69°
Gam. = -0.3508
Dur. = 07m25s

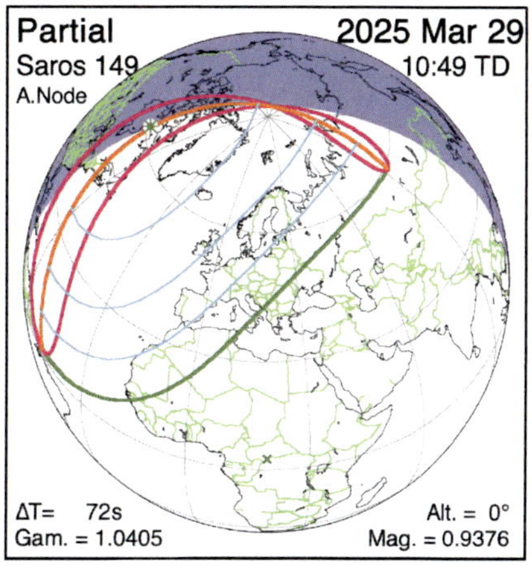
Partial
2025 Mar 29
Saros 149
10:49 TD
A.Node
ΔT= 72s
Alt. = 0°
Gam. = 1.0405
Mag. = 0.9376

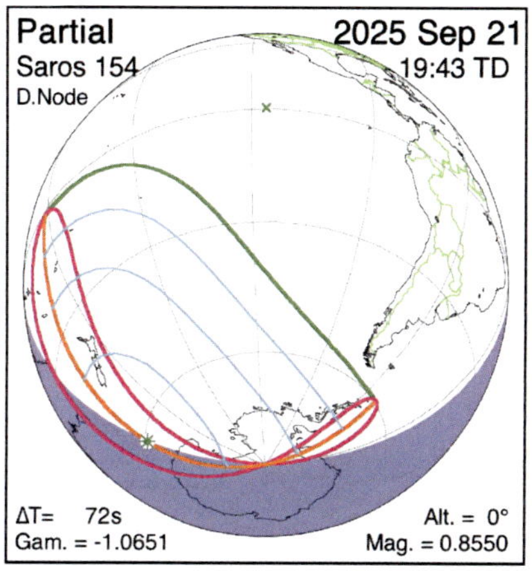
Partial
2025 Sep 21
Saros 154
19:43 TD
D.Node
ΔT= 72s
Alt. = 0°
Gam. = -1.0651
Mag. = 0.8550

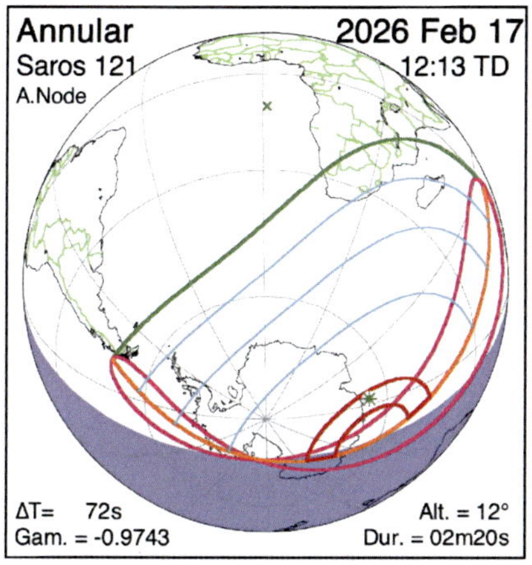
Annular
2026 Feb 17
Saros 121
12:13 TD
A.Node
ΔT= 72s
Alt. = 12°
Gam. = -0.9743
Dur. = 02m20s

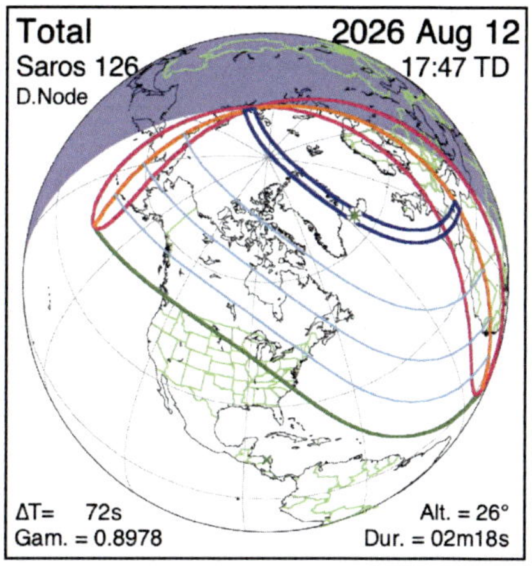
Total
2026 Aug 12
Saros 126
17:47 TD
D.Node
ΔT= 72s
Alt. = 26°
Gam. = 0.8978
Dur. = 02m18s

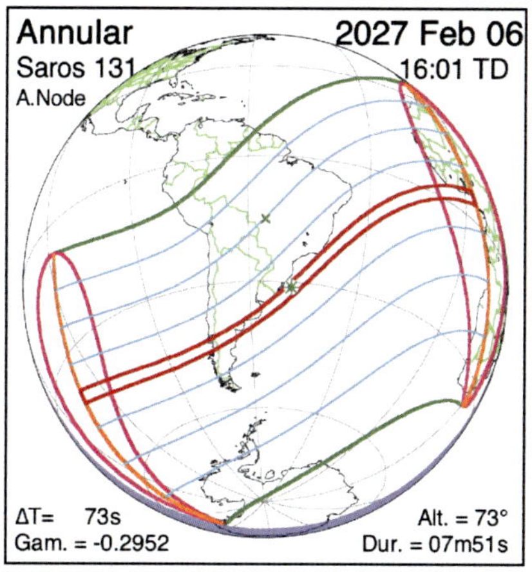
Annular
2027 Feb 06
Saros 131
16:01 TD
A.Node
ΔT= 73s
Alt. = 73°
Gam. = -0.2952
Dur. = 07m51s

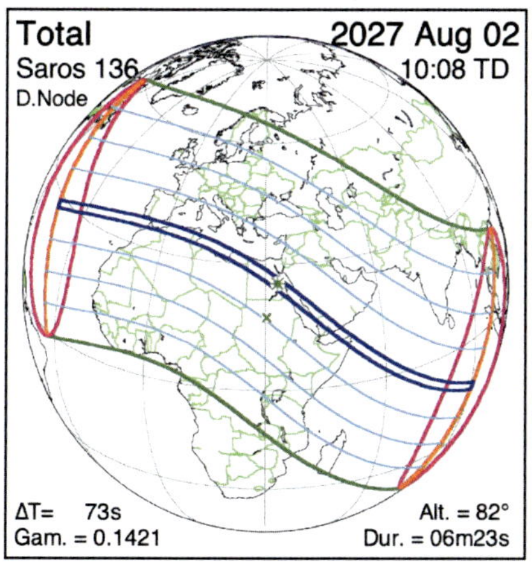
Total
2027 Aug 02
Saros 136
10:08 TD
D.Node
ΔT= 73s
Alt. = 82°
Gam. = 0.1421
Dur. = 06m23s

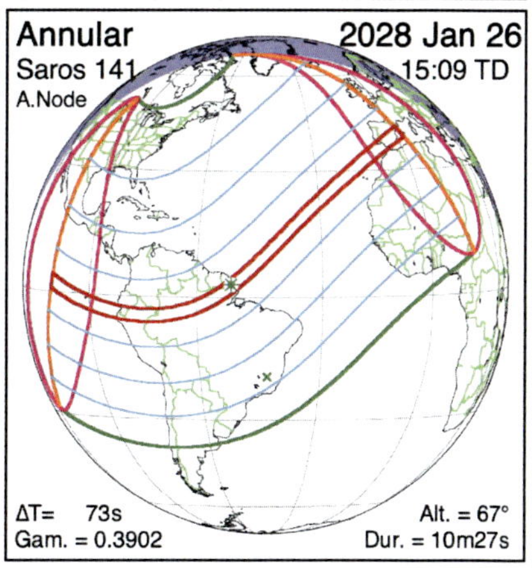
Annular
2028 Jan 26
Saros 141
15:09 TD
A.Node
ΔT= 73s
Alt. = 67°
Gam. = 0.3902
Dur. = 10m27s

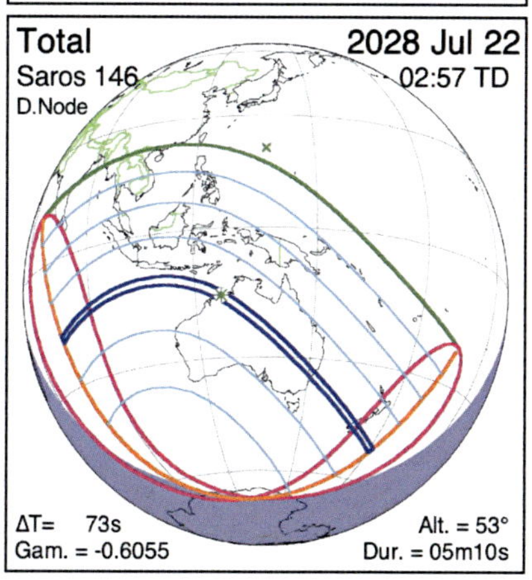
Total
2028 Jul 22
Saros 146
02:57 TD
D.Node
ΔT= 73s
Alt. = 53°
Gam. = -0.6055
Dur. = 05m10s

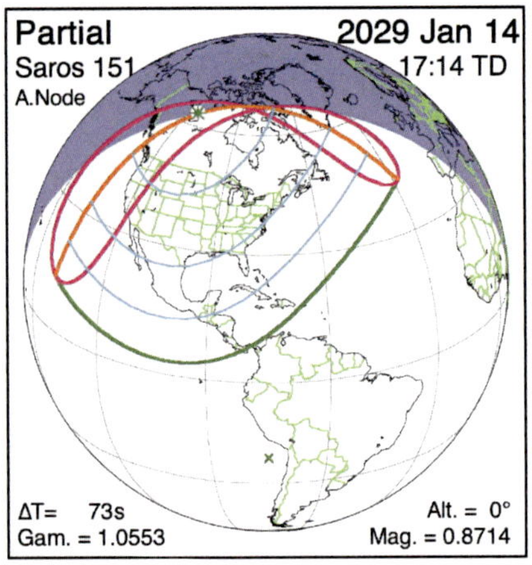
Partial
2029 Jan 14
Saros 151
17:14 TD
A.Node
ΔT= 73s
Alt. = 0°
Gam. = 1.0553
Mag. = 0.8714

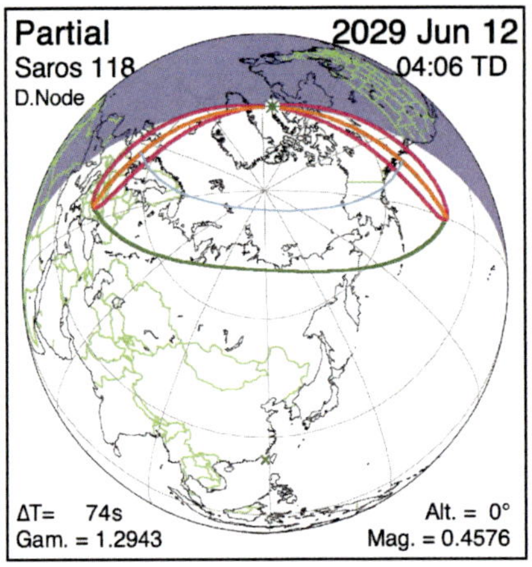
Partial
2029 Jun 12
Saros 118
04:06 TD
D.Node
ΔT= 74s
Alt. = 0°
Gam. = 1.2943
Mag. = 0.4576

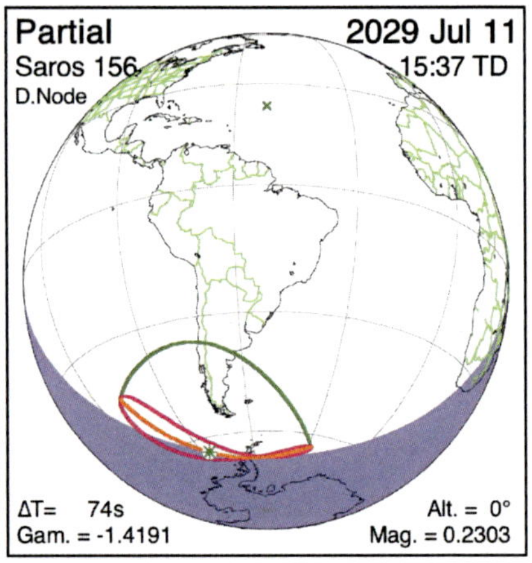
Partial
2029 Jul 11
Saros 156
15:37 TD
D.Node
ΔT= 74s
Alt. = 0°
Gam. = -1.4191
Mag. = 0.2303

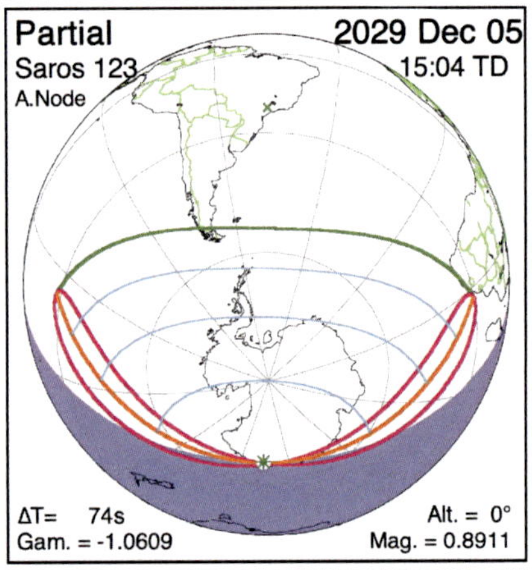
Partial
2029 Dec 05
Saros 123
15:04 TD
A.Node
ΔT= 74s
Alt. = 0°
Gam. = -1.0609
Mag. = 0.8911

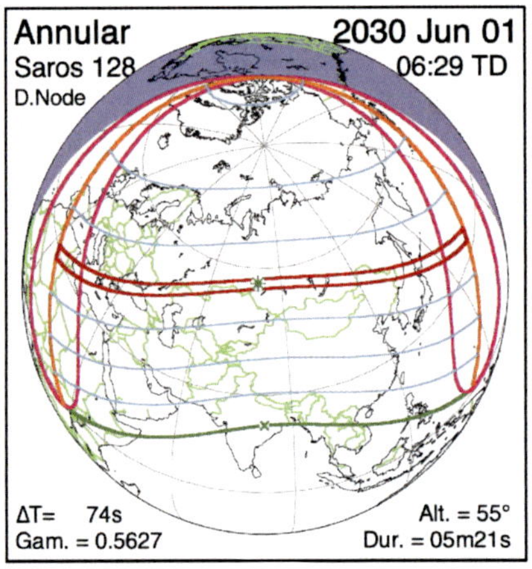
Annular
2030 Jun 01
Saros 128
06:29 TD
D.Node
ΔT= 74s
Alt. = 55°
Gam. = 0.5627
Dur. = 05m21s

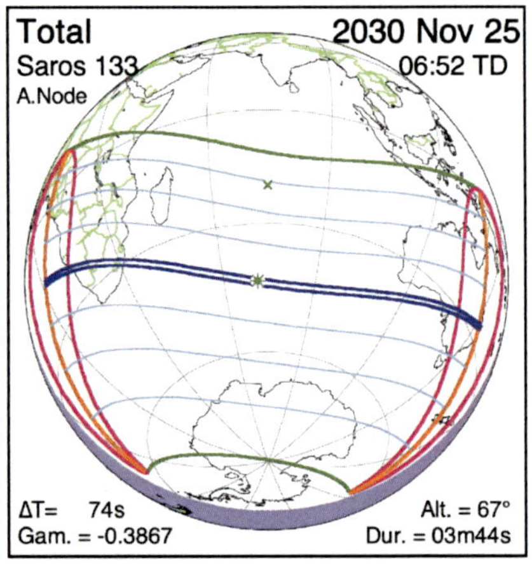
Total
2030 Nov 25
Saros 133
06:52 TD
A.Node
ΔT= 74s
Alt. = 67°
Gam. = -0.3867
Dur. = 03m44s

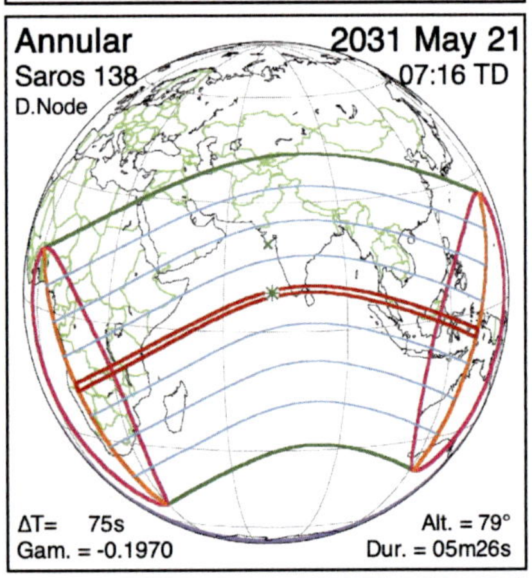
Annular
2031 May 21
Saros 138
07:16 TD
D.Node
ΔT= 75s
Alt. = 79°
Gam. = -0.1970
Dur. = 05m26s

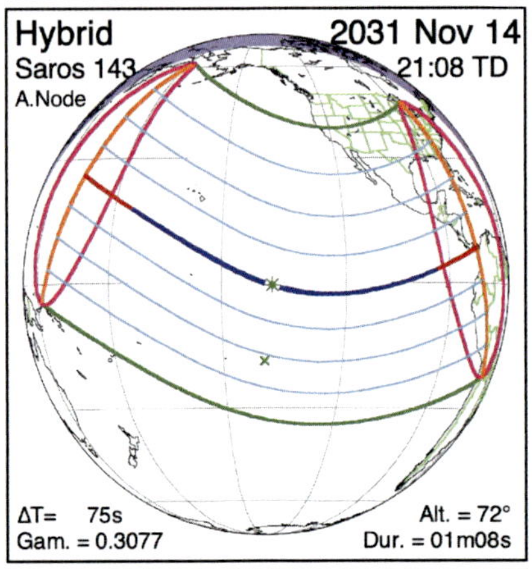
Hybrid
2031 Nov 14
Saros 143
21:08 TD
A.Node
ΔT= 75s
Alt. = 72°
Gam. = 0.3077
Dur. = 01m08s

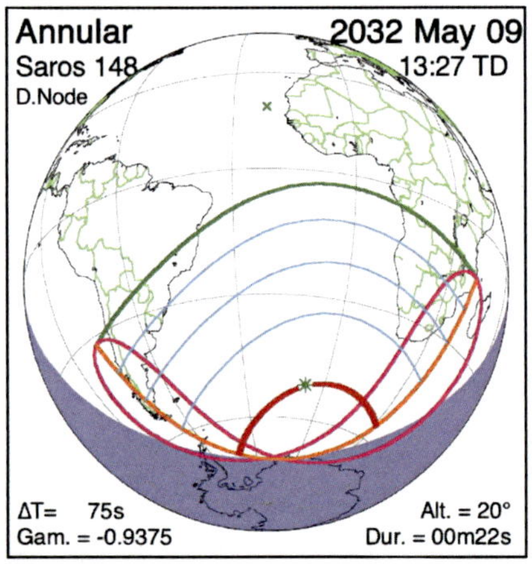
Annular
2032 May 09
Saros 148
13:27 TD
D.Node
ΔT= 75s
Alt. = 20°
Gam. = -0.9375
Dur. = 00m22s

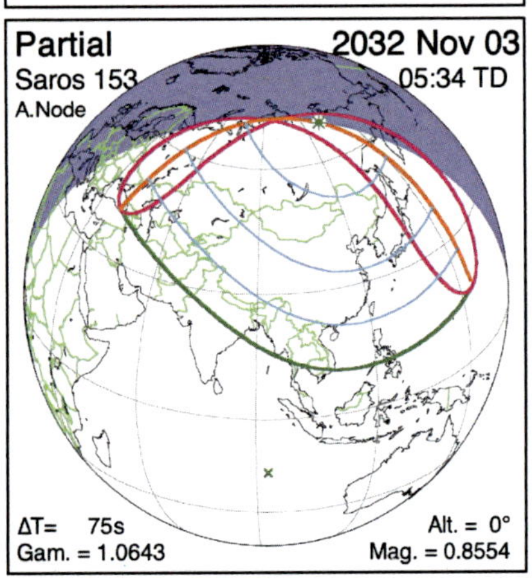
Partial
2032 Nov 03
Saros 153
05:34 TD
A.Node
ΔT= 75s
Alt. = 0°
Gam. = 1.0643
Mag. = 0.8554

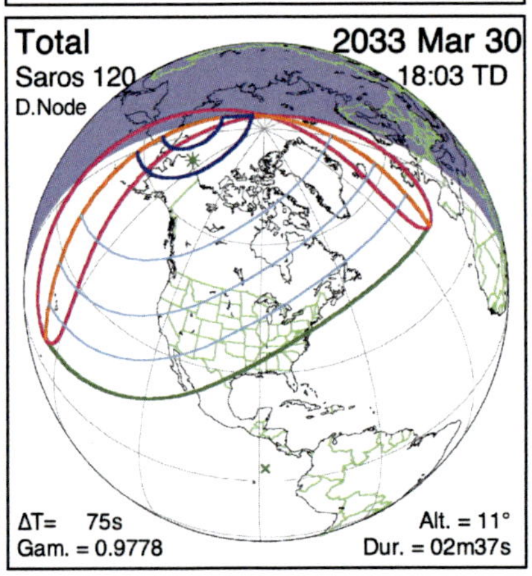
Total
2033 Mar 30
Saros 120
18:03 TD
D.Node
ΔT= 75s
Alt. = 11°
Gam. = 0.9778
Dur. = 02m37s

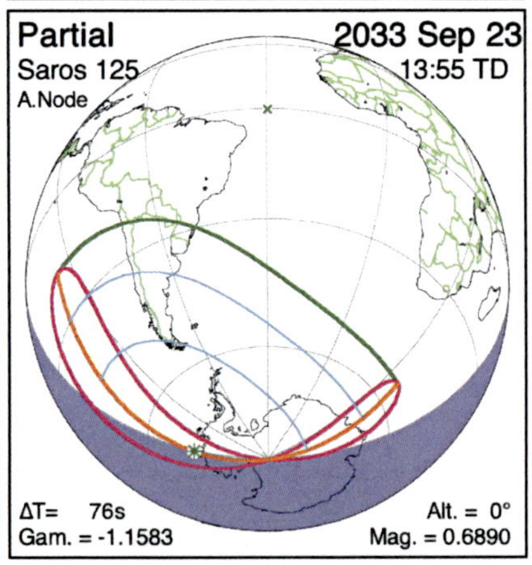
Partial
2033 Sep 23
Saros 125
13:55 TD
A.Node
ΔT= 76s
Alt. = 0°
Gam. = -1.1583
Mag. = 0.6890

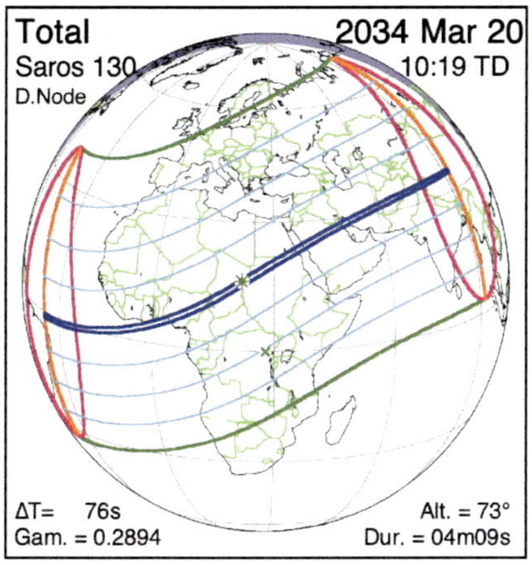
Total
2034 Mar 20
Saros 130
10:19 TD
D.Node
ΔT= 76s
Alt. = 73°
Gam. = 0.2894
Dur. = 04m09s

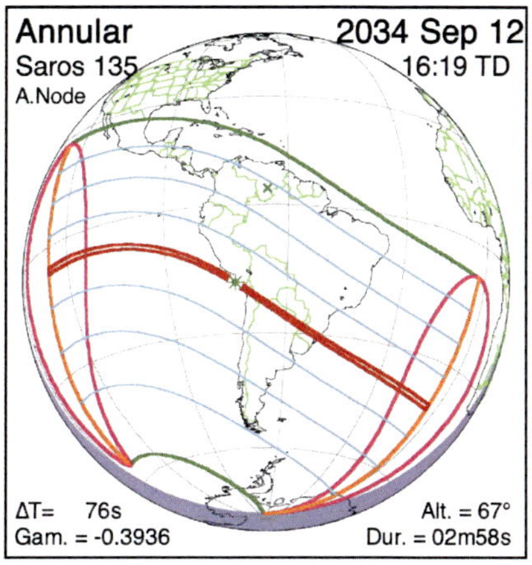
Annular
2034 Sep 12
Saros 135
16:19 TD
A.Node
ΔT= 76s
Alt. = 67°
Gam. = -0.3936
Dur. = 02m58s

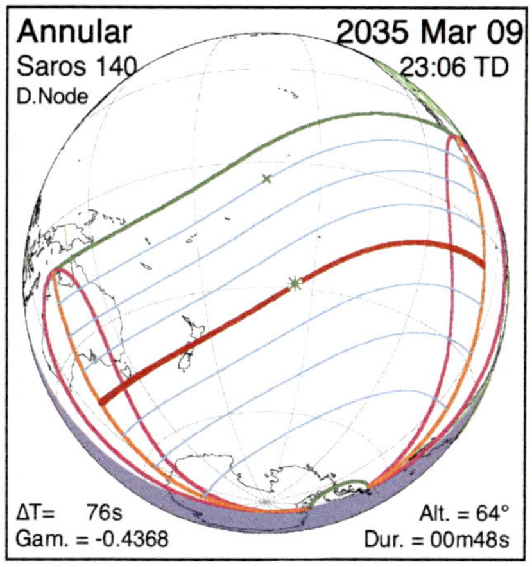
Annular
2035 Mar 09
Saros 140
23:06 TD
D.Node
ΔT= 76s
Alt. = 64°
Gam. = -0.4368
Dur. = 00m48s

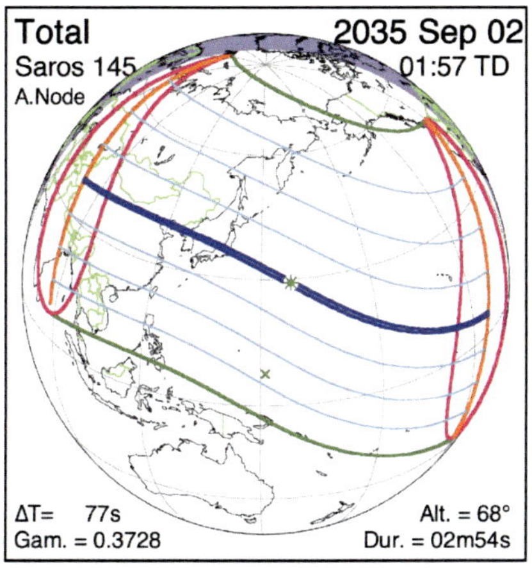
Total
2035 Sep 02
Saros 145
01:57 TD
A.Node
ΔT= 77s
Alt. = 68°
Gam. = 0.3728
Dur. = 02m54s

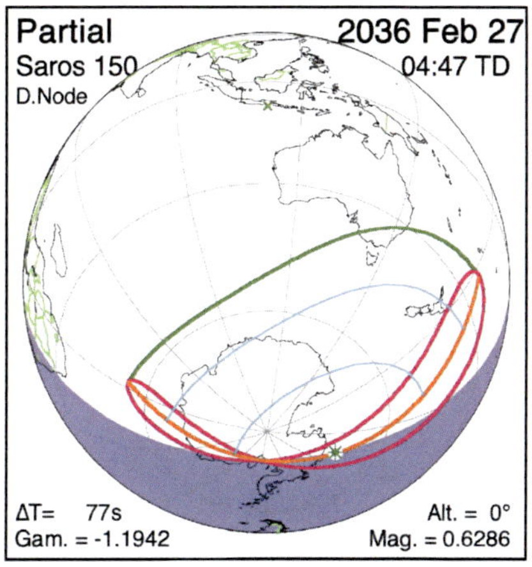
Partial
2036 Feb 27
Saros 150
04:47 TD
D.Node
ΔT= 77s
Alt. = 0°
Gam. = -1.1942
Mag. = 0.6286

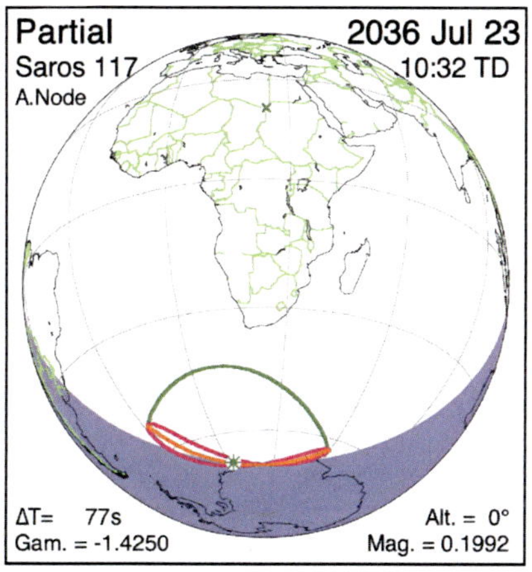
Partial
2036 Jul 23
Saros 117
10:32 TD
A.Node
ΔT= 77s
Alt. = 0°
Gam. = -1.4250
Mag. = 0.1992

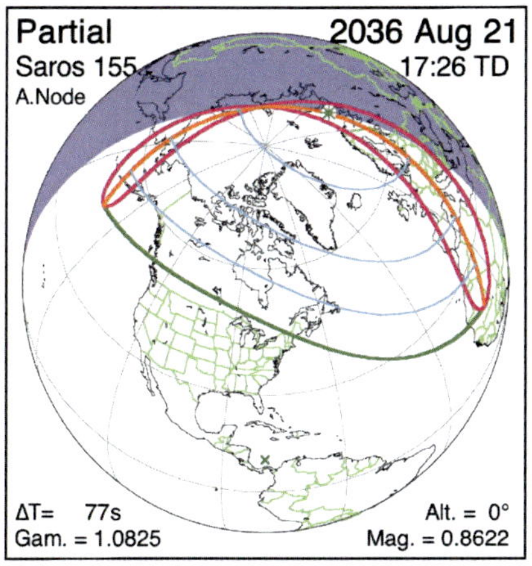
Partial
2036 Aug 21
Saros 155
17:26 TD
A.Node
ΔT= 77s
Alt. = 0°
Gam. = 1.0825
Mag. = 0.8622

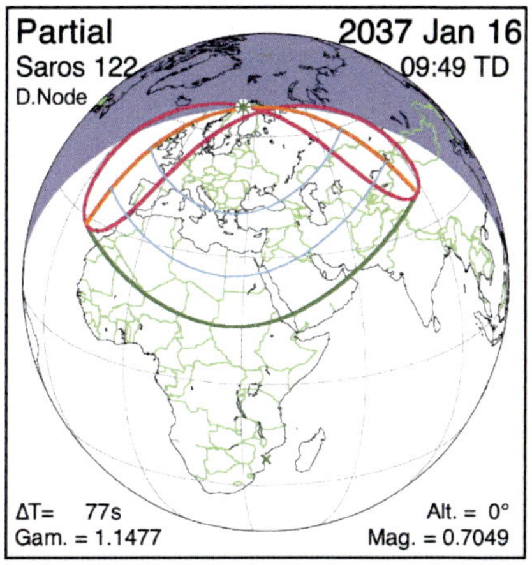
Partial
2037 Jan 16
Saros 122
09:49 TD
D.Node
ΔT= 77s
Alt. = 0°
Gam. = 1.1477
Mag. = 0.7049

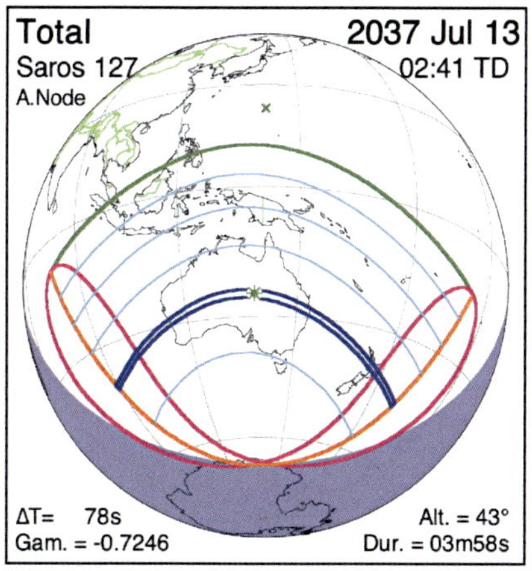
Total
2037 Jul 13
Saros 127
02:41 TD
A.Node
ΔT= 78s
Alt. = 43°
Gam. = -0.7246
Dur. = 03m58s

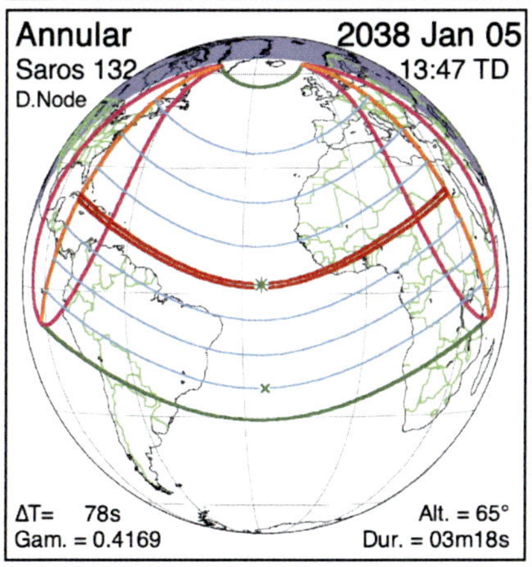
Annular
2038 Jan 05
Saros 132
13:47 TD
D.Node
ΔT= 78s
Alt. = 65°
Gam. = 0.4169
Dur. = 03m18s

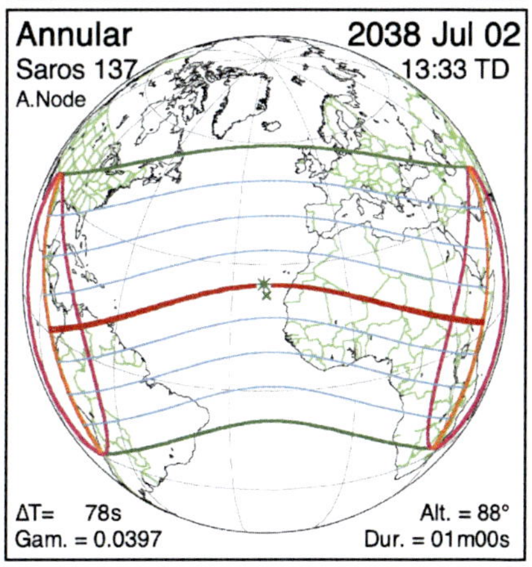
Annular
2038 Jul 02
Saros 137
13:33 TD
A.Node
ΔT= 78s
Alt. = 88°
Gam. = 0.0397
Dur. = 01m00s

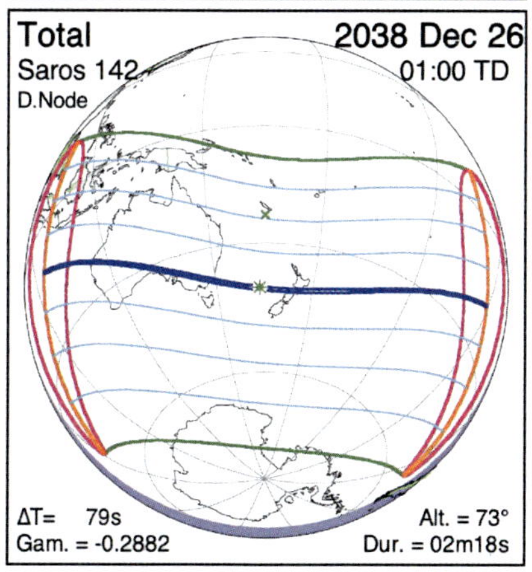
Total
2038 Dec 26
Saros 142
01:00 TD
D.Node
ΔT= 79s
Alt. = 73°
Gam. = -0.2882
Dur. = 02m18s

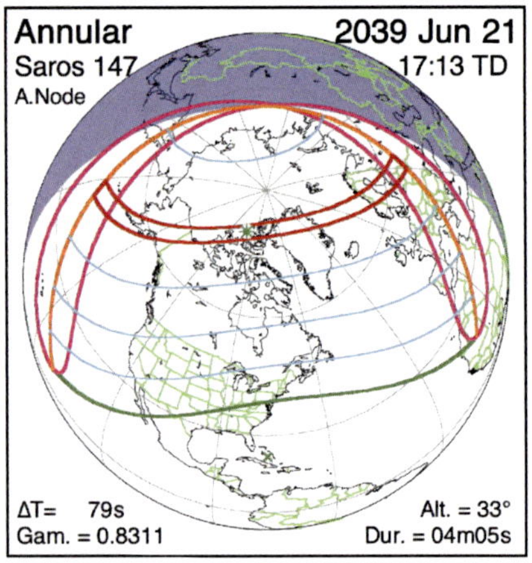
Annular
2039 Jun 21
Saros 147
17:13 TD
A.Node
ΔT= 79s
Alt. = 33°
Gam. = 0.8311
Dur. = 04m05s

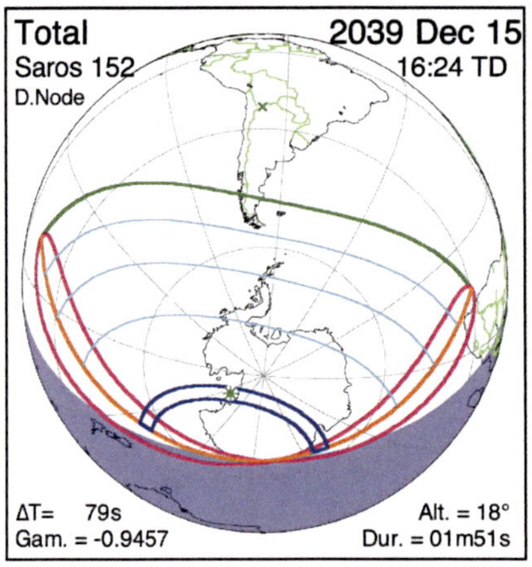
Total
2039 Dec 15
Saros 152
16:24 TD
D.Node
ΔT= 79s
Alt. = 18°
Gam. = -0.9457
Dur. = 01m51s

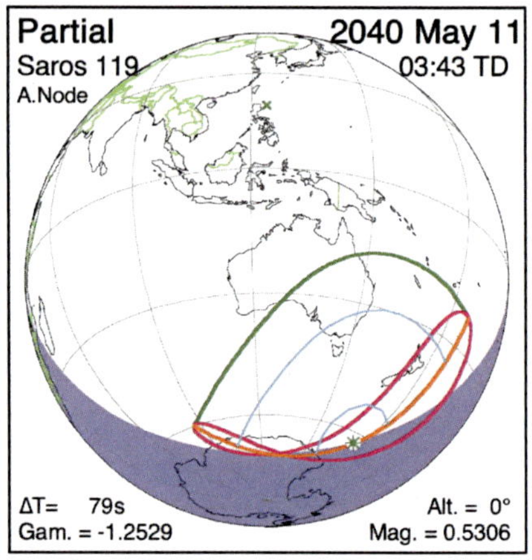
Partial
2040 May 11
Saros 119
03:43 TD
A.Node
ΔT= 79s
Alt. = 0°
Gam. = -1.2529
Mag. = 0.5306

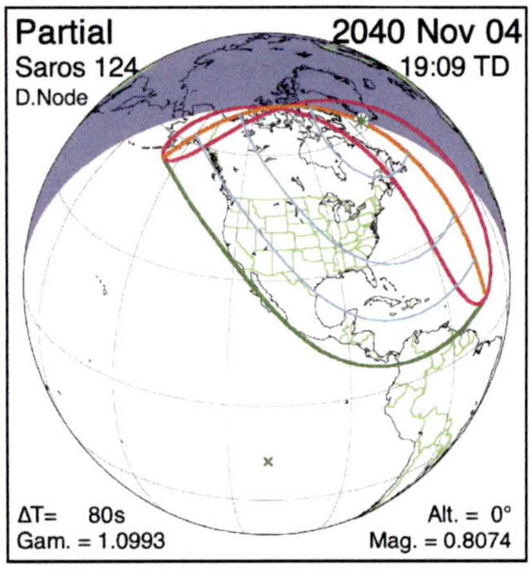
Partial
2040 Nov 04
Saros 124
19:09 TD
D.Node
ΔT= 80s
Alt. = 0°
Gam. = 1.0993
Mag. = 0.8074

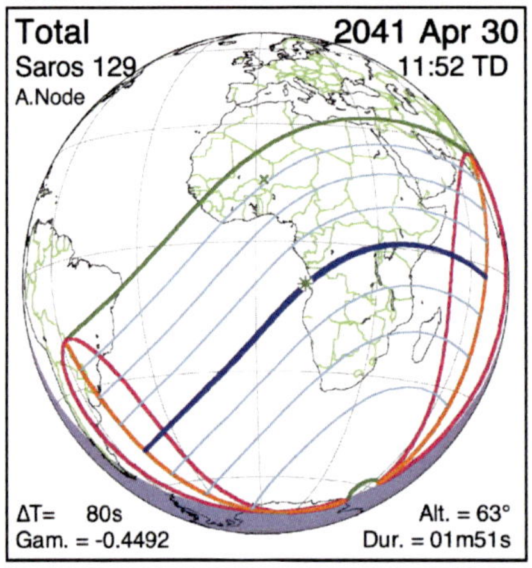
Total
2041 Apr 30
Saros 129
11:52 TD
A.Node
ΔT= 80s
Alt. = 63°
Gam. = -0.4492
Dur. = 01m51s

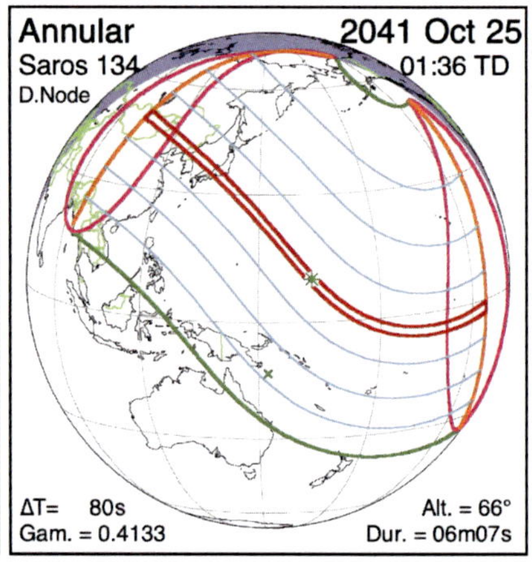
Annular
2041 Oct 25
Saros 134
01:36 TD
D.Node
ΔT= 80s
Alt. = 66°
Gam. = 0.4133
Dur. = 06m07s

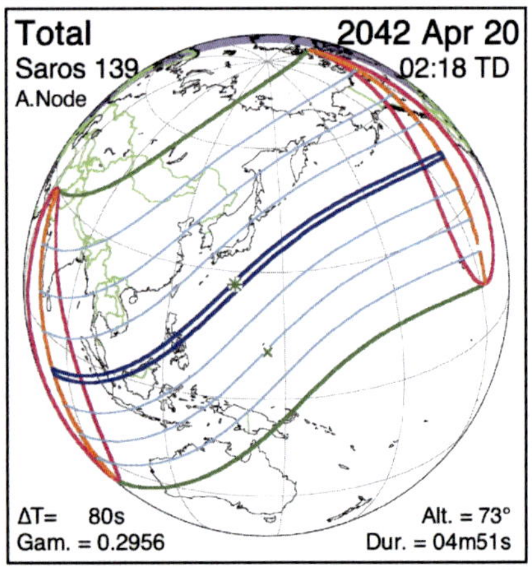
Total
2042 Apr 20
Saros 139
02:18 TD
A.Node
ΔT= 80s
Alt. = 73°
Gam. = 0.2956
Dur. = 04m51s

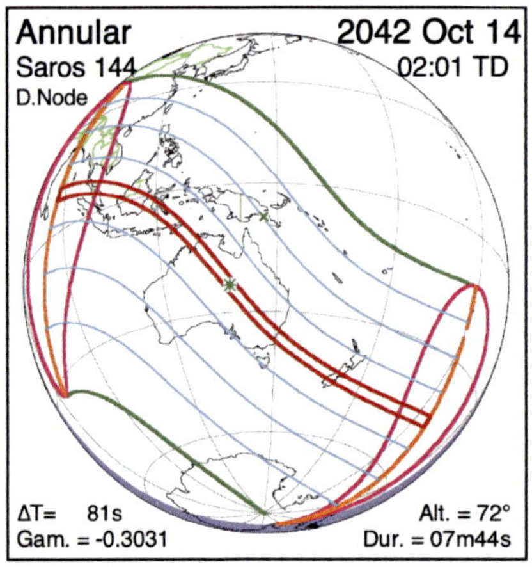
Annular
2042 Oct 14
Saros 144
02:01 TD
D.Node
ΔT= 81s
Alt. = 72°
Gam. = -0.3031
Dur. = 07m44s

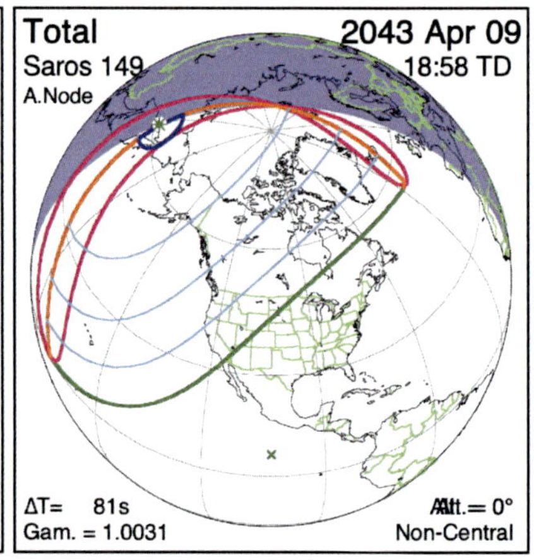
Total
2043 Apr 09
Saros 149
18:58 TD
A.Node
ΔT= 81s
Alt. = 0°
Gam. = 1.0031
Non-Central

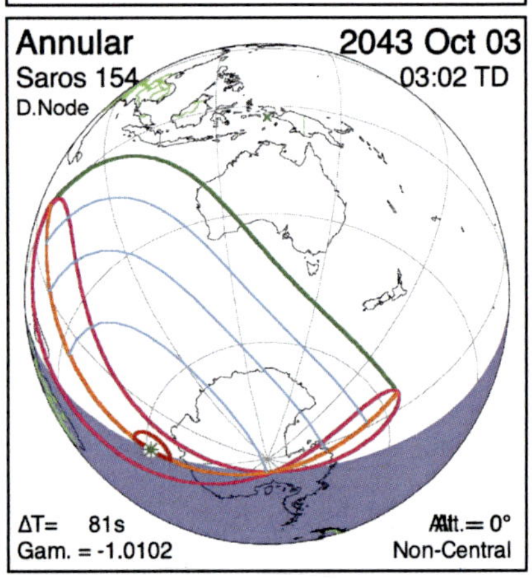
Annular
2043 Oct 03
Saros 154
03:02 TD
D.Node
ΔT= 81s
Alt. = 0°
Gam. = -1.0102
Non-Central

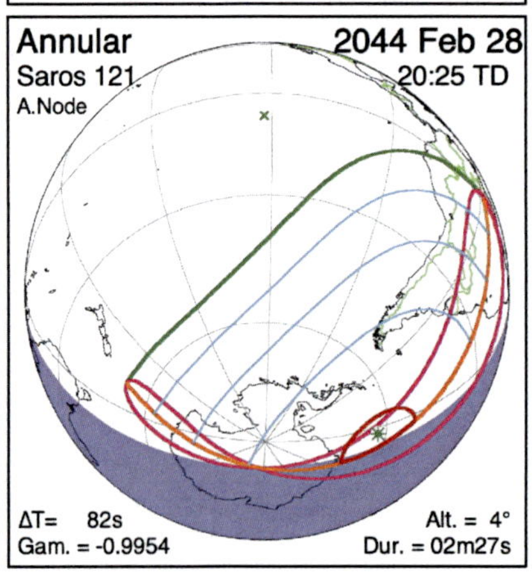
Annular
2044 Feb 28
Saros 121
20:25 TD
A.Node
ΔT= 82s
Alt. = 4°
Gam. = -0.9954
Dur. = 02m27s

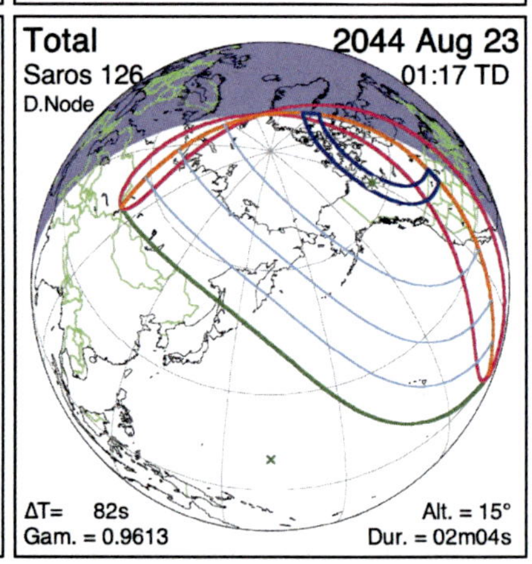
Total
2044 Aug 23
Saros 126
01:17 TD
D.Node
ΔT= 82s
Alt. = 15°
Gam. = 0.9613
Dur. = 02m04s

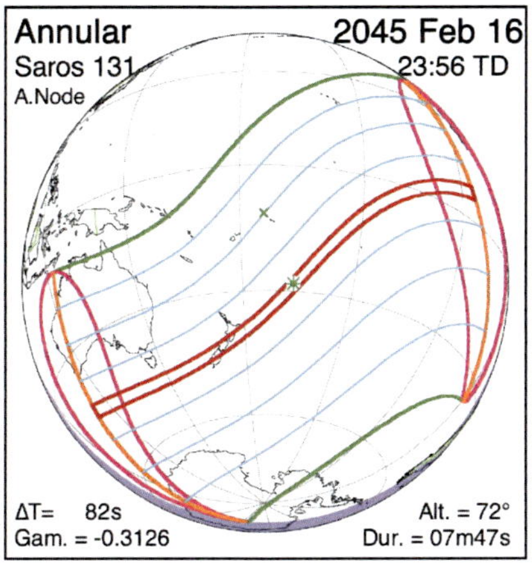
Annular
2045 Feb 16
Saros 131
23:56 TD
A.Node
ΔT= 82s
Alt. = 72°
Gam. = -0.3126
Dur. = 07m47s

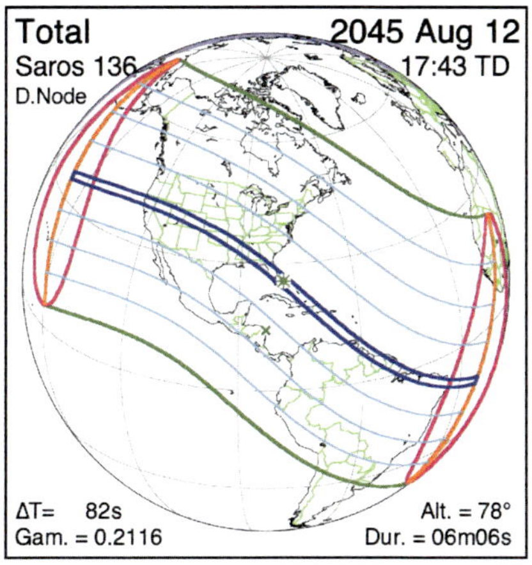
Total
2045 Aug 12
Saros 136
17:43 TD
D.Node
ΔT= 82s
Alt. = 78°
Gam. = 0.2116
Dur. = 06m06s

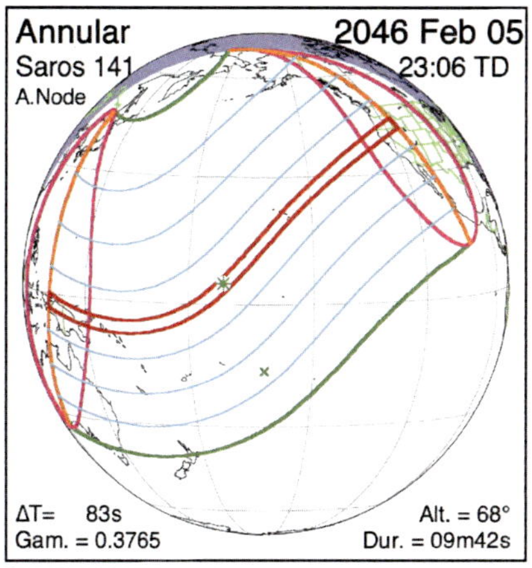
Annular
2046 Feb 05
Saros 141
23:06 TD
A.Node
ΔT= 83s
Alt. = 68°
Gam. = 0.3765
Dur. = 09m42s

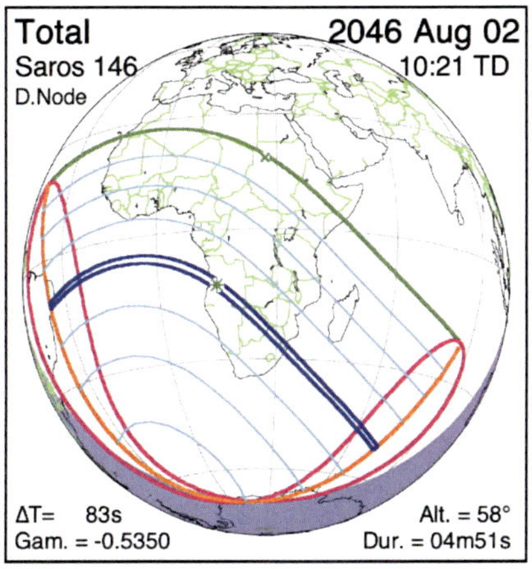
Total
2046 Aug 02
Saros 146
10:21 TD
D.Node
ΔT= 83s
Alt. = 58°
Gam. = -0.5350
Dur. = 04m51s

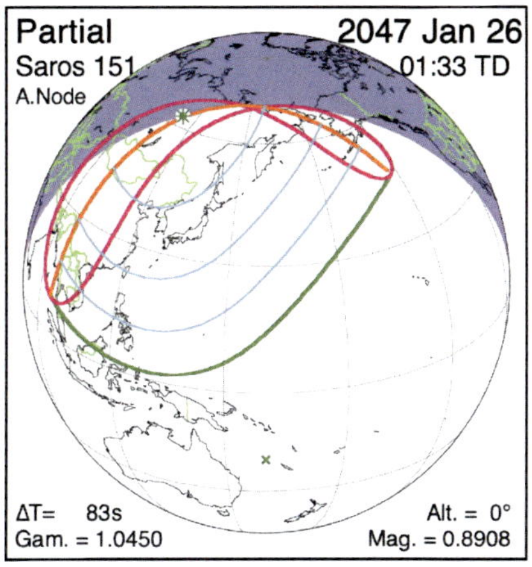
Partial
2047 Jan 26
Saros 151
01:33 TD
A.Node
ΔT= 83s
Alt. = 0°
Gam. = 1.0450
Mag. = 0.8908

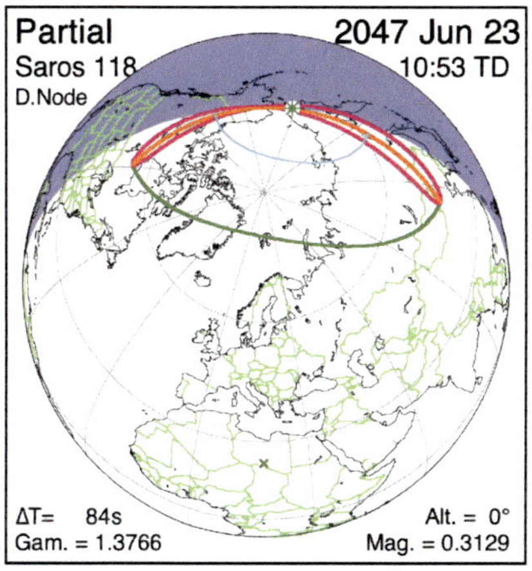
Partial
2047 Jun 23
Saros 118
10:53 TD
D.Node
ΔT= 84s
Alt. = 0°
Gam. = 1.3766
Mag. = 0.3129

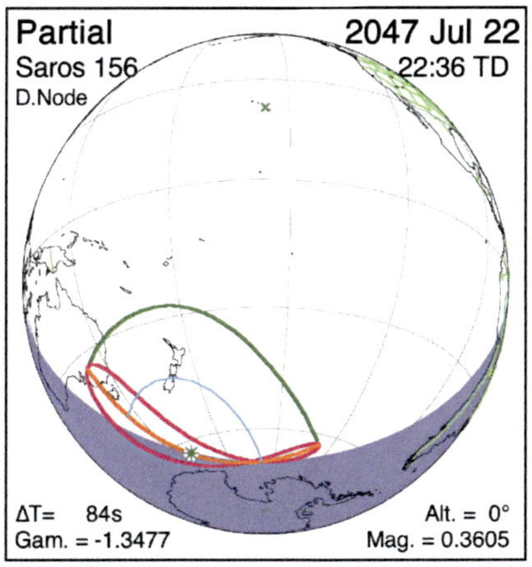
Partial
2047 Jul 22
Saros 156
22:36 TD
D.Node
ΔT= 84s
Alt. = 0°
Gam. = -1.3477
Mag. = 0.3605

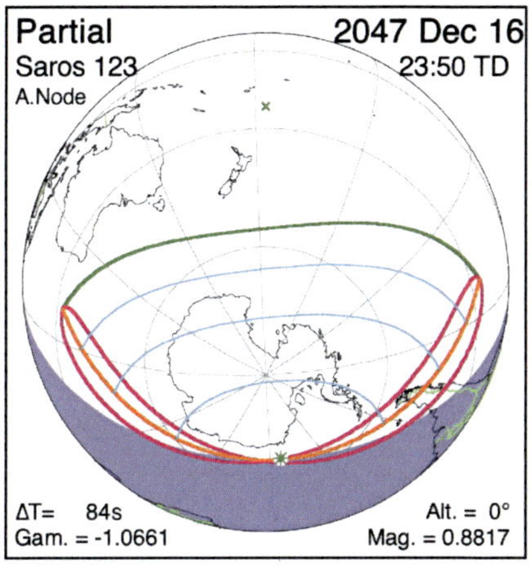
Partial
2047 Dec 16
Saros 123
23:50 TD
A.Node
ΔT= 84s
Alt. = 0°
Gam. = -1.0661
Mag. = 0.8817

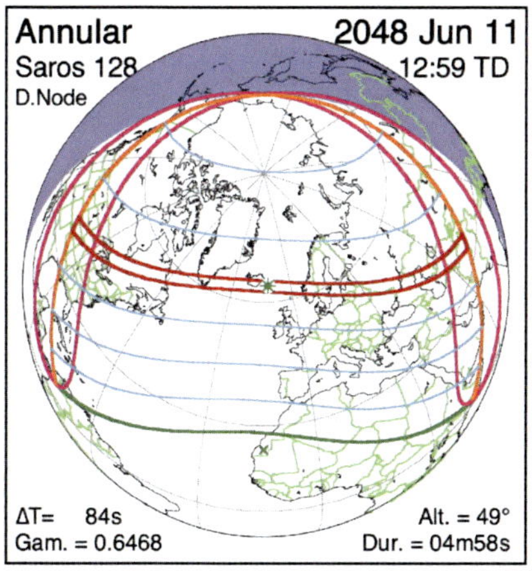
Annular
2048 Jun 11
Saros 128
12:59 TD
D.Node
ΔT= 84s
Alt. = 49°
Gam. = 0.6468
Dur. = 04m58s

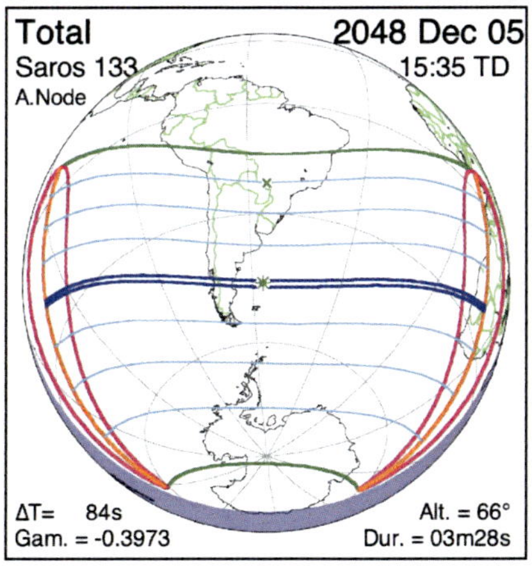
Total
2048 Dec 05
Saros 133
15:35 TD
A.Node
ΔT= 84s
Alt. = 66°
Gam. = -0.3973
Dur. = 03m28s

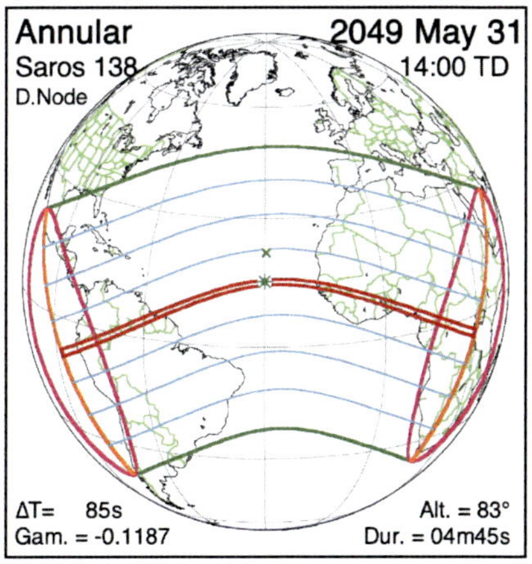
Annular
2049 May 31
Saros 138
14:00 TD
D.Node
ΔT= 85s
Alt. = 83°
Gam. = -0.1187
Dur. = 04m45s

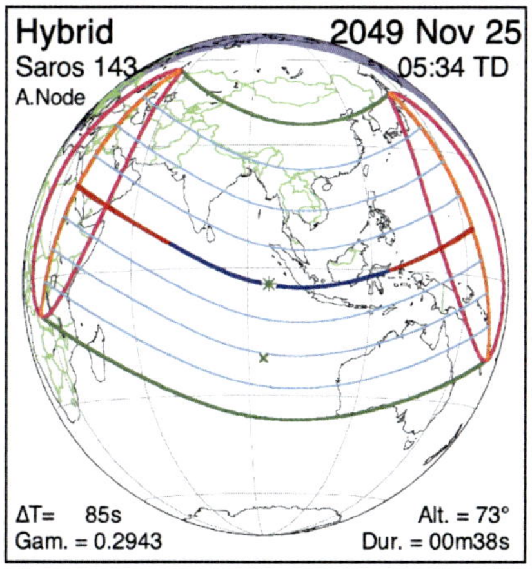
Hybrid
2049 Nov 25
Saros 143
05:34 TD
A.Node
ΔT= 85s
Alt. = 73°
Gam. = 0.2943
Dur. = 00m38s

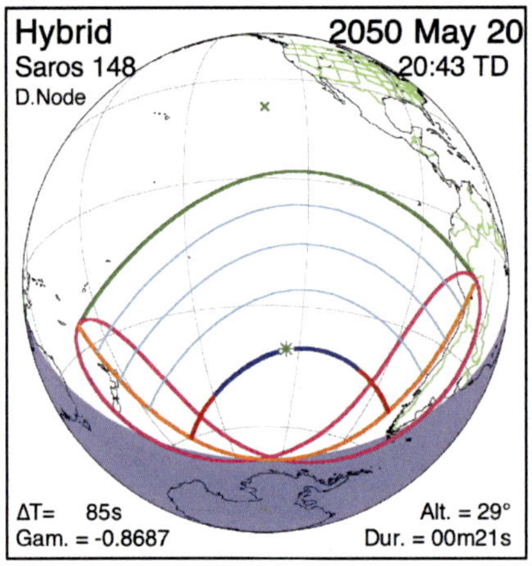
Hybrid
2050 May 20
Saros 148
20:43 TD
D.Node
ΔT= 85s
Alt. = 29°
Gam. = -0.8687
Dur. = 00m21s

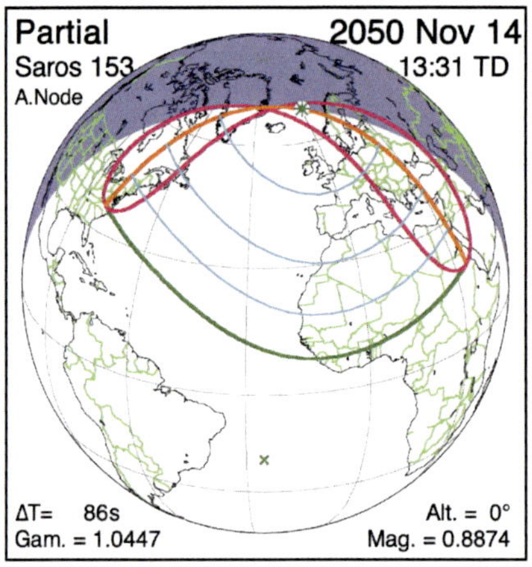
Partial
2050 Nov 14
Saros 153
13:31 TD
A.Node
ΔT= 86s
Alt. = 0°
Gam. = 1.0447
Mag. = 0.8874

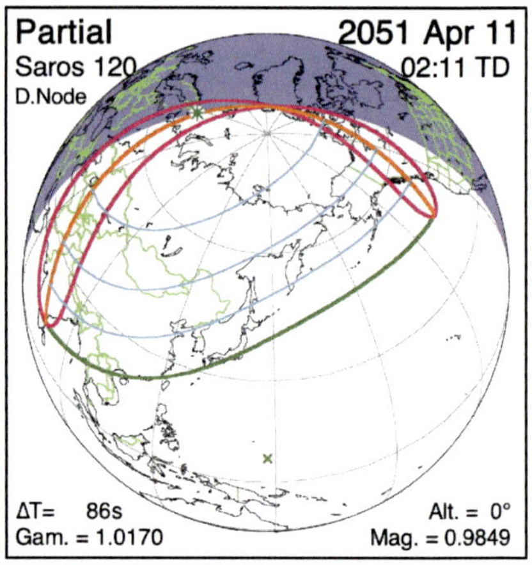
Partial
2051 Apr 11
Saros 120
02:11 TD
D.Node
ΔT= 86s
Alt. = 0°
Gam. = 1.0170
Mag. = 0.9849

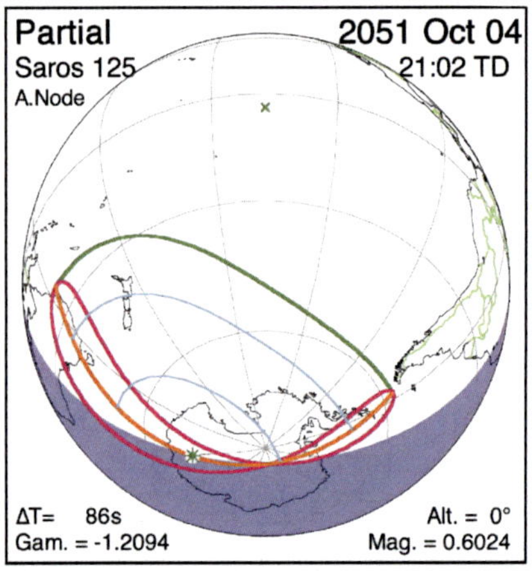
Partial
2051 Oct 04
Saros 125
21:02 TD
A.Node
ΔT= 86s
Alt. = 0°
Gam. = -1.2094
Mag. = 0.6024

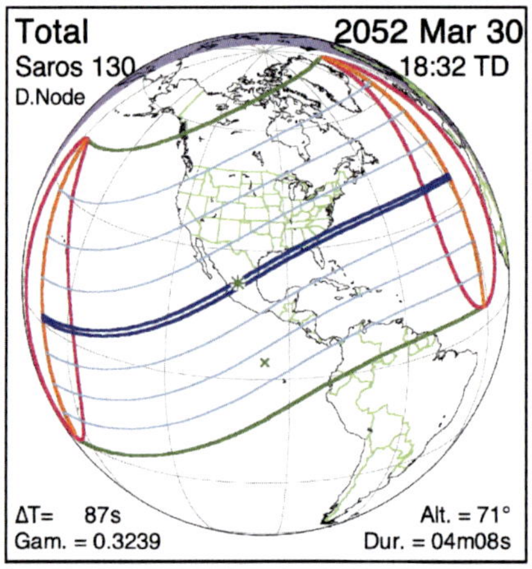
Total
2052 Mar 30
Saros 130
18:32 TD
D.Node
ΔT= 87s
Alt. = 71°
Gam. = 0.3239
Dur. = 04m08s

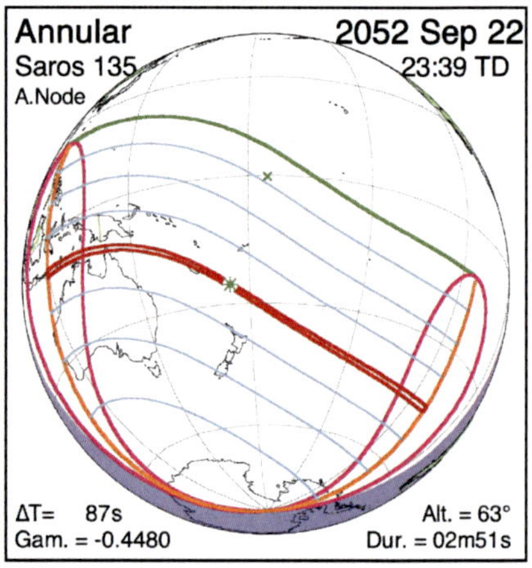
Annular
2052 Sep 22
Saros 135
23:39 TD
A.Node
ΔT= 87s
Alt. = 63°
Gam. = -0.4480
Dur. = 02m51s

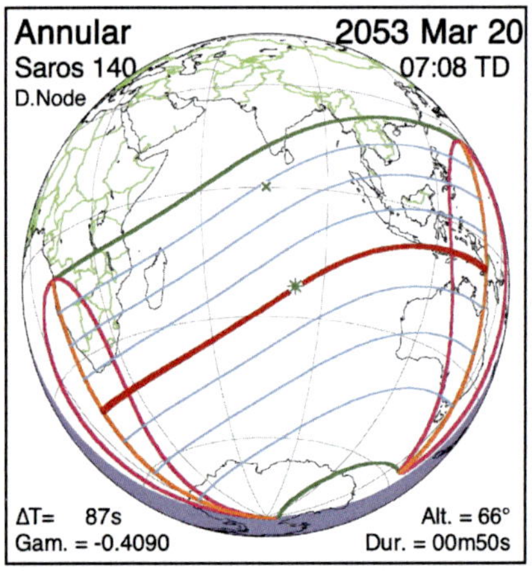
Annular
2053 Mar 20
Saros 140
07:08 TD
D.Node
ΔT= 87s
Alt. = 66°
Gam. = -0.4090
Dur. = 00m50s

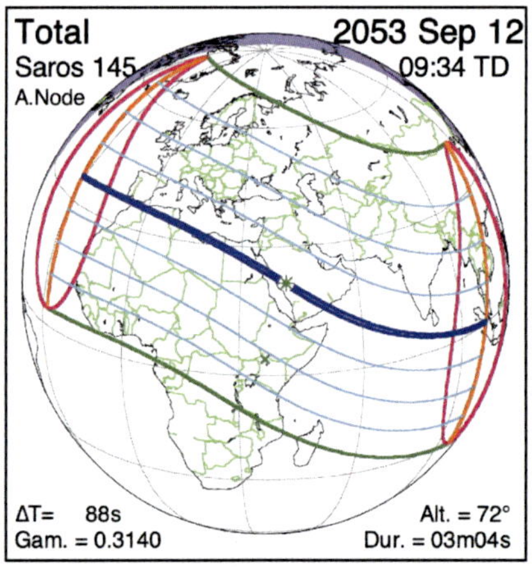
Total
2053 Sep 12
Saros 145
09:34 TD
A.Node
ΔT= 88s
Alt. = 72°
Gam. = 0.3140
Dur. = 03m04s

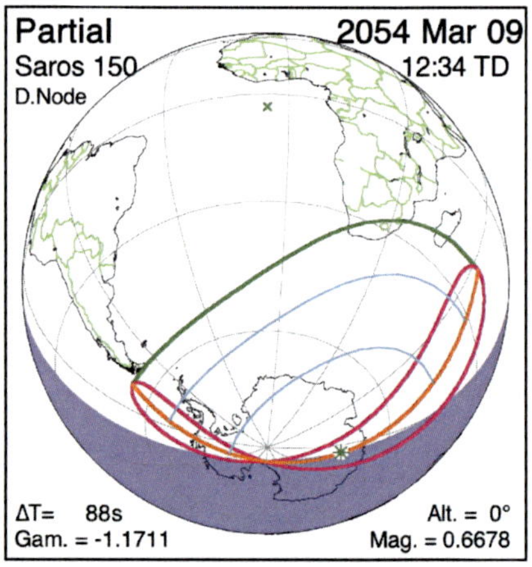
Partial
2054 Mar 09
Saros 150
12:34 TD
D.Node
ΔT= 88s
Alt. = 0°
Gam. = -1.1711
Mag. = 0.6678

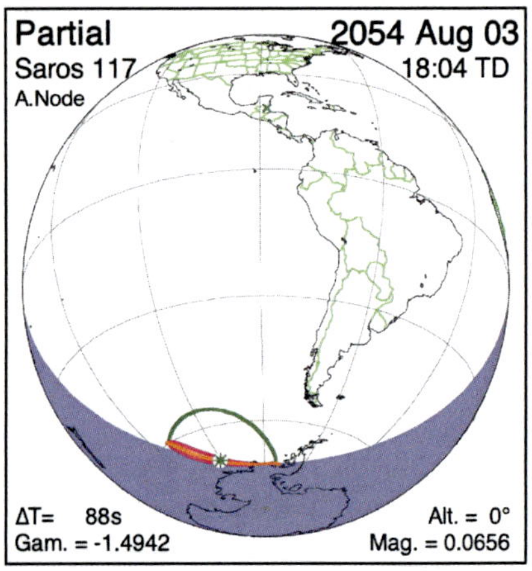
Partial
2054 Aug 03
Saros 117
18:04 TD
A.Node
ΔT= 88s
Alt. = 0°
Gam. = -1.4942
Mag. = 0.0656

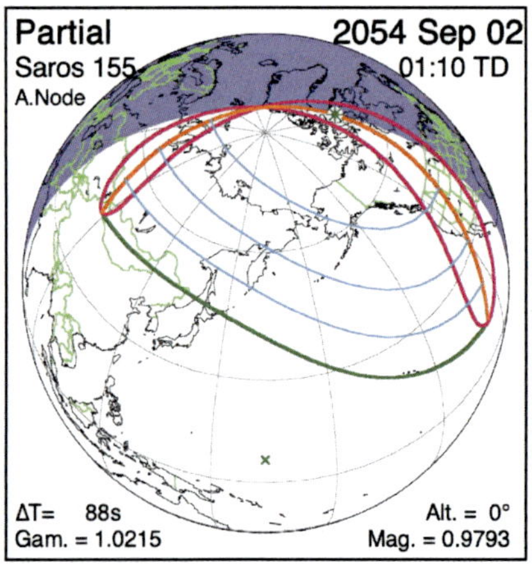
Partial
2054 Sep 02
Saros 155
01:10 TD
A.Node
ΔT= 88s
Alt. = 0°
Gam. = 1.0215
Mag. = 0.9793

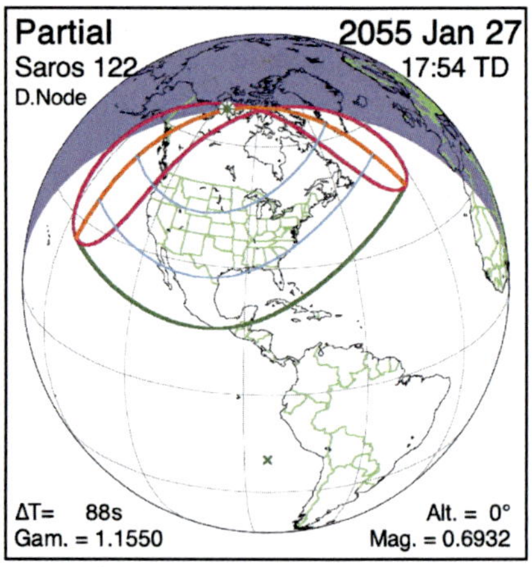
Partial
2055 Jan 27
Saros 122
17:54 TD
D.Node
ΔT= 88s
Alt. = 0°
Gam. = 1.1550
Mag. = 0.6932

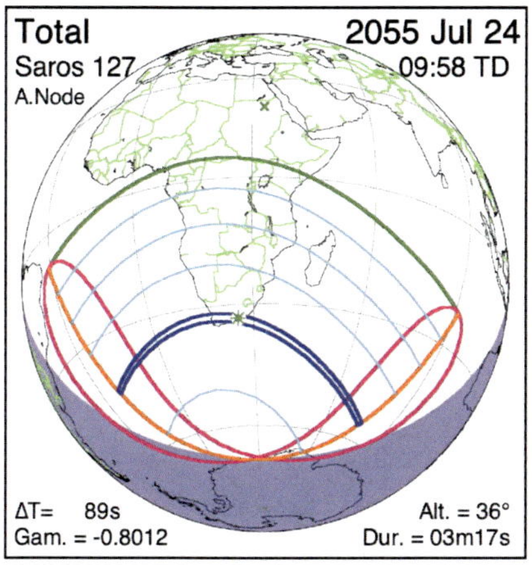
Total
2055 Jul 24
Saros 127
09:58 TD
A.Node
ΔT= 89s
Alt. = 36°
Gam. = -0.8012
Dur. = 03m17s

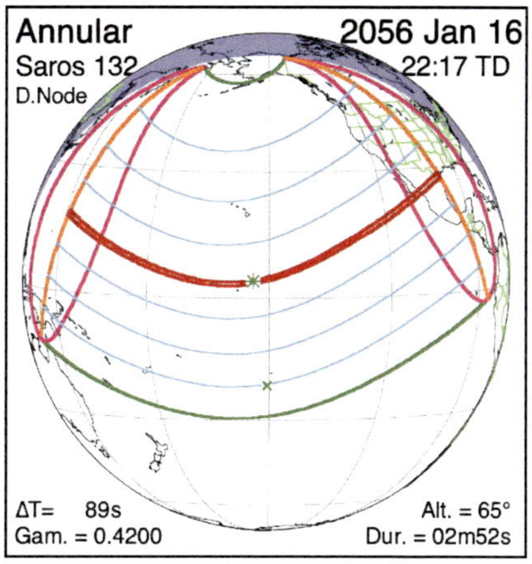
Annular
2056 Jan 16
Saros 132
22:17 TD
D.Node
ΔT= 89s
Alt. = 65°
Gam. = 0.4200
Dur. = 02m52s

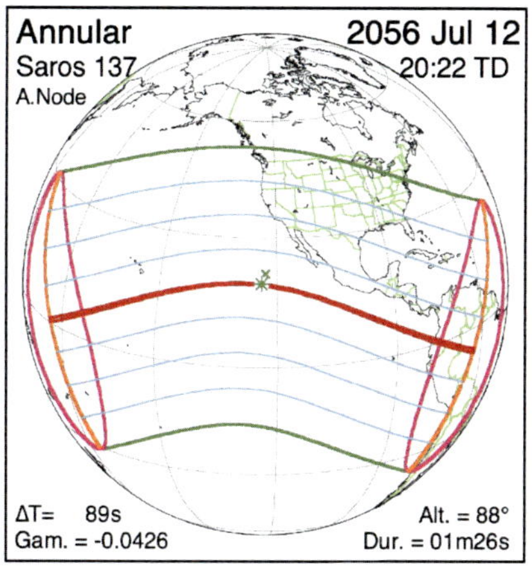
Annular
2056 Jul 12
Saros 137
20:22 TD
A.Node
ΔT= 89s
Alt. = 88°
Gam. = -0.0426
Dur. = 01m26s

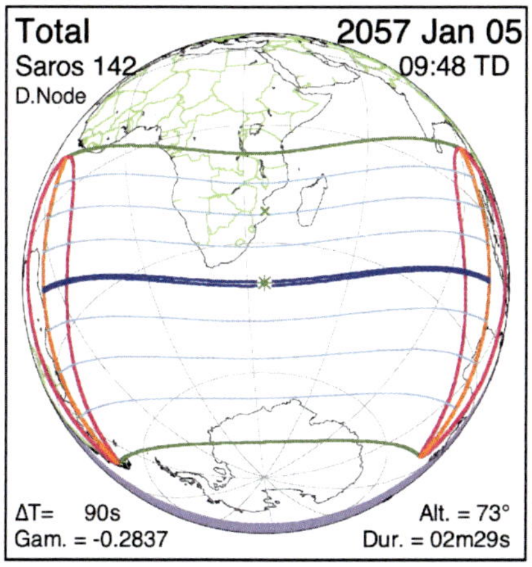
Total
2057 Jan 05
Saros 142
09:48 TD
D.Node
ΔT= 90s
Alt. = 73°
Gam. = -0.2837
Dur. = 02m29s

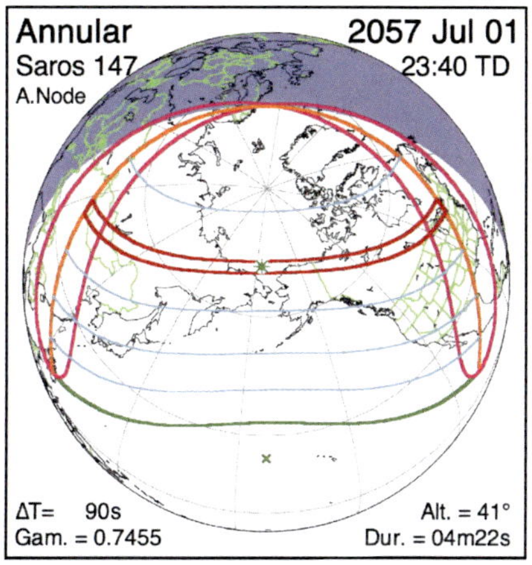
Annular
2057 Jul 01
Saros 147
23:40 TD
A.Node
ΔT= 90s
Alt. = 41°
Gam. = 0.7455
Dur. = 04m22s

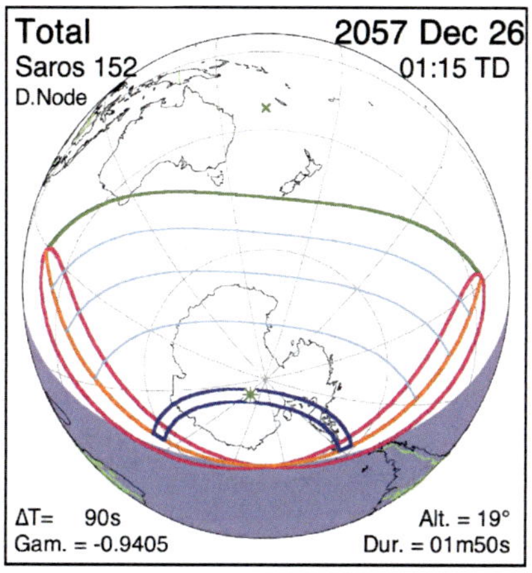
Total
2057 Dec 26
Saros 152
01:15 TD
D.Node
ΔT= 90s
Alt. = 19°
Gam. = -0.9405
Dur. = 01m50s

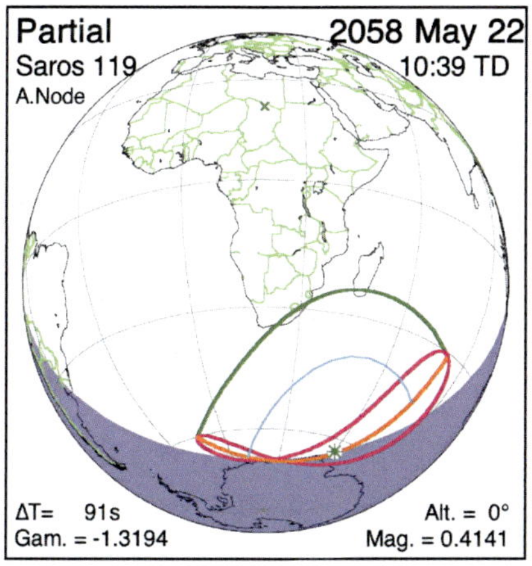
Partial
2058 May 22
Saros 119
10:39 TD
A.Node
ΔT= 91s
Alt. = 0°
Gam. = -1.3194
Mag. = 0.4141

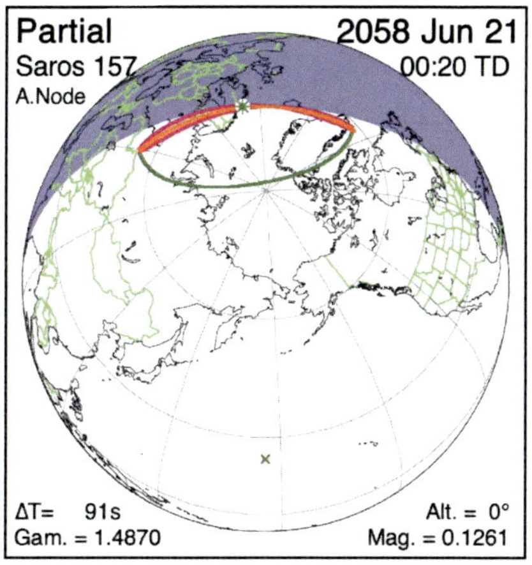
Partial
2058 Jun 21
Saros 157
00:20 TD
A.Node
ΔT= 91s
Alt. = 0°
Gam. = 1.4870
Mag. = 0.1261

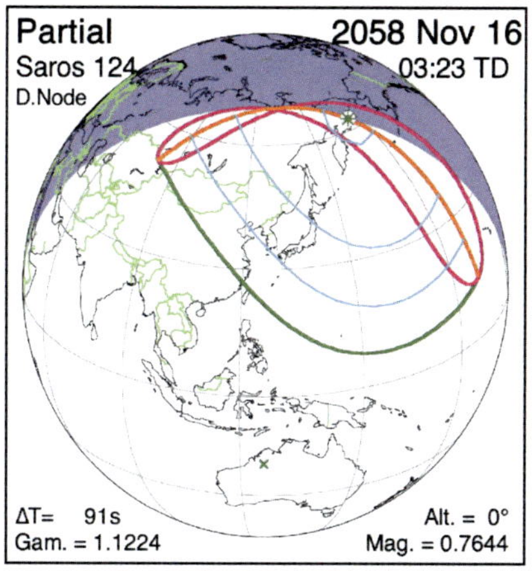
Partial
2058 Nov 16
Saros 124
03:23 TD
D.Node
ΔT= 91s
Alt. = 0°
Gam. = 1.1224
Mag. = 0.7644

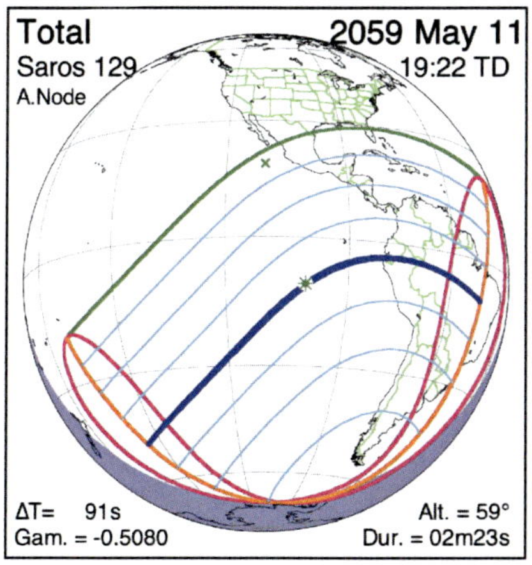
Total
2059 May 11
Saros 129
19:22 TD
A.Node
ΔT= 91s
Alt. = 59°
Gam. = -0.5080
Dur. = 02m23s

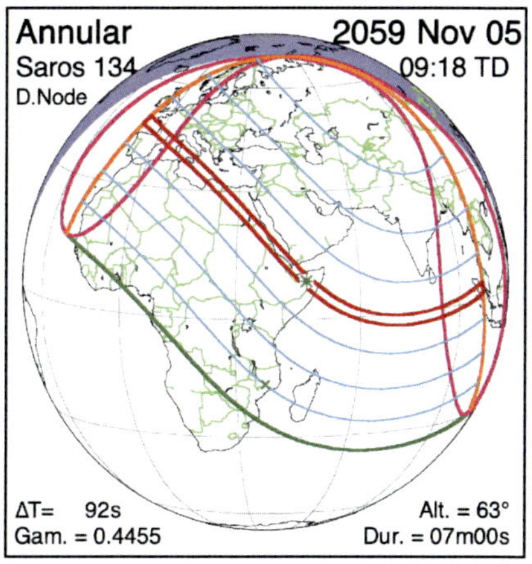
Annular
2059 Nov 05
Saros 134
09:18 TD
D.Node
ΔT= 92s
Alt. = 63°
Gam. = 0.4455
Dur. = 07m00s

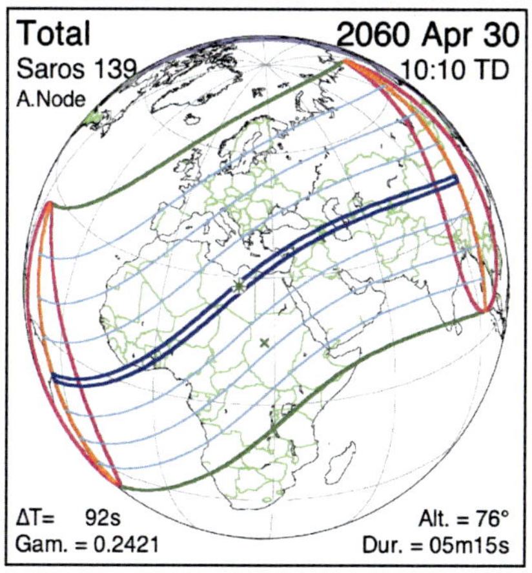
Total
2060 Apr 30
Saros 139
10:10 TD
A.Node
ΔT= 92s
Alt. = 76°
Gam. = 0.2421
Dur. = 05m15s

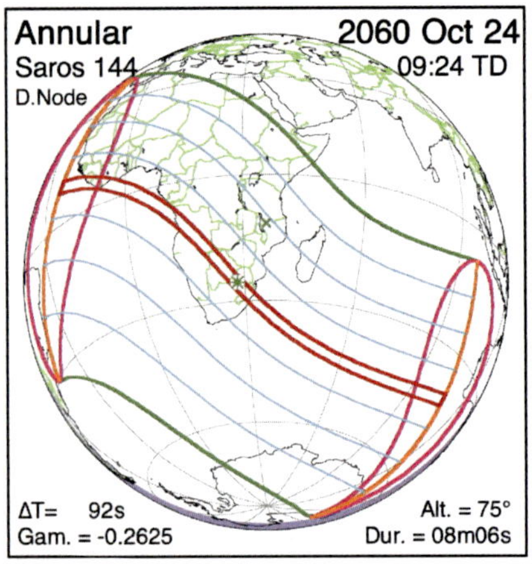
Annular
2060 Oct 24
Saros 144
09:24 TD
D.Node
ΔT= 92s
Alt. = 75°
Gam. = -0.2625
Dur. = 08m06s

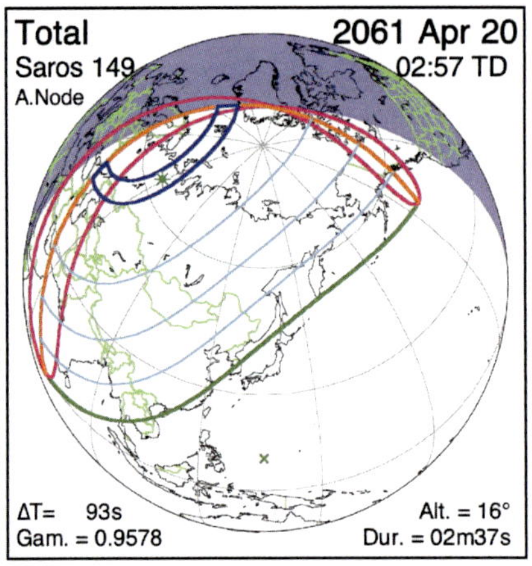
Total
2061 Apr 20
Saros 149
02:57 TD
A.Node
ΔT= 93s
Alt. = 16°
Gam. = 0.9578
Dur. = 02m37s

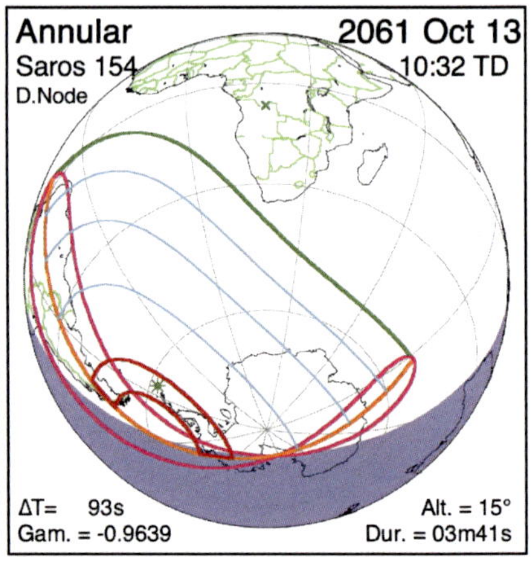
Annular
2061 Oct 13
Saros 154
10:32 TD
D.Node
ΔT= 93s
Alt. = 15°
Gam. = -0.9639
Dur. = 03m41s

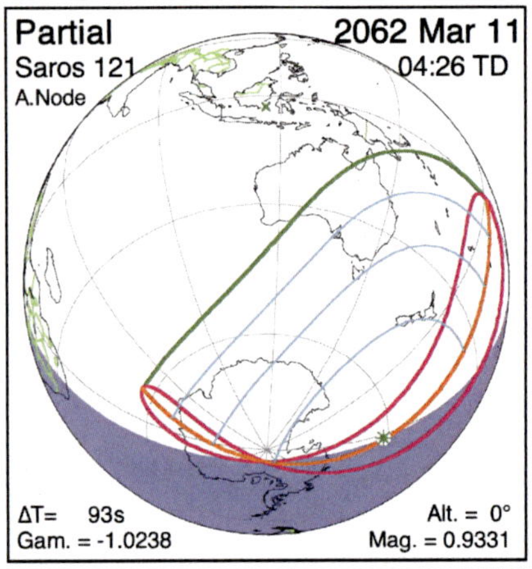
Partial
2062 Mar 11
Saros 121
04:26 TD
A.Node
ΔT= 93s
Alt. = 0°
Gam. = -1.0238
Mag. = 0.9331

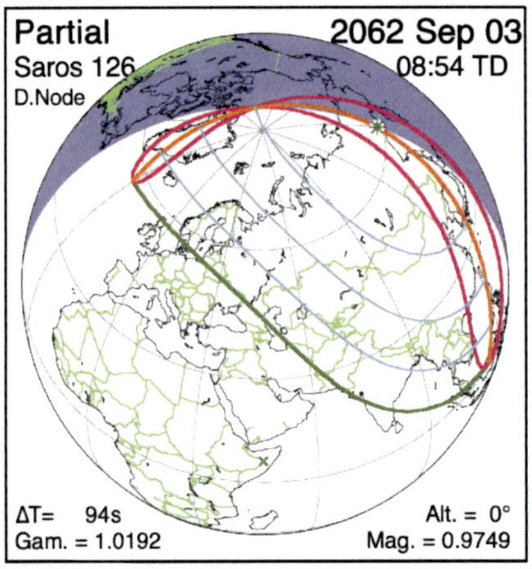
Partial
2062 Sep 03
Saros 126
08:54 TD
D.Node
ΔT= 94s
Alt. = 0°
Gam. = 1.0192
Mag. = 0.9749

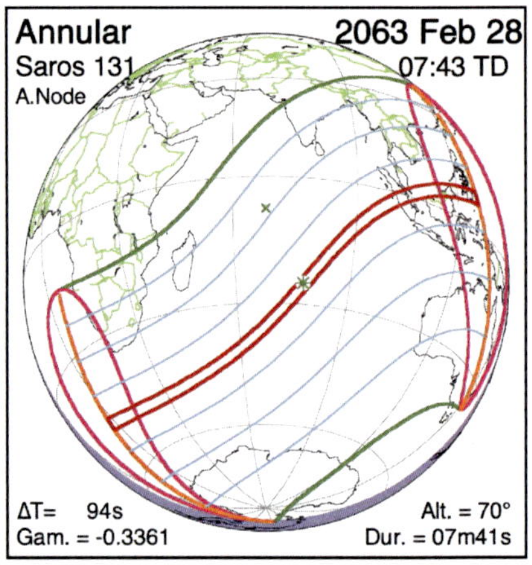
Annular
2063 Feb 28
Saros 131
07:43 TD
A.Node
ΔT= 94s
Alt. = 70°
Gam. = -0.3361
Dur. = 07m41s

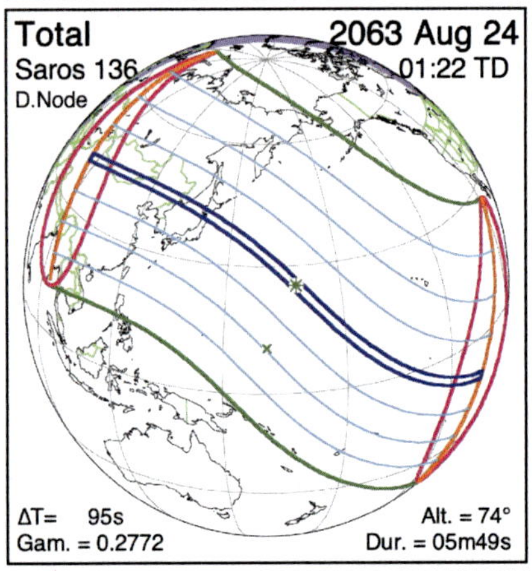
Total
2063 Aug 24
Saros 136
01:22 TD
D.Node
ΔT= 95s
Alt. = 74°
Gam. = 0.2772
Dur. = 05m49s

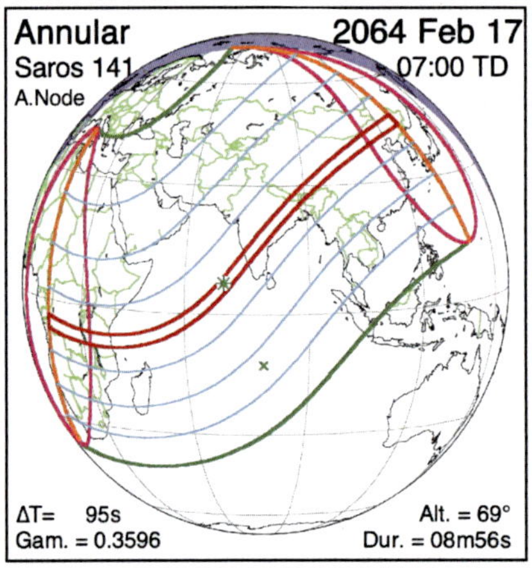
Annular
2064 Feb 17
Saros 141
07:00 TD
A.Node
ΔT= 95s
Alt. = 69°
Gam. = 0.3596
Dur. = 08m56s

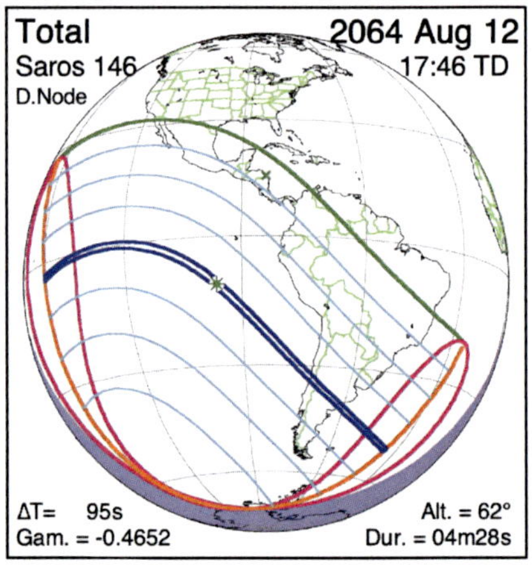
Total
2064 Aug 12
Saros 146
17:46 TD
D.Node
ΔT= 95s
Alt. = 62°
Gam. = -0.4652
Dur. = 04m28s

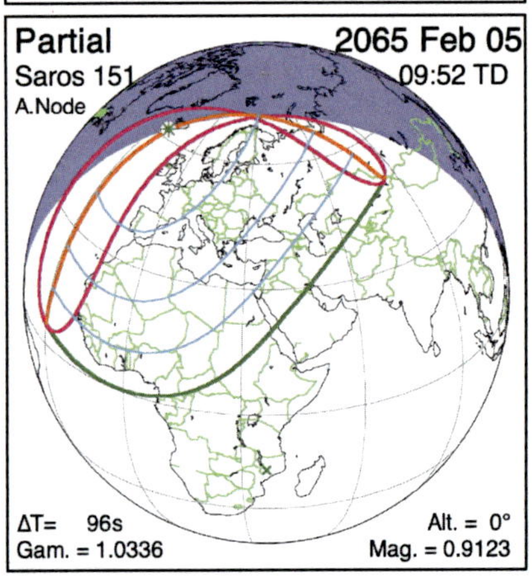
Partial
2065 Feb 05
Saros 151
09:52 TD
A.Node
ΔT= 96s
Alt. = 0°
Gam. = 1.0336
Mag. = 0.9123

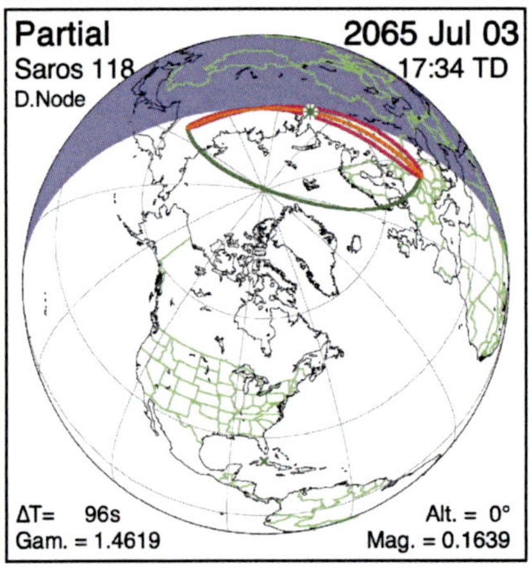
Partial
2065 Jul 03
Saros 118
17:34 TD
D.Node
ΔT= 96s
Alt. = 0°
Gam. = 1.4619
Mag. = 0.1639

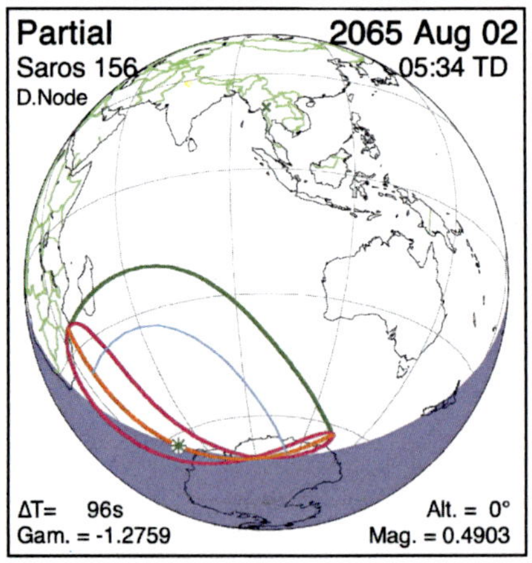
Partial
2065 Aug 02
Saros 156
05:34 TD
D.Node
ΔT= 96s
Alt. = 0°
Gam. = -1.2759
Mag. = 0.4903

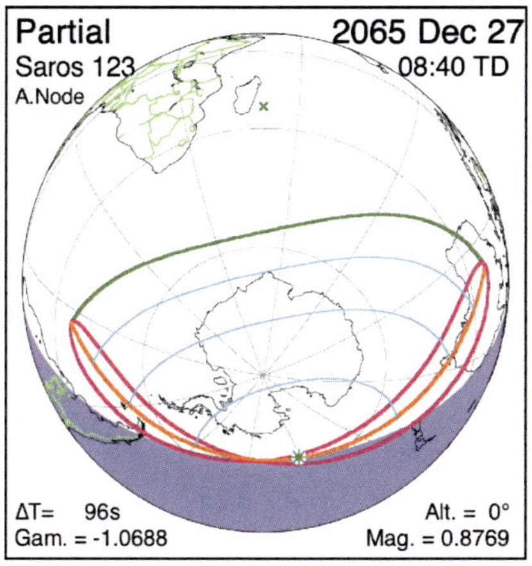
Partial
2065 Dec 27
Saros 123
08:40 TD
A.Node
ΔT= 96s
Alt. = 0°
Gam. = -1.0688
Mag. = 0.8769

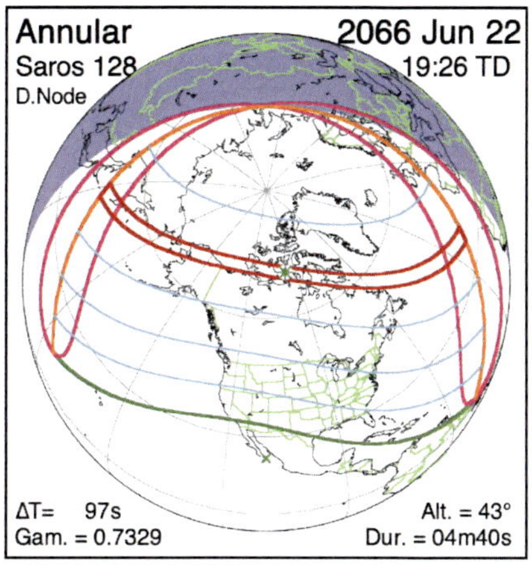
Annular
2066 Jun 22
Saros 128
19:26 TD
D.Node
ΔT= 97s
Alt. = 43°
Gam. = 0.7329
Dur. = 04m40s

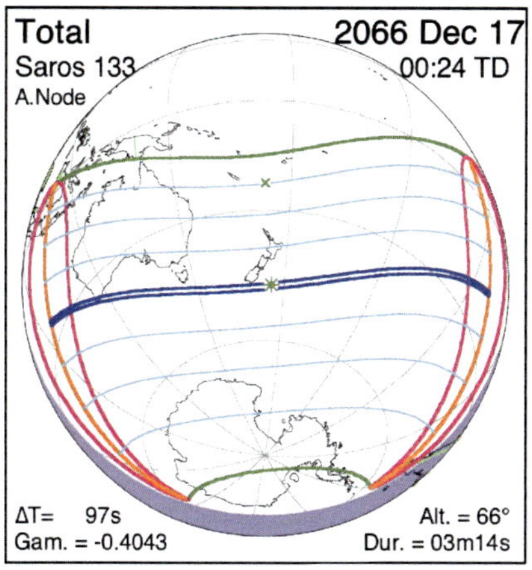
Total
2066 Dec 17
Saros 133
00:24 TD
A.Node
ΔT= 97s
Alt. = 66°
Gam. = -0.4043
Dur. = 03m14s

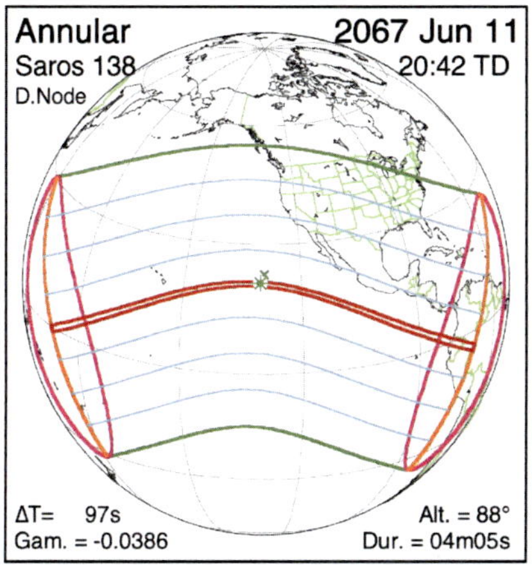
Annular
2067 Jun 11
Saros 138
20:42 TD
D.Node
ΔT= 97s
Alt. = 88°
Gam. = -0.0386
Dur. = 04m05s

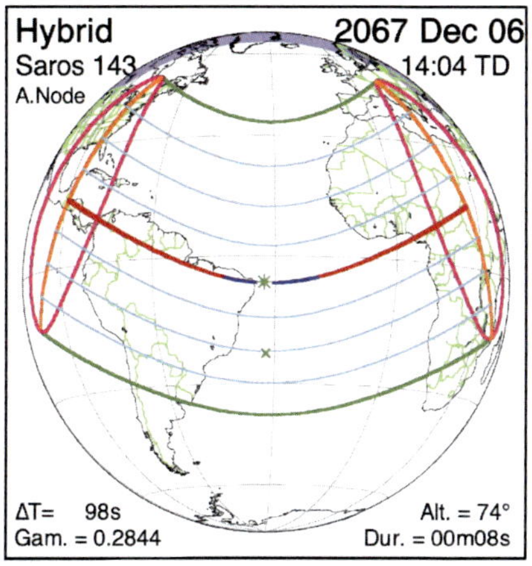
Hybrid
2067 Dec 06
Saros 143
14:04 TD
A.Node
ΔT= 98s
Alt. = 74°
Gam. = 0.2844
Dur. = 00m08s

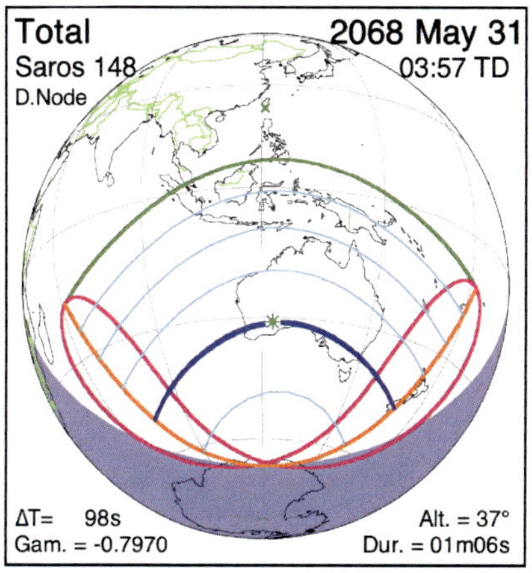
Total
2068 May 31
Saros 148
03:57 TD
D.Node
ΔT= 98s
Alt. = 37°
Gam. = -0.7970
Dur. = 01m06s

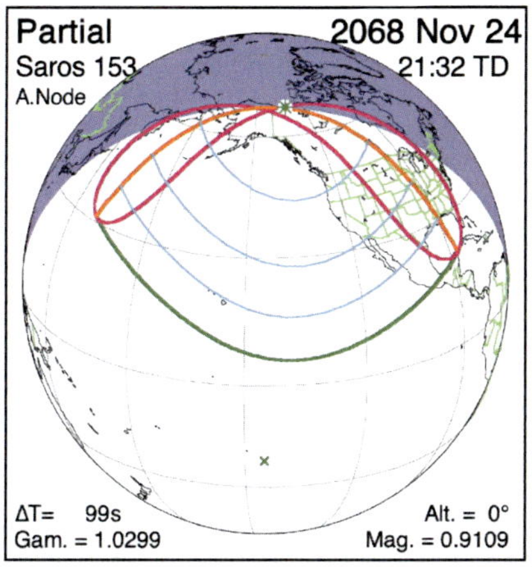
Partial
2068 Nov 24
Saros 153
21:32 TD
A.Node
ΔT= 99s
Alt. = 0°
Gam. = 1.0299
Mag. = 0.9109

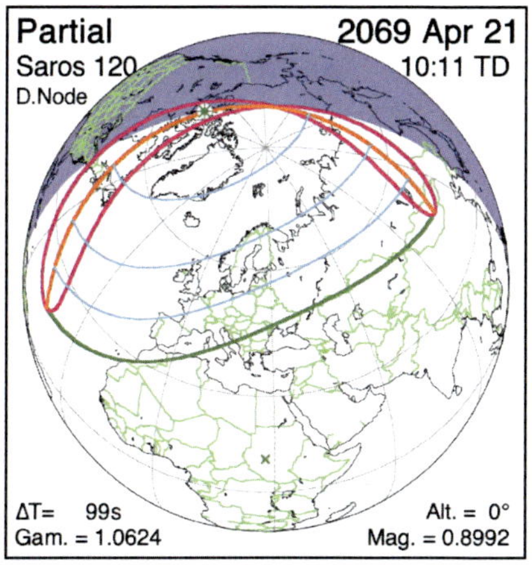
Partial
2069 Apr 21
Saros 120
10:11 TD
D.Node
ΔT= 99s
Alt. = 0°
Gam. = 1.0624
Mag. = 0.8992

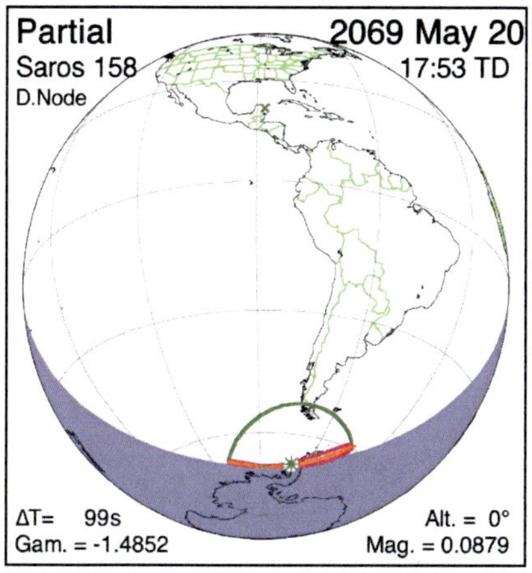
Partial
2069 May 20
Saros 158
17:53 TD
D.Node
ΔT= 99s
Alt. = 0°
Gam. = -1.4852
Mag. = 0.0879

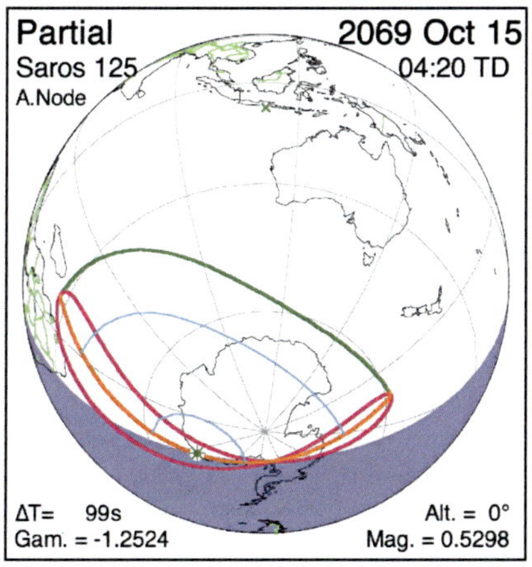
Partial
2069 Oct 15
Saros 125
04:20 TD
A.Node
ΔT= 99s
Alt. = 0°
Gam. = -1.2524
Mag. = 0.5298

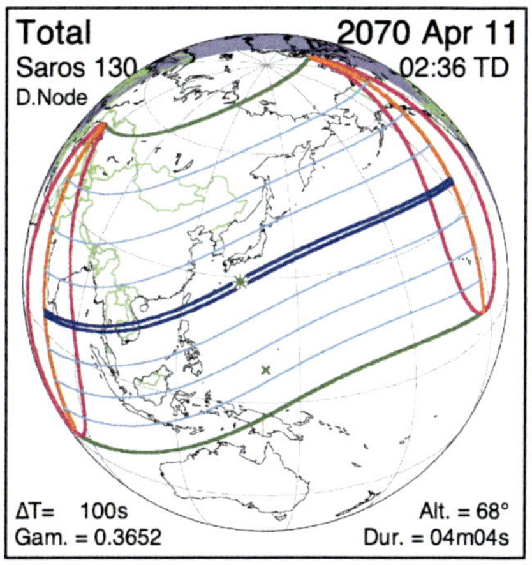
Total
2070 Apr 11
Saros 130
02:36 TD
D.Node
ΔT= 100s
Alt. = 68°
Gam. = 0.3652
Dur. = 04m04s

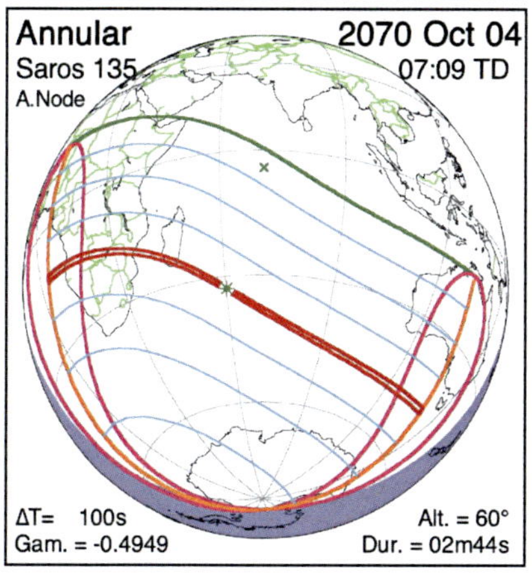
Annular
2070 Oct 04
Saros 135
07:09 TD
A.Node
ΔT= 100s
Alt. = 60°
Gam. = -0.4949
Dur. = 02m44s

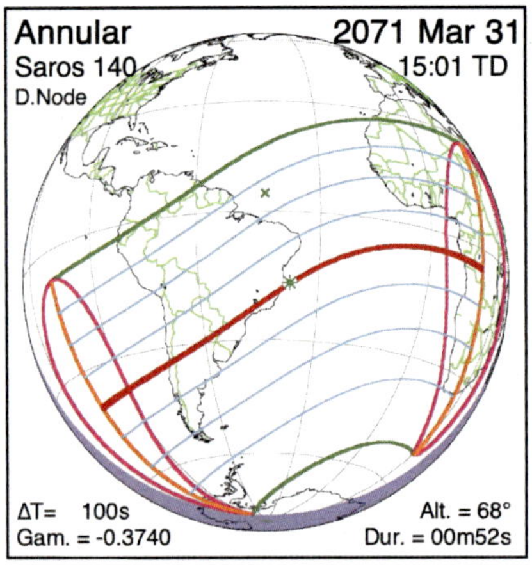
Annular
2071 Mar 31
Saros 140
15:01 TD
D.Node
ΔT= 100s
Alt. = 68°
Gam. = -0.3740
Dur. = 00m52s

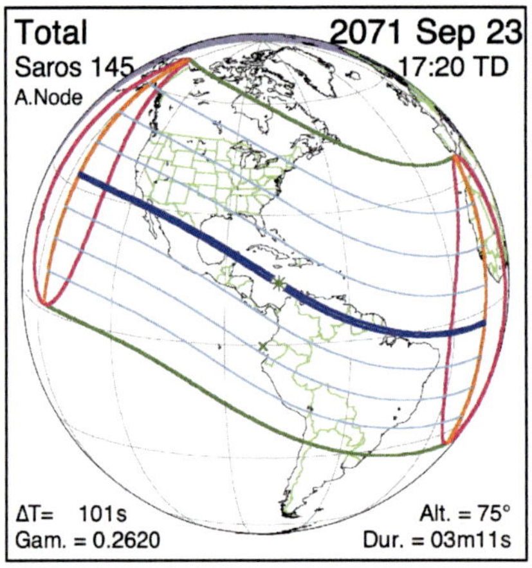
Total
2071 Sep 23
Saros 145
17:20 TD
A.Node
ΔT= 101s
Alt. = 75°
Gam. = 0.2620
Dur. = 03m11s

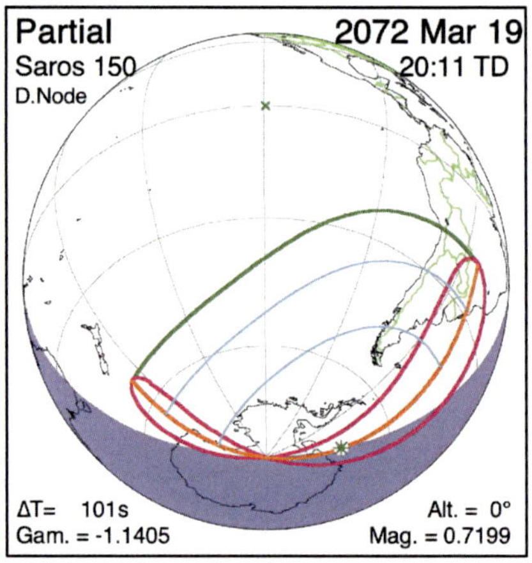
Partial
2072 Mar 19
Saros 150
20:11 TD
D.Node
ΔT= 101s
Alt. = 0°
Gam. = -1.1405
Mag. = 0.7199

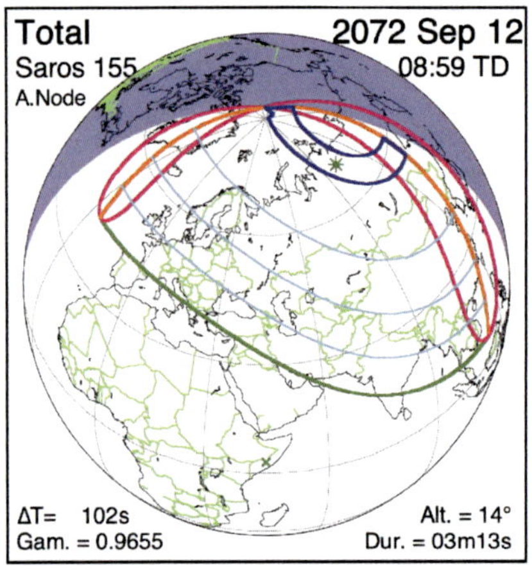
Total
2072 Sep 12
Saros 155
08:59 TD
A.Node
ΔT= 102s
Alt. = 14°
Gam. = 0.9655
Dur. = 03m13s

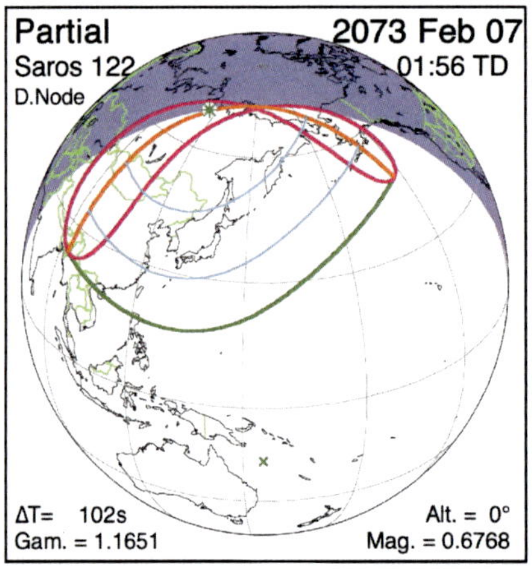
Partial
2073 Feb 07
Saros 122
01:56 TD
D.Node
ΔT= 102s
Alt. = 0°
Gam. = 1.1651
Mag. = 0.6768

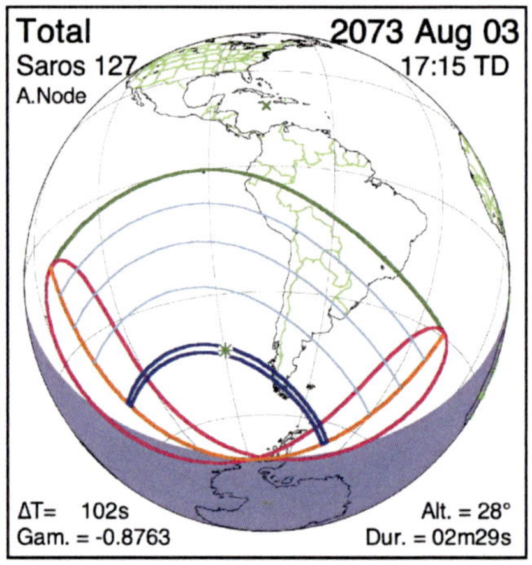
Total
2073 Aug 03
Saros 127
17:15 TD
A.Node
ΔT= 102s
Alt. = 28°
Gam. = -0.8763
Dur. = 02m29s

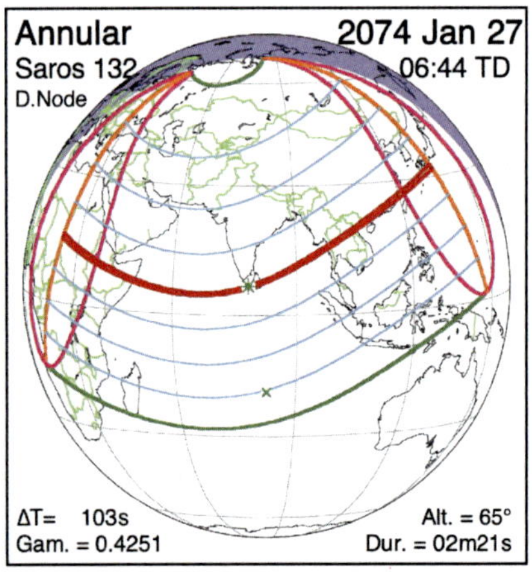
Annular
2074 Jan 27
Saros 132
06:44 TD
D.Node
ΔT= 103s
Alt. = 65°
Gam. = 0.4251
Dur. = 02m21s

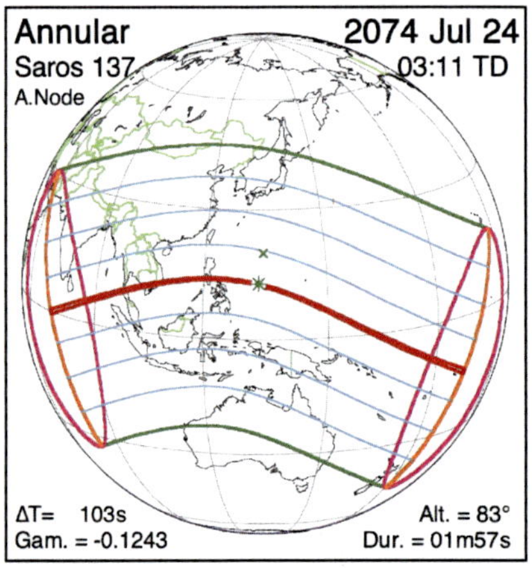
Annular
2074 Jul 24
Saros 137
03:11 TD
A.Node
ΔT= 103s
Alt. = 83°
Gam. = -0.1243
Dur. = 01m57s

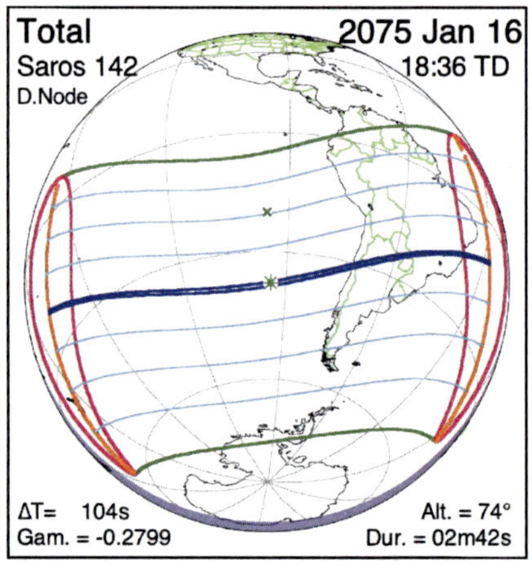
Total
2075 Jan 16
Saros 142
18:36 TD
D.Node
ΔT= 104s
Alt. = 74°
Gam. = -0.2799
Dur. = 02m42s

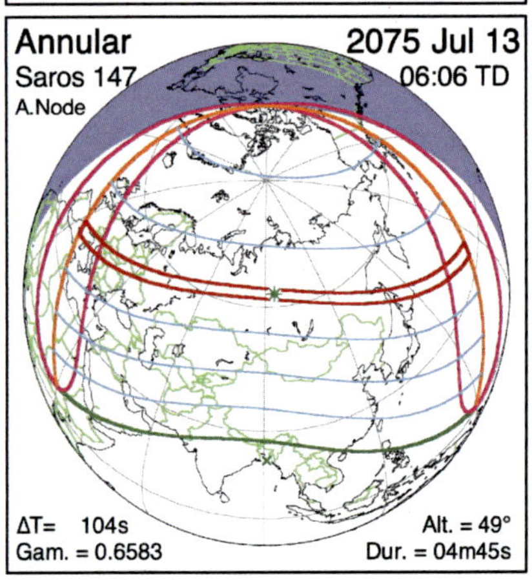
Annular
2075 Jul 13
Saros 147
06:06 TD
A.Node
ΔT= 104s
Alt. = 49°
Gam. = 0.6583
Dur. = 04m45s

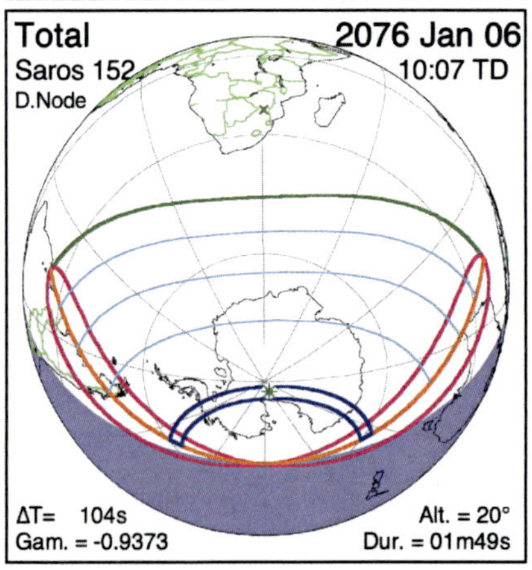
Total
2076 Jan 06
Saros 152
10:07 TD
D.Node
ΔT= 104s
Alt. = 20°
Gam. = -0.9373
Dur. = 01m49s

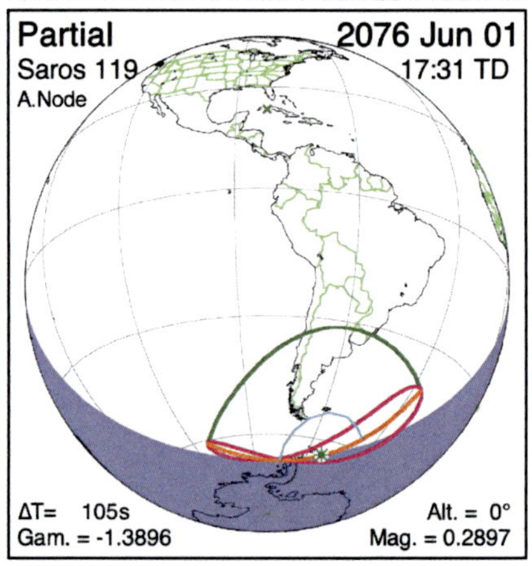
Partial
2076 Jun 01
Saros 119
17:31 TD
A.Node
ΔT= 105s
Alt. = 0°
Gam. = -1.3896
Mag. = 0.2897

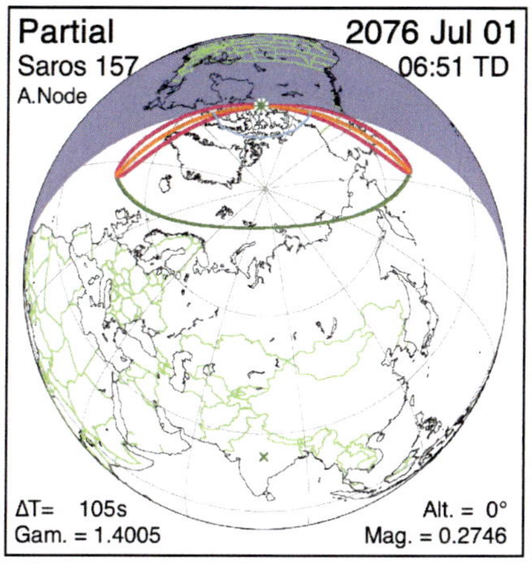
Partial
2076 Jul 01
Saros 157
06:51 TD
A.Node
ΔT= 105s
Alt. = 0°
Gam. = 1.4005
Mag. = 0.2746

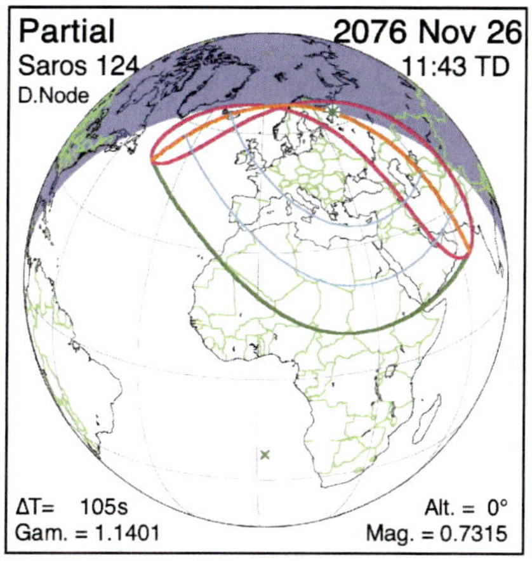
Partial
2076 Nov 26
Saros 124
11:43 TD
D.Node
ΔT= 105s
Alt. = 0°
Gam. = 1.1401
Mag. = 0.7315

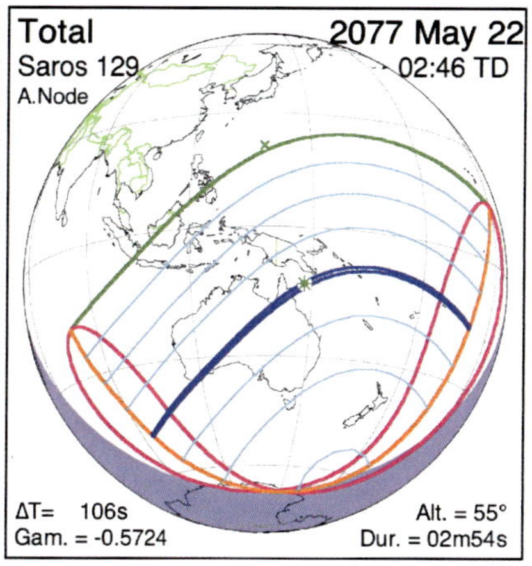
Total
2077 May 22
Saros 129
02:46 TD
A.Node
ΔT= 106s
Alt. = 55°
Gam. = -0.5724
Dur. = 02m54s

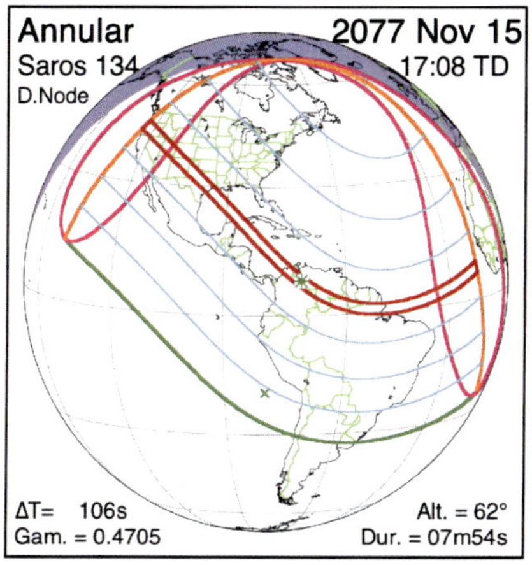
Annular
2077 Nov 15
Saros 134
17:08 TD
D.Node
ΔT= 106s
Alt. = 62°
Gam. = 0.4705
Dur. = 07m54s

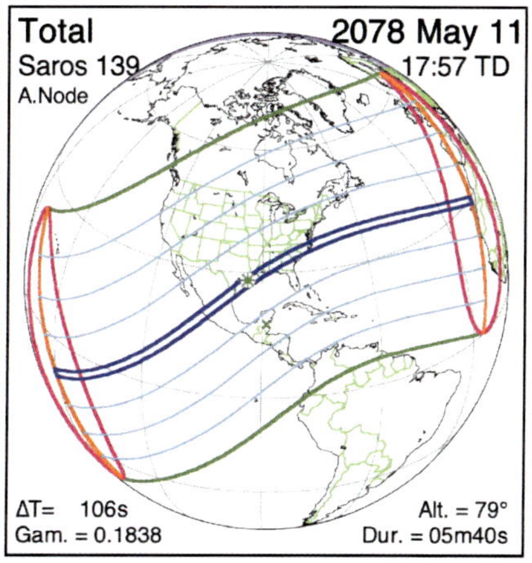
Total
2078 May 11
Saros 139
17:57 TD
A.Node
ΔT= 106s
Alt. = 79°
Gam. = 0.1838
Dur. = 05m40s

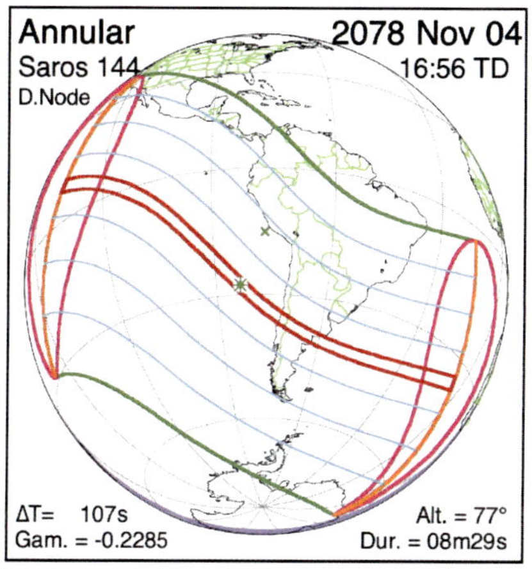
Annular
2078 Nov 04
Saros 144
16:56 TD
D.Node
ΔT= 107s
Alt. = 77°
Gam. = -0.2285
Dur. = 08m29s

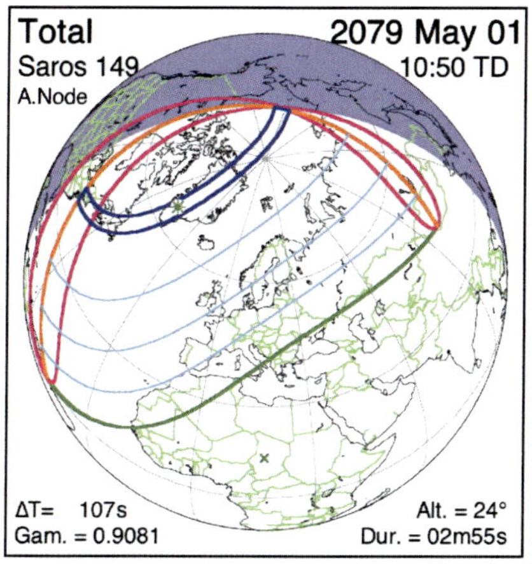
Total
2079 May 01
Saros 149
10:50 TD
A.Node
ΔT= 107s
Alt. = 24°
Gam. = 0.9081
Dur. = 02m55s

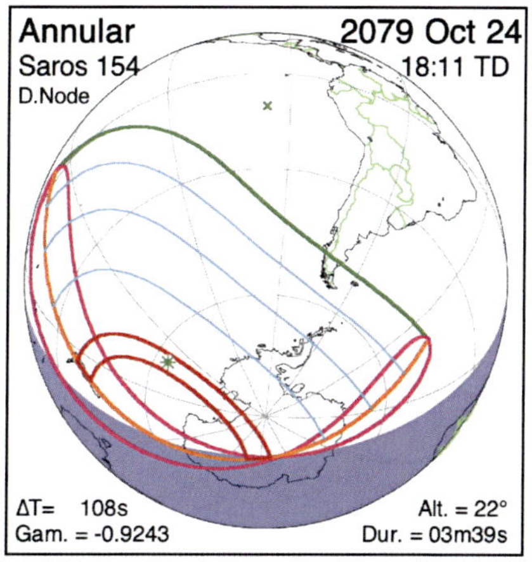
Annular
2079 Oct 24
Saros 154
18:11 TD
D.Node
ΔT= 108s
Alt. = 22°
Gam. = -0.9243
Dur. = 03m39s

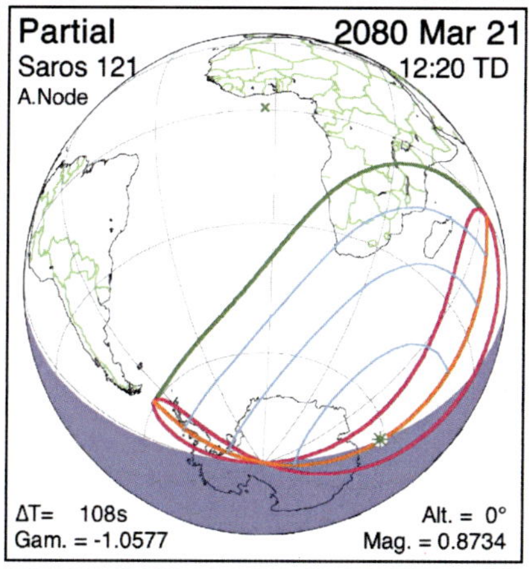
Partial
2080 Mar 21
Saros 121
12:20 TD
A.Node
ΔT= 108s
Alt. = 0°
Gam. = -1.0577
Mag. = 0.8734

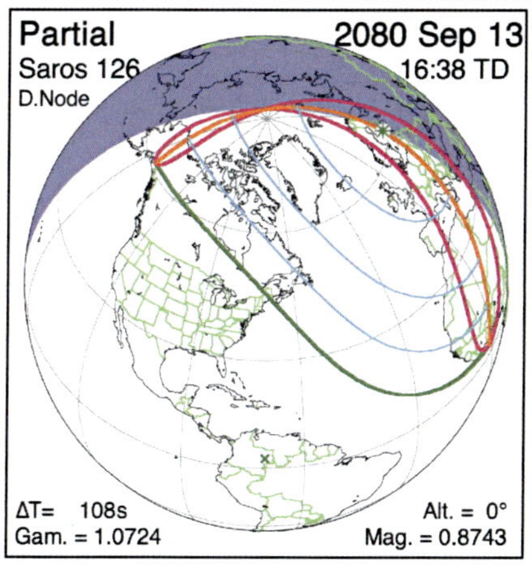
Partial
2080 Sep 13
Saros 126
16:38 TD
D.Node
ΔT= 108s
Alt. = 0°
Gam. = 1.0724
Mag. = 0.8743

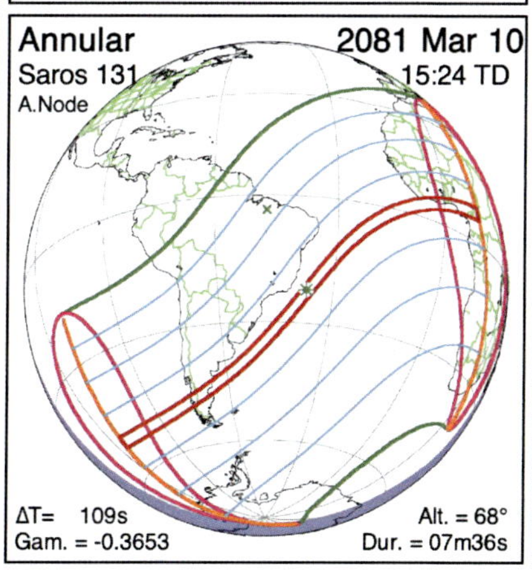
Annular
2081 Mar 10
Saros 131
15:24 TD
A.Node
ΔT= 109s
Alt. = 68°
Gam. = -0.3653
Dur. = 07m36s

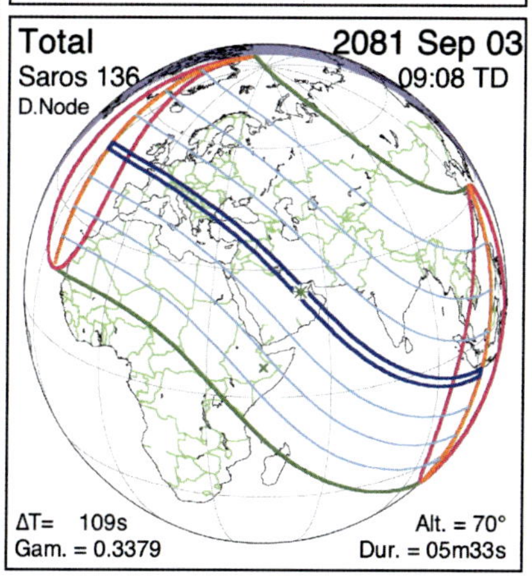
Total
2081 Sep 03
Saros 136
09:08 TD
D.Node
ΔT= 109s
Alt. = 70°
Gam. = 0.3379
Dur. = 05m33s

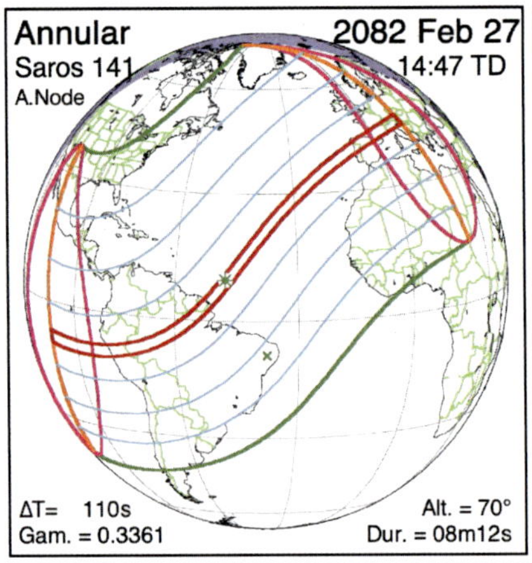
Annular
2082 Feb 27
Saros 141
14:47 TD
A.Node
ΔT= 110s
Alt. = 70°
Gam. = 0.3361
Dur. = 08m12s

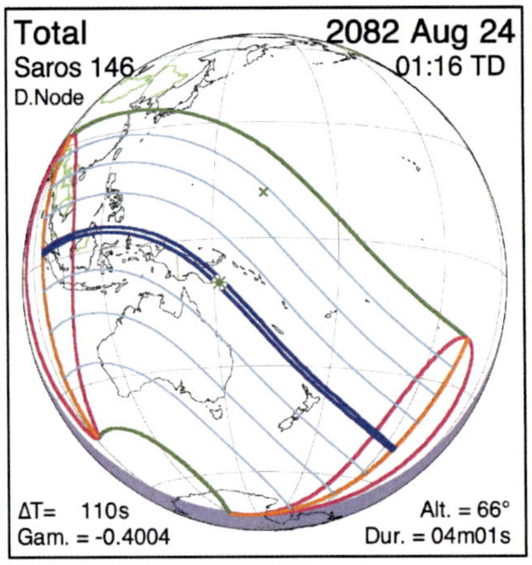
Total
2082 Aug 24
Saros 146
01:16 TD
D.Node
ΔT= 110s
Alt. = 66°
Gam. = -0.4004
Dur. = 04m01s

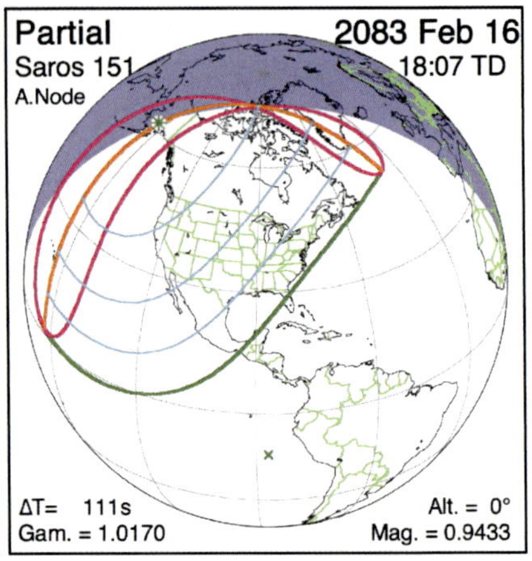
Partial
2083 Feb 16
Saros 151
18:07 TD
A.Node
ΔT= 111s
Alt. = 0°
Gam. = 1.0170
Mag. = 0.9433

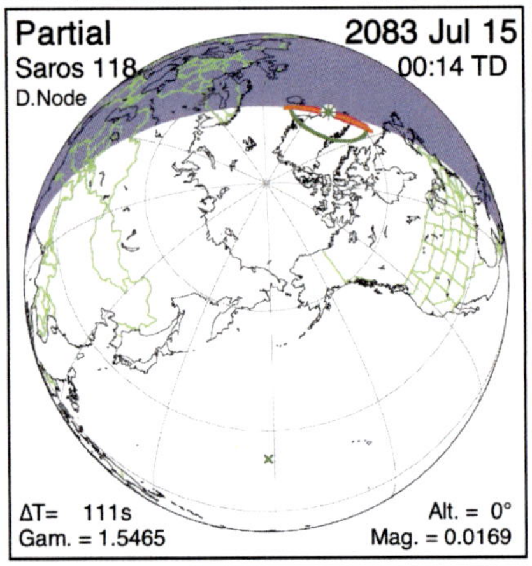
Partial
2083 Jul 15
Saros 118
00:14 TD
D.Node
ΔT= 111s
Alt. = 0°
Gam. = 1.5465
Mag. = 0.0169

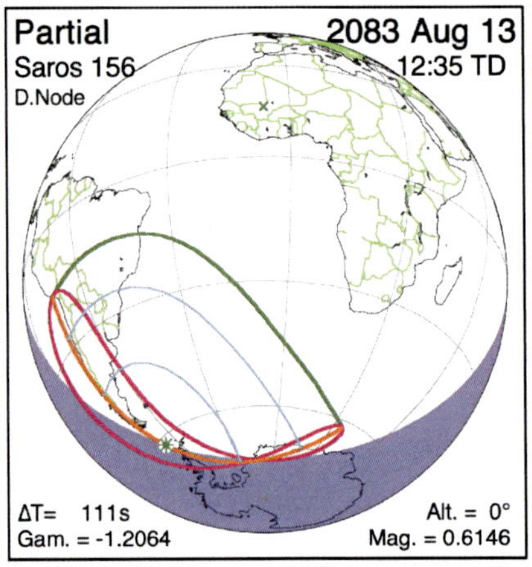
Partial
2083 Aug 13
Saros 156
12:35 TD
D.Node
ΔT= 111s
Alt. = 0°
Gam. = -1.2064
Mag. = 0.6146

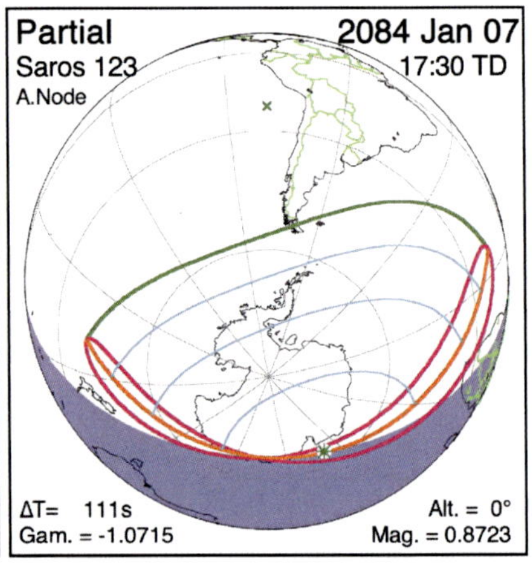
Partial
2084 Jan 07
Saros 123
17:30 TD
A.Node
ΔT= 111s
Alt. = 0°
Gam. = -1.0715
Mag. = 0.8723

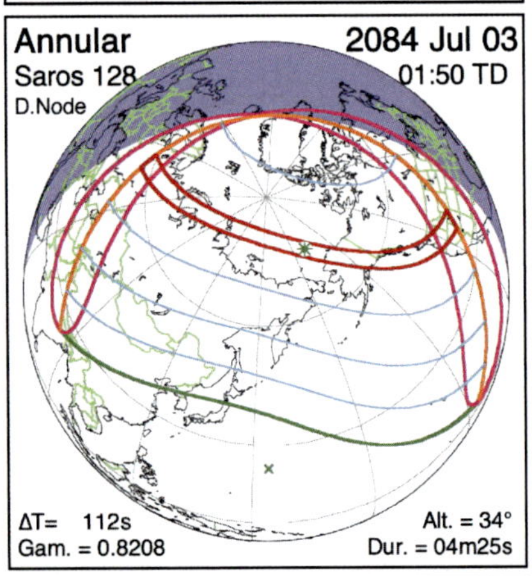
Annular
2084 Jul 03
Saros 128
01:50 TD
D.Node
ΔT= 112s
Alt. = 34°
Gam. = 0.8208
Dur. = 04m25s

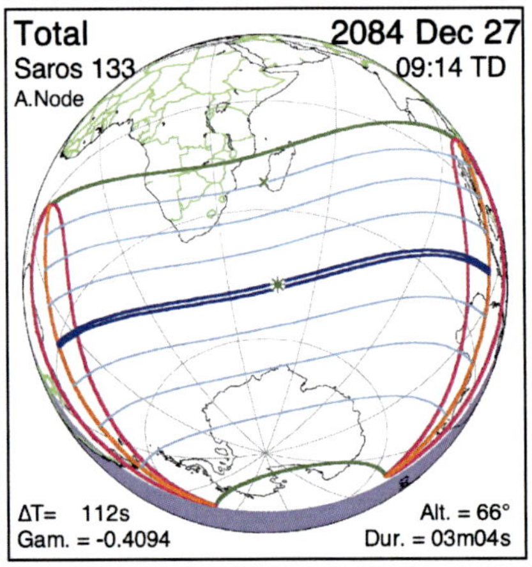
Total
2084 Dec 27
Saros 133
09:14 TD
A.Node
ΔT= 112s
Alt. = 66°
Gam. = -0.4094
Dur. = 03m04s

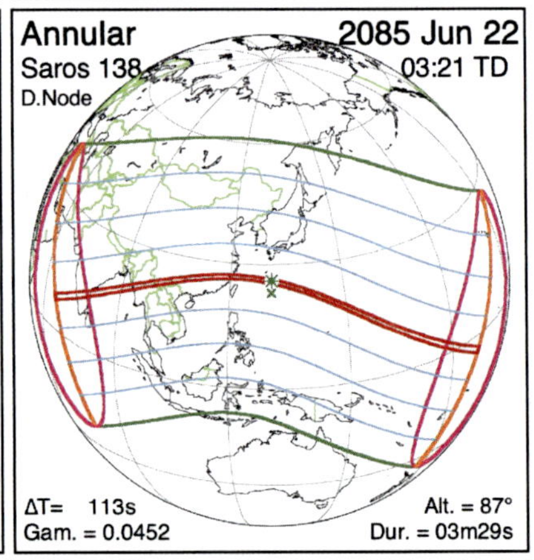
Annular
2085 Jun 22
Saros 138
03:21 TD
D.Node
ΔT= 113s
Alt. = 87°
Gam. = 0.0452
Dur. = 03m29s

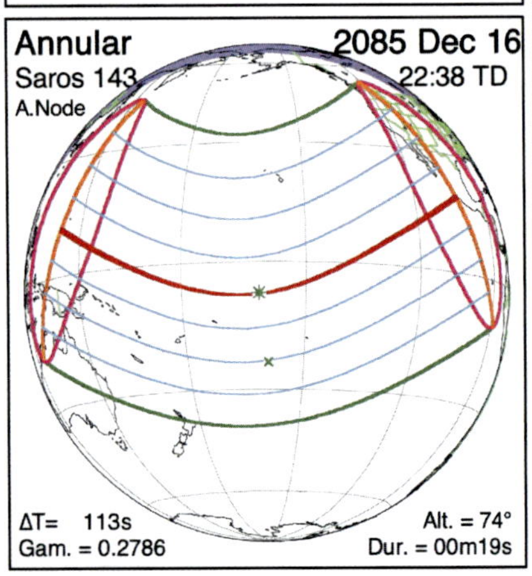
Annular
2085 Dec 16
Saros 143
22:38 TD
A.Node
ΔT= 113s
Alt. = 74°
Gam. = 0.2786
Dur. = 00m19s

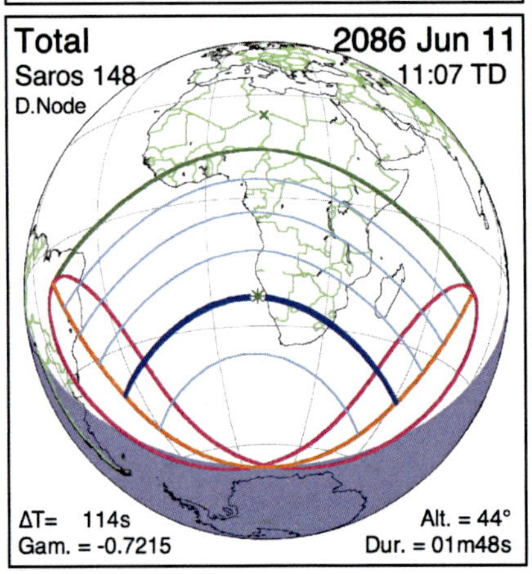
Total
2086 Jun 11
Saros 148
11:07 TD
D.Node
ΔT= 114s
Alt. = 44°
Gam. = -0.7215
Dur. = 01m48s

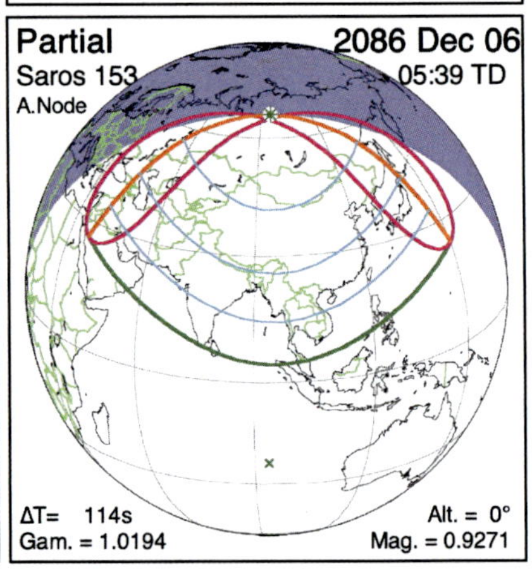
Partial
2086 Dec 06
Saros 153
05:39 TD
A.Node
ΔT= 114s
Alt. = 0°
Gam. = 1.0194
Mag. = 0.9271

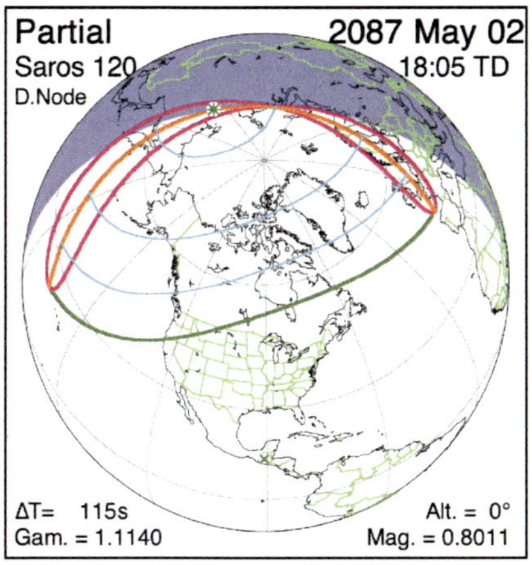
Partial
2087 May 02
Saros 120
18:05 TD
D.Node
ΔT= 115s
Alt. = 0°
Gam. = 1.1140
Mag. = 0.8011

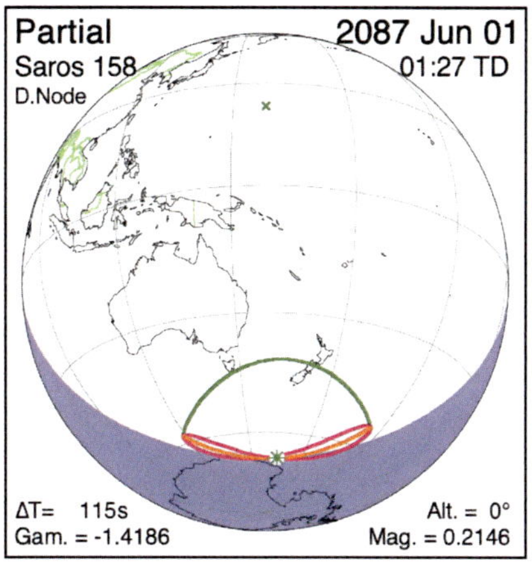
Partial
2087 Jun 01
Saros 158
01:27 TD
D.Node
ΔT= 115s
Alt. = 0°
Gam. = -1.4186
Mag. = 0.2146

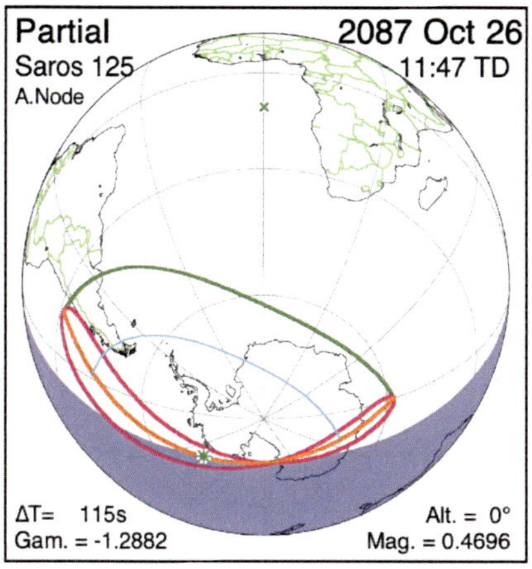
Partial
2087 Oct 26
Saros 125
11:47 TD
A.Node
ΔT= 115s
Alt. = 0°
Gam. = -1.2882
Mag. = 0.4696

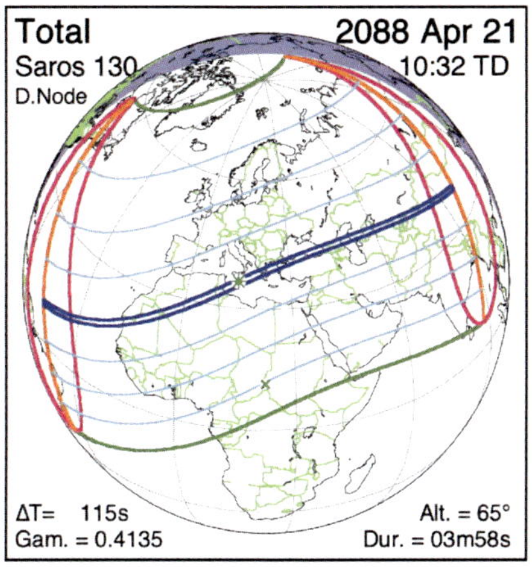
Total
2088 Apr 21
Saros 130
10:32 TD
D.Node
ΔT= 115s
Alt. = 65°
Gam. = 0.4135
Dur. = 03m58s

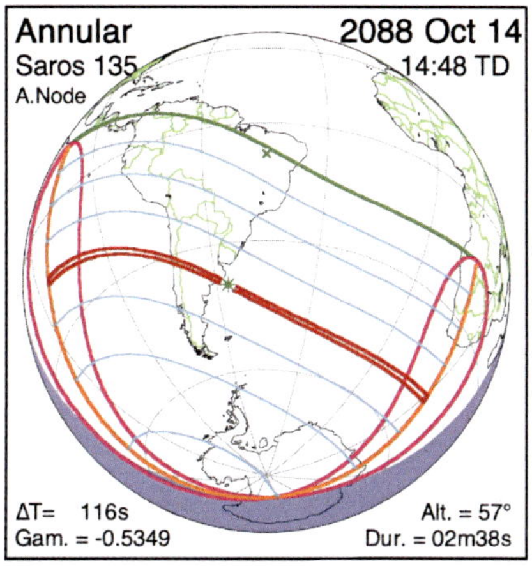
Annular
2088 Oct 14
Saros 135
14:48 TD
A.Node
ΔT= 116s
Alt. = 57°
Gam. = -0.5349
Dur. = 02m38s

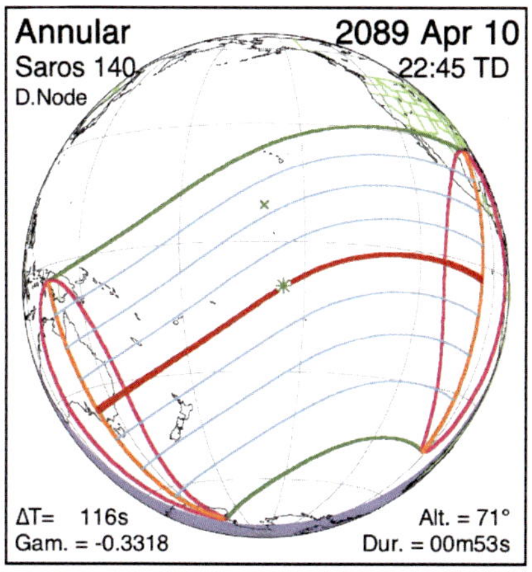
Annular
2089 Apr 10
Saros 140
22:45 TD
D.Node
ΔT= 116s
Alt. = 71°
Gam. = -0.3318
Dur. = 00m53s

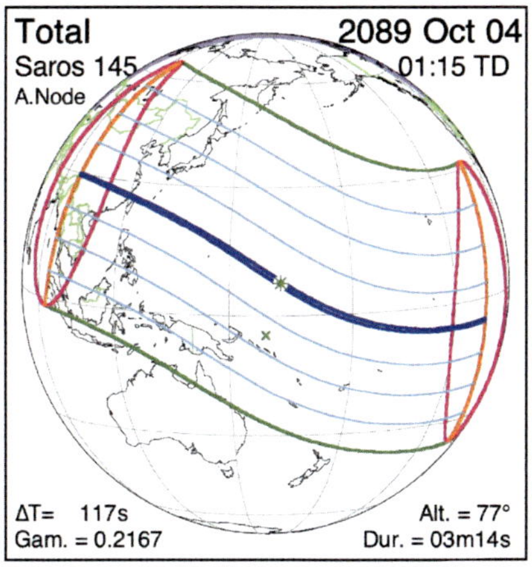
Total
2089 Oct 04
Saros 145
01:15 TD
A.Node
ΔT= 117s
Alt. = 77°
Gam. = 0.2167
Dur. = 03m14s

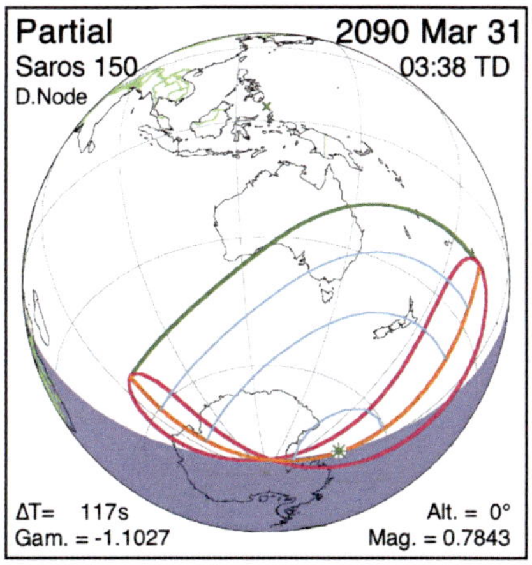
Partial
2090 Mar 31
Saros 150
03:38 TD
D.Node
ΔT= 117s
Alt. = 0°
Gam. = -1.1027
Mag. = 0.7843

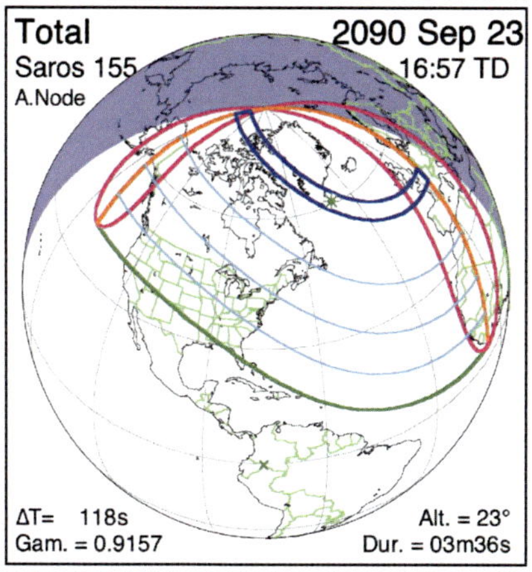
Total
2090 Sep 23
Saros 155
16:57 TD
A.Node
ΔT= 118s
Alt. = 23°
Gam. = 0.9157
Dur. = 03m36s

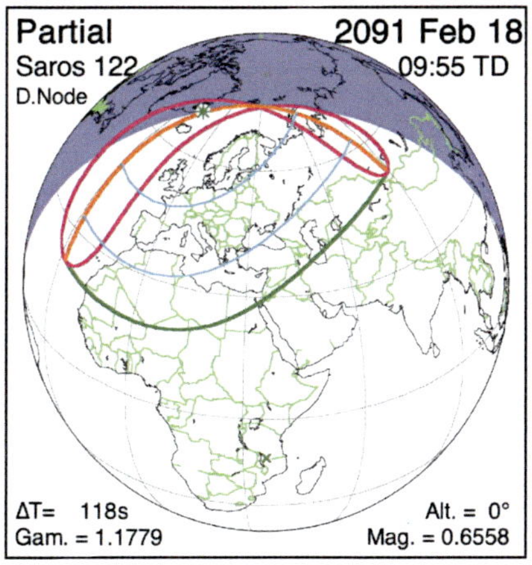
Partial
2091 Feb 18
Saros 122
09:55 TD
D.Node
ΔT= 118s
Alt. = 0°
Gam. = 1.1779
Mag. = 0.6558

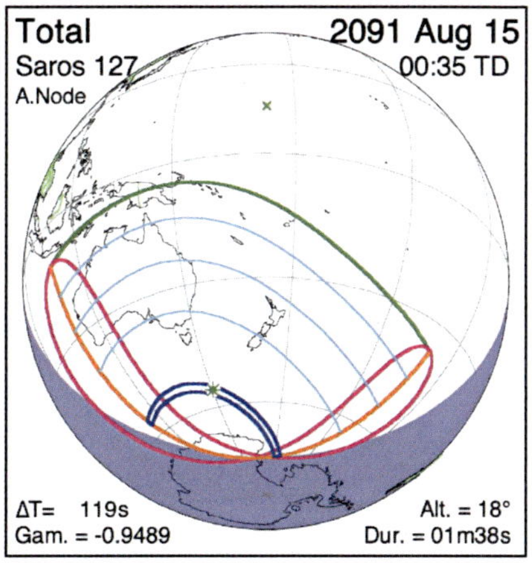
Total
2091 Aug 15
Saros 127
00:35 TD
A.Node
ΔT= 119s
Alt. = 18°
Gam. = -0.9489
Dur. = 01m38s

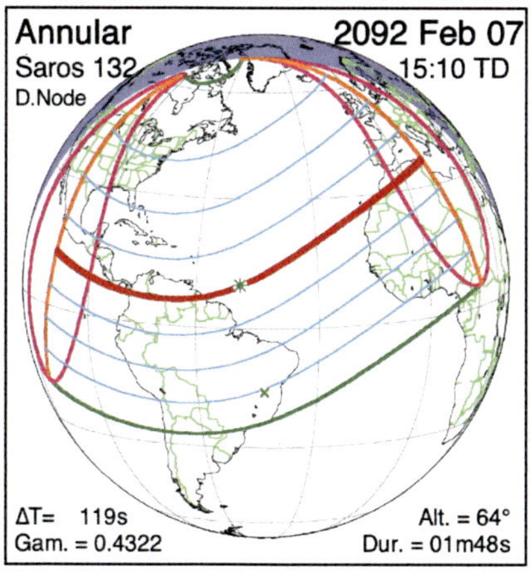
Annular
2092 Feb 07
Saros 132
15:10 TD
D.Node
ΔT= 119s
Alt. = 64°
Gam. = 0.4322
Dur. = 01m48s

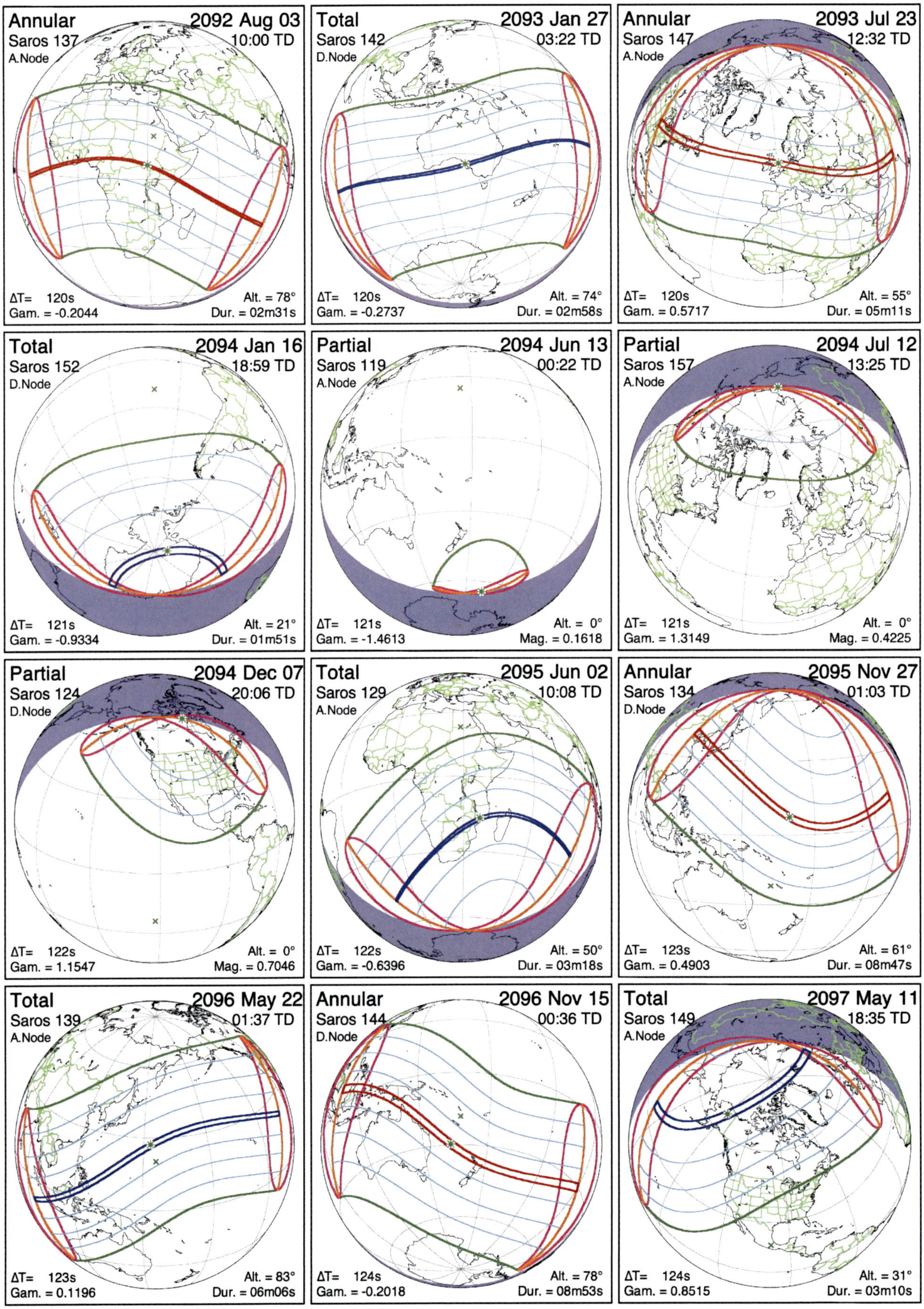
Annular 2092 Aug 03
Saros 137 10:00 TD
A.Node
ΔT= 120s Alt. = 78°
Gam. = -0.2044 Dur. = 02m31s
Total 2093 Jan 27
Saros 142 03:22 TD
D.Node
ΔT= 120s Alt. = 74°
Gam. = -0.2737 Dur. = 02m58s
Annular 2093 Jul 23
Saros 147 12:32 TD
A.Node
ΔT= 120s Alt. = 55°
Gam. = 0.5717 Dur. = 05m11s
Total 2094 Jan 16
Saros 152 18:59 TD
D.Node
ΔT= 121s Alt. = 21°
Gam. = -0.9334 Dur. = 01m51s
Partial 2094 Jun 13
Saros 119 00:22 TD
A.Node
ΔT= 121s Alt. = 0°
Gam. = -1.4613 Mag. = 0.1618
Partial 2094 Jul 12
Saros 157 13:25 TD
A.Node
ΔT= 121s Alt. = 0°
Gam. = 1.3149 Mag. = 0.4225
Partial 2094 Dec 07
Saros 124 20:06 TD
D.Node
ΔT= 122s Alt. = 0°
Gam. = 1.1547 Mag. = 0.7046
Total 2095 Jun 02
Saros 129 10:08 TD
A.Node
ΔT= 122s Alt. = 50°
Gam. = -0.6396 Dur. = 03m18s
Annular 2095 Nov 27
Saros 134 01:03 TD
D.Node
ΔT= 123s Alt. = 61°
Gam. = 0.4903 Dur. = 08m47s
Total 2096 May 22
Saros 139 01:37 TD
A.Node
ΔT= 123s Alt. = 83°
Gam. = 0.1196 Dur. = 06m06s
Annular 2096 Nov 15
Saros 144 00:36 TD
D.Node
ΔT= 124s Alt. = 78°
Gam. = -0.2018 Dur. = 08m53s
Total 2097 May 11
Saros 149 18:35 TD
A.Node
ΔT= 124s Alt. = 31°
Gam. = 0.8515 Dur. = 03m10s

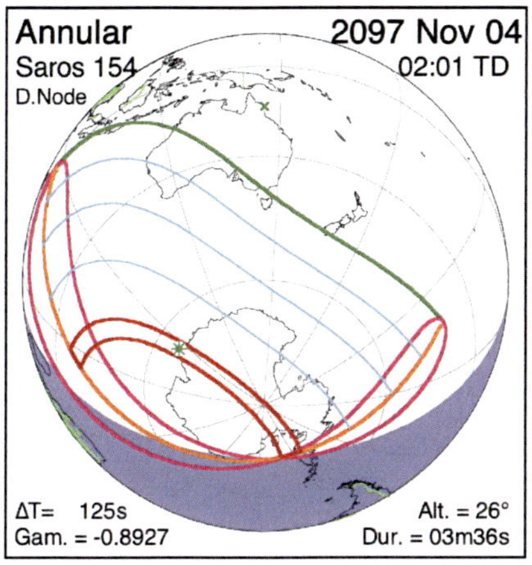
Annular
2097 Nov 04
Saros 154
02:01 TD
D.Node
ΔT= 125s
Alt. = 26°
Gam. = -0.8927
Dur. = 03m36s

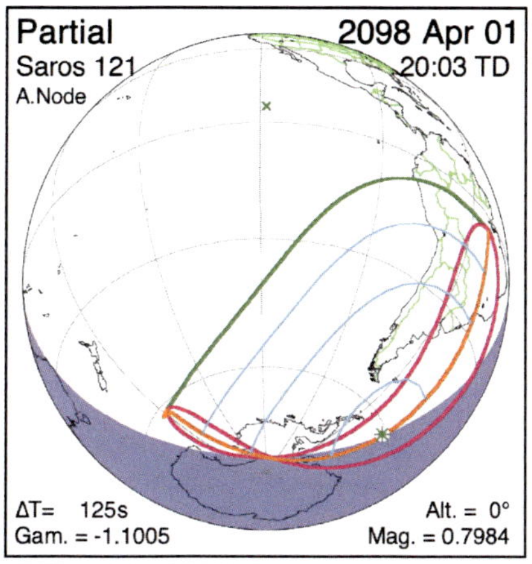
Partial
2098 Apr 01
Saros 121
20:03 TD
A.Node
ΔT= 125s
Alt. = 0°
Gam. = -1.1005
Mag. = 0.7984

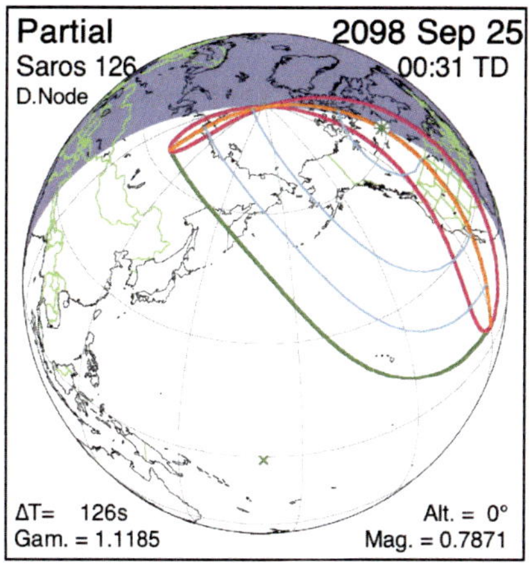
Partial
2098 Sep 25
Saros 126
00:31 TD
D.Node
ΔT= 126s
Alt. = 0°
Gam. = 1.1185
Mag. = 0.7871

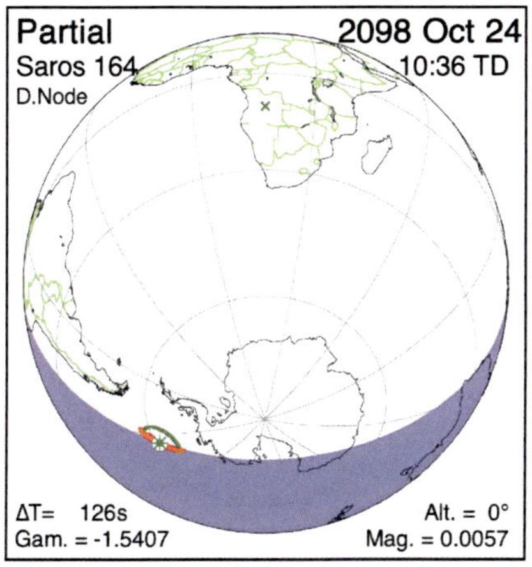
Partial
2098 Oct 24
Saros 164
10:36 TD
D.Node
ΔT= 126s
Alt. = 0°
Gam. = -1.5407
Mag. = 0.0057

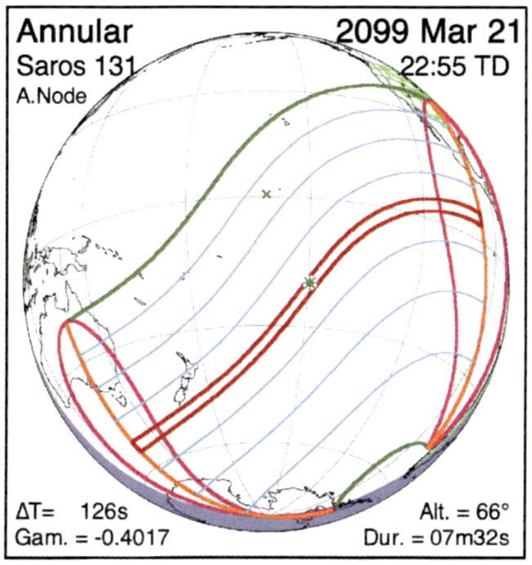
Annular
2099 Mar 21
Saros 131
22:55 TD
A.Node
ΔT= 126s
Alt. = 66°
Gam. = -0.4017
Dur. = 07m32s

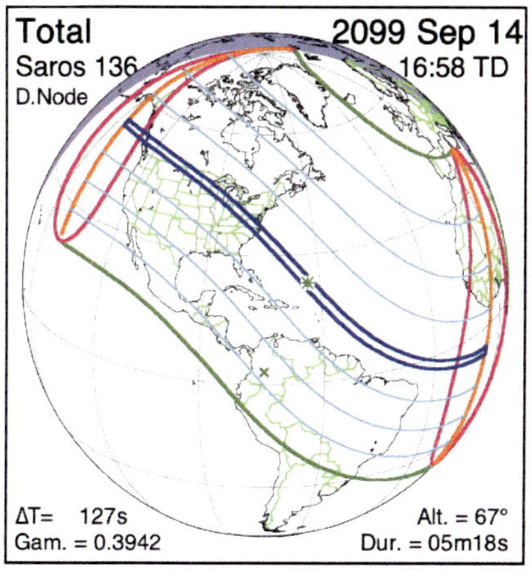
Total
2099 Sep 14
Saros 136
16:58 TD
D.Node
ΔT= 127s
Alt. = 67°
Gam. = 0.3942
Dur. = 05m18s

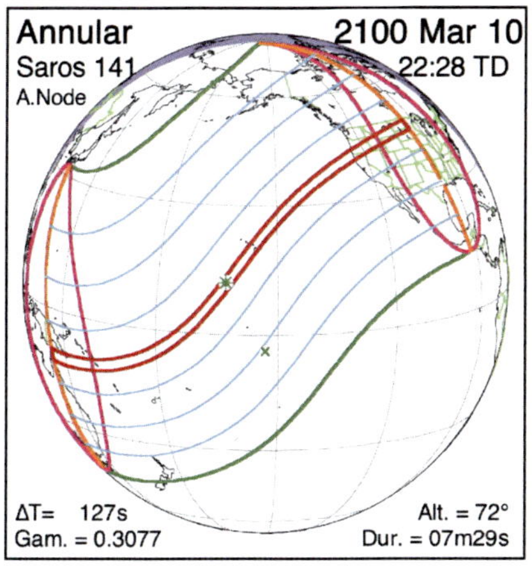
Annular
2100 Mar 10
Saros 141
22:28 TD
A.Node
ΔT= 127s
Alt. = 72°
Gam. = 0.3077
Dur. = 07m29s

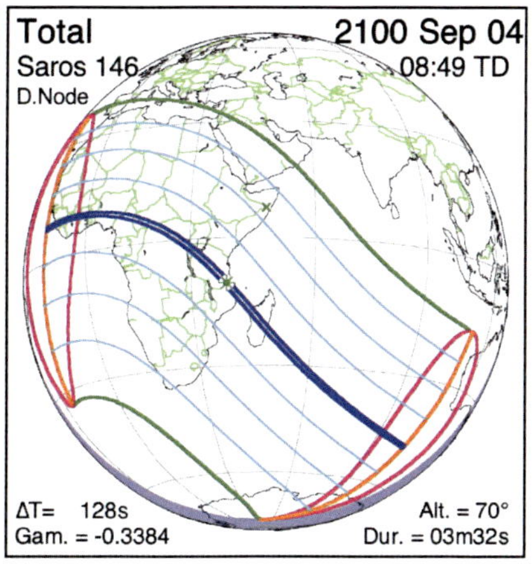
Total
2100 Sep 04
Saros 146
08:49 TD
D.Node
ΔT= 128s
Alt. = 70°
Gam. = -0.3384
Dur. = 03m32s

Appendix C

Large Global Solar Eclipse Maps: 2001 to 2100

Key to Large Global Solar Eclipse Maps

Figure 5—2: Key to Large Global Solar Eclipse Maps (Appendix C)

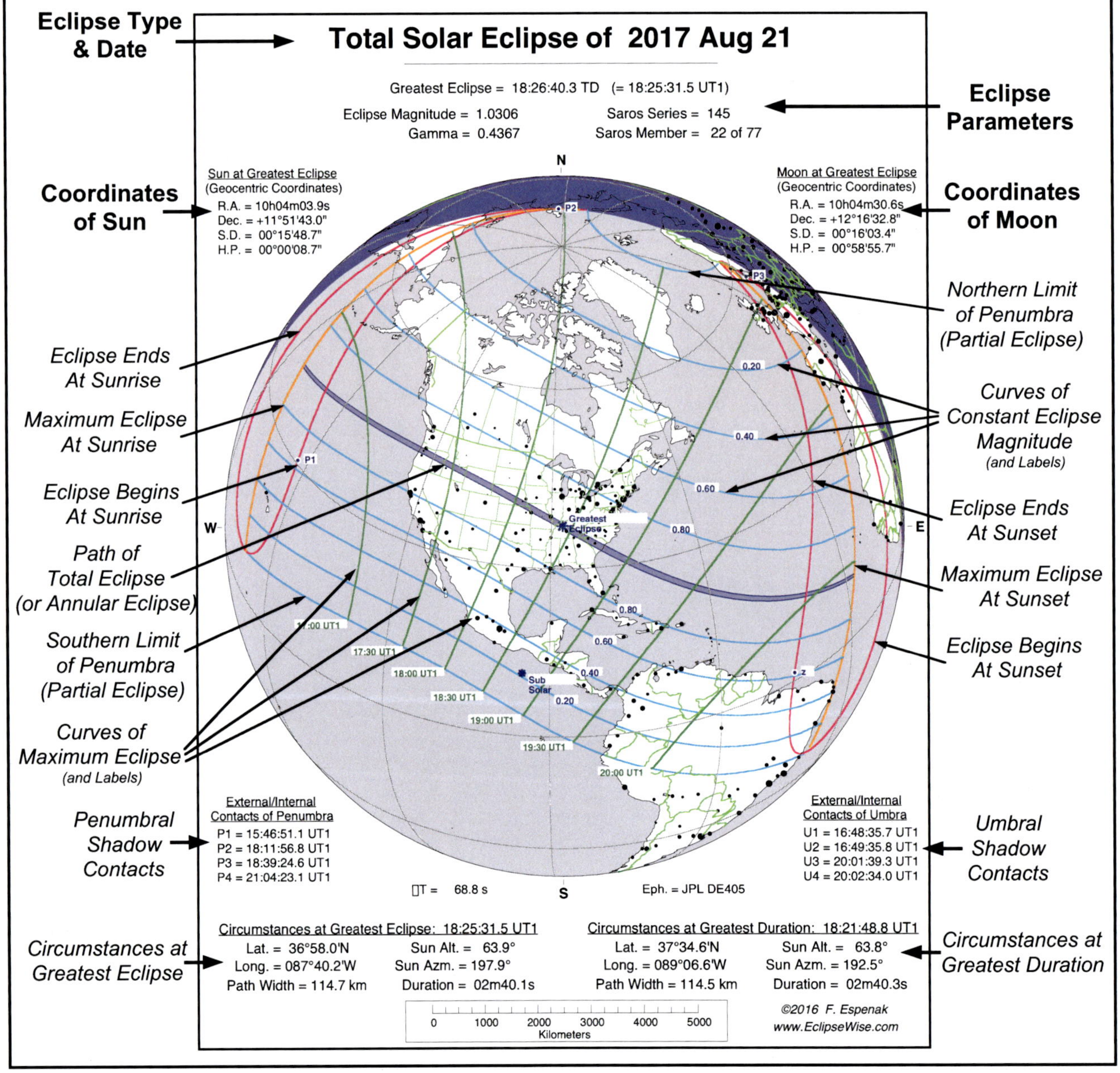

Total Solar Eclipse of 2001 Jun 21

Greatest Eclipse = 12:04:46.3 TD (= 12:03:42.2 UT1)

Eclipse Magnitude = 1.0495 Saros Series = 127

Gamma = -0.5701 Saros Member = 57 of 82

Sun at Greatest Eclipse
(Geocentric Coordinates)
R.A. = 06h00m46.1s
Dec. = +23°26'18.2"
S.D. = 00°15'44.3"
H.P. = 00°00'08.7"

Moon at Greatest Eclipse
(Geocentric Coordinates)
R.A. = 06h01m00.5s
Dec. = +22°52'27.2"
S.D. = 00°16'17.6"
H.P. = 00°59'47.9"

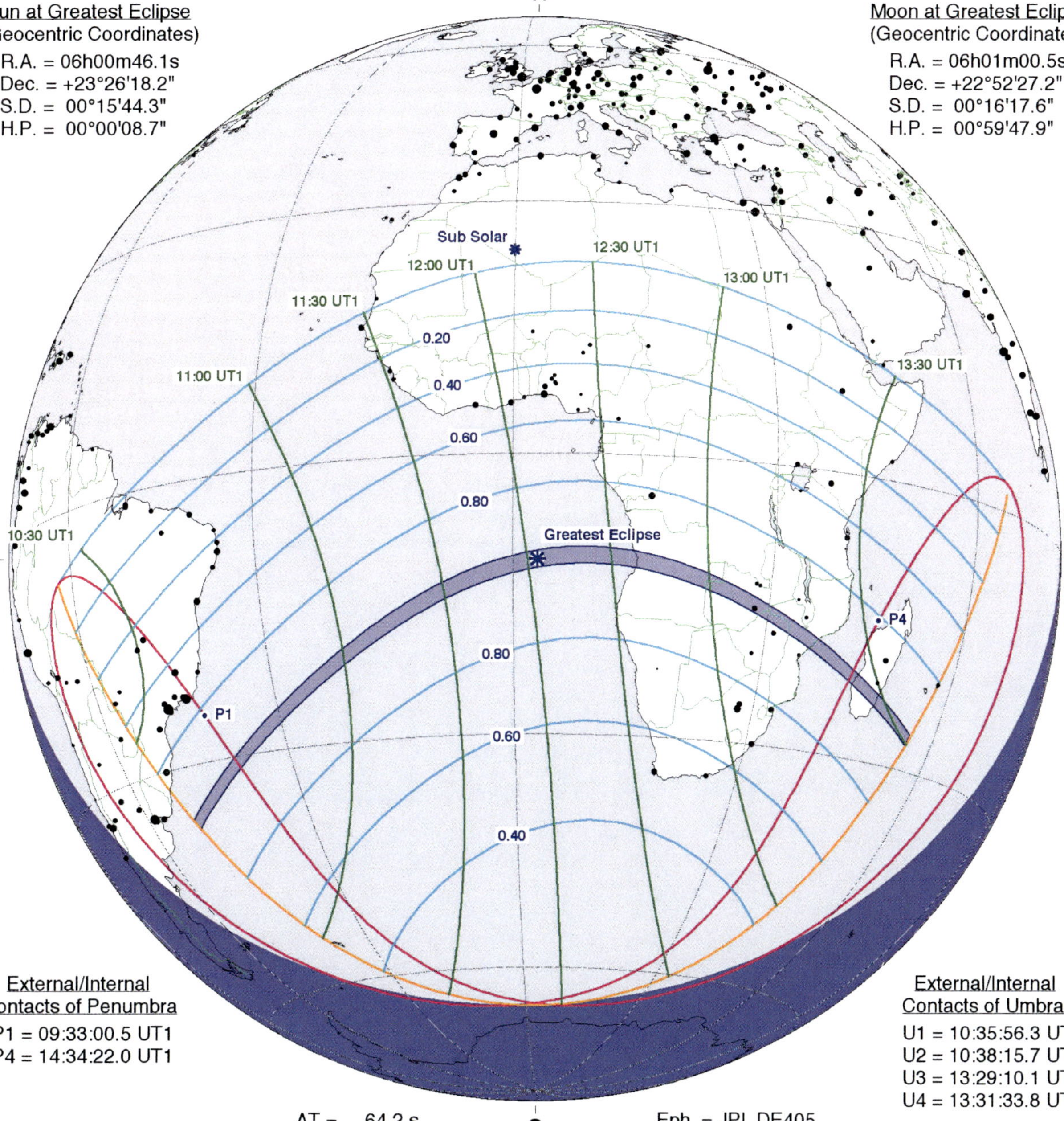

External/Internal
Contacts of Penumbra
P1 = 09:33:00.5 UT1
P4 = 14:34:22.0 UT1

External/Internal
Contacts of Umbra
U1 = 10:35:56.3 UT1
U2 = 10:38:15.7 UT1
U3 = 13:29:10.1 UT1
U4 = 13:31:33.8 UT1

ΔT = 64.2 s Eph. = JPL DE405

Circumstances at Greatest Eclipse: 12:03:42.2 UT1		Circumstances at Greatest Duration: 12:06:07.4 UT1	
Lat. = 11°15.2'S	Sun Alt. = 55.2°	Lat. = 11°08.5'S	Sun Alt. = 55.1°
Long. = 002°44.8'E	Sun Azm. = 354.8°	Long. = 003°28.5'E	Sun Azm. = 352.7°
Path Width = 200.0 km	Duration = 04m56.5s	Path Width = 200.3 km	Duration = 04m56.7s

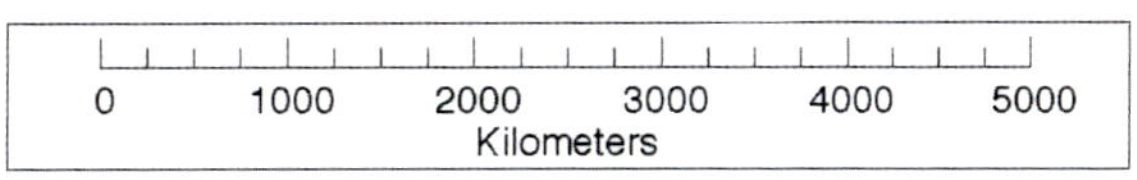

Annular Solar Eclipse of 2001 Dec 14

Greatest Eclipse = 20:53:01.4 TD (= 20:51:57.1 UT1)

Eclipse Magnitude = 0.9681 Saros Series = 132

Gamma = 0.4089 Saros Member = 45 of 71

Sun at Greatest Eclipse
(Geocentric Coordinates)

R.A. = 17h29m14.2s
Dec. = -23°15'02.0"
S.D. = 00°16'14.9"
H.P. = 00°00'08.9"

Moon at Greatest Eclipse
(Geocentric Coordinates)

R.A. = 17h29m29.5s
Dec. = -22°52'05.9"
S.D. = 00°15'30.3"
H.P. = 00°56'54.2"

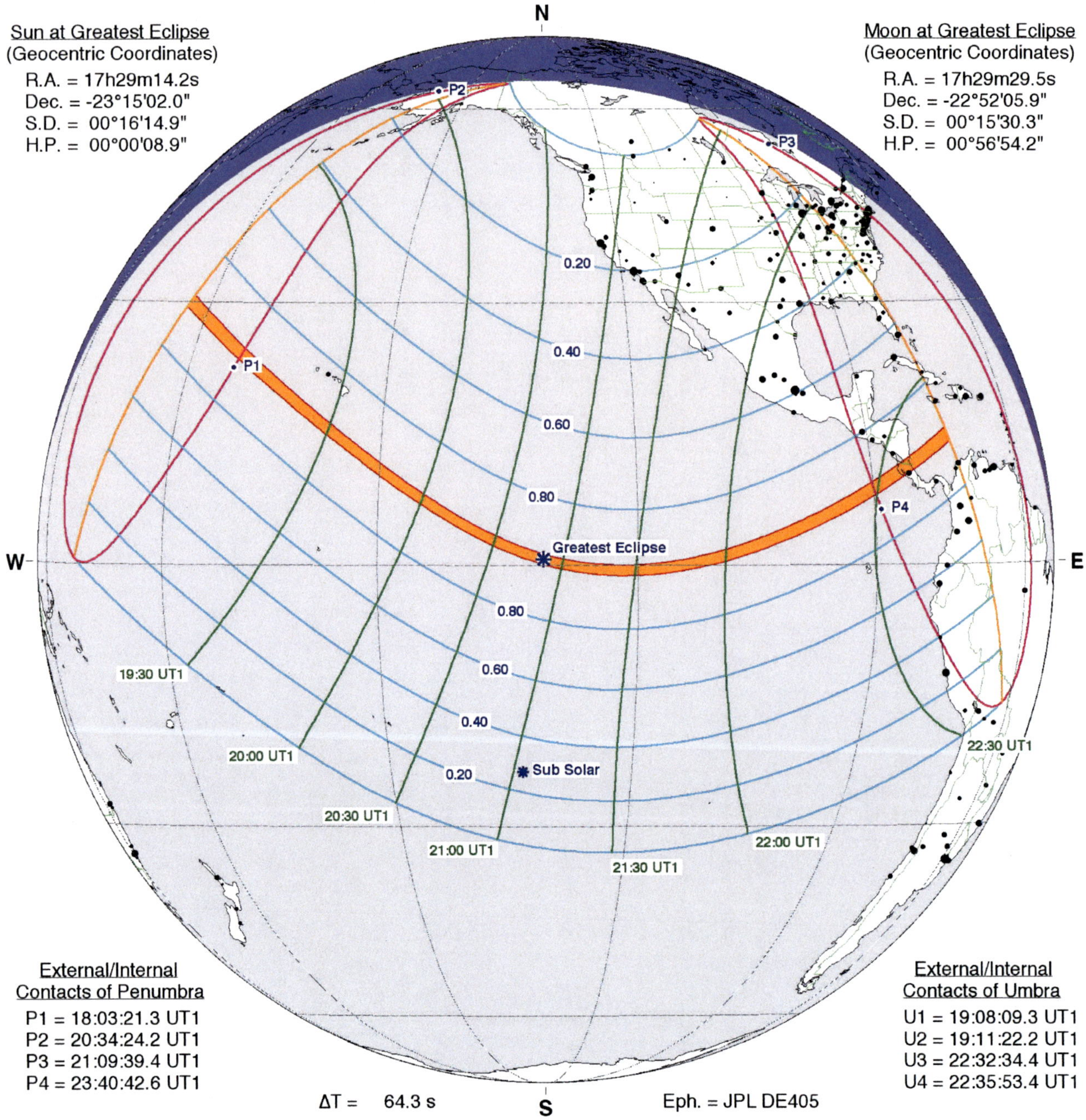

External/Internal
Contacts of Penumbra

P1 = 18:03:21.3 UT1
P2 = 20:34:24.2 UT1
P3 = 21:09:39.4 UT1
P4 = 23:40:42.6 UT1

External/Internal
Contacts of Umbra

U1 = 19:08:09.3 UT1
U2 = 19:11:22.2 UT1
U3 = 22:32:34.4 UT1
U4 = 22:35:53.4 UT1

ΔT = 64.3 s Eph. = JPL DE405

Circumstances at Greatest Eclipse: 20:51:57.1 UT1

Lat. = 00°38.1'N Sun Alt. = 65.9°
Long. = 130°42.2'W Sun Azm. = 188.0°
Path Width = 125.6 km Duration = 03m53.0s

Circumstances at Greatest Duration: 21:07:23.8 UT1

Lat. = 00°14.2'S Sun Alt. = 64.4°
Long. = 126°41.3'W Sun Azm. = 205.0°
Path Width = 127.4 km Duration = 03m54.1s

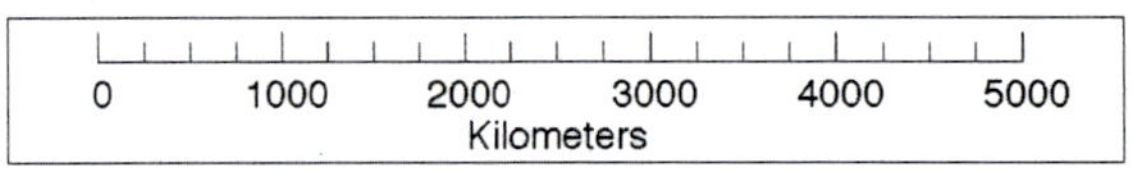

Annular Solar Eclipse of 2002 Jun 10

Greatest Eclipse = 23:45:22.2 TD (= 23:44:17.9 UT1)

Eclipse Magnitude = 0.9962 Saros Series = 137

Gamma = 0.1993 Saros Member = 35 of 70

Sun at Greatest Eclipse
(Geocentric Coordinates)

R.A. = 05h16m04.1s
Dec. = +23°03'18.9"
S.D. = 00°15'45.1"
H.P. = 00°00'08.7"

Moon at Greatest Eclipse
(Geocentric Coordinates)

R.A. = 05h15m55.6s
Dec. = +23°14'25.0"
S.D. = 00°15'27.1"
H.P. = 00°56'42.5"

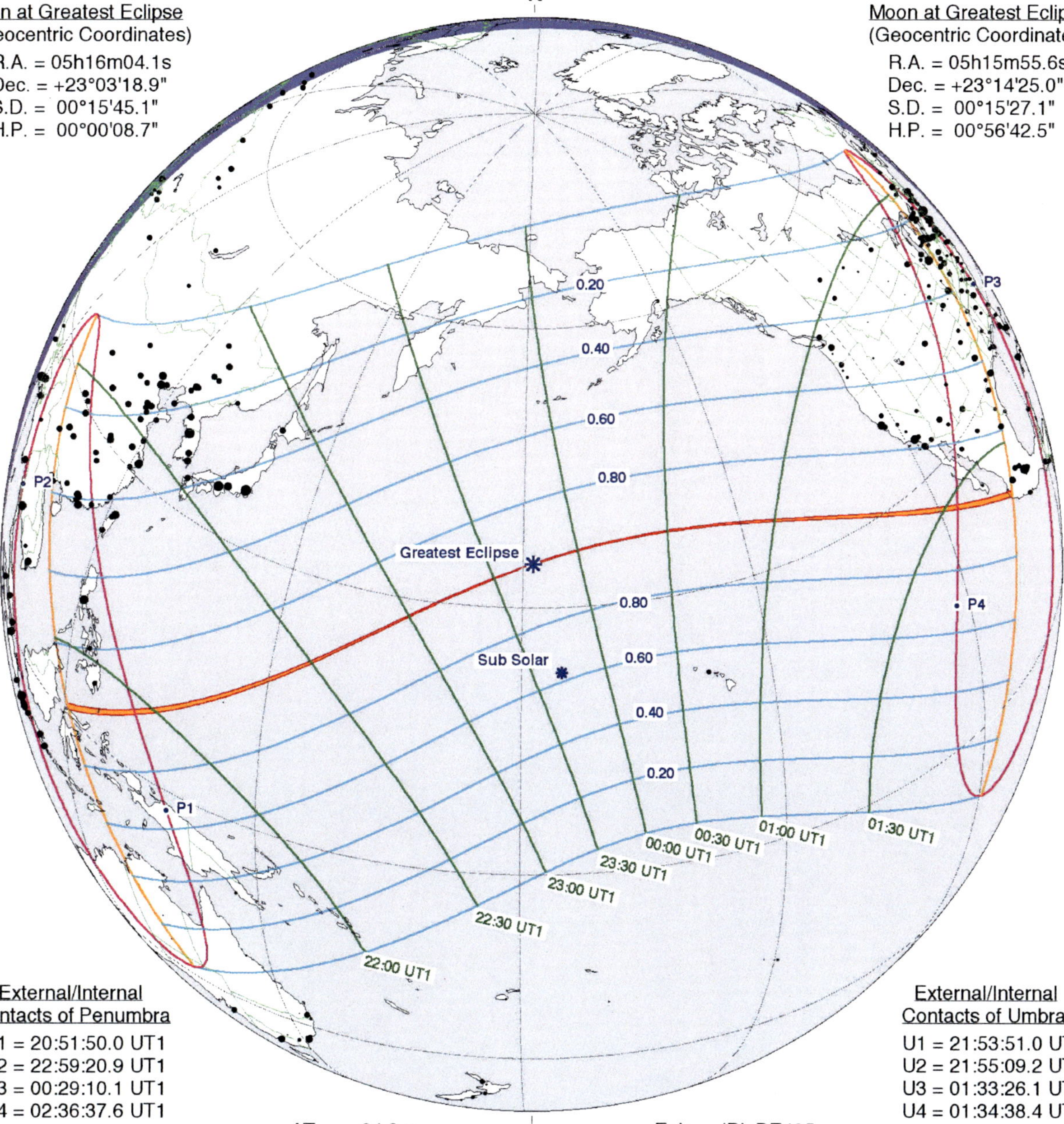

External/Internal
Contacts of Penumbra

P1 = 20:51:50.0 UT1
P2 = 22:59:20.9 UT1
P3 = 00:29:10.1 UT1
P4 = 02:36:37.6 UT1

External/Internal
Contacts of Umbra

U1 = 21:53:51.0 UT1
U2 = 21:55:09.2 UT1
U3 = 01:33:26.1 UT1
U4 = 01:34:38.4 UT1

ΔT = 64.3 s Eph. = JPL DE405

Circumstances at Greatest Eclipse: 23:44:17.9 UT1

Lat. = 34°33.0'N	Sun Alt. = 78.3°
Long. = 178°37.2'W	Sun Azm. = 169.0°
Path Width = 13.5 km	Duration = 00m22.8s

Circumstances at Greatest Duration: 21:54:30.1 UT1

Lat. = 01°19.8'N	Sun Alt. = 0.0°
Long. = 120°40.5'E	Sun Azm. = 66.9°
Path Width = 78.3 km	Duration = 01m12.6s

0 1000 2000 3000 4000 5000
Kilometers

Total Solar Eclipse of 2002 Dec 04

Greatest Eclipse = 07:32:15.7 TD (= 07:31:11.3 UT1)

Eclipse Magnitude = 1.0244 | Saros Series = 142

Gamma = -0.3020 | Saros Member = 22 of 72

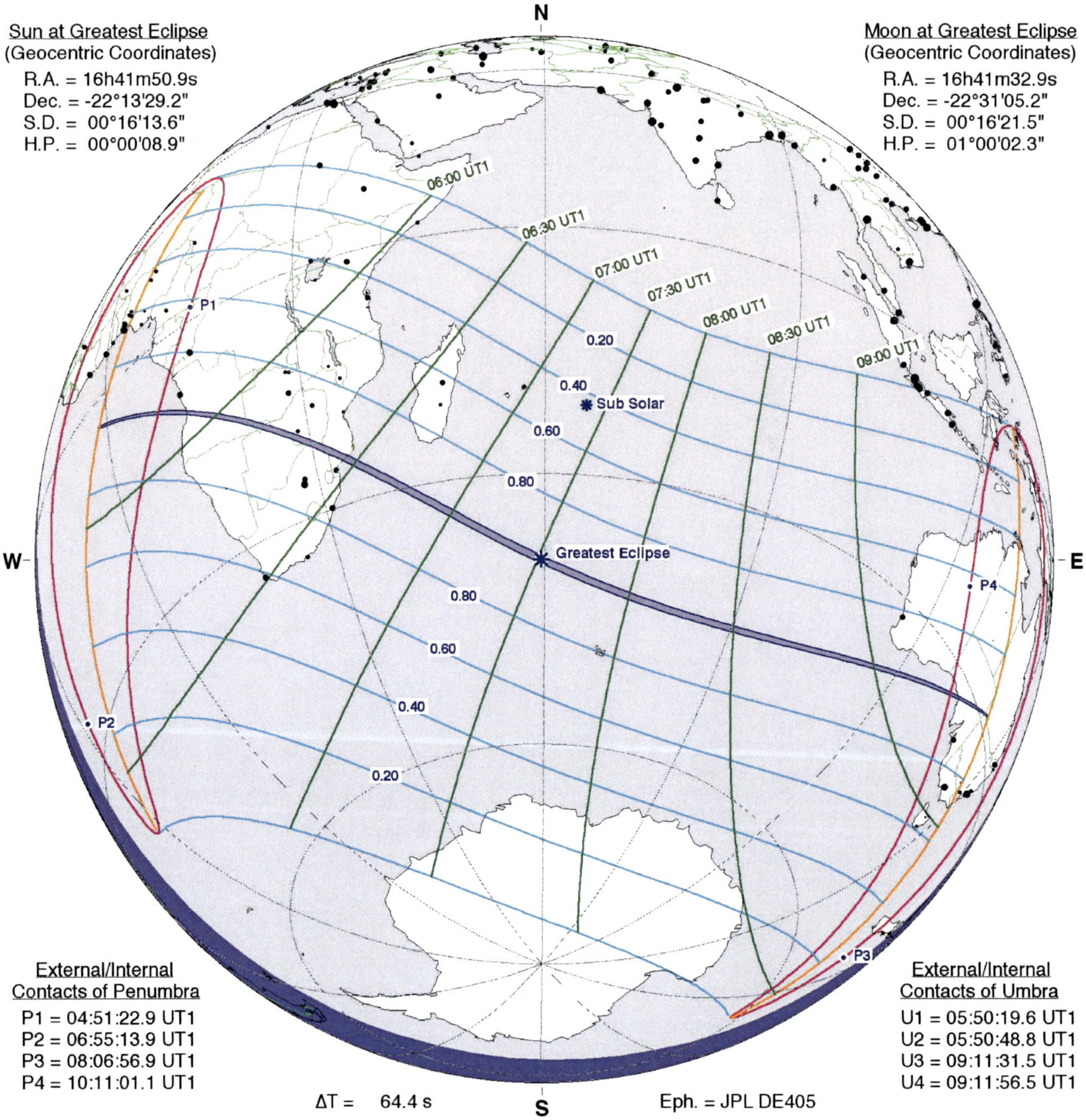

Circumstances at Greatest Eclipse: 07:31:11.3 UT1	
Lat. = 39°27.4'S	Sun Alt. = 72.2°
Long. = 059°33.3'E	Sun Azm. = 15.8°
Path Width = 87.0 km	Duration = 02m03.8s

Circumstances at Greatest Duration: 07:31:56.7 UT1	
Lat. = 39°34.2'S	Sun Alt. = 72.2°
Long. = 059°52.6'E	Sun Azm. = 14.2°
Path Width = 86.9 km	Duration = 02m03.8s

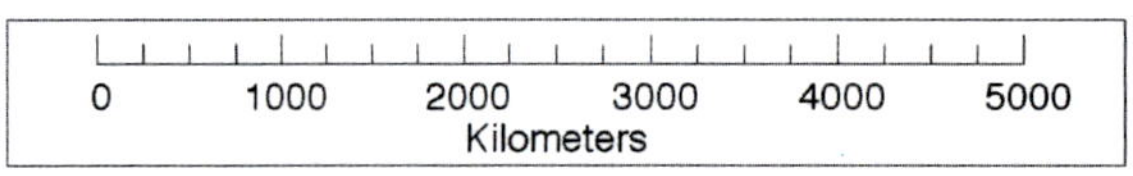

Annular Solar Eclipse of 2003 May 31

Greatest Eclipse = 04:09:22.5 TD (= 04:08:18.0 UT1)

Eclipse Magnitude = 0.9384
Gamma = 0.9960

Saros Series = 147
Saros Member = 22 of 80

Sun at Greatest Eclipse
(Geocentric Coordinates)
R.A. = 04h30m33.5s
Dec. = +21°50'57.1"
S.D. = 00°15'46.5"
H.P. = 00°00'08.7"

Moon at Greatest Eclipse
(Geocentric Coordinates)
R.A. = 04h29m35.5s
Dec. = +22°43'13.1"
S.D. = 00°14'48.1"
H.P. = 00°54'19.3"

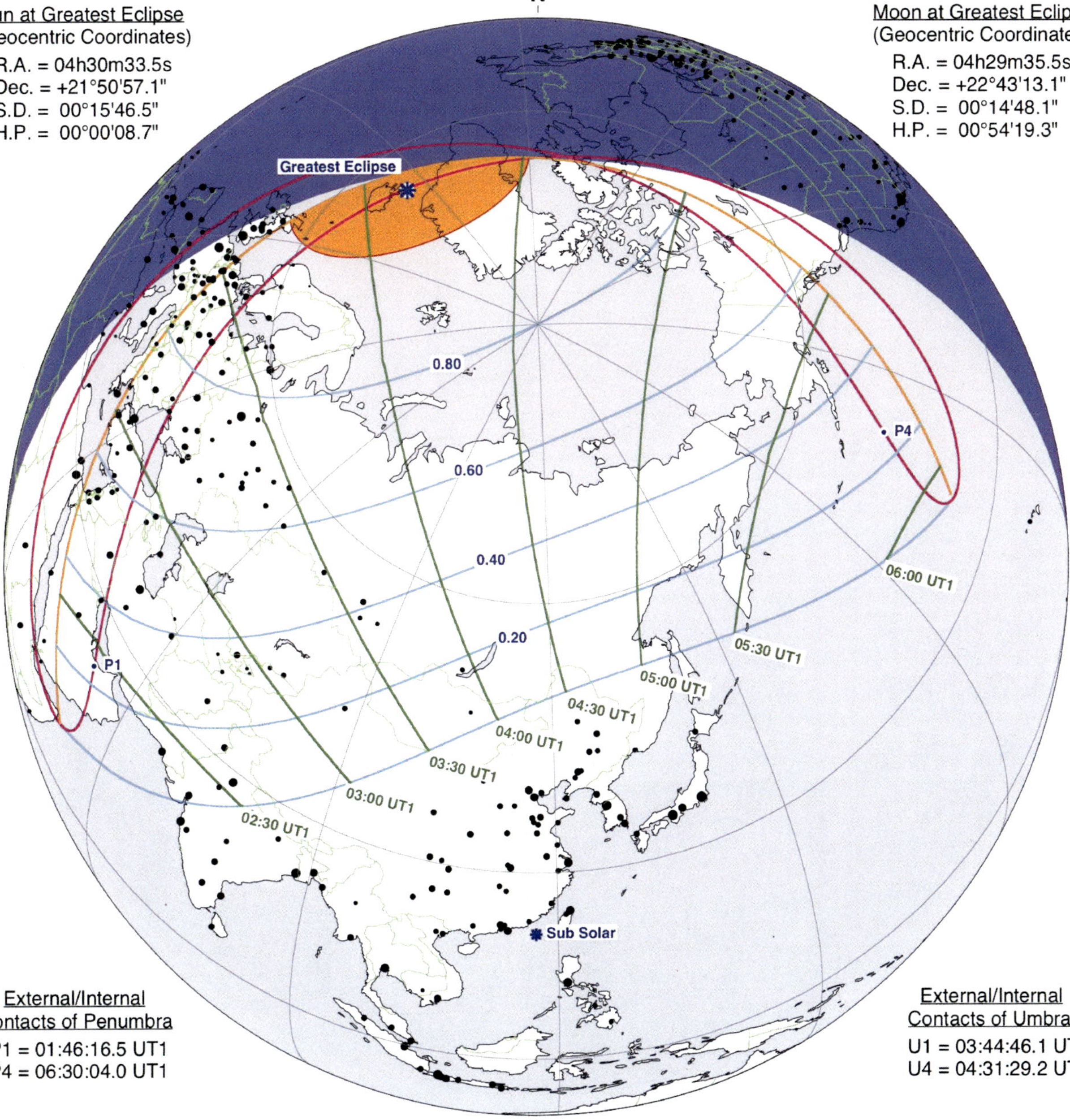

External/Internal
Contacts of Penumbra
P1 = 01:46:16.5 UT1
P4 = 06:30:04.0 UT1

External/Internal
Contacts of Umbra
U1 = 03:44:46.1 UT1
U4 = 04:31:29.2 UT1

ΔT = 64.5 s

Eph. = JPL DE405

Circumstances at Greatest Eclipse: 04:08:18.0 UT1
Lat. = 66°33.4'N
Long. = 024°29.5'W
Path Width =4638.4 km
Sun Alt. = 2.9°
Sun Azm. = 35.1°
Duration = 03m36.8s

Circumstances at Greatest Duration: 04:07:54.1 UT1
Lat. = 66°26.6'N
Long. = 023°59.7'W
Path Width =4659.3 km
Sun Alt. = 2.9°
Sun Azm. = 35.4°
Duration = 03m36.8s

0 1000 2000 3000 4000 5000
Kilometers

Total Solar Eclipse of 2003 Nov 23

Greatest Eclipse = 22:50:21.7 TD (= 22:49:17.2 UT1)

Eclipse Magnitude = 1.0379 Saros Series = 152
Gamma = -0.9638 Saros Member = 12 of 70

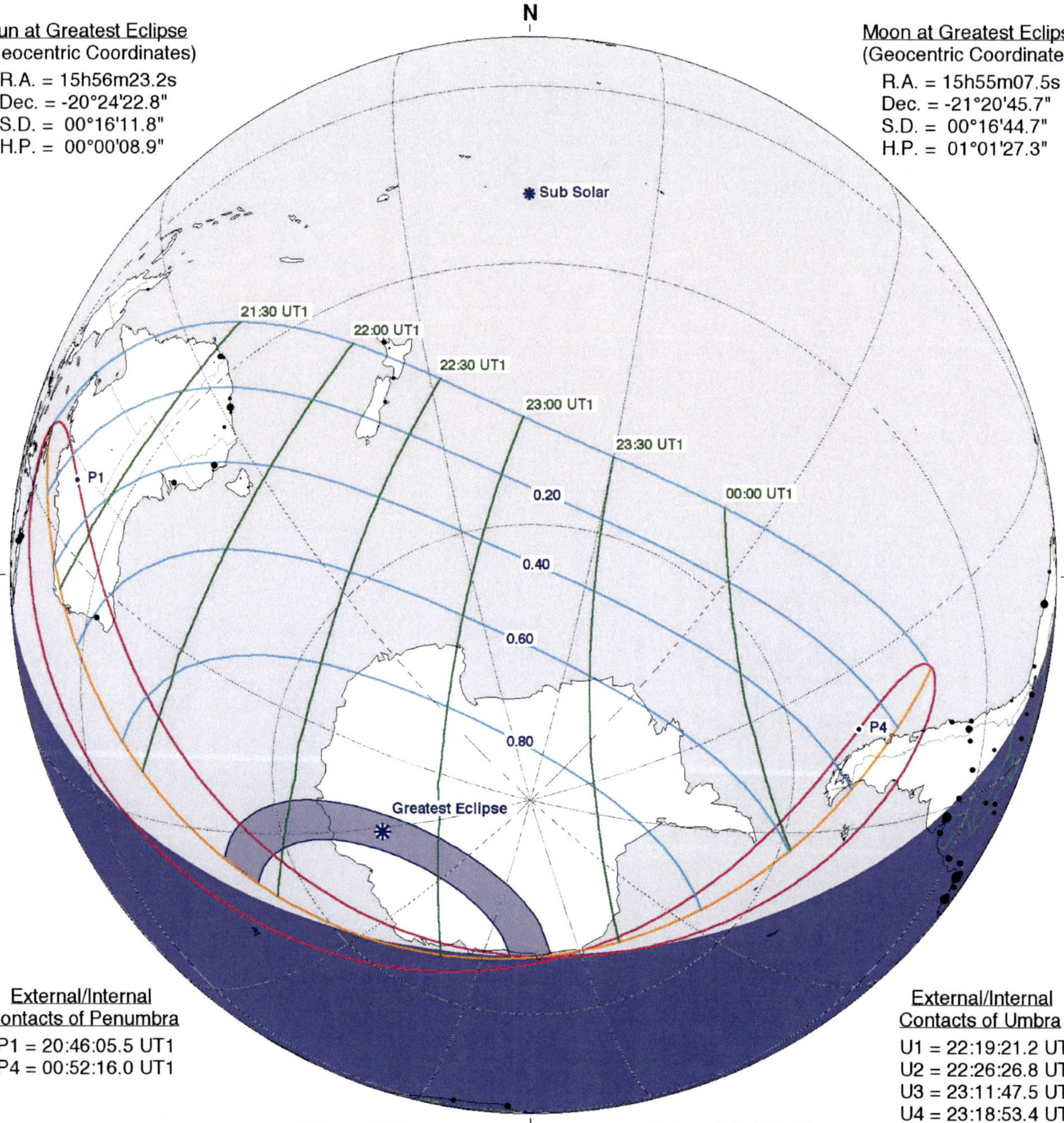

Sun at Greatest Eclipse
(Geocentric Coordinates)
R.A. = 15h56m23.2s
Dec. = -20°24'22.8"
S.D. = 00°16'11.8"
H.P. = 00°00'08.9"

Moon at Greatest Eclipse
(Geocentric Coordinates)
R.A. = 15h55m07.5s
Dec. = -21°20'45.7"
S.D. = 00°16'44.7"
H.P. = 01°01'27.3"

External/Internal
Contacts of Penumbra
P1 = 20:46:05.5 UT1
P4 = 00:52:16.0 UT1

External/Internal
Contacts of Umbra
U1 = 22:19:21.2 UT1
U2 = 22:26:26.8 UT1
U3 = 23:11:47.5 UT1
U4 = 23:18:53.4 UT1

ΔT = 64.5 s Eph. = JPL DE405

Circumstances at Greatest Eclipse: 22:49:17.2 UT1

Lat. = 72°40.0'S Sun Alt. = 14.9°
Long. = 088°23.7'E Sun Azm. = 111.2°
Path Width = 495.5 km Duration = 01m57.3s

Circumstances at Greatest Duration: 22:49:14.2 UT1

Lat. = 72°38.5'S Sun Alt. = 14.9°
Long. = 088°26.5'E Sun Azm. = 111.1°
Path Width = 495.6 km Duration = 01m57.3s

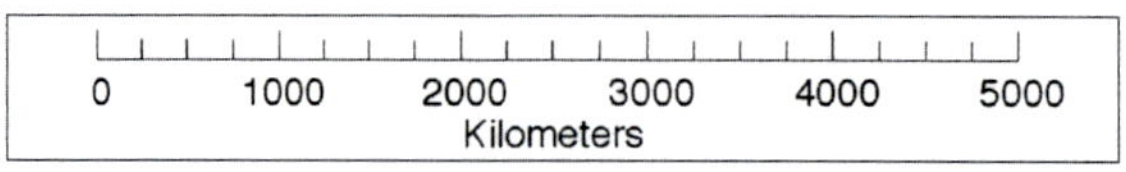

Partial Solar Eclipse of 2004 Apr 19

Greatest Eclipse = 13:35:05.3 TD (= 13:34:00.7 UT1)

Eclipse Magnitude = 0.7367
Gamma = -1.1335

Saros Series = 119
Saros Member = 65 of 71

Sun at Greatest Eclipse
(Geocentric Coordinates)
R.A. = 01h50m58.6s
Dec. = +11°24'41.2"
S.D. = 00°15'55.2"
H.P. = 00°00'08.8"

Moon at Greatest Eclipse
(Geocentric Coordinates)
R.A. = 01h52m50.5s
Dec. = +10°28'42.9"
S.D. = 00°15'01.7"
H.P. = 00°55'09.3"

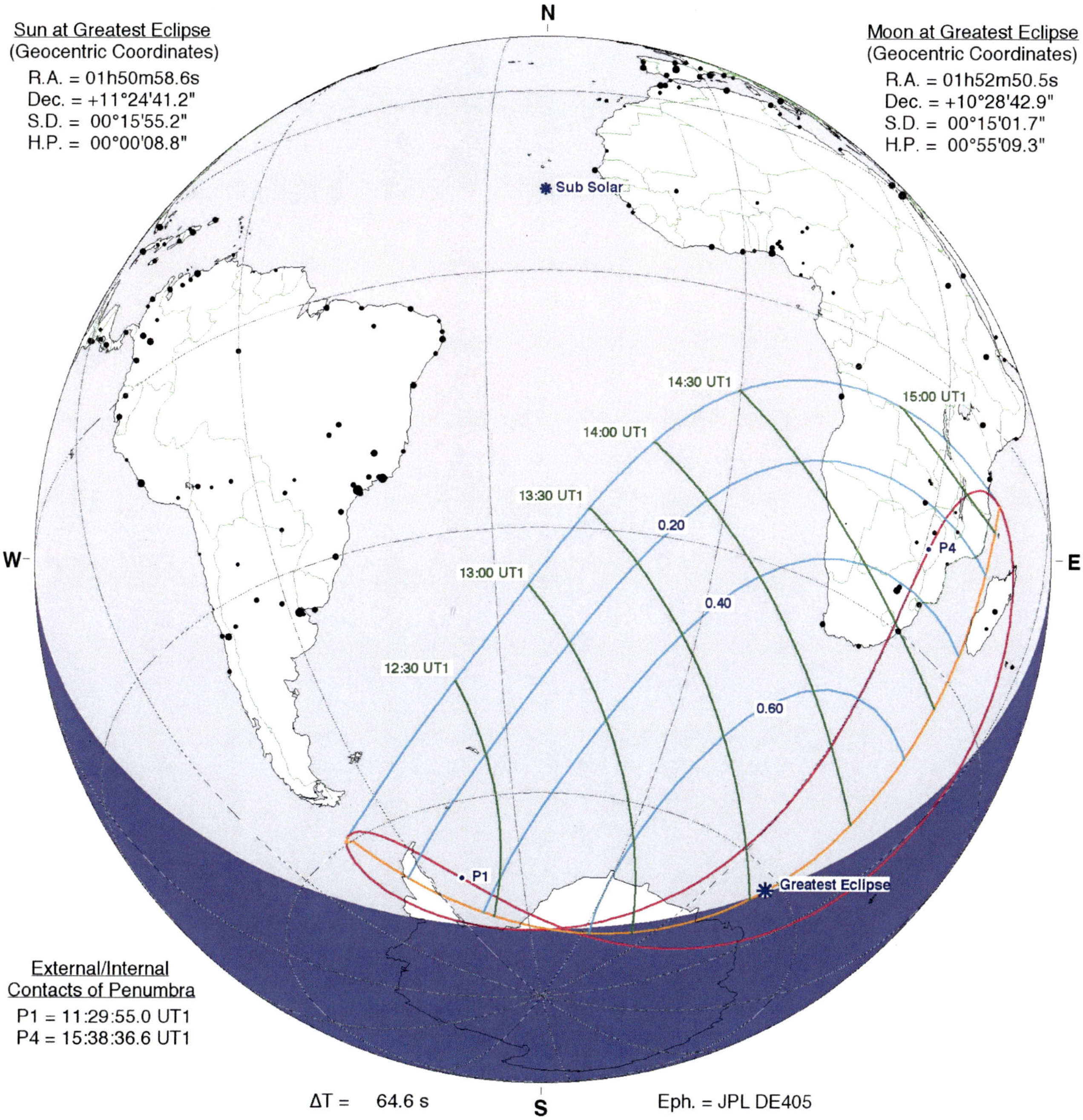

External/Internal
Contacts of Penumbra
P1 = 11:29:55.0 UT1
P4 = 15:38:36.6 UT1

ΔT = 64.6 s

Eph. = JPL DE405

Circumstances at Greatest Eclipse: 13:34:00.7 UT1

Lat. = 61°35.6'S
Long. = 044°15.6'E

Sun Alt. = 0.0°
Sun Azm. = 294.6°

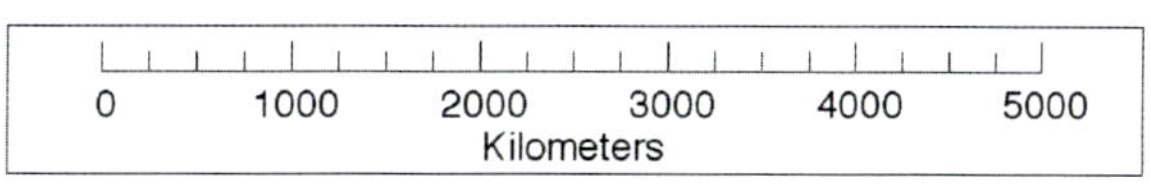

Partial Solar Eclipse of 2004 Oct 14

Greatest Eclipse = 03:00:23.0 TD (= 02:59:18.4 UT1)

Eclipse Magnitude = 0.9283
Gamma = 1.0348

Saros Series = 124
Saros Member = 54 of 73

Sun at Greatest Eclipse
(Geocentric Coordinates)
R.A. = 13h18m00.5s
Dec. = -08°14'10.7"
S.D. = 00°16'02.2"
H.P. = 00°00'08.8"

Moon at Greatest Eclipse
(Geocentric Coordinates)
R.A. = 13h19m53.4s
Dec. = -07°20'43.5"
S.D. = 00°15'55.4"
H.P. = 00°58'26.4"

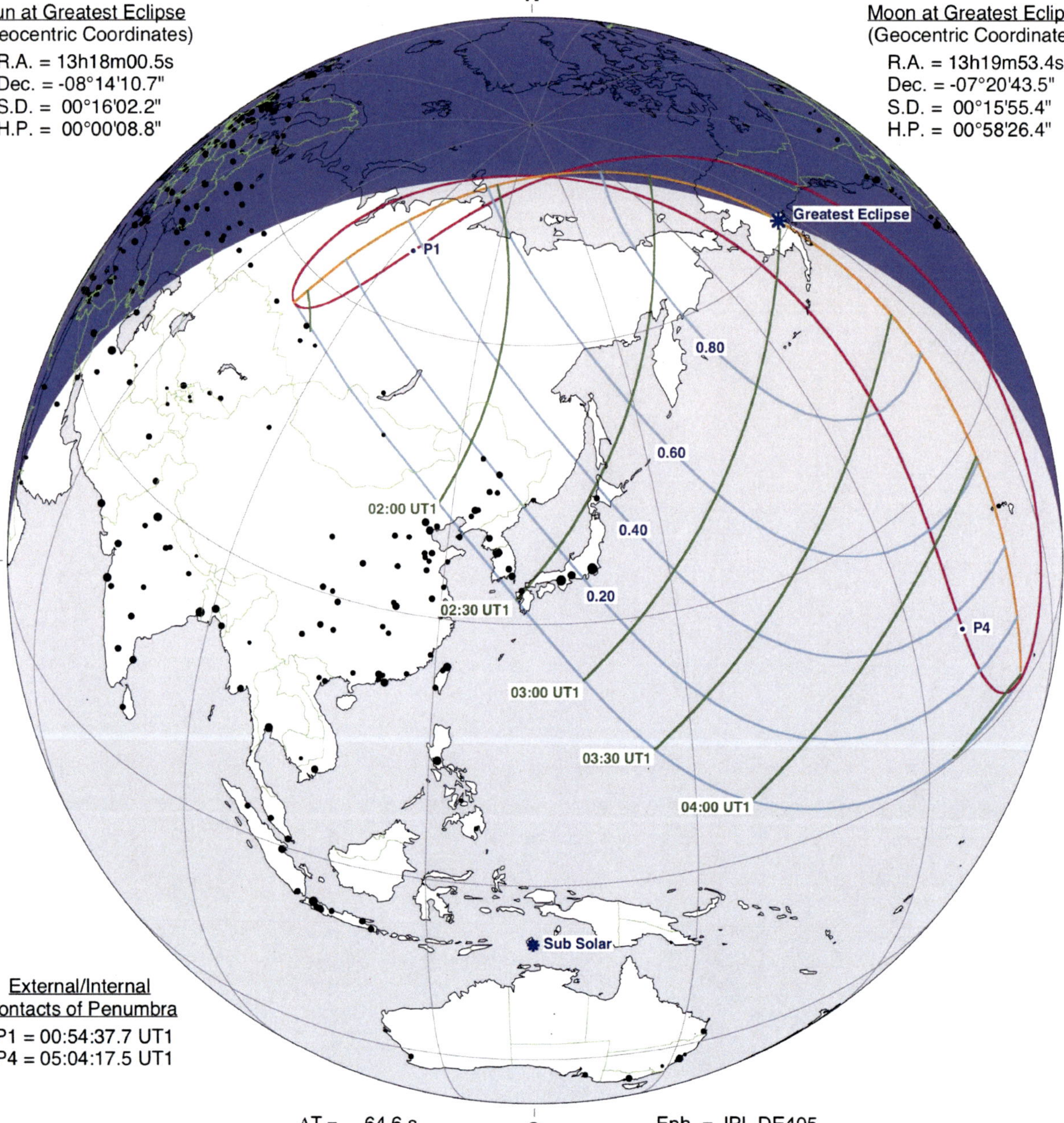

External/Internal
Contacts of Penumbra
P1 = 00:54:37.7 UT1
P4 = 05:04:17.5 UT1

ΔT = 64.6 s

Eph. = JPL DE405

Circumstances at Greatest Eclipse: 02:59:18.4 UT1

Lat. = 61°15.0'N
Long. = 153°42.1'W

Sun Alt. = 0.0°
Sun Azm. = 252.7°

0 1000 2000 3000 4000 5000
Kilometers

Hybrid Solar Eclipse of 2005 Apr 08

Greatest Eclipse = 20:36:50.5 TD (= 20:35:45.9 UT1)

Eclipse Magnitude = 1.0074
Gamma = -0.3473

Saros Series = 129
Saros Member = 51 of 80

Sun at Greatest Eclipse
(Geocentric Coordinates)
R.A. = 01h10m30.3s
Dec. = +07°28'47.2"
S.D. = 00°15'58.1"
H.P. = 00°00'08.8"

Moon at Greatest Eclipse
(Geocentric Coordinates)
R.A. = 01h11m08.3s
Dec. = +07°10'59.2"
S.D. = 00°15'50.7"
H.P. = 00°58'09.0"

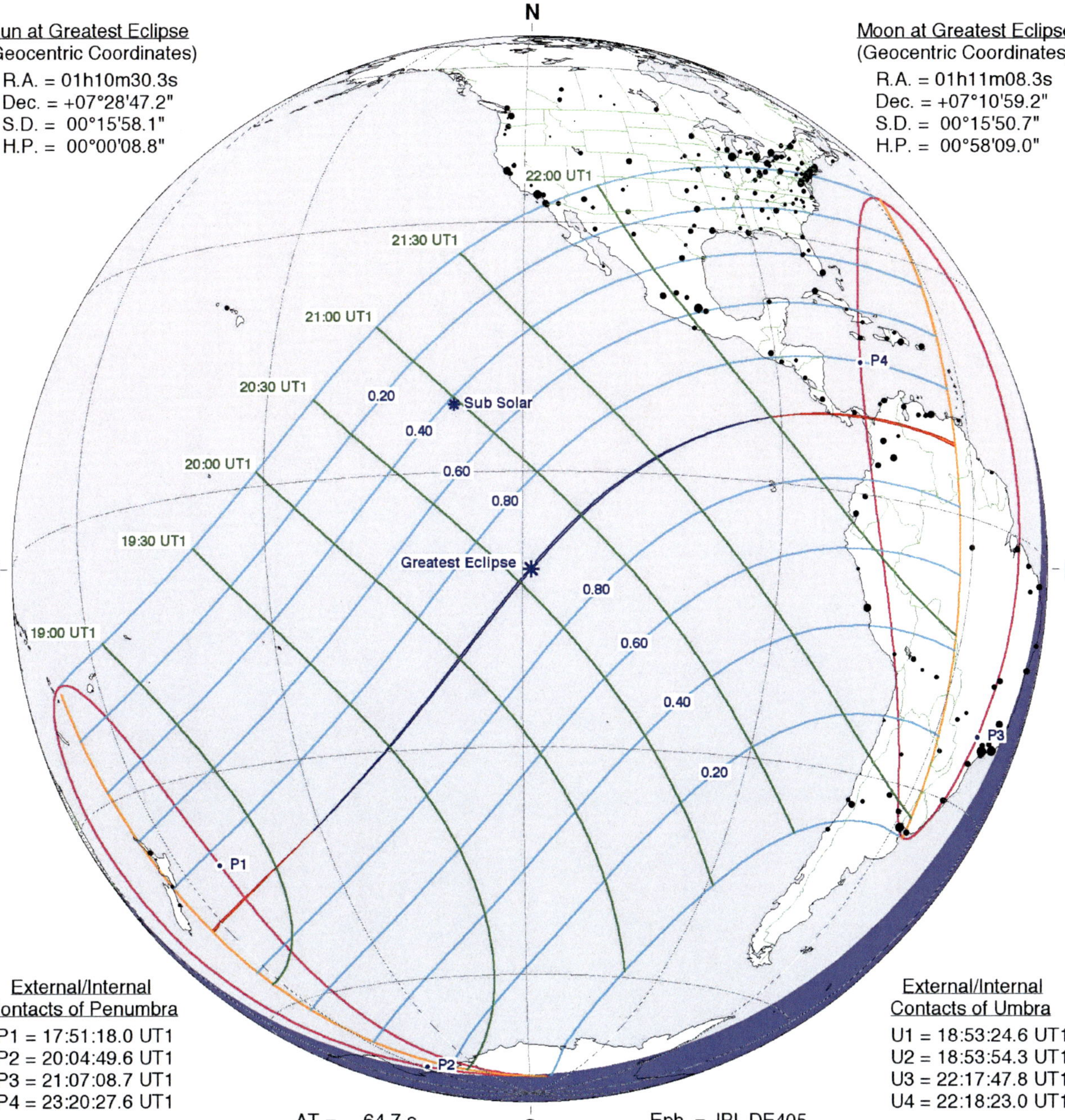

External/Internal
Contacts of Penumbra
P1 = 17:51:18.0 UT1
P2 = 20:04:49.6 UT1
P3 = 21:07:08.7 UT1
P4 = 23:20:27.6 UT1

External/Internal
Contacts of Umbra
U1 = 18:53:24.6 UT1
U2 = 18:53:54.3 UT1
U3 = 22:17:47.8 UT1
U4 = 22:18:23.0 UT1

ΔT = 64.7 s

Eph. = JPL DE405

Circumstances at Greatest Eclipse: 20:35:45.9 UT1

Lat. = 10°34.0'S
Long. = 118°59.1'W
Path Width = 27.0 km
Sun Alt. = 69.6°
Sun Azm. = 331.9°
Duration = 00m42.1s

Circumstances at Greatest Duration: 20:33:14.1 UT1

Lat. = 11°12.1'S
Long. = 119°32.6'W
Path Width = 27.0 km
Sun Alt. = 69.6°
Sun Azm. = 335.7°
Duration = 00m42.1s

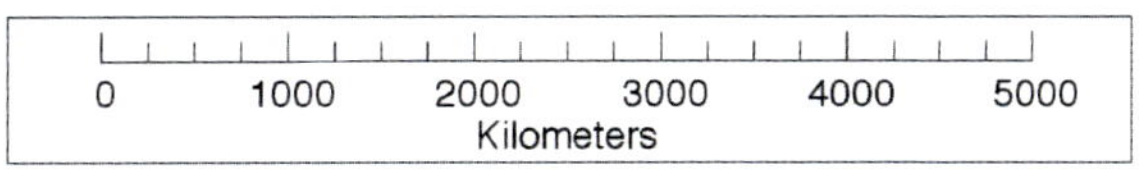

Annular Solar Eclipse of 2005 Oct 03

Greatest Eclipse = 10:32:47.3 TD (= 10:31:42.6 UT1)

Eclipse Magnitude = 0.9576
Gamma = 0.3306

Saros Series = 134
Saros Member = 43 of 71

Sun at Greatest Eclipse
(Geocentric Coordinates)
R.A. = 12h37m55.0s
Dec. = -04°05'04.2"
S.D. = 00°15'59.1"
H.P. = 00°00'08.8"

Moon at Greatest Eclipse
(Geocentric Coordinates)
R.A. = 12h38m30.3s
Dec. = -03°49'04.7"
S.D. = 00°15'05.2"
H.P. = 00°55'22.1"

N

W

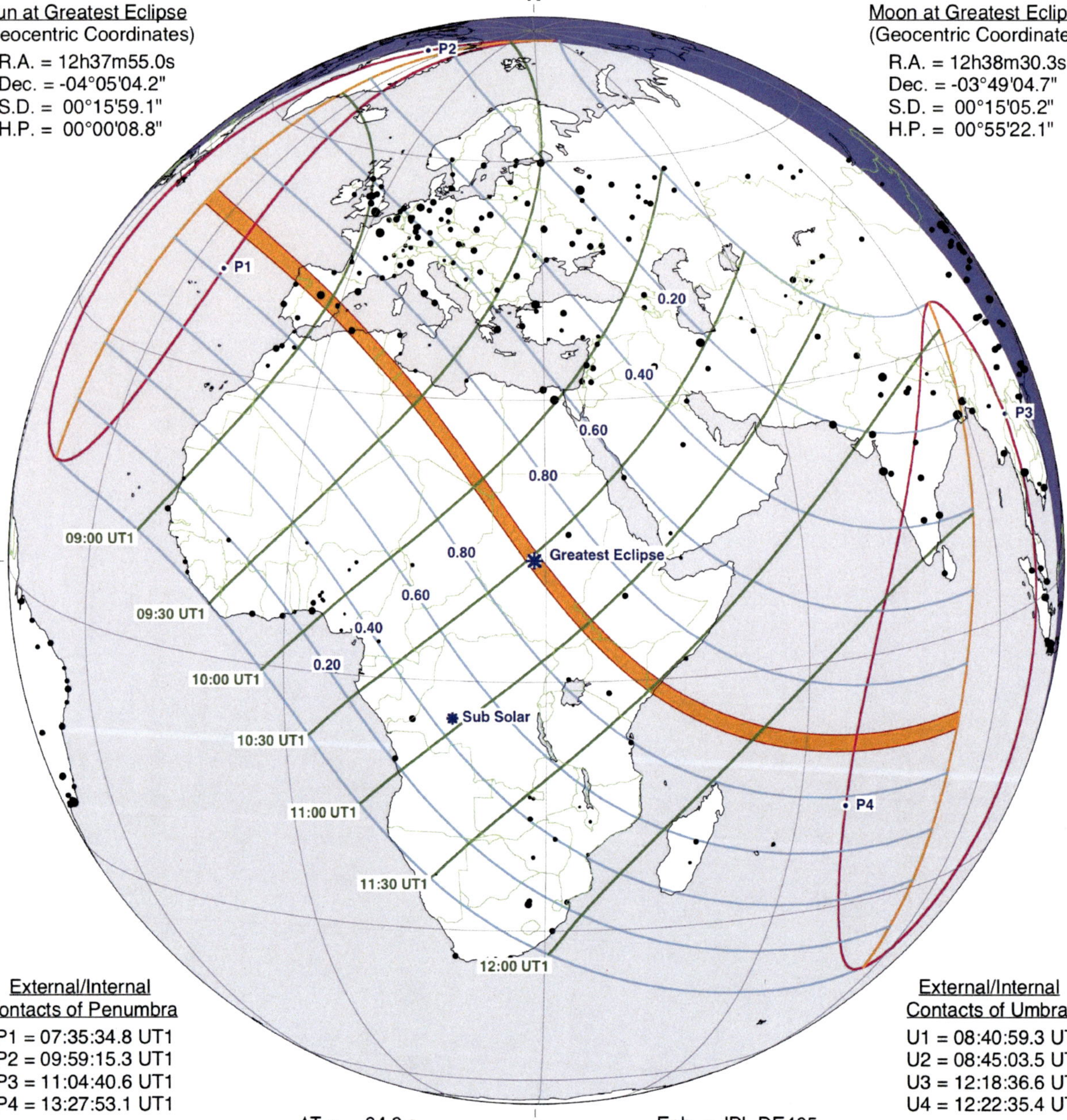

E

S

External/Internal
Contacts of Penumbra
P1 = 07:35:34.8 UT1
P2 = 09:59:15.3 UT1
P3 = 11:04:40.6 UT1
P4 = 13:27:53.1 UT1

External/Internal
Contacts of Umbra
U1 = 08:40:59.3 UT1
U2 = 08:45:03.5 UT1
U3 = 12:18:36.6 UT1
U4 = 12:22:35.4 UT1

ΔT = 64.8 s

Eph. = JPL DE405

Circumstances at Greatest Eclipse: 10:31:42.6 UT1
Lat. = 12°53.4'N Sun Alt. = 70.6°
Long. = 028°44.1'E Sun Azm. = 209.4°
Path Width = 162.2 km Duration = 04m31.5s

Circumstances at Greatest Duration: 10:37:00.5 UT1
Lat. = 11°35.3'N Sun Alt. = 70.4°
Long. = 029°45.4'E Sun Azm. = 217.4°
Path Width = 163.1 km Duration = 04m31.6s

0 1000 2000 3000 4000 5000
Kilometers

Total Solar Eclipse of 2006 Mar 29

Greatest Eclipse = 10:12:22.7 TD (= 10:11:17.8 UT1)

Eclipse Magnitude = 1.0515   Saros Series = 139

Gamma = 0.3843   Saros Member = 29 of 71

Sun at Greatest Eclipse
(Geocentric Coordinates)
R.A. = 00h31m31.7s
Dec. = +03°24'10.3"
S.D. = 00°16'01.1"
H.P. = 00°00'08.8"

Moon at Greatest Eclipse
(Geocentric Coordinates)
R.A. = 00h30m46.6s
Dec. = +03°44'36.2"
S.D. = 00°16'34.9"
H.P. = 01°00'51.4"

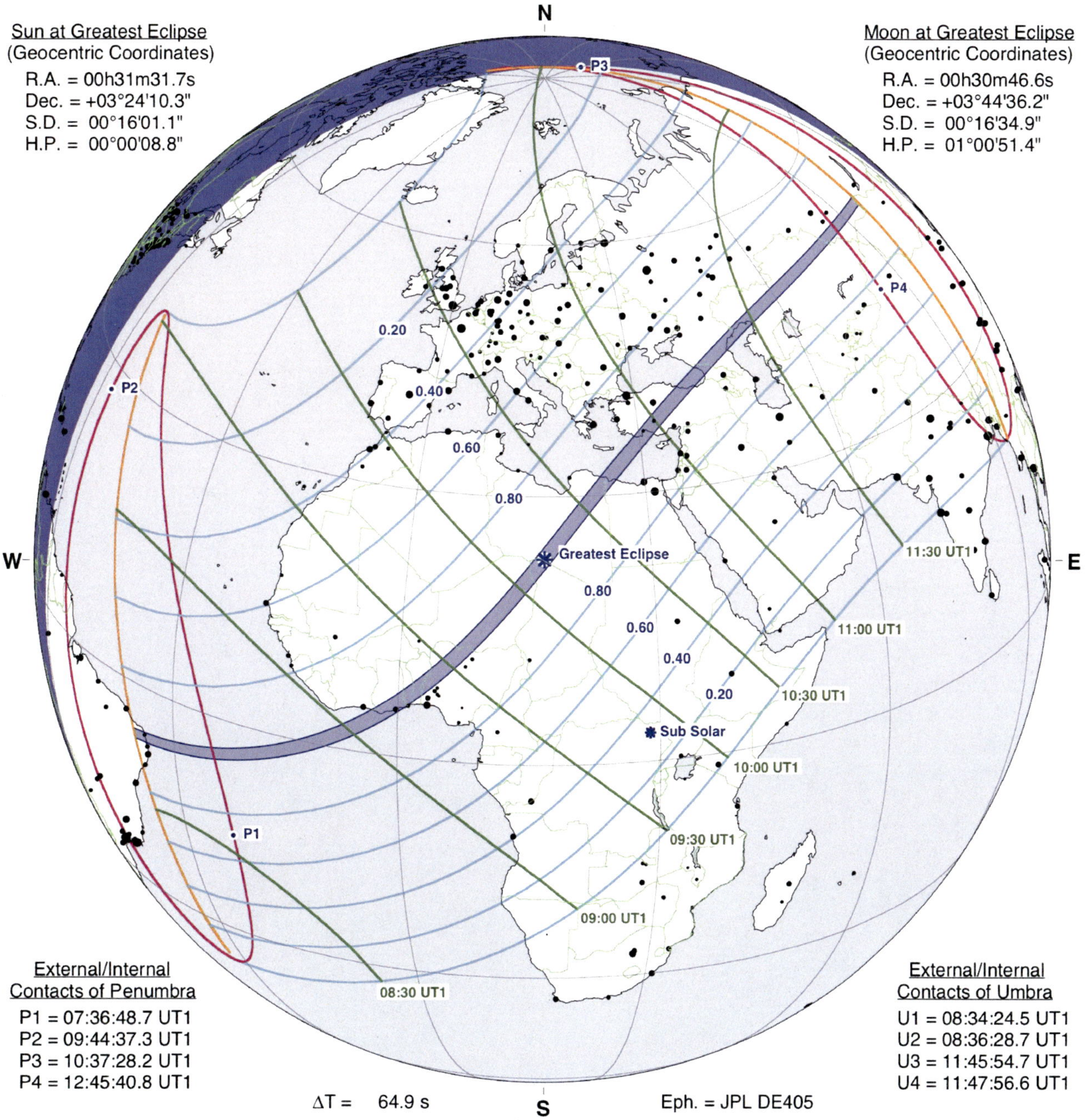

External/Internal
Contacts of Penumbra
P1 = 07:36:48.7 UT1
P2 = 09:44:37.3 UT1
P3 = 10:37:28.2 UT1
P4 = 12:45:40.8 UT1

External/Internal
Contacts of Umbra
U1 = 08:34:24.5 UT1
U2 = 08:36:28.7 UT1
U3 = 11:45:54.7 UT1
U4 = 11:47:56.6 UT1

ΔT = 64.9 s   Eph. = JPL DE405

Circumstances at Greatest Eclipse: 10:11:17.8 UT1
Lat. = 23°09.1'N   Sun Alt. = 67.3°
Long. = 016°44.8'E   Sun Azm. = 148.6°
Path Width = 183.5 km   Duration = 04m06.7s

Circumstances at Greatest Duration: 10:11:40.5 UT1
Lat. = 23°15.7'N   Sun Alt. = 67.3°
Long. = 016°50.8'E   Sun Azm. = 149.2°
Path Width = 183.5 km   Duration = 04m06.7s

0 1000 2000 3000 4000 5000
Kilometers

Annular Solar Eclipse of 2006 Sep 22

Greatest Eclipse = 11:41:16.4 TD (= 11:40:11.4 UT1)

Eclipse Magnitude = 0.9352 Saros Series = 144

Gamma = -0.4062 Saros Member = 16 of 70

Sun at Greatest Eclipse
(Geocentric Coordinates)
R.A. = 11h57m32.9s
Dec. = +00°15'56.9"
S.D. = 00°15'56.1"
H.P. = 00°00'08.8"

Moon at Greatest Eclipse
(Geocentric Coordinates)
R.A. = 11h56m50.2s
Dec. = -00°03'07.8"
S.D. = 00°14'41.9"
H.P. = 00°53'56.7"

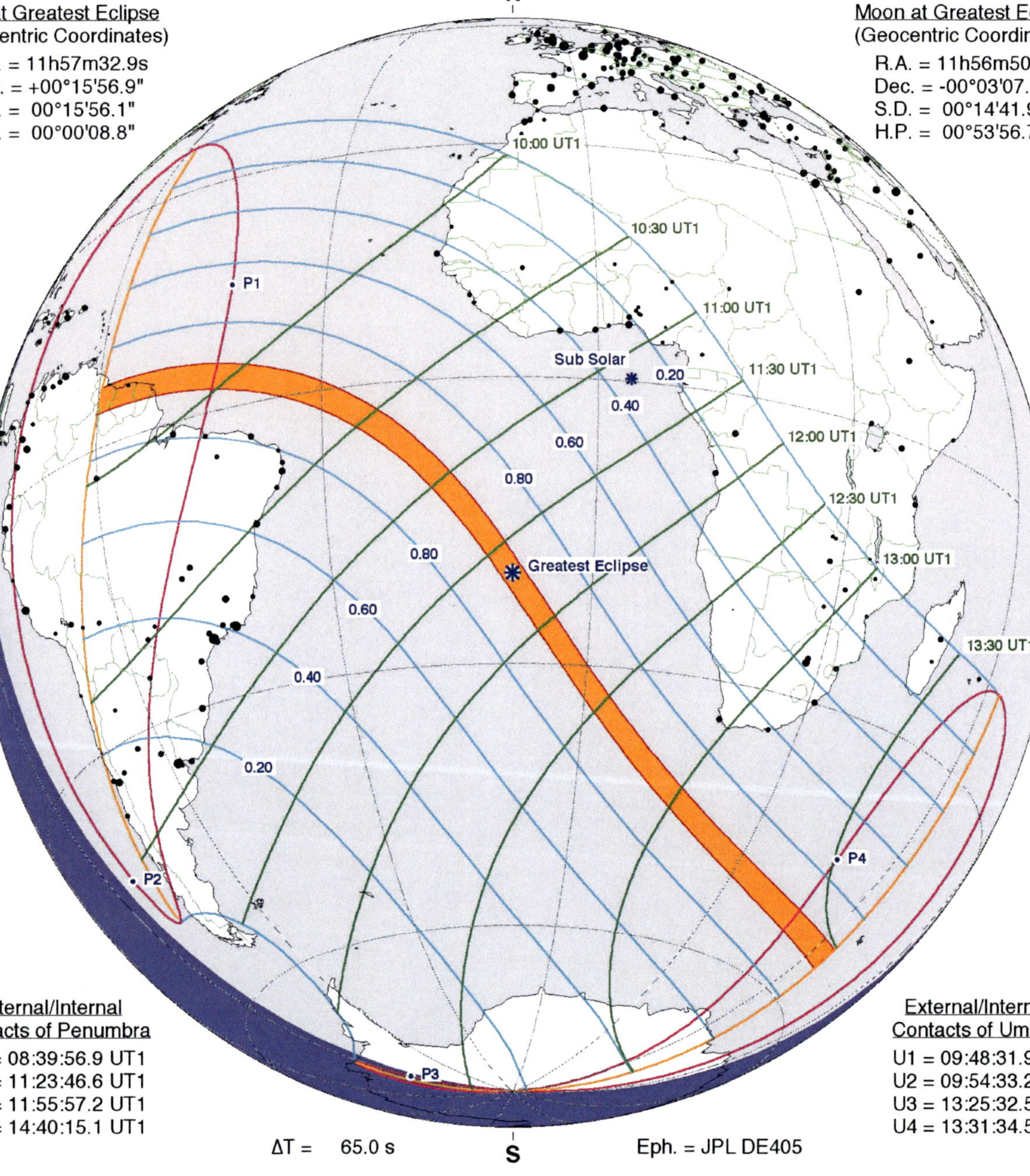

External/Internal Contacts of Penumbra
P1 = 08:39:56.9 UT1
P2 = 11:23:46.6 UT1
P3 = 11:55:57.2 UT1
P4 = 14:40:15.1 UT1

External/Internal Contacts of Umbra
U1 = 09:48:31.9 UT1
U2 = 09:54:33.2 UT1
U3 = 13:25:32.5 UT1
U4 = 13:31:34.5 UT1

ΔT = 65.0 s Eph. = JPL DE405

Circumstances at Greatest Eclipse: 11:40:11.4 UT1

Lat. = 20°38.8'S Sun Alt. = 65.9°
Long. = 009°04.6'W Sun Azm. = 31.2°
Path Width = 261.0 km Duration = 07m09.3s

Circumstances at Greatest Duration: 11:39:23.5 UT1

Lat. = 20°26.7'S Sun Alt. = 65.9°
Long. = 009°13.7'W Sun Azm. = 32.2°
Path Width = 261.3 km Duration = 07m09.3s

0 1000 2000 3000 4000 5000
Kilometers

Partial Solar Eclipse of 2007 Mar 19

Greatest Eclipse = 02:32:57.5 TD (= 02:31:52.3 UT1)

Eclipse Magnitude = 0.8756 Saros Series = 149
Gamma = 1.0728 Saros Member = 20 of 71

Sun at Greatest Eclipse
(Geocentric Coordinates)
R.A. = 23h53m04.0s
Dec. = -00°45'04.8"
S.D. = 00°16'04.0"
H.P. = 00°00'08.8"

Moon at Greatest Eclipse
(Geocentric Coordinates)
R.A. = 23h50m57.2s
Dec. = +00°12'14.7"
S.D. = 00°16'40.7"
H.P. = 01°01'12.5"

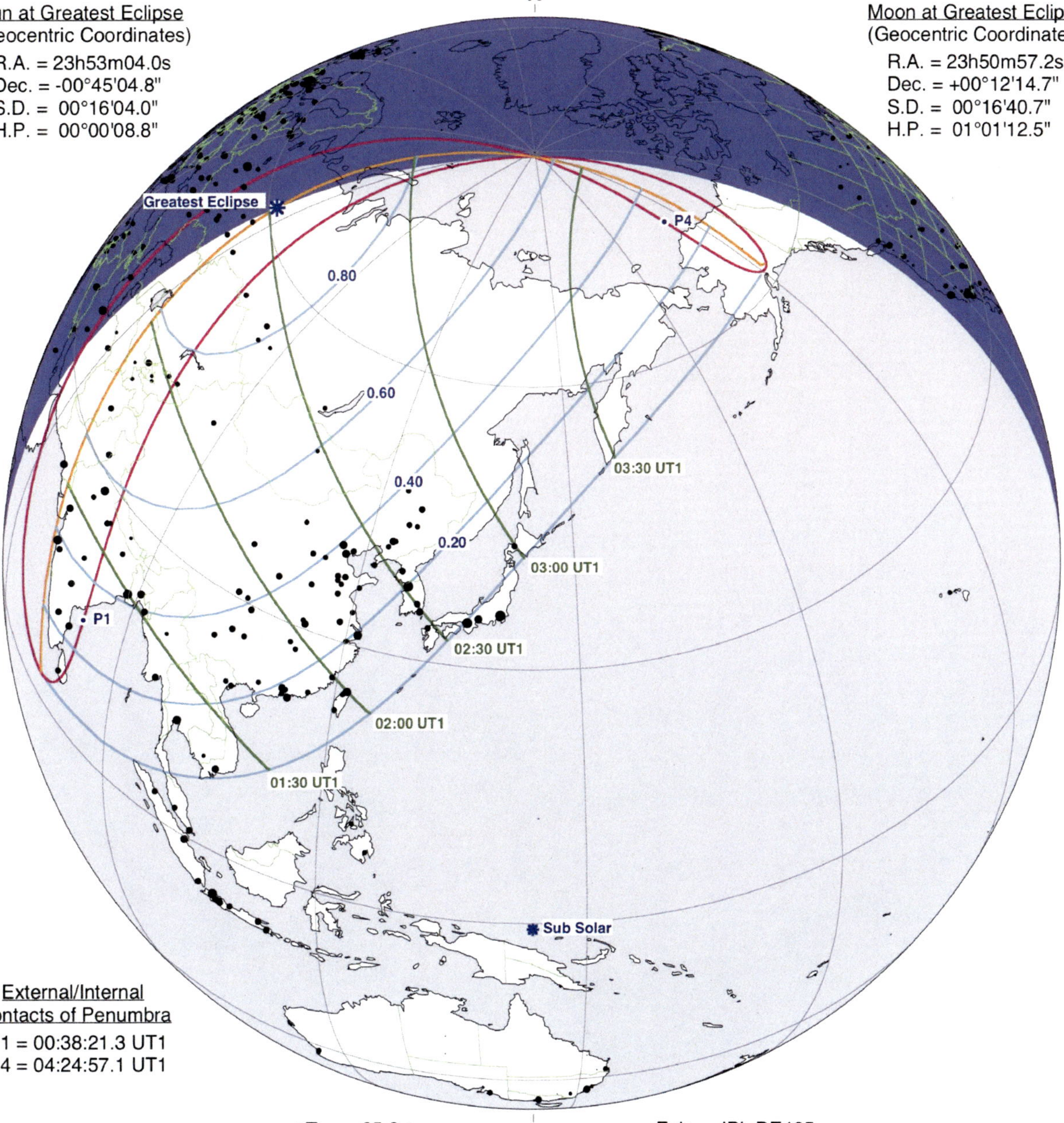

External/Internal
Contacts of Penumbra
P1 = 00:38:21.3 UT1
P4 = 04:24:57.1 UT1

ΔT = 65.2 s Eph. = JPL DE405

Circumstances at Greatest Eclipse: 02:31:52.3 UT1
Lat. = 61°02.9'N Sun Alt. = 0.0°
Long. = 055°28.1'E Sun Azm. = 91.6°

0	1000	2000	3000	4000	5000
Kilometers					

Partial Solar Eclipse of 2007 Sep 11

Greatest Eclipse = 12:32:24.5 TD (= 12:31:19.2 UT1)

Eclipse Magnitude = 0.7507 | Saros Series = 154

Gamma = -1.1255 | Saros Member = 6 of 71

<u>Sun at Greatest Eclipse</u>
(Geocentric Coordinates)
R.A. = 11h17m20.8s
Dec. = +04°35'13.3"
S.D. = 00°15'53.2"
H.P. = 00°00'08.7"

<u>Moon at Greatest Eclipse</u>
(Geocentric Coordinates)
R.A. = 11h15m21.9s
Dec. = +03°40'57.3"
S.D. = 00°15'00.5"
H.P. = 00°55'05.0"

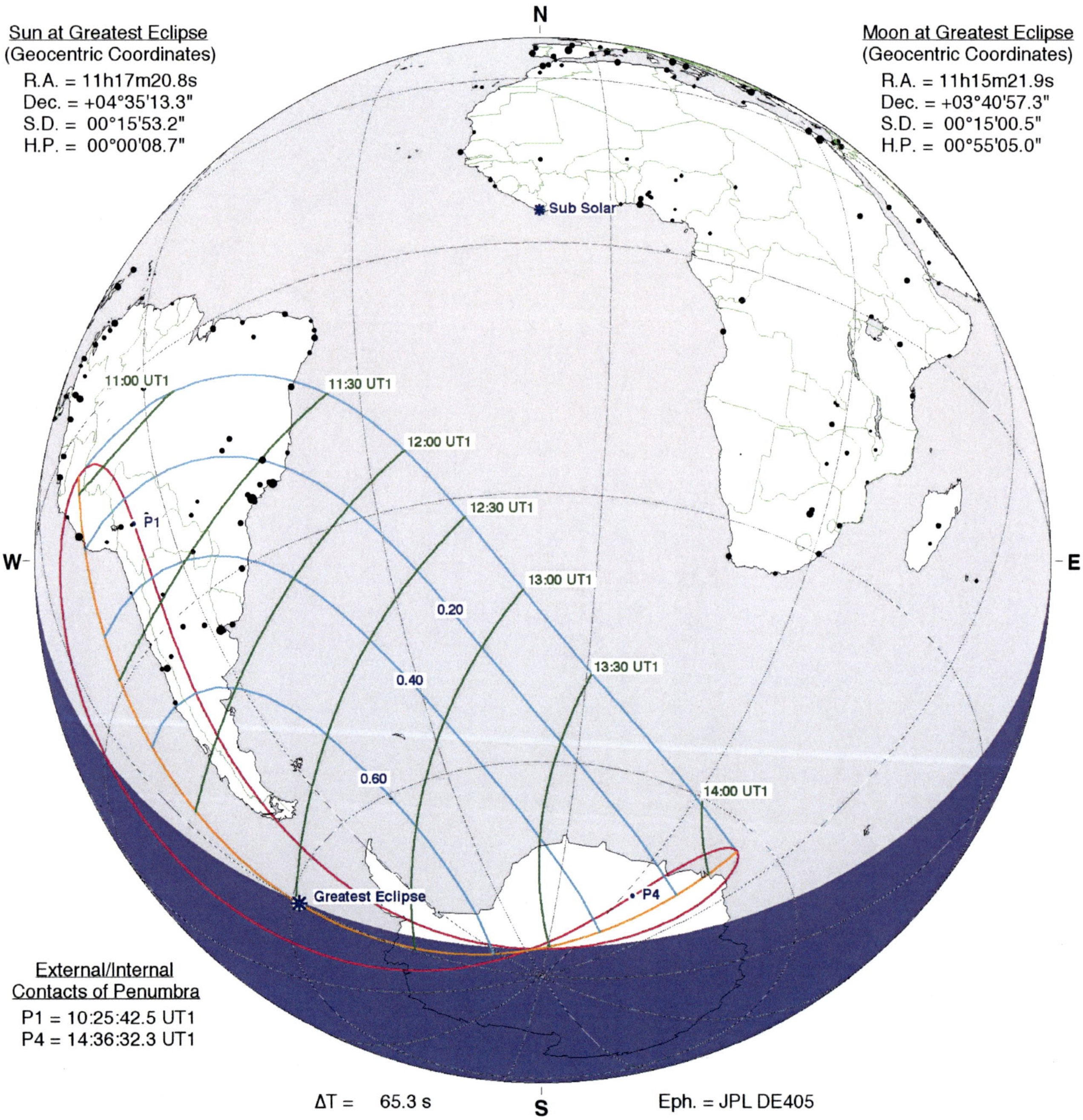

<u>External/Internal</u>
<u>Contacts of Penumbra</u>
P1 = 10:25:42.5 UT1
P4 = 14:36:32.3 UT1

ΔT = 65.3 s | Eph. = JPL DE405

<u>Circumstances at Greatest Eclipse: 12:31:19.2 UT1</u>

Lat. = 61°00.3'S | Sun Alt. = 0.0°

Long. = 090°14.1'W | Sun Azm. = 80.5°

0 1000 2000 3000 4000 5000
Kilometers

Annular Solar Eclipse of 2008 Feb 07

Greatest Eclipse = 03:56:10.5 TD (= 03:55:05.0 UT1)

Eclipse Magnitude = 0.9650 Saros Series = 121

Gamma = -0.9570 Saros Member = 60 of 71

<u>Sun at Greatest Eclipse</u>
(Geocentric Coordinates)
R.A. = 21h20m44.7s
Dec. = -15°30'56.2"
S.D. = 00°16'13.1"
H.P. = 00°00'08.9"

<u>Moon at Greatest Eclipse</u>
(Geocentric Coordinates)
R.A. = 21h22m15.3s
Dec. = -16°21'00.5"
S.D. = 00°15'35.2"
H.P. = 00°57'12.3"

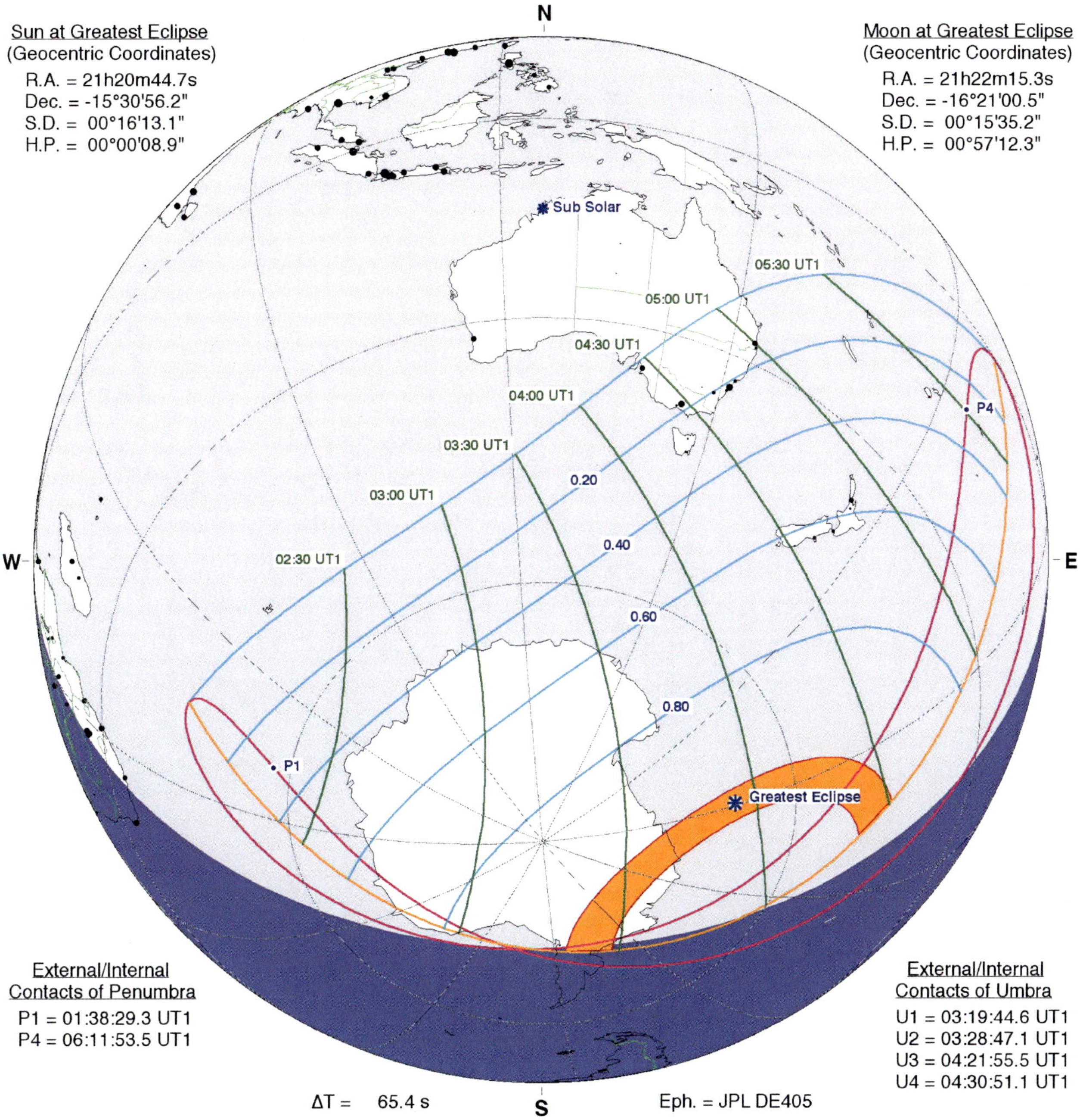

<u>External/Internal Contacts of Penumbra</u>
P1 = 01:38:29.3 UT1
P4 = 06:11:53.5 UT1

<u>External/Internal Contacts of Umbra</u>
U1 = 03:19:44.6 UT1
U2 = 03:28:47.1 UT1
U3 = 04:21:55.5 UT1
U4 = 04:30:51.1 UT1

ΔT = 65.4 s Eph. = JPL DE405

<u>Circumstances at Greatest Eclipse: 03:55:05.0 UT1</u>		<u>Circumstances at Greatest Duration: 03:23:57.9 UT1</u>	
Lat. = 67°34.6'S	Sun Alt. = 16.3°	Lat. = 72°57.4'S	Sun Alt. = 0.0°
Long. = 150°30.1'W	Sun Azm. = 268.8°	Long. = 072°31.9'W	Sun Azm. = 204.1°
Path Width = 444.4 km	Duration = 02m11.7s	Path Width = 581.2 km	Duration = 02m14.1s

0 1000 2000 3000 4000 5000
Kilometers

Total Solar Eclipse of 2008 Aug 01

Greatest Eclipse = 10:22:12.3 TD (= 10:21:06.7 UT1)

Eclipse Magnitude = 1.0394   Saros Series = 126

Gamma = 0.8307   Saros Member = 47 of 72

Sun at Greatest Eclipse
(Geocentric Coordinates)
R.A. = 08h47m54.1s
Dec. = +17°51'56.4"
S.D. = 00°15'45.5"
H.P. = 00°00'08.7"

Moon at Greatest Eclipse
(Geocentric Coordinates)
R.A. = 08h49m08.8s
Dec. = +18°38'01.6"
S.D. = 00°16'14.1"
H.P. = 00°59'34.8"

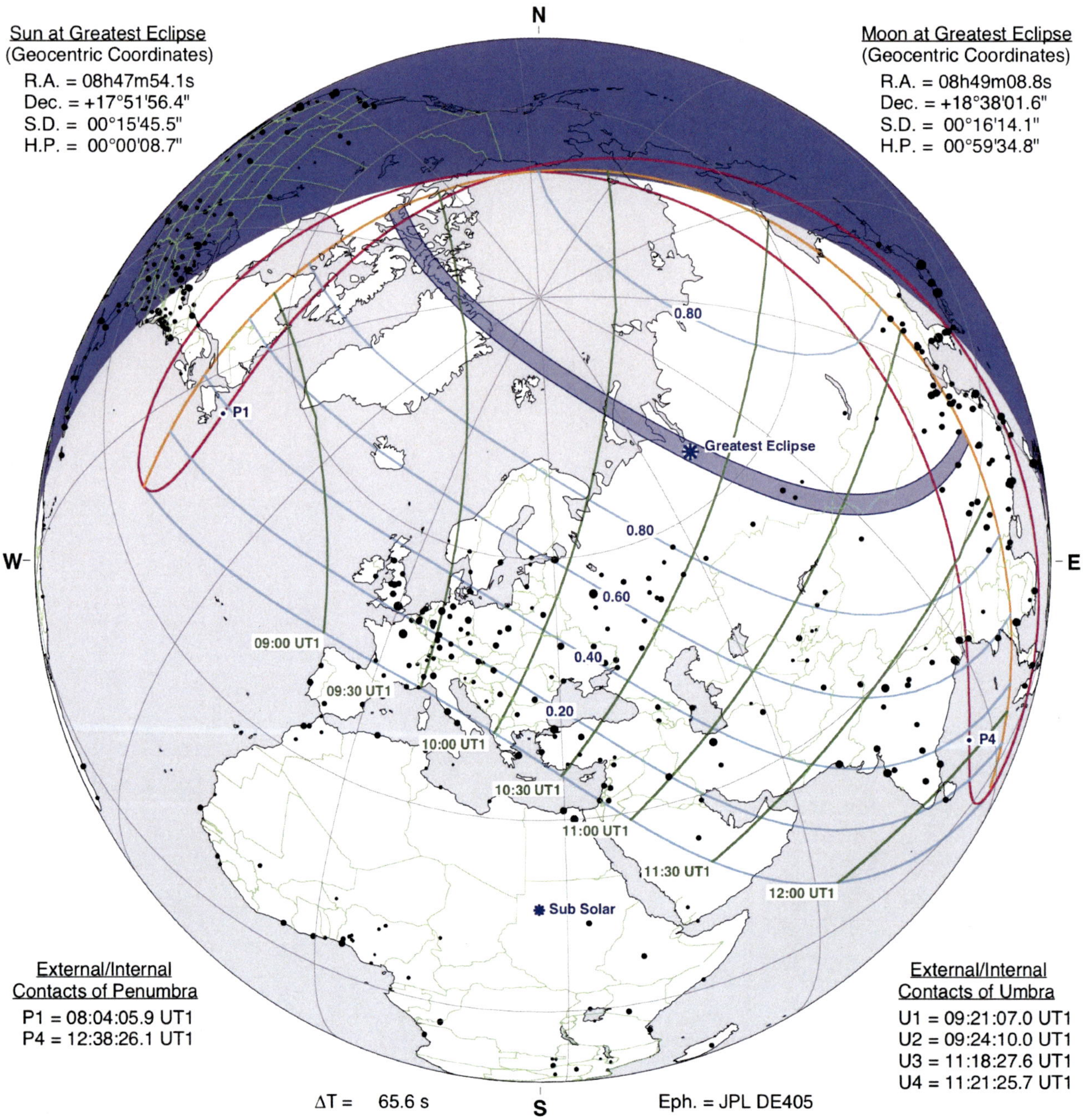

External/Internal
Contacts of Penumbra
P1 = 08:04:05.9 UT1
P4 = 12:38:26.1 UT1

External/Internal
Contacts of Umbra
U1 = 09:21:07.0 UT1
U2 = 09:24:10.0 UT1
U3 = 11:18:27.6 UT1
U4 = 11:21:25.7 UT1

ΔT = 65.6 s   Eph. = JPL DE405

Circumstances at Greatest Eclipse: 10:21:06.7 UT1

Lat. = 65°39.2'N	Sun Alt. = 33.5°
Long. = 072°17.8'E	Sun Azm. = 235.2°
Path Width = 236.9 km	Duration = 02m27.1s

Circumstances at Greatest Duration: 10:19:11.5 UT1

Lat. = 66°30.6'N	Sun Alt. = 33.5°
Long. = 071°20.0'E	Sun Azm. = 233.2°
Path Width = 236.1 km	Duration = 02m27.2s

0 1000 2000 3000 4000 5000
Kilometers

Annular Solar Eclipse of 2009 Jan 26

Greatest Eclipse = 07:59:44.5 TD (= 07:58:38.8 UT1)

Eclipse Magnitude = 0.9282 | Saros Series = 131

Gamma = -0.2820 | Saros Member = 50 of 70

Sun at Greatest Eclipse
(Geocentric Coordinates)

R.A. = 20h35m32.8s
Dec. = -18°38'55.0"
S.D. = 00°16'14.6"
H.P. = 00°00'08.9"

Moon at Greatest Eclipse
(Geocentric Coordinates)

R.A. = 20h35m55.2s
Dec. = -18°53'18.2"
S.D. = 00°14'51.6"
H.P. = 00°54'32.2"

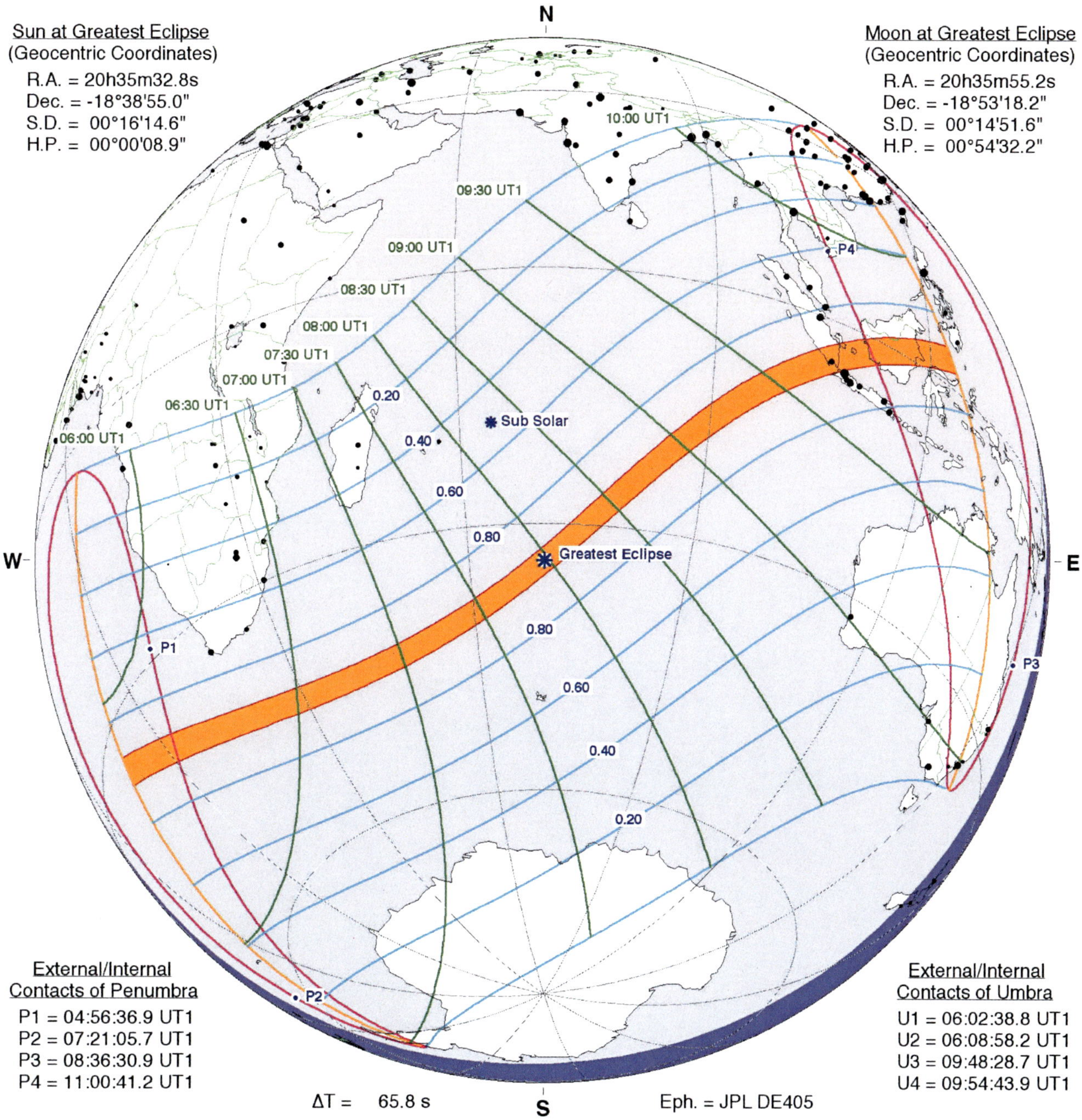

External/Internal
Contacts of Penumbra

P1 = 04:56:36.9 UT1
P2 = 07:21:05.7 UT1
P3 = 08:36:30.9 UT1
P4 = 11:00:41.2 UT1

External/Internal
Contacts of Umbra

U1 = 06:02:38.8 UT1
U2 = 06:08:58.2 UT1
U3 = 09:48:28.7 UT1
U4 = 09:54:43.9 UT1

ΔT = 65.8 s

Eph. = JPL DE405

Circumstances at Greatest Eclipse: 07:58:38.8 UT1

Lat. = 34°04.5'S	Sun Alt. = 73.4°
Long. = 070°13.9'E	Sun Azm. = 337.0°
Path Width = 280.2 km	Duration = 07m53.6s

Circumstances at Greatest Duration: 07:42:17.9 UT1

Lat. = 37°03.7'S	Sun Alt. = 71.5°
Long. = 065°12.8'E	Sun Azm. = 7.0°
Path Width = 279.0 km	Duration = 07m56.0s

0 1000 2000 3000 4000 5000
Kilometers

Total Solar Eclipse of 2009 Jul 22

Greatest Eclipse = 02:36:24.6 TD (= 02:35:18.7 UT1)

Eclipse Magnitude = 1.0799

Gamma = 0.0698

Saros Series = 136

Saros Member = 37 of 71

Sun at Greatest Eclipse
(Geocentric Coordinates)

R.A. = 08h06m24.1s
Dec. = +20°16'03.0"
S.D. = 00°15'44.5"
H.P. = 00°00'08.7"

Moon at Greatest Eclipse
(Geocentric Coordinates)

R.A. = 08h06m29.6s
Dec. = +20°20'07.0"
S.D. = 00°16'42.7"
H.P. = 01°01'19.8"

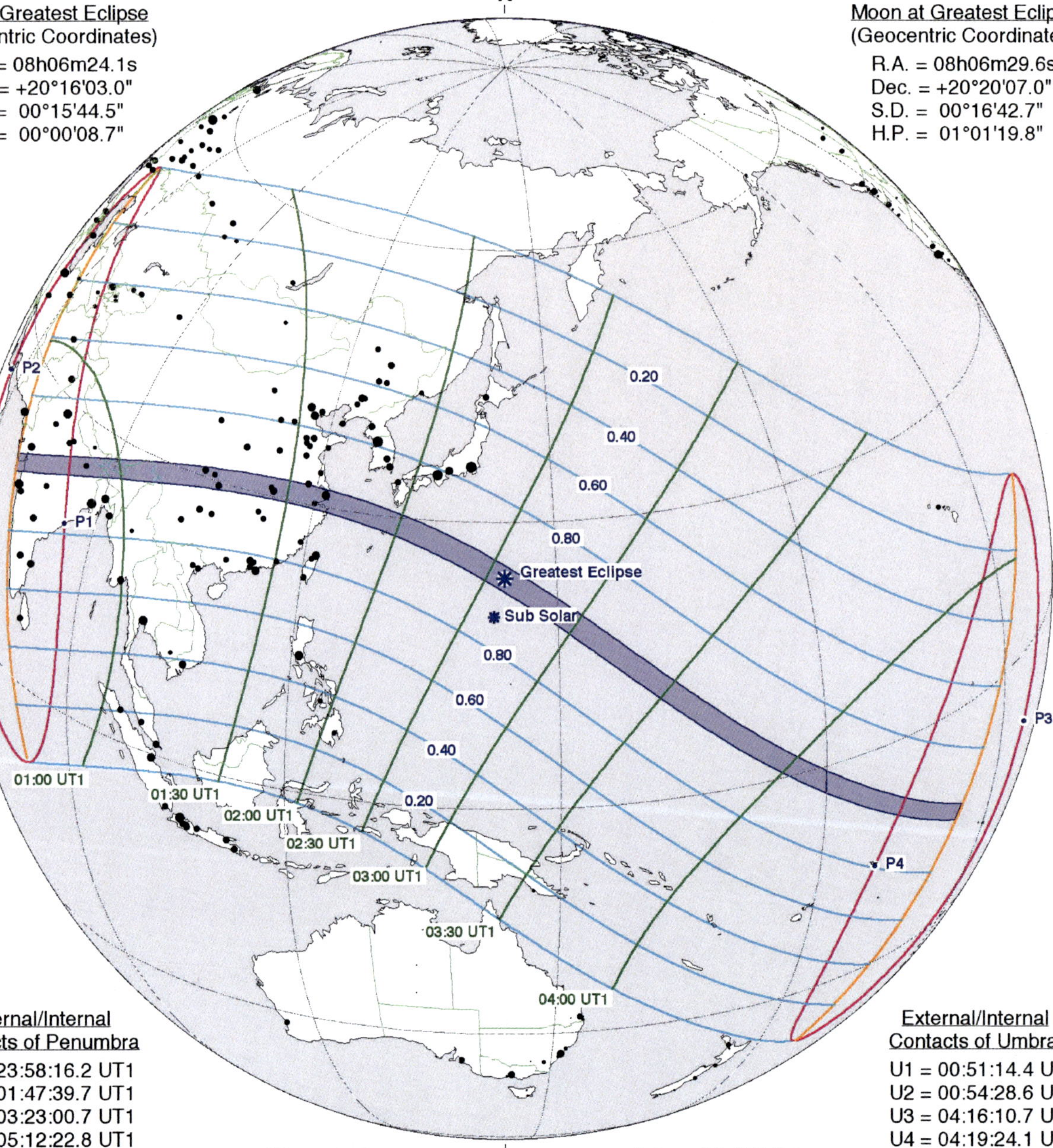

External/Internal Contacts of Penumbra

P1 = 23:58:16.2 UT1
P2 = 01:47:39.7 UT1
P3 = 03:23:00.7 UT1
P4 = 05:12:22.8 UT1

External/Internal Contacts of Umbra

U1 = 00:51:14.4 UT1
U2 = 00:54:28.6 UT1
U3 = 04:16:10.7 UT1
U4 = 04:19:24.1 UT1

ΔT = 65.9 s

Eph. = JPL DE405

Circumstances at Greatest Eclipse: 02:35:18.7 UT1

Lat. = 24°13.2'N	Sun Alt. = 85.9°
Long. = 144°06.9'E	Sun Azm. = 197.6°
Path Width = 258.5 km	Duration = 06m38.9s

Circumstances at Greatest Duration: 02:29:16.6 UT1

Lat. = 25°14.5'N	Sun Alt. = 84.6°
Long. = 142°05.5'E	Sun Azm. = 157.4°
Path Width = 258.2 km	Duration = 06m39.5s

0 1000 2000 3000 4000 5000
Kilometers

Annular Solar Eclipse of 2010 Jan 15

Greatest Eclipse = 07:07:39.2 TD (= 07:06:33.2 UT1)

Eclipse Magnitude = 0.9190 Saros Series = 141
Gamma = 0.4002 Saros Member = 23 of 70

Sun at Greatest Eclipse
(Geocentric Coordinates)
R.A. = 19h47m51.0s
Dec. = -21°07'38.7"
S.D. = 00°16'15.5"
H.P. = 00°00'08.9"

Moon at Greatest Eclipse
(Geocentric Coordinates)
R.A. = 19h47m25.3s
Dec. = -20°46'54.8"
S.D. = 00°14'44.3"
H.P. = 00°54'05.4"

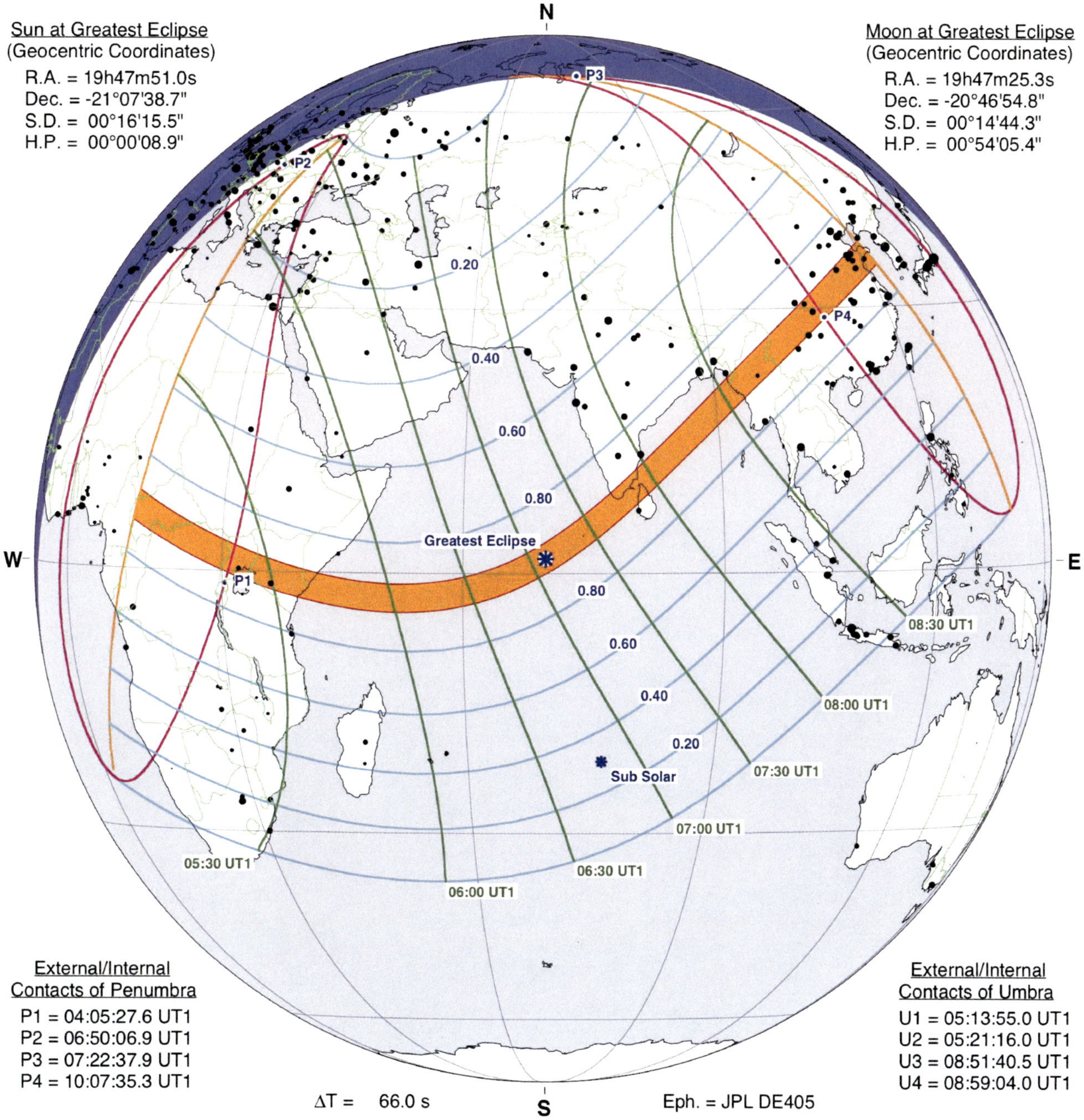

External/Internal
Contacts of Penumbra
P1 = 04:05:27.6 UT1
P2 = 06:50:06.9 UT1
P3 = 07:22:37.9 UT1
P4 = 10:07:35.3 UT1

External/Internal
Contacts of Umbra
U1 = 05:13:55.0 UT1
U2 = 05:21:16.0 UT1
U3 = 08:51:40.5 UT1
U4 = 08:59:04.0 UT1

ΔT = 66.0 s Eph. = JPL DE405

Circumstances at Greatest Eclipse: 07:06:33.2 UT1
Lat. = 01°37.4'N Sun Alt. = 66.4°
Long. = 069°17.4'E Sun Azm. = 164.9°
Path Width = 333.1 km Duration = 11m07.8s

Circumstances at Greatest Duration: 06:54:29.6 UT1
Lat. = 00°13.2'N Sun Alt. = 65.6°
Long. = 066°39.1'E Sun Azm. = 151.8°
Path Width = 337.2 km Duration = 11m10.7s

0 1000 2000 3000 4000 5000
Kilometers

Total Solar Eclipse of 2010 Jul 11

Greatest Eclipse = 19:34:37.9 TD (= 19:33:31.7 UT1)

Eclipse Magnitude = 1.0580 Saros Series = 146

Gamma = -0.6788 Saros Member = 27 of 76

Sun at Greatest Eclipse
(Geocentric Coordinates)
R.A. = 07h23m57.6s
Dec. = +22°02'11.0"
S.D. = 00°15'43.9"
H.P. = 00°00'08.7"

Moon at Greatest Eclipse
(Geocentric Coordinates)
R.A. = 07h23m15.8s
Dec. = +21°22'29.3"
S.D. = 00°16'26.6"
H.P. = 01°00'20.9"

N

W

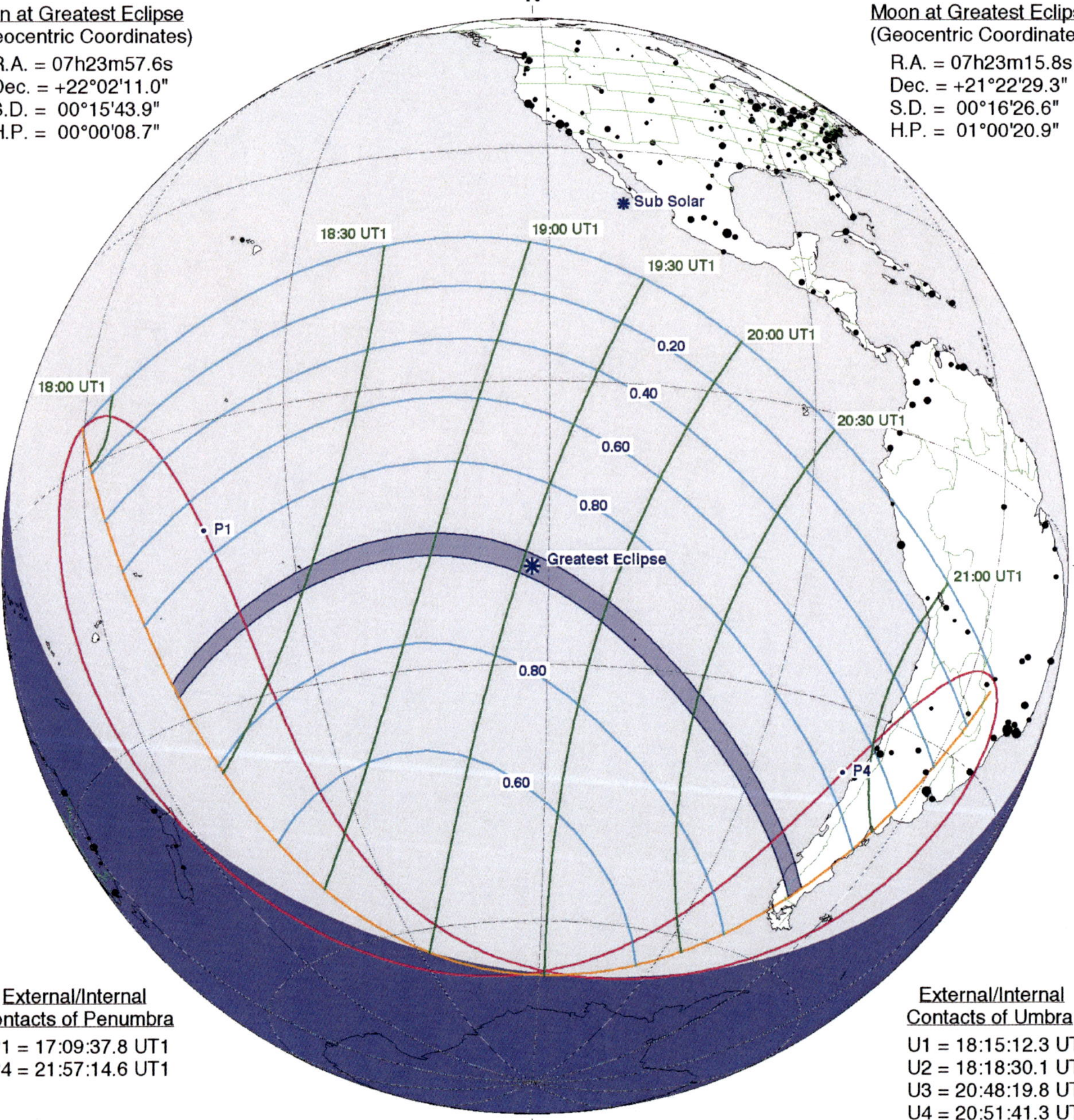

E

S

External/Internal
Contacts of Penumbra
P1 = 17:09:37.8 UT1
P4 = 21:57:14.6 UT1

External/Internal
Contacts of Umbra
U1 = 18:15:12.3 UT1
U2 = 18:18:30.1 UT1
U3 = 20:48:19.8 UT1
U4 = 20:51:41.3 UT1

ΔT = 66.2 s Eph. = JPL DE405

Circumstances at Greatest Eclipse: 19:33:31.7 UT1
Lat. = 19°44.8'S Sun Alt. = 47.1°
Long. = 121°52.5'W Sun Azm. = 13.5°
Path Width = 258.6 km Duration = 05m20.2s

Circumstances at Greatest Duration: 19:31:26.0 UT1
Lat. = 19°29.0'S Sun Alt. = 47.1°
Long. = 122°32.2'W Sun Azm. = 15.1°
Path Width = 259.5 km Duration = 05m20.3s

0 1000 2000 3000 4000 5000
Kilometers

Partial Solar Eclipse of 2011 Jan 04

Greatest Eclipse = 08:51:42.4 TD (= 08:50:36.1 UT1)

Eclipse Magnitude = 0.8576
Gamma = 1.0626

Saros Series = 151
Saros Member = 14 of 72

Sun at Greatest Eclipse
(Geocentric Coordinates)
R.A. = 18h59m14.9s
Dec. = -22°44'21.1"
S.D. = 00°16'15.9"
H.P. = 00°00'08.9"

Moon at Greatest Eclipse
(Geocentric Coordinates)
R.A. = 18h58m23.8s
Dec. = -21°46'01.2"
S.D. = 00°15'18.1"
H.P. = 00°56'09.6"

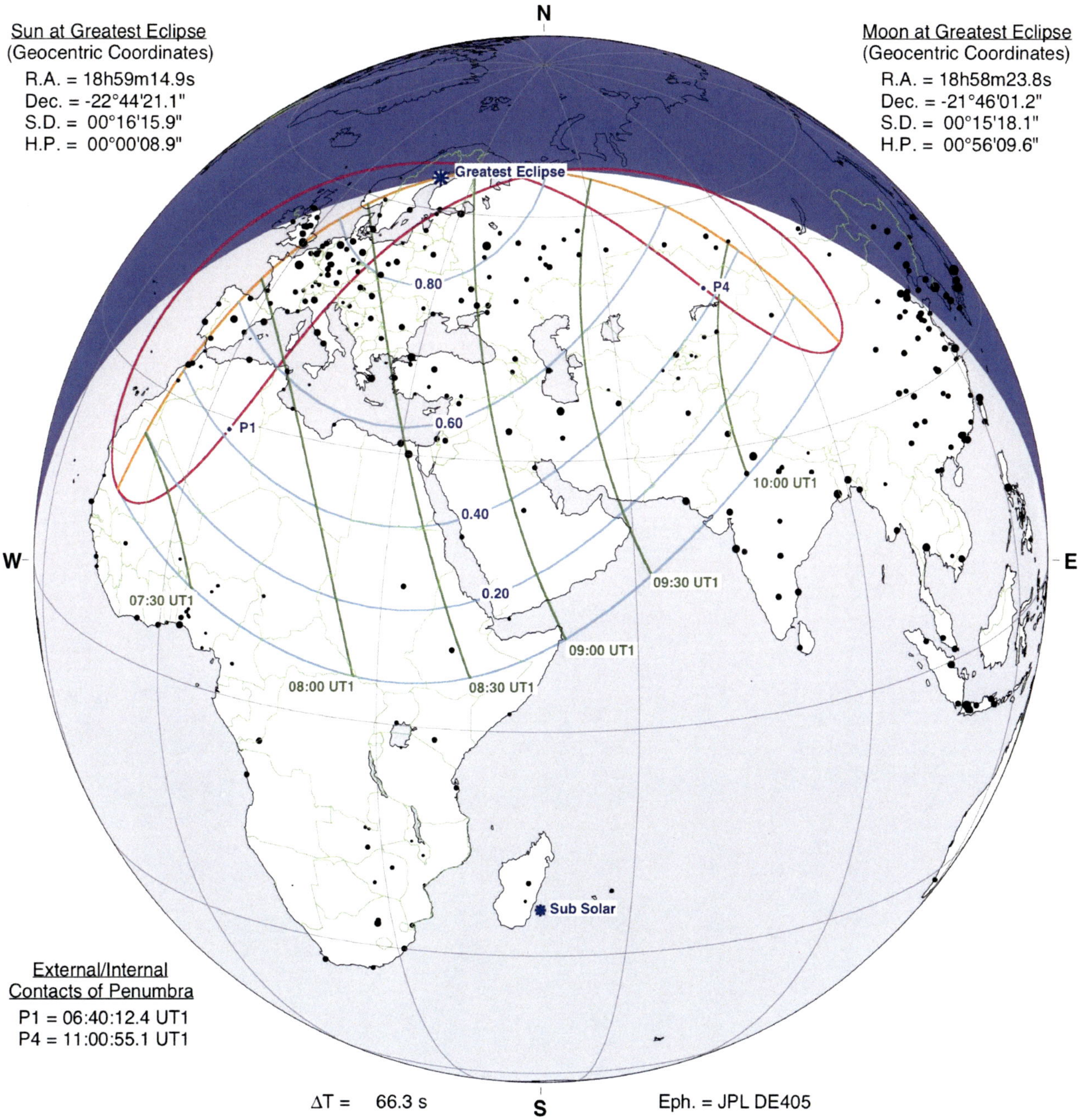

External/Internal
Contacts of Penumbra
P1 = 06:40:12.4 UT1
P4 = 11:00:55.1 UT1

ΔT = 66.3 s

Eph. = JPL DE405

Circumstances at Greatest Eclipse: 08:50:36.1 UT1

Lat. = 64°39.1'N
Long. = 020°47.9'E

Sun Alt. = 0.0°
Sun Azm. = 154.5°

0 1000 2000 3000 4000 5000
Kilometers

Partial Solar Eclipse of 2011 Jun 01

Greatest Eclipse = 21:17:18.4 TD (= 21:16:12.0 UT1)

Eclipse Magnitude = 0.6011 Saros Series = 118

Gamma = 1.2130 Saros Member = 68 of 72

Sun at Greatest Eclipse
(Geocentric Coordinates)
R.A. = 04h37m53.4s
Dec. = +22°05'47.3"
S.D. = 00°15'46.3"
H.P. = 00°00'08.7"

Moon at Greatest Eclipse
(Geocentric Coordinates)
R.A. = 04h37m41.0s
Dec. = +23°13'19.4"
S.D. = 00°15'13.4"
H.P. = 00°55'52.1"

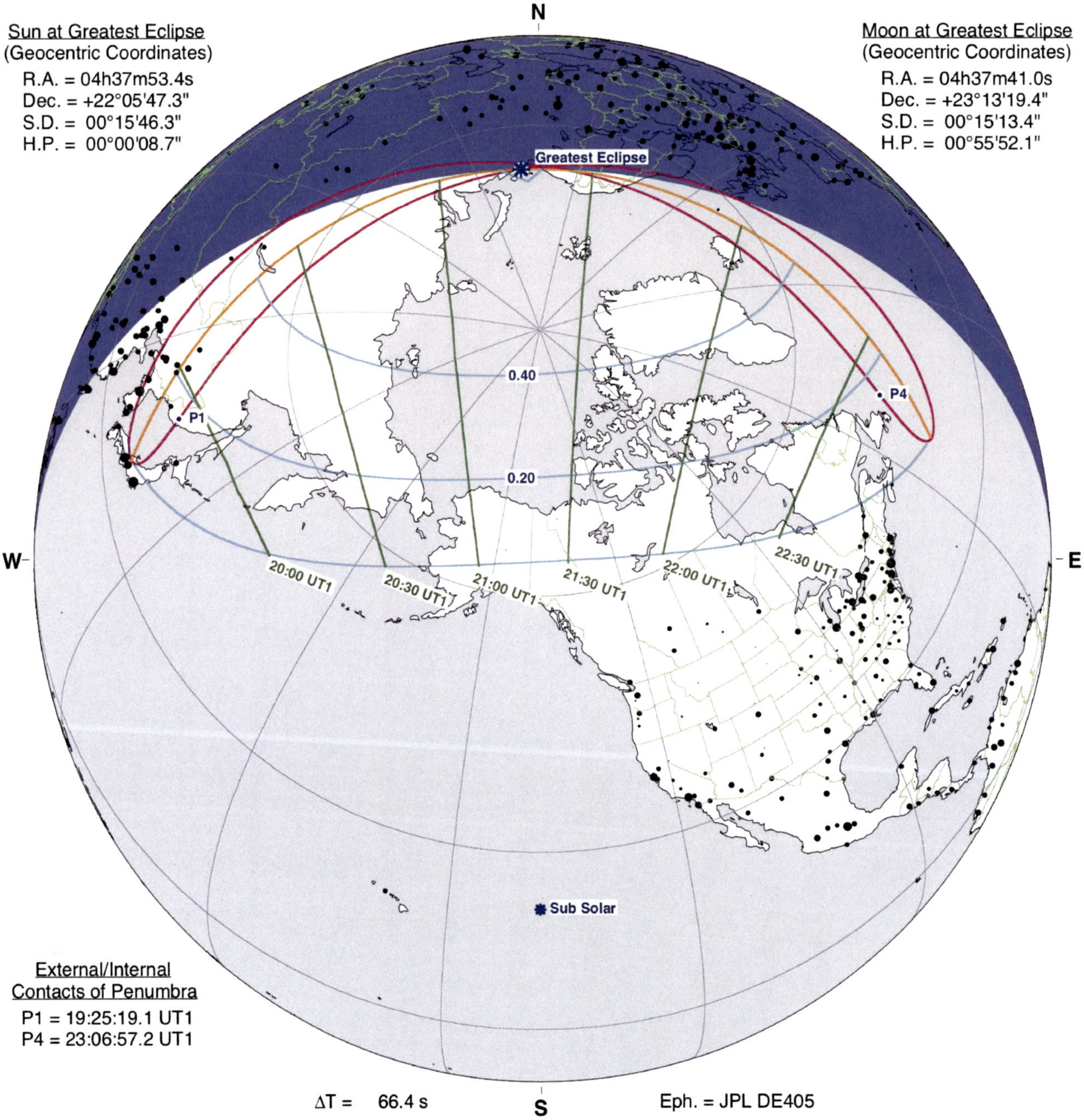

External/Internal
Contacts of Penumbra
P1 = 19:25:19.1 UT1
P4 = 23:06:57.2 UT1

ΔT = 66.4 s Eph. = JPL DE405

Circumstances at Greatest Eclipse: 21:16:12.0 UT1

Lat. = 67°47.0'N Sun Alt. = 0.0°

Long. = 046°47.2'E Sun Azm. = 5.9°

0	1000	2000	3000	4000	5000

Kilometers

Partial Solar Eclipse of 2011 Jul 01

Greatest Eclipse = 08:39:30.3 TD (= 08:38:23.9 UT1)

Eclipse Magnitude = 0.0971 **Saros Series = 156**

Gamma = -1.4917 **Saros Member = 1 of 69**

Sun at Greatest Eclipse
(Geocentric Coordinates)

R.A. = 06h40m01.7s
Dec. = +23°07'05.9"
S.D. = 00°15'43.9"
H.P. = 00°00'08.6"

Moon at Greatest Eclipse
(Geocentric Coordinates)

R.A. = 06h39m02.0s
Dec. = +21°42'47.5"
S.D. = 00°15'38.6"
H.P. = 00°57'24.6"

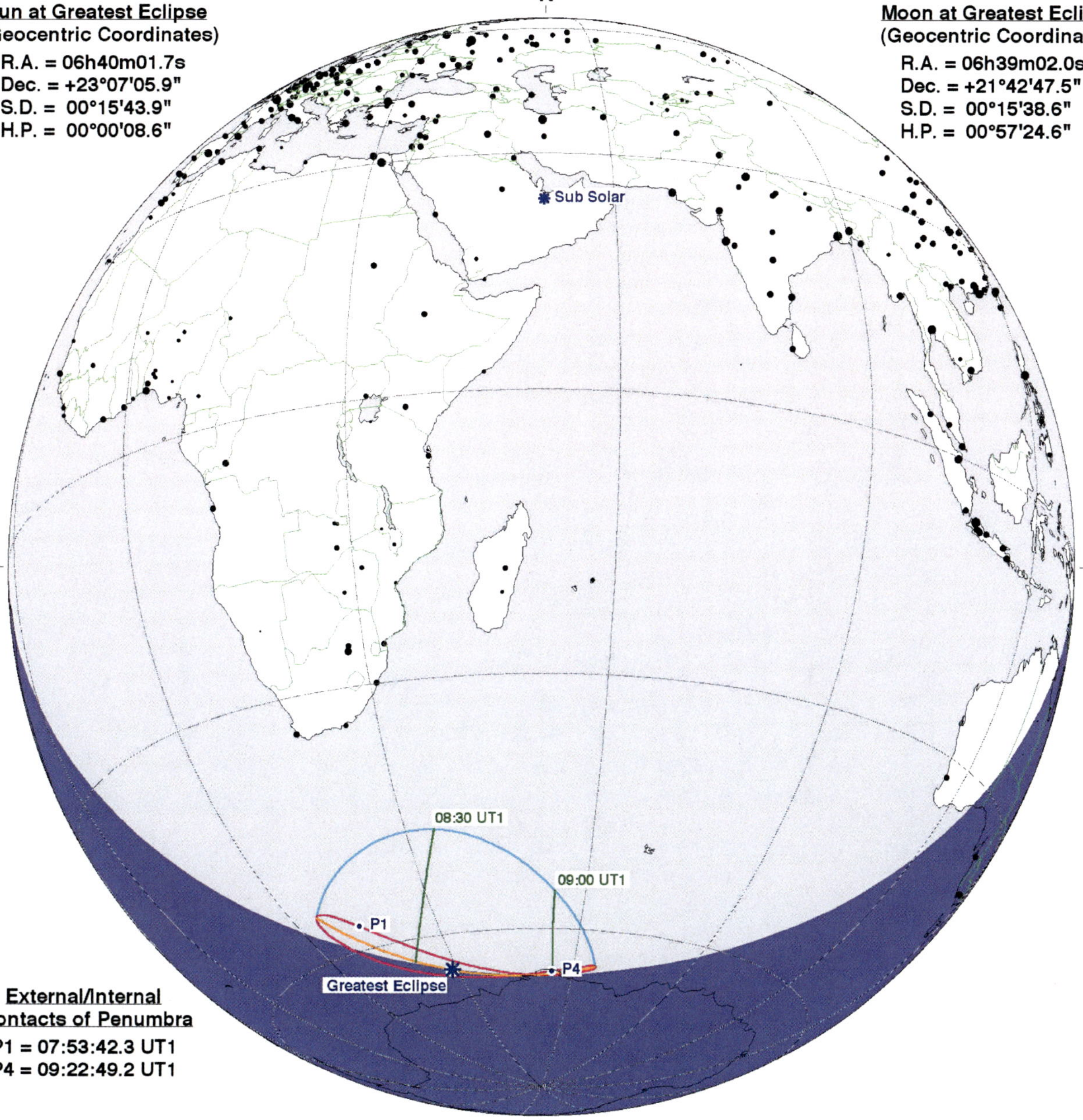

External/Internal Contacts of Penumbra

P1 = 07:53:42.3 UT1
P4 = 09:22:49.2 UT1

ΔT = 66.4 s

Eph. = JPL DE405

Circumstances at Greatest Eclipse: 08:38:23.9 UT1

Lat. = 65°09.5'S Sun Alt. = 0.0°
Long. = 028°38.2'E Sun Azm. = 20.8°

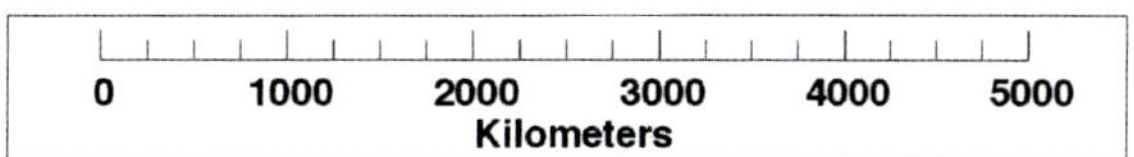

Partial Solar Eclipse of 2011 Nov 25

Greatest Eclipse = 06:21:24.5 TD (= 06:20:18.0 UT1)

Eclipse Magnitude = 0.9047 Saros Series = 123

Gamma = -1.0536 Saros Member = 53 of 70

Sun at Greatest Eclipse
(Geocentric Coordinates)
R.A. = 16h02m13.7s
Dec. = -20°40'56.2"
S.D. = 00°16'12.1"
H.P. = 00°00'08.9"

Moon at Greatest Eclipse
(Geocentric Coordinates)
R.A. = 16h01m46.2s
Dec. = -21°44'25.4"
S.D. = 00°16'32.6"
H.P. = 01°00'42.7"

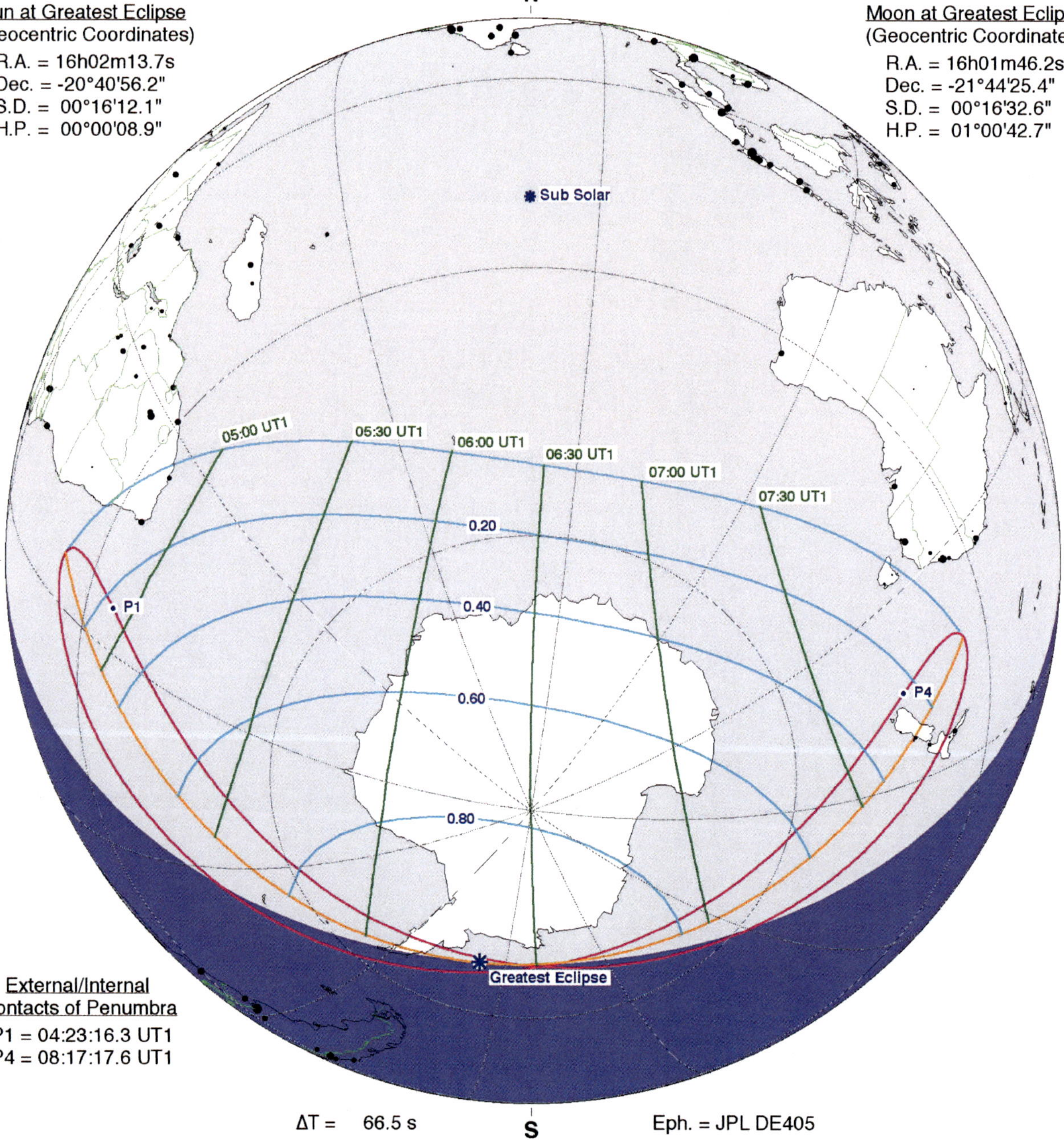

External/Internal Contacts of Penumbra
P1 = 04:23:16.3 UT1
P4 = 08:17:17.6 UT1

ΔT = 66.5 s Eph. = JPL DE405

Circumstances at Greatest Eclipse: 06:20:18.0 UT1

Lat. = 68°34.1'S Sun Alt. = 0.0°

Long. = 082°25.9'W Sun Azm. = 165.1°

0 1000 2000 3000 4000 5000
Kilometers

Annular Solar Eclipse of 2012 May 20

Greatest Eclipse = 23:53:53.6 TD (= 23:52:47.0 UT1)

Eclipse Magnitude = 0.9439 Saros Series = 128

Gamma = 0.4828 Saros Member = 58 of 73

Sun at Greatest Eclipse
(Geocentric Coordinates)
R.A. = 03h52m43.0s
Dec. = +20°13'15.1"
S.D. = 00°15'48.1"
H.P. = 00°00'08.7"

Moon at Greatest Eclipse
(Geocentric Coordinates)
R.A. = 03h52m30.7s
Dec. = +20°39'06.3"
S.D. = 00°14'43.3"
H.P. = 00°54'01.7"

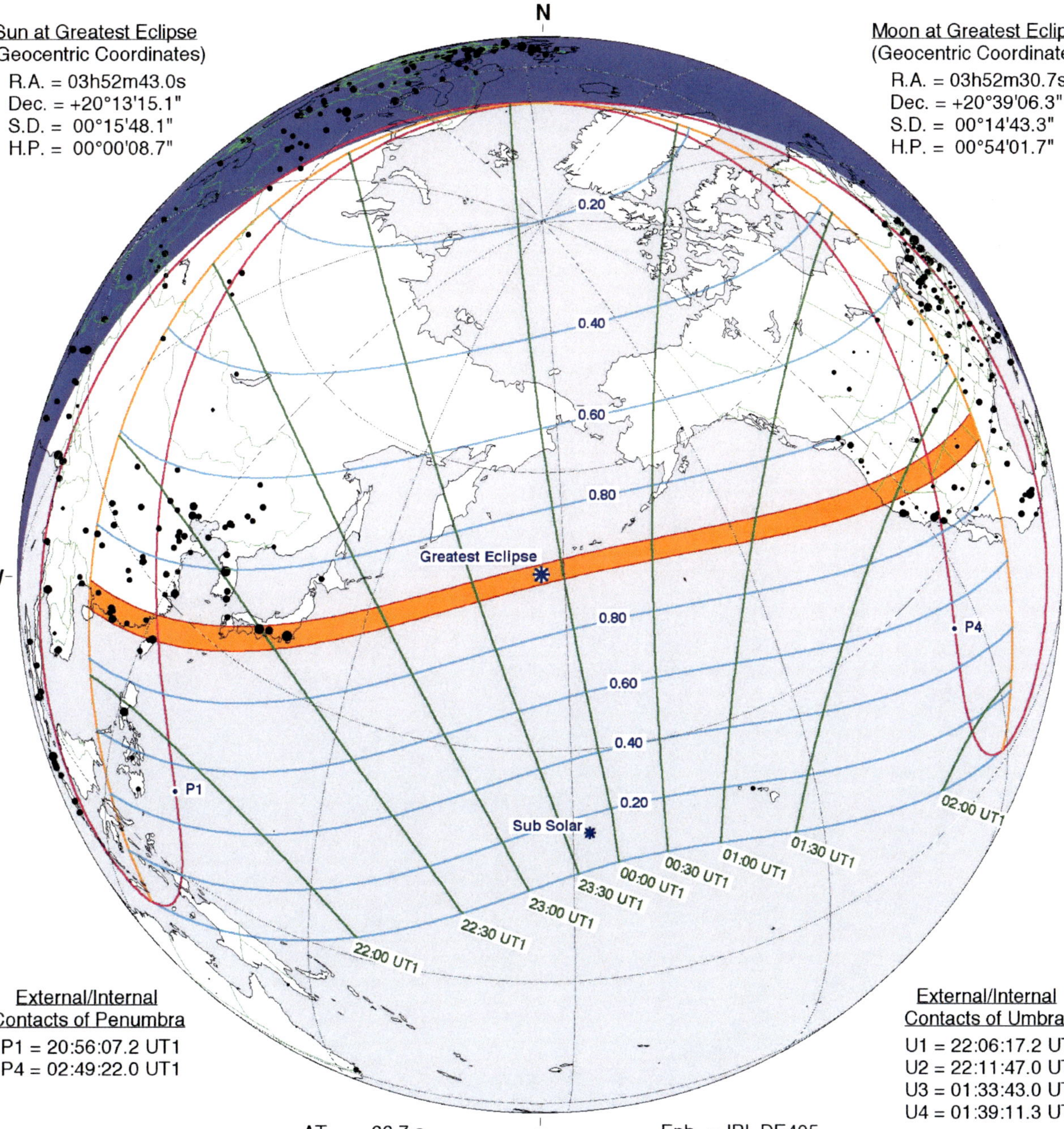

External/Internal Contacts of Penumbra
P1 = 20:56:07.2 UT1
P4 = 02:49:22.0 UT1

External/Internal Contacts of Umbra
U1 = 22:06:17.2 UT1
U2 = 22:11:47.0 UT1
U3 = 01:33:43.0 UT1
U4 = 01:39:11.3 UT1

ΔT = 66.7 s Eph. = JPL DE405

Circumstances at Greatest Eclipse: 23:52:47.0 UT1

Lat. = 49°05.6'N	Sun Alt. = 60.9°
Long. = 176°16.4'E	Sun Azm. = 171.0°
Path Width = 236.9 km	Duration = 05m46.3s

Circumstances at Greatest Duration: 23:55:53.1 UT1

Lat. = 49°19.3'N	Sun Alt. = 60.8°
Long. = 177°45.8'E	Sun Azm. = 175.4°
Path Width = 236.7 km	Duration = 05m46.3s

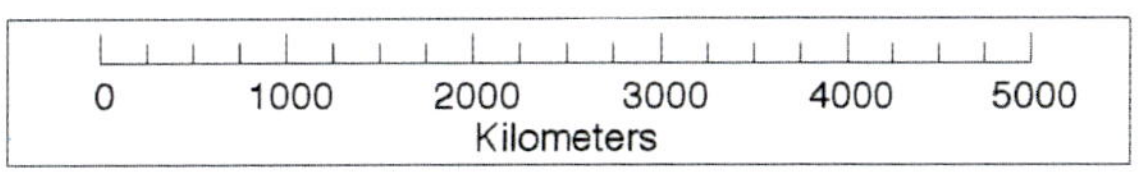

Total Solar Eclipse of 2012 Nov 13

Greatest Eclipse = 22:12:55.2 TD (= 22:11:48.4 UT1)

Eclipse Magnitude = 1.0500 Saros Series = 133

Gamma = -0.3719 Saros Member = 45 of 72

Sun at Greatest Eclipse
(Geocentric Coordinates)
R.A. = 15h18m06.7s
Dec. = -18°15'02.6"
S.D. = 00°16'09.9"
H.P. = 00°00'08.9"

Moon at Greatest Eclipse
(Geocentric Coordinates)
R.A. = 15h17m51.2s
Dec. = -18°37'29.5"
S.D. = 00°16'42.4"
H.P. = 01°01'19.0"

N

W

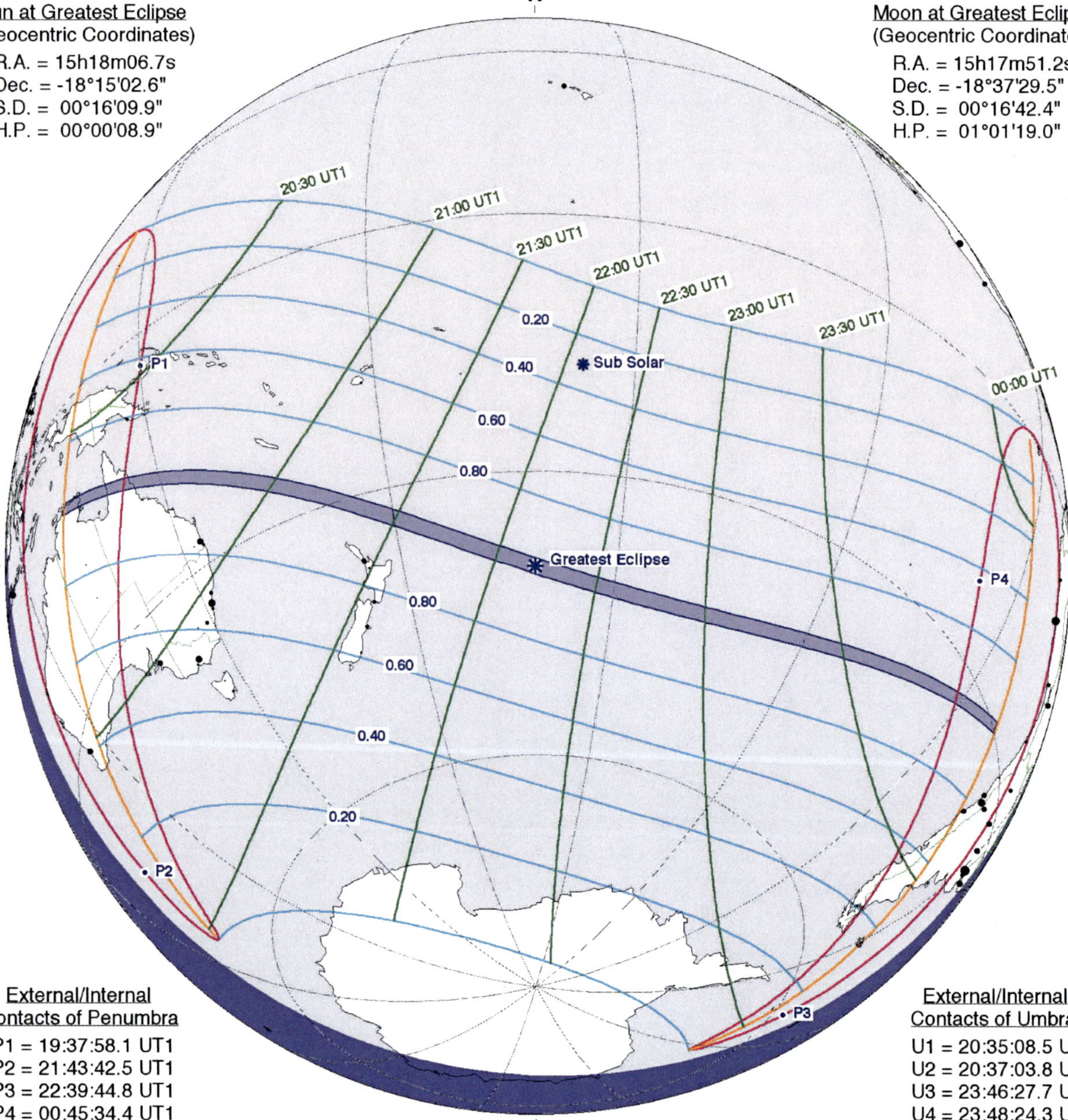

E

External/Internal
Contacts of Penumbra
P1 = 19:37:58.1 UT1
P2 = 21:43:42.5 UT1
P3 = 22:39:44.8 UT1
P4 = 00:45:34.4 UT1

External/Internal
Contacts of Umbra
U1 = 20:35:08.5 UT1
U2 = 20:37:03.8 UT1
U3 = 23:46:27.7 UT1
U4 = 23:48:24.3 UT1

ΔT = 66.8 s S Eph. = JPL DE405

Circumstances at Greatest Eclipse: 22:11:48.4 UT1

Lat. = 39°57.4'S Sun Alt. = 68.0°
Long. = 161°20.2'W Sun Azm. = 11.4°
Path Width = 179.0 km Duration = 04m02.2s

Circumstances at Greatest Duration: 22:14:00.1 UT1

Lat. = 40°11.7'S Sun Alt. = 67.9°
Long. = 160°19.4'W Sun Azm. = 7.4°
Path Width = 178.8 km Duration = 04m02.2s

0 1000 2000 3000 4000 5000
Kilometers

Annular Solar Eclipse of 2013 May 10

Greatest Eclipse = 00:26:20.3 TD (= 00:25:13.3 UT1)

Eclipse Magnitude = 0.9544   Saros Series = 138

Gamma = -0.2694   Saros Member = 31 of 70

Sun at Greatest Eclipse
(Geocentric Coordinates)
R.A. = 03h08m17.4s
Dec. = +17°36'34.3"
S.D. = 00°15'50.4"
H.P. = 00°00'08.7"

Moon at Greatest Eclipse
(Geocentric Coordinates)
R.A. = 03h08m28.1s
Dec. = +17°22'06.3"
S.D. = 00°14'53.8"
H.P. = 00°54'40.4"

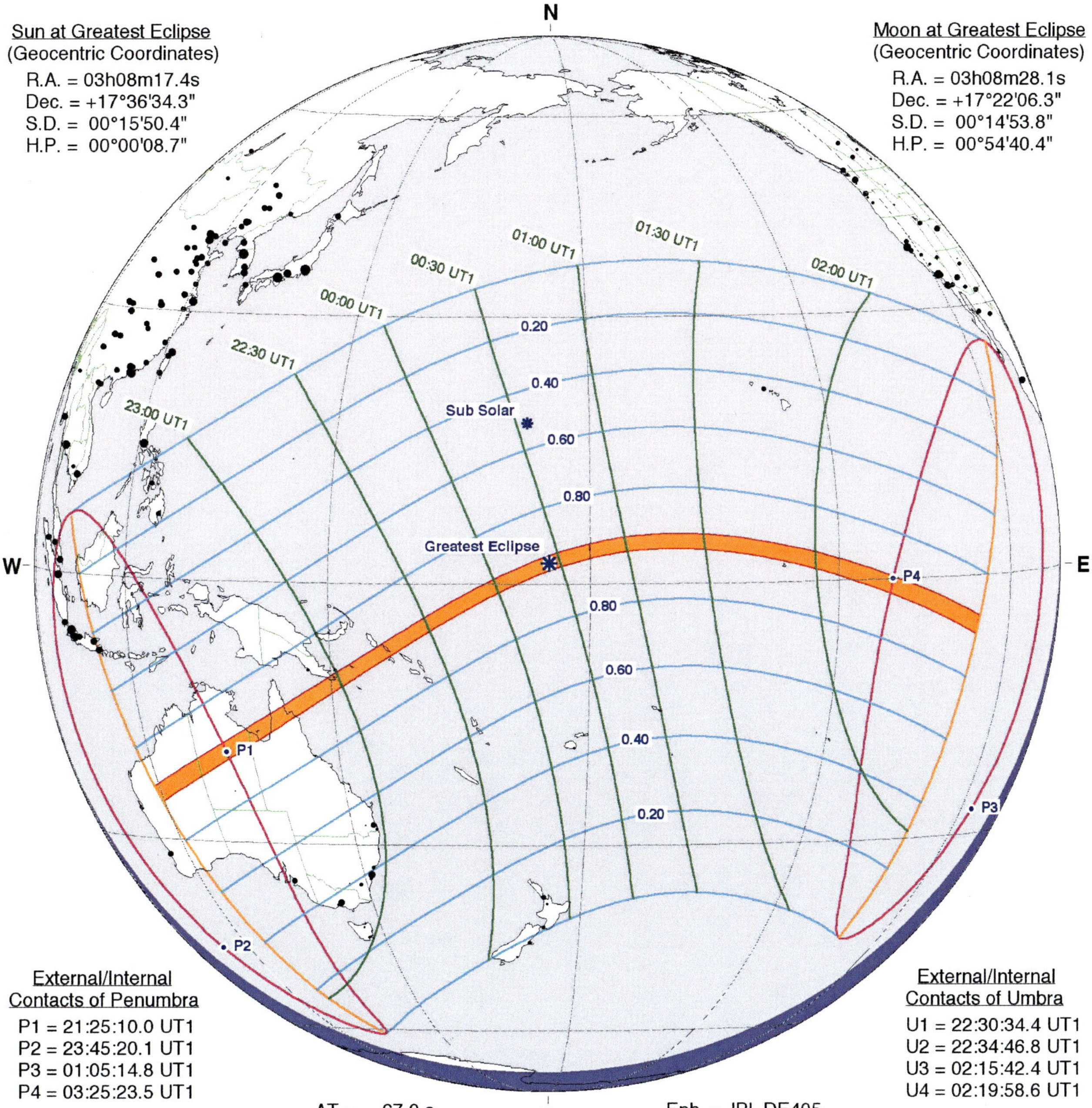

External/Internal
Contacts of Penumbra
P1 = 21:25:10.0 UT1
P2 = 23:45:20.1 UT1
P3 = 01:05:14.8 UT1
P4 = 03:25:23.5 UT1

External/Internal
Contacts of Umbra
U1 = 22:30:34.4 UT1
U2 = 22:34:46.8 UT1
U3 = 02:15:42.4 UT1
U4 = 02:19:58.6 UT1

ΔT = 67.0 s   Eph. = JPL DE405

Circumstances at Greatest Eclipse: 00:25:13.3 UT1

Lat. = 02°13.2'N   Sun Alt. = 74.4°
Long. = 175°27.9'E   Sun Azm. = 350.5°
Path Width = 172.6 km   Duration = 06m03.4s

Circumstances at Greatest Duration: 00:35:20.7 UT1

Lat. = 02°58.7'N   Sun Alt. = 73.6°
Long. = 177°50.9'E   Sun Azm. = 333.6°
Path Width = 173.6 km   Duration = 06m04.3s

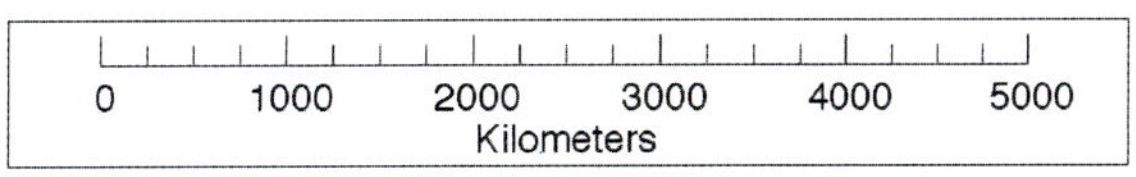

Hybrid Solar Eclipse of 2013 Nov 03

Greatest Eclipse = 12:47:36.1 TD (= 12:46:28.9 UT1)

Eclipse Magnitude = 1.0159 | Saros Series = 143

Gamma = 0.3272 | Saros Member = 23 of 72

Sun at Greatest Eclipse
(Geocentric Coordinates)
R.A. = 14h35m19.9s
Dec. = -15°12'22.5"
S.D. = 00°16'07.4"
H.P. = 00°00'08.9"

Moon at Greatest Eclipse
(Geocentric Coordinates)
R.A. = 14h35m37.0s
Dec. = -14°53'30.7"
S.D. = 00°16'07.6"
H.P. = 00°59'11.0"

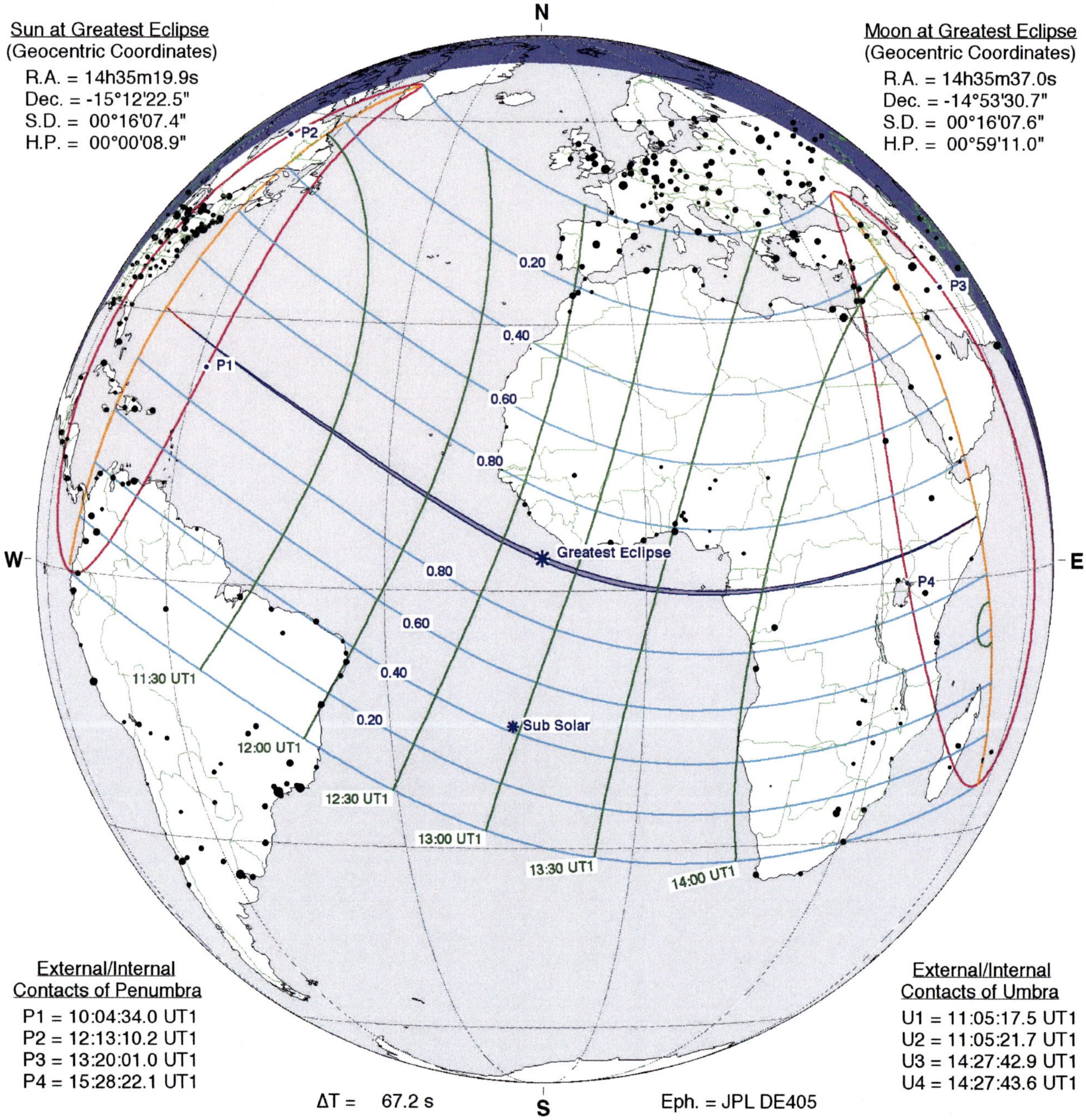

External/Internal
Contacts of Penumbra
P1 = 10:04:34.0 UT1
P2 = 12:13:10.2 UT1
P3 = 13:20:01.0 UT1
P4 = 15:28:22.1 UT1

External/Internal
Contacts of Umbra
U1 = 11:05:17.5 UT1
U2 = 11:05:21.7 UT1
U3 = 14:27:42.9 UT1
U4 = 14:27:43.6 UT1

ΔT = 67.2 s

Eph. = JPL DE405

Circumstances at Greatest Eclipse: 12:46:28.9 UT1

Lat. = 03°29.4'N | Sun Alt. = 70.9°
Long. = 011°41.9'W | Sun Azm. = 192.0°
Path Width = 57.5 km | Duration = 01m39.6s

Circumstances at Greatest Duration: 12:50:51.1 UT1

Lat. = 03°00.9'N | Sun Alt. = 70.7°
Long. = 010°29.9'W | Sun Azm. = 198.8°
Path Width = 57.7 km | Duration = 01m39.7s

0 1000 2000 3000 4000 5000
Kilometers

Annular Solar Eclipse of 2014 Apr 29

Greatest Eclipse = 06:04:32.9 TD (= 06:03:25.5 UT1)

Eclipse Magnitude = 0.9868 | Saros Series = 148

Gamma = -1.0000 | Saros Member = 21 of 75

Sun at Greatest Eclipse
(Geocentric Coordinates)
R.A. = 02h25m52.9s
Dec. = +14°26'54.2"
S.D. = 00°15'52.9"
H.P. = 00°00'08.7"

Moon at Greatest Eclipse
(Geocentric Coordinates)
R.A. = 02h26m46.0s
Dec. = +13°31'06.8"
S.D. = 00°15'38.4"
H.P. = 00°57'24.1"

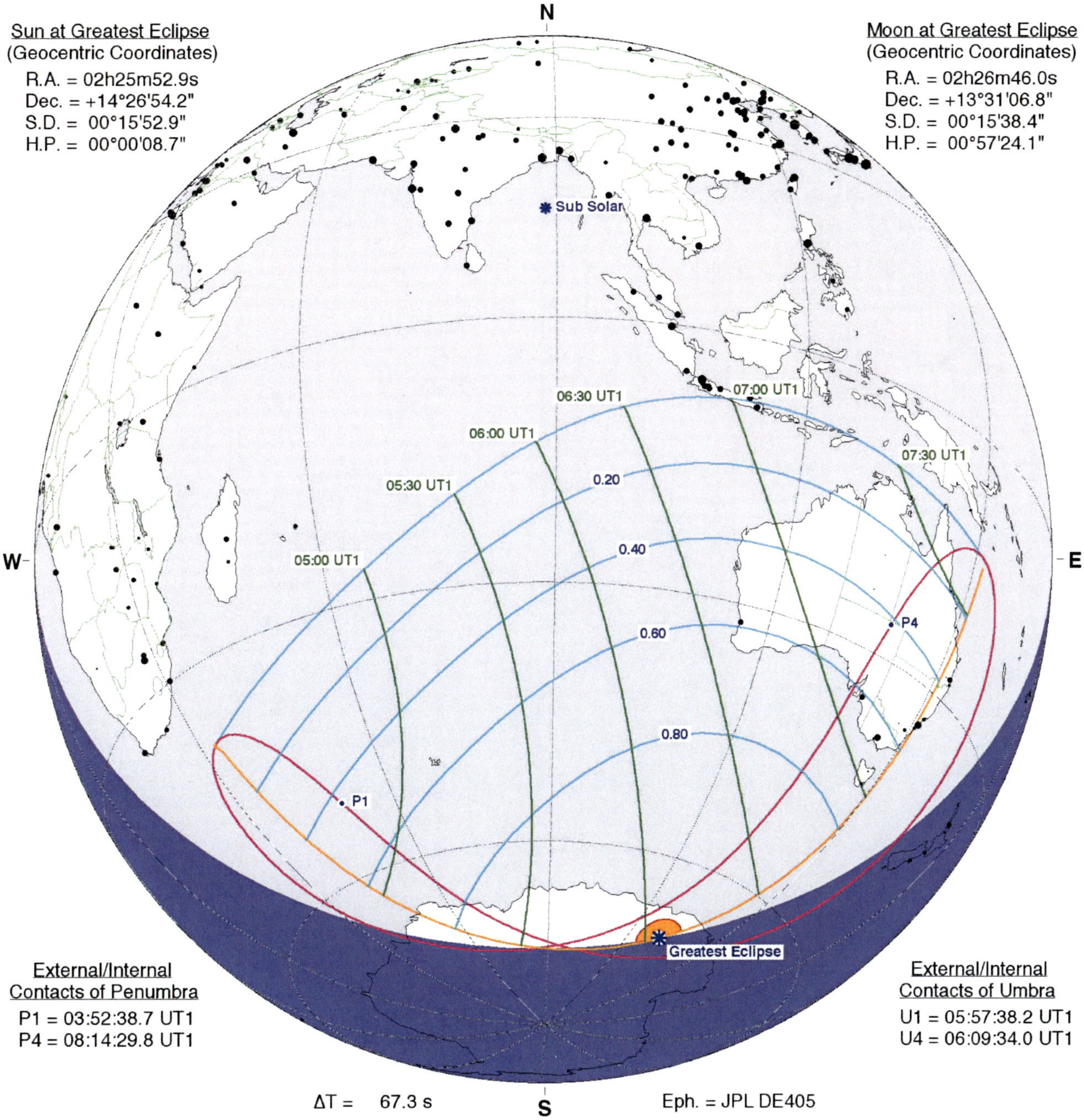

External/Internal Contacts of Penumbra
P1 = 03:52:38.7 UT1
P4 = 08:14:29.8 UT1

External/Internal Contacts of Umbra
U1 = 05:57:38.2 UT1
U4 = 06:09:34.0 UT1

ΔT = 67.3 s

Eph. = JPL DE405

Circumstances at Greatest Eclipse: 06:03:25.5 UT1

Lat. = 70°38.7'S | Sun Alt. = 0.0°
Long. = 131°15.4'E | Sun Azm. = 318.8°
Path Width = 0.0 km | Duration = 00m00.0s

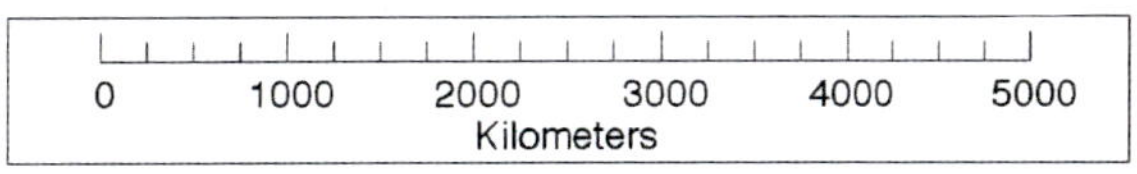

Partial Solar Eclipse of 2014 Oct 23

Greatest Eclipse = 21:45:39.2 TD (= 21:44:31.7 UT1)

Eclipse Magnitude = 0.8114
Gamma = 1.0908

Saros Series = 153
Saros Member = 9 of 70

Sun at Greatest Eclipse
(Geocentric Coordinates)
R.A. = 13h53m11.9s
Dec. = -11°36'45.1"
S.D. = 00°16'04.6"
H.P. = 00°00'08.8"

Moon at Greatest Eclipse
(Geocentric Coordinates)
R.A. = 13h54m15.8s
Dec. = -10°37'52.6"
S.D. = 00°15'15.5"
H.P. = 00°55'59.9"

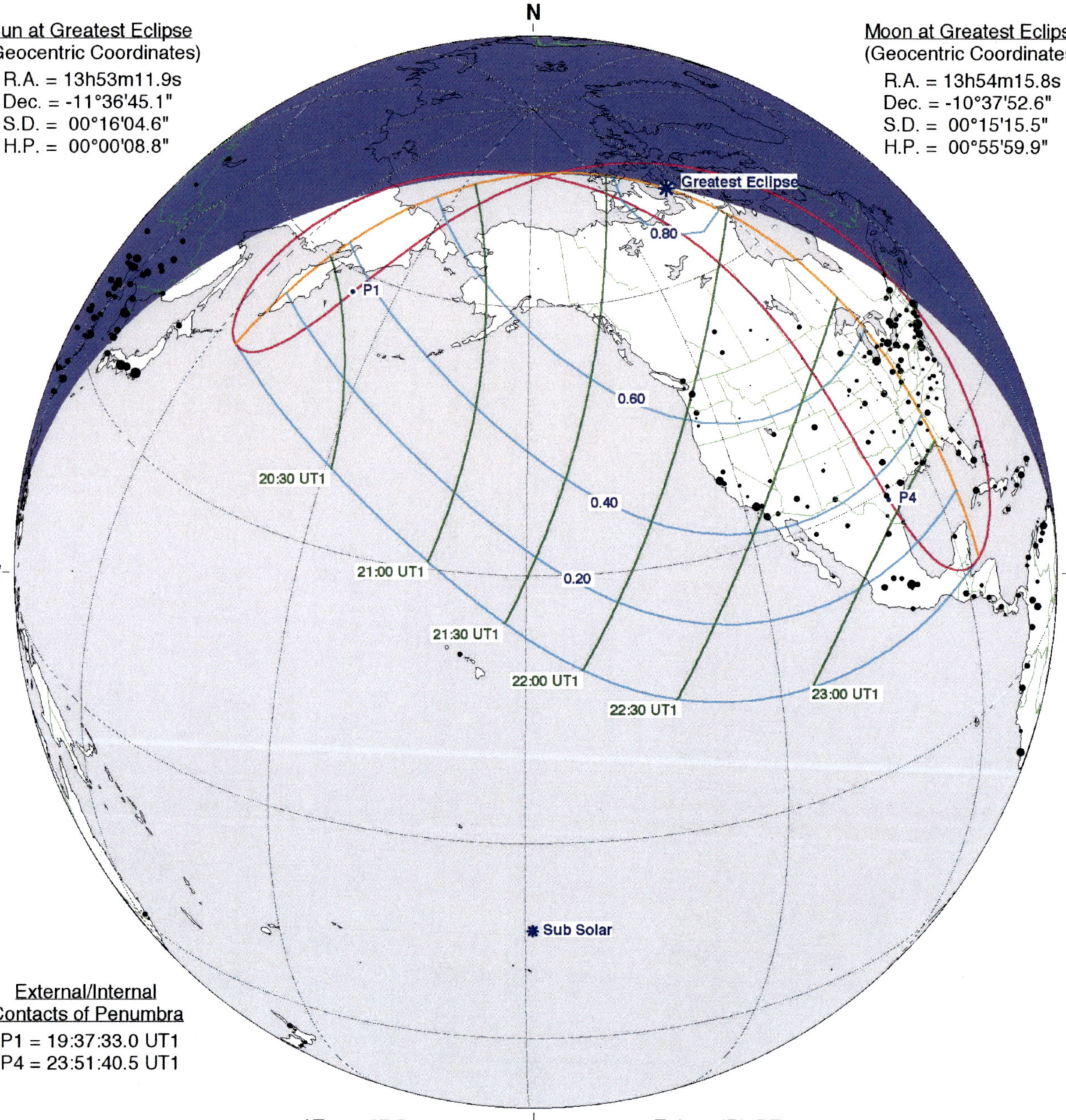

External/Internal
Contacts of Penumbra
P1 = 19:37:33.0 UT1
P4 = 23:51:40.5 UT1

ΔT = 67.5 s

Eph. = JPL DE405

Circumstances at Greatest Eclipse: 21:44:31.7 UT1

Lat. = 71°10.1'N
Long. = 097°10.3'W

Sun Alt. = 0.0°
Sun Azm. = 231.4°

0 1000 2000 3000 4000 5000
Kilometers

Total Solar Eclipse of 2015 Mar 20

Greatest Eclipse = 09:46:46.8 TD (= 09:45:39.2 UT1)

Eclipse Magnitude = 1.0446 Saros Series = 120

Gamma = 0.9454 Saros Member = 61 of 71

Sun at Greatest Eclipse
(Geocentric Coordinates)
R.A. = 23h58m01.5s
Dec. = -00°12'50.4"
S.D. = 00°16'03.7"
H.P. = 00°00'08.8"

Moon at Greatest Eclipse
(Geocentric Coordinates)
R.A. = 23h56m50.5s
Dec. = +00°42'08.7"
S.D. = 00°16'41.6"
H.P. = 01°01'15.8"

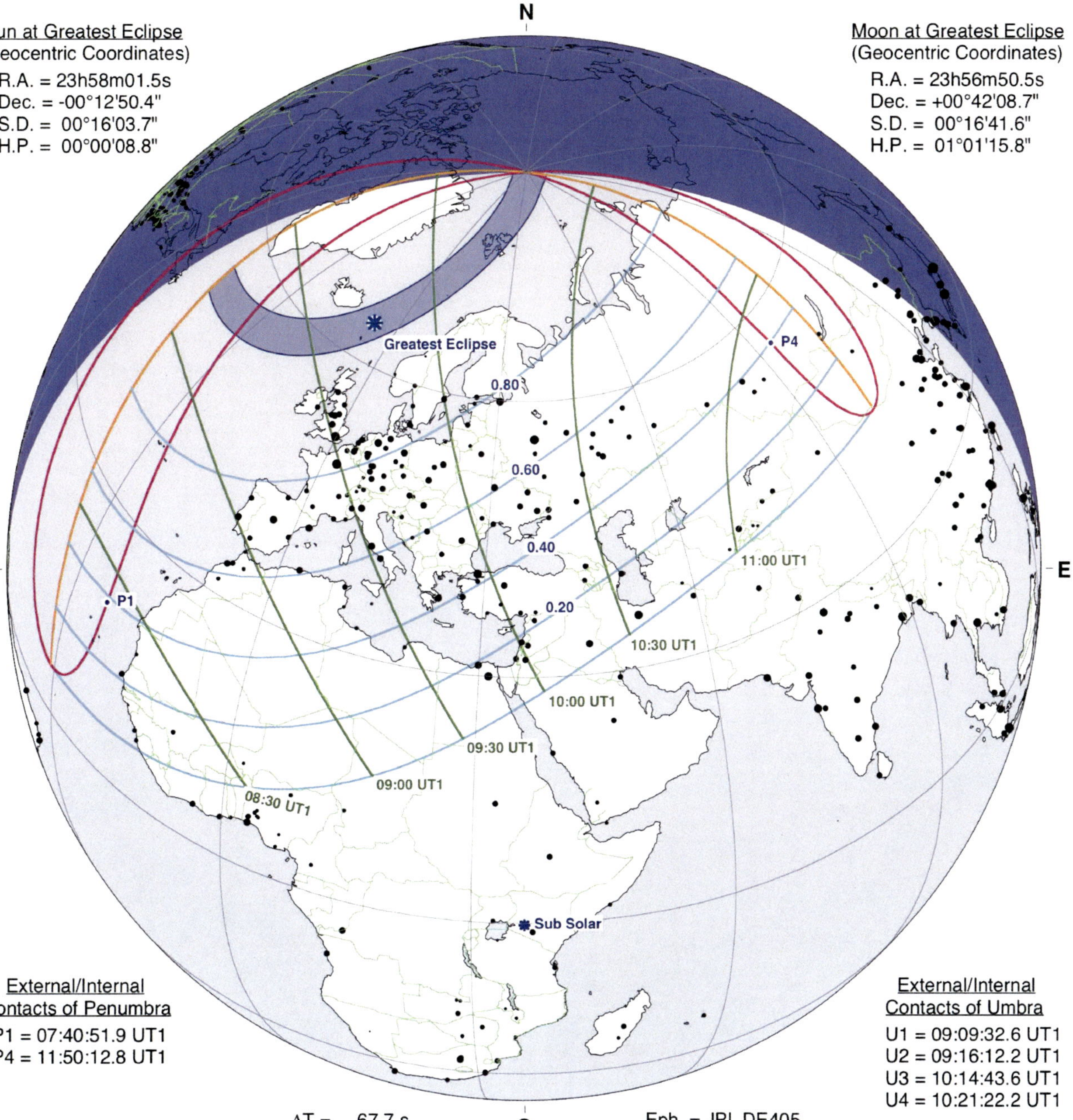

External/Internal
Contacts of Penumbra
P1 = 07:40:51.9 UT1
P4 = 11:50:12.8 UT1

External/Internal
Contacts of Umbra
U1 = 09:09:32.6 UT1
U2 = 09:16:12.2 UT1
U3 = 10:14:43.6 UT1
U4 = 10:21:22.2 UT1

ΔT = 67.7 s

Eph. = JPL DE405

Circumstances at Greatest Eclipse: 09:45:39.2 UT1

Lat. = 64°25.9'N Sun Alt. = 18.5°
Long. = 006°38.8'W Sun Azm. = 135.0°
Path Width = 462.6 km Duration = 02m46.9s

Circumstances at Greatest Duration: 09:45:16.5 UT1

Lat. = 64°17.0'N Sun Alt. = 18.5°
Long. = 006°53.6'W Sun Azm. = 134.6°
Path Width = 463.3 km Duration = 02m46.9s

0 1000 2000 3000 4000 5000
Kilometers

Partial Solar Eclipse of 2015 Sep 13

Greatest Eclipse = 06:55:19.2 TD (= 06:54:11.4 UT1)

Eclipse Magnitude = 0.7875 Saros Series = 125
Gamma = -1.1004 Saros Member = 54 of 73

Sun at Greatest Eclipse
(Geocentric Coordinates)
R.A. = 11h23m54.6s
Dec. = +03°53'20.1"
S.D. = 00°15'53.6"
H.P. = 00°00'08.7"

Moon at Greatest Eclipse
(Geocentric Coordinates)
R.A. = 11h22m43.3s
Dec. = +02°56'47.8"
S.D. = 00°14'43.0"
H.P. = 00°54'00.6"

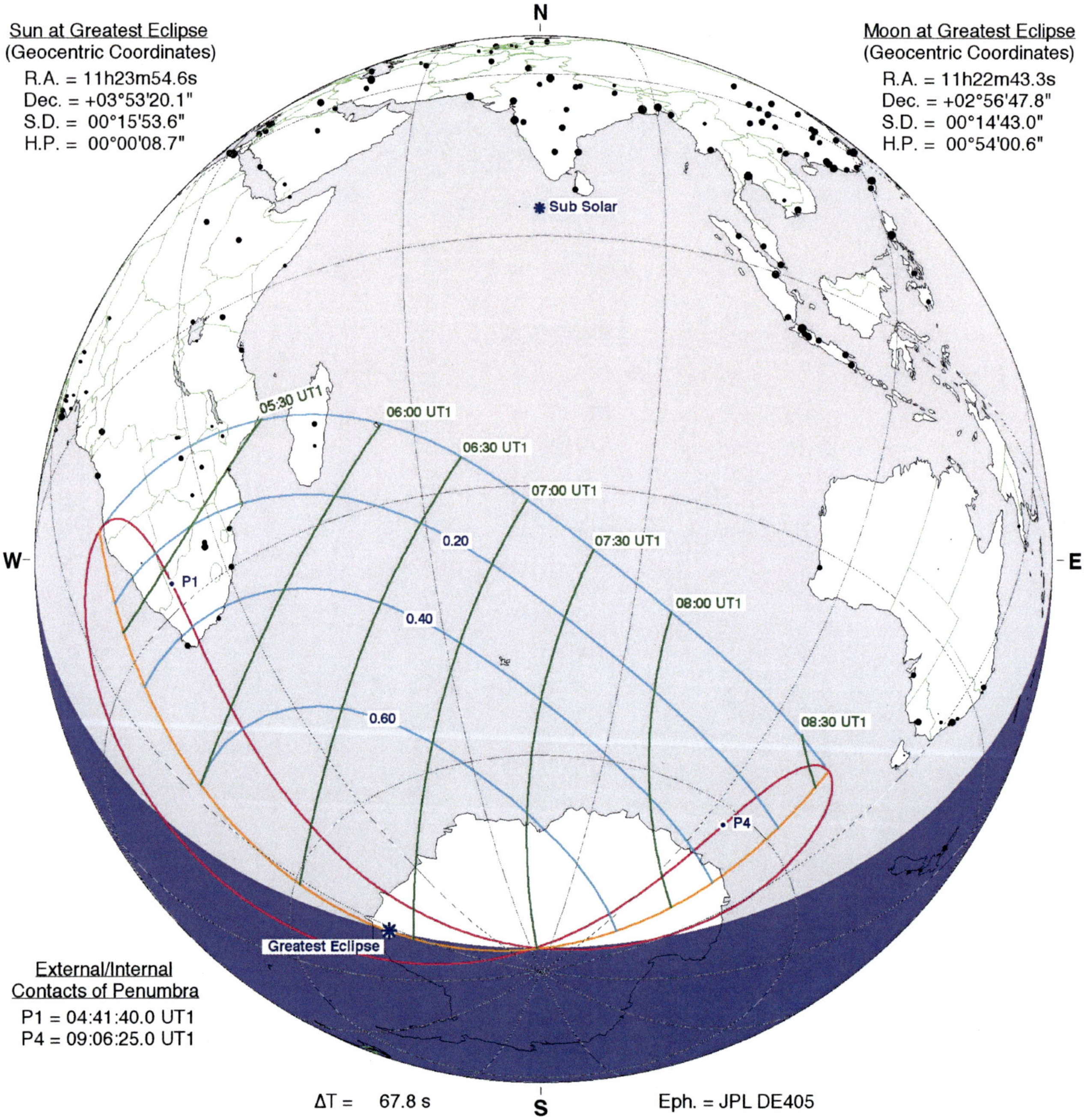

External/Internal
Contacts of Penumbra
P1 = 04:41:40.0 UT1
P4 = 09:06:25.0 UT1

ΔT = 67.8 s Eph. = JPL DE405

Circumstances at Greatest Eclipse: 06:54:11.4 UT1
Lat. = 72°06.4'S Sun Alt. = 0.0°
Long. = 002°18.2'W Sun Azm. = 77.2°

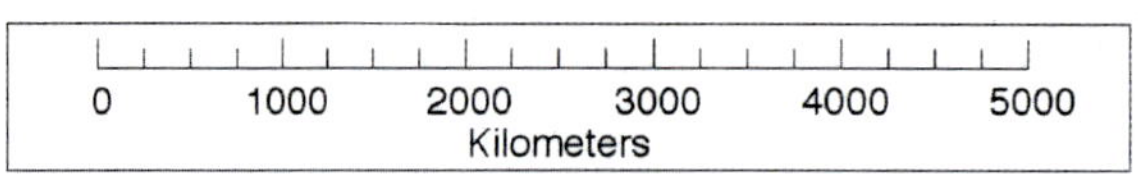

Total Solar Eclipse of 2016 Mar 09

Greatest Eclipse = 01:58:19.5 TD (= 01:57:11.4 UT1)

Eclipse Magnitude = 1.0450 Saros Series = 130

Gamma = 0.2609 Saros Member = 52 of 73

Sun at Greatest Eclipse
(Geocentric Coordinates)
R.A. = 23h19m17.6s
Dec. = -04°22'46.4"
S.D. = 00°16'06.5"
H.P. = 00°00'08.9"

Moon at Greatest Eclipse
(Geocentric Coordinates)
R.A. = 23h18m58.7s
Dec. = -04°07'40.6"
S.D. = 00°16'33.5"
H.P. = 01°00'46.2"

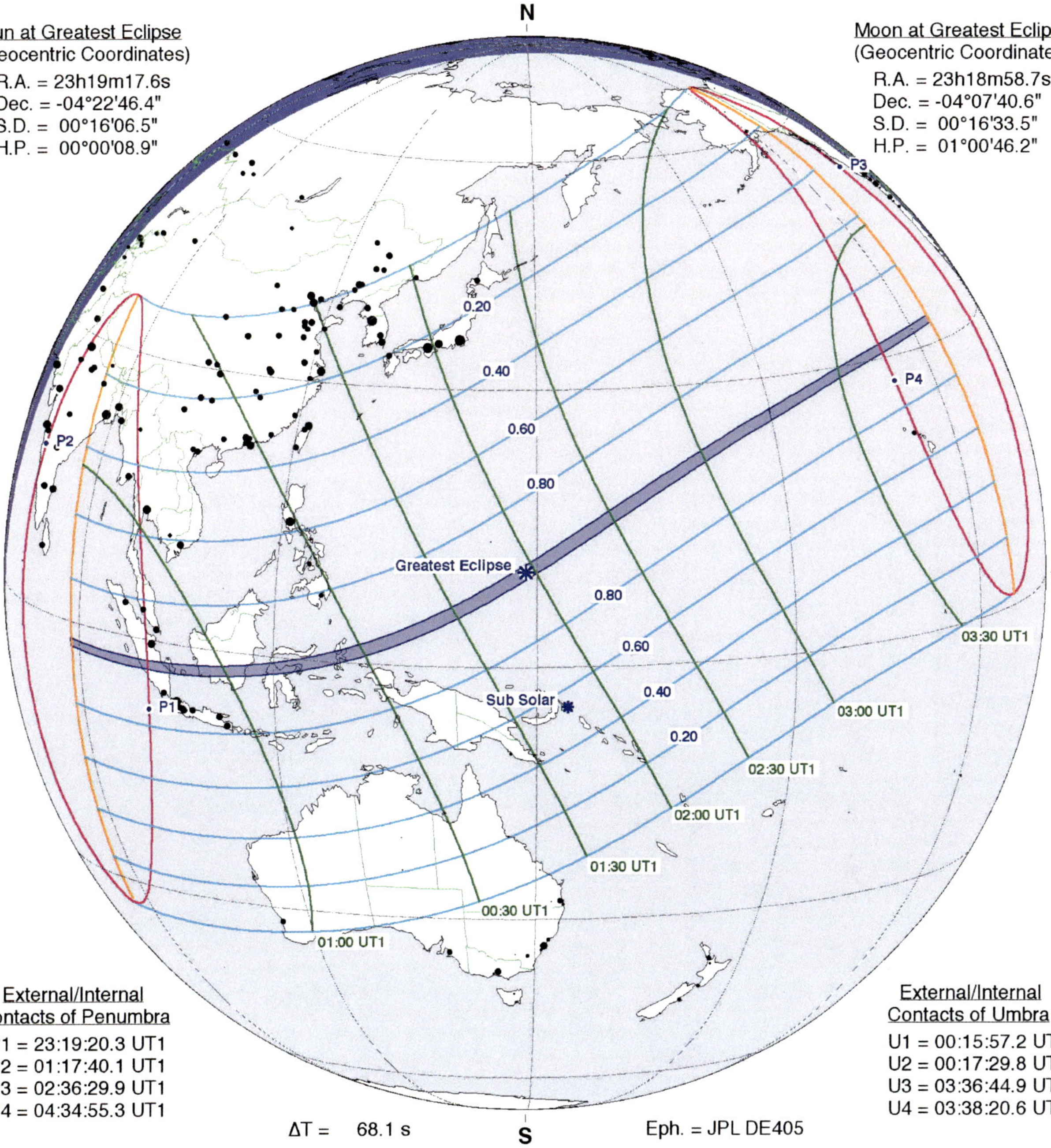

External/Internal
Contacts of Penumbra
P1 = 23:19:20.3 UT1
P2 = 01:17:40.1 UT1
P3 = 02:36:29.9 UT1
P4 = 04:34:55.3 UT1

External/Internal
Contacts of Umbra
U1 = 00:15:57.2 UT1
U2 = 00:17:29.8 UT1
U3 = 03:36:44.9 UT1
U4 = 03:38:20.6 UT1

ΔT = 68.1 s Eph. = JPL DE405

Circumstances at Greatest Eclipse: 01:57:11.4 UT1

Lat. = 10°07.3'N Sun Alt. = 74.8°
Long. = 148°47.6'E Sun Azm. = 162.5°
Path Width = 155.1 km Duration = 04m09.5s

Circumstances at Greatest Duration: 01:56:51.8 UT1

Lat. = 10°04.0'N Sun Alt. = 74.8°
Long. = 148°42.1'E Sun Azm. = 161.8°
Path Width = 155.1 km Duration = 04m09.5s

0 1000 2000 3000 4000 5000
Kilometers

Annular Solar Eclipse of 2016 Sep 01

Greatest Eclipse = 09:08:02.0 TD (= 09:06:53.7 UT1)

Eclipse Magnitude = 0.9736 Saros Series = 135

Gamma = -0.3330 Saros Member = 39 of 71

Sun at Greatest Eclipse
(Geocentric Coordinates)
R.A. = 10h43m43.3s
Dec. = +08°03'38.0"
S.D. = 00°15'51.0"
H.P. = 00°00'08.7"

Moon at Greatest Eclipse
(Geocentric Coordinates)
R.A. = 10h43m22.2s
Dec. = +07°45'51.0"
S.D. = 00°15'12.4"
H.P. = 00°55'48.6"

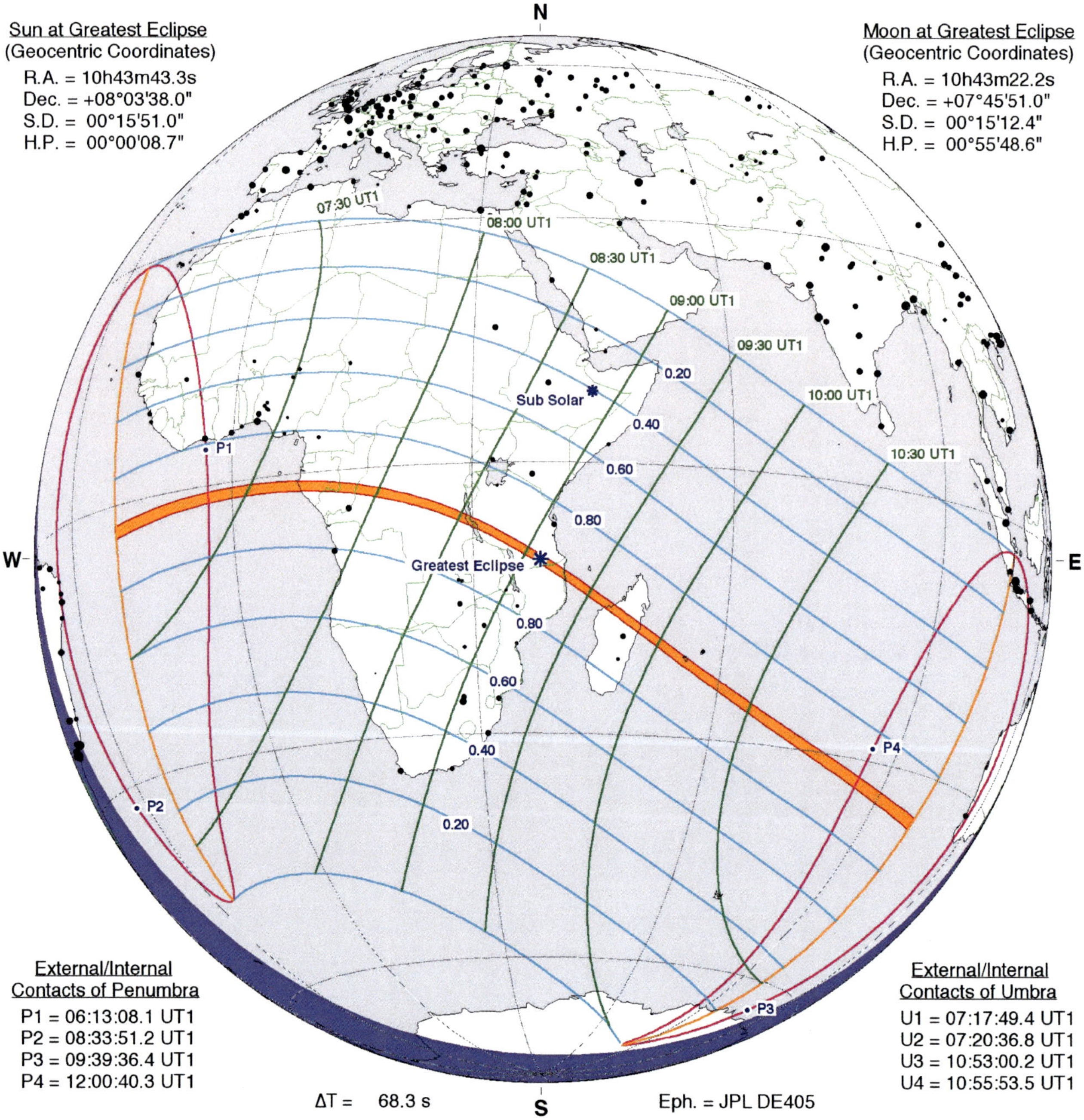

External/Internal Contacts of Penumbra
P1 = 06:13:08.1 UT1
P2 = 08:33:51.2 UT1
P3 = 09:39:36.4 UT1
P4 = 12:00:40.3 UT1

External/Internal Contacts of Umbra
U1 = 07:17:49.4 UT1
U2 = 07:20:36.8 UT1
U3 = 10:53:00.2 UT1
U4 = 10:55:53.5 UT1

ΔT = 68.3 s Eph. = JPL DE405

Circumstances at Greatest Eclipse: 09:06:53.7 UT1

Lat. = 10°40.9'S Sun Alt. = 70.5°
Long. = 037°45.7'E Sun Azm. = 16.4°
Path Width = 99.7 km Duration = 03m05.6s

Circumstances at Greatest Duration: 09:05:09.4 UT1

Lat. = 10°25.9'S Sun Alt. = 70.5°
Long. = 037°20.7'E Sun Azm. = 19.1°
Path Width = 99.8 km Duration = 03m05.6s

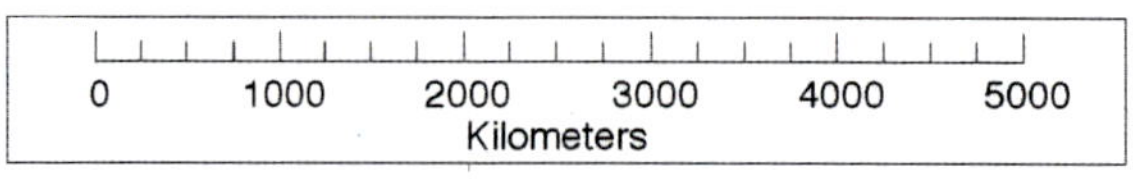

Annular Solar Eclipse of 2017 Feb 26

Greatest Eclipse = 14:54:32.8 TD (= 14:53:24.3 UT1)

Eclipse Magnitude = 0.9922
Gamma = -0.4578

Saros Series = 140
Saros Member = 29 of 71

Sun at Greatest Eclipse
(Geocentric Coordinates)
R.A. = 22h39m23.1s
Dec. = -08°29'38.8"
S.D. = 00°16'09.0"
H.P. = 00°00'08.9"

Moon at Greatest Eclipse
(Geocentric Coordinates)
R.A. = 22h39m53.2s
Dec. = -08°55'03.6"
S.D. = 00°15'47.8"
H.P. = 00°57'58.6"

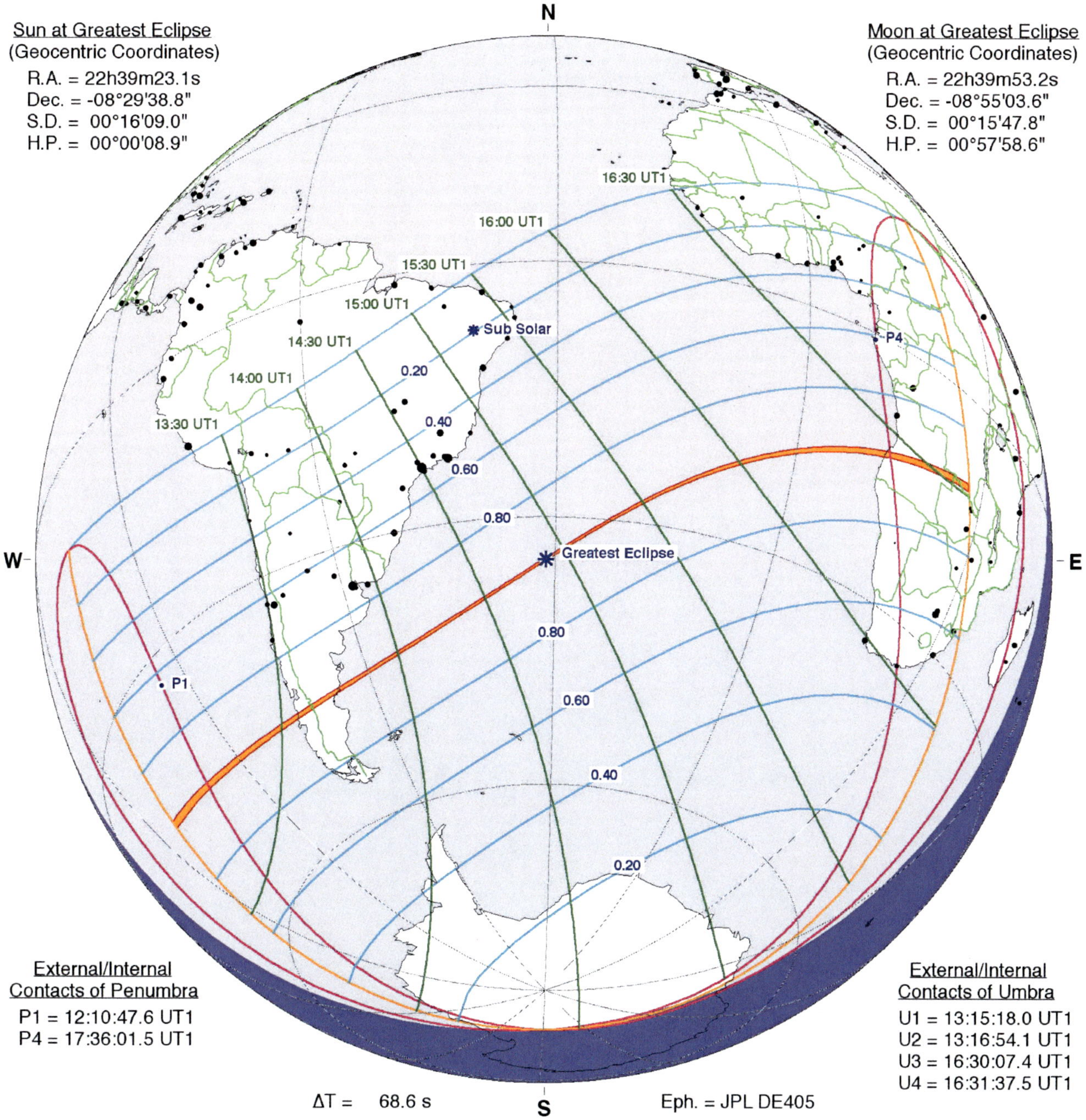

External/Internal
Contacts of Penumbra
P1 = 12:10:47.6 UT1
P4 = 17:36:01.5 UT1

External/Internal
Contacts of Umbra
U1 = 13:15:18.0 UT1
U2 = 13:16:54.1 UT1
U3 = 16:30:07.4 UT1
U4 = 16:31:37.5 UT1

ΔT = 68.6 s

Eph. = JPL DE405

Circumstances at Greatest Eclipse: 14:53:24.3 UT1	
Lat. = 34°40.8'S	Sun Alt. = 62.6°
Long. = 031°11.4'W	Sun Azm. = 340.5°
Path Width = 30.6 km	Duration = 00m44.0s

Circumstances at Greatest Duration: 13:16:06.0 UT1	
Lat. = 43°07.5'S	Sun Alt. = 0.0°
Long. = 113°52.9'W	Sun Azm. = 101.7°
Path Width = 96.3 km	Duration = 01m22.4s

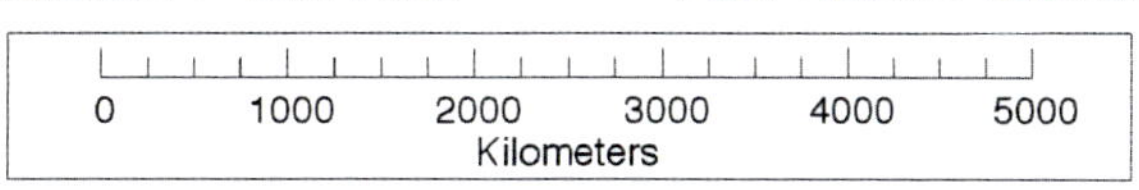

Total Solar Eclipse of 2017 Aug 21

Greatest Eclipse = 18:26:40.3 TD (= 18:25:31.5 UT1)

Eclipse Magnitude = 1.0306 Saros Series = 145

Gamma = 0.4367 Saros Member = 22 of 77

Sun at Greatest Eclipse
(Geocentric Coordinates)

R.A. = 10h04m03.9s
Dec. = +11°51'43.0"
S.D. = 00°15'48.7"
H.P. = 00°00'08.7"

Moon at Greatest Eclipse
(Geocentric Coordinates)

R.A. = 10h04m30.6s
Dec. = +12°16'32.8"
S.D. = 00°16'03.4"
H.P. = 00°58'55.7"

N

W

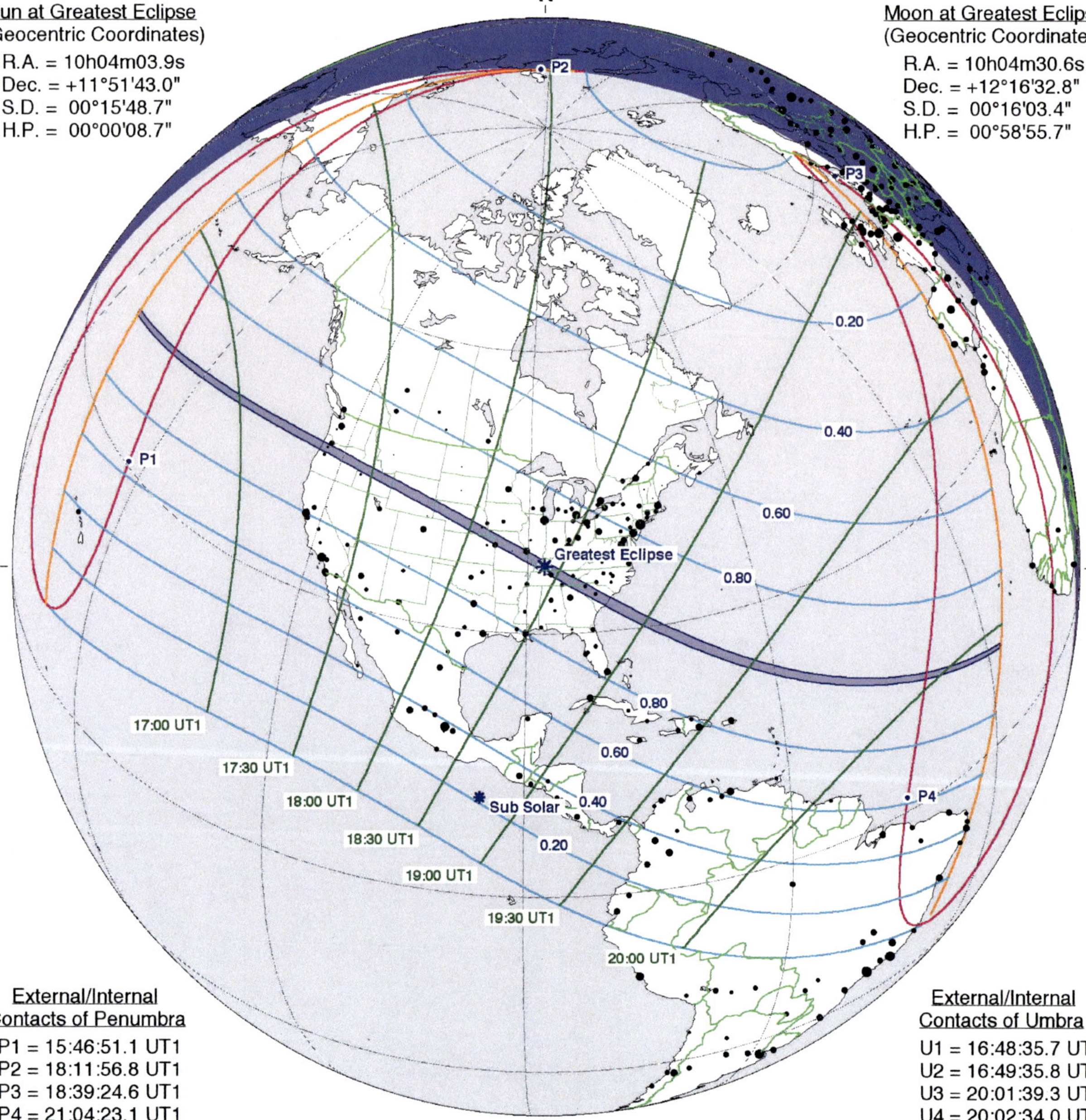

E

S

External/Internal
Contacts of Penumbra

P1 = 15:46:51.1 UT1
P2 = 18:11:56.8 UT1
P3 = 18:39:24.6 UT1
P4 = 21:04:23.1 UT1

External/Internal
Contacts of Umbra

U1 = 16:48:35.7 UT1
U2 = 16:49:35.8 UT1
U3 = 20:01:39.3 UT1
U4 = 20:02:34.0 UT1

ΔT = 68.8 s Eph. = JPL DE405

Circumstances at Greatest Eclipse: 18:25:31.5 UT1

Lat. = 36°58.0'N Sun Alt. = 63.9°
Long. = 087°40.2'W Sun Azm. = 197.9°
Path Width = 114.7 km Duration = 02m40.1s

Circumstances at Greatest Duration: 18:21:48.8 UT1

Lat. = 37°34.6'N Sun Alt. = 63.8°
Long. = 089°06.6'W Sun Azm. = 192.5°
Path Width = 114.5 km Duration = 02m40.3s

0 1000 2000 3000 4000 5000
Kilometers

Partial Solar Eclipse of 2018 Feb 15

Greatest Eclipse = 20:52:33.3 TD (= 20:51:24.3 UT1)

Eclipse Magnitude = 0.5991 Saros Series = 150

Gamma = -1.2116 Saros Member = 17 of 71

Sun at Greatest Eclipse
(Geocentric Coordinates)
R.A. = 21h57m18.8s
Dec. = -12°28'07.3"
S.D. = 00°16'11.4"
H.P. = 00°00'08.9"

Moon at Greatest Eclipse
(Geocentric Coordinates)
R.A. = 21h58m26.9s
Dec. = -13°32'29.9"
S.D. = 00°14'59.4"
H.P. = 00°55'00.9"

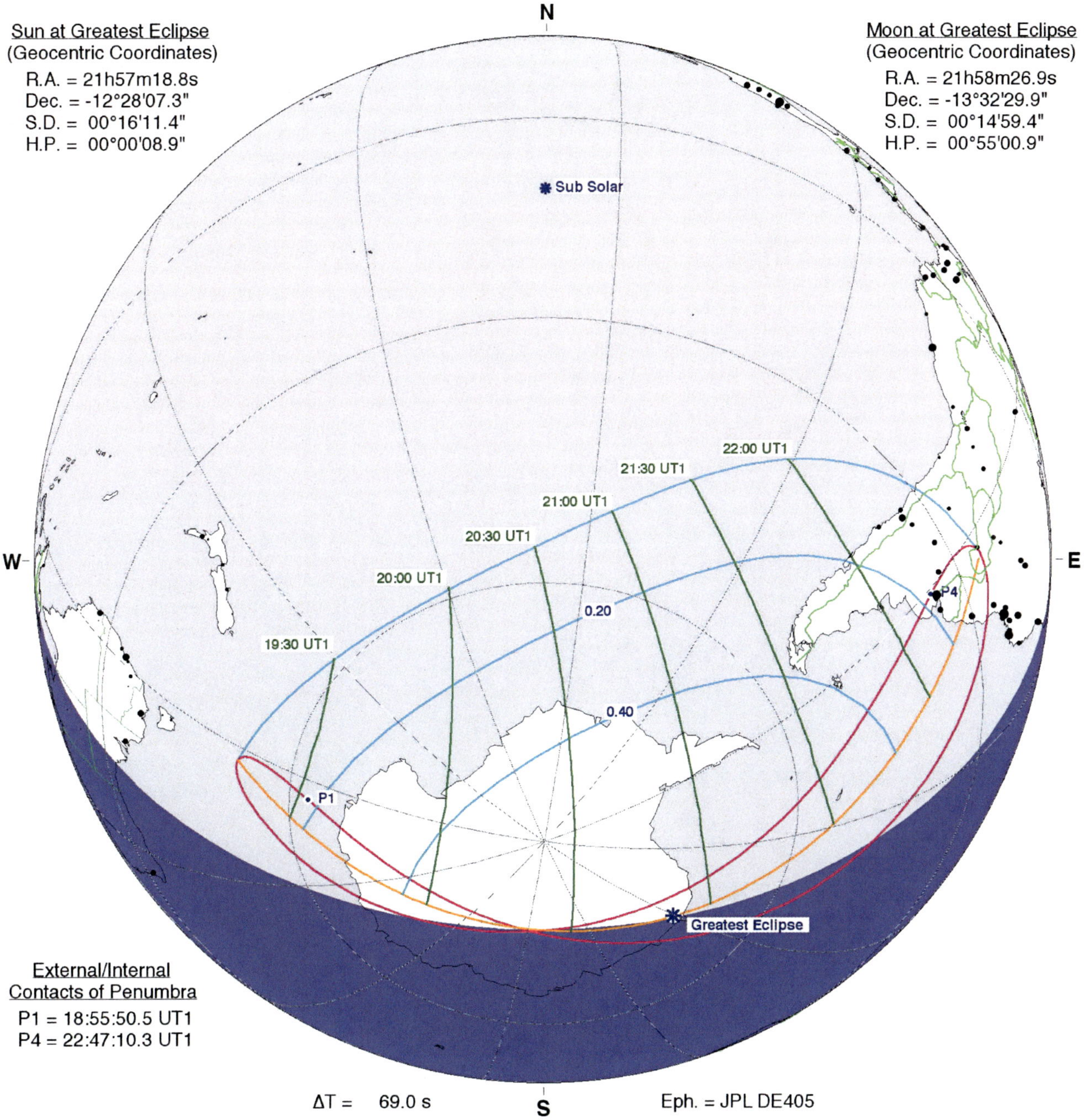

External/Internal
Contacts of Penumbra
P1 = 18:55:50.5 UT1
P4 = 22:47:10.3 UT1

ΔT = 69.0 s Eph. = JPL DE405

Circumstances at Greatest Eclipse: 20:51:24.3 UT1

Lat. = 71°01.6'S Sun Alt. = 0.0°

Long. = 000°38.5'E Sun Azm. = 228.4°

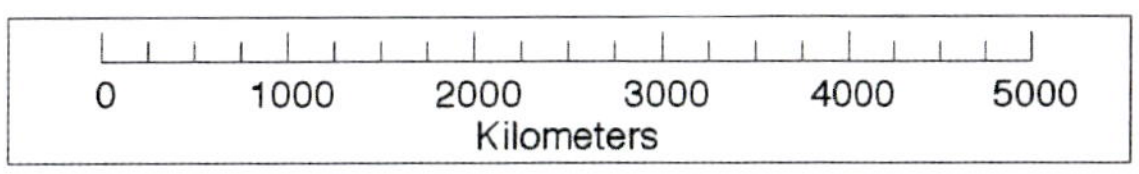

Partial Solar Eclipse of 2018 Jul 13

Greatest Eclipse = 03:02:16.1 TD (= 03:01:07.0 UT1)

Eclipse Magnitude = 0.3365 Saros Series = 117

Gamma = -1.3542 Saros Member = 69 of 71

Sun at Greatest Eclipse
(Geocentric Coordinates)

R.A. = 07h29m31.1s
Dec. = +21°50'30.6"
S.D. = 00°15'44.0"
H.P. = 00°00'08.7"

Moon at Greatest Eclipse
(Geocentric Coordinates)

R.A. = 07h29m10.9s
Dec. = +20°27'46.1"
S.D. = 00°16'42.8"
H.P. = 01°01'20.4"

N

W

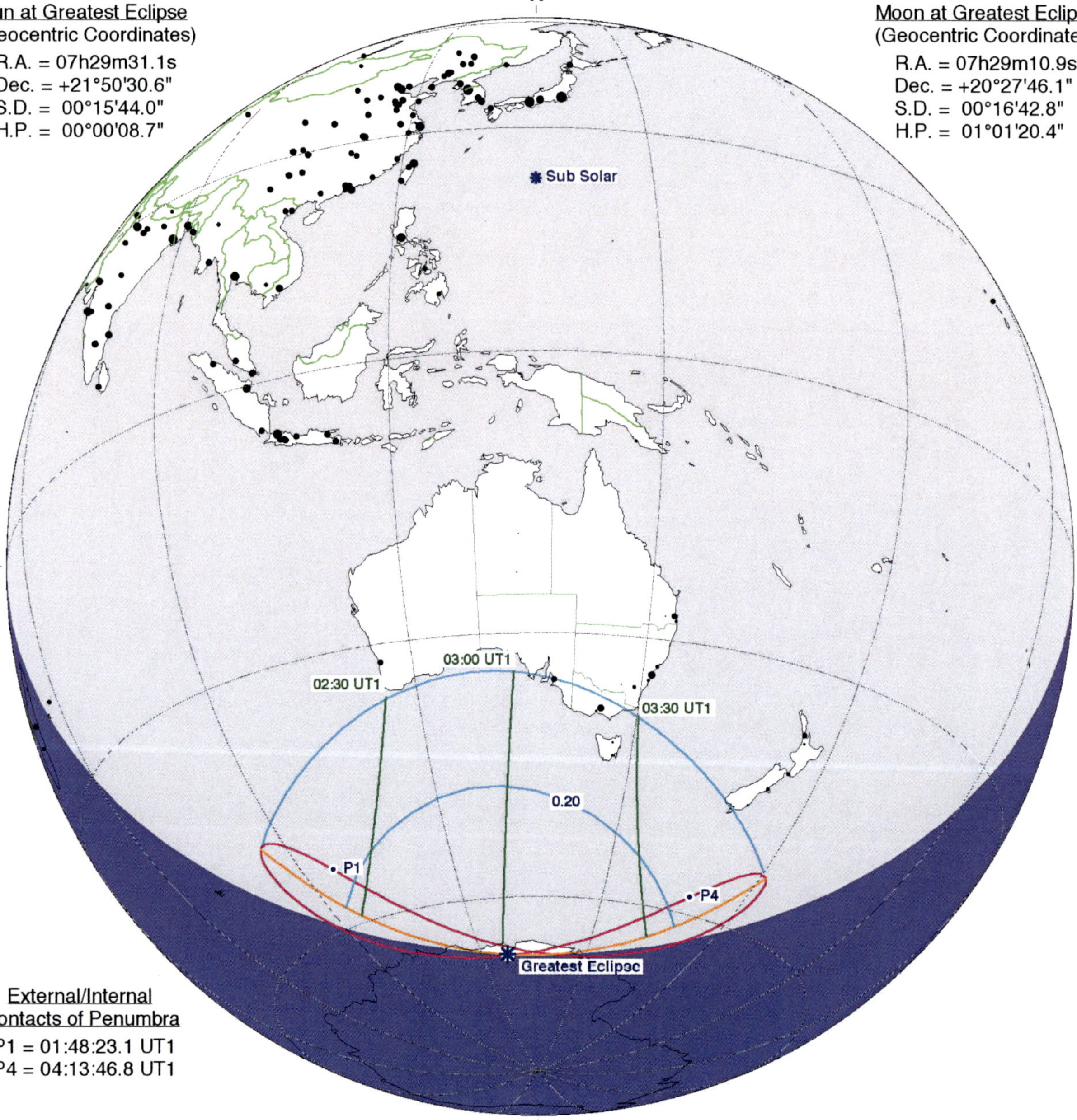

E

S

External/Internal
Contacts of Penumbra

P1 = 01:48:23.1 UT1
P4 = 04:13:46.8 UT1

ΔT = 69.2 s

Eph. = JPL DE405

Circumstances at Greatest Eclipse: 03:01:07.0 UT1

Lat. = 67°55.4'S Sun Alt. = 0.0°

Long. = 127°25.8'E Sun Azm. = 8.1°

0 1000 2000 3000 4000 5000
Kilometers

Partial Solar Eclipse of 2018 Aug 11

Greatest Eclipse = 09:47:28.0 TD (= 09:46:18.8 UT1)

Eclipse Magnitude = 0.7368
Gamma = 1.1476

Saros Series = 155
Saros Member = 6 of 71

Sun at Greatest Eclipse
(Geocentric Coordinates)
R.A. = 09h24m28.1s
Dec. = +15°13'19.1"
S.D. = 00°15'46.8"
H.P. = 00°00'08.7"

Moon at Greatest Eclipse
(Geocentric Coordinates)
R.A. = 09h25m31.3s
Dec. = +16°21'40.4"
S.D. = 00°16'40.0"
H.P. = 01°01'10.1"

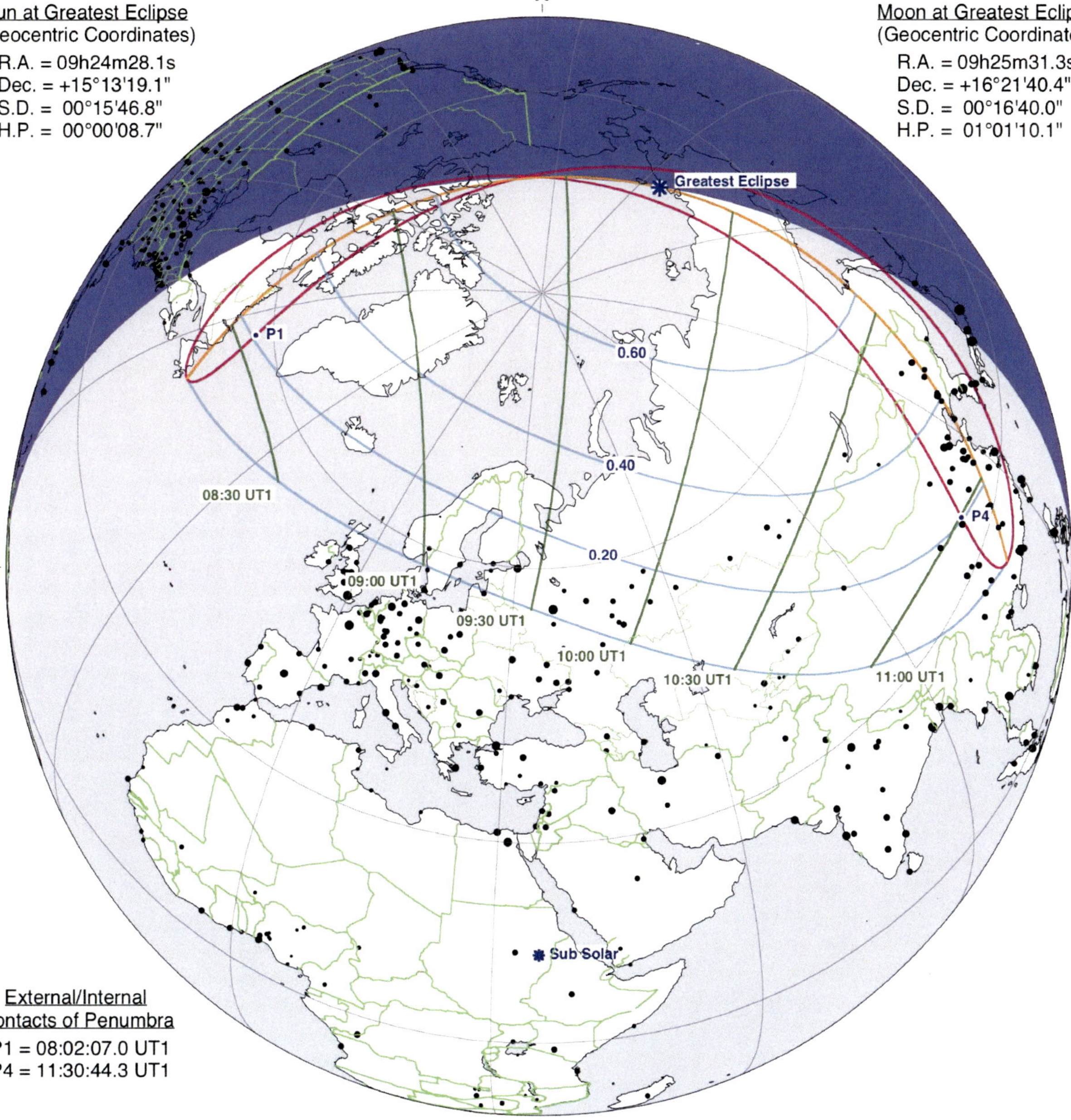

External/Internal
Contacts of Penumbra
P1 = 08:02:07.0 UT1
P4 = 11:30:44.3 UT1

ΔT = 69.2 s

Eph. = JPL DE405

Circumstances at Greatest Eclipse: 09:46:18.8 UT1

Lat. = 70°23.3'N
Long. = 174°28.1'E

Sun Alt. = 0.0°
Sun Azm. = 321.5°

0	1000	2000	3000	4000	5000

Kilometers

Partial Solar Eclipse of 2019 Jan 06

Greatest Eclipse = 01:42:37.7 TD (= 01:41:28.3 UT1)

Eclipse Magnitude = 0.7146
Gamma = 1.1417

Saros Series = 122
Saros Member = 58 of 70

Sun at Greatest Eclipse
(Geocentric Coordinates)
R.A. = 19h06m57.4s
Dec. = -22°32'36.6"
S.D. = 00°16'15.9"
H.P. = 00°00'08.9"

Moon at Greatest Eclipse
(Geocentric Coordinates)
R.A. = 19h06m53.0s
Dec. = -21°30'36.4"
S.D. = 00°14'50.4"
H.P. = 00°54'27.6"

N

W

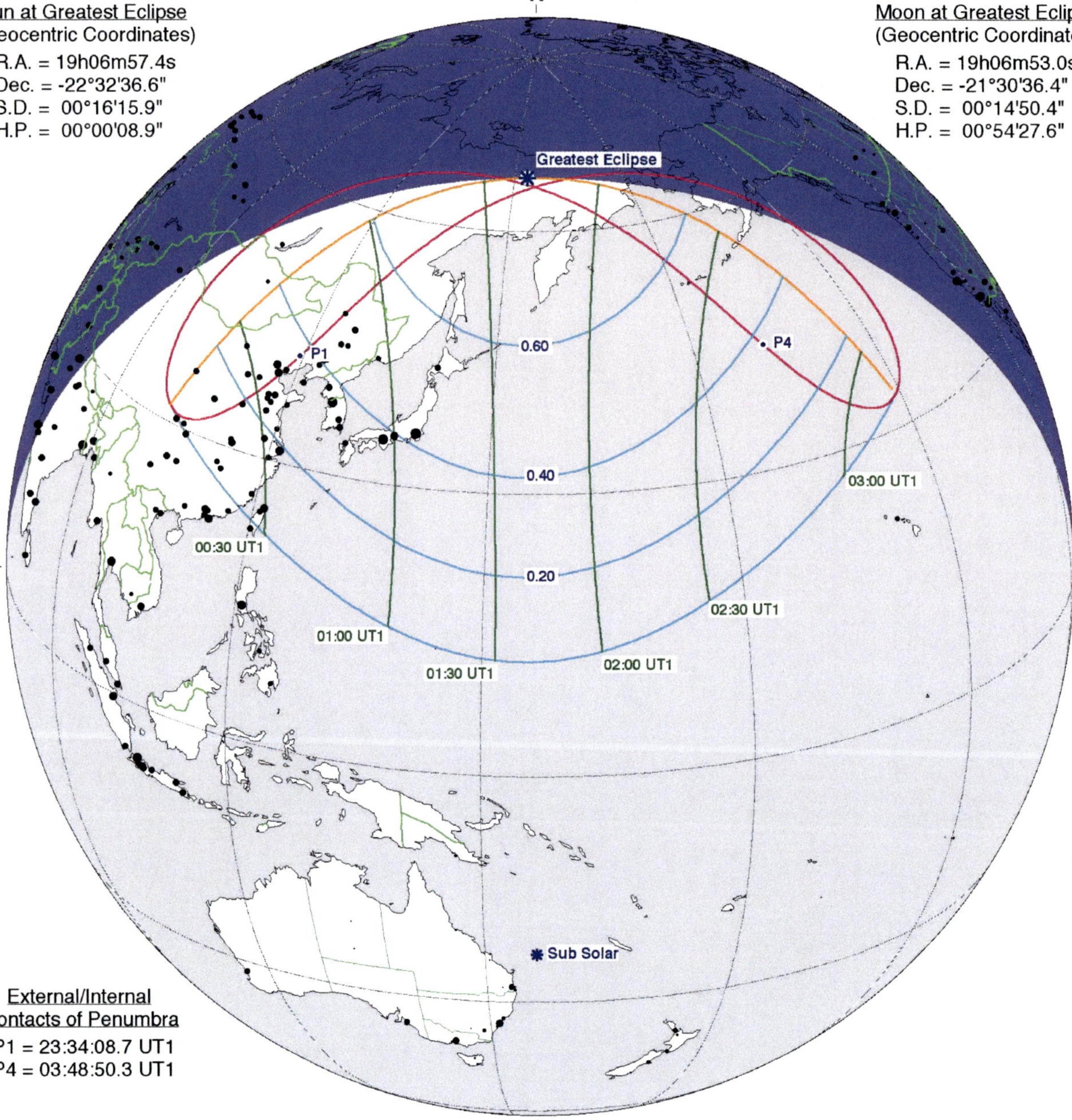

E

S

External/Internal
Contacts of Penumbra
P1 = 23:34:08.7 UT1
P4 = 03:48:50.3 UT1

ΔT = 69.4 s

Eph. = JPL DE405

Circumstances at Greatest Eclipse: 01:41:28.3 UT1

Lat. = 67°26.1'N
Long. = 153°33.7'E

Sun Alt. = 0.0°
Sun Azm. = 177.7°

0 1000 2000 3000 4000 5000
Kilometers

Total Solar Eclipse of 2019 Jul 02

Greatest Eclipse = 19:24:07.5 TD (= 19:22:57.9 UT1)

Eclipse Magnitude = 1.0459 | Saros Series = 127

Gamma = -0.6466 | Saros Member = 58 of 82

Sun at Greatest Eclipse
(Geocentric Coordinates)
R.A. = 06h46m14.8s
Dec. = +23°00'36.4"
S.D. = 00°15'43.8"
H.P. = 00°00'08.6"

Moon at Greatest Eclipse
(Geocentric Coordinates)
R.A. = 06h46m17.9s
Dec. = +22°22'09.2"
S.D. = 00°16'14.9"
H.P. = 00°59'37.8"

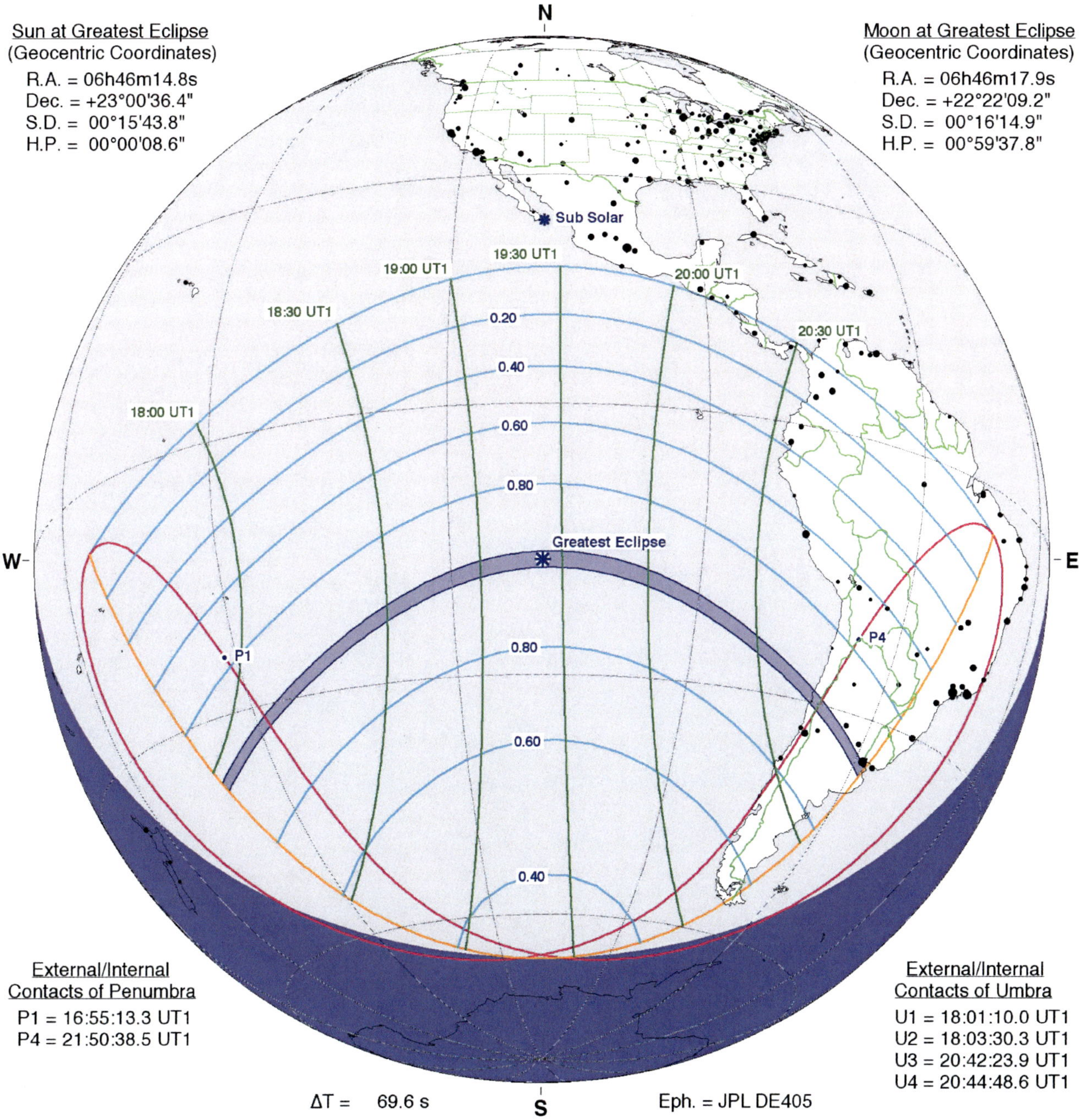

External/Internal Contacts of Penumbra
P1 = 16:55:13.3 UT1
P4 = 21:50:38.5 UT1

External/Internal Contacts of Umbra
U1 = 18:01:10.0 UT1
U2 = 18:03:30.3 UT1
U3 = 20:42:23.9 UT1
U4 = 20:44:48.6 UT1

ΔT = 69.6 s | Eph. = JPL DE405

Circumstances at Greatest Eclipse: 19:22:57.9 UT1

Lat. =	17°23.3'S	Sun Alt. =	49.6°
Long. =	108°59.9'W	Sun Azm. =	359.0°
Path Width =	200.6 km	Duration =	04m32.8s

Circumstances at Greatest Duration: 19:24:09.3 UT1

Lat. =	17°22.8'S	Sun Alt. =	49.6°
Long. =	108°37.3'W	Sun Azm. =	358.0°
Path Width =	200.7 km	Duration =	04m32.8s

0 1000 2000 3000 4000 5000
Kilometers

Annular Solar Eclipse of 2019 Dec 26

Greatest Eclipse = 05:18:53.1 TD (= 05:17:43.3 UT1)

Eclipse Magnitude = 0.9701
Gamma = 0.4135

Saros Series = 132
Saros Member = 46 of 71

Sun at Greatest Eclipse
(Geocentric Coordinates)
R.A. = 18h17m56.7s
Dec. = -23°22'19.2"
S.D. = 00°16'15.7"
H.P. = 00°00'08.9"

Moon at Greatest Eclipse
(Geocentric Coordinates)
R.A. = 18h18m03.7s
Dec. = -22°58'50.4"
S.D. = 00°15'33.0"
H.P. = 00°57'04.0"

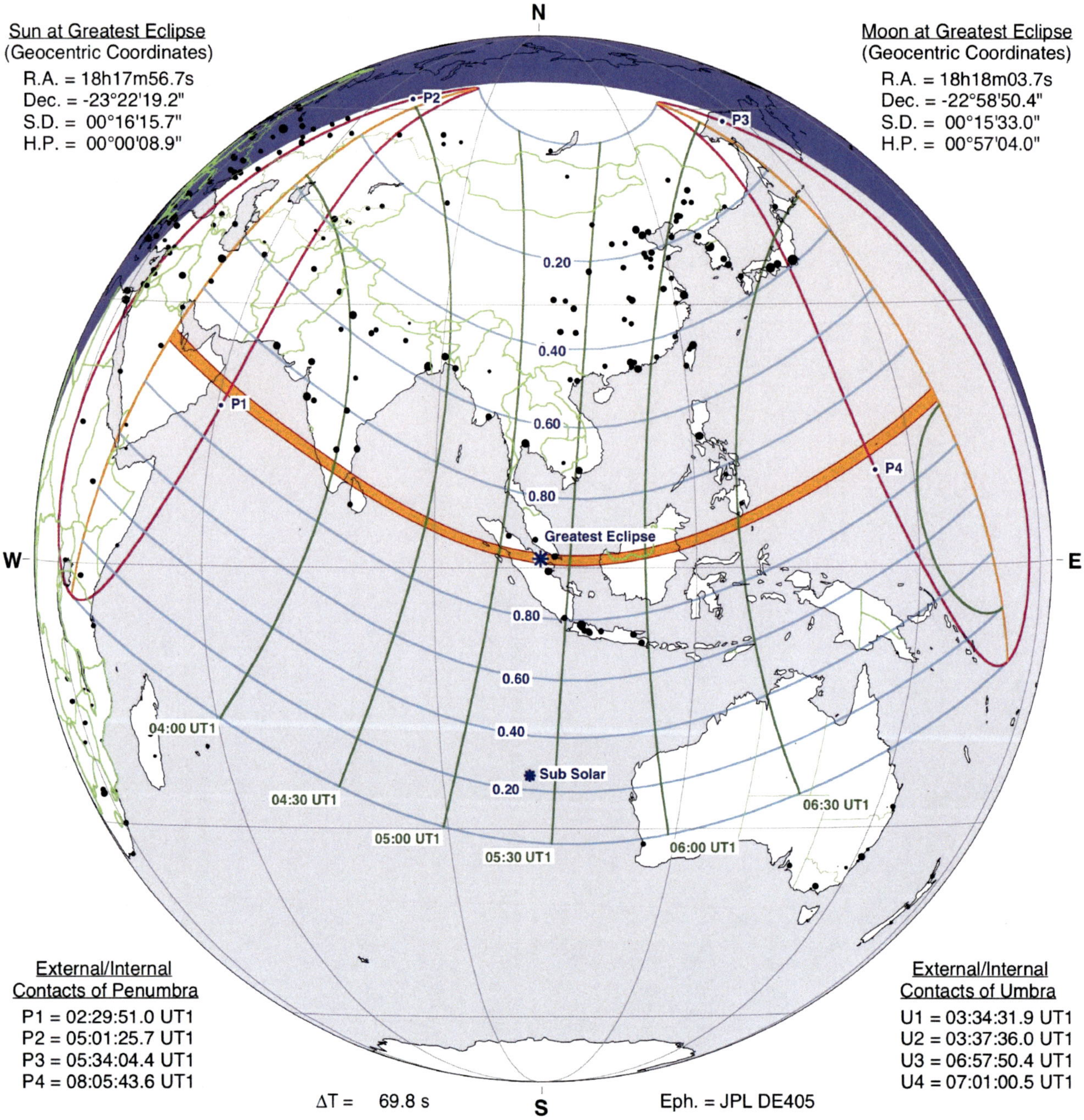

External/Internal
Contacts of Penumbra
P1 = 02:29:51.0 UT1
P2 = 05:01:25.7 UT1
P3 = 05:34:04.4 UT1
P4 = 08:05:43.6 UT1

External/Internal
Contacts of Umbra
U1 = 03:34:31.9 UT1
U2 = 03:37:36.0 UT1
U3 = 06:57:50.4 UT1
U4 = 07:01:00.5 UT1

ΔT = 69.8 s

Eph. = JPL DE405

Circumstances at Greatest Eclipse: 05:17:43.3 UT1
Lat. = 01°00.5'N
Long. = 102°14.9'E
Path Width = 117.9 km
Sun Alt. = 65.6°
Sun Azm. = 183.6°
Duration = 03m39.5s

Circumstances at Greatest Duration: 05:28:29.5 UT1
Lat. = 00°46.7'N
Long. = 105°05.1'E
Path Width = 118.6 km
Sun Alt. = 64.9°
Sun Azm. = 195.6°
Duration = 03m40.0s

0 1000 2000 3000 4000 5000
Kilometers

Annular Solar Eclipse of 2020 Jun 21

Greatest Eclipse = 06:41:15.4 TD (= 06:40:05.4 UT1)

Eclipse Magnitude = 0.9940 Saros Series = 137

Gamma = 0.1209 Saros Member = 36 of 70

Sun at Greatest Eclipse
(Geocentric Coordinates)
R.A. = 06h01m33.0s
Dec. = +23°26'09.7"
S.D. = 00°15'44.2"
H.P. = 00°00'08.7"

Moon at Greatest Eclipse
(Geocentric Coordinates)
R.A. = 06h01m30.2s
Dec. = +23°32'56.7"
S.D. = 00°15'24.0"
H.P. = 00°56'31.1"

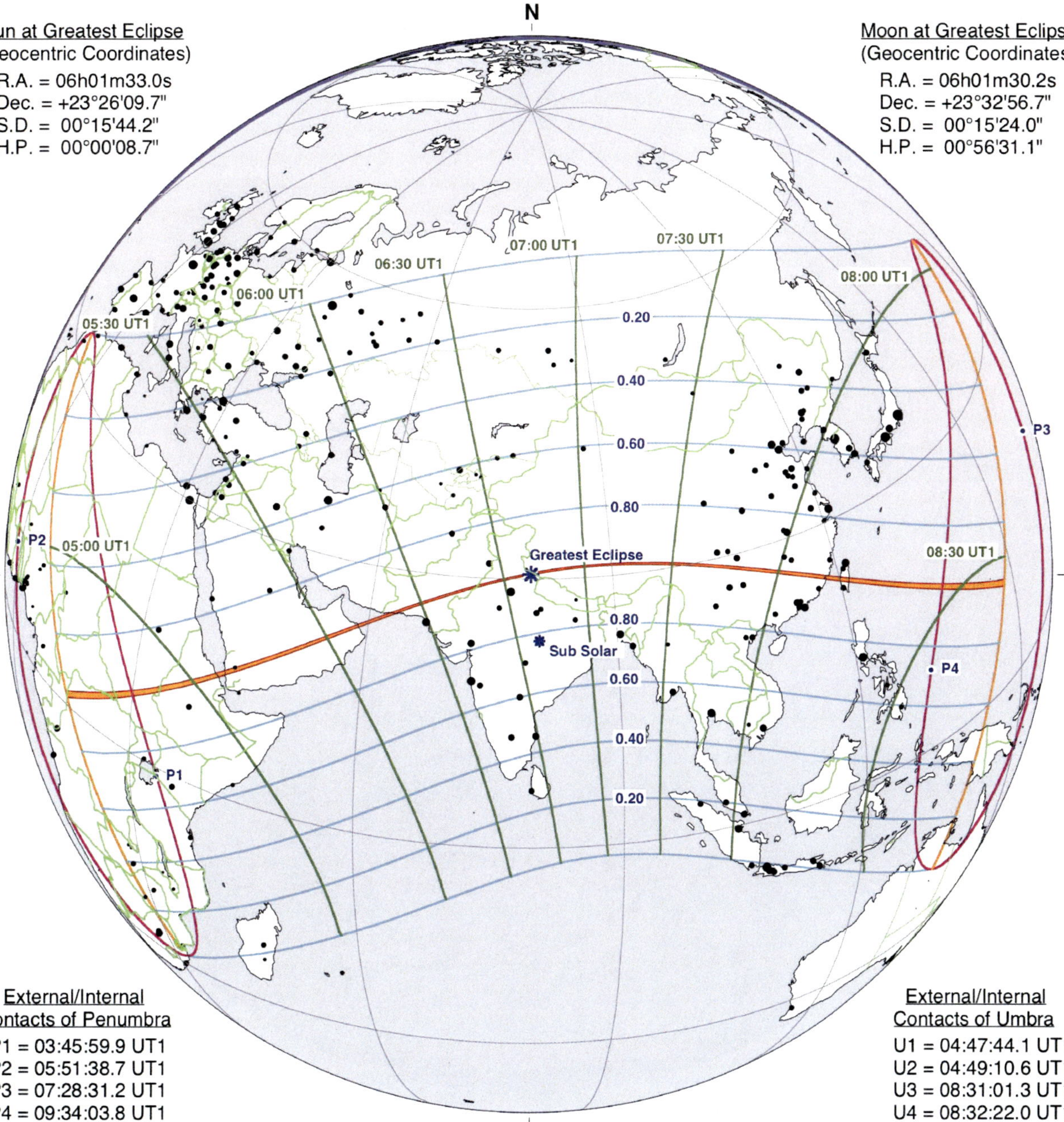

External/Internal
Contacts of Penumbra
P1 = 03:45:59.9 UT1
P2 = 05:51:38.7 UT1
P3 = 07:28:31.2 UT1
P4 = 09:34:03.8 UT1

External/Internal
Contacts of Umbra
U1 = 04:47:44.1 UT1
U2 = 04:49:10.6 UT1
U3 = 08:31:01.3 UT1
U4 = 08:32:22.0 UT1

ΔT = 70.0 s

Eph. = JPL DE405

Circumstances at Greatest Eclipse: 06:40:05.4 UT1

Lat. = 30°31.2'N	Sun Alt. = 82.9°
Long. = 079°40.0'E	Sun Azm. = 174.2°
Path Width = 21.2 km	Duration = 00m38.2s

Circumstances at Greatest Duration: 04:48:27.4 UT1

Lat. = 01°16.1'N	Sun Alt. = 0.0°
Long. = 017°47.9'E	Sun Azm. = 66.6°
Path Width = 85.2 km	Duration = 01m22.4s

0 1000 2000 3000 4000 5000
Kilometers

Total Solar Eclipse of 2020 Dec 14

Greatest Eclipse = 16:14:39.4 TD (= 16:13:29.1 UT1)

Eclipse Magnitude = 1.0254
Gamma = -0.2939

Saros Series = 142
Saros Member = 23 of 72

Sun at Greatest Eclipse
(Geocentric Coordinates)
R.A. = 17h30m05.9s
Dec. = -23°15'32.3"
S.D. = 00°16'14.9"
H.P. = 00°00'08.9"

Moon at Greatest Eclipse
(Geocentric Coordinates)
R.A. = 17h29m54.3s
Dec. = -23°32'58.8"
S.D. = 00°16'23.7"
H.P. = 01°00'10.4"

N

W

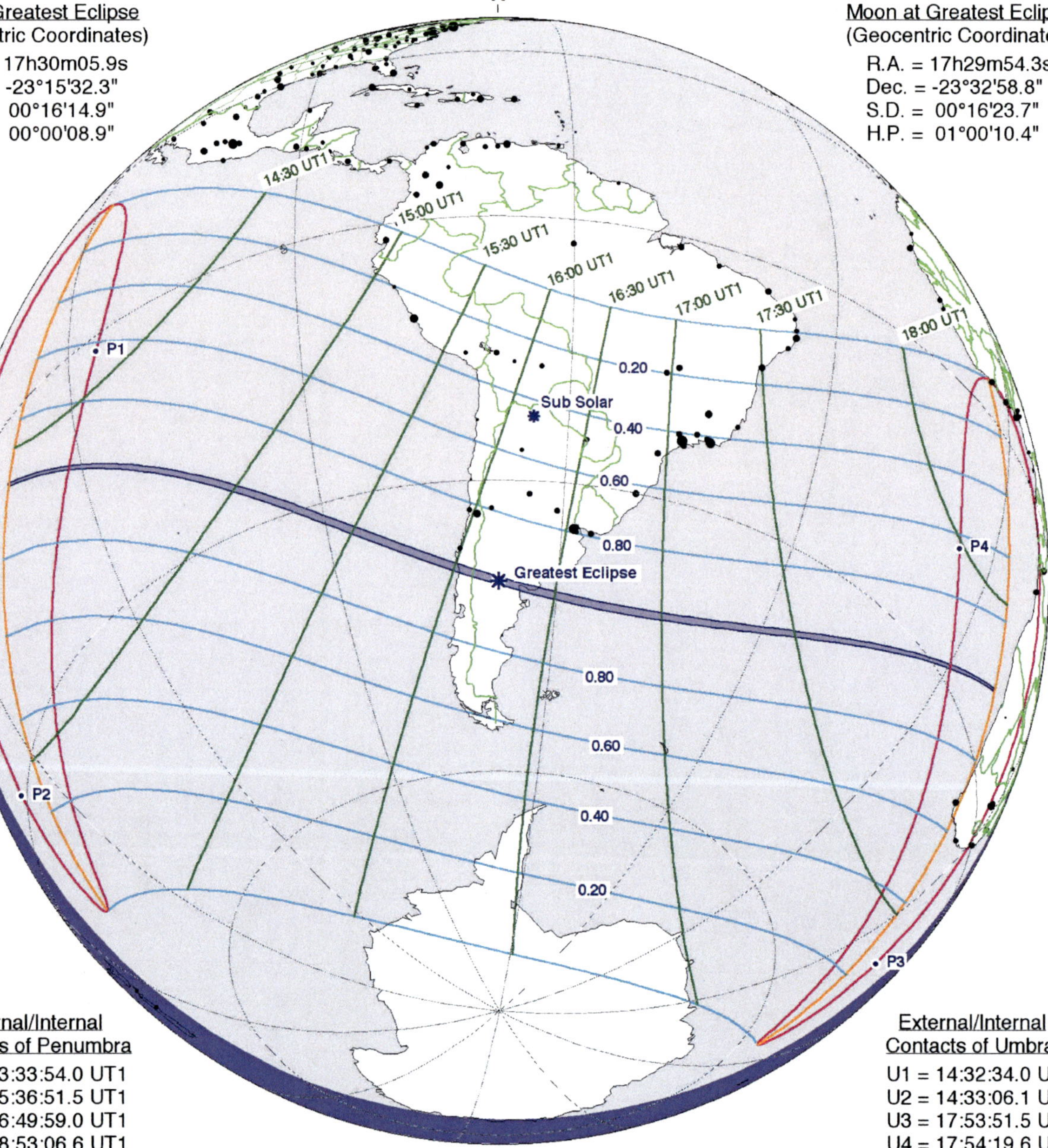

E

S

External/Internal
Contacts of Penumbra
P1 = 13:33:54.0 UT1
P2 = 15:36:51.5 UT1
P3 = 16:49:59.0 UT1
P4 = 18:53:06.6 UT1

External/Internal
Contacts of Umbra
U1 = 14:32:34.0 UT1
U2 = 14:33:06.1 UT1
U3 = 17:53:51.5 UT1
U4 = 17:54:19.6 UT1

ΔT = 70.2 s

Eph. = JPL DE405

Circumstances at Greatest Eclipse: 16:13:29.1 UT1

Lat. = 40°20.1'S	Sun Alt. = 72.7°
Long. = 067°57.4'W	Sun Azm. = 10.3°
Path Width = 90.2 km	Duration = 02m09.7s

Circumstances at Greatest Duration: 16:13:34.9 UT1

Lat. = 40°20.7'S	Sun Alt. = 72.7°
Long. = 067°54.8'W	Sun Azm. = 10.1°
Path Width = 90.2 km	Duration = 02m09.7s

0 1000 2000 3000 4000 5000
Kilometers

Annular Solar Eclipse of 2021 Jun 10

Greatest Eclipse = 10:43:06.7 TD (= 10:41:56.3 UT1)

Eclipse Magnitude = 0.9435
Gamma = 0.9152

Saros Series = 147
Saros Member = 23 of 80

Sun at Greatest Eclipse
(Geocentric Coordinates)
R.A. = 05h15m31.4s
Dec. = +23°02'37.1"
S.D. = 00°15'45.2"
H.P. = 00°00'08.7"

Moon at Greatest Eclipse
(Geocentric Coordinates)
R.A. = 05h14m53.6s
Dec. = +23°51'21.6"
S.D. = 00°14'46.8"
H.P. = 00°54'14.5"

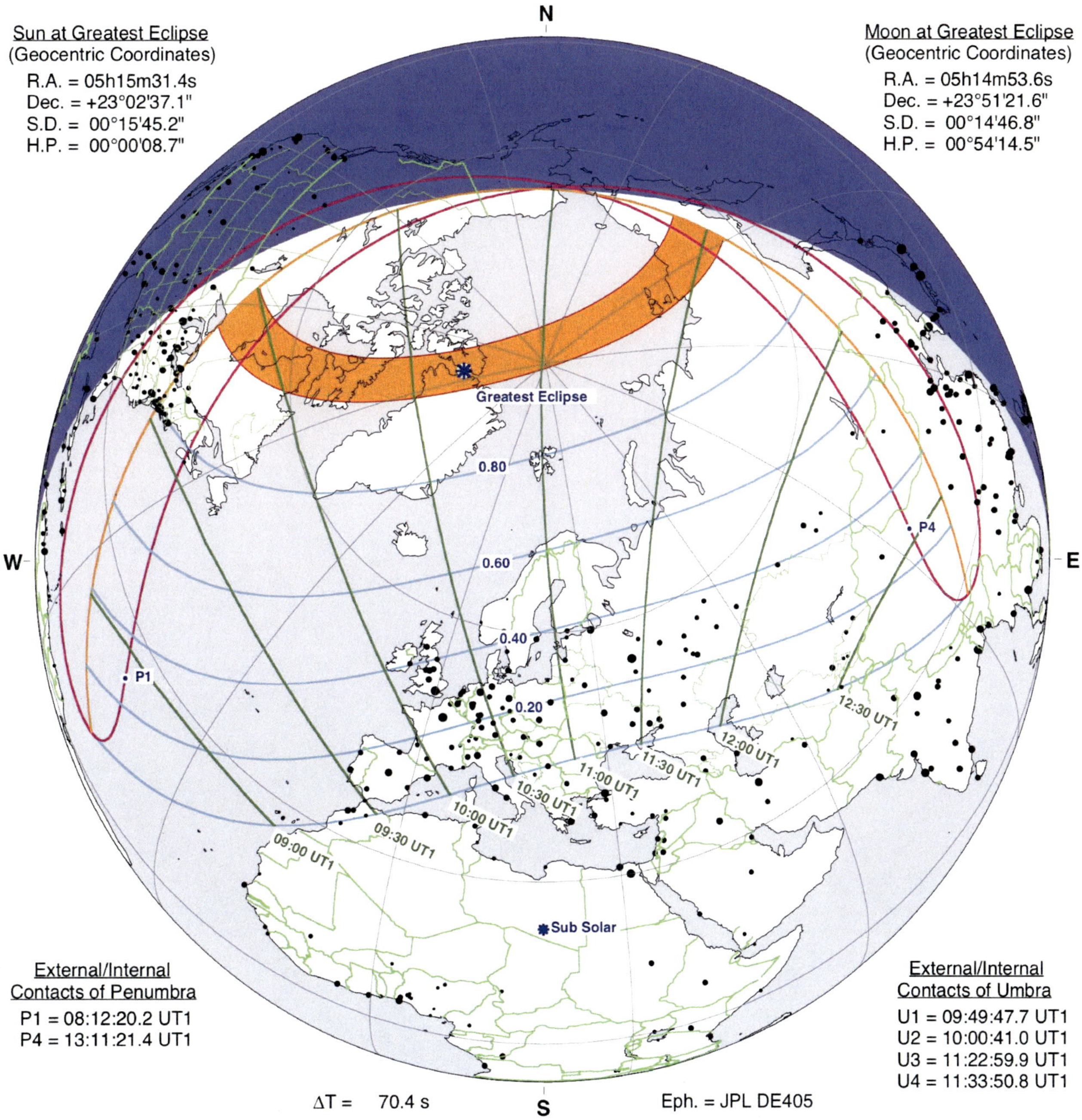

External/Internal
Contacts of Penumbra
P1 = 08:12:20.2 UT1
P4 = 13:11:21.4 UT1

External/Internal
Contacts of Umbra
U1 = 09:49:47.7 UT1
U2 = 10:00:41.0 UT1
U3 = 11:22:59.9 UT1
U4 = 11:33:50.8 UT1

ΔT = 70.4 s

Eph. = JPL DE405

Circumstances at Greatest Eclipse: 10:41:56.3 UT1
Lat. = 80°48.9'N   Sun Alt. = 23.3°
Long. = 066°46.1'W   Sun Azm. = 89.9°
Path Width = 526.8 km   Duration = 03m51.2s

Circumstances at Greatest Duration: 10:41:57.4 UT1
Lat. = 80°49.4'N   Sun Alt. = 23.3°
Long. = 066°46.4'W   Sun Azm. = 89.9°
Path Width = 526.8 km   Duration = 03m51.2s

0 1000 2000 3000 4000 5000
Kilometers

Total Solar Eclipse of 2021 Dec 04

Greatest Eclipse = 07:34:37.9 TD (= 07:33:27.3 UT1)

Eclipse Magnitude = 1.0367 Saros Series = 152

Gamma = -0.9526 Saros Member = 13 of 70

N

Sun at Greatest Eclipse
(Geocentric Coordinates)
R.A. = 16h43m32.4s
Dec. = -22°16'29.4"
S.D. = 00°16'13.6"
H.P. = 00°00'08.9"

Moon at Greatest Eclipse
(Geocentric Coordinates)
R.A. = 16h42m35.0s
Dec. = -23°13'22.3"
S.D. = 00°16'44.7"
H.P. = 01°01'27.3"

W

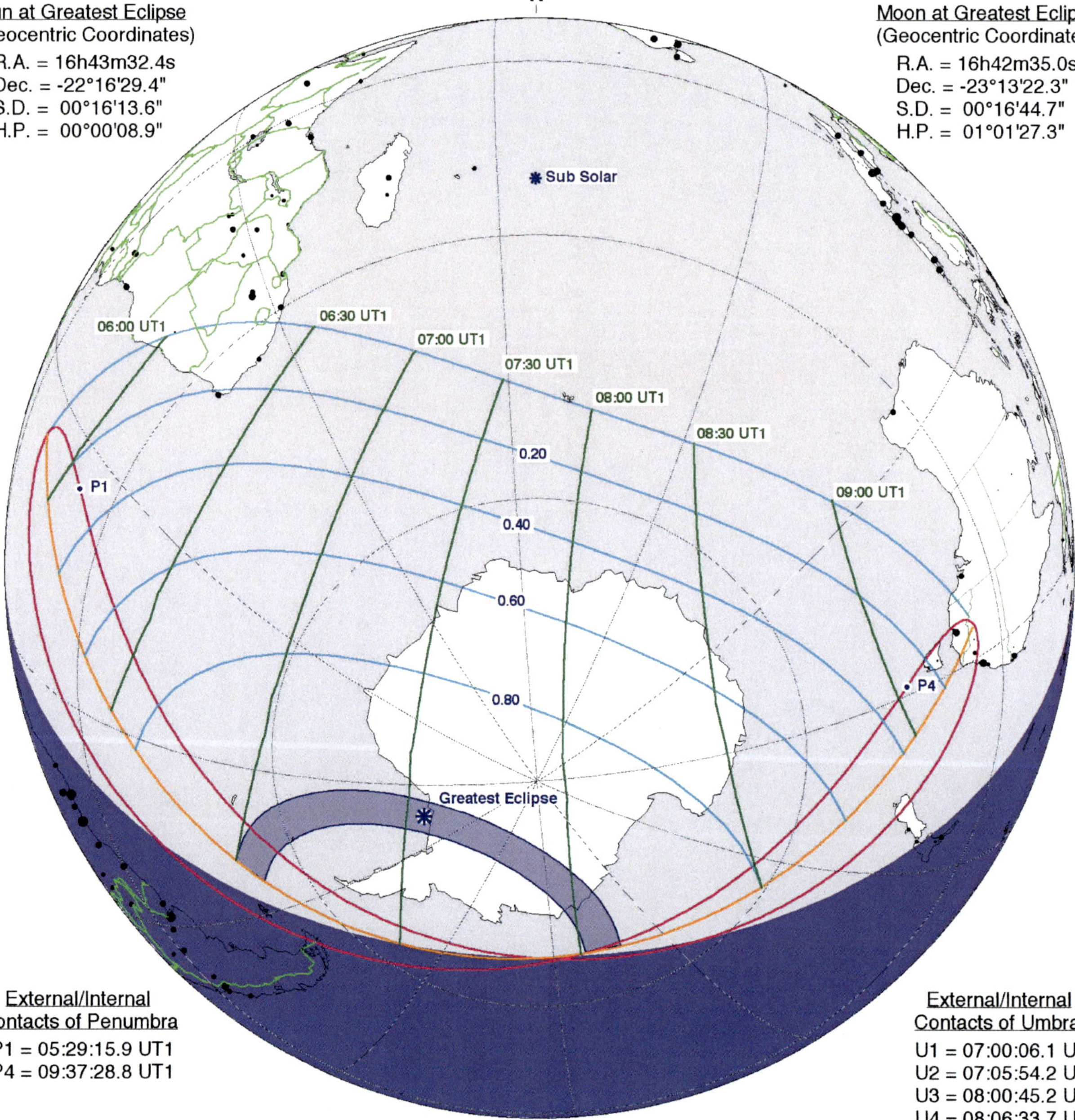

E

External/Internal Contacts of Penumbra
P1 = 05:29:15.9 UT1
P4 = 09:37:28.8 UT1

External/Internal Contacts of Umbra
U1 = 07:00:06.1 UT1
U2 = 07:05:54.2 UT1
U3 = 08:00:45.2 UT1
U4 = 08:06:33.7 UT1

ΔT = 70.6 s S Eph. = JPL DE405

Circumstances at Greatest Eclipse: 07:33:27.3 UT1

Lat. = 76°46.6'S	Sun Alt. = 17.2°
Long. = 046°13.7'W	Sun Azm. = 114.8°
Path Width = 418.8 km	Duration = 01m54.4s

Circumstances at Greatest Duration: 07:33:29.6 UT1

Lat. = 76°47.7'S	Sun Alt. = 17.2°
Long. = 046°16.7'W	Sun Azm. = 114.8°
Path Width = 418.7 km	Duration = 01m54.4s

0 1000 2000 3000 4000 5000
Kilometers

Partial Solar Eclipse of 2022 Apr 30

Greatest Eclipse = 20:42:36.5 TD (= 20:41:25.8 UT1)

Eclipse Magnitude = 0.6396
Gamma = -1.1901

Saros Series = 119
Saros Member = 66 of 71

Sun at Greatest Eclipse
(Geocentric Coordinates)
R.A. = 02h32m15.6s
Dec. = +14°57'53.5"
S.D. = 00°15'52.6"
H.P. = 00°00'08.7"

Moon at Greatest Eclipse
(Geocentric Coordinates)
R.A. = 02h34m04.8s
Dec. = +13°57'48.8"
S.D. = 00°15'04.0"
H.P. = 00°55'17.7"

N

W

E

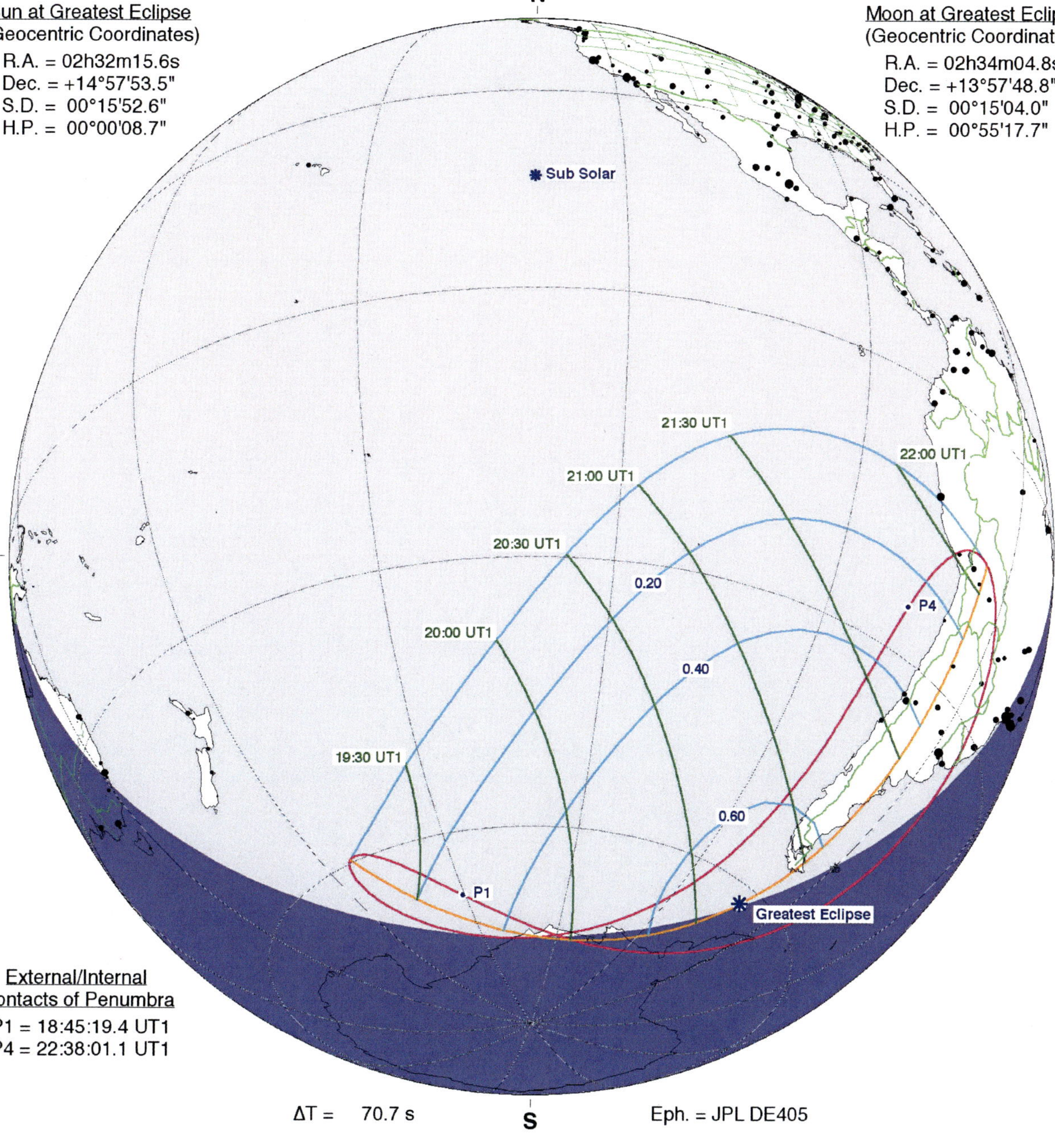

S

External/Internal
Contacts of Penumbra
P1 = 18:45:19.4 UT1
P4 = 22:38:01.1 UT1

ΔT = 70.7 s

Eph. = JPL DE405

Circumstances at Greatest Eclipse: 20:41:25.8 UT1

Lat. = 62°07.0'S
Long. = 071°28.7'W

Sun Alt. = 0.0°
Sun Azm. = 303.5°

0 1000 2000 3000 4000 5000
Kilometers

Partial Solar Eclipse of 2022 Oct 25

Greatest Eclipse = 11:01:20.0 TD (= 11:00:09.1 UT1)

Eclipse Magnitude = 0.8619 Saros Series = 124

Gamma = 1.0701 Saros Member = 55 of 73

Sun at Greatest Eclipse
(Geocentric Coordinates)
R.A. = 13h59m20.5s
Dec. = -12°10'17.0"
S.D. = 00°16'05.0"
H.P. = 00°00'08.8"

Moon at Greatest Eclipse
(Geocentric Coordinates)
R.A. = 14h01m10.9s
Dec. = -11°14'16.0"
S.D. = 00°15'52.6"
H.P. = 00°58'16.0"

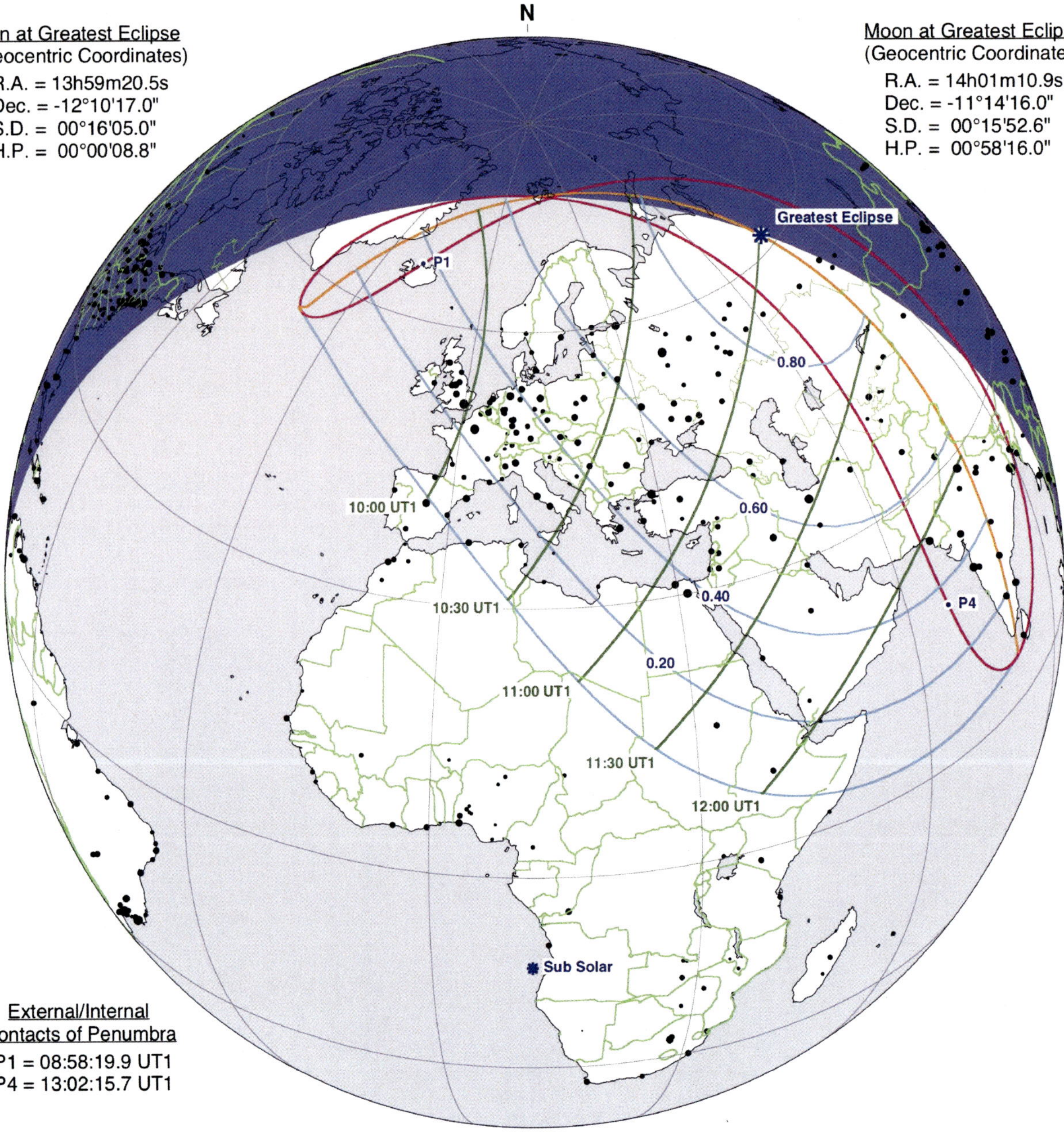

External/Internal
Contacts of Penumbra
P1 = 08:58:19.9 UT1
P4 = 13:02:15.7 UT1

ΔT = 70.9 s Eph. = JPL DE405

Circumstances at Greatest Eclipse: 11:00:09.1 UT1

Lat. = 61°38.9'N Sun Alt. = 0.0°

Long. = 077°20.8'E Sun Azm. = 243.6°

0 1000 2000 3000 4000 5000
Kilometers

Hybrid Solar Eclipse of 2023 Apr 20

Greatest Eclipse = 04:17:56.0 TD (= 04:16:44.9 UT1)

Eclipse Magnitude = 1.0132
Gamma = -0.3952

Saros Series = 129
Saros Member = 52 of 80

Sun at Greatest Eclipse
(Geocentric Coordinates)
R.A. = 01h51m01.7s
Dec. = +11°24'54.1"
S.D. = 00°15'55.4"
H.P. = 00°00'08.8"

Moon at Greatest Eclipse
(Geocentric Coordinates)
R.A. = 01h51m43.2s
Dec. = +11°04'16.7"
S.D. = 00°15'53.6"
H.P. = 00°58'19.9"

N

W

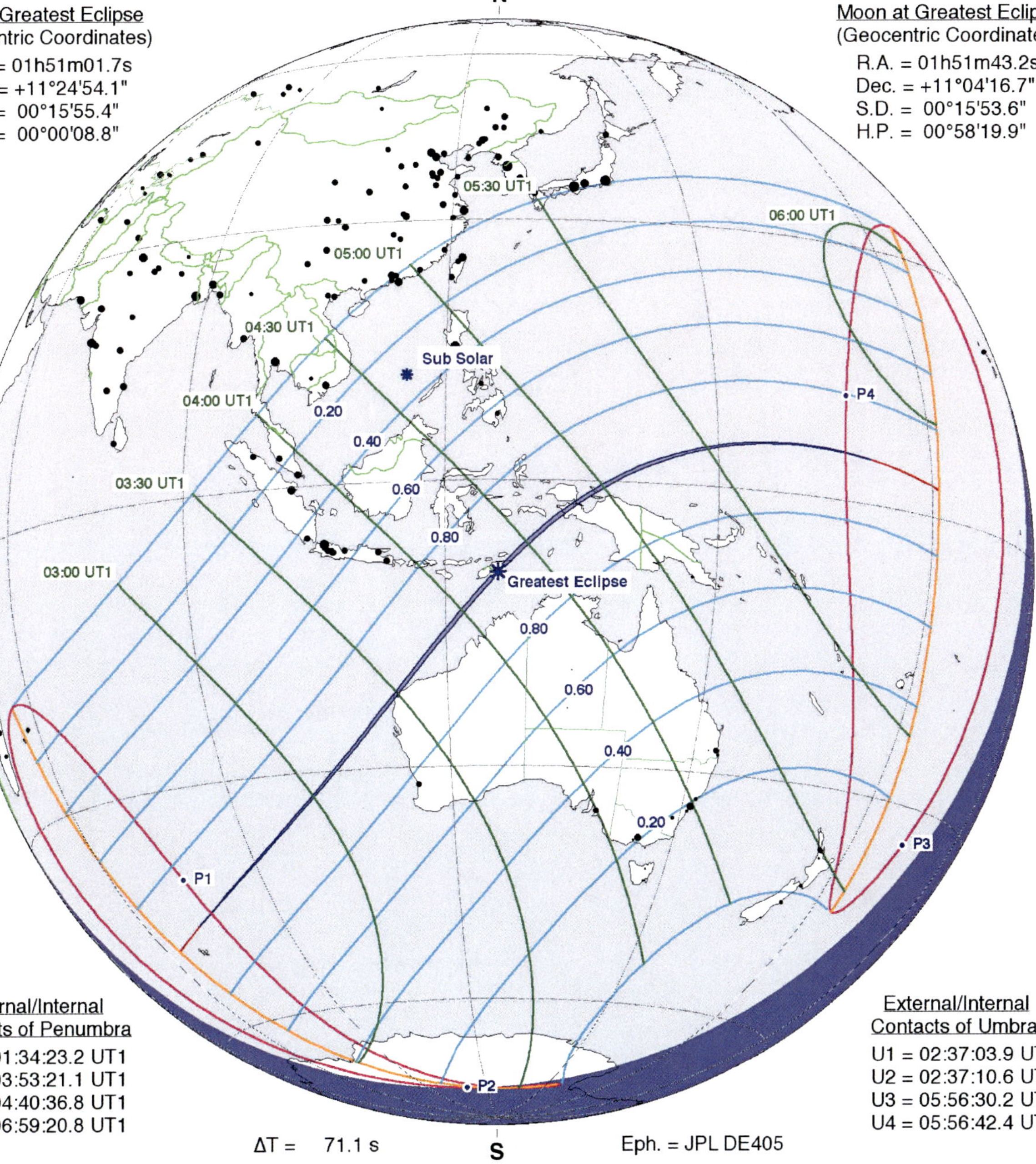

E

S

External/Internal
Contacts of Penumbra
P1 = 01:34:23.2 UT1
P2 = 03:53:21.1 UT1
P3 = 04:40:36.8 UT1
P4 = 06:59:20.8 UT1

External/Internal
Contacts of Umbra
U1 = 02:37:03.9 UT1
U2 = 02:37:10.6 UT1
U3 = 05:56:30.2 UT1
U4 = 05:56:42.4 UT1

ΔT = 71.1 s

Eph. = JPL DE405

Circumstances at Greatest Eclipse: 04:16:44.9 UT1
Lat. = 09°35.7'S
Long. = 125°46.8'E
Path Width = 49.0 km
Sun Alt. = 66.7°
Sun Azm. = 334.0°
Duration = 01m16.1s

Circumstances at Greatest Duration: 04:16:15.6 UT1
Lat. = 09°42.5'S
Long. = 125°40.1'E
Path Width = 48.9 km
Sun Alt. = 66.7°
Sun Azm. = 334.6°
Duration = 01m16.1s

0 1000 2000 3000 4000 5000
Kilometers

Annular Solar Eclipse of 2023 Oct 14

Greatest Eclipse = 18:00:40.6 TD (= 17:59:29.3 UT1)

Eclipse Magnitude = 0.9520 | Saros Series = 134
Gamma = 0.3753 | Saros Member = 44 of 71

Sun at Greatest Eclipse
(Geocentric Coordinates)
R.A. = 13h18m05.4s
Dec. = -08°14'36.7"
S.D. = 00°16'02.0"
H.P. = 00°00'08.8"

Moon at Greatest Eclipse
(Geocentric Coordinates)
R.A. = 13h18m44.3s
Dec. = -07°56'18.9"
S.D. = 00°15'02.9"
H.P. = 00°55'13.8"

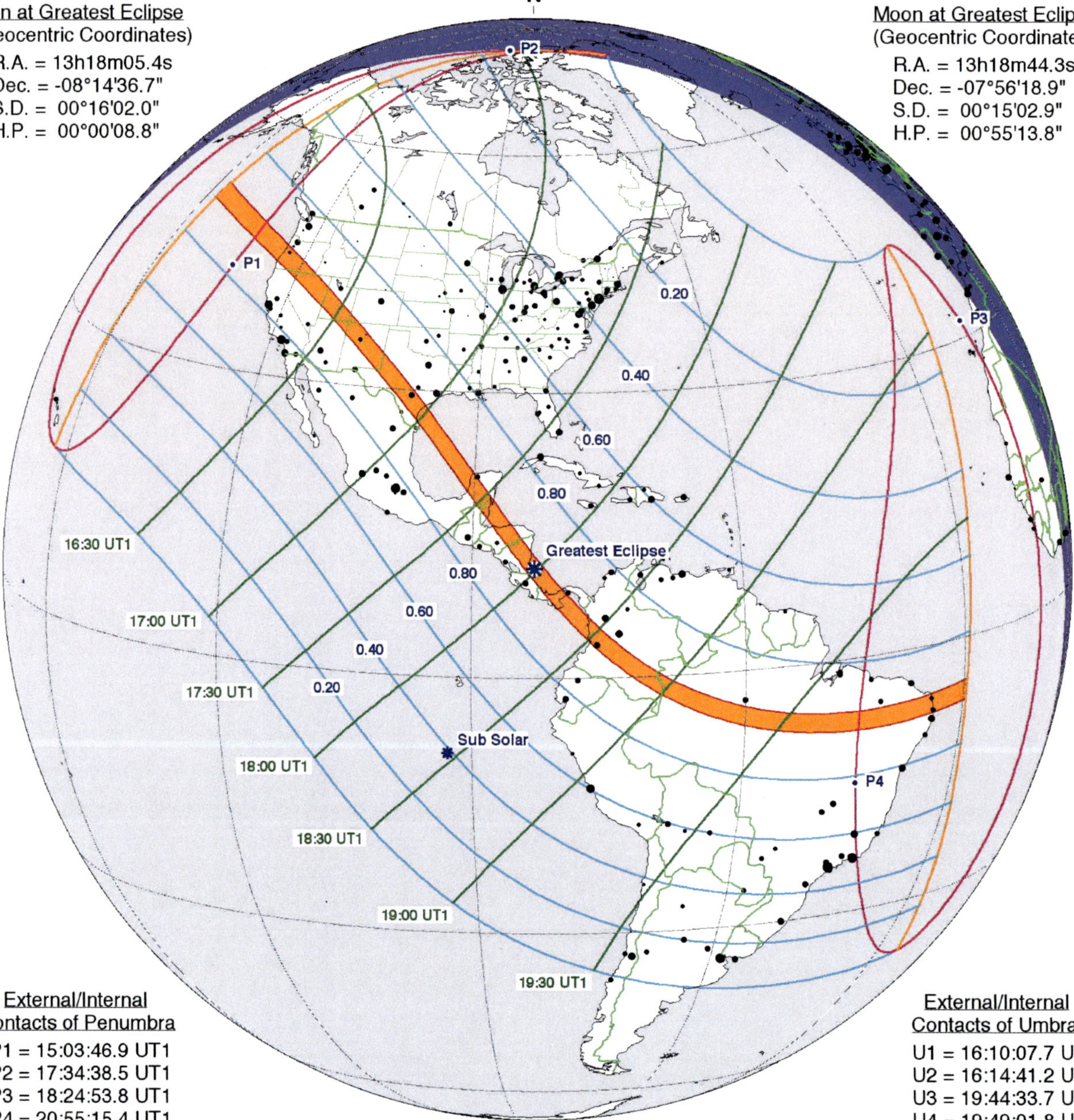

External/Internal Contacts of Penumbra
P1 = 15:03:46.9 UT1
P2 = 17:34:38.5 UT1
P3 = 18:24:53.8 UT1
P4 = 20:55:15.4 UT1

External/Internal Contacts of Umbra
U1 = 16:10:07.7 UT1
U2 = 16:14:41.2 UT1
U3 = 19:44:33.7 UT1
U4 = 19:49:01.8 UT1

ΔT = 71.3 s | Eph. = JPL DE405

Circumstances at Greatest Eclipse: 17:59:29.3 UT1
Lat. = 11°22.1'N | Sun Alt. = 67.9°
Long. = 083°06.1'W | Sun Azm. = 208.0°
Path Width = 187.4 km | Duration = 05m17.2s

Circumstances at Greatest Duration: 18:13:09.2 UT1
Lat. = 08°14.6'N | Sun Alt. = 66.8°
Long. = 080°24.1'W | Sun Azm. = 225.1°
Path Width = 191.1 km | Duration = 05m17.8s

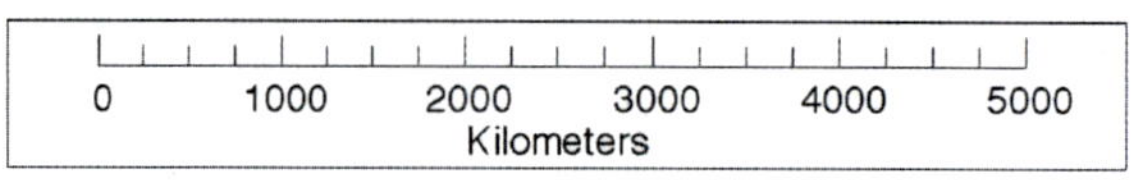

Total Solar Eclipse of 2024 Apr 08

Greatest Eclipse = 18:18:29.4 TD (= 18:17:17.9 UT1)

Eclipse Magnitude = 1.0566 Saros Series = 139

Gamma = 0.3431 Saros Member = 30 of 71

Sun at Greatest Eclipse
(Geocentric Coordinates)
R.A. = 01h11m36.9s
Dec. = +07°35'29.4"
S.D. = 00°15'58.2"
H.P. = 00°00'08.8"

Moon at Greatest Eclipse
(Geocentric Coordinates)
R.A. = 01h10m57.5s
Dec. = +07°53'55.5"
S.D. = 00°16'36.3"
H.P. = 01°00'56.6"

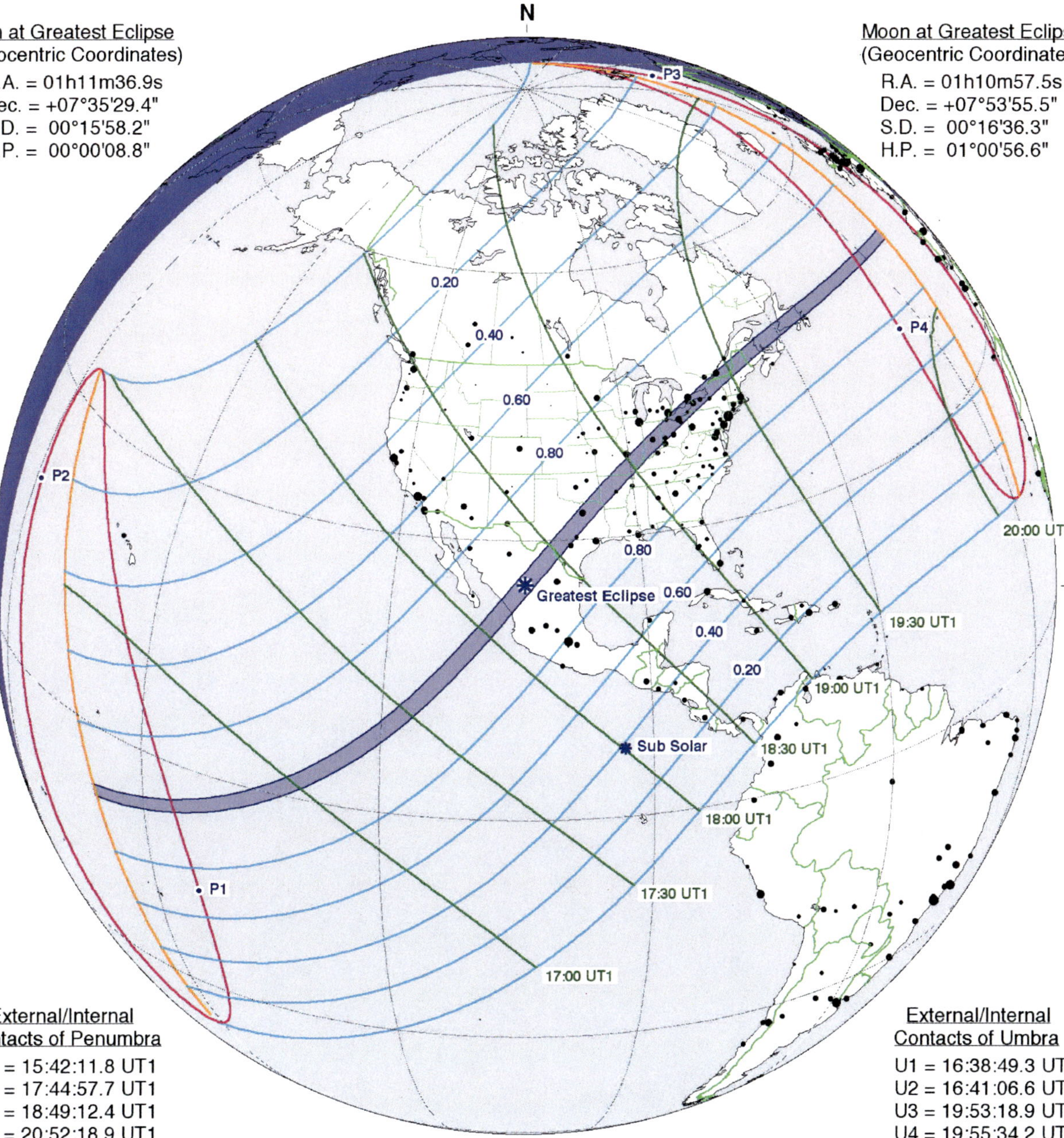

External/Internal
Contacts of Penumbra
P1 = 15:42:11.8 UT1
P2 = 17:44:57.7 UT1
P3 = 18:49:12.4 UT1
P4 = 20:52:18.9 UT1

External/Internal
Contacts of Umbra
U1 = 16:38:49.3 UT1
U2 = 16:41:06.6 UT1
U3 = 19:53:18.9 UT1
U4 = 19:55:34.2 UT1

ΔT = 71.5 s

Eph. = JPL DE405

Circumstances at Greatest Eclipse: 18:17:17.9 UT1

Lat. = 25°17.4'N Sun Alt. = 69.8°
Long. = 104°08.3'W Sun Azm. = 149.4°
Path Width = 197.5 km Duration = 04m28.1s

Circumstances at Greatest Duration: 18:19:32.7 UT1

Lat. = 25°55.8'N Sun Alt. = 69.7°
Long. = 103°31.1'W Sun Azm. = 153.3°
Path Width = 197.0 km Duration = 04m28.2s

0 1000 2000 3000 4000 5000
Kilometers

Annular Solar Eclipse of 2024 Oct 02

Greatest Eclipse = 18:46:13.3 TD (= 18:45:01.6 UT1)

Eclipse Magnitude = 0.9326 Saros Series = 144

Gamma = -0.3509 Saros Member = 17 of 70

<u>Sun at Greatest Eclipse</u>
(Geocentric Coordinates)
R.A. = 12h36m58.9s
Dec. = -03°59'03.9"
S.D. = 00°15'58.9"
H.P. = 00°00'08.8"

<u>Moon at Greatest Eclipse</u>
(Geocentric Coordinates)
R.A. = 12h36m22.3s
Dec. = -04°15'35.4"
S.D. = 00°14'41.8"
H.P. = 00°53'56.4"

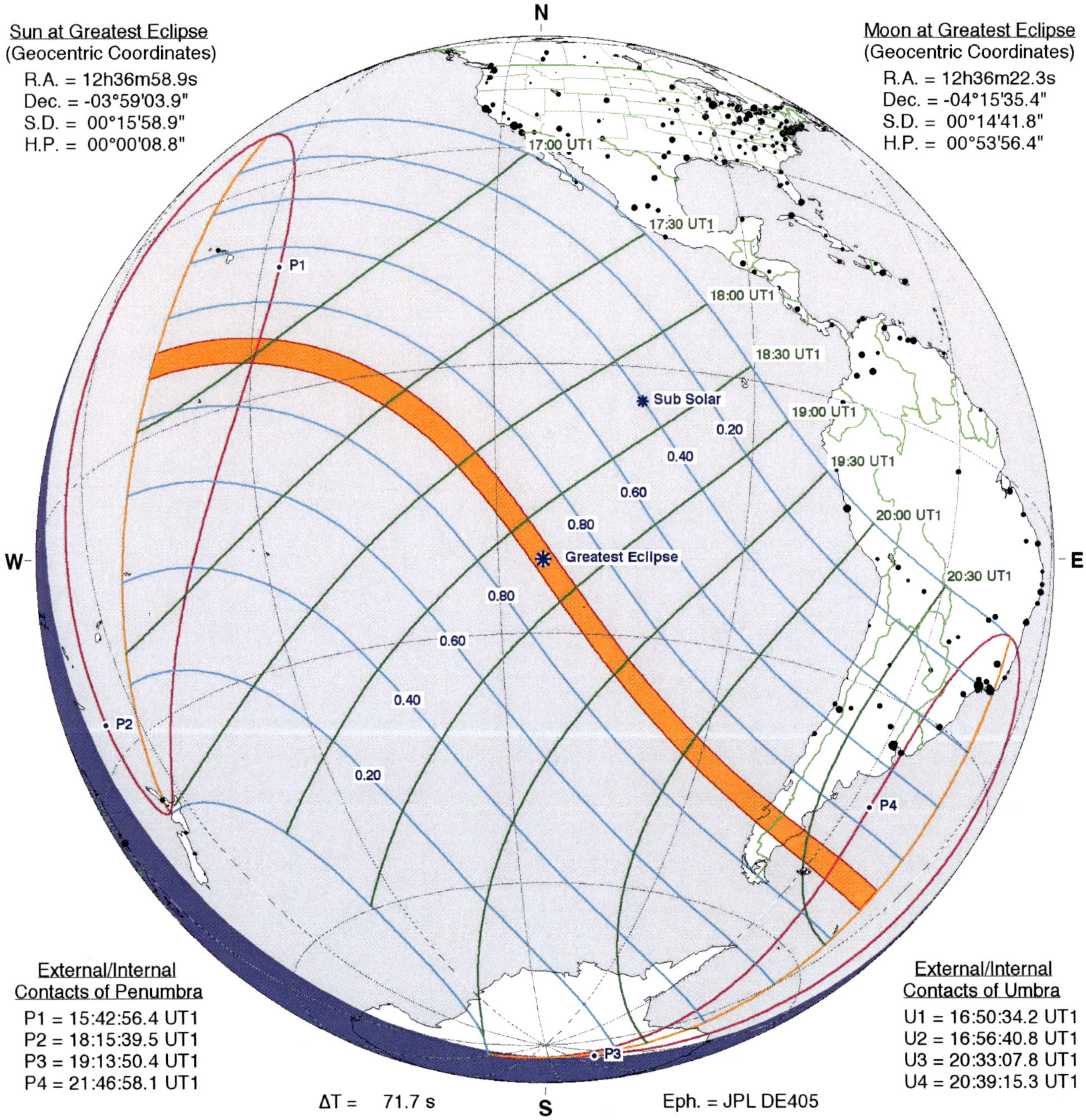

<u>External/Internal Contacts of Penumbra</u>
P1 = 15:42:56.4 UT1
P2 = 18:15:39.5 UT1
P3 = 19:13:50.4 UT1
P4 = 21:46:58.1 UT1

<u>External/Internal Contacts of Umbra</u>
U1 = 16:50:34.2 UT1
U2 = 16:56:40.8 UT1
U3 = 20:33:07.8 UT1
U4 = 20:39:15.3 UT1

ΔT = 71.7 s Eph. = JPL DE405

<u>Circumstances at Greatest Eclipse: 18:45:01.6 UT1</u>

Lat. = 21°57.2'S Sun Alt. = 69.3°
Long. = 114°30.5'W Sun Azm. = 31.1°
Path Width = 266.5 km Duration = 07m25.1s

<u>Circumstances at Greatest Duration: 18:53:00.1 UT1</u>

Lat. = 23°57.8'S Sun Alt. = 68.9°
Long. = 112°56.3'W Sun Azm. = 19.6°
Path Width = 264.4 km Duration = 07m25.4s

0 1000 2000 3000 4000 5000
Kilometers

Partial Solar Eclipse of 2025 Mar 29

Greatest Eclipse = 10:48:36.1 TD (= 10:47:24.2 UT1)

Eclipse Magnitude = 0.9376 Saros Series = 149

Gamma = 1.0405 Saros Member = 21 of 71

Sun at Greatest Eclipse
(Geocentric Coordinates)
R.A. = 00h33m03.1s
Dec. = +03°33'55.0"
S.D. = 00°16'01.1"
H.P. = 00°00'08.8"

Moon at Greatest Eclipse
(Geocentric Coordinates)
R.A. = 00h31m00.8s
Dec. = +04°29'34.1"
S.D. = 00°16'39.4"
H.P. = 01°01'07.8"

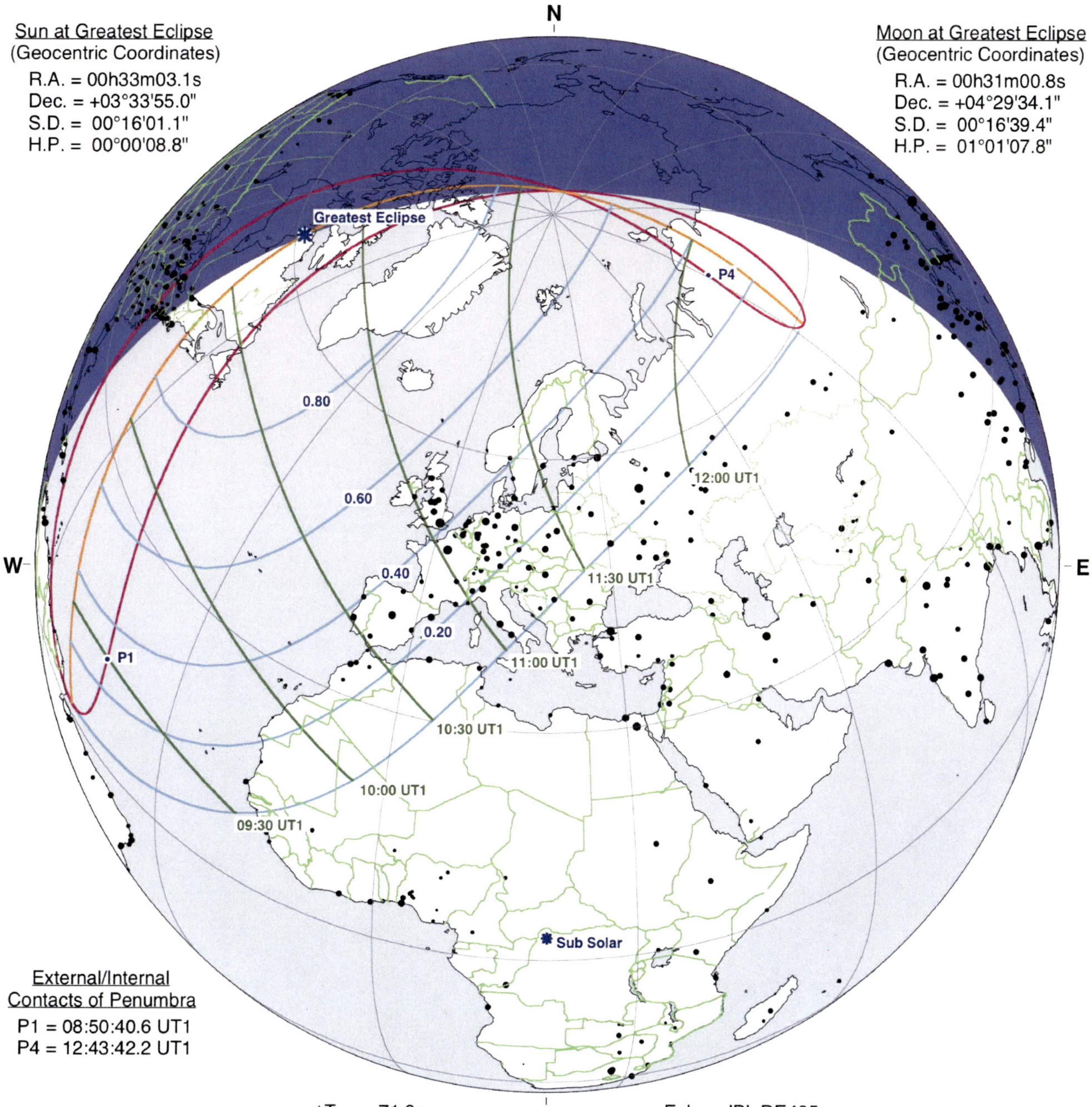

External/Internal
Contacts of Penumbra
P1 = 08:50:40.6 UT1
P4 = 12:43:42.2 UT1

ΔT = 71.9 s Eph. = JPL DE405

Circumstances at Greatest Eclipse: 10:47:24.2 UT1

Lat. = 61°06.0'N Sun Alt. = 0.0°

Long. = 077°05.1'W Sun Azm. = 82.6°

0 1000 2000 3000 4000 5000
Kilometers

Partial Solar Eclipse of 2025 Sep 21

Greatest Eclipse = 19:43:04.2 TD (= 19:41:52.2 UT1)

Eclipse Magnitude = 0.8550   Saros Series = 154

Gamma = -1.0651   Saros Member = 7 of 71

Sun at Greatest Eclipse
(Geocentric Coordinates)

R.A. = 11h56m36.9s
Dec. = +00°22'00.7"
S.D. = 00°15'55.9"
H.P. = 00°00'08.8"

Moon at Greatest Eclipse
(Geocentric Coordinates)

R.A. = 11h54m42.8s
Dec. = -00°29'14.7"
S.D. = 00°15'02.8"
H.P. = 00°55'13.2"

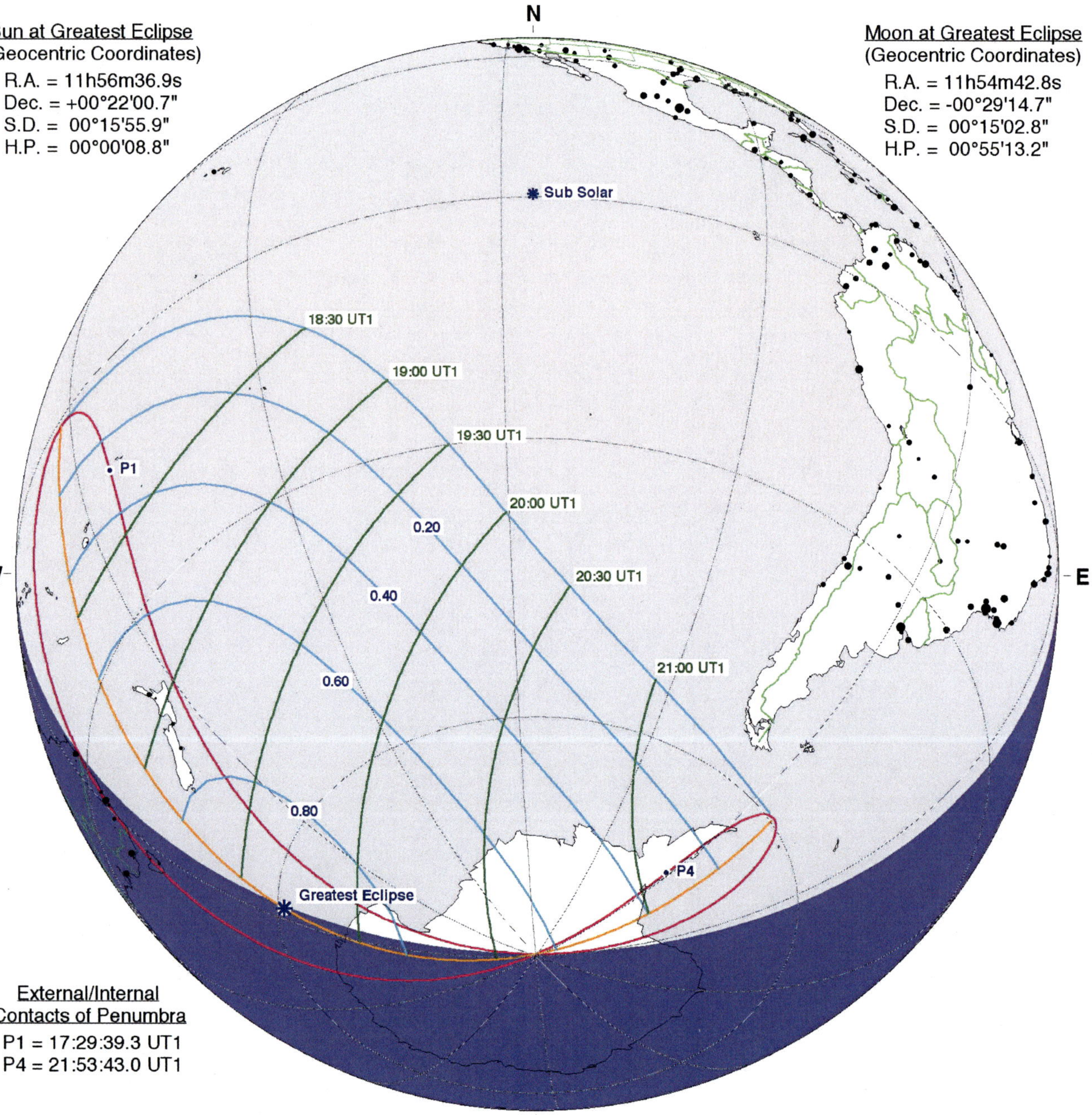

External/Internal
Contacts of Penumbra

P1 = 17:29:39.3 UT1
P4 = 21:53:43.0 UT1

ΔT = 72.1 s   Eph. = JPL DE405

Circumstances at Greatest Eclipse: 19:41:52.2 UT1

Lat. = 60°54.1'S   Sun Alt. = 0.0°
Long. = 153°29.9'E   Sun Azm. = 89.2°

0 1000 2000 3000 4000 5000
Kilometers

Annular Solar Eclipse of 2026 Feb 17

Greatest Eclipse = 12:13:05.8 TD (= 12:11:53.6 UT1)

Eclipse Magnitude = 0.9630 Saros Series = 121
Gamma = -0.9743 Saros Member = 61 of 71

<u>Sun at Greatest Eclipse</u>
(Geocentric Coordinates)
R.A. = 22h03m54.3s
Dec. = -11°52'42.3"
S.D. = 00°16'11.1"
H.P. = 00°00'08.9"

<u>Moon at Greatest Eclipse</u>
(Geocentric Coordinates)
R.A. = 22h05m34.0s
Dec. = -12°42'29.5"
S.D. = 00°15'32.4"
H.P. = 00°57'02.0"

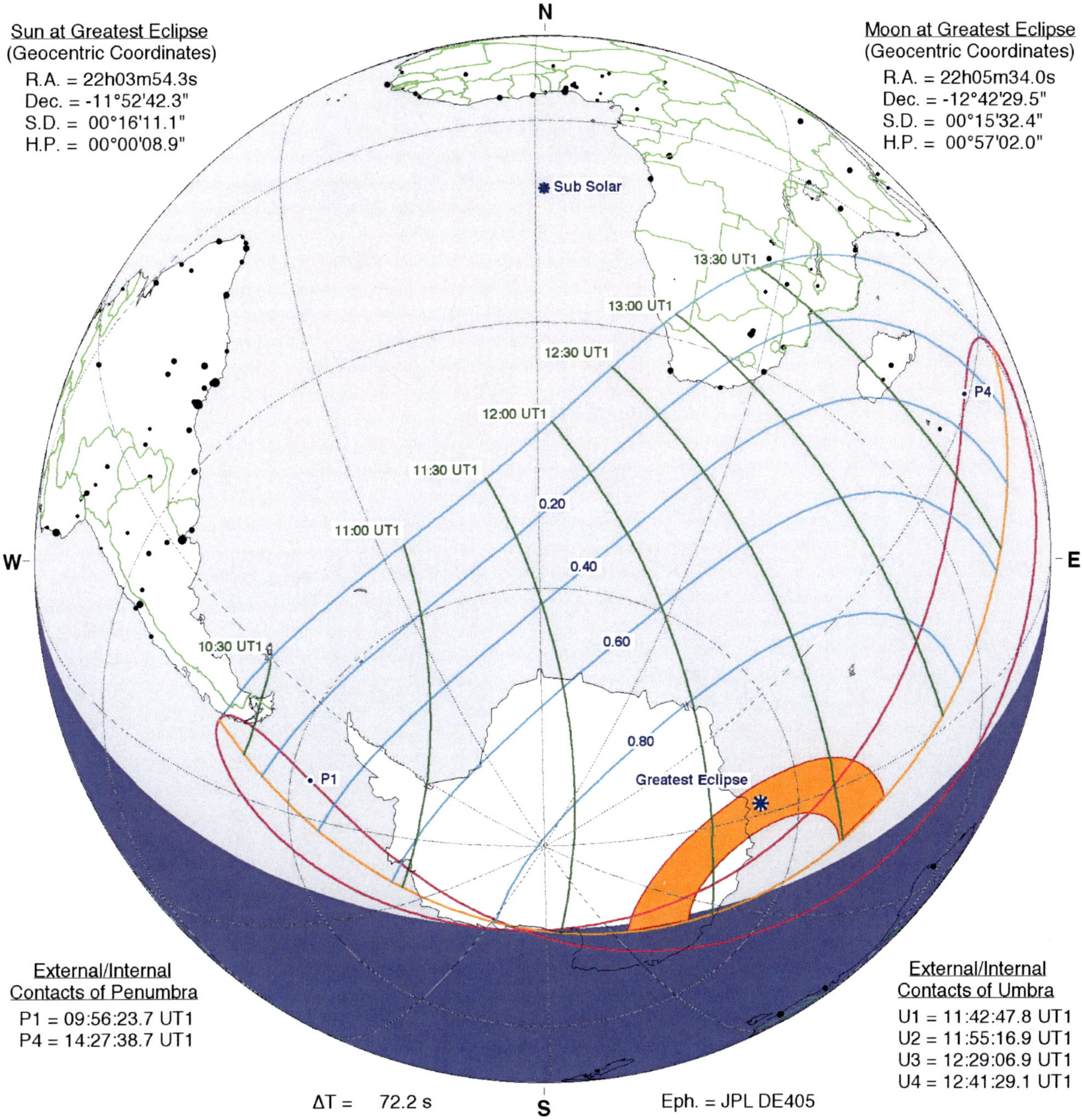

<u>External/Internal Contacts of Penumbra</u>
P1 = 09:56:23.7 UT1
P4 = 14:27:38.7 UT1

<u>External/Internal Contacts of Umbra</u>
U1 = 11:42:47.8 UT1
U2 = 11:55:16.9 UT1
U3 = 12:29:06.9 UT1
U4 = 12:41:29.1 UT1

ΔT = 72.2 s Eph. = JPL DE405

<u>Circumstances at Greatest Eclipse: 12:11:53.6 UT1</u>
Lat. = 64°43.1'S Sun Alt. = 12.3°
Long. = 086°44.4'E Sun Azm. = 268.3°
Path Width = 616.2 km Duration = 02m19.6s

<u>Circumstances at Greatest Duration: 11:48:15.2 UT1</u>
Lat. = 71°57.6'S Sun Alt. = 0.0°
Long. = 136°39.8'E Sun Azm. = 228.4°
Path Width = 765.7 km Duration = 02m20.9s

0 1000 2000 3000 4000 5000
Kilometers

Total Solar Eclipse of 2026 Aug 12

Greatest Eclipse = 17:47:05.8 TD (= 17:45:53.3 UT1)

Eclipse Magnitude = 1.0386 Saros Series = 126

Gamma = 0.8977 Saros Member = 48 of 72

Sun at Greatest Eclipse
(Geocentric Coordinates)
R.A. = 09h29m47.3s
Dec. = +14°48'04.5"
S.D. = 00°15'47.0"
H.P. = 00°00'08.7"

Moon at Greatest Eclipse
(Geocentric Coordinates)
R.A. = 09h31m17.3s
Dec. = +15°36'58.5"
S.D. = 00°16'16.9"
H.P. = 00°59'45.1"

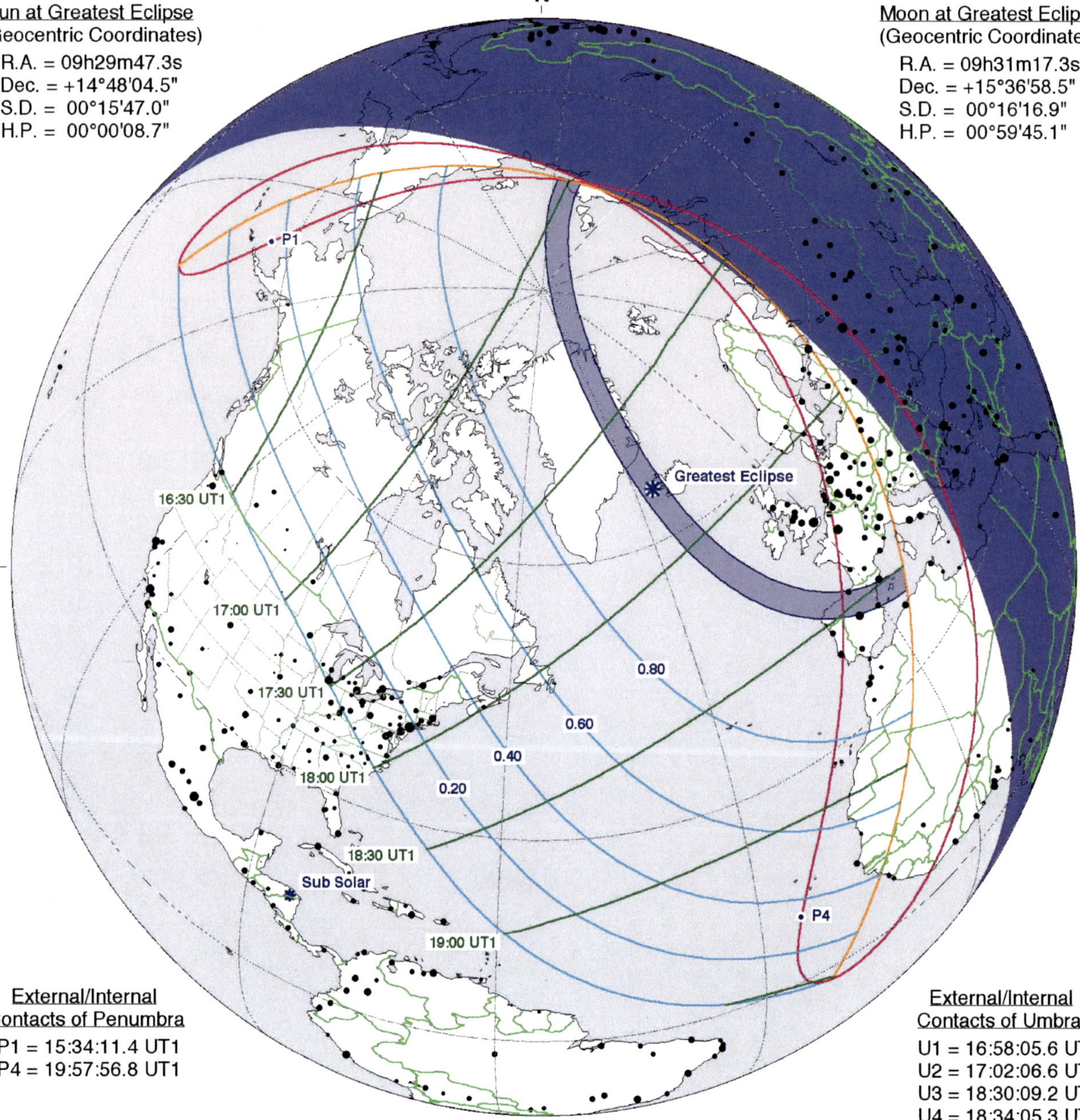

External/Internal
Contacts of Penumbra
P1 = 15:34:11.4 UT1
P4 = 19:57:56.8 UT1

External/Internal
Contacts of Umbra
U1 = 16:58:05.6 UT1
U2 = 17:02:06.6 UT1
U3 = 18:30:09.2 UT1
U4 = 18:34:05.3 UT1

ΔT = 72.4 s Eph. = JPL DE405

Circumstances at Greatest Eclipse: 17:45:53.3 UT1

Lat. = 65°13.4'N Sun Alt. = 25.8°
Long. = 025°13.7'W Sun Azm. = 248.4°
Path Width = 294.0 km Duration = 02m18.2s

Circumstances at Greatest Duration: 17:44:41.5 UT1

Lat. = 65°49.8'N Sun Alt. = 25.7°
Long. = 025°29.1'W Sun Azm. = 247.6°
Path Width = 292.9 km Duration = 02m18.2s

0 1000 2000 3000 4000 5000
Kilometers

Annular Solar Eclipse of 2027 Feb 06

Greatest Eclipse = 16:00:47.7 TD (= 15:59:35.1 UT1)

Eclipse Magnitude = 0.9281 Saros Series = 131

Gamma = -0.2952 Saros Member = 51 of 70

N

Sun at Greatest Eclipse
(Geocentric Coordinates)

R.A. = 21h20m17.6s
Dec. = -15°32'54.5"
S.D. = 00°16'13.1"
H.P. = 00°00'08.9"

Moon at Greatest Eclipse
(Geocentric Coordinates)

R.A. = 21h20m44.2s
Dec. = -15°47'36.0"
S.D. = 00°14'50.2"
H.P. = 00°54'27.0"

W

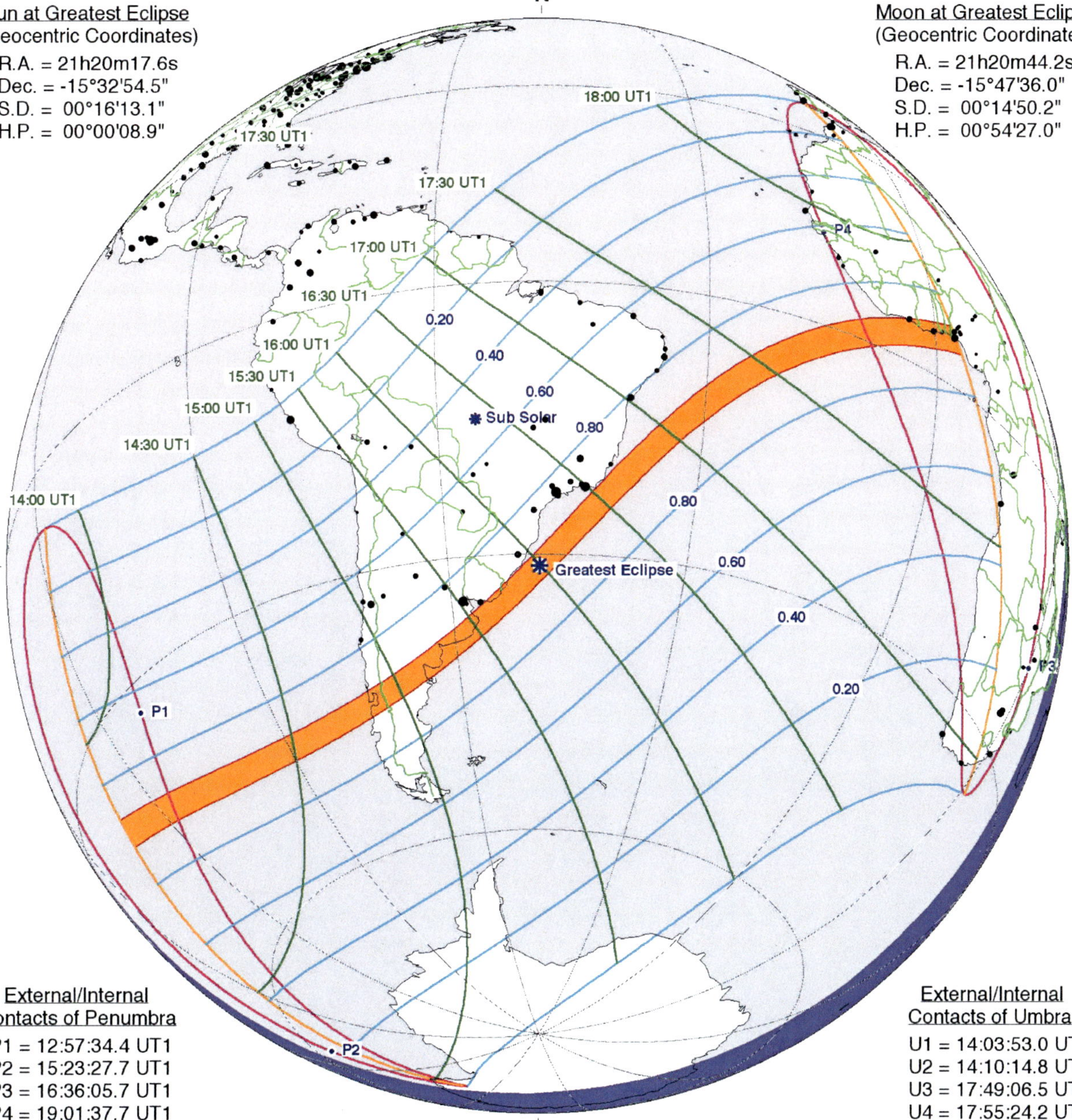

E

External/Internal
Contacts of Penumbra

P1 = 12:57:34.4 UT1
P2 = 15:23:27.7 UT1
P3 = 16:36:05.7 UT1
P4 = 19:01:37.7 UT1

External/Internal
Contacts of Umbra

U1 = 14:03:53.0 UT1
U2 = 14:10:14.8 UT1
U3 = 17:49:06.5 UT1
U4 = 17:55:24.2 UT1

ΔT = 72.6 s S Eph. = JPL DE405

Circumstances at Greatest Eclipse: 15:59:35.1 UT1

Lat. = 31°18.2'S Sun Alt. = 72.7°
Long. = 048°28.0'W Sun Azm. = 333.5°
Path Width = 281.5 km Duration = 07m50.9s

Circumstances at Greatest Duration: 15:41:47.6 UT1

Lat. = 35°04.2'S Sun Alt. = 70.4°
Long. = 053°22.3'W Sun Azm. = 4.1°
Path Width = 279.7 km Duration = 07m53.5s

0 1000 2000 3000 4000 5000
Kilometers

Total Solar Eclipse of 2027 Aug 02

Greatest Eclipse = 10:07:50.2 TD (= 10:06:37.4 UT1)

Eclipse Magnitude = 1.0790 | Saros Series = 136

Gamma = 0.1421 | Saros Member = 38 of 71

Sun at Greatest Eclipse
(Geocentric Coordinates)
R.A. = 08h49m26.9s
Dec. = +17°45'41.3"
S.D. = 00°15'45.5"
H.P. = 00°00'08.7"

Moon at Greatest Eclipse
(Geocentric Coordinates)
R.A. = 08h49m40.1s
Dec. = +17°53'47.8"
S.D. = 00°16'43.1"
H.P. = 01°01'21.4"

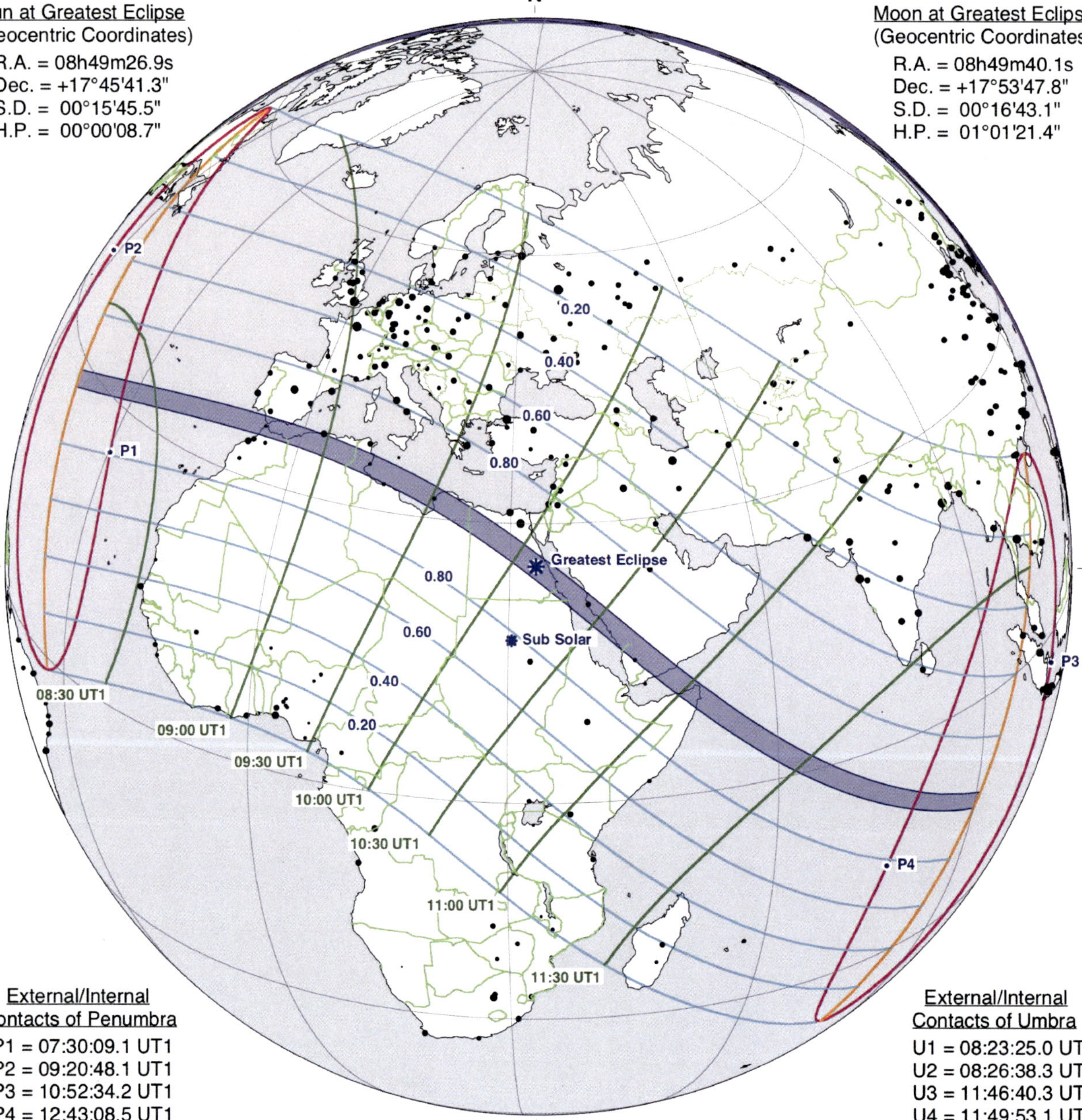

External/Internal Contacts of Penumbra
P1 = 07:30:09.1 UT1
P2 = 09:20:48.1 UT1
P3 = 10:52:34.2 UT1
P4 = 12:43:08.5 UT1

External/Internal Contacts of Umbra
U1 = 08:23:25.0 UT1
U2 = 08:26:38.3 UT1
U3 = 11:46:40.3 UT1
U4 = 11:49:53.1 UT1

ΔT = 72.8 s

Eph. = JPL DE405

Circumstances at Greatest Eclipse: 10:06:37.4 UT1

Lat. = 25°30.2'N | Sun Alt. = 81.7°
Long. = 033°11.0'E | Sun Azm. = 202.0°
Path Width = 257.7 km | Duration = 06m22.6s

Circumstances at Greatest Duration: 10:00:20.9 UT1

Lat. = 26°48.9'N | Sun Alt. = 80.9°
Long. = 031°07.9'E | Sun Azm. = 177.8°
Path Width = 257.2 km | Duration = 06m23.2s

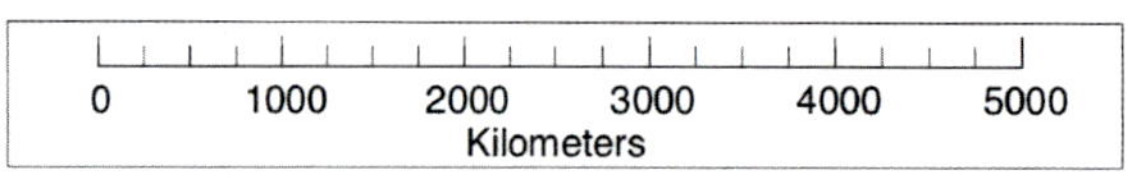

Annular Solar Eclipse of 2028 Jan 26

Greatest Eclipse = 15:08:58.8 TD (= 15:07:45.8 UT1)

Eclipse Magnitude = 0.9208 Saros Series = 141

Gamma = 0.3901 Saros Member = 24 of 70

Sun at Greatest Eclipse
(Geocentric Coordinates)
R.A. = 20h34m14.2s
Dec. = -18°43'33.0"
S.D. = 00°16'14.6"
H.P. = 00°00'08.9"

Moon at Greatest Eclipse
(Geocentric Coordinates)
R.A. = 20h33m43.7s
Dec. = -18°23'46.3"
S.D. = 00°14'45.1"
H.P. = 00°54'08.3"

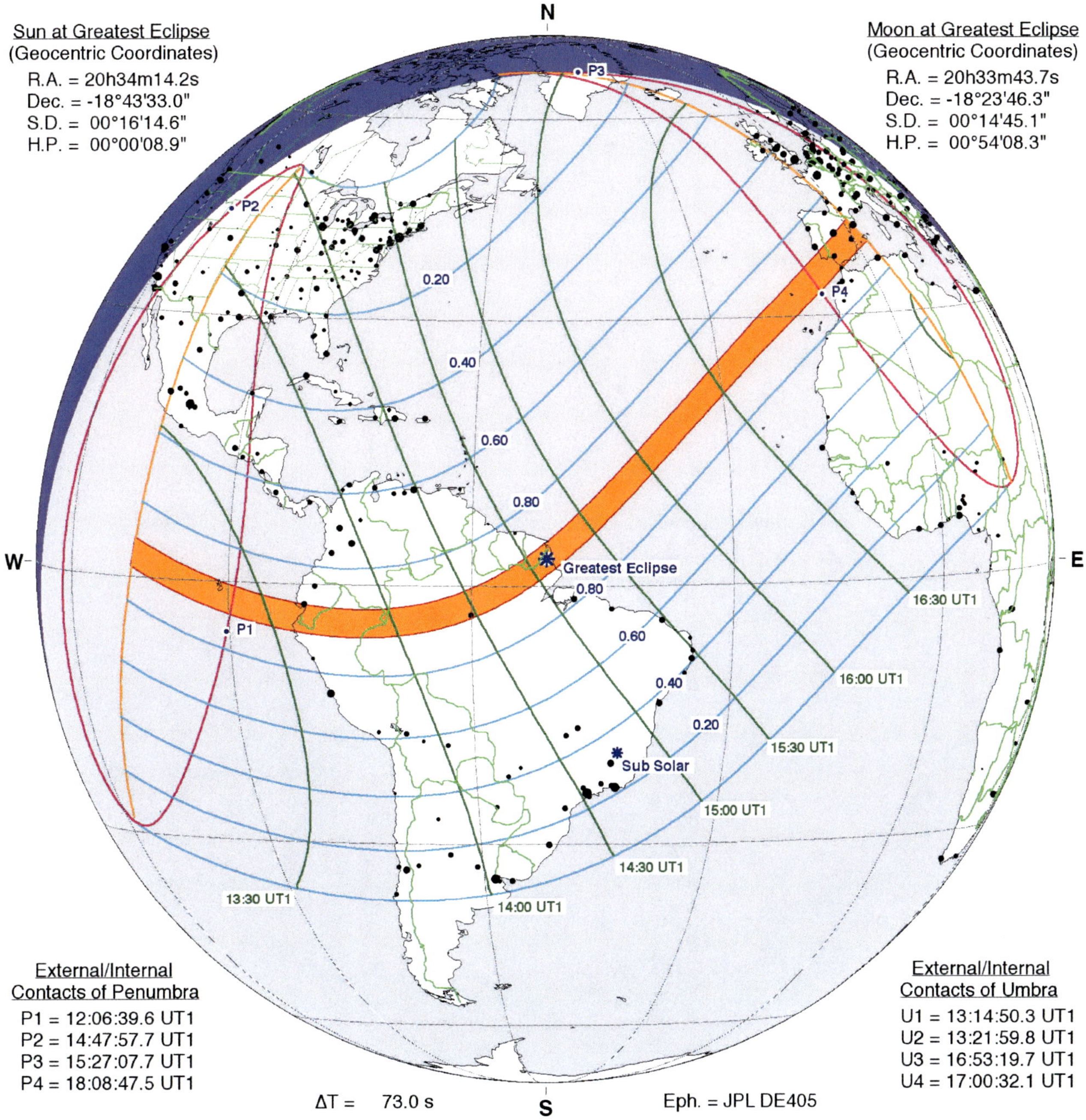

External/Internal Contacts of Penumbra
P1 = 12:06:39.6 UT1
P2 = 14:47:57.7 UT1
P3 = 15:27:07.7 UT1
P4 = 18:08:47.5 UT1

External/Internal Contacts of Umbra
U1 = 13:14:50.3 UT1
U2 = 13:21:59.8 UT1
U3 = 16:53:19.7 UT1
U4 = 17:00:32.1 UT1

ΔT = 73.0 s Eph. = JPL DE405

Circumstances at Greatest Eclipse: 15:07:45.8 UT1

Lat. = 02°57.6'N Sun Alt. = 67.0°
Long. = 051°33.6'W Sun Azm. = 161.0°
Path Width = 323.0 km Duration = 10m27.1s

Circumstances at Greatest Duration: 14:53:07.6 UT1

Lat. = 00°47.6'N Sun Alt. = 65.8°
Long. = 054°39.0'W Sun Azm. = 144.6°
Path Width = 328.7 km Duration = 10m30.6s

0 1000 2000 3000 4000 5000
Kilometers

Total Solar Eclipse of 2028 Jul 22

Greatest Eclipse = 02:56:39.6 TD (= 02:55:26.4 UT1)

Eclipse Magnitude = 1.0560 Saros Series = 146
Gamma = -0.6056 Saros Member = 28 of 76

Sun at Greatest Eclipse
(Geocentric Coordinates)
R.A. = 08h08m03.8s
Dec. = +20°10'53.0"
S.D. = 00°15'44.5"
H.P. = 00°00'08.7"

Moon at Greatest Eclipse
(Geocentric Coordinates)
R.A. = 08h07m16.7s
Dec. = +19°36'14.4"
S.D. = 00°16'24.3"
H.P. = 01°00'12.3"

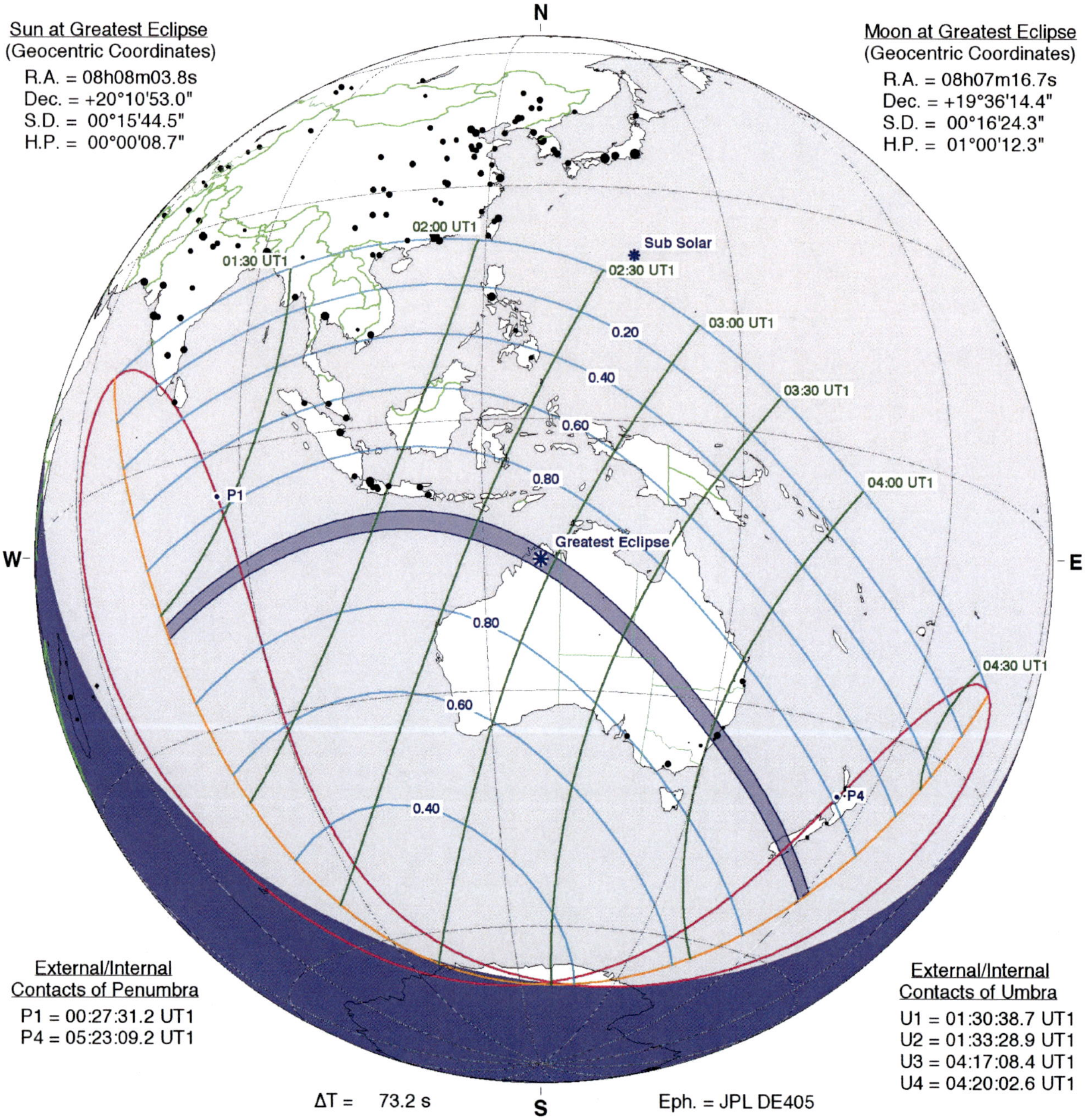

External/Internal
Contacts of Penumbra
P1 = 00:27:31.2 UT1
P4 = 05:23:09.2 UT1

External/Internal
Contacts of Umbra
U1 = 01:30:38.7 UT1
U2 = 01:33:28.9 UT1
U3 = 04:17:08.4 UT1
U4 = 04:20:02.6 UT1

ΔT = 73.2 s Eph. = JPL DE405

Circumstances at Greatest Eclipse: 02:55:26.4 UT1

Lat. = 15°34.7'S	Sun Alt. = 52.6°
Long. = 126°42.4'E	Sun Azm. = 17.2°
Path Width = 230.2 km	Duration = 05m09.7s

Circumstances at Greatest Duration: 02:52:17.7 UT1

Lat. = 15°04.6'S	Sun Alt. = 52.6°
Long. = 125°47.3'E	Sun Azm. = 19.9°
Path Width = 231.4 km	Duration = 05m09.9s

0 1000 2000 3000 4000 5000
Kilometers

Partial Solar Eclipse of 2029 Jan 14

Greatest Eclipse = 17:13:47.5 TD (= 17:12:34.1 UT1)

Eclipse Magnitude = 0.8714
Gamma = 1.0553

Saros Series = 151
Saros Member = 15 of 72

Sun at Greatest Eclipse
(Geocentric Coordinates)
R.A. = 19h47m03.1s
Dec. = -21°09'31.8"
S.D. = 00°16'15.6"
H.P. = 00°00'08.9"

Moon at Greatest Eclipse
(Geocentric Coordinates)
R.A. = 19h45m53.5s
Dec. = -20°12'32.3"
S.D. = 00°15'20.6"
H.P. = 00°56'18.7"

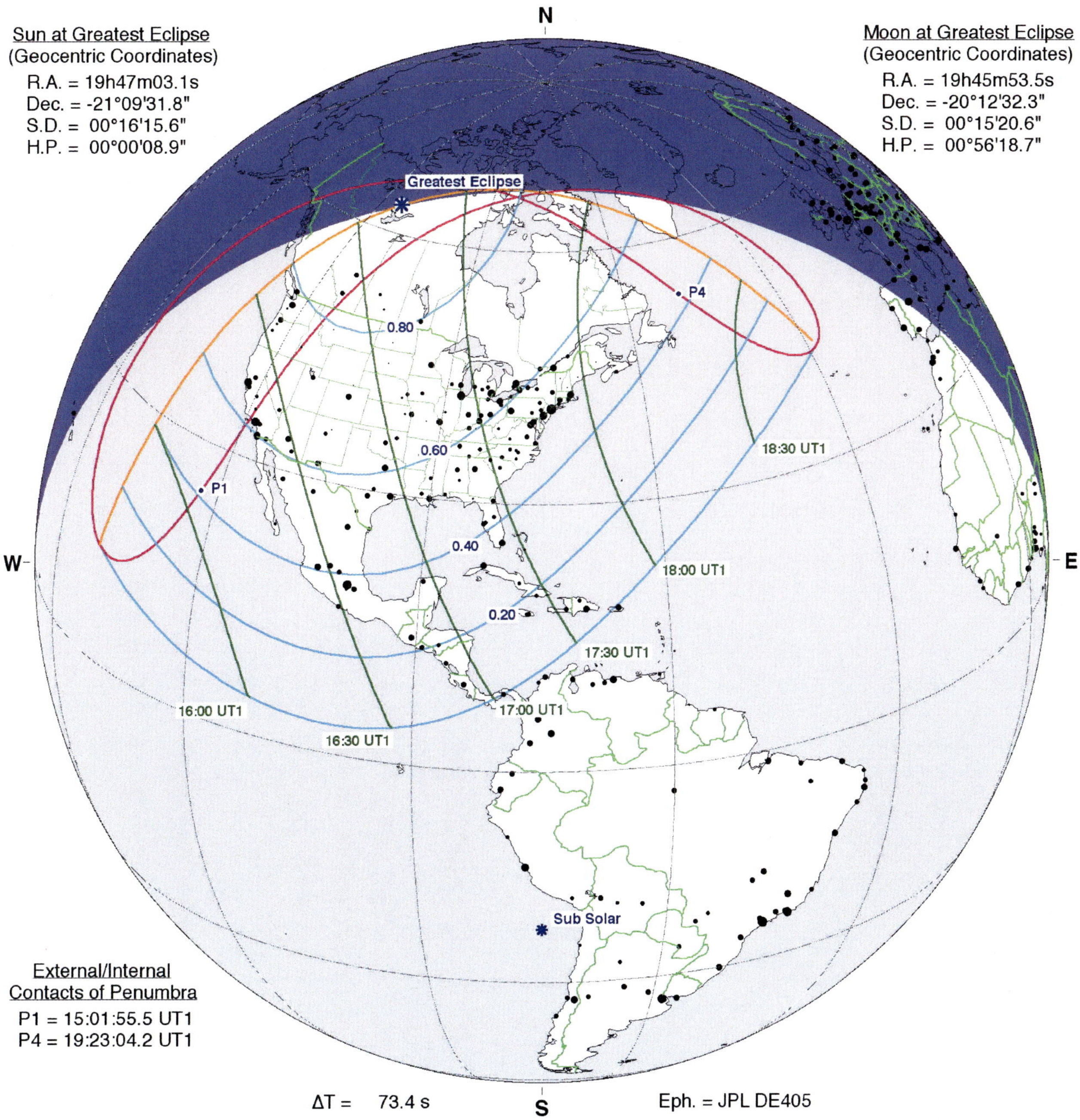

External/Internal Contacts of Penumbra
P1 = 15:01:55.5 UT1
P4 = 19:23:04.2 UT1

ΔT = 73.4 s

Eph. = JPL DE405

Circumstances at Greatest Eclipse: 17:12:34.1 UT1

Lat. = 63°42.1'N
Long. = 114°13.5'W

Sun Alt. = 0.0°
Sun Azm. = 144.6°

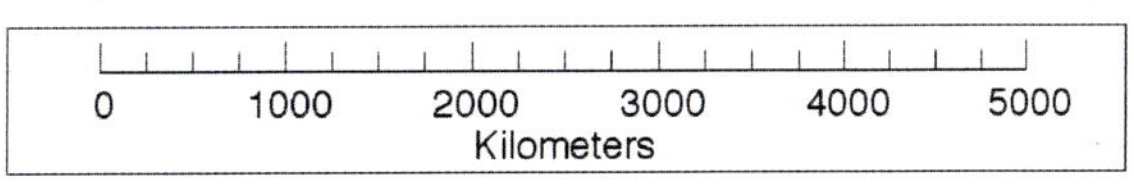

Partial Solar Eclipse of 2029 Jun 12

Greatest Eclipse = 04:06:13.0 TD (= 04:04:59.5 UT1)

Eclipse Magnitude = 0.4576 Saros Series = 118

Gamma = 1.2943 Saros Member = 69 of 72

Sun at Greatest Eclipse
(Geocentric Coordinates)
R.A. = 05h22m58.2s
Dec. = +23°09'45.7"
S.D. = 00°15'45.0"
H.P. = 00°00'08.7"

Moon at Greatest Eclipse
(Geocentric Coordinates)
R.A. = 05h23m08.9s
Dec. = +24°21'37.7"
S.D. = 00°15'10.6"
H.P. = 00°55'42.0"

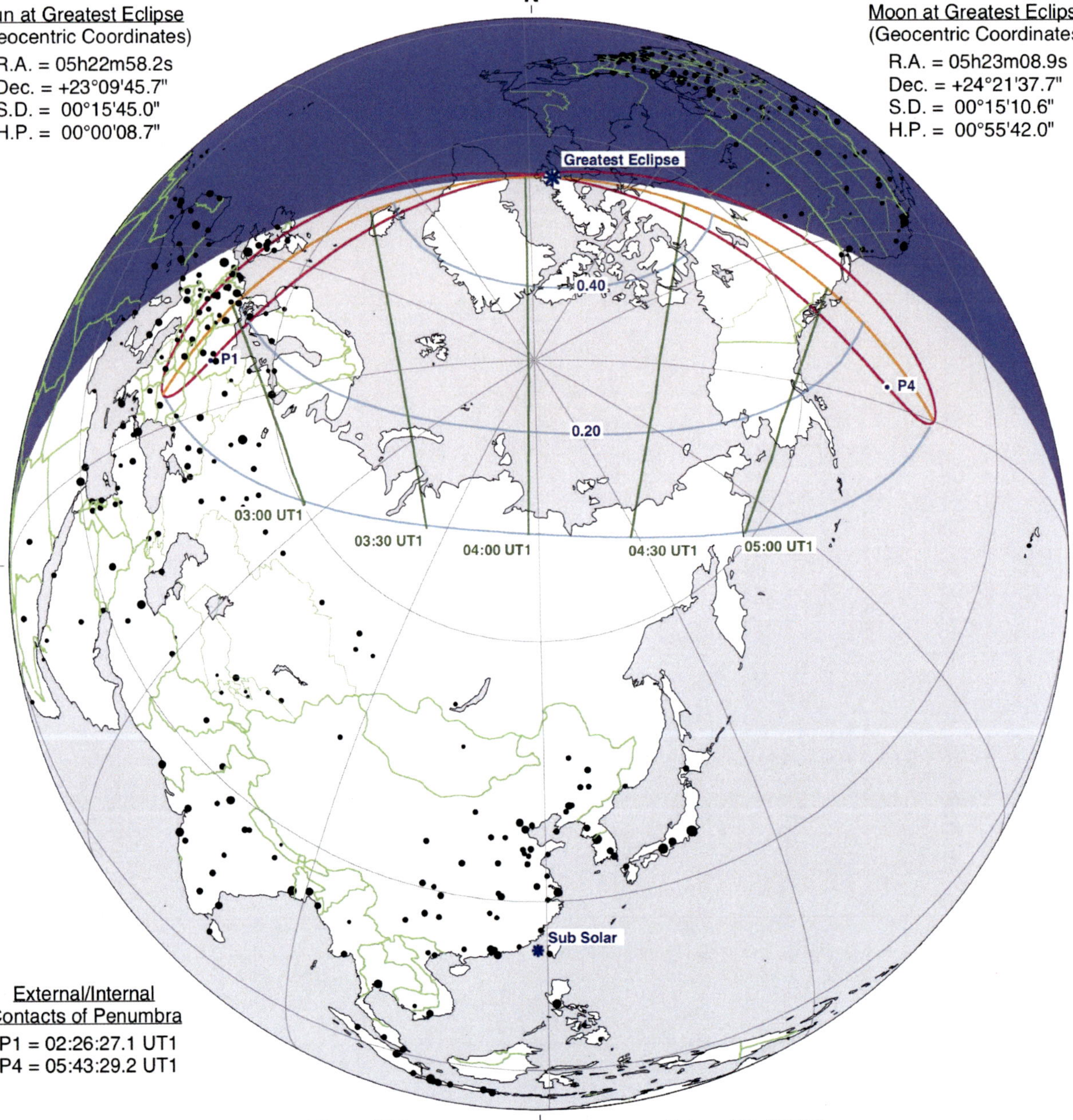

External/Internal
Contacts of Penumbra
P1 = 02:26:27.1 UT1
P4 = 05:43:29.2 UT1

ΔT = 73.6 s Eph. = JPL DE405

Circumstances at Greatest Eclipse: 04:04:59.5 UT1

Lat. = 66°45.9'N Sun Alt. = 0.0°

Long. = 066°11.0'W Sun Azm. = 355.5°

0 1000 2000 3000 4000 5000
Kilometers

Partial Solar Eclipse of 2029 Jul 11

Greatest Eclipse = 15:37:18.9 TD (= 15:36:05.3 UT1)

Eclipse Magnitude = 0.2303
Gamma = -1.4191

Saros Series = 156
Saros Member = 2 of 69

Sun at Greatest Eclipse
(Geocentric Coordinates)
R.A. = 07h24m55.6s
Dec. = +22°00'04.3"
S.D. = 00°15'43.9"
H.P. = 00°00'08.7"

Moon at Greatest Eclipse
(Geocentric Coordinates)
R.A. = 07h23m33.7s
Dec. = +20°41'22.0"
S.D. = 00°15'35.3"
H.P. = 00°57'12.6"

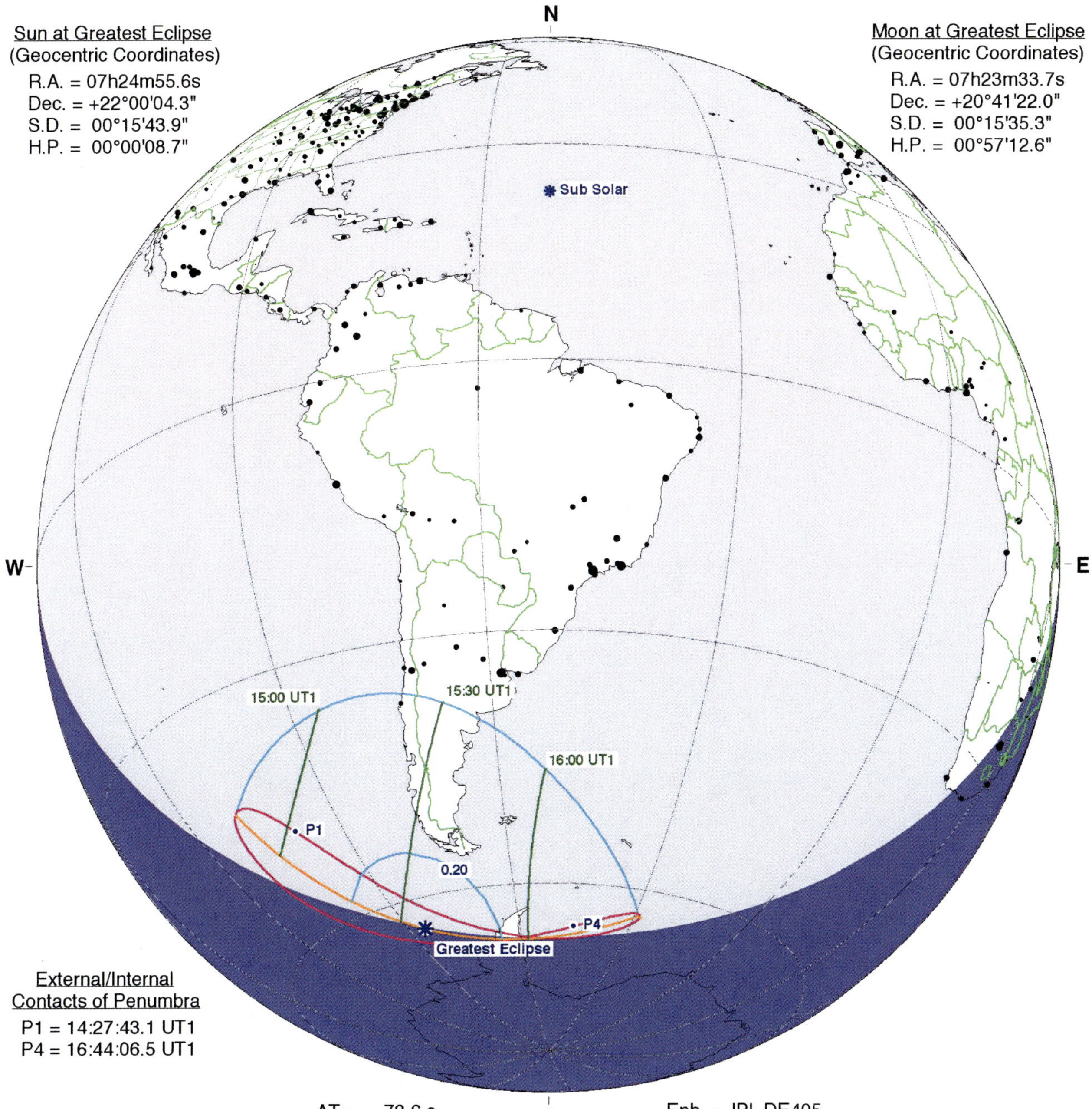

External/Internal
Contacts of Penumbra
P1 = 14:27:43.1 UT1
P4 = 16:44:06.5 UT1

ΔT = 73.6 s

Eph. = JPL DE405

Circumstances at Greatest Eclipse: 15:36:05.3 UT1

Lat. = 64°15.8'S
Long. = 085°37.4'W

Sun Alt. = 0.0°
Sun Azm. = 30.4°

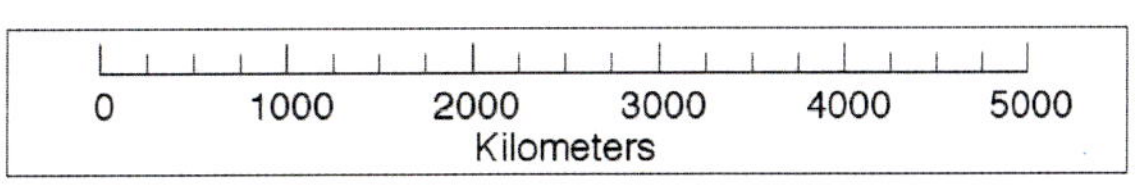

Partial Solar Eclipse of 2029 Dec 05

Greatest Eclipse = 15:03:58.0 TD (= 15:02:44.2 UT1)

Eclipse Magnitude = 0.8911 Saros Series = 123

Gamma = -1.0609 Saros Member = 54 of 70

Sun at Greatest Eclipse
(Geocentric Coordinates)

R.A. = 16h49m34.2s
Dec. = -22°26'54.3"
S.D. = 00°16'13.8"
H.P. = 00°00'08.9"

Moon at Greatest Eclipse
(Geocentric Coordinates)

R.A. = 16h49m27.4s
Dec. = -23°31'15.0"
S.D. = 00°16'34.3"
H.P. = 01°00'49.1"

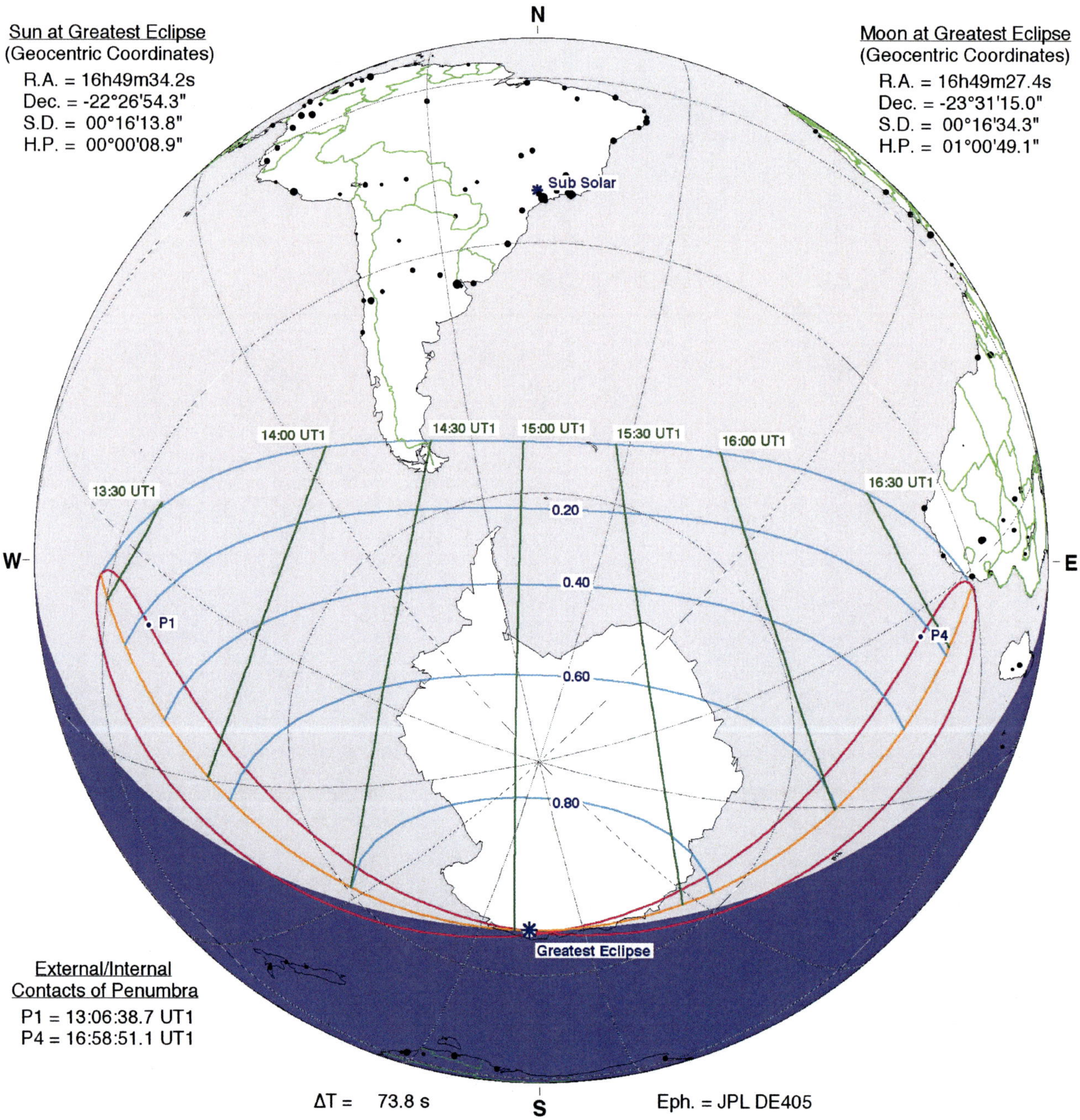

External/Internal
Contacts of Penumbra

P1 = 13:06:38.7 UT1
P4 = 16:58:51.1 UT1

ΔT = 73.8 s Eph. = JPL DE405

Circumstances at Greatest Eclipse: 15:02:44.2 UT1

Lat. = 67°30.8'S Sun Alt. = 0.0°

Long. = 135°38.7'E Sun Azm. = 176.6°

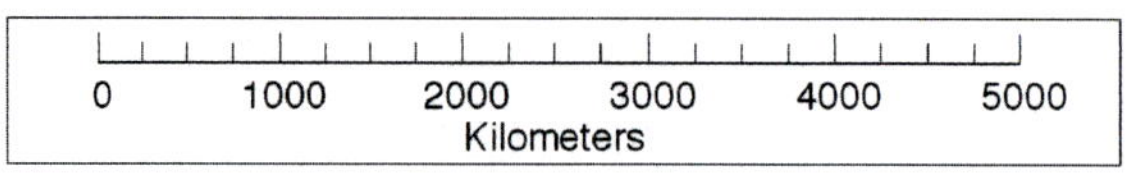

Annular Solar Eclipse of 2030 Jun 01

Greatest Eclipse = 06:29:12.9 TD (= 06:27:58.8 UT1)

Eclipse Magnitude = 0.9443 Saros Series = 128

Gamma = 0.5626 Saros Member = 59 of 73

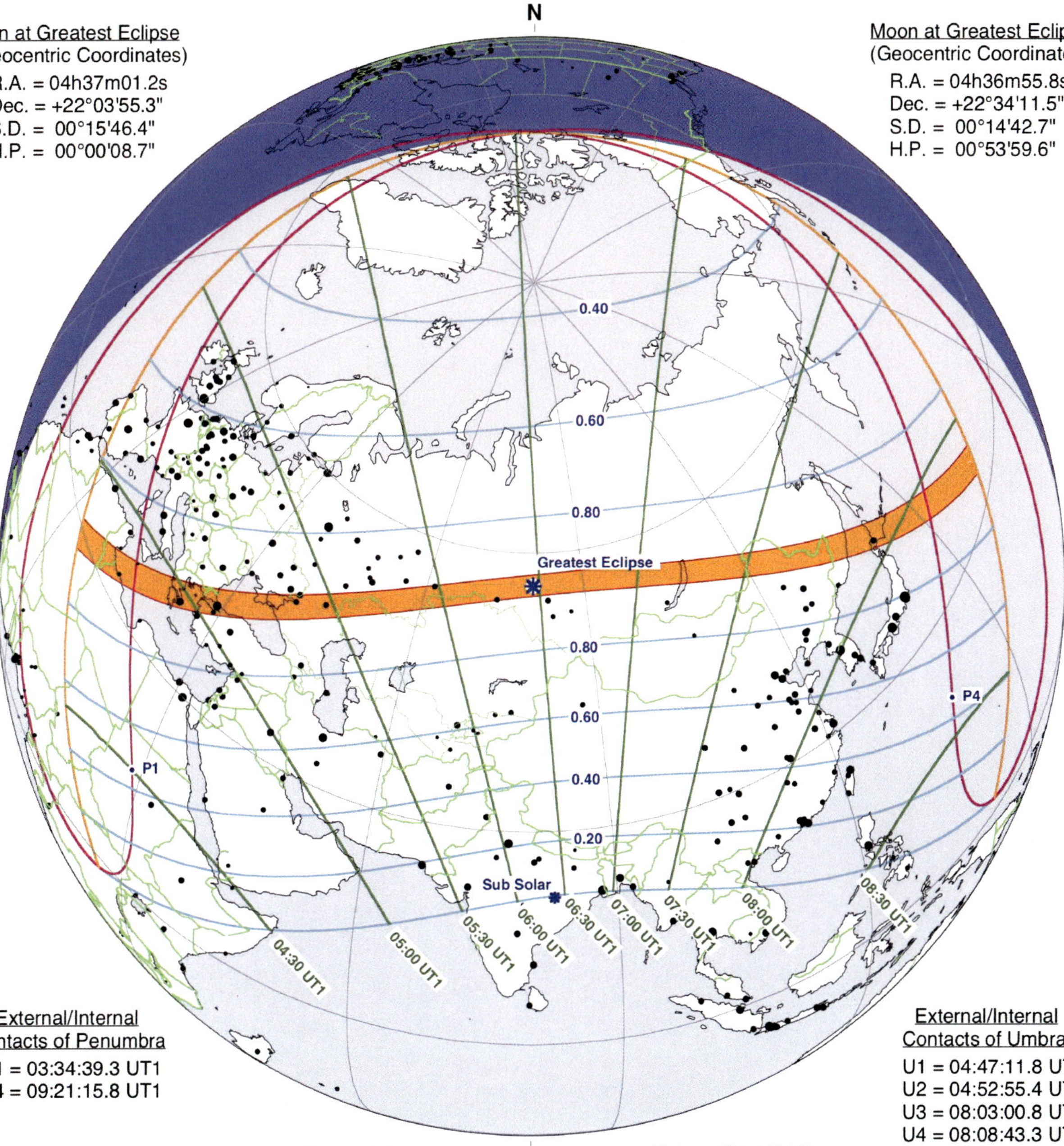

Sun at Greatest Eclipse
(Geocentric Coordinates)
R.A. = 04h37m01.2s
Dec. = +22°03'55.3"
S.D. = 00°15'46.4"
H.P. = 00°00'08.7"

Moon at Greatest Eclipse
(Geocentric Coordinates)
R.A. = 04h36m55.8s
Dec. = +22°34'11.5"
S.D. = 00°14'42.7"
H.P. = 00°53'59.6"

External/Internal
Contacts of Penumbra
P1 = 03:34:39.3 UT1
P4 = 09:21:15.8 UT1

External/Internal
Contacts of Umbra
U1 = 04:47:11.8 UT1
U2 = 04:52:55.4 UT1
U3 = 08:03:00.8 UT1
U4 = 08:08:43.3 UT1

ΔT = 74.0 s Eph. = JPL DE405

Circumstances at Greatest Eclipse: 06:27:58.8 UT1

Lat. = 56°31.5'N Sun Alt. = 55.5°
Long. = 080°03.7'E Sun Azm. = 176.1°
Path Width = 249.6 km Duration = 05m20.8s

Circumstances at Greatest Duration: 06:28:41.2 UT1

Lat. = 56°32.9'N Sun Alt. = 55.5°
Long. = 080°30.1'E Sun Azm. = 177.1°
Path Width = 249.6 km Duration = 05m20.8s

0 1000 2000 3000 4000 5000
Kilometers

Total Solar Eclipse of 2030 Nov 25

Greatest Eclipse = 06:51:36.9 TD (= 06:50:22.7 UT1)

Eclipse Magnitude = 1.0468 Saros Series = 133

Gamma = -0.3867 Saros Member = 46 of 72

Sun at Greatest Eclipse
(Geocentric Coordinates)

R.A. = 16h03m58.7s
Dec. = -20°45'39.0"
S.D. = 00°16'12.1"
H.P. = 00°00'08.9"

Moon at Greatest Eclipse
(Geocentric Coordinates)

R.A. = 16h03m49.1s
Dec. = -21°09'10.6"
S.D. = 00°16'41.7"
H.P. = 01°01'16.4"

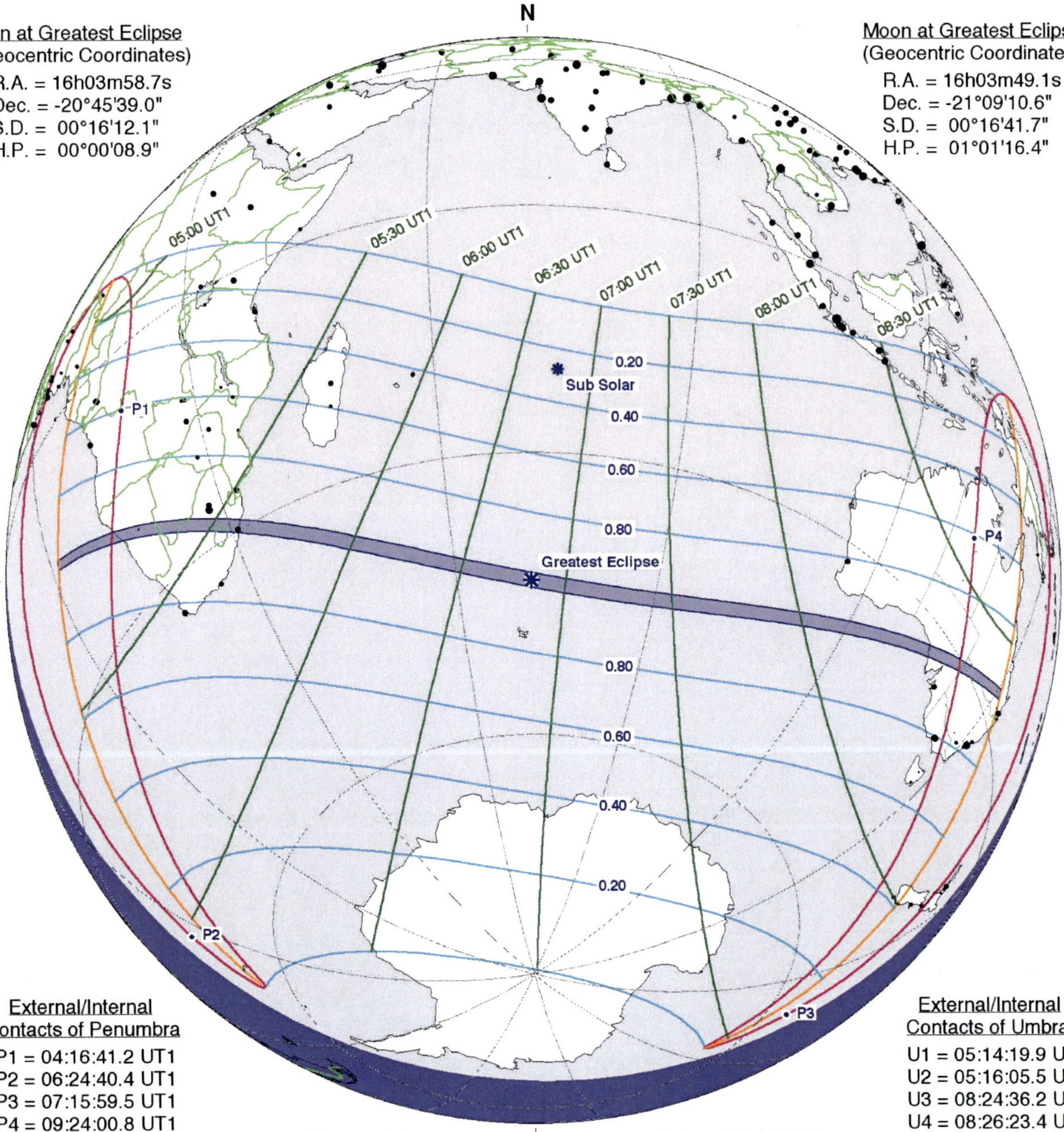

External/Internal Contacts of Penumbra

P1 = 04:16:41.2 UT1
P2 = 06:24:40.4 UT1
P3 = 07:15:59.5 UT1
P4 = 09:24:00.8 UT1

External/Internal Contacts of Umbra

U1 = 05:14:19.9 UT1
U2 = 05:16:05.5 UT1
U3 = 08:24:36.2 UT1
U4 = 08:26:23.4 UT1

ΔT = 74.2 s Eph. = JPL DE405

Circumstances at Greatest Eclipse: 06:50:22.7 UT1		Circumstances at Greatest Duration: 06:51:56.0 UT1	
Lat. = 43°36.6'S	Sun Alt. = 67.0°	Lat. = 43°42.7'S	Sun Alt. = 67.0°
Long. = 071°13.7'E	Sun Azm. = 7.0°	Long. = 072°01.4'E	Sun Azm. = 4.1°
Path Width = 169.3 km	Duration = 03m43.5s	Path Width = 169.2 km	Duration = 03m43.6s

0 1000 2000 3000 4000 5000
Kilometers

Annular Solar Eclipse of 2031 May 21

Greatest Eclipse = 07:16:04.3 TD (= 07:14:49.9 UT1)

Eclipse Magnitude = 0.9589 Saros Series = 138

Gamma = -0.1970 Saros Member = 32 of 70

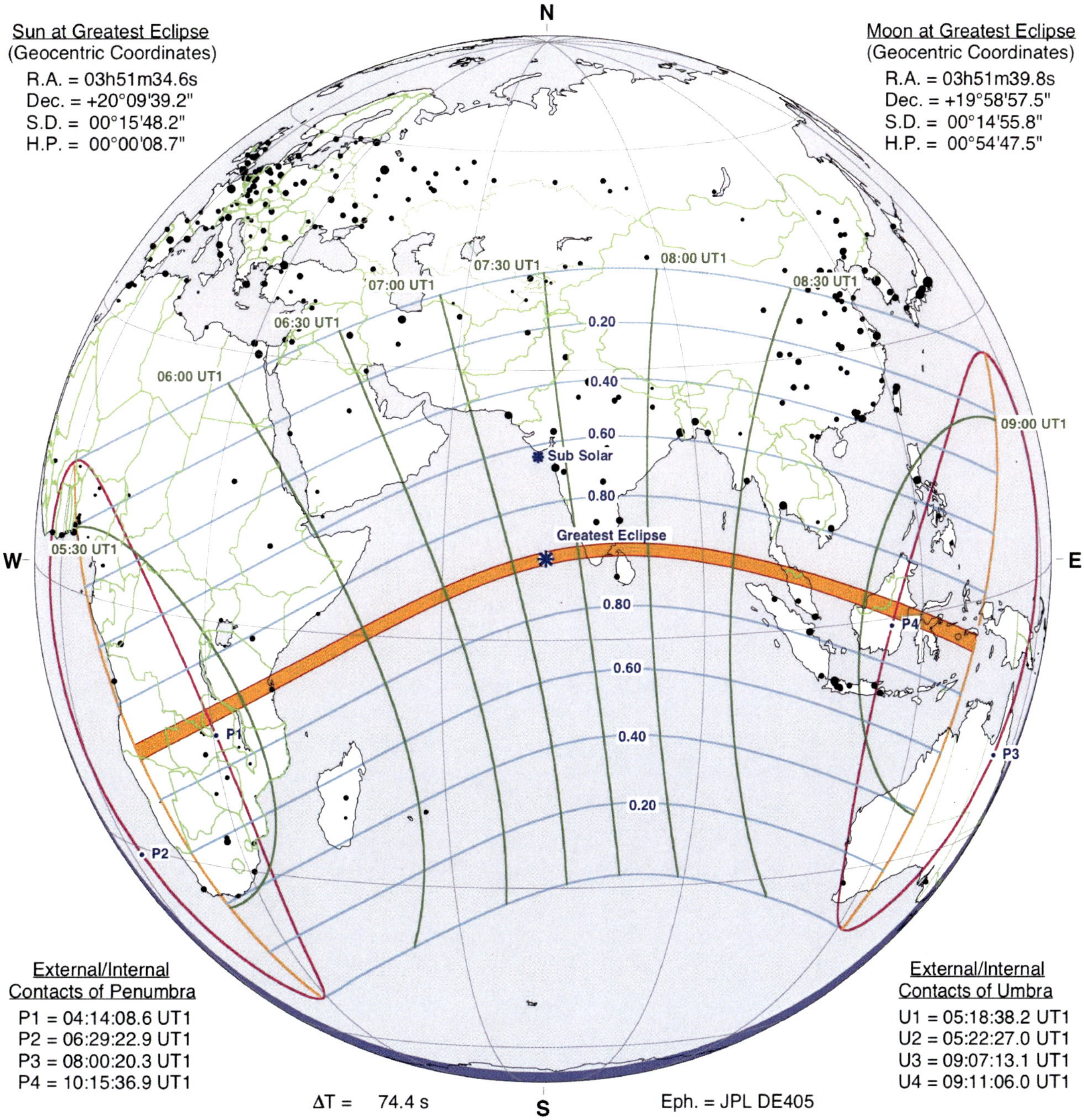

Sun at Greatest Eclipse
(Geocentric Coordinates)
R.A. = 03h51m34.6s
Dec. = +20°09'39.2"
S.D. = 00°15'48.2"
H.P. = 00°00'08.7"

Moon at Greatest Eclipse
(Geocentric Coordinates)
R.A. = 03h51m39.8s
Dec. = +19°58'57.5"
S.D. = 00°14'55.8"
H.P. = 00°54'47.5"

External/Internal Contacts of Penumbra
P1 = 04:14:08.6 UT1
P2 = 06:29:22.9 UT1
P3 = 08:00:20.3 UT1
P4 = 10:15:36.9 UT1

External/Internal Contacts of Umbra
U1 = 05:18:38.2 UT1
U2 = 05:22:27.0 UT1
U3 = 09:07:13.1 UT1
U4 = 09:11:06.0 UT1

ΔT = 74.4 s

Eph. = JPL DE405

Circumstances at Greatest Eclipse: 07:14:49.9 UT1

Lat. = 08°55.6'N Sun Alt. = 78.7°
Long. = 071°43.6'E Sun Azm. = 353.8°
Path Width = 152.2 km Duration = 05m25.5s

Circumstances at Greatest Duration: 07:23:14.4 UT1

Lat. = 09°19.1'N Sun Alt. = 77.9°
Long. = 073°47.6'E Sun Azm. = 334.7°
Path Width = 152.7 km Duration = 05m26.1s

0	1000	2000	3000	4000	5000

Kilometers

Hybrid Solar Eclipse of 2031 Nov 14

Greatest Eclipse = 21:07:30.7 TD (= 21:06:16.0 UT1)

Eclipse Magnitude = 1.0106
Gamma = 0.3078

Saros Series = 143
Saros Member = 24 of 72

Sun at Greatest Eclipse
(Geocentric Coordinates)
R.A. = 15h19m31.2s
Dec. = -18°20'14.5"
S.D. = 00°16'09.9"
H.P. = 00°00'08.9"

Moon at Greatest Eclipse
(Geocentric Coordinates)
R.A. = 15h19m43.3s
Dec. = -18°02'21.3"
S.D. = 00°16'05.0"
H.P. = 00°59'01.4"

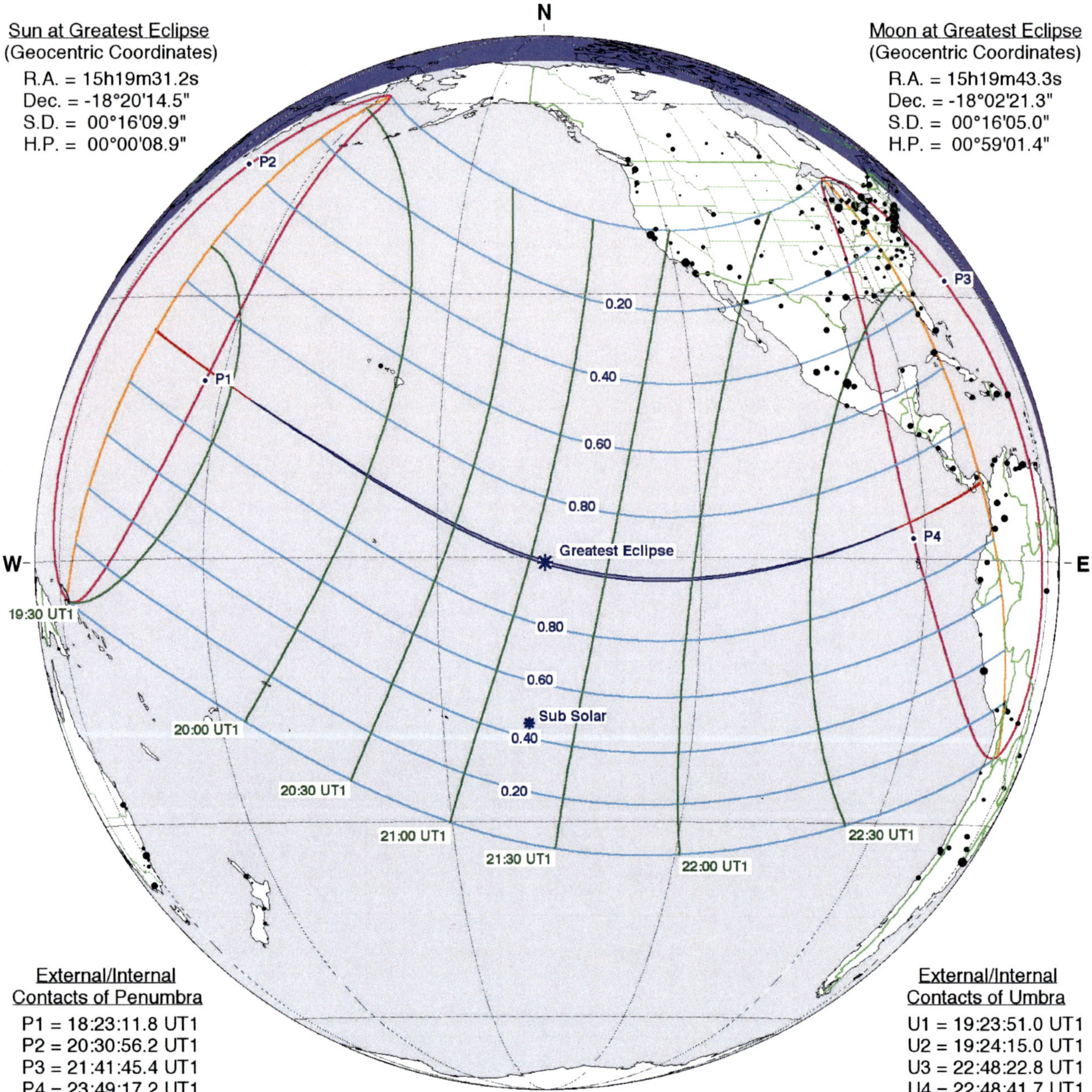

External/Internal
Contacts of Penumbra
P1 = 18:23:11.8 UT1
P2 = 20:30:56.2 UT1
P3 = 21:41:45.4 UT1
P4 = 23:49:17.2 UT1

External/Internal
Contacts of Umbra
U1 = 19:23:51.0 UT1
U2 = 19:24:15.0 UT1
U3 = 22:48:22.8 UT1
U4 = 22:48:41.7 UT1

ΔT = 74.7 s S Eph. = JPL DE405

Circumstances at Greatest Eclipse: 21:06:16.0 UT1

Lat. = 00°37.9'S	Sun Alt. = 72.1°
Long. = 137°38.7'W	Sun Azm. = 188.7°
Path Width = 38.3 km	Duration = 01m08.3s

Circumstances at Greatest Duration: 21:10:29.2 UT1

Lat. = 00°57.7'S	Sun Alt. = 71.9°
Long. = 136°28.4'W	Sun Azm. = 195.6°
Path Width = 38.4 km	Duration = 01m08.4s

0 1000 2000 3000 4000 5000
Kilometers

Annular Solar Eclipse of 2032 May 09

Greatest Eclipse = 13:26:42.4 TD (= 13:25:27.5 UT1)

Eclipse Magnitude = 0.9957 Saros Series = 148

Gamma = -0.9375 Saros Member = 22 of 75

Sun at Greatest Eclipse
(Geocentric Coordinates)

R.A. = 03h08m06.7s
Dec. = +17°35'43.5"
S.D. = 00°15'50.4"
H.P. = 00°00'08.7"

Moon at Greatest Eclipse
(Geocentric Coordinates)

R.A. = 03h08m46.1s
Dec. = +16°42'42.0"
S.D. = 00°15'41.5"
H.P. = 00°57'35.4"

N

W

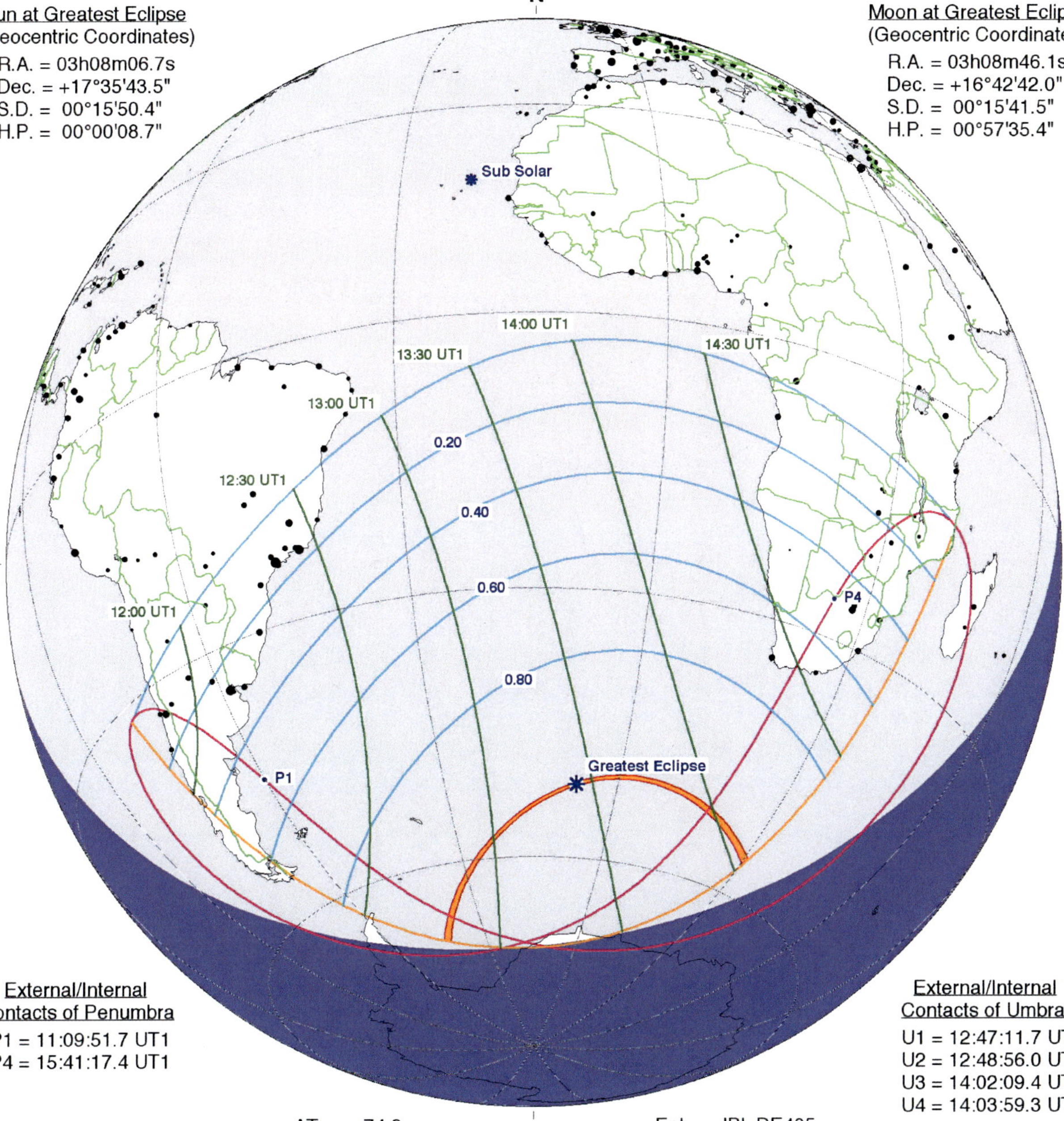

E

External/Internal Contacts of Penumbra

P1 = 11:09:51.7 UT1
P4 = 15:41:17.4 UT1

External/Internal Contacts of Umbra

U1 = 12:47:11.7 UT1
U2 = 12:48:56.0 UT1
U3 = 14:02:09.4 UT1
U4 = 14:03:59.3 UT1

ΔT = 74.9 s

S

Eph. = JPL DE405

Circumstances at Greatest Eclipse: 13:25:27.5 UT1

Lat. =	51°15.6'S	Sun Alt. =	19.9°
Long. =	007°04.4'W	Sun Azm. =	344.6°
Path Width =	43.7 km	Duration =	00m22.0s

Circumstances at Greatest Duration: 12:48:03.3 UT1

Lat. =	69°57.7'S	Sun Alt. =	0.0°
Long. =	042°31.9'W	Sun Azm. =	28.1°
Path Width =	86.7 km	Duration =	00m40.8s

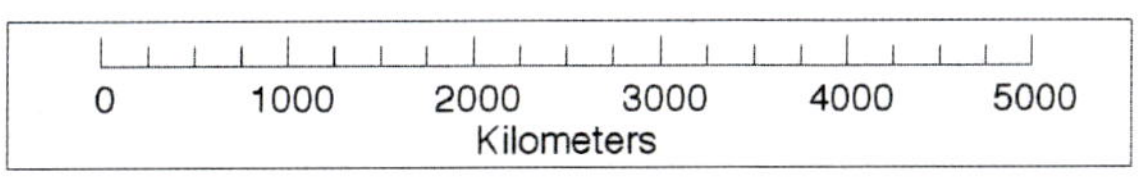

Partial Solar Eclipse of 2032 Nov 03

Greatest Eclipse = 05:34:12.9 TD (= 05:32:57.8 UT1)

Eclipse Magnitude = 0.8554 Saros Series = 153

Gamma = 1.0643 Saros Member = 10 of 70

Sun at Greatest Eclipse
(Geocentric Coordinates)
R.A. = 14h35m40.9s
Dec. = -15°13'54.9"
S.D. = 00°16'07.4"
H.P. = 00°00'08.9"

Moon at Greatest Eclipse
(Geocentric Coordinates)
R.A. = 14h36m33.6s
Dec. = -14°16'01.1"
S.D. = 00°15'13.0"
H.P. = 00°55'50.8"

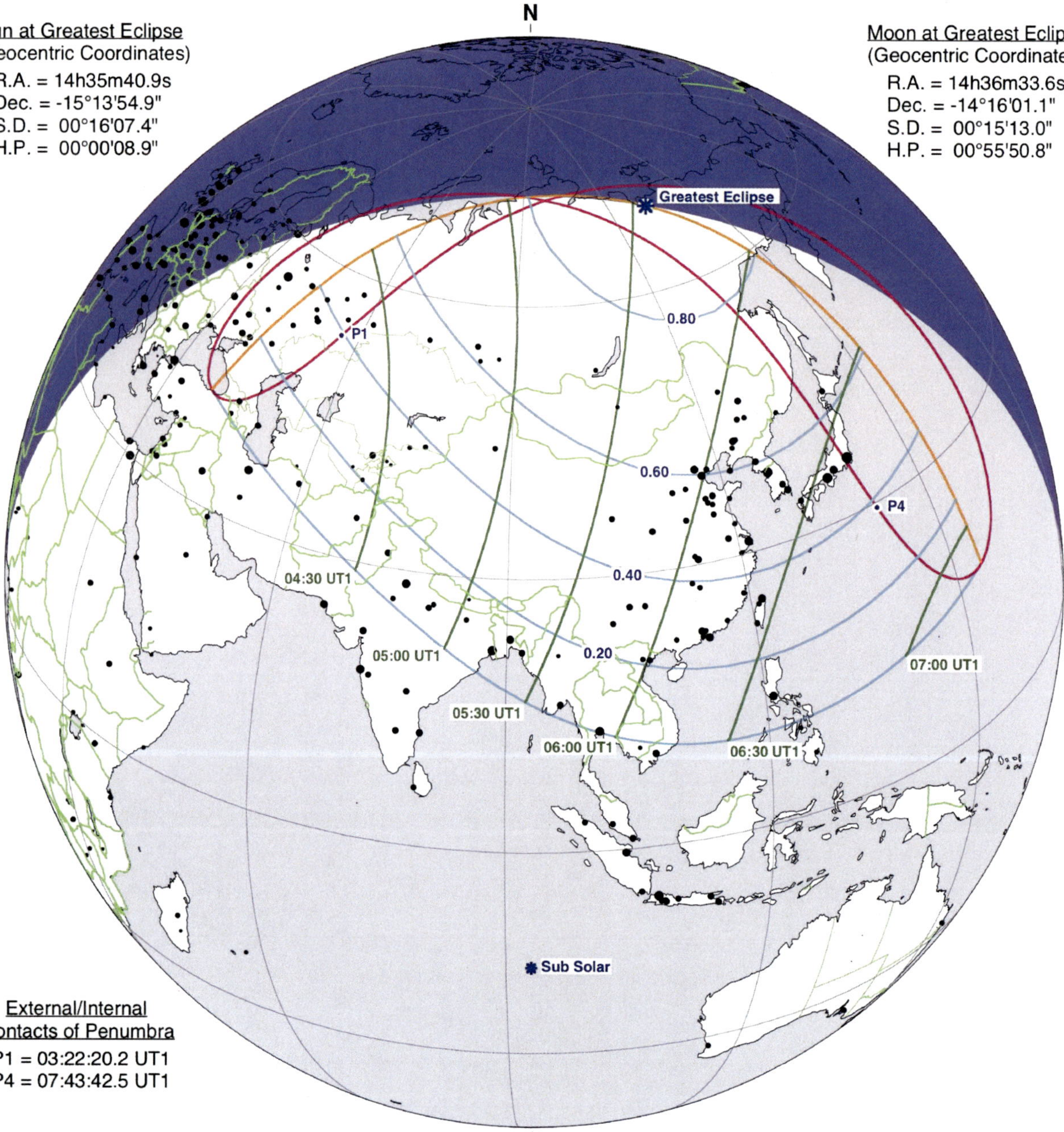

External/Internal
Contacts of Penumbra
P1 = 03:22:20.2 UT1
P4 = 07:43:42.5 UT1

ΔT = 75.1 s Eph. = JPL DE405

Circumstances at Greatest Eclipse: 05:32:57.8 UT1

Lat. = 70°25.7'N Sun Alt. = 0.0°

Long. = 132°36.5'E Sun Azm. = 218.3°

0 1000 2000 3000 4000 5000
Kilometers

Total Solar Eclipse of 2033 Mar 30

Greatest Eclipse = 18:02:35.7 TD (= 18:01:20.4 UT1)

Eclipse Magnitude = 1.0462 | Saros Series = 120
Gamma = 0.9778 | Saros Member = 62 of 71

Sun at Greatest Eclipse
(Geocentric Coordinates)
R.A. = 00h38m02.8s
Dec. = +04°05'47.8"
S.D. = 00°16'00.8"
H.P. = 00°00'08.8"

Moon at Greatest Eclipse
(Geocentric Coordinates)
R.A. = 00h36m50.4s
Dec. = +05°02'48.6"
S.D. = 00°16'42.2"
H.P. = 01°01'18.3"

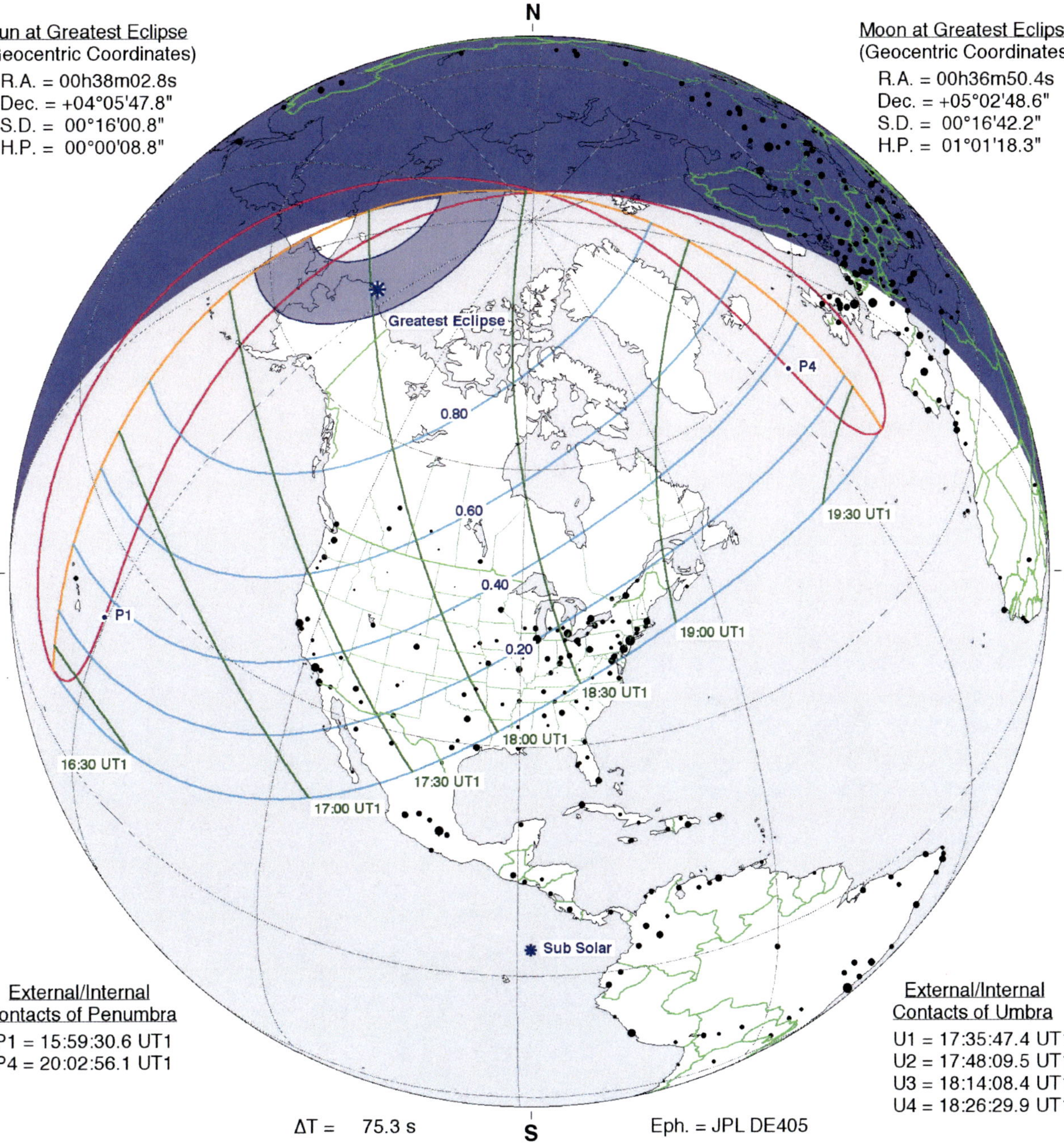

External/Internal Contacts of Penumbra
P1 = 15:59:30.6 UT1
P4 = 20:02:56.1 UT1

External/Internal Contacts of Umbra
U1 = 17:35:47.4 UT1
U2 = 17:48:09.5 UT1
U3 = 18:14:08.4 UT1
U4 = 18:26:29.9 UT1

ΔT = 75.3 s

Eph. = JPL DE405

Circumstances at Greatest Eclipse: 18:01:20.4 UT1

Lat. = 71°18.9'N	Sun Alt. = 11.2°
Long. = 155°50.1'W	Sun Azm. = 111.1°
Path Width = 781.2 km	Duration = 02m37.0s

Circumstances at Greatest Duration: 18:01:04.2 UT1

Lat. = 71°10.5'N	Sun Alt. = 11.2°
Long. = 155°55.9'W	Sun Azm. = 110.9°
Path Width = 782.6 km	Duration = 02m37.0s

0 1000 2000 3000 4000 5000
Kilometers

Partial Solar Eclipse of 2033 Sep 23

Greatest Eclipse = 13:54:31.2 TD (= 13:53:15.7 UT1)

Eclipse Magnitude = 0.6890
Gamma = -1.1583

Saros Series = 125
Saros Member = 55 of 73

Sun at Greatest Eclipse
(Geocentric Coordinates)
R.A. = 12h03m08.9s
Dec. = -00°20'27.7"
S.D. = 00°15'56.3"
H.P. = 00°00'08.8"

Moon at Greatest Eclipse
(Geocentric Coordinates)
R.A. = 12h01m52.5s
Dec. = -01°19'54.7"
S.D. = 00°14'43.6"
H.P. = 00°54'03.0"

N

W

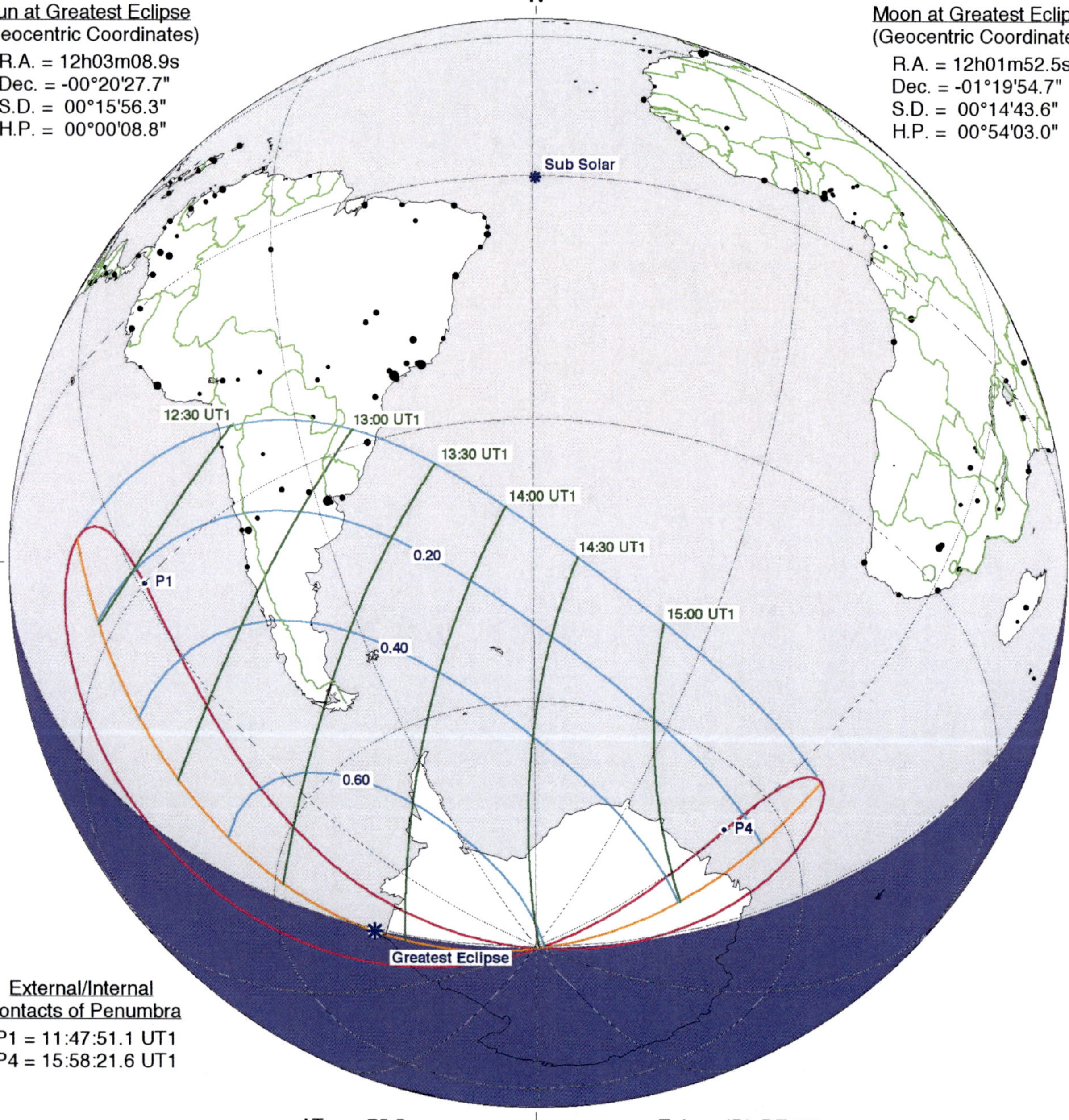

E

S

External/Internal
Contacts of Penumbra
P1 = 11:47:51.1 UT1
P4 = 15:58:21.6 UT1

ΔT = 75.5 s

Eph. = JPL DE405

Circumstances at Greatest Eclipse: 13:53:15.7 UT1

Lat. = 72°12.2'S
Long. = 121°15.2'W

Sun Alt. = 0.0°
Sun Azm. = 91.1°

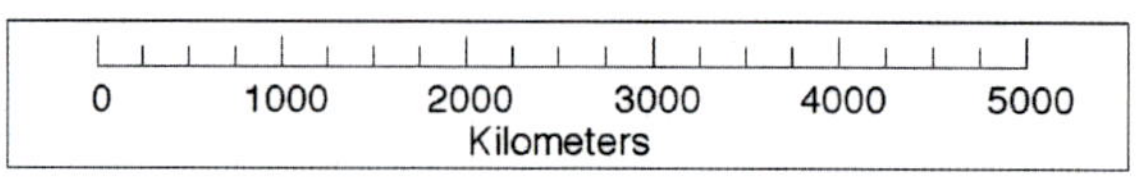

Total Solar Eclipse of 2034 Mar 20

Greatest Eclipse = 10:18:45.2 TD (= 10:17:29.5 UT1)

Eclipse Magnitude = 1.0458 Saros Series = 130

Gamma = 0.2894 Saros Member = 53 of 73

Sun at Greatest Eclipse
(Geocentric Coordinates)
R.A. = 23h59m32.7s
Dec. = -00°02'58.0"
S.D. = 00°16'03.7"
H.P. = 00°00'08.8"

Moon at Greatest Eclipse
(Geocentric Coordinates)
R.A. = 23h59m11.3s
Dec. = +00°13'42.6"
S.D. = 00°16'31.6"
H.P. = 01°00'39.3"

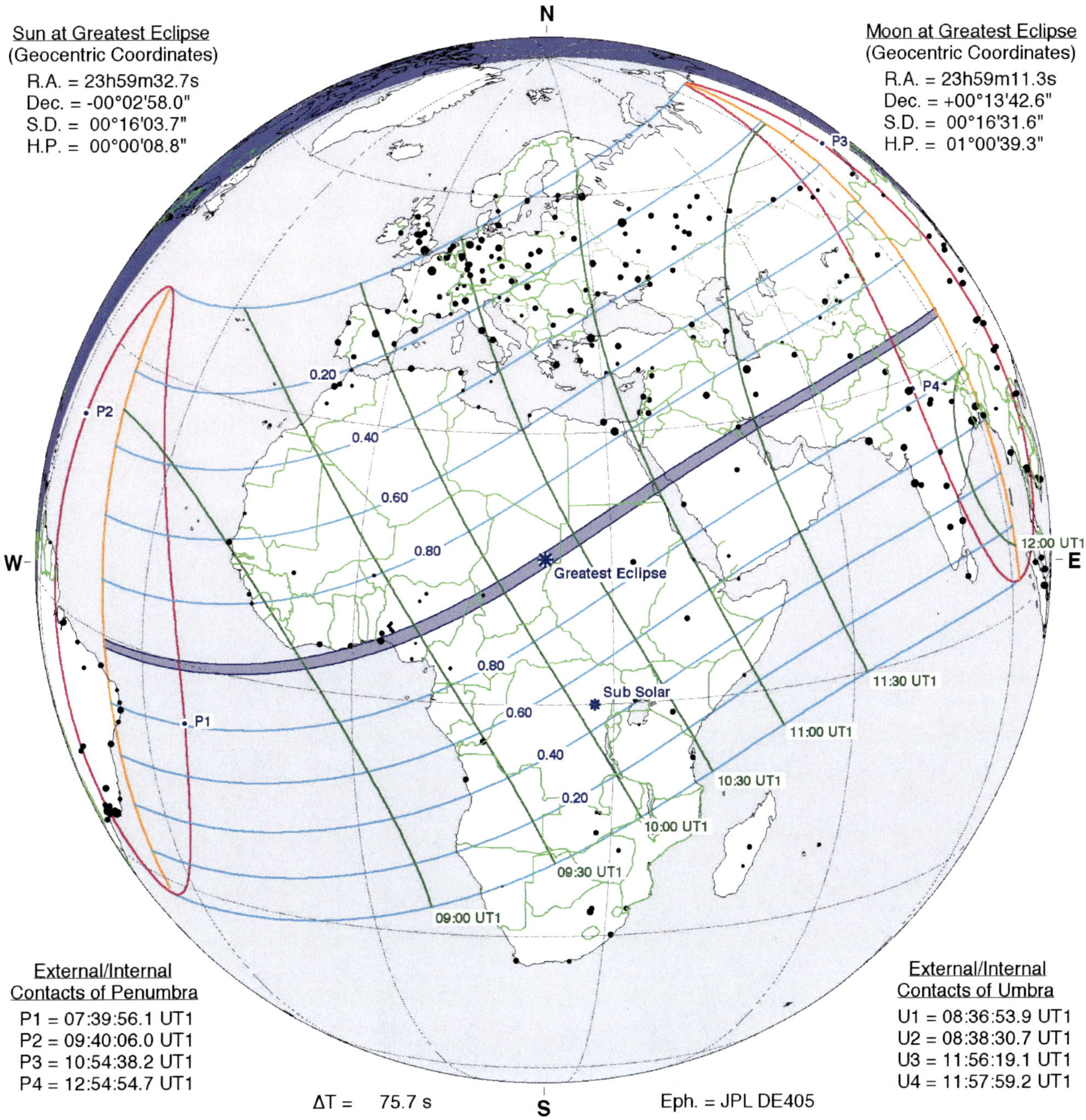

External/Internal Contacts of Penumbra
P1 = 07:39:56.1 UT1
P2 = 09:40:06.0 UT1
P3 = 10:54:38.2 UT1
P4 = 12:54:54.7 UT1

External/Internal Contacts of Umbra
U1 = 08:36:53.9 UT1
U2 = 08:38:30.7 UT1
U3 = 11:56:19.1 UT1
U4 = 11:57:59.2 UT1

ΔT = 75.7 s Eph. = JPL DE405

Circumstances at Greatest Eclipse: 10:17:29.5 UT1

Lat. = 16°03.3'N Sun Alt. = 73.1°
Long. = 022°13.1'E Sun Azm. = 161.6°
Path Width = 159.1 km Duration = 04m09.3s

Circumstances at Greatest Duration: 10:18:26.1 UT1

Lat. = 16°13.3'N Sun Alt. = 73.1°
Long. = 022°29.6'E Sun Azm. = 163.5°
Path Width = 159.0 km Duration = 04m09.4s

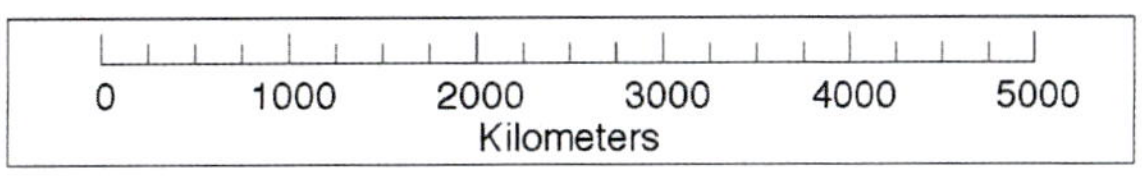

Annular Solar Eclipse of 2034 Sep 12

Greatest Eclipse = 16:19:27.5 TD (= 16:18:11.6 UT1)

Eclipse Magnitude = 0.9736 Saros Series = 135

Gamma = -0.3936 Saros Member = 40 of 71

<u>Sun at Greatest Eclipse</u>
(Geocentric Coordinates)
R.A. = 11h23m10.9s
Dec. = +03°57'57.5"
S.D. = 00°15'53.5"
H.P. = 00°00'08.7"

<u>Moon at Greatest Eclipse</u>
(Geocentric Coordinates)
R.A. = 11h22m44.5s
Dec. = +03°36'59.6"
S.D. = 00°15'15.1"
H.P. = 00°55'58.6"

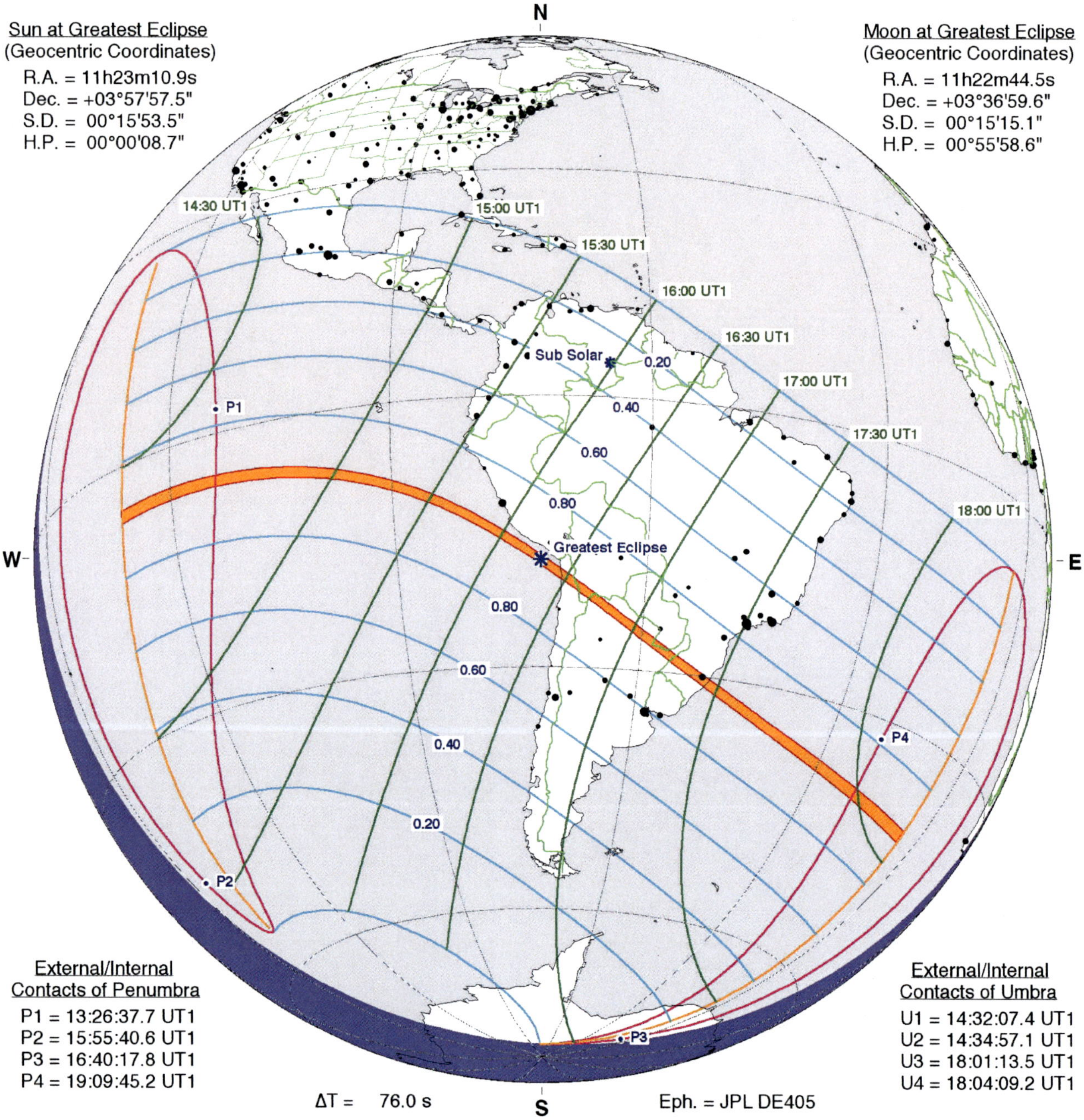

<u>External/Internal Contacts of Penumbra</u>
P1 = 13:26:37.7 UT1
P2 = 15:55:40.6 UT1
P3 = 16:40:17.8 UT1
P4 = 19:09:45.2 UT1

<u>External/Internal Contacts of Umbra</u>
U1 = 14:32:07.4 UT1
U2 = 14:34:57.1 UT1
U3 = 18:01:13.5 UT1
U4 = 18:04:09.2 UT1

ΔT = 76.0 s Eph. = JPL DE405

<u>Circumstances at Greatest Eclipse: 16:18:11.6 UT1</u>

Lat. = 18°14.5'S	Sun Alt. = 66.7°
Long. = 072°37.4'W	Sun Azm. = 18.2°
Path Width = 102.1 km	Duration = 02m57.7s

<u>Circumstances at Greatest Duration: 16:29:28.2 UT1</u>

Lat. = 20°04.3'S	Sun Alt. = 65.9°
Long. = 069°44.5'W	Sun Azm. = 3.5°
Path Width = 101.7 km	Duration = 02m57.8s

0 1000 2000 3000 4000 5000
Kilometers

Annular Solar Eclipse of 2035 Mar 09

Greatest Eclipse = 23:05:53.6 TD (= 23:04:37.4 UT1)

Eclipse Magnitude = 0.9919 Saros Series = 140
Gamma = -0.4368 Saros Member = 30 of 71

Sun at Greatest Eclipse
(Geocentric Coordinates)
R.A. = 23h20m17.6s
Dec. = -04°16'22.2"
S.D. = 00°16'06.5"
H.P. = 00°00'08.9"

Moon at Greatest Eclipse
(Geocentric Coordinates)
R.A. = 23h20m47.9s
Dec. = -04°40'23.8"
S.D. = 00°15'44.9"
H.P. = 00°57'47.9"

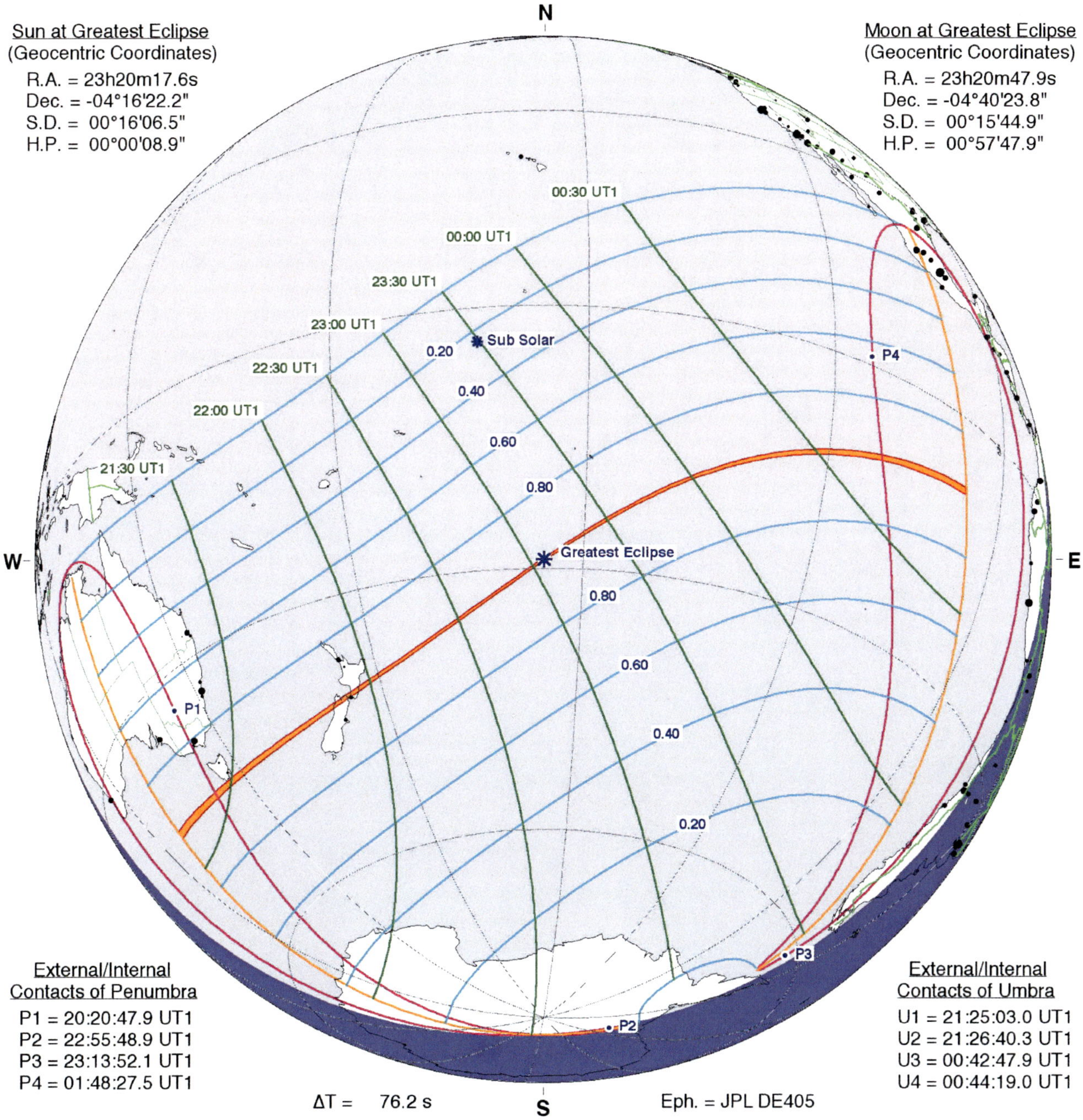

External/Internal
Contacts of Penumbra
P1 = 20:20:47.9 UT1
P2 = 22:55:48.9 UT1
P3 = 23:13:52.1 UT1
P4 = 01:48:27.5 UT1

External/Internal
Contacts of Umbra
U1 = 21:25:03.0 UT1
U2 = 21:26:40.3 UT1
U3 = 00:42:47.9 UT1
U4 = 00:44:19.0 UT1

ΔT = 76.2 s

Eph. = JPL DE405

Circumstances at Greatest Eclipse: 23:04:37.4 UT1
Lat. = 29°02.6'S Sun Alt. = 63.9°
Long. = 154°58.1'W Sun Azm. = 340.2°
Path Width = 31.5 km Duration = 00m47.5s

Circumstances at Greatest Duration: 21:25:51.6 UT1
Lat. = 43°22.2'S Sun Alt. = 0.0°
Long. = 127°04.5'E Sun Azm. = 95.9°
Path Width = 95.5 km Duration = 01m25.9s

0 1000 2000 3000 4000 5000
Kilometers

Total Solar Eclipse of 2035 Sep 02

Greatest Eclipse = 01:56:46.3 TD (= 01:55:29.9 UT1)

Eclipse Magnitude = 1.0320 Saros Series = 145

Gamma = 0.3727 Saros Member = 23 of 77

Sun at Greatest Eclipse
(Geocentric Coordinates)
R.A. = 10h44m07.3s
Dec. = +08°01'09.8"
S.D. = 00°15'50.9"
H.P. = 00°00'08.7"

Moon at Greatest Eclipse
(Geocentric Coordinates)
R.A. = 10h44m32.4s
Dec. = +08°22'14.7"
S.D. = 00°16'06.4"
H.P. = 00°59'06.9"

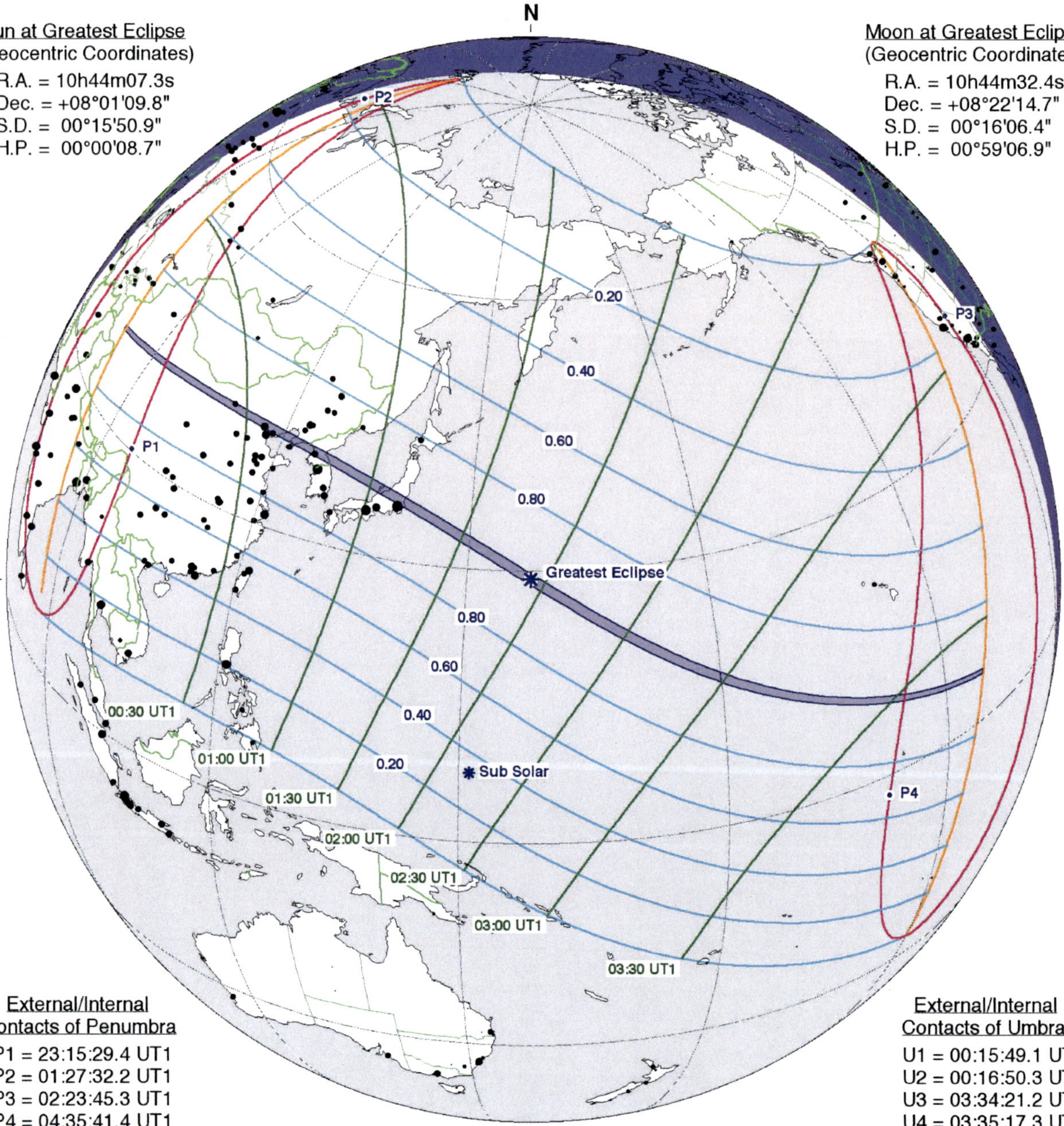

External/Internal Contacts of Penumbra
P1 = 23:15:29.4 UT1
P2 = 01:27:32.2 UT1
P3 = 02:23:45.3 UT1
P4 = 04:35:41.4 UT1

External/Internal Contacts of Umbra
U1 = 00:15:49.1 UT1
U2 = 00:16:50.3 UT1
U3 = 03:34:21.2 UT1
U4 = 03:35:17.3 UT1

ΔT = 76.4 s Eph. = JPL DE405

Circumstances at Greatest Eclipse: 01:55:29.9 UT1

Lat. = 29°05.6'N Sun Alt. = 67.9°
Long. = 158°00.9'E Sun Azm. = 198.5°
Path Width = 116.3 km Duration = 02m54.2s

Circumstances at Greatest Duration: 01:52:01.1 UT1

Lat. = 29°41.4'N Sun Alt. = 67.8°
Long. = 156°51.6'E Sun Azm. = 192.9°
Path Width = 116.1 km Duration = 02m54.4s

0 1000 2000 3000 4000 5000
Kilometers

Partial Solar Eclipse of 2036 Feb 27

Greatest Eclipse = 04:46:49.0 TD (= 04:45:32.4 UT1)

Eclipse Magnitude = 0.6286
Gamma = -1.1942

Saros Series = 150
Saros Member = 18 of 71

Sun at Greatest Eclipse
(Geocentric Coordinates)
R.A. = 22h39m15.4s
Dec. = -08°30'21.2"
S.D. = 00°16'09.1"
H.P. = 00°00'08.9"

Moon at Greatest Eclipse
(Geocentric Coordinates)
R.A. = 22h40m29.9s
Dec. = -09°33'05.6"
S.D. = 00°14'57.5"
H.P. = 00°54'53.9"

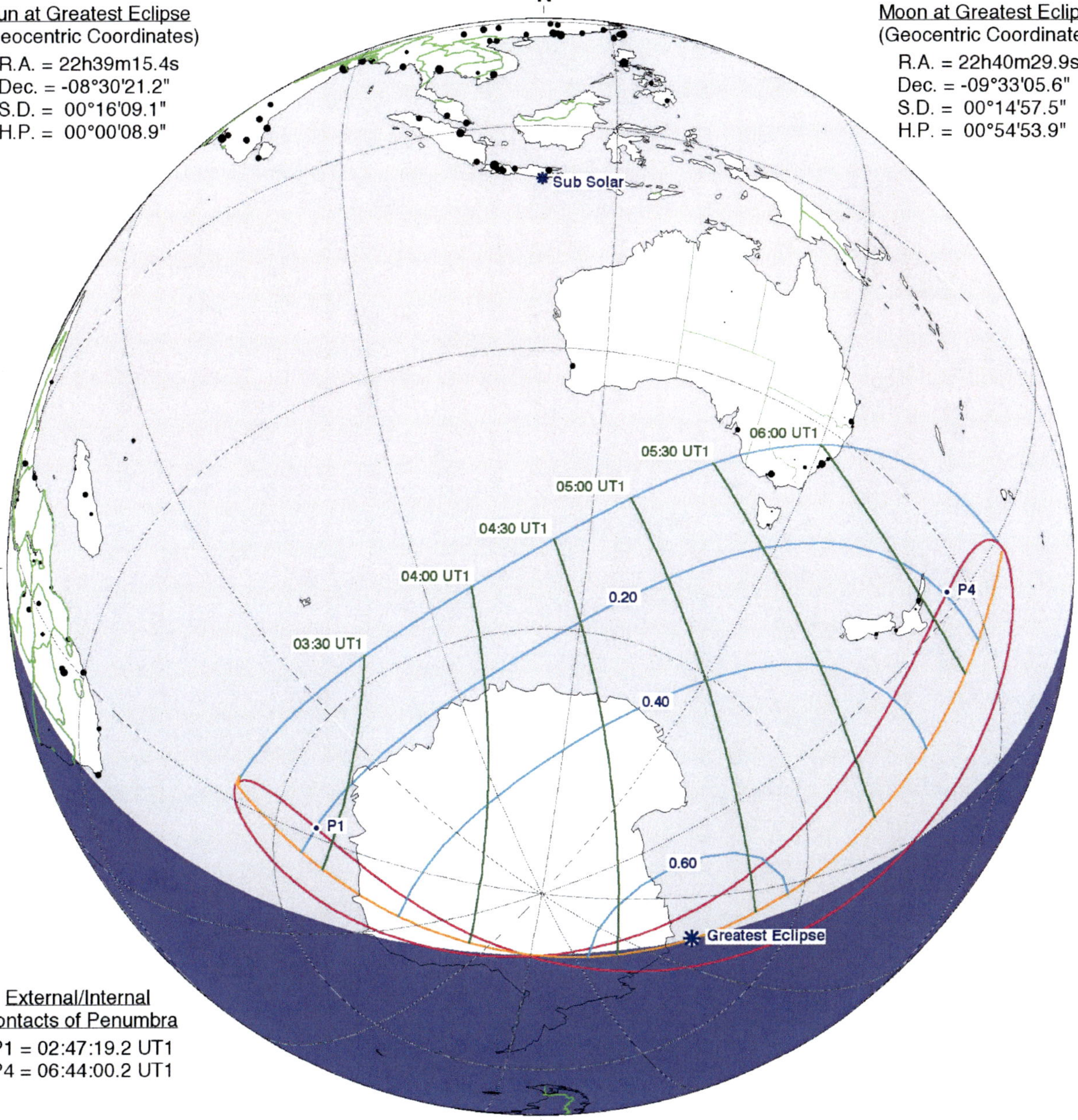

External/Internal
Contacts of Penumbra
P1 = 02:47:19.2 UT1
P4 = 06:44:00.2 UT1

ΔT = 76.7 s

Eph. = JPL DE405

Circumstances at Greatest Eclipse: 04:45:32.4 UT1

Lat. = 71°38.9'S
Long. = 131°27.3'W

Sun Alt. = 0.0°
Sun Azm. = 242.0°

0 1000 2000 3000 4000 5000
Kilometers

Partial Solar Eclipse of 2036 Jul 23

Greatest Eclipse = 10:32:06.5 TD (= 10:30:49.6 UT1)

Eclipse Magnitude = 0.1992 Saros Series = 117

Gamma = -1.4250 Saros Member = 70 of 71

Sun at Greatest Eclipse
(Geocentric Coordinates)
R.A. = 08h13m32.5s
Dec. = +19°53'41.2"
S.D. = 00°15'44.6"
H.P. = 00°00'08.7"

Moon at Greatest Eclipse
(Geocentric Coordinates)
R.A. = 08h12m46.3s
Dec. = +18°27'12.2"
S.D. = 00°16'42.4"
H.P. = 01°01'18.7"

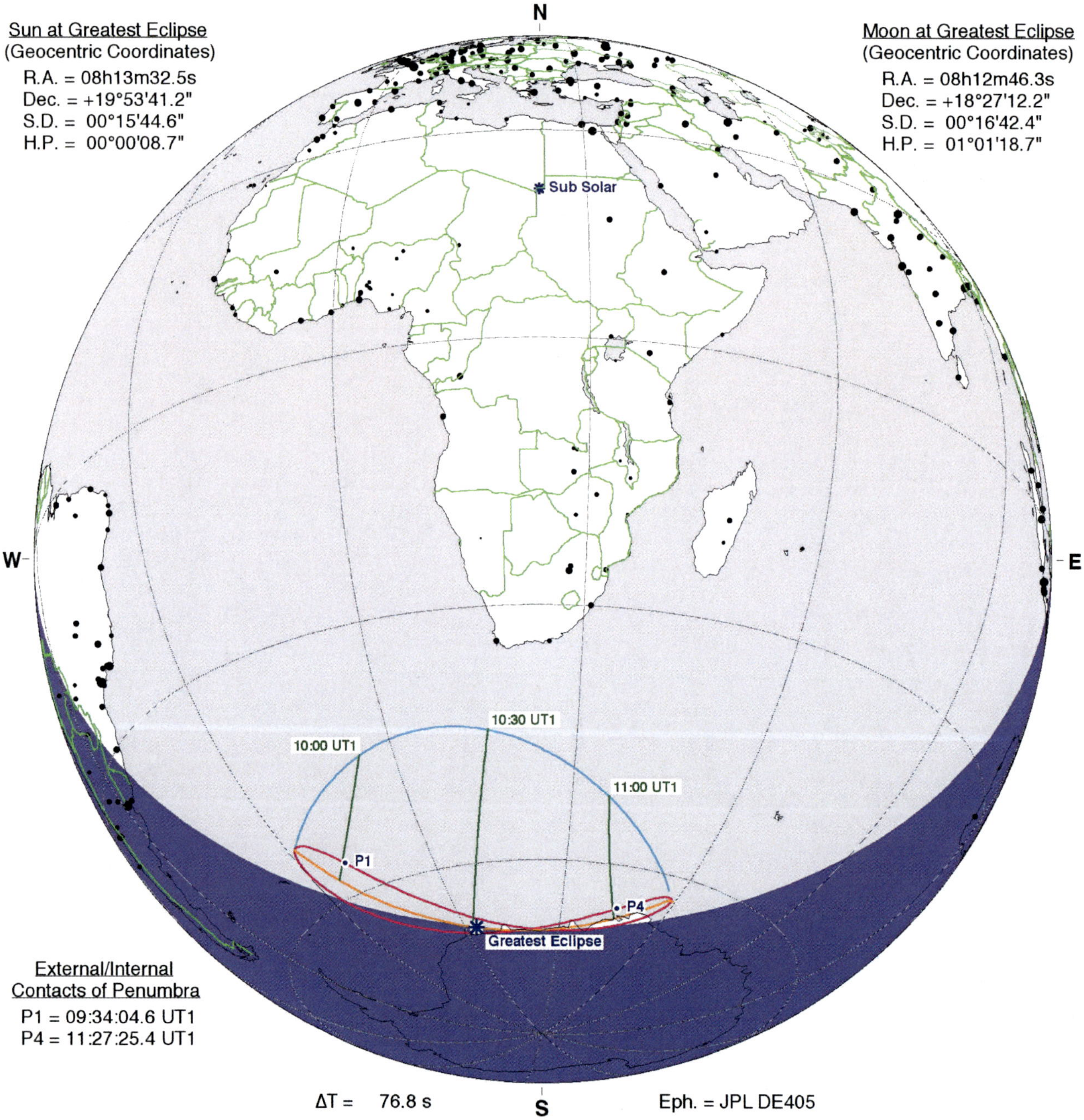

External/Internal
Contacts of Penumbra
P1 = 09:34:04.6 UT1
P4 = 11:27:25.4 UT1

ΔT = 76.8 s Eph. = JPL DE405

Circumstances at Greatest Eclipse: 10:30:49.6 UT1

Lat. = 68°53.0'S Sun Alt. = 0.0°

Long. = 003°32.3'E Sun Azm. = 19.1°

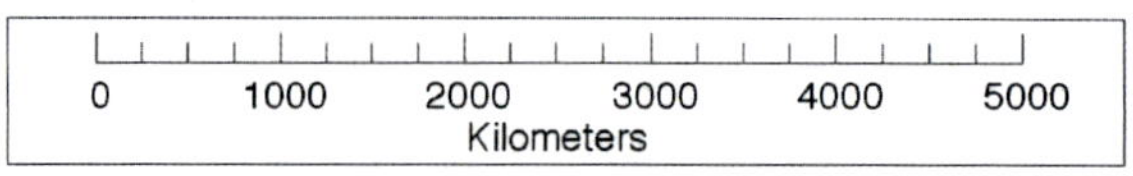

Partial Solar Eclipse of 2036 Aug 21

Greatest Eclipse = 17:25:45.4 TD (= 17:24:28.5 UT1)

Eclipse Magnitude = 0.8622 Saros Series = 155

Gamma = 1.0825 Saros Member = 7 of 71

Sun at Greatest Eclipse
(Geocentric Coordinates)

R.A. = 10h05m24.9s
Dec. = +11°44'16.4"
S.D. = 00°15'48.7"
H.P. = 00°00'08.7"

Moon at Greatest Eclipse
(Geocentric Coordinates)

R.A. = 10h06m34.6s
Dec. = +12°48'10.2"
S.D. = 00°16'41.1"
H.P. = 01°01'14.1"

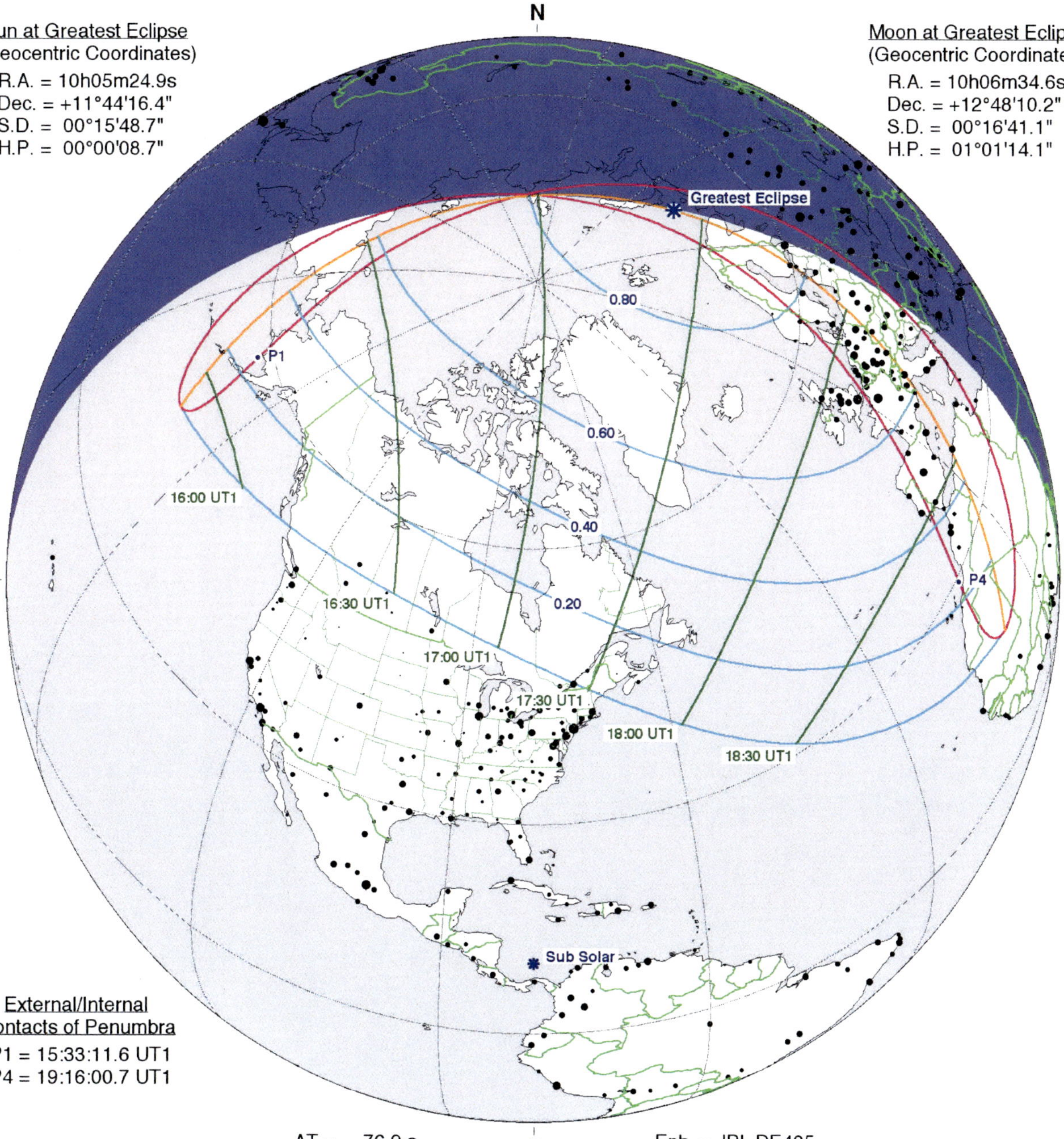

External/Internal
Contacts of Penumbra

P1 = 15:33:11.6 UT1
P4 = 19:16:00.7 UT1

ΔT = 76.9 s Eph. = JPL DE405

Circumstances at Greatest Eclipse: 17:24:28.5 UT1

Lat. = 71°07.2'N Sun Alt. = 0.0°
Long. = 046°59.1'E Sun Azm. = 308.9°

0 1000 2000 3000 4000 5000
Kilometers

Partial Solar Eclipse of 2037 Jan 16

Greatest Eclipse = 09:48:55.1 TD (= 09:47:38.0 UT1)

Eclipse Magnitude = 0.7049 Saros Series = 122
Gamma = 1.1477 Saros Member = 59 of 70

Sun at Greatest Eclipse
(Geocentric Coordinates)
R.A. = 19h54m30.1s
Dec. = -20°49'43.5"
S.D. = 00°16'15.5"
H.P. = 00°00'08.9"

Moon at Greatest Eclipse
(Geocentric Coordinates)
R.A. = 19h54m05.4s
Dec. = -19°47'33.7"
S.D. = 00°14'51.8"
H.P. = 00°54'32.8"

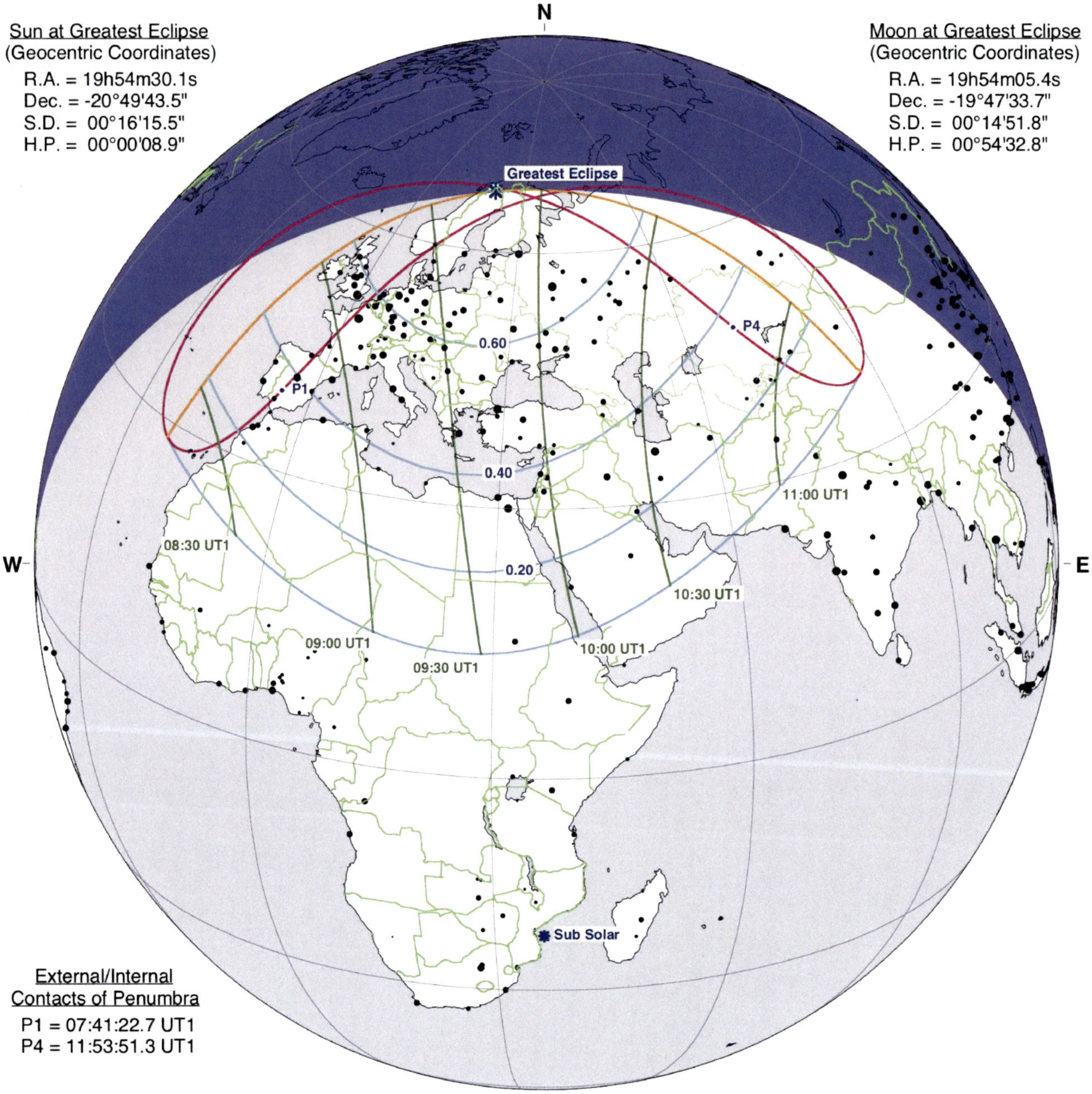

External/Internal
Contacts of Penumbra
P1 = 07:41:22.7 UT1
P4 = 11:53:51.3 UT1

ΔT = 77.1 s Eph. = JPL DE405

Circumstances at Greatest Eclipse: 09:47:38.0 UT1
Lat. = 68°31.3'N Sun Alt. = 0.0°
Long. = 020°47.6'E Sun Azm. = 166.2°

0 1000 2000 3000 4000 5000
Kilometers

Total Solar Eclipse of 2037 Jul 13

Greatest Eclipse = 02:40:35.9 TD (= 02:39:18.5 UT1)

Eclipse Magnitude = 1.0413 Saros Series = 127

Gamma = -0.7246 Saros Member = 59 of 82

Sun at Greatest Eclipse
(Geocentric Coordinates)

R.A. = 07h31m06.7s
Dec. = +21°46'57.5"
S.D. = 00°15'44.0"
H.P. = 00°00'08.7"

Moon at Greatest Eclipse
(Geocentric Coordinates)

R.A. = 07h30m56.4s
Dec. = +21°04'03.1"
S.D. = 00°16'12.0"
H.P. = 00°59'27.3"

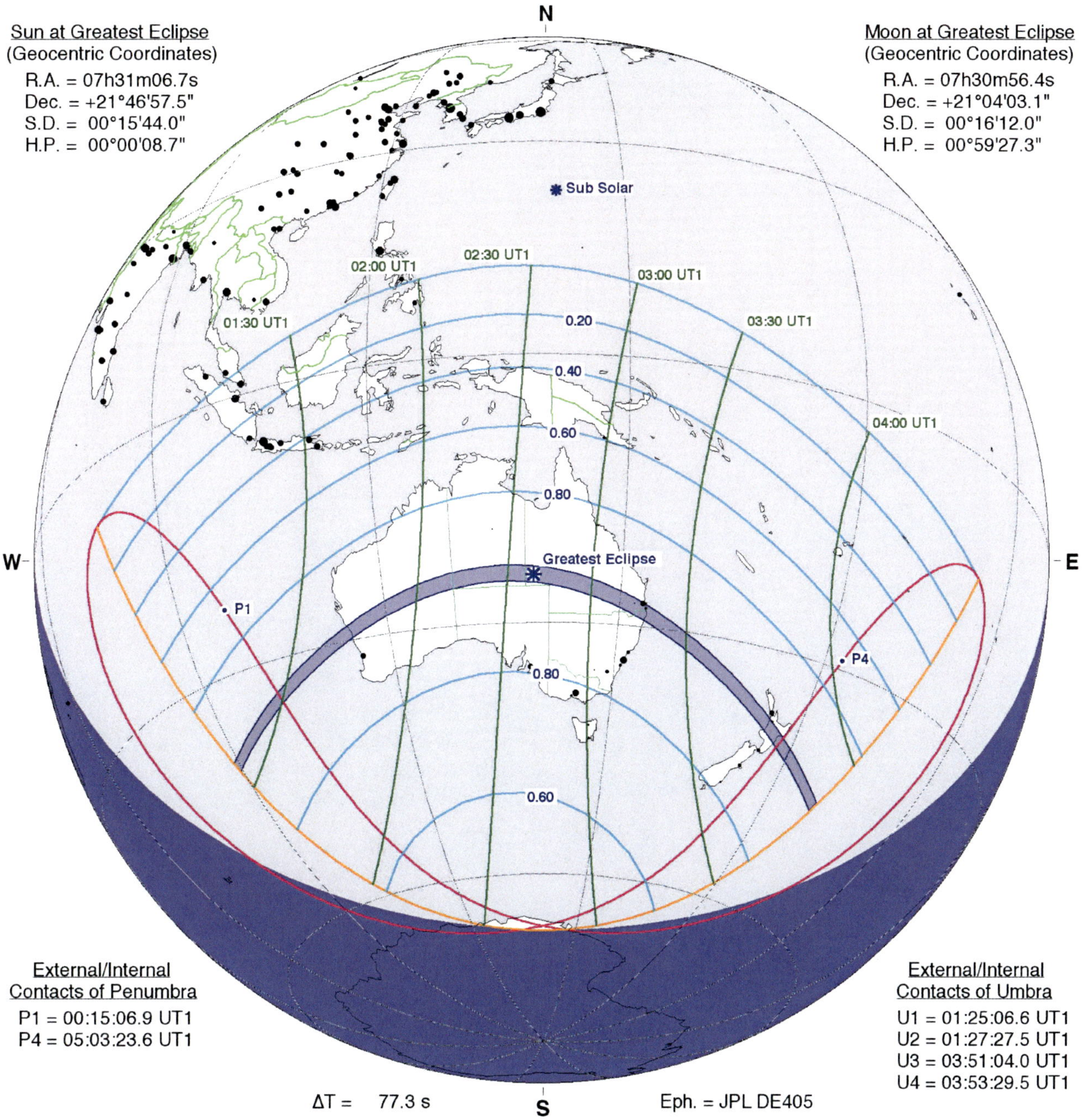

External/Internal
Contacts of Penumbra

P1 = 00:15:06.9 UT1
P4 = 05:03:23.6 UT1

External/Internal
Contacts of Umbra

U1 = 01:25:06.6 UT1
U2 = 01:27:27.5 UT1
U3 = 03:51:04.0 UT1
U4 = 03:53:29.5 UT1

ΔT = 77.3 s Eph. = JPL DE405

Circumstances at Greatest Eclipse: 02:39:18.5 UT1

Lat. = 24°45.8'S Sun Alt. = 43.4°
Long. = 139°03.2'E Sun Azm. = 3.3°
Path Width = 200.7 km Duration = 03m58.4s

Circumstances at Greatest Duration: 02:39:45.9 UT1

Lat. = 24°46.7'S Sun Alt. = 43.4°
Long. = 139°12.5'E Sun Azm. = 2.9°
Path Width = 200.6 km Duration = 03m58.4s

0 1000 2000 3000 4000 5000
Kilometers

Annular Solar Eclipse of 2038 Jan 05

Greatest Eclipse = 13:47:10.9 TD (= 13:45:53.4 UT1)

Eclipse Magnitude = 0.9728	Saros Series = 132
Gamma = 0.4169	Saros Member = 47 of 71

Sun at Greatest Eclipse
(Geocentric Coordinates)

R.A. = 19h06m27.4s
Dec. = -22°33'17.3"
S.D. = 00°16'15.9"
H.P. = 00°00'08.9"

Moon at Greatest Eclipse
(Geocentric Coordinates)

R.A. = 19h06m25.9s
Dec. = -22°09'29.7"
S.D. = 00°15'35.7"
H.P. = 00°57'13.9"

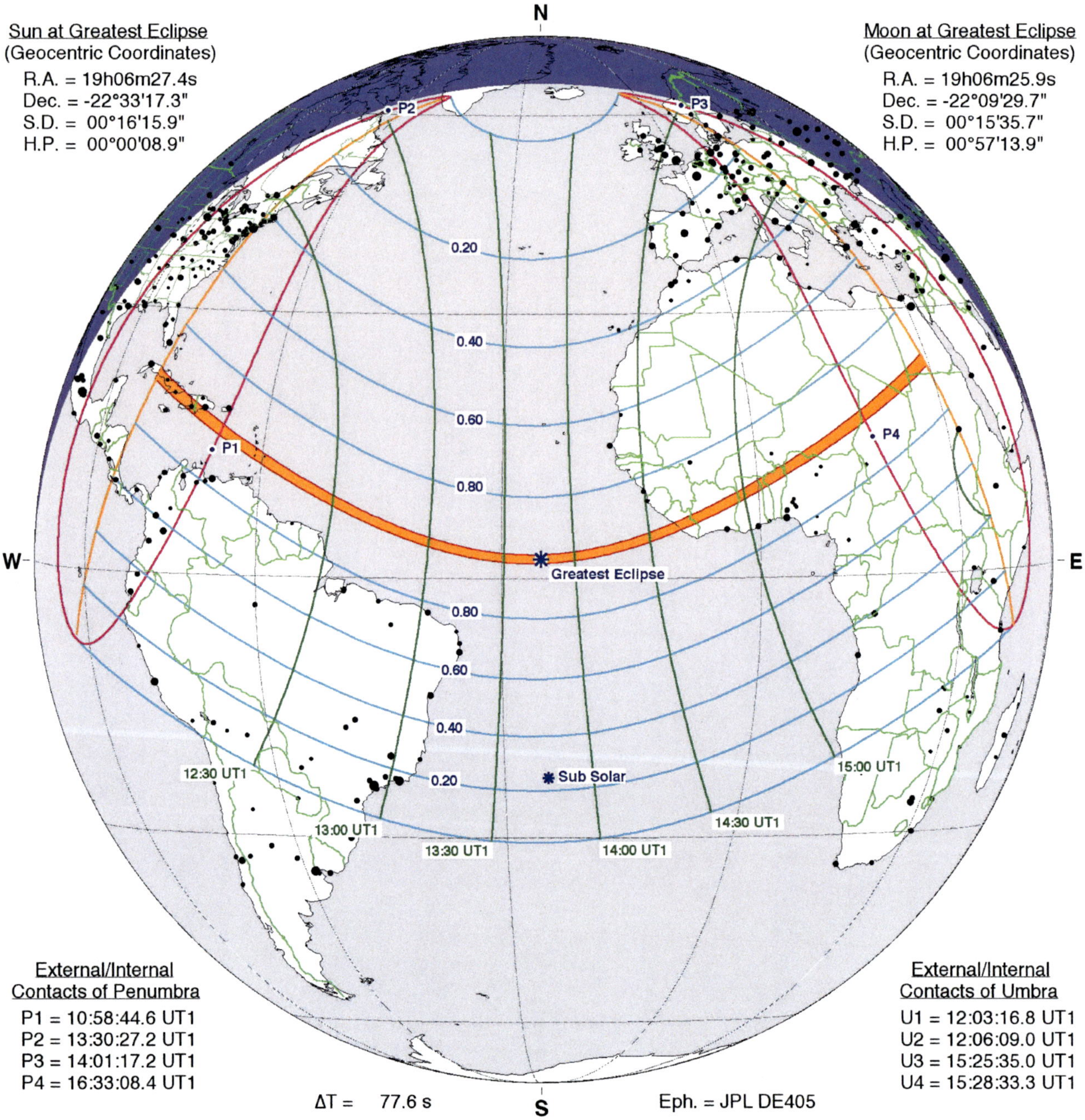

External/Internal Contacts of Penumbra

P1 = 10:58:44.6 UT1
P2 = 13:30:27.2 UT1
P3 = 14:01:17.2 UT1
P4 = 16:33:08.4 UT1

External/Internal Contacts of Umbra

U1 = 12:03:16.8 UT1
U2 = 12:06:09.0 UT1
U3 = 15:25:35.0 UT1
U4 = 15:28:33.3 UT1

ΔT = 77.6 s

Eph. = JPL DE405

Circumstances at Greatest Eclipse: 13:45:53.4 UT1

Lat. = 02°05.7'N	Sun Alt. = 65.3°
Long. = 025°28.0'W	Sun Azm. = 179.2°
Path Width = 107.2 km	Duration = 03m18.5s

Circumstances at Greatest Duration: 13:52:36.3 UT1

Lat. = 02°11.2'N	Sun Alt. = 65.1°
Long. = 023°41.3'W	Sun Azm. = 186.8°
Path Width = 107.4 km	Duration = 03m18.6s

0 1000 2000 3000 4000 5000
Kilometers

www.EclipseWise.com

Annular Solar Eclipse of 2038 Jul 02

Greatest Eclipse = 13:32:55.0 TD (= 13:31:37.2 UT1)

Eclipse Magnitude = 0.9911 Saros Series = 137

Gamma = 0.0398 Saros Member = 37 of 70

Sun at Greatest Eclipse
(Geocentric Coordinates)
R.A. = 06h46m55.4s
Dec. = +22°59'44.2"
S.D. = 00°15'43.9"
H.P. = 00°00'08.6"

Moon at Greatest Eclipse
(Geocentric Coordinates)
R.A. = 06h46m55.2s
Dec. = +23°01'58.2"
S.D. = 00°15'20.9"
H.P. = 00°56'19.9"

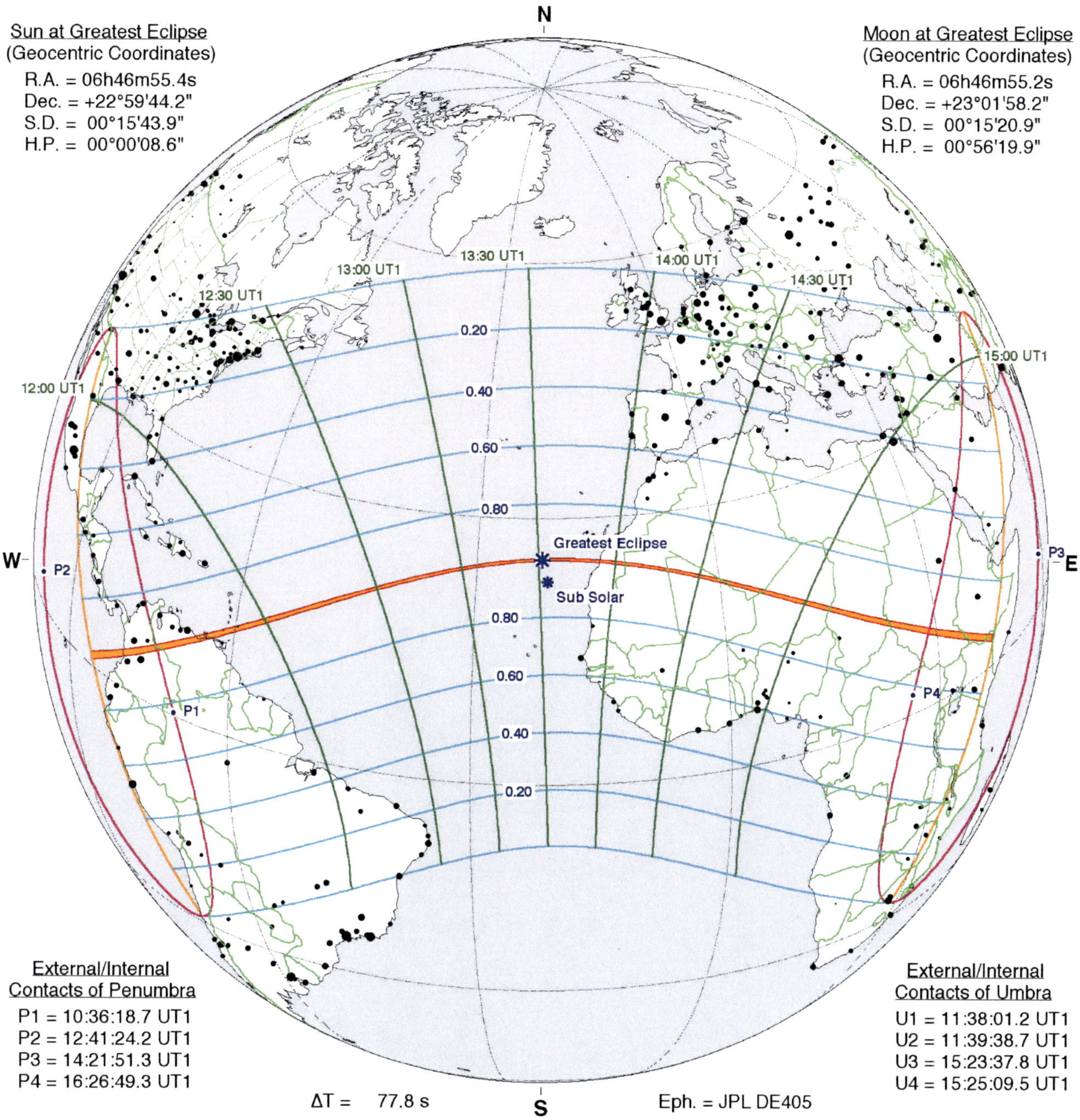

External/Internal Contacts of Penumbra
P1 = 10:36:18.7 UT1
P2 = 12:41:24.2 UT1
P3 = 14:21:51.3 UT1
P4 = 16:26:49.3 UT1

External/Internal Contacts of Umbra
U1 = 11:38:01.2 UT1
U2 = 11:39:38.7 UT1
U3 = 15:23:37.8 UT1
U4 = 15:25:09.5 UT1

ΔT = 77.8 s Eph. = JPL DE405

Circumstances at Greatest Eclipse: 13:31:37.2 UT1

Lat. = 25°25.4'N Sun Alt. = 87.6°
Long. = 021°54.8'W Sun Azm. = 179.0°
Path Width = 31.2 km Duration = 00m59.7s

Circumstances at Greatest Duration: 11:38:50.0 UT1

Lat. = 01°05.8'N Sun Alt. = 0.0°
Long. = 084°08.2'W Sun Azm. = 67.0°
Path Width = 94.2 km Duration = 01m34.9s

0 1000 2000 3000 4000 5000
Kilometers

Total Solar Eclipse of 2038 Dec 26

Greatest Eclipse = 01:00:09.7 TD (= 00:58:51.6 UT1)

Eclipse Magnitude = 1.0269 Saros Series = 142

Gamma = -0.2881 Saros Member = 24 of 72

Sun at Greatest Eclipse
(Geocentric Coordinates)
R.A. = 18h18m51.7s
Dec. = -23°21'47.8"
S.D. = 00°16'15.7"
H.P. = 00°00'08.9"

Moon at Greatest Eclipse
(Geocentric Coordinates)
R.A. = 18h18m46.7s
Dec. = -23°39'05.4"
S.D. = 00°16'25.8"
H.P. = 01°00'18.1"

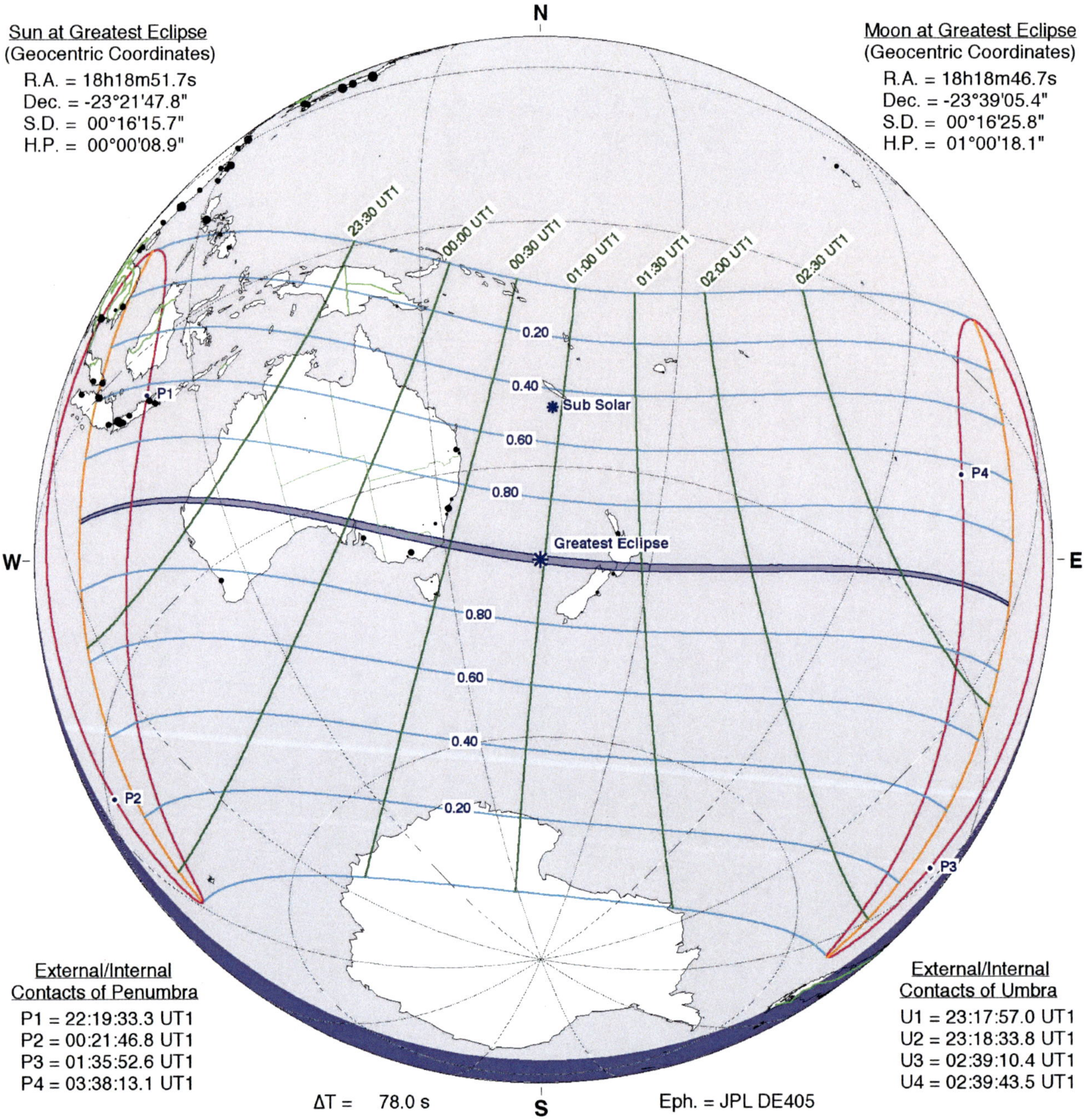

External/Internal
Contacts of Penumbra
P1 = 22:19:33.3 UT1
P2 = 00:21:46.8 UT1
P3 = 01:35:52.6 UT1
P4 = 03:38:13.1 UT1

External/Internal
Contacts of Umbra
U1 = 23:17:57.0 UT1
U2 = 23:18:33.8 UT1
U3 = 02:39:10.4 UT1
U4 = 02:39:43.5 UT1

ΔT = 78.0 s Eph. = JPL DE405

Circumstances at Greatest Eclipse: 00:58:51.6 UT1	
Lat. = 40°16.7'S	Sun Alt. = 73.0°
Long. = 163°55.6'E	Sun Azm. = 4.5°
Path Width = 95.3 km	Duration = 02m18.2s

Circumstances at Greatest Duration: 00:58:08.2 UT1	
Lat. = 40°14.8'S	Sun Alt. = 73.0°
Long. = 163°35.6'E	Sun Azm. = 6.2°
Path Width = 95.3 km	Duration = 02m18.2s

0 1000 2000 3000 4000 5000
Kilometers

Annular Solar Eclipse of 2039 Jun 21

Greatest Eclipse = 17:12:53.8 TD (= 17:11:35.5 UT1)

Eclipse Magnitude = 0.9454
Gamma = 0.8312

Saros Series = 147
Saros Member = 24 of 80

Sun at Greatest Eclipse
(Geocentric Coordinates)
R.A. = 06h00m54.5s
Dec. = +23°26'03.6"
S.D. = 00°15'44.3"
H.P. = 00°00'08.7"

Moon at Greatest Eclipse
(Geocentric Coordinates)
R.A. = 06h00m35.3s
Dec. = +24°10'44.9"
S.D. = 00°14'45.6"
H.P. = 00°54'10.2"

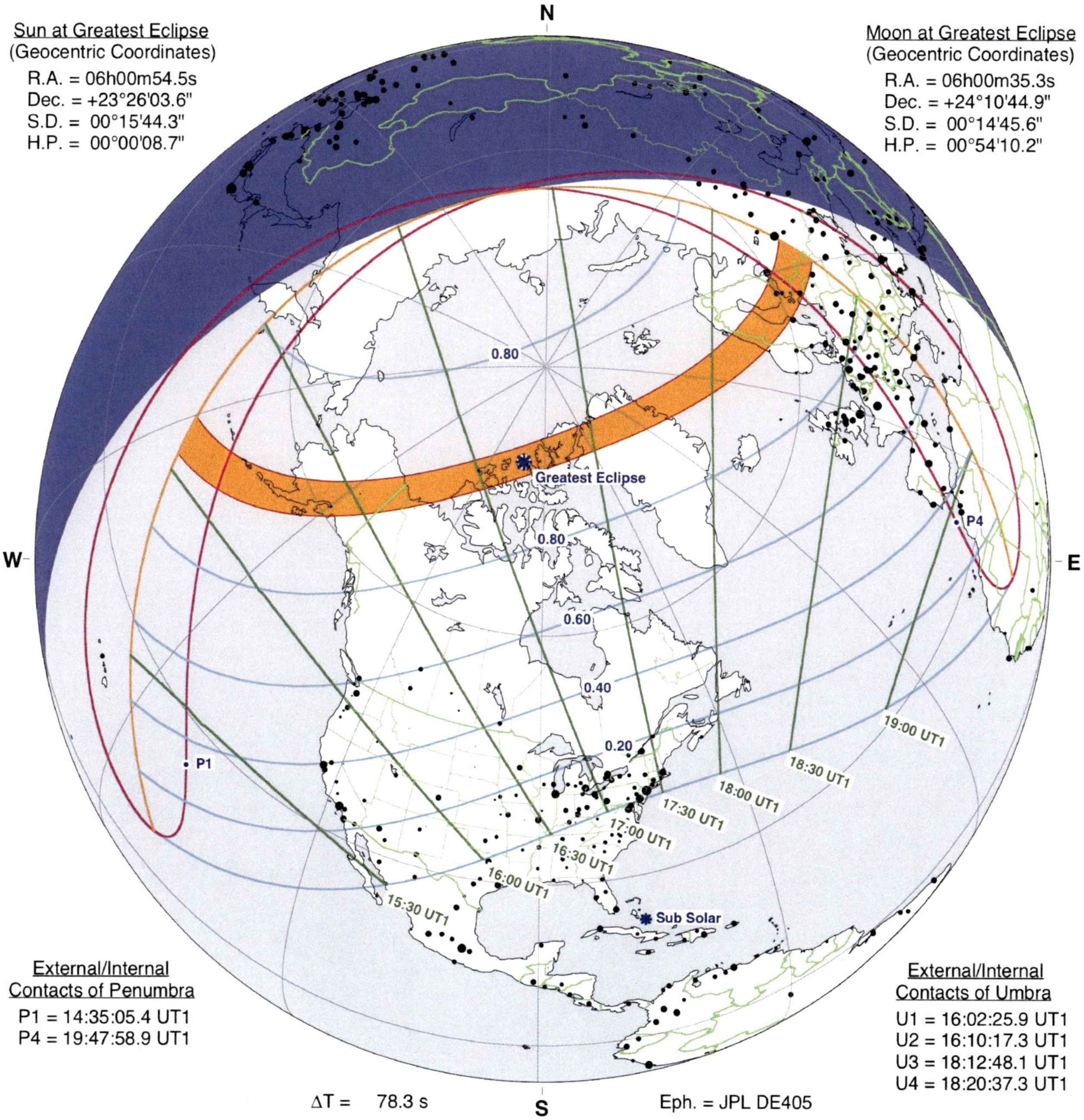

External/Internal
Contacts of Penumbra
P1 = 14:35:05.4 UT1
P4 = 19:47:58.9 UT1

External/Internal
Contacts of Umbra
U1 = 16:02:25.9 UT1
U2 = 16:10:17.3 UT1
U3 = 18:12:48.1 UT1
U4 = 18:20:37.3 UT1

ΔT = 78.3 s

Eph. = JPL DE405

Circumstances at Greatest Eclipse: 17:11:35.5 UT1

Lat. = 78°52.5'N	Sun Alt. = 33.4°
Long. = 102°07.9'W	Sun Azm. = 152.6°
Path Width = 365.3 km	Duration = 04m04.9s

Circumstances at Greatest Duration: 17:11:45.6 UT1

Lat. = 78°54.8'N	Sun Alt. = 33.4°
Long. = 101°47.9'W	Sun Azm. = 153.1°
Path Width = 365.2 km	Duration = 04m04.9s

0 1000 2000 3000 4000 5000
Kilometers

Total Solar Eclipse of 2039 Dec 15

Greatest Eclipse = 16:23:45.9 TD (= 16:22:27.4 UT1)

Eclipse Magnitude = 1.0356 Saros Series = 152

Gamma = -0.9458 Saros Member = 14 of 70

Sun at Greatest Eclipse
(Geocentric Coordinates)
R.A. = 17h31m51.4s
Dec. = -23°16'37.6"
S.D. = 00°16'14.9"
H.P. = 00°00'08.9"

Moon at Greatest Eclipse
(Geocentric Coordinates)
R.A. = 17h31m14.4s
Dec. = -24°13'58.8"
S.D. = 00°16'44.6"
H.P. = 01°01'26.8"

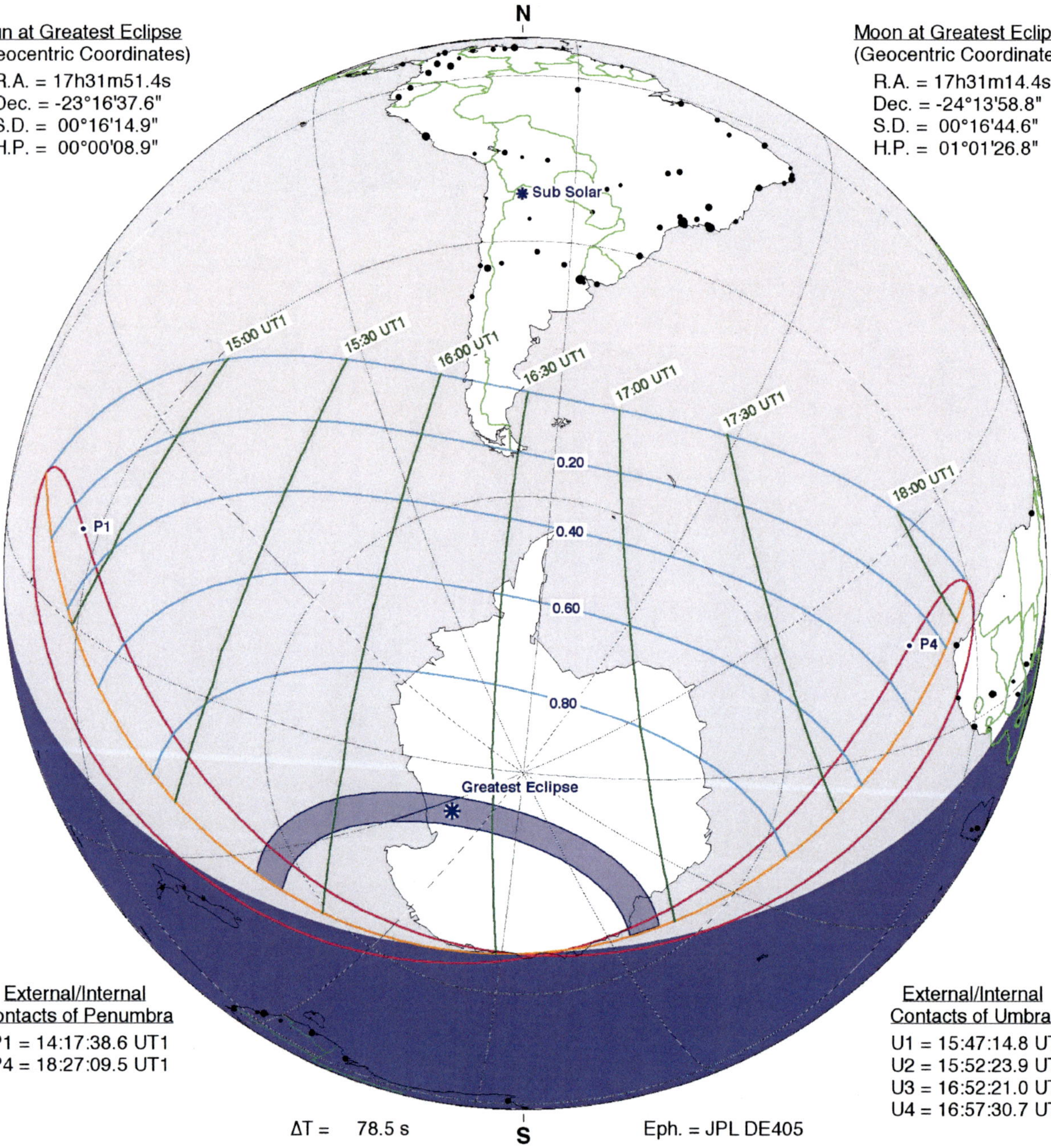

External/Internal
Contacts of Penumbra
P1 = 14:17:38.6 UT1
P4 = 18:27:09.5 UT1

External/Internal
Contacts of Umbra
U1 = 15:47:14.8 UT1
U2 = 15:52:23.9 UT1
U3 = 16:52:21.0 UT1
U4 = 16:57:30.7 UT1

ΔT = 78.5 s Eph. = JPL DE405

Circumstances at Greatest Eclipse: 16:22:27.4 UT1

Lat. = 80°51.4'S Sun Alt. = 18.4°
Long. = 172°44.9'E Sun Azm. = 123.4°
Path Width = 379.8 km Duration = 01m51.4s

Circumstances at Greatest Duration: 16:22:32.9 UT1

Lat. = 80°54.0'S Sun Alt. = 18.4°
Long. = 172°32.7'E Sun Azm. = 123.5°
Path Width = 379.7 km Duration = 01m51.4s

0 1000 2000 3000 4000 5000
Kilometers

Partial Solar Eclipse of 2040 May 11

Greatest Eclipse = 03:43:02.1 TD (= 03:41:43.4 UT1)

Eclipse Magnitude = 0.5306 Saros Series = 119

Gamma = -1.2529 Saros Member = 67 of 71

Sun at Greatest Eclipse
(Geocentric Coordinates)
R.A. = 03h14m33.6s
Dec. = +18°01'19.7"
S.D. = 00°15'50.1"
H.P. = 00°00'08.7"

Moon at Greatest Eclipse
(Geocentric Coordinates)
R.A. = 03h16m16.3s
Dec. = +16°56'30.8"
S.D. = 00°15'06.4"
H.P. = 00°55'26.7"

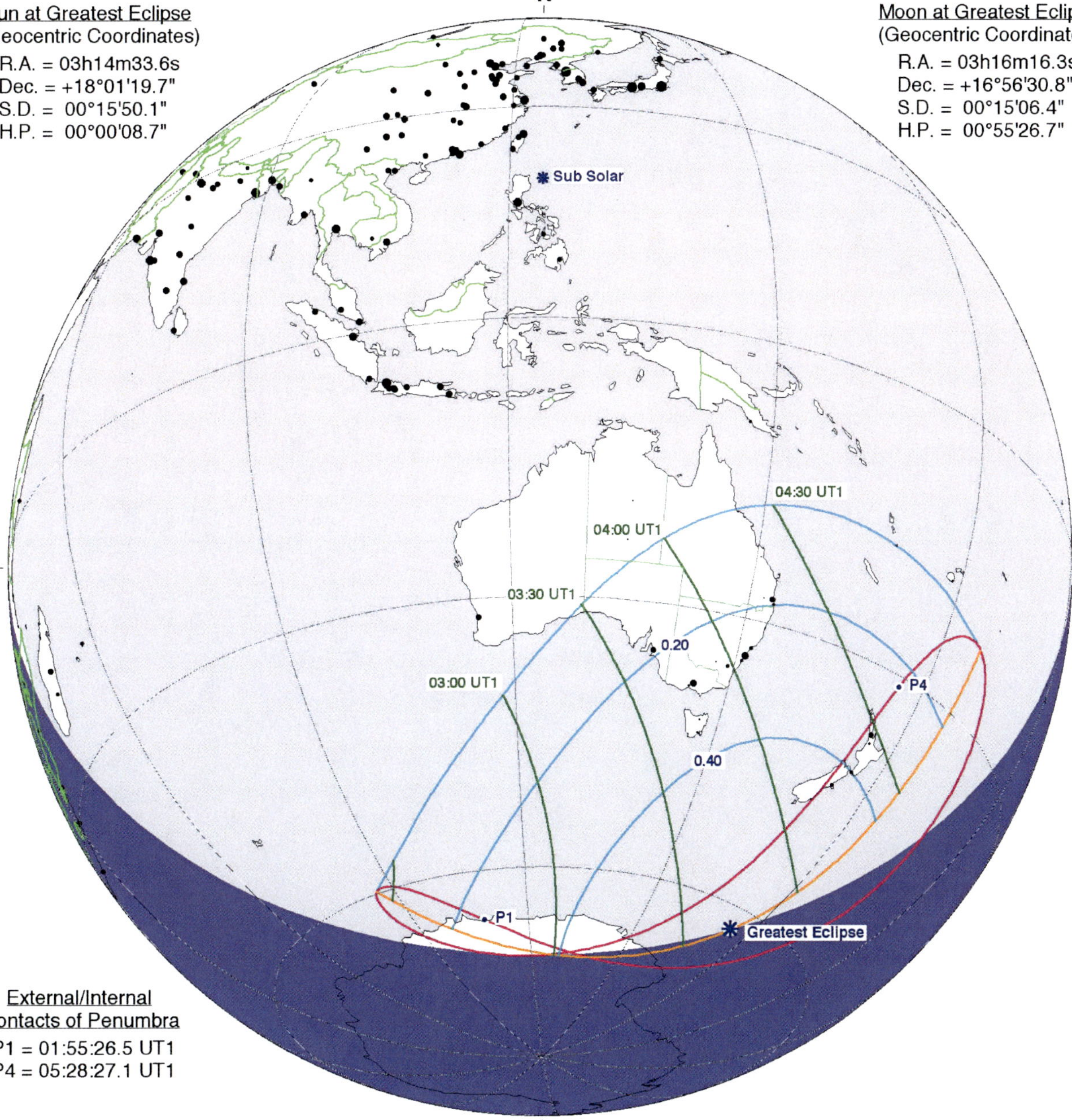

External/Internal
Contacts of Penumbra
P1 = 01:55:26.5 UT1
P4 = 05:28:27.1 UT1

ΔT = 78.8 s Eph. = JPL DE405

Circumstances at Greatest Eclipse: 03:41:43.4 UT1

Lat. = 62°46.1'S Sun Alt. = 0.0°
Long. = 174°23.3'E Sun Azm. = 312.5°

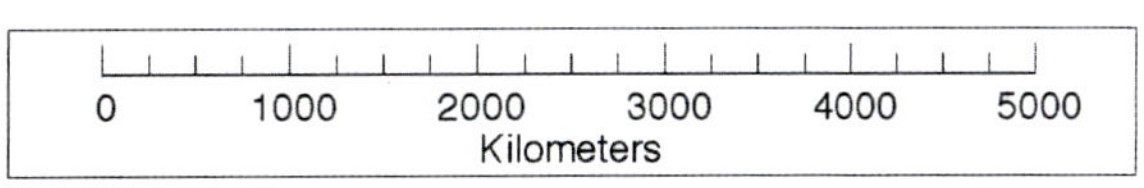

Partial Solar Eclipse of 2040 Nov 04

Greatest Eclipse = 19:09:02.0 TD (= 19:07:43.0 UT1)

Eclipse Magnitude = 0.8074 | Saros Series = 124
Gamma = 1.0993 | Saros Member = 56 of 73

Sun at Greatest Eclipse
(Geocentric Coordinates)
R.A. = 14h42m06.9s
Dec. = -15°43'53.8"
S.D. = 00°16'07.7"
H.P. = 00°00'08.9"

Moon at Greatest Eclipse
(Geocentric Coordinates)
R.A. = 14h43m50.8s
Dec. = -14°45'19.8"
S.D. = 00°15'49.8"
H.P. = 00°58'05.7"

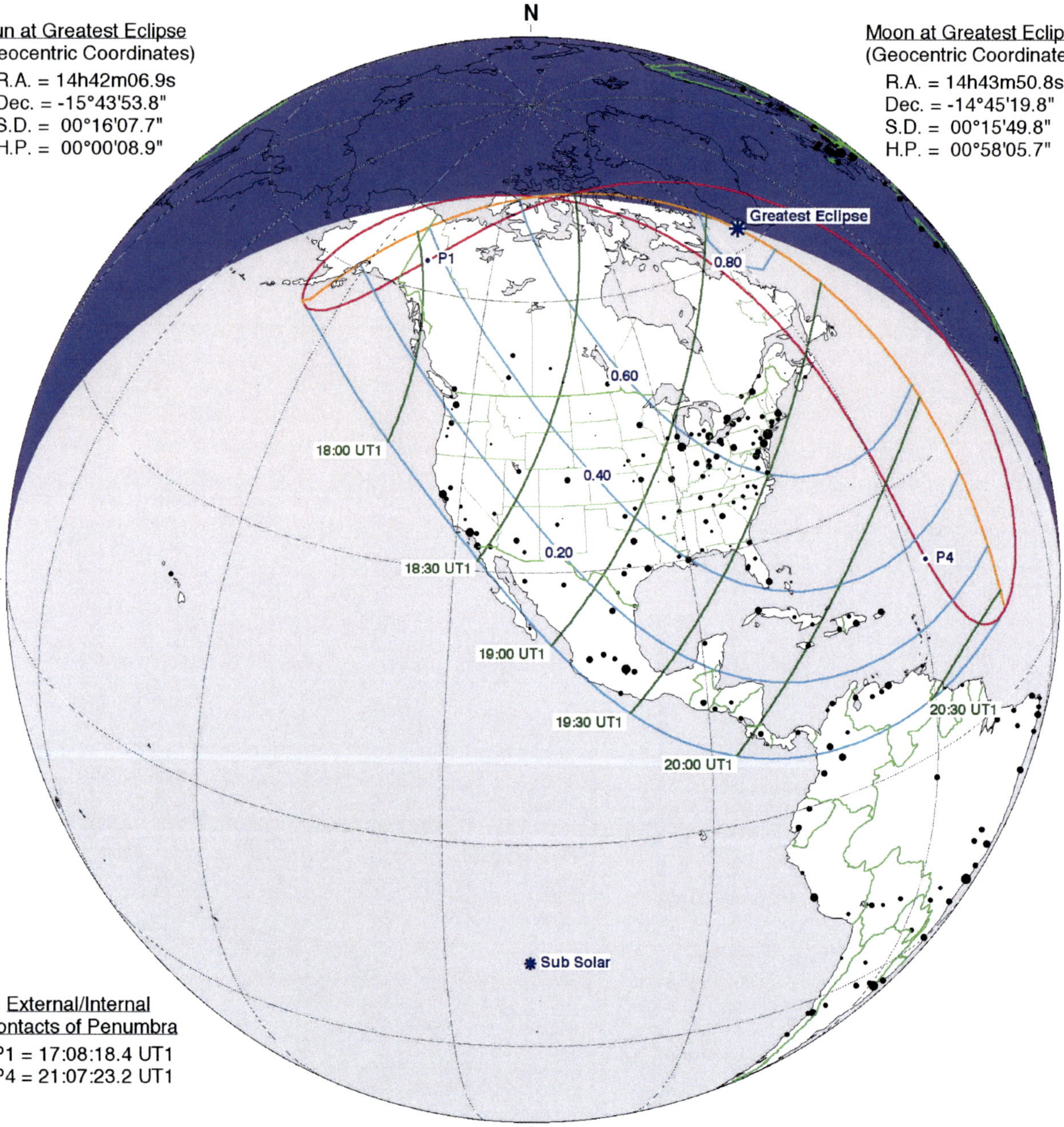

External/Internal
Contacts of Penumbra
P1 = 17:08:18.4 UT1
P4 = 21:07:23.2 UT1

ΔT = 79.0 s | Eph. = JPL DE405

Circumstances at Greatest Eclipse: 19:07:43.0 UT1
Lat. = 62°12.1'N | Sun Alt. = 0.0°
Long. = 053°24.2'W | Sun Azm. = 234.4°

0 1000 2000 3000 4000 5000
Kilometers

Total Solar Eclipse of 2041 Apr 30

Greatest Eclipse = 11:52:20.8 TD (= 11:51:01.5 UT1)

Eclipse Magnitude = 1.0189   Saros Series = 129
Gamma = -0.4492   Saros Member = 53 of 80

Sun at Greatest Eclipse
(Geocentric Coordinates)
R.A. = 02h32m22.2s
Dec. = +14°58'18.8"
S.D. = 00°15'52.6"
H.P. = 00°00'08.7"

Moon at Greatest Eclipse
(Geocentric Coordinates)
R.A. = 02h33m06.0s
Dec. = +14°34'20.1"
S.D. = 00°15'56.6"
H.P. = 00°58'30.8"

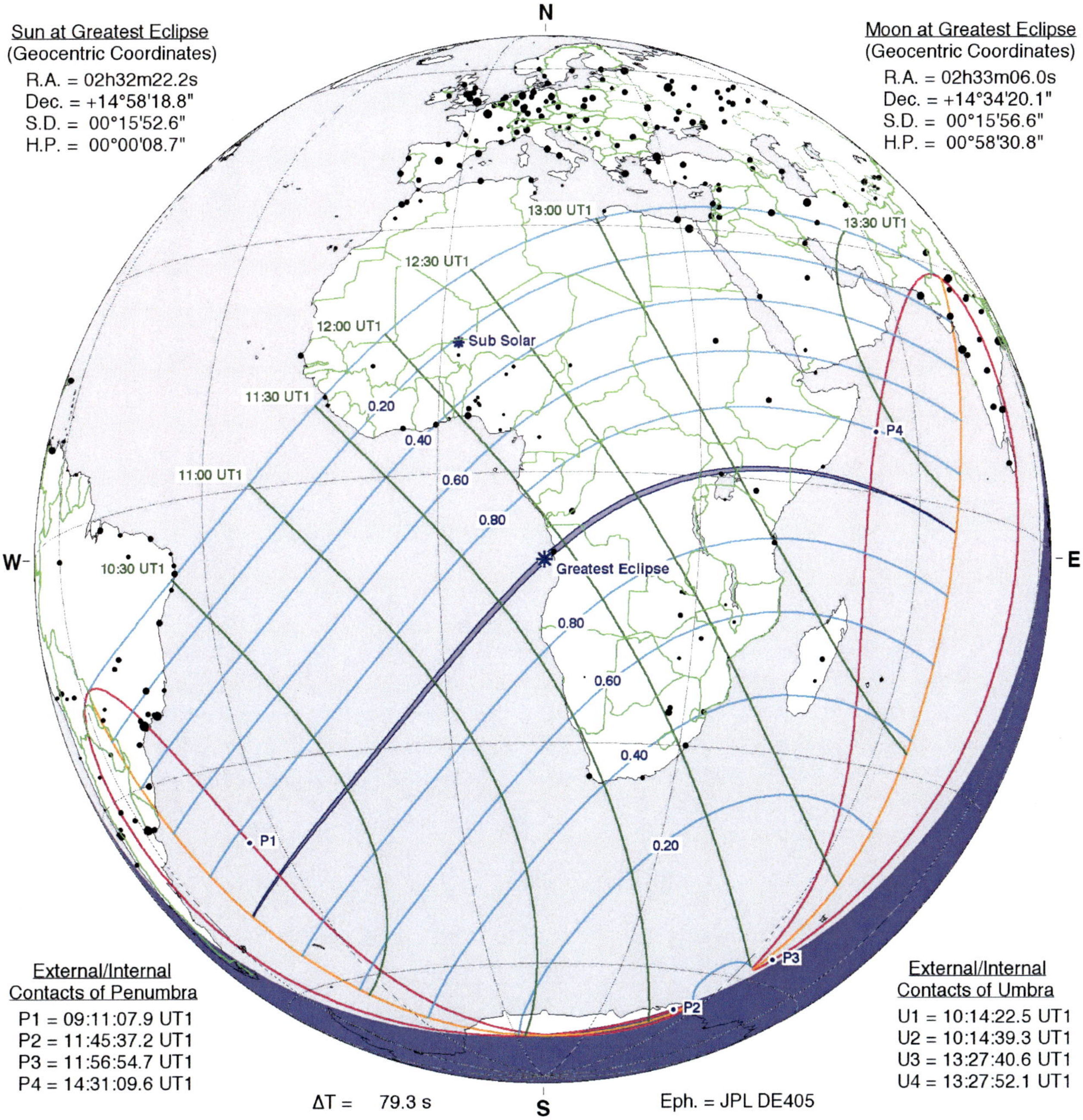

External/Internal Contacts of Penumbra
P1 = 09:11:07.9 UT1
P2 = 11:45:37.2 UT1
P3 = 11:56:54.7 UT1
P4 = 14:31:09.6 UT1

External/Internal Contacts of Umbra
U1 = 10:14:22.5 UT1
U2 = 10:14:39.3 UT1
U3 = 13:27:40.6 UT1
U4 = 13:27:52.1 UT1

ΔT = 79.3 s   Eph. = JPL DE405

Circumstances at Greatest Eclipse: 11:51:01.5 UT1

Lat. = 09°37.2'S   Sun Alt. = 63.2°
Long. = 012°09.7'E   Sun Azm. = 336.7°
Path Width = 71.7 km   Duration = 01m50.6s

Circumstances at Greatest Duration: 11:52:09.5 UT1

Lat. = 09°22.9'S   Sun Alt. = 63.2°
Long. = 012°26.2'E   Sun Azm. = 335.4°
Path Width = 71.7 km   Duration = 01m50.6s

0 1000 2000 3000 4000 5000
Kilometers

Annular Solar Eclipse of 2041 Oct 25

Greatest Eclipse = 01:36:21.7 TD (= 01:35:02.2 UT1)

Eclipse Magnitude = 0.9467 Saros Series = 134

Gamma = 0.4133 Saros Member = 45 of 71

Sun at Greatest Eclipse
(Geocentric Coordinates)

R.A. = 13h59m22.0s
Dec. = -12°10'20.1"
S.D. = 00°16'04.9"
H.P. = 00°00'08.8"

Moon at Greatest Eclipse
(Geocentric Coordinates)

R.A. = 14h00m02.5s
Dec. = -11°49'54.3"
S.D. = 00°15'00.8"
H.P. = 00°55'06.0"

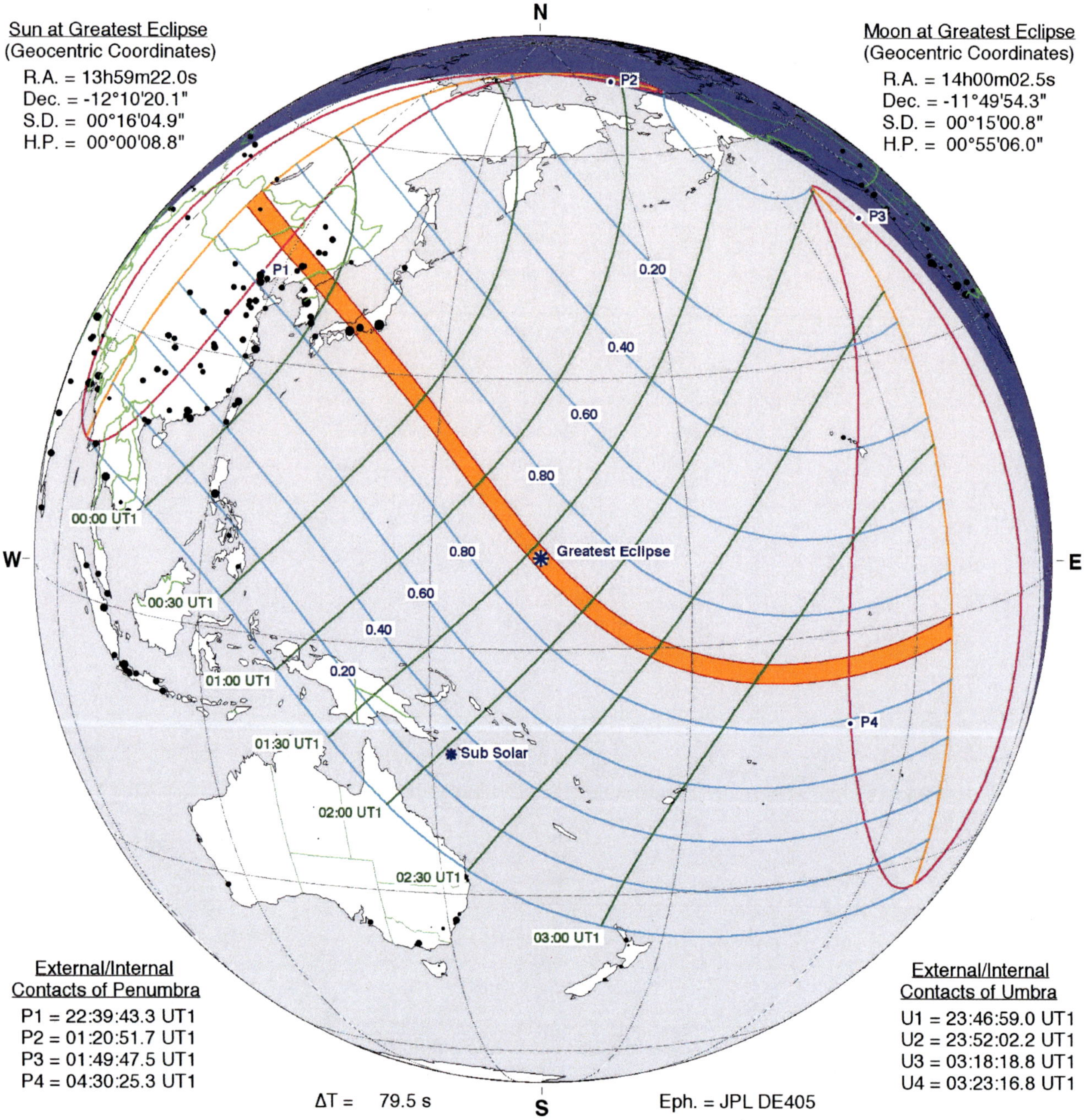

External/Internal
Contacts of Penumbra

P1 = 22:39:43.3 UT1
P2 = 01:20:51.7 UT1
P3 = 01:49:47.5 UT1
P4 = 04:30:25.3 UT1

External/Internal
Contacts of Umbra

U1 = 23:46:59.0 UT1
U2 = 23:52:02.2 UT1
U3 = 03:18:18.8 UT1
U4 = 03:23:16.8 UT1

ΔT = 79.5 s Eph. = JPL DE405

Circumstances at Greatest Eclipse: 01:35:02.2 UT1

Lat. = 09°56.0'N Sun Alt. = 65.5°
Long. = 162°49.8'E Sun Azm. = 205.6°
Path Width = 213.1 km Duration = 06m07.1s

Circumstances at Greatest Duration: 01:50:35.9 UT1

Lat. = 06°42.8'N Sun Alt. = 64.2°
Long. = 166°01.4'E Sun Azm. = 223.0°
Path Width = 218.3 km Duration = 06m08.4s

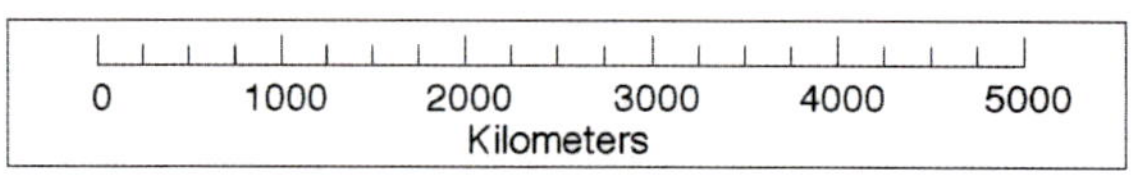

www.EclipseWise.com

Total Solar Eclipse of 2042 Apr 20

Greatest Eclipse = 02:17:30.1 TD (= 02:16:10.4 UT1)

Eclipse Magnitude = 1.0614 Saros Series = 139

Gamma = 0.2956 Saros Member = 31 of 71

Sun at Greatest Eclipse
(Geocentric Coordinates)

R.A. = 01h52m12.4s
Dec. = +11°31'19.4"
S.D. = 00°15'55.3"
H.P. = 00°00'08.8"

Moon at Greatest Eclipse
(Geocentric Coordinates)

R.A. = 01h51m39.9s
Dec. = +11°47'27.9"
S.D. = 00°16'37.6"
H.P. = 01°01'01.4"

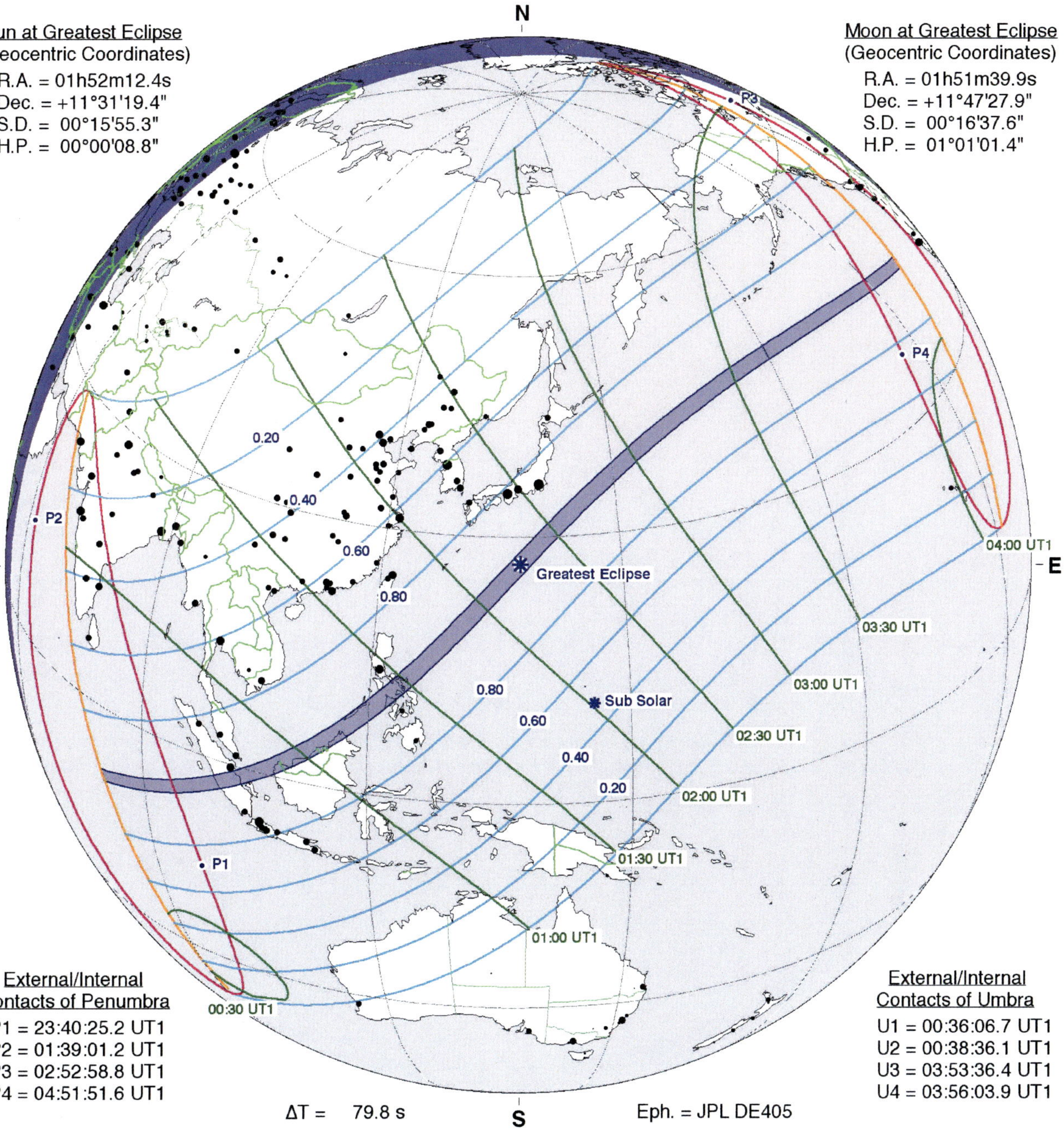

External/Internal Contacts of Penumbra

P1 = 23:40:25.2 UT1
P2 = 01:39:01.2 UT1
P3 = 02:52:58.8 UT1
P4 = 04:51:51.6 UT1

External/Internal Contacts of Umbra

U1 = 00:36:06.7 UT1
U2 = 00:38:36.1 UT1
U3 = 03:53:36.4 UT1
U4 = 03:56:03.9 UT1

ΔT = 79.8 s Eph. = JPL DE405

Circumstances at Greatest Eclipse: 02:16:10.4 UT1

Lat. = 26°57.2'N	Sun Alt. = 72.7°
Long. = 137°17.0'E	Sun Azm. = 151.2°
Path Width = 210.4 km	Duration = 04m51.0s

Circumstances at Greatest Duration: 02:20:00.4 UT1

Lat. = 27°59.1'N	Sun Alt. = 72.5°
Long. = 138°25.2'E	Sun Azm. = 159.0°
Path Width = 209.7 km	Duration = 04m51.2s

0 1000 2000 3000 4000 5000
Kilometers

Annular Solar Eclipse of 2042 Oct 14

Greatest Eclipse = 02:00:41.9 TD (= 01:59:21.8 UT1)

Eclipse Magnitude = 0.9301 Saros Series = 144

Gamma = -0.3030 Saros Member = 18 of 70

Sun at Greatest Eclipse
(Geocentric Coordinates)

R.A. = 13h17m05.8s
Dec. = -08°08'35.1"
S.D. = 00°16'01.9"
H.P. = 00°00'08.8"

Moon at Greatest Eclipse
(Geocentric Coordinates)

R.A. = 13h16m35.0s
Dec. = -08°23'00.1"
S.D. = 00°14'41.9"
H.P. = 00°53'56.6"

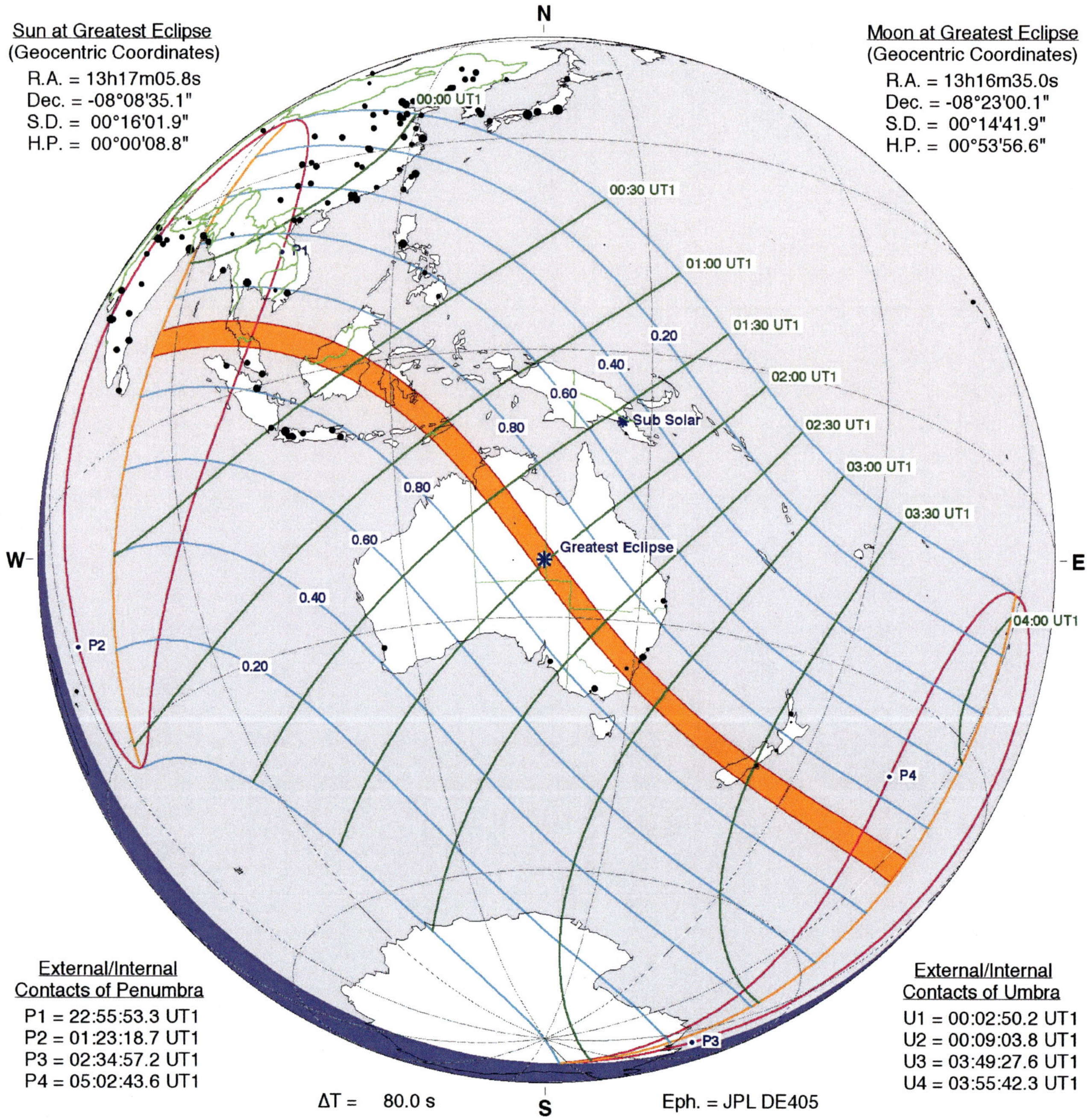

External/Internal
Contacts of Penumbra

P1 = 22:55:53.3 UT1
P2 = 01:23:18.7 UT1
P3 = 02:34:57.2 UT1
P4 = 05:02:43.6 UT1

External/Internal
Contacts of Umbra

U1 = 00:02:50.2 UT1
U2 = 00:09:03.8 UT1
U3 = 03:49:27.6 UT1
U4 = 03:55:42.3 UT1

ΔT = 80.0 s Eph. = JPL DE405

Circumstances at Greatest Eclipse: 01:59:21.8 UT1

Lat. = 23°44.7'S Sun Alt. = 72.2°
Long. = 137°47.4'E Sun Azm. = 30.1°
Path Width = 273.3 km Duration = 07m44.2s

Circumstances at Greatest Duration: 02:13:39.4 UT1

Lat. = 27°13.1'S Sun Alt. = 70.8°
Long. = 140°49.9'E Sun Azm. = 6.8°
Path Width = 270.8 km Duration = 07m45.4s

0 1000 2000 3000 4000 5000
Kilometers

Total Solar Eclipse of 2043 Apr 09

Greatest Eclipse = 18:57:49.4 TD (= 18:56:29.1 UT1)

Eclipse Magnitude = 1.0096 Saros Series = 149

Gamma = 1.0031 Saros Member = 22 of 71

<u>Sun at Greatest Eclipse</u>
(Geocentric Coordinates)
R.A. = 01h13m12.2s
Dec. = +07°45'05.1"
S.D. = 00°15'58.1"
H.P. = 00°00'08.8"

<u>Moon at Greatest Eclipse</u>
(Geocentric Coordinates)
R.A. = 01h11m17.3s
Dec. = +08°39'09.1"
S.D. = 00°16'38.0"
H.P. = 01°01'02.7"

N

W

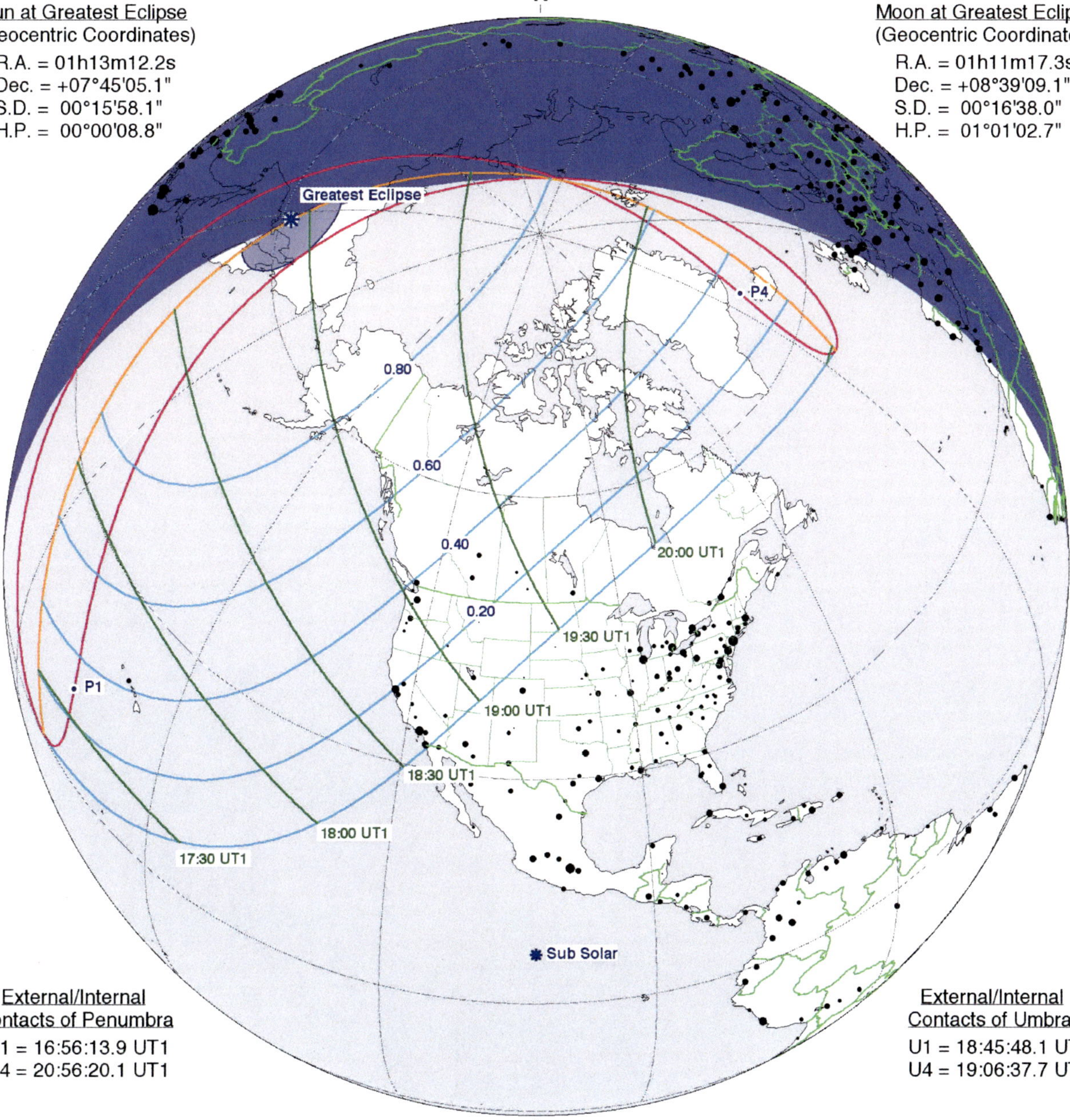

E

<u>External/Internal Contacts of Penumbra</u>
P1 = 16:56:13.9 UT1
P4 = 20:56:20.1 UT1

<u>External/Internal Contacts of Umbra</u>
U1 = 18:45:48.1 UT1
U4 = 19:06:37.7 UT1

ΔT = 80.3 s S Eph. = JPL DE405

<u>Circumstances at Greatest Eclipse: 18:56:29.1 UT1</u>

Lat. = 61°18.7'N Sun Alt. = 0.0°
Long. = 151°56.2'E Sun Azm. = 73.7°
Path Width = 0.0 km Duration = 00m00.0s

0 1000 2000 3000 4000 5000
Kilometers

Annular Solar Eclipse of 2043 Oct 03

Greatest Eclipse = 03:01:48.9 TD (= 03:00:28.4 UT1)

Eclipse Magnitude = 0.9497 Saros Series = 154

Gamma = -1.0102 Saros Member = 8 of 71

Sun at Greatest Eclipse
(Geocentric Coordinates)

R.A. = 12h36m02.9s
Dec. = -03°53'04.6"
S.D. = 00°15'58.8"
H.P. = 00°00'08.8"

Moon at Greatest Eclipse
(Geocentric Coordinates)

R.A. = 12h34m15.0s
Dec. = -04°41'56.9"
S.D. = 00°15'05.1"
H.P. = 00°55'21.7"

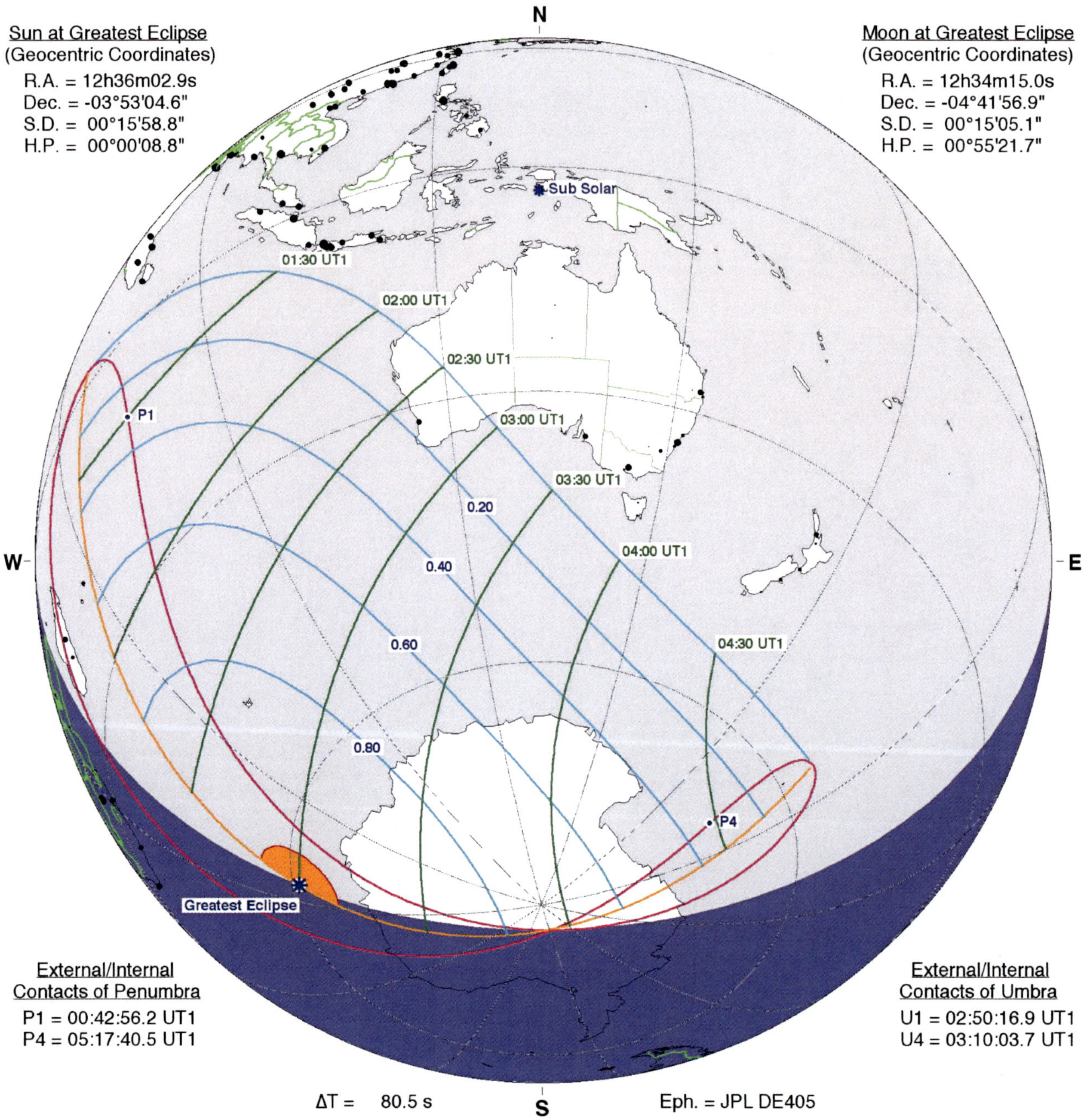

External/Internal
Contacts of Penumbra

P1 = 00:42:56.2 UT1
P4 = 05:17:40.5 UT1

External/Internal
Contacts of Umbra

U1 = 02:50:16.9 UT1
U4 = 03:10:03.7 UT1

ΔT = 80.5 s Eph. = JPL DE405

Circumstances at Greatest Eclipse: 03:00:28.4 UT1

Lat. = 60°57.6'S Sun Alt. = 0.0°
Long. = 035°14.0'E Sun Azm. = 98.0°
Path Width = 0.0 km Duration = 00m00.0s

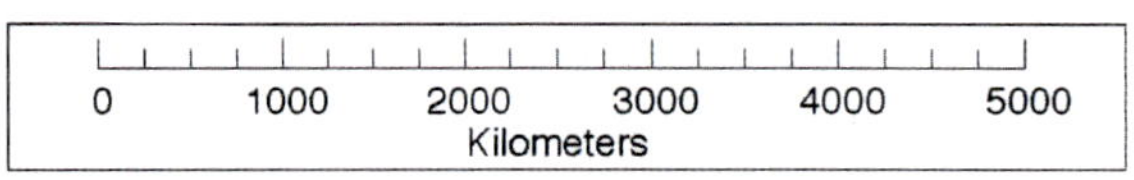

Annular Solar Eclipse of 2044 Feb 28

Greatest Eclipse = 20:24:39.5 TD (= 20:23:18.7 UT1)

Eclipse Magnitude = 0.9600 | Saros Series = 121

Gamma = -0.9954 | Saros Member = 62 of 71

Sun at Greatest Eclipse
(Geocentric Coordinates)

R.A. = 22h45m44.1s
Dec. = -07°51'30.6"
S.D. = 00°16'08.8"
H.P. = 00°00'08.9"

Moon at Greatest Eclipse
(Geocentric Coordinates)

R.A. = 22h47m30.6s
Dec. = -08°41'25.7"
S.D. = 00°15'29.6"
H.P. = 00°56'51.8"

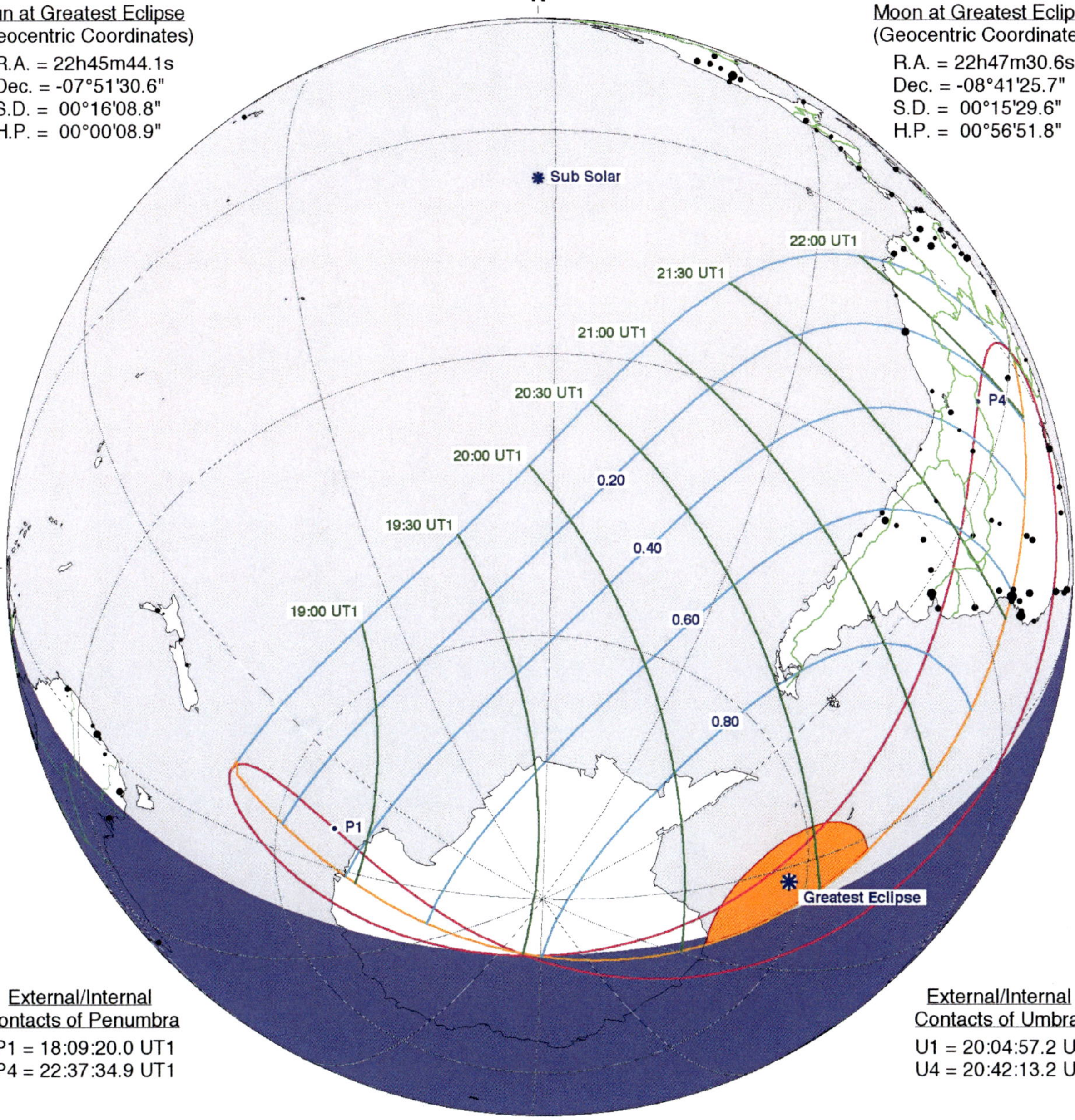

External/Internal
Contacts of Penumbra

P1 = 18:09:20.0 UT1
P4 = 22:37:34.9 UT1

External/Internal
Contacts of Umbra

U1 = 20:04:57.2 UT1
U4 = 20:42:13.2 UT1

ΔT = 80.8 s | Eph. = JPL DE405

Circumstances at Greatest Eclipse: 20:23:18.7 UT1

Lat. = 62°10.9'S | Sun Alt. = 3.7°
Long. = 025°38.6'W | Sun Azm. = 260.1°
Path Width =2237.9 km | Duration = 02m27.2s

Circumstances at Greatest Duration: 20:16:25.2 UT1

Lat. = 64°41.7'S | Sun Alt. = 0.0°
Long. = 014°00.7'W | Sun Azm. = 251.4°
Path Width =2582.2 km | Duration = 02m27.4s

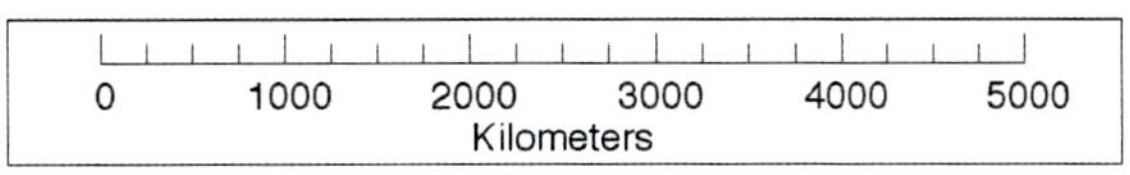

Total Solar Eclipse of 2044 Aug 23

Greatest Eclipse = 01:17:01.7 TD (= 01:15:40.6 UT1)

Eclipse Magnitude = 1.0364
Gamma = 0.9613

Saros Series = 126
Saros Member = 49 of 72

Sun at Greatest Eclipse
(Geocentric Coordinates)
R.A. = 10h10m33.4s
Dec. = +11°16'02.2"
S.D. = 00°15'48.9"
H.P. = 00°00'08.7"

Moon at Greatest Eclipse
(Geocentric Coordinates)
R.A. = 10h12m17.2s
Dec. = +12°07'34.4"
S.D. = 00°16'19.6"
H.P. = 00°59'55.1"

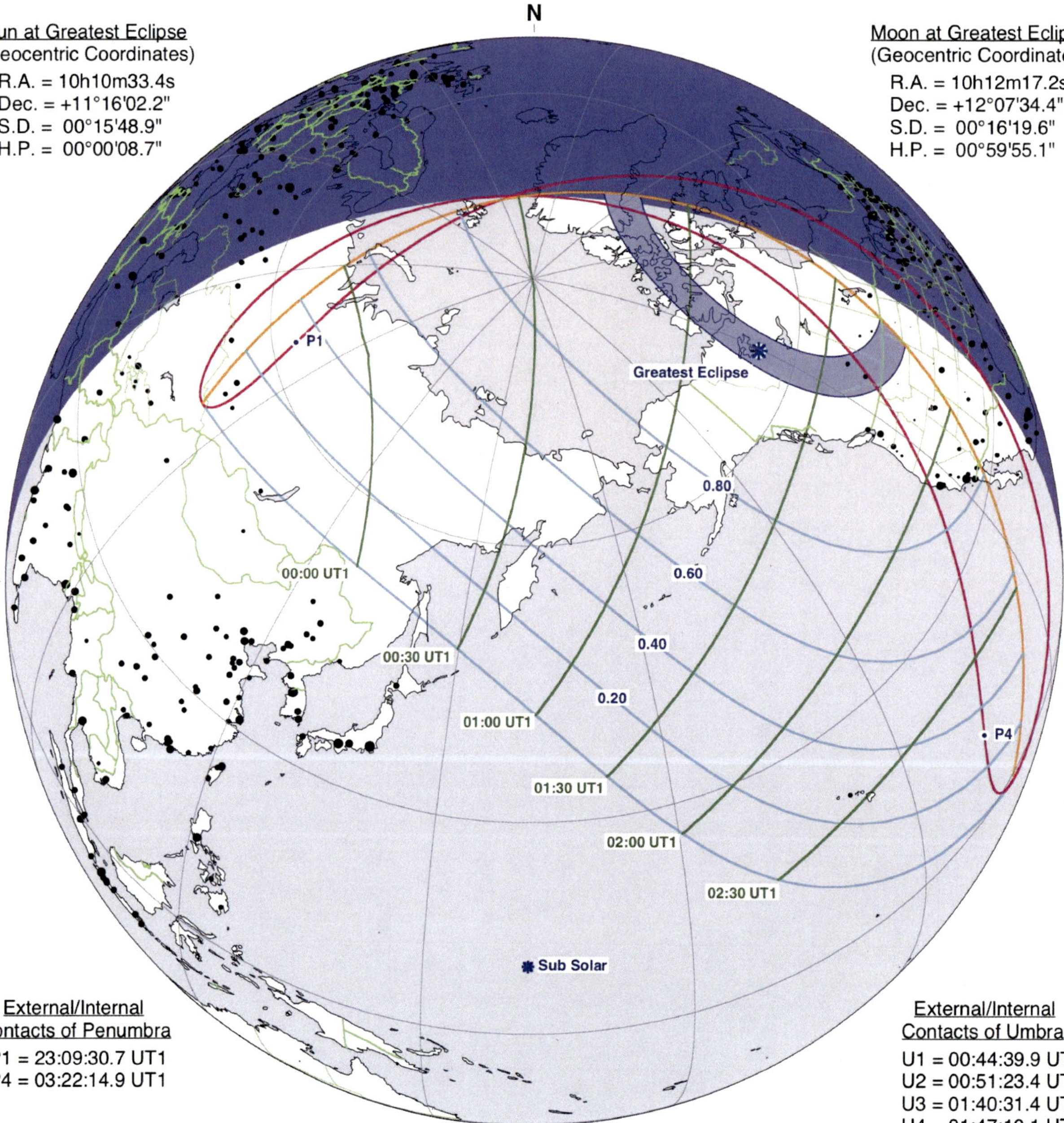

External/Internal
Contacts of Penumbra
P1 = 23:09:30.7 UT1
P4 = 03:22:14.9 UT1

External/Internal
Contacts of Umbra
U1 = 00:44:39.9 UT1
U2 = 00:51:23.4 UT1
U3 = 01:40:31.4 UT1
U4 = 01:47:10.1 UT1

ΔT = 81.0 s

Eph. = JPL DE405

Circumstances at Greatest Eclipse: 01:15:40.6 UT1

Lat. = 64°20.1'N
Long. = 120°28.6'W
Path Width = 452.7 km
Sun Alt. = 15.4°
Sun Azm. = 263.9°
Duration = 02m03.9s

Circumstances at Greatest Duration: 01:15:14.8 UT1

Lat. = 64°34.1'N
Long. = 120°25.0'W
Path Width = 451.6 km
Sun Alt. = 15.4°
Sun Azm. = 263.8°
Duration = 02m03.9s

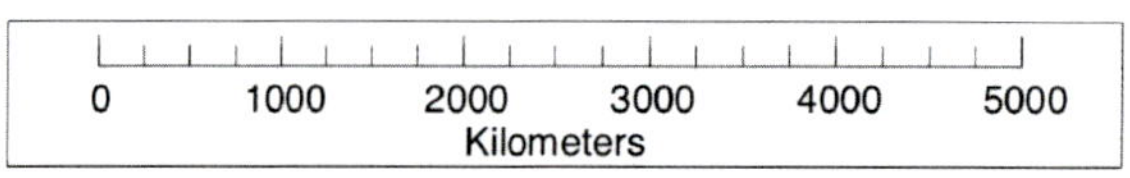

Annular Solar Eclipse of 2045 Feb 16

Greatest Eclipse = 23:56:06.6 TD (= 23:54:45.3 UT1)

Eclipse Magnitude = 0.9285

Gamma = -0.3125

Saros Series = 131

Saros Member = 52 of 70

Sun at Greatest Eclipse
(Geocentric Coordinates)

R.A. = 22h03m27.1s
Dec. = -11°55'04.8"
S.D. = 00°16'11.2"
H.P. = 00°00'08.9"

Moon at Greatest Eclipse
(Geocentric Coordinates)

R.A. = 22h03m57.6s
Dec. = -12°10'17.7"
S.D. = 00°14'48.9"
H.P. = 00°54'22.2"

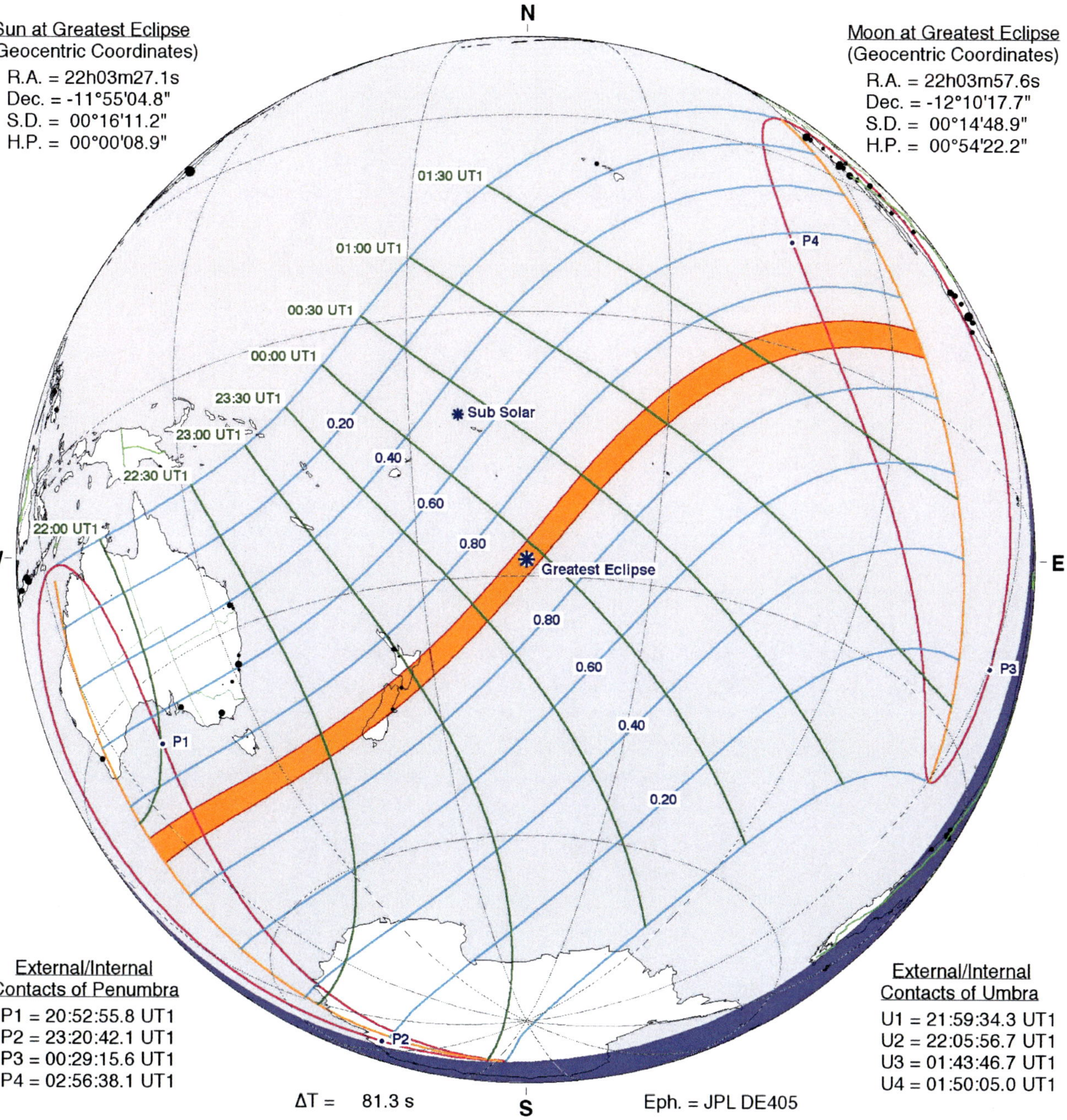

External/Internal Contacts of Penumbra

P1 = 20:52:55.8 UT1
P2 = 23:20:42.1 UT1
P3 = 00:29:15.6 UT1
P4 = 02:56:38.1 UT1

External/Internal Contacts of Umbra

U1 = 21:59:34.3 UT1
U2 = 22:05:56.7 UT1
U3 = 01:43:46.7 UT1
U4 = 01:50:05.0 UT1

ΔT = 81.3 s

Eph. = JPL DE405

Circumstances at Greatest Eclipse: 23:54:45.3 UT1

Lat. = 28°15.4'S	Sun Alt. = 71.6°
Long. = 166°13.8'W	Sun Azm. = 331.0°
Path Width = 281.2 km	Duration = 07m46.5s

Circumstances at Greatest Duration: 23:37:15.9 UT1

Lat. = 32°20.8'S	Sun Alt. = 69.6°
Long. = 170°31.5'W	Sun Azm. = 359.1°
Path Width = 278.7 km	Duration = 07m48.7s

0 1000 2000 3000 4000 5000
Kilometers

Total Solar Eclipse of 2045 Aug 12

Greatest Eclipse = 17:42:39.1 TD (= 17:41:17.5 UT1)

Eclipse Magnitude = 1.0774 Saros Series = 136

Gamma = 0.2116 Saros Member = 39 of 71

Sun at Greatest Eclipse
(Geocentric Coordinates)
R.A. = 09h31m17.7s
Dec. = +14°40'40.5"
S.D. = 00°15'47.0"
H.P. = 00°00'08.7"

Moon at Greatest Eclipse
(Geocentric Coordinates)
R.A. = 09h31m39.7s
Dec. = +14°52'29.9"
S.D. = 00°16'43.3"
H.P. = 01°01'22.3"

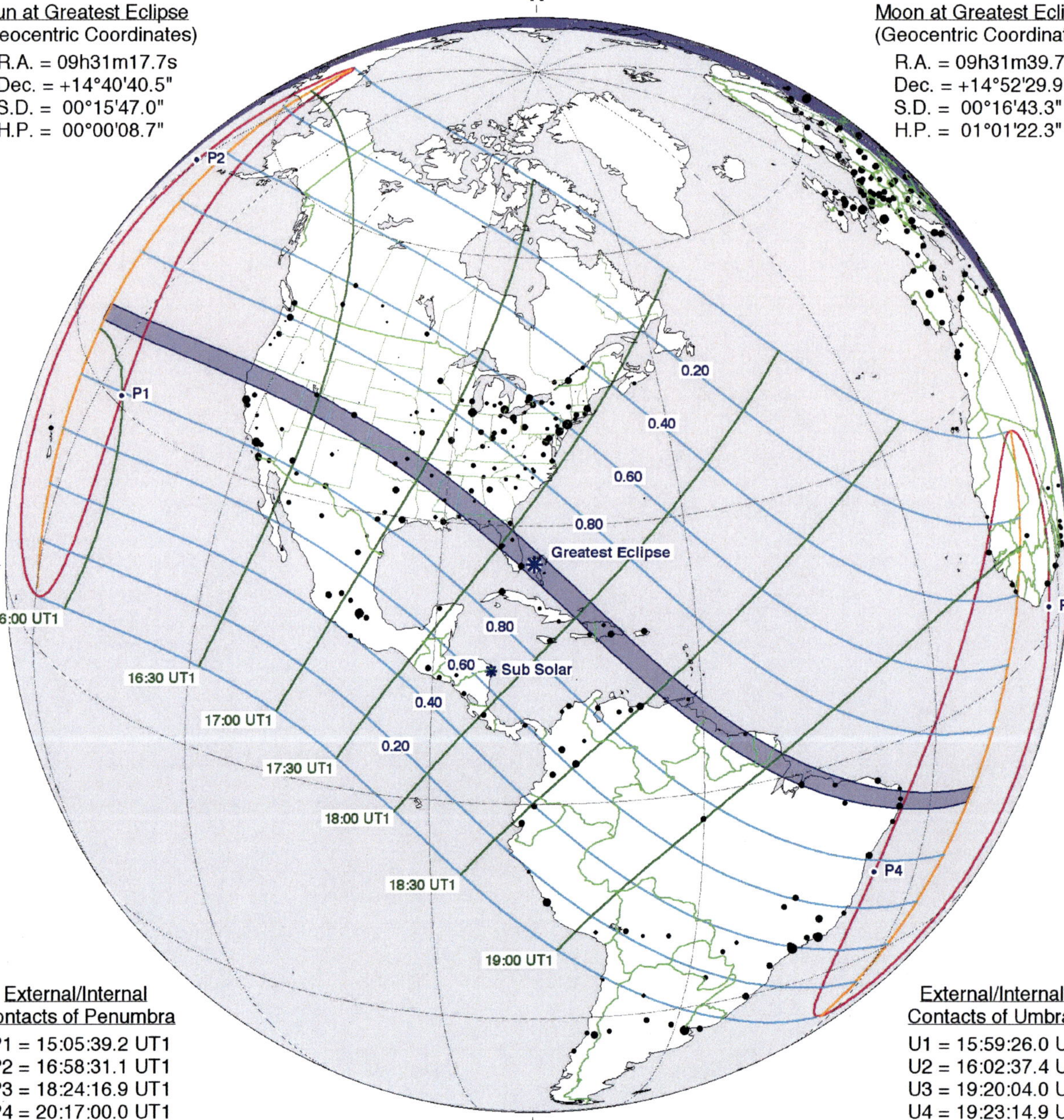

External/Internal
Contacts of Penumbra
P1 = 15:05:39.2 UT1
P2 = 16:58:31.1 UT1
P3 = 18:24:16.9 UT1
P4 = 20:17:00.0 UT1

External/Internal
Contacts of Umbra
U1 = 15:59:26.0 UT1
U2 = 16:02:37.4 UT1
U3 = 19:20:04.0 UT1
U4 = 19:23:14.9 UT1

ΔT = 81.6 s

Eph. = JPL DE405

Circumstances at Greatest Eclipse: 17:41:17.5 UT1

Lat. = 25°54.6'N Sun Alt. = 77.6°
Long. = 078°34.1'W Sun Azm. = 205.7°
Path Width = 255.6 km Duration = 06m05.7s

Circumstances at Greatest Duration: 17:35:29.1 UT1

Lat. = 27°19.3'N Sun Alt. = 77.2°
Long. = 080°22.9'W Sun Azm. = 189.9°
Path Width = 254.8 km Duration = 06m06.2s

0	1000	2000	3000	4000	5000

Kilometers

Annular Solar Eclipse of 2046 Feb 05

Greatest Eclipse = 23:06:26.2 TD (= 23:05:04.3 UT1)

Eclipse Magnitude = 0.9232 Saros Series = 141
Gamma = 0.3765 Saros Member = 25 of 70

Sun at Greatest Eclipse
(Geocentric Coordinates)
R.A. = 21h19m00.8s
Dec. = -15°38'42.4"
S.D. = 00°16'13.2"
H.P. = 00°00'08.9"

Moon at Greatest Eclipse
(Geocentric Coordinates)
R.A. = 21h18m27.2s
Dec. = -15°20'02.1"
S.D. = 00°14'46.0"
H.P. = 00°54'11.7"

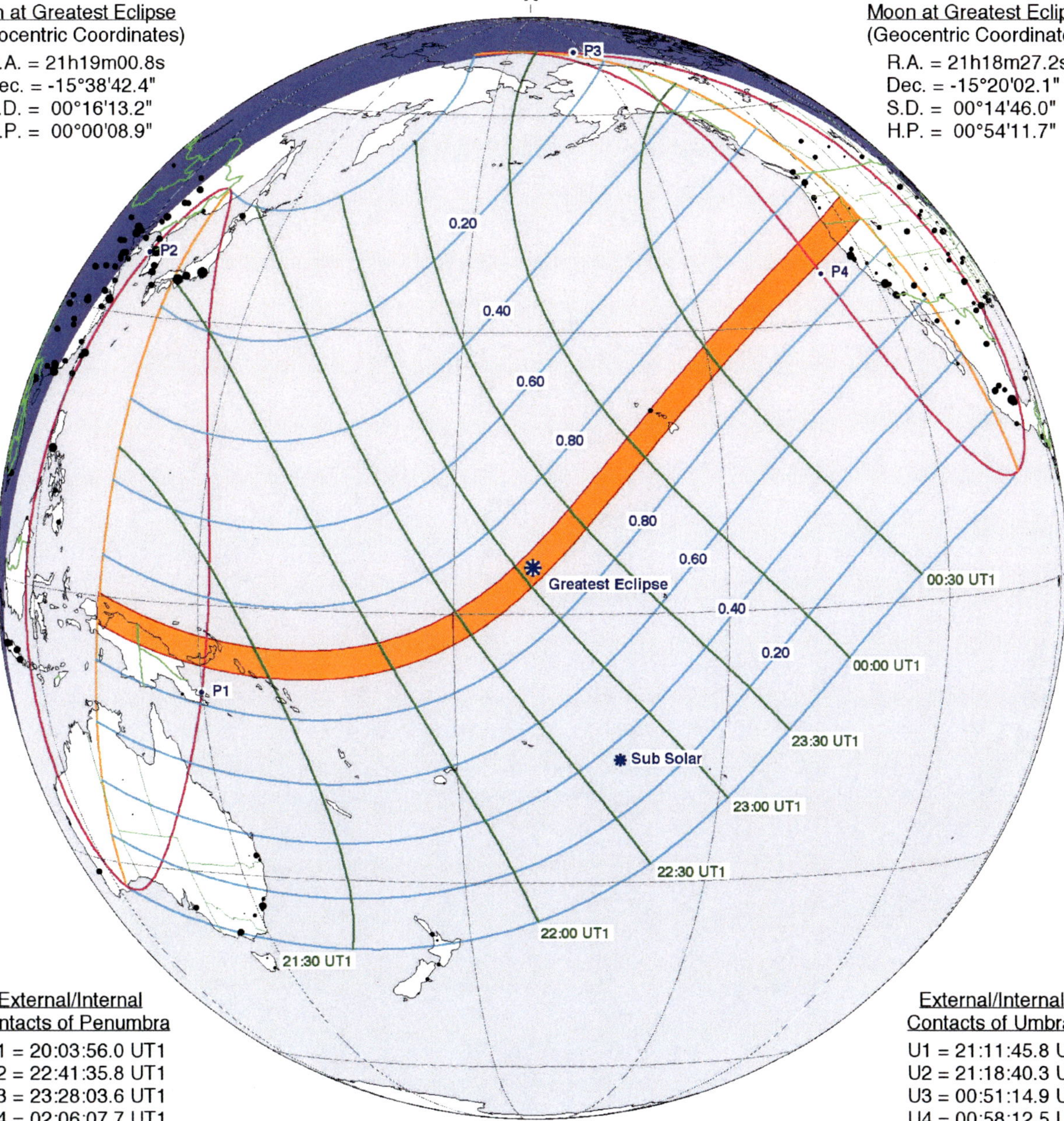

External/Internal
Contacts of Penumbra
P1 = 20:03:56.0 UT1
P2 = 22:41:35.8 UT1
P3 = 23:28:03.6 UT1
P4 = 02:06:07.7 UT1

External/Internal
Contacts of Umbra
U1 = 21:11:45.8 UT1
U2 = 21:18:40.3 UT1
U3 = 00:51:14.9 UT1
U4 = 00:58:12.5 UT1

ΔT = 81.8 s Eph. = JPL DE405

Circumstances at Greatest Eclipse: 23:05:04.3 UT1
Lat. = 04°46.9'N Sun Alt. = 67.9°
Long. = 171°25.6'W Sun Azm. = 157.4°
Path Width = 310.1 km Duration = 09m42.4s

Circumstances at Greatest Duration: 22:49:00.6 UT1
Lat. = 01°56.1'N Sun Alt. = 66.4°
Long. = 174°40.8'W Sun Azm. = 138.7°
Path Width = 316.7 km Duration = 09m45.6s

0 1000 2000 3000 4000 5000
Kilometers

Total Solar Eclipse of 2046 Aug 02

Greatest Eclipse = 10:21:13.4 TD (= 10:19:51.3 UT1)

Eclipse Magnitude = 1.0531 Saros Series = 146

Gamma = -0.5350 Saros Member = 29 of 76

Sun at Greatest Eclipse
(Geocentric Coordinates)
R.A. = 08h51m04.7s
Dec. = +17°39'03.1"
S.D. = 00°15'45.5"
H.P. = 00°00'08.7"

Moon at Greatest Eclipse
(Geocentric Coordinates)
R.A. = 08h50m16.1s
Dec. = +17°09'10.7"
S.D. = 00°16'21.8"
H.P. = 01°00'03.4"

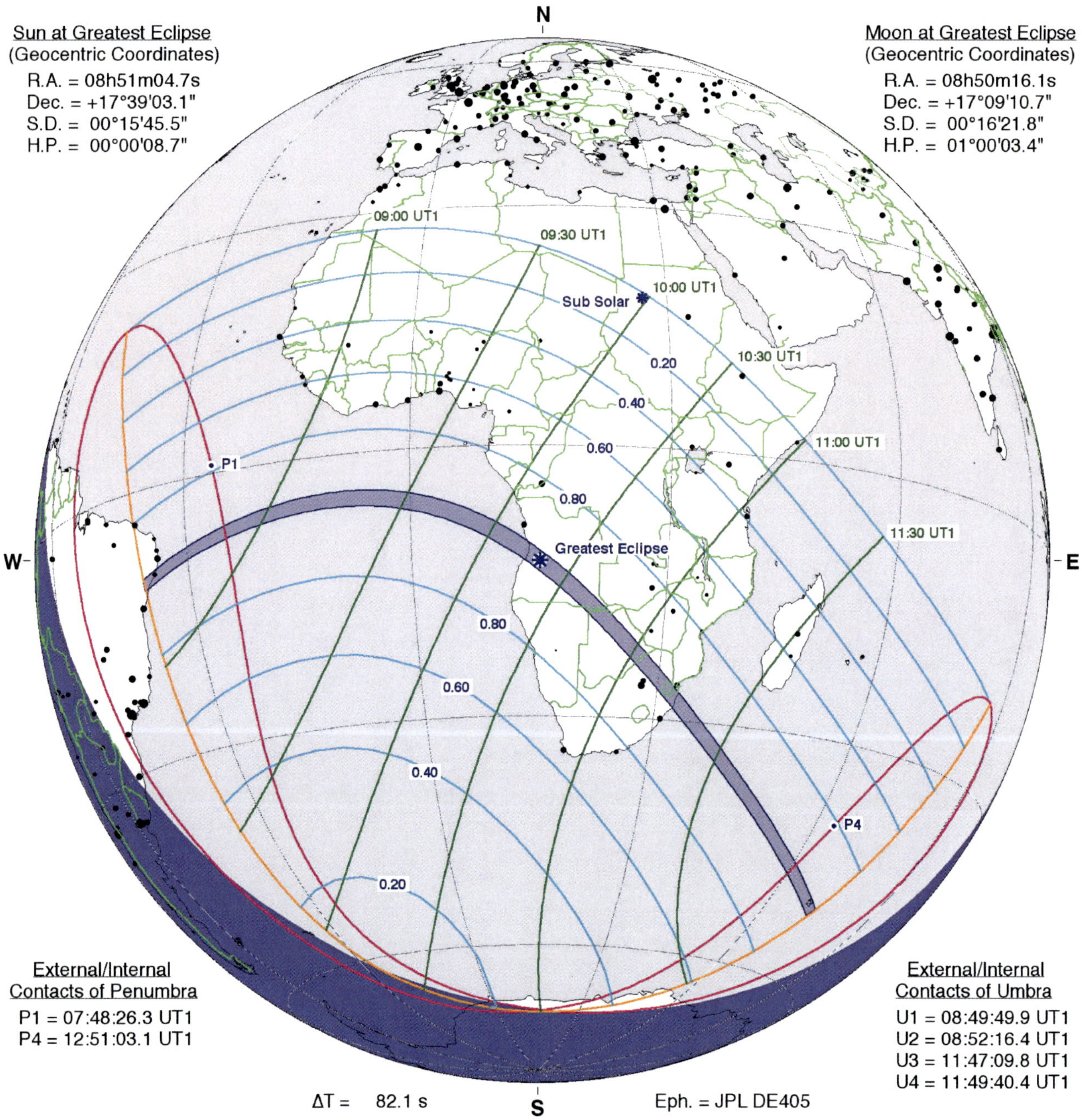

External/Internal Contacts of Penumbra
P1 = 07:48:26.3 UT1
P4 = 12:51:03.1 UT1

External/Internal Contacts of Umbra
U1 = 08:49:49.9 UT1
U2 = 08:52:16.4 UT1
U3 = 11:47:09.8 UT1
U4 = 11:49:40.4 UT1

ΔT = 82.1 s Eph. = JPL DE405

Circumstances at Greatest Eclipse: 10:19:51.3 UT1

Lat. = 12°43.8'S Sun Alt. = 57.6°
Long. = 015°08.8'E Sun Azm. = 20.7°
Path Width = 206.0 km Duration = 04m51.4s

Circumstances at Greatest Duration: 10:16:03.2 UT1

Lat. = 12°00.3'S Sun Alt. = 57.5°
Long. = 014°06.4'E Sun Azm. = 24.4°
Path Width = 207.2 km Duration = 04m51.6s

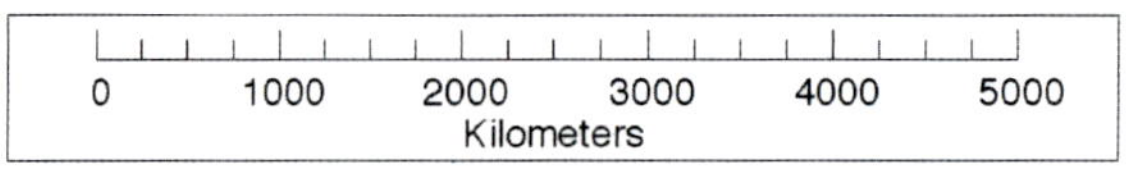

Partial Solar Eclipse of 2047 Jan 26

Greatest Eclipse = 01:33:17.8 TD (= 01:31:55.4 UT1)

Eclipse Magnitude = 0.8908 Saros Series = 151

Gamma = 1.0450 Saros Member = 16 of 72

Sun at Greatest Eclipse
(Geocentric Coordinates)
R.A. = 20h33m28.4s
Dec. = -18°46'10.9"
S.D. = 00°16'14.7"
H.P. = 00°00'08.9"

Moon at Greatest Eclipse
(Geocentric Coordinates)
R.A. = 20h32m04.0s
Dec. = -17°50'50.8"
S.D. = 00°15'23.2"
H.P. = 00°56'28.0"

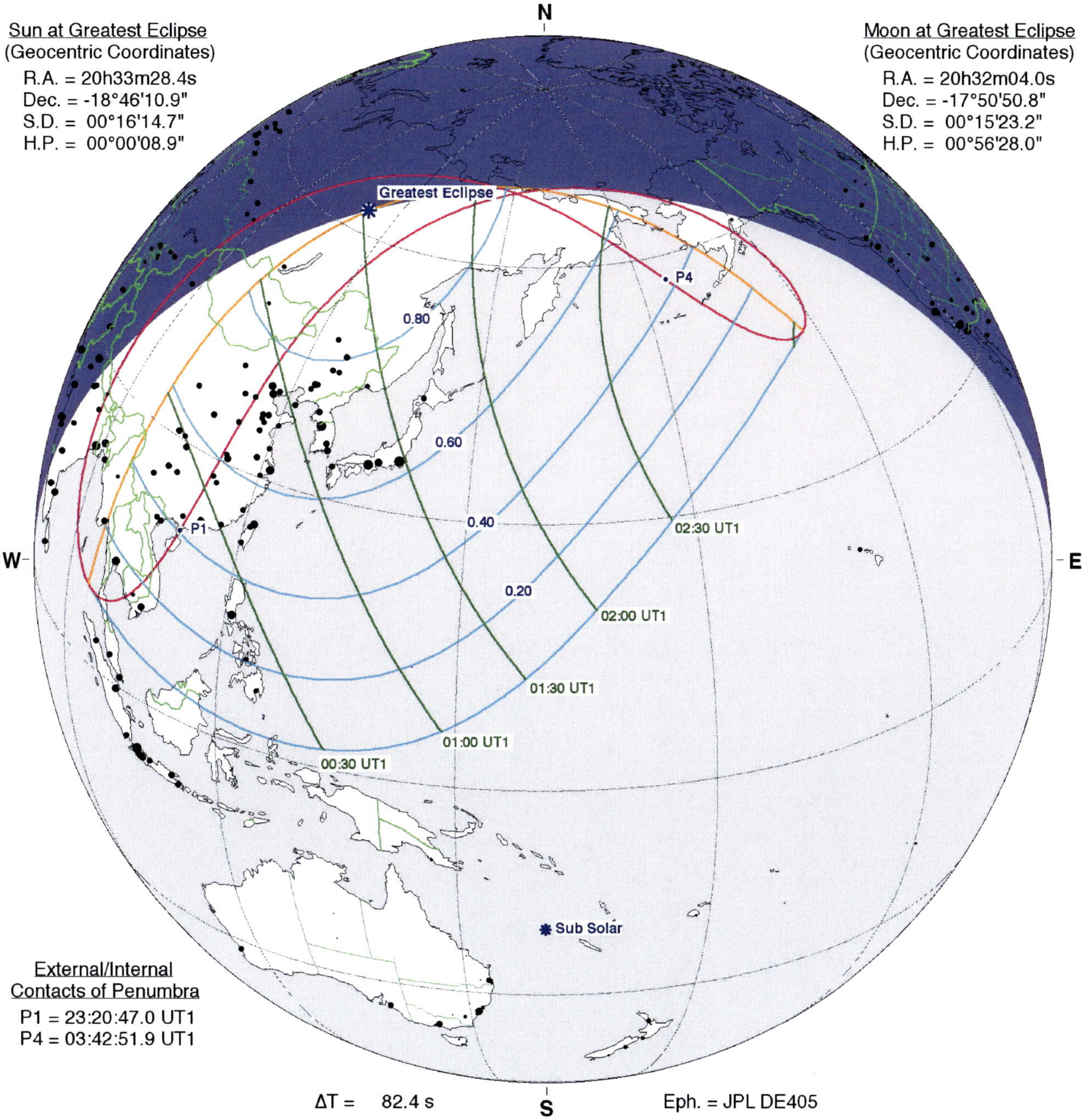

External/Internal
Contacts of Penumbra
P1 = 23:20:47.0 UT1
P4 = 03:42:51.9 UT1

ΔT = 82.4 s Eph. = JPL DE405

Circumstances at Greatest Eclipse: 01:31:55.4 UT1

Lat. = 62°52.0'N Sun Alt. = 0.0°

Long. = 111°42.3'E Sun Azm. = 134.9°

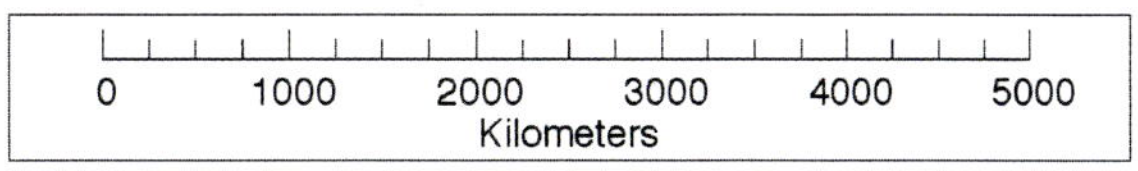

Partial Solar Eclipse of 2047 Jun 23

Greatest Eclipse = 10:52:30.6 TD (= 10:51:08.0 UT1)

Eclipse Magnitude = 0.3129
Gamma = 1.3766

Saros Series = 118
Saros Member = 70 of 72

Sun at Greatest Eclipse
(Geocentric Coordinates)
R.A. = 06h08m27.7s
Dec. = +23°25'10.2"
S.D. = 00°15'44.2"
H.P. = 00°00'08.7"

Moon at Greatest Eclipse
(Geocentric Coordinates)
R.A. = 06h09m05.2s
Dec. = +24°40'56.6"
S.D. = 00°15'07.9"
H.P. = 00°55'32.1"

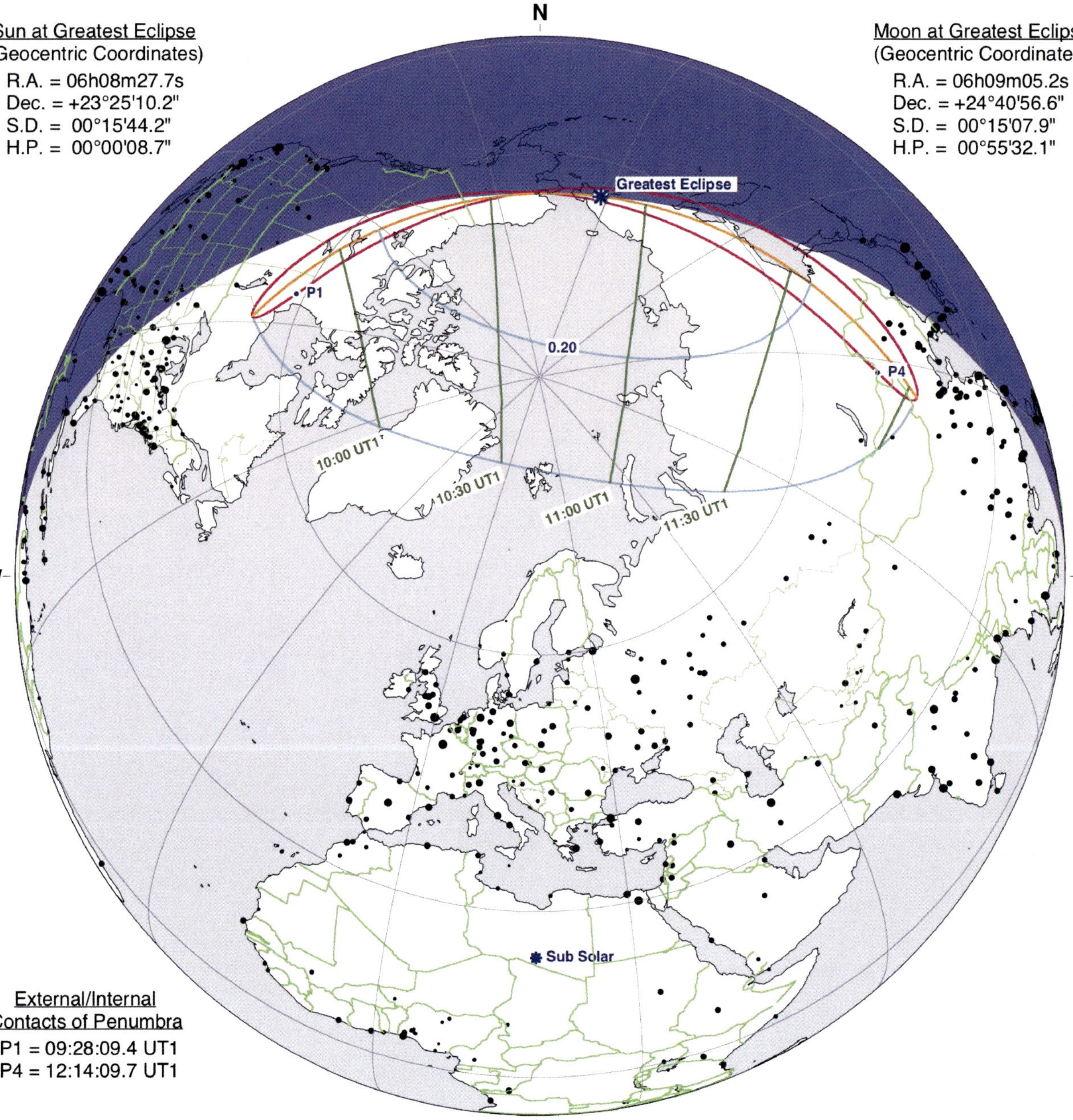

External/Internal
Contacts of Penumbra
P1 = 09:28:09.4 UT1
P4 = 12:14:09.7 UT1

ΔT = 82.6 s

Eph. = JPL DE405

Circumstances at Greatest Eclipse: 10:51:08.0 UT1

Lat. = 65°46.0'N
Long. = 178°02.4'W

Sun Alt. = 0.0°
Sun Azm. = 345.5°

0	1000	2000	3000	4000	5000

Kilometers

Partial Solar Eclipse of 2047 Jul 22

Greatest Eclipse = 22:36:17.4 TD (= 22:34:54.7 UT1)

Eclipse Magnitude = 0.3605 Saros Series = 156

Gamma = -1.3477 Saros Member = 3 of 69

Sun at Greatest Eclipse
(Geocentric Coordinates)
R.A. = 08h08m59.7s
Dec. = +20°07'53.9"
S.D. = 00°15'44.5"
H.P. = 00°00'08.7"

Moon at Greatest Eclipse
(Geocentric Coordinates)
R.A. = 08h07m21.2s
Dec. = +18°54'51.1"
S.D. = 00°15'32.1"
H.P. = 00°57'00.9"

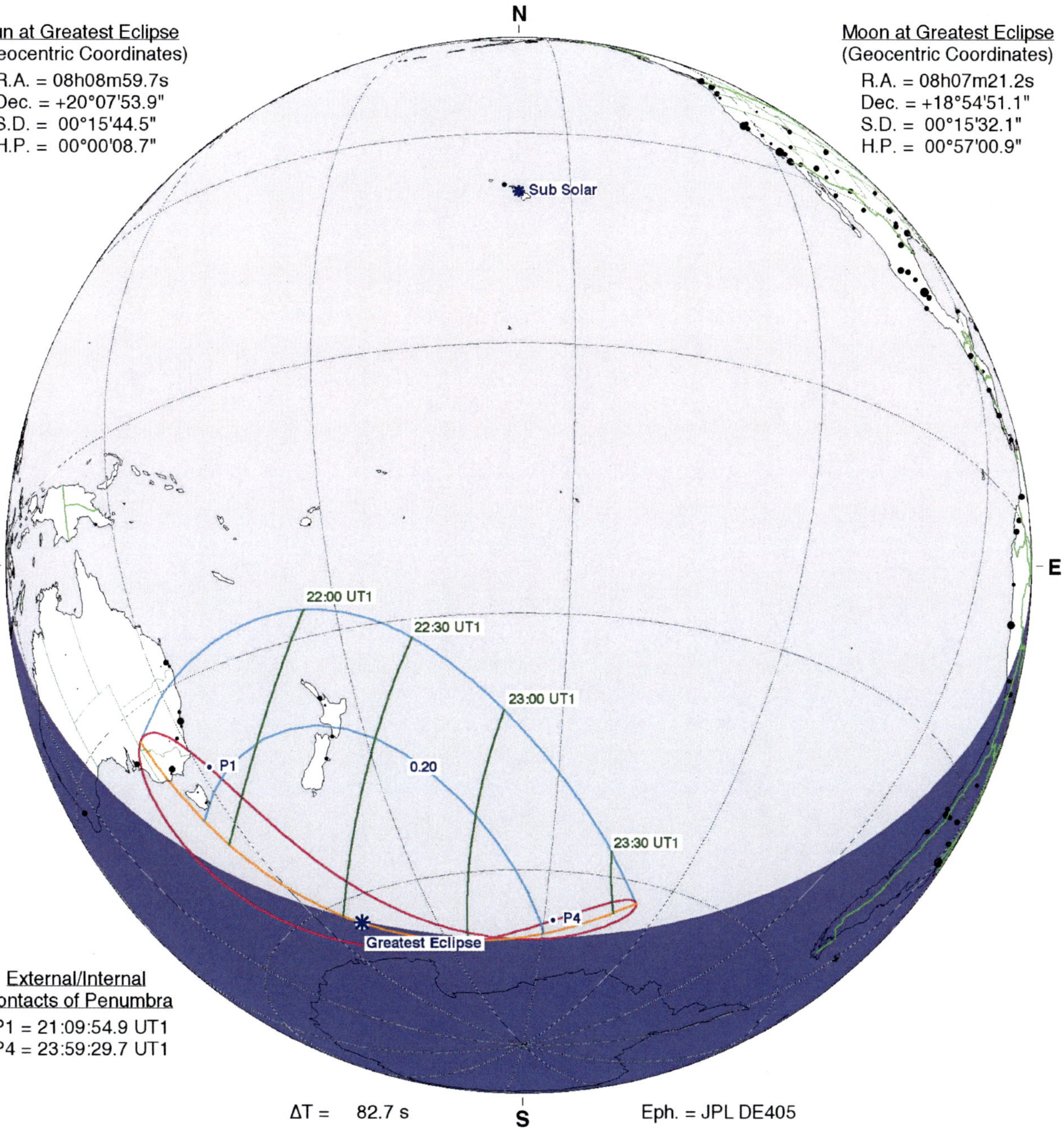

External/Internal
Contacts of Penumbra
P1 = 21:09:54.9 UT1
P4 = 23:59:29.7 UT1

ΔT = 82.7 s Eph. = JPL DE405

Circumstances at Greatest Eclipse: 22:34:54.7 UT1

Lat. = 63°26.5'S Sun Alt. = 0.0°
Long. = 160°08.3'E Sun Azm. = 39.7°

0 1000 2000 3000 4000 5000
Kilometers

Partial Solar Eclipse of 2047 Dec 16

Greatest Eclipse = 23:50:12.3 TD (= 23:48:49.4 UT1)

Eclipse Magnitude = 0.8817
Gamma = -1.0661

Saros Series = 123
Saros Member = 55 of 70

Sun at Greatest Eclipse
(Geocentric Coordinates)
R.A. = 17h37m56.6s
Dec. = -23°20'10.9"
S.D. = 00°16'15.0"
H.P. = 00°00'08.9"

Moon at Greatest Eclipse
(Geocentric Coordinates)
R.A. = 17h38m13.1s
Dec. = -24°24'51.1"
S.D. = 00°16'35.9"
H.P. = 01°00'54.9"

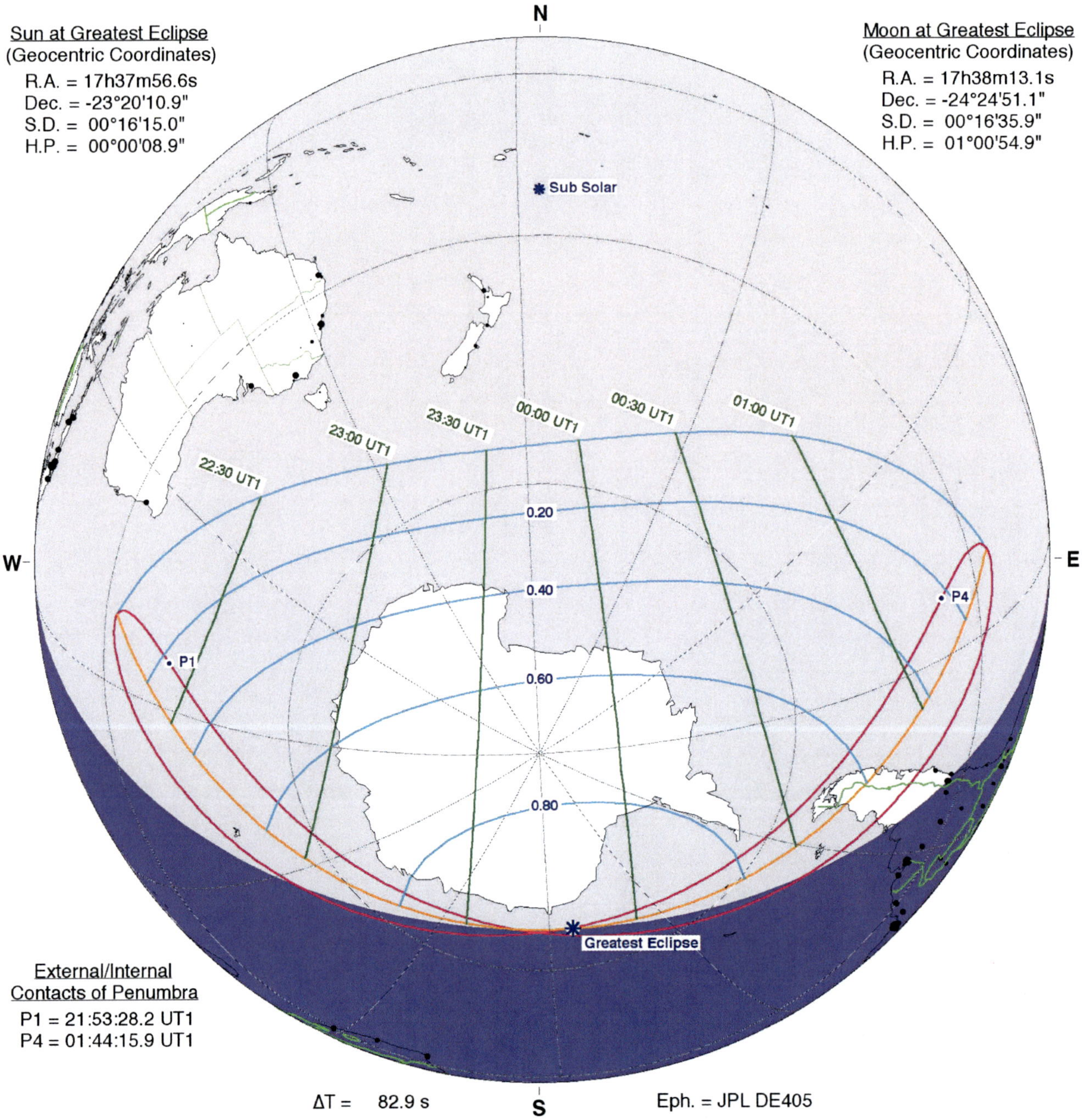

External/Internal Contacts of Penumbra
P1 = 21:53:28.2 UT1
P4 = 01:44:15.9 UT1

ΔT = 82.9 s

Eph. = JPL DE405

Circumstances at Greatest Eclipse: 23:48:49.4 UT1

Lat. = 66°26.6'S
Long. = 006°38.2'W

Sun Alt. = 0.0°
Sun Azm. = 187.7°

0 1000 2000 3000 4000 5000
Kilometers

Annular Solar Eclipse of 2048 Jun 11

Greatest Eclipse = 12:58:52.8 TD (= 12:57:29.6 UT1)

Eclipse Magnitude = 0.9441 | Saros Series = 128
Gamma = 0.6468 | Saros Member = 60 of 73

Sun at Greatest Eclipse
(Geocentric Coordinates)
R.A. = 05h22m03.9s
Dec. = +23°08'47.0"
S.D. = 00°15'45.1"
H.P. = 00°00'08.7"

Moon at Greatest Eclipse
(Geocentric Coordinates)
R.A. = 05h22m09.1s
Dec. = +23°43'34.6"
S.D. = 00°14'42.3"
H.P. = 00°53'58.0"

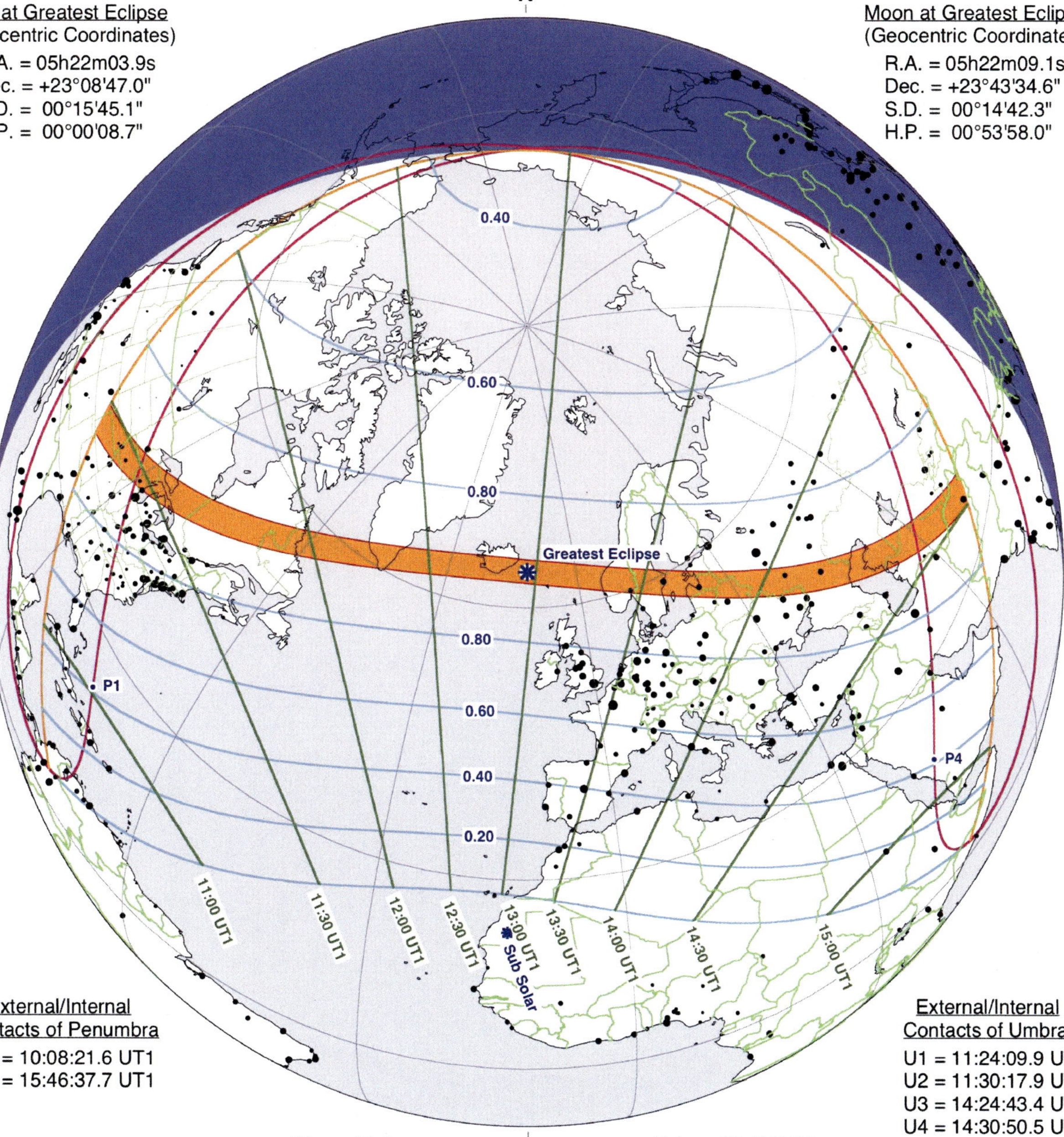

External/Internal Contacts of Penumbra
P1 = 10:08:21.6 UT1
P4 = 15:46:37.7 UT1

External/Internal Contacts of Umbra
U1 = 11:24:09.9 UT1
U2 = 11:30:17.9 UT1
U3 = 14:24:43.4 UT1
U4 = 14:30:50.5 UT1

ΔT = 83.2 s

Eph. = JPL DE405

Circumstances at Greatest Eclipse: 12:57:29.6 UT1
Lat. = 63°41.3'N | Sun Alt. = 49.4°
Long. = 011°32.5'W | Sun Azm. = 184.0°
Path Width = 271.5 km | Duration = 04m58.3s

Circumstances at Greatest Duration: 12:56:04.3 UT1
Lat. = 63°43.9'N | Sun Alt. = 49.4°
Long. = 012°43.9'W | Sun Azm. = 181.9°
Path Width = 271.5 km | Duration = 04m58.3s

0 1000 2000 3000 4000 5000
Kilometers

Total Solar Eclipse of 2048 Dec 05

Greatest Eclipse = 15:35:26.7 TD (= 15:34:03.2 UT1)

Eclipse Magnitude = 1.0440 Saros Series = 133

Gamma = -0.3973 Saros Member = 47 of 72

Sun at Greatest Eclipse
(Geocentric Coordinates)
R.A. = 16h51m20.5s
Dec. = -22°29'40.9"
S.D. = 00°16'13.8"
H.P. = 00°00'08.9"

Moon at Greatest Eclipse
(Geocentric Coordinates)
R.A. = 16h51m18.6s
Dec. = -22°53'56.4"
S.D. = 00°16'40.9"
H.P. = 01°01'13.3"

N

W

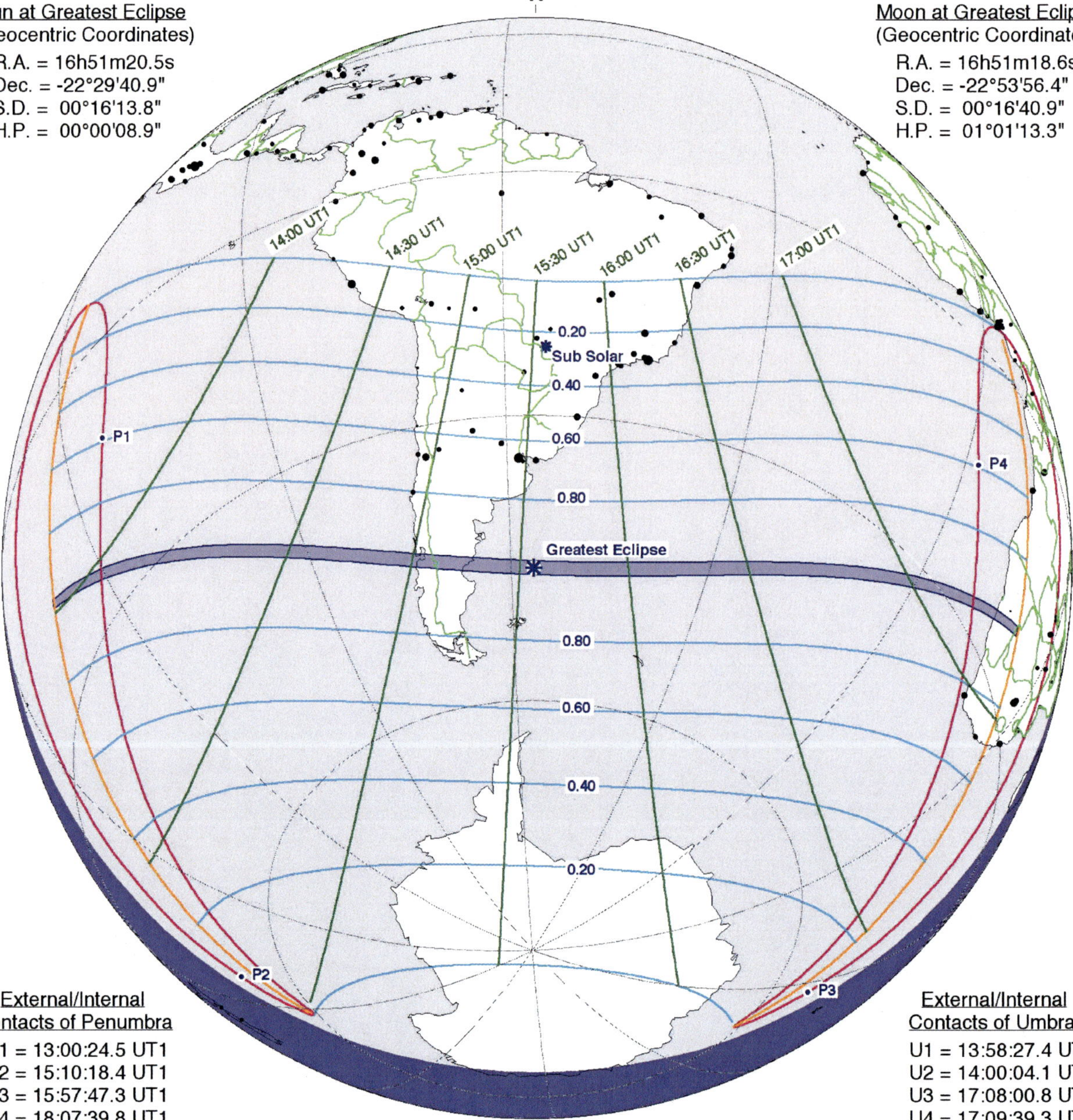

E

S

External/Internal
Contacts of Penumbra
P1 = 13:00:24.5 UT1
P2 = 15:10:18.4 UT1
P3 = 15:57:47.3 UT1
P4 = 18:07:39.8 UT1

External/Internal
Contacts of Umbra
U1 = 13:58:27.4 UT1
U2 = 14:00:04.1 UT1
U3 = 17:08:00.8 UT1
U4 = 17:09:39.3 UT1

ΔT = 83.5 s

Eph. = JPL DE405

Circumstances at Greatest Eclipse: 15:34:03.2 UT1

Lat. = 46°07.9'S Sun Alt. = 66.4°
Long. = 056°24.4'W Sun Azm. = 1.4°
Path Width = 160.3 km Duration = 03m27.6s

Circumstances at Greatest Duration: 15:34:47.8 UT1

Lat. = 46°08.5'S Sun Alt. = 66.4°
Long. = 056°00.0'W Sun Azm. = 0.0°
Path Width = 160.3 km Duration = 03m27.6s

0 1000 2000 3000 4000 5000
Kilometers

Annular Solar Eclipse of 2049 May 31

Greatest Eclipse = 13:59:58.8 TD (= 13:58:35.0 UT1)

Eclipse Magnitude = 0.9631 Saros Series = 138

Gamma = -0.1187 Saros Member = 33 of 70

Sun at Greatest Eclipse
(Geocentric Coordinates)
R.A. = 04h35m51.4s
Dec. = +22°01'26.4"
S.D. = 00°15'46.5"
H.P. = 00°00'08.7"

Moon at Greatest Eclipse
(Geocentric Coordinates)
R.A. = 04h35m52.6s
Dec. = +21°54'56.7"
S.D. = 00°14'57.8"
H.P. = 00°54'55.1"

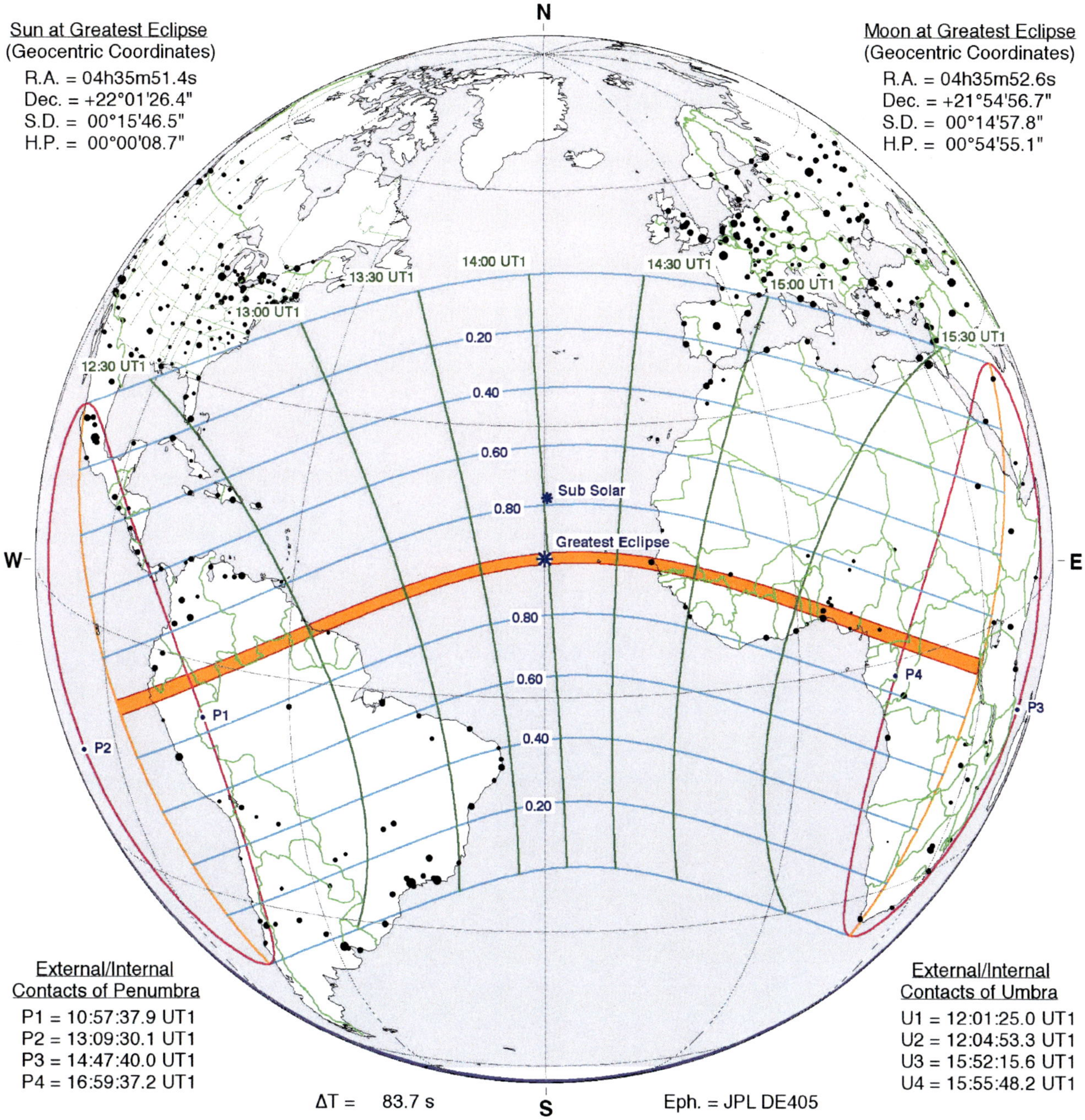

External/Internal
Contacts of Penumbra
P1 = 10:57:37.9 UT1
P2 = 13:09:30.1 UT1
P3 = 14:47:40.0 UT1
P4 = 16:59:37.2 UT1

External/Internal
Contacts of Umbra
U1 = 12:01:25.0 UT1
U2 = 12:04:53.3 UT1
U3 = 15:52:15.6 UT1
U4 = 15:55:48.2 UT1

ΔT = 83.7 s Eph. = JPL DE405

Circumstances at Greatest Eclipse: 13:58:35.0 UT1

Lat. = 15°18.6'N Sun Alt. = 83.3°
Long. = 029°53.3'W Sun Azm. = 357.6°
Path Width = 134.4 km Duration = 04m45.3s

Circumstances at Greatest Duration: 14:04:03.2 UT1

Lat. = 15°23.8'N Sun Alt. = 82.8°
Long. = 028°27.6'W Sun Azm. = 336.6°
Path Width = 134.6 km Duration = 04m45.5s

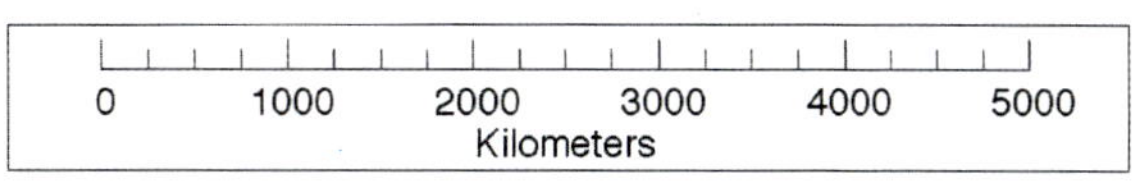

Hybrid Solar Eclipse of 2049 Nov 25

Greatest Eclipse = 05:33:47.9 TD (= 05:32:23.9 UT1)

Eclipse Magnitude = 1.0057
Gamma = 0.2943

Saros Series = 143
Saros Member = 25 of 72

Sun at Greatest Eclipse
(Geocentric Coordinates)
R.A. = 16h05m24.9s
Dec. = -20°49'25.8"
S.D. = 00°16'12.0"
H.P. = 00°00'08.9"

Moon at Greatest Eclipse
(Geocentric Coordinates)
R.A. = 16h05m31.7s
Dec. = -20°32'13.6"
S.D. = 00°16'02.3"
H.P. = 00°58'51.9"

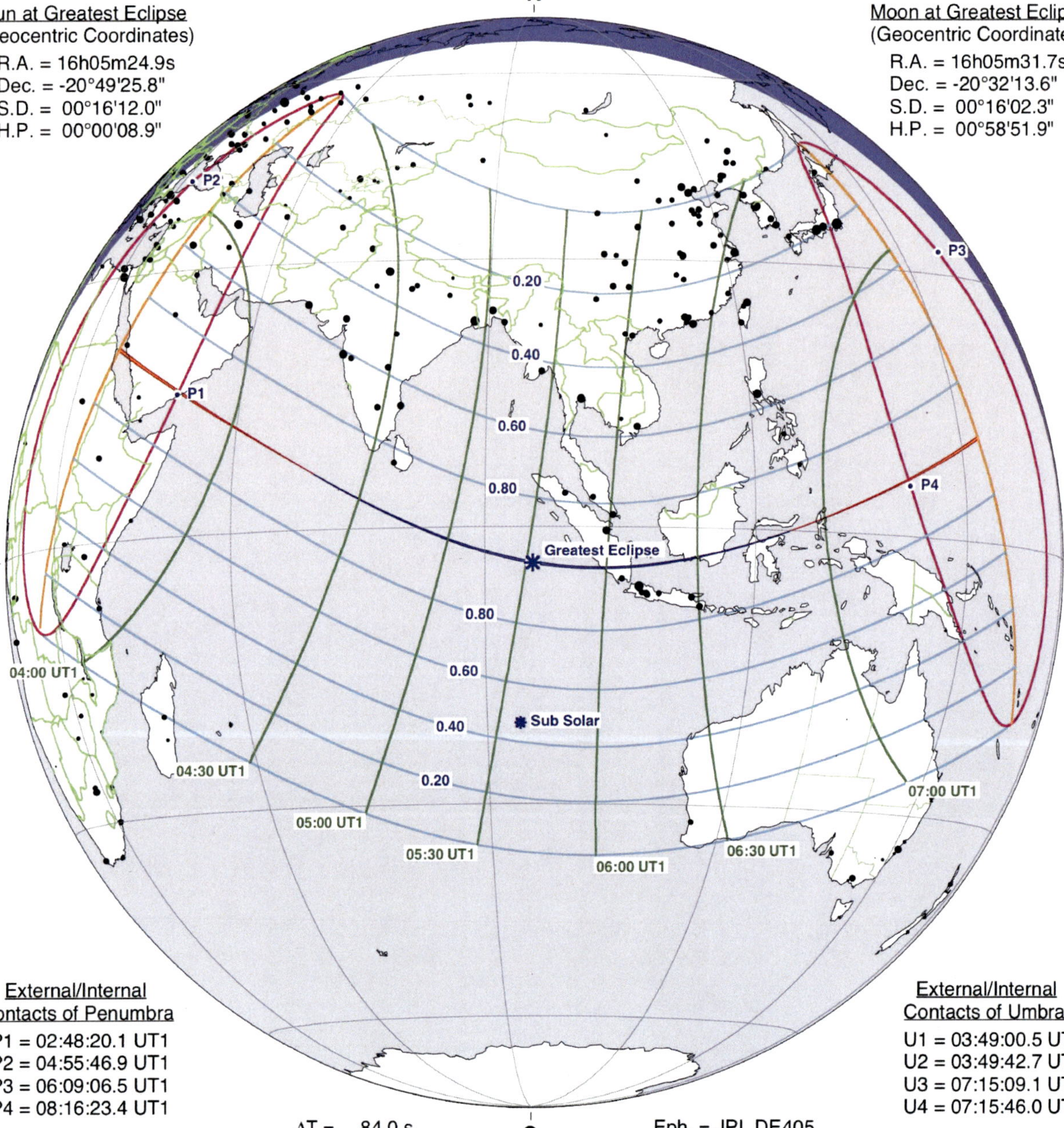

External/Internal
Contacts of Penumbra
P1 = 02:48:20.1 UT1
P2 = 04:55:46.9 UT1
P3 = 06:09:06.5 UT1
P4 = 08:16:23.4 UT1

External/Internal
Contacts of Umbra
U1 = 03:49:00.5 UT1
U2 = 03:49:42.7 UT1
U3 = 07:15:09.1 UT1
U4 = 07:15:46.0 UT1

ΔT = 84.0 s

Eph. = JPL DE405

Circumstances at Greatest Eclipse: 05:32:23.9 UT1

Lat. = 03°48.2'S
Long. = 095°12.7'E
Path Width = 20.7 km
Sun Alt. = 72.9°
Sun Azm. = 185.0°
Duration = 00m37.5s

Circumstances at Greatest Duration: 03:49:21.6 UT1

Lat. = 21°02.0'N
Long. = 037°48.4'E
Path Width = 40.0 km
Sun Alt. = 0.0°
Sun Azm. = 112.4°
Duration = 00m41.8s

0 1000 2000 3000 4000 5000
Kilometers

Hybrid Solar Eclipse of 2050 May 20

Greatest Eclipse = 20:42:50.2 TD (= 20:41:25.9 UT1)

Eclipse Magnitude = 1.0038 Saros Series = 148

Gamma = -0.8688 Saros Member = 23 of 75

Sun at Greatest Eclipse
(Geocentric Coordinates)
R.A. = 03h51m25.4s
Dec. = +20°09'01.9"
S.D. = 00°15'48.3"
H.P. = 00°00'08.7"

Moon at Greatest Eclipse
(Geocentric Coordinates)
R.A. = 03h51m49.6s
Dec. = +19°19'17.1"
S.D. = 00°15'44.7"
H.P. = 00°57'47.0"

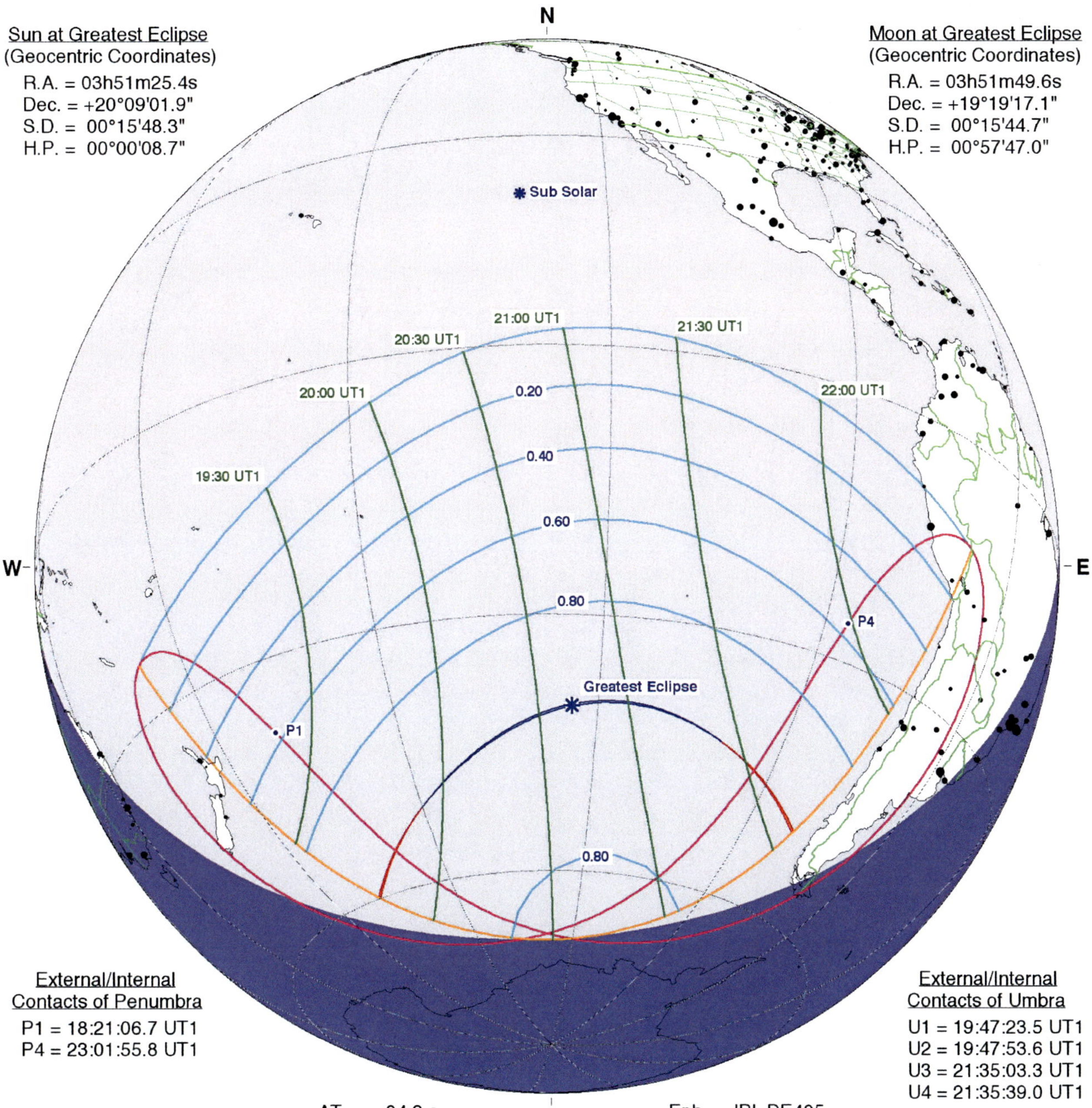

External/Internal
Contacts of Penumbra
P1 = 18:21:06.7 UT1
P4 = 23:01:55.8 UT1

External/Internal
Contacts of Umbra
U1 = 19:47:23.5 UT1
U2 = 19:47:53.6 UT1
U3 = 21:35:03.3 UT1
U4 = 21:35:39.0 UT1

ΔT = 84.3 s Eph. = JPL DE405

Circumstances at Greatest Eclipse: 20:41:25.9 UT1

Lat. = 40°05.5'S Sun Alt. = 29.4°
Long. = 123°45.1'W Sun Azm. = 352.0°
Path Width = 26.7 km Duration = 00m21.3s

Circumstances at Greatest Duration: 20:39:25.9 UT1

Lat. = 40°14.9'S Sun Alt. = 29.3°
Long. = 124°36.1'W Sun Azm. = 353.4°
Path Width = 26.6 km Duration = 00m21.4s

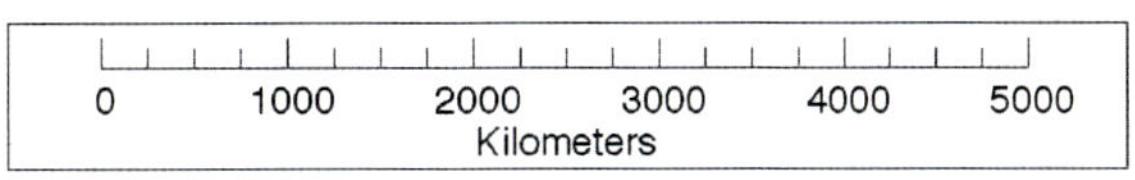

Partial Solar Eclipse of 2050 Nov 14

Greatest Eclipse = 13:30:52.8 TD (= 13:29:28.2 UT1)

Eclipse Magnitude = 0.8874 Saros Series = 153

Gamma = 1.0447 Saros Member = 11 of 70

Sun at Greatest Eclipse
(Geocentric Coordinates)
R.A. = 15h19m50.5s
Dec. = -18°21'19.2"
S.D. = 00°16'09.8"
H.P. = 00°00'08.9"

Moon at Greatest Eclipse
(Geocentric Coordinates)
R.A. = 15h20m29.5s
Dec. = -17°24'01.9"
S.D. = 00°15'10.6"
H.P. = 00°55'41.9"

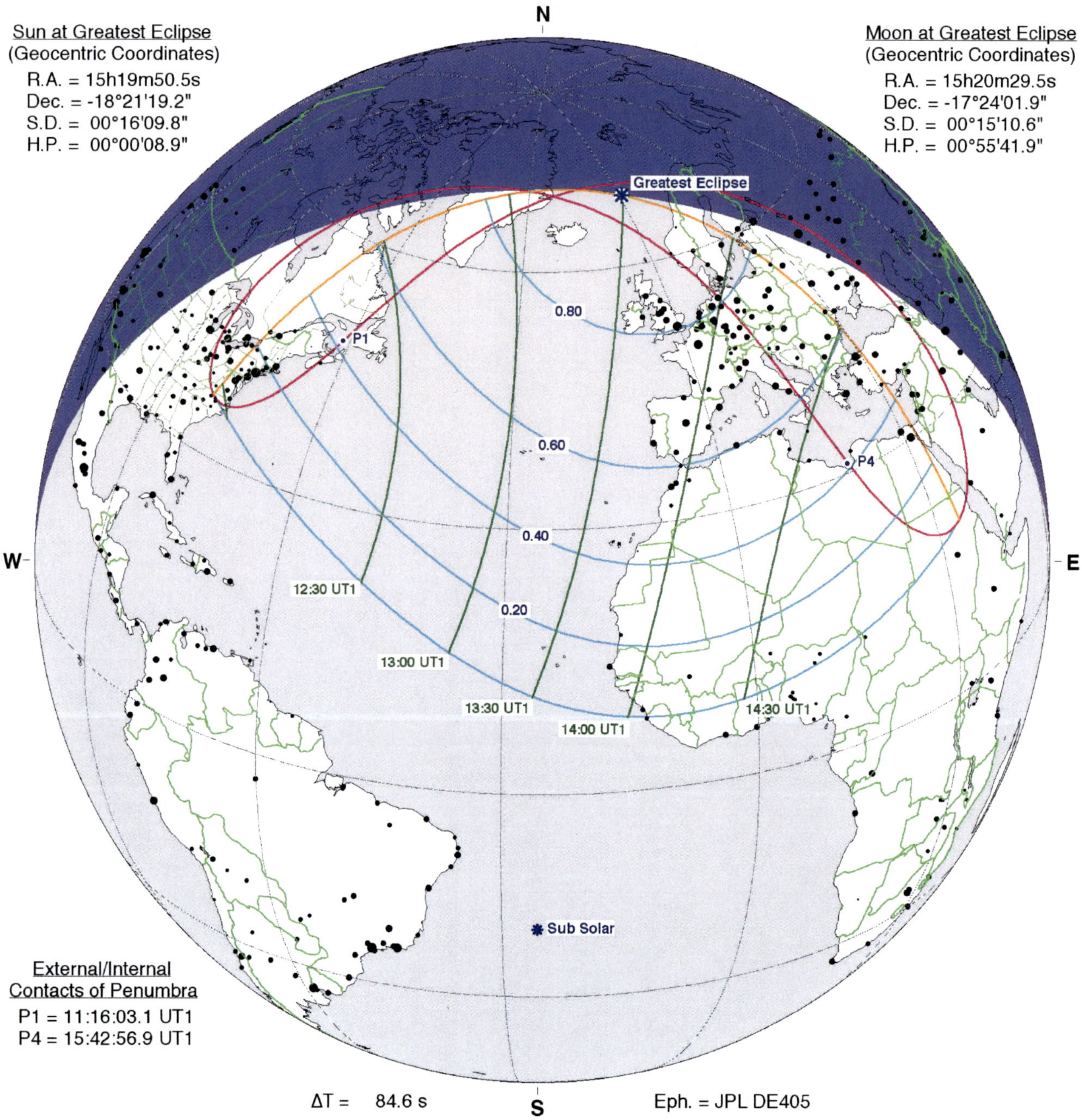

External/Internal Contacts of Penumbra
P1 = 11:16:03.1 UT1
P4 = 15:42:56.9 UT1

ΔT = 84.6 s Eph. = JPL DE405

Circumstances at Greatest Eclipse: 13:29:28.2 UT1

Lat. = 69°31.7'N Sun Alt. = 0.0°

Long. = 000°59.0'E Sun Azm. = 205.8°

0 1000 2000 3000 4000 5000
Kilometers

Partial Solar Eclipse of 2051 Apr 11

Greatest Eclipse = 02:10:38.6 TD (= 02:09:13.8 UT1)

Eclipse Magnitude = 0.9849 Saros Series = 120
Gamma = 1.0169 Saros Member = 63 of 71

Sun at Greatest Eclipse
(Geocentric Coordinates)
R.A. = 01h18m13.3s
Dec. = +08°15'12.8"
S.D. = 00°15'57.8"
H.P. = 00°00'08.8"

Moon at Greatest Eclipse
(Geocentric Coordinates)
R.A. = 01h17m01.7s
Dec. = +09°14'52.8"
S.D. = 00°16'42.8"
H.P. = 01°01'20.2"

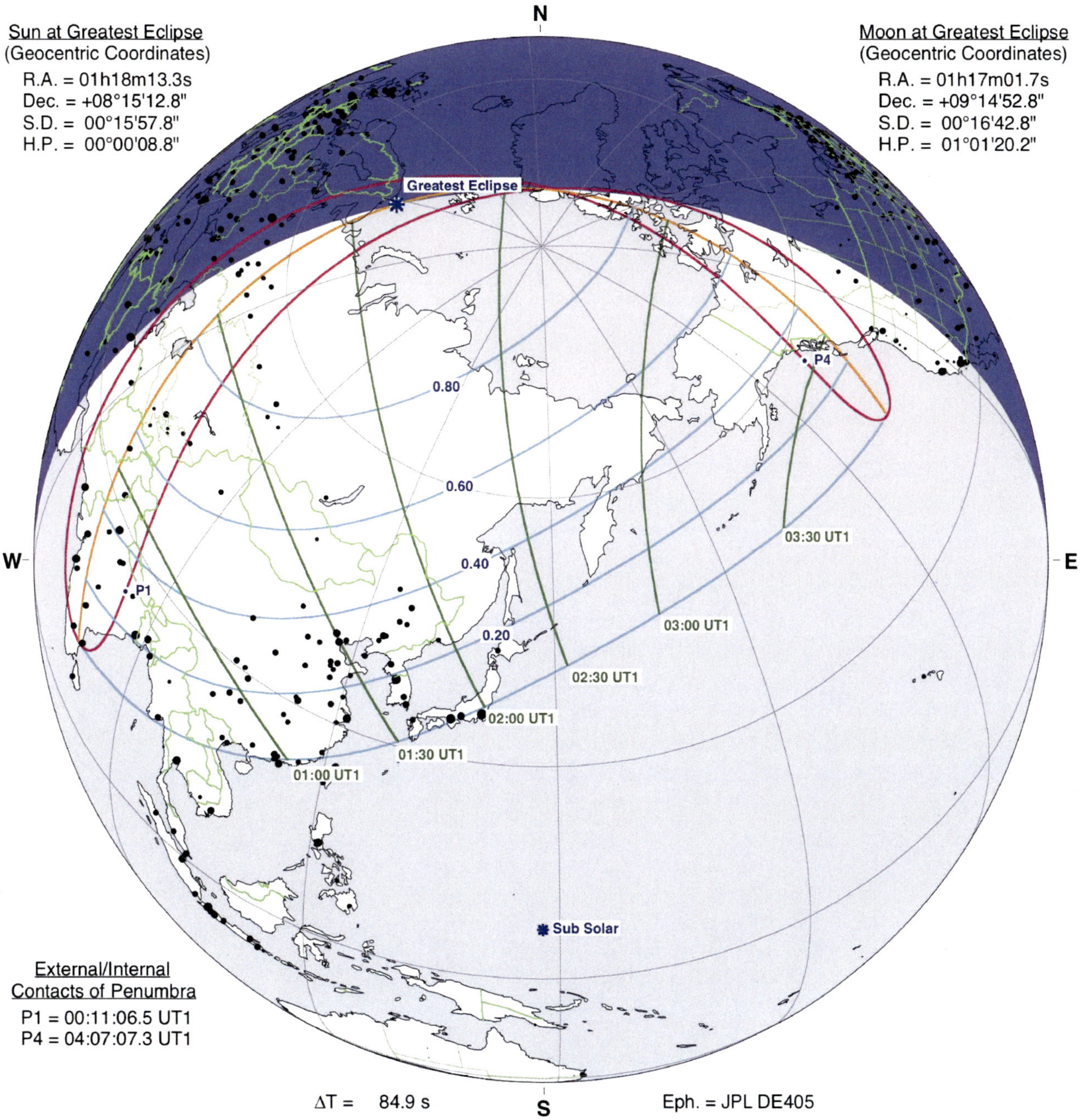

External/Internal
Contacts of Penumbra
P1 = 00:11:06.5 UT1
P4 = 04:07:07.3 UT1

ΔT = 84.9 s Eph. = JPL DE405

Circumstances at Greatest Eclipse: 02:09:13.8 UT1

Lat. = 71°37.6'N Sun Alt. = 0.0°
Long. = 032°08.7'E Sun Azm. = 62.9°

0 1000 2000 3000 4000 5000
Kilometers

Partial Solar Eclipse of 2051 Oct 04

Greatest Eclipse = 21:02:14.5 TD (= 21:00:49.4 UT1)

Eclipse Magnitude = 0.6024 Saros Series = 125

Gamma = -1.2094 Saros Member = 56 of 73

Sun at Greatest Eclipse
(Geocentric Coordinates)
R.A. = 12h42m39.3s
Dec. = -04°35'05.4"
S.D. = 00°15'59.2"
H.P. = 00°00'08.8"

Moon at Greatest Eclipse
(Geocentric Coordinates)
R.A. = 12h41m20.9s
Dec. = -05°37'21.2"
S.D. = 00°14'44.4"
H.P. = 00°54'05.8"

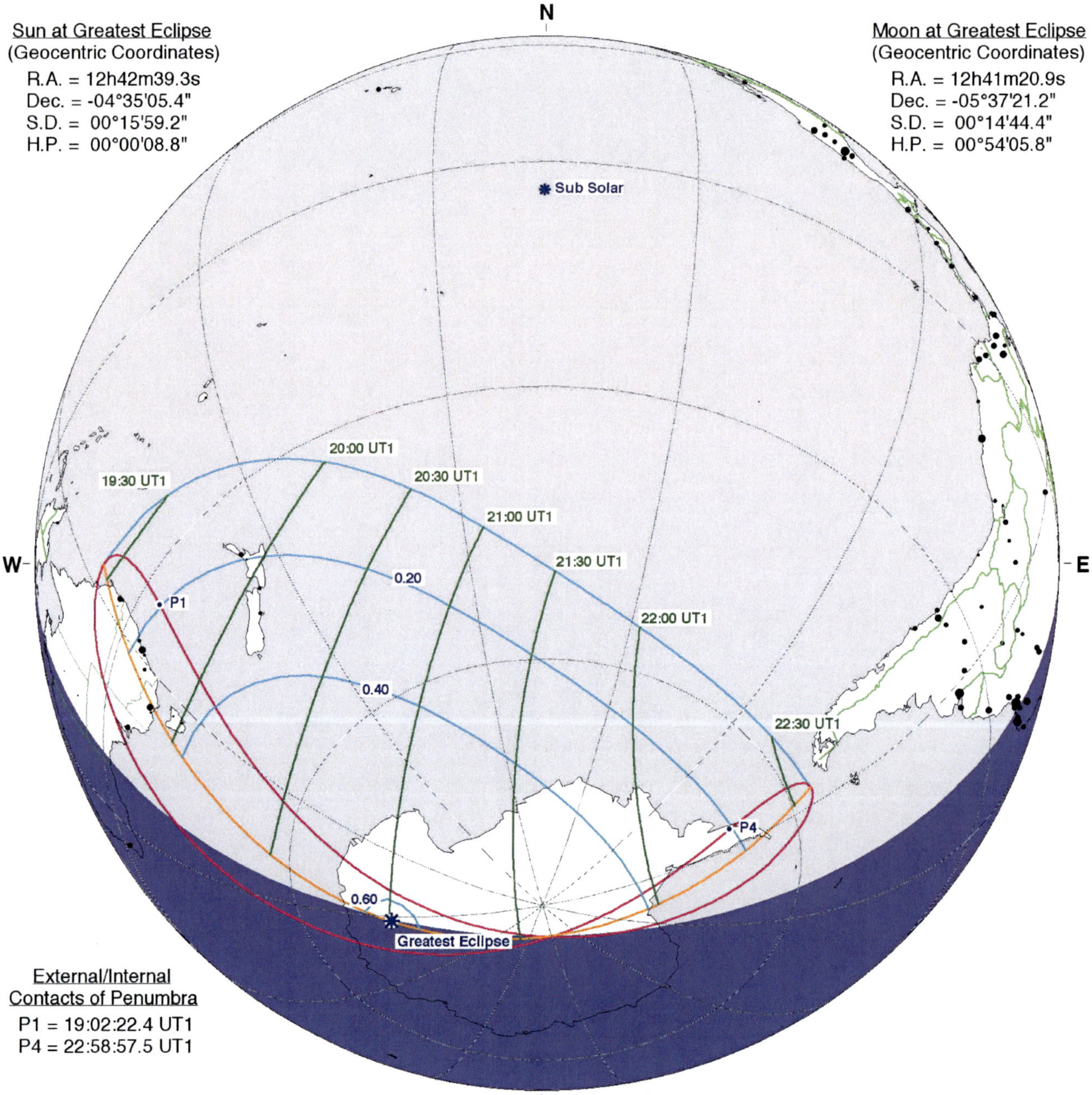

External/Internal
Contacts of Penumbra
P1 = 19:02:22.4 UT1
P4 = 22:58:57.5 UT1

ΔT = 85.1 s Eph. = JPL DE405

Circumstances at Greatest Eclipse: 21:00:49.4 UT1

Lat. = 72°02.7'S Sun Alt. = 0.0°

Long. = 117°41.2'E Sun Azm. = 105.0°

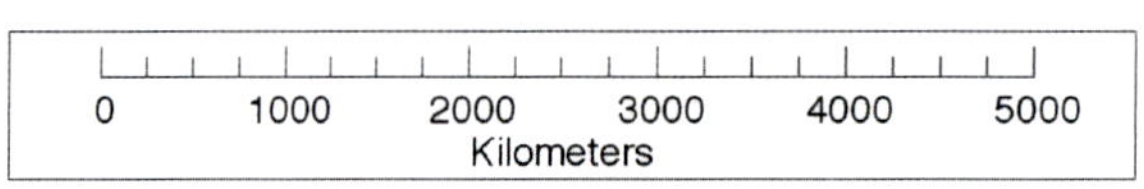

Total Solar Eclipse of 2052 Mar 30

Greatest Eclipse = 18:31:52.9 TD (= 18:30:27.4 UT1)

Eclipse Magnitude =	1.0466	Saros Series =	130
Gamma =	0.3239	Saros Member =	54 of 73

Sun at Greatest Eclipse
(Geocentric Coordinates)

R.A. = 00h39m33.8s
Dec. = +04°15'25.9"
S.D. = 00°16'00.7"
H.P. = 00°00'08.8"

Moon at Greatest Eclipse
(Geocentric Coordinates)

R.A. = 00h39m10.3s
Dec. = +04°34'05.5"
S.D. = 00°16'29.6"
H.P. = 01°00'31.8"

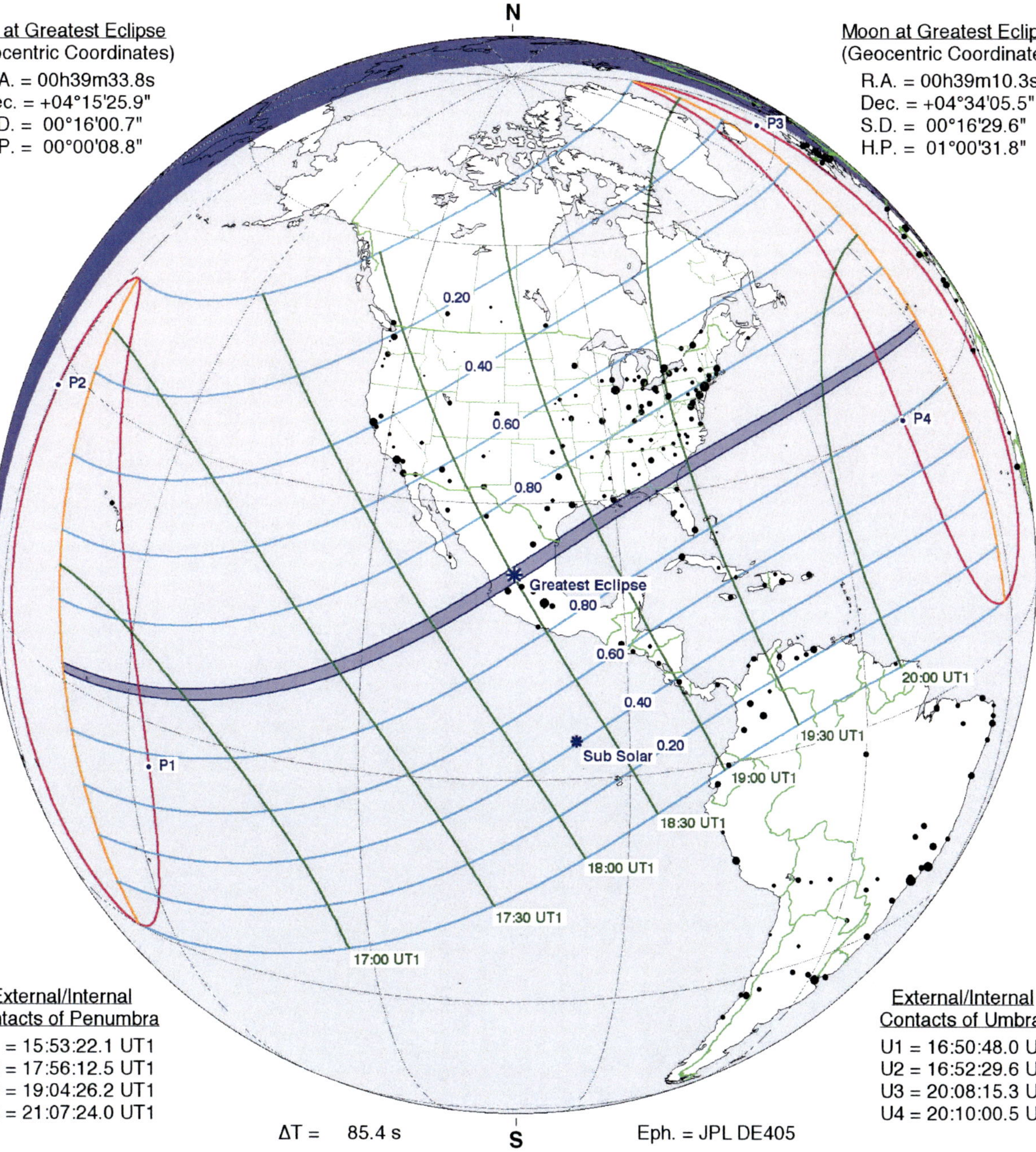

External/Internal
Contacts of Penumbra

P1 = 15:53:22.1 UT1
P2 = 17:56:12.5 UT1
P3 = 19:04:26.2 UT1
P4 = 21:07:24.0 UT1

External/Internal
Contacts of Umbra

U1 = 16:50:48.0 UT1
U2 = 16:52:29.6 UT1
U3 = 20:08:15.3 UT1
U4 = 20:10:00.5 UT1

ΔT = 85.4 s

Eph. = JPL DE405

Circumstances at Greatest Eclipse: 18:30:27.4 UT1

Lat. = 22°22.8'N	Sun Alt. = 71.0°
Long. = 102°34.9'W	Sun Azm. = 161.3°
Path Width = 163.7 km	Duration = 04m08.2s

Circumstances at Greatest Duration: 18:32:33.3 UT1

Lat. = 22°45.2'N	Sun Alt. = 70.9°
Long. = 101°55.6'W	Sun Azm. = 165.1°
Path Width = 163.5 km	Duration = 04m08.2s

0 1000 2000 3000 4000 5000
Kilometers

Annular Solar Eclipse of 2052 Sep 22

Greatest Eclipse = 23:39:09.7 TD (= 23:37:44.0 UT1)

Eclipse Magnitude = 0.9734   Saros Series = 135

Gamma = -0.4480   Saros Member = 41 of 71

Sun at Greatest Eclipse
(Geocentric Coordinates)

R.A. = 12h02m27.0s
Dec. = -00°15'55.5"
S.D. = 00°15'56.2"
H.P. = 00°00'08.8"

Moon at Greatest Eclipse
(Geocentric Coordinates)

R.A. = 12h01m56.4s
Dec. = -00°39'49.3"
S.D. = 00°15'17.9"
H.P. = 00°56'08.8"

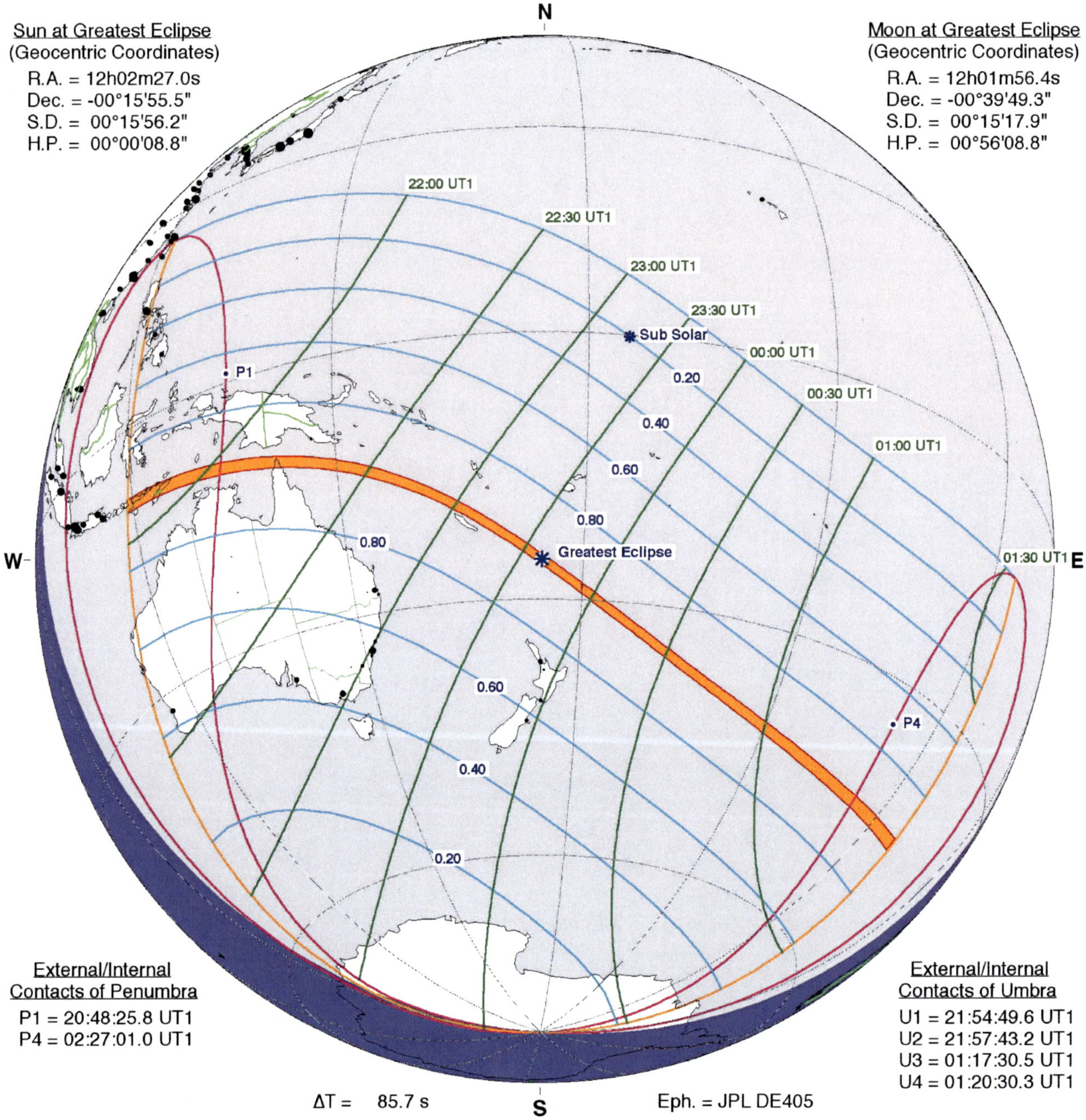

External/Internal Contacts of Penumbra

P1 = 20:48:25.8 UT1
P4 = 02:27:01.0 UT1

External/Internal Contacts of Umbra

U1 = 21:54:49.6 UT1
U2 = 21:57:43.2 UT1
U3 = 01:17:30.5 UT1
U4 = 01:20:30.3 UT1

ΔT = 85.7 s   Eph. = JPL DE405

Circumstances at Greatest Eclipse: 23:37:44.0 UT1

Lat. = 25°41.3'S   Sun Alt. = 63.2°
Long. = 174°56.0'E   Sun Azm. = 19.6°
Path Width = 106.0 km   Duration = 02m50.9s

Circumstances at Greatest Duration: 00:04:03.2 UT1

Lat. = 30°17.6'S   Sun Alt. = 59.5°
Long. = 177°20.4'W   Sun Azm. = 348.9°
Path Width = 106.1 km   Duration = 02m51.5s

0 1000 2000 3000 4000 5000
Kilometers

Annular Solar Eclipse of 2053 Mar 20

Greatest Eclipse = 07:08:19.4 TD (= 07:06:53.4 UT1)

Eclipse Magnitude = 0.9919 Saros Series = 140

Gamma = -0.4089 Saros Member = 31 of 71

Sun at Greatest Eclipse
(Geocentric Coordinates)
R.A. = 00h00m30.3s
Dec. = +00°03'17.2"
S.D. = 00°16'03.6"
H.P. = 00°00'08.8"

Moon at Greatest Eclipse
(Geocentric Coordinates)
R.A. = 00h00m59.0s
Dec. = -00°19'05.6"
S.D. = 00°15'41.9"
H.P. = 00°57'37.0"

N

W

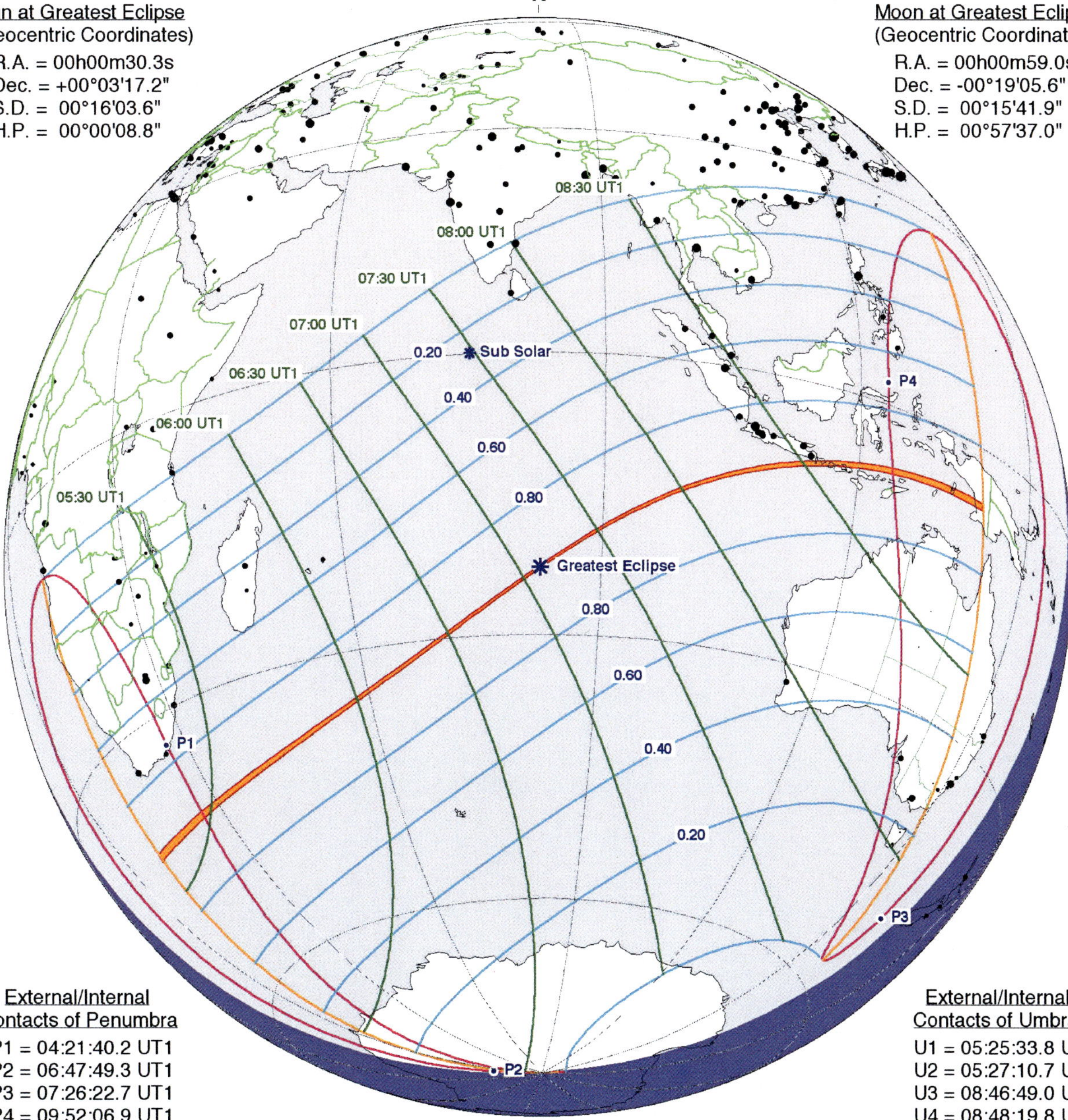

E

External/Internal
Contacts of Penumbra
P1 = 04:21:40.2 UT1
P2 = 06:47:49.3 UT1
P3 = 07:26:22.7 UT1
P4 = 09:52:06.9 UT1

External/Internal
Contacts of Umbra
U1 = 05:25:33.8 UT1
U2 = 05:27:10.7 UT1
U3 = 08:46:49.0 UT1
U4 = 08:48:19.8 UT1

ΔT = 86.0 s

S

Eph. = JPL DE405

Circumstances at Greatest Eclipse: 07:06:53.4 UT1	
Lat. = 23°00.9'S	Sun Alt. = 65.7°
Long. = 082°54.1'E	Sun Azm. = 340.8°
Path Width = 31.1 km	Duration = 00m49.5s

Circumstances at Greatest Duration: 05:26:22.2 UT1	
Lat. = 42°07.6'S	Sun Alt. = 0.0°
Long. = 010°16.4'E	Sun Azm. = 90.0°
Path Width = 93.4 km	Duration = 01m28.3s

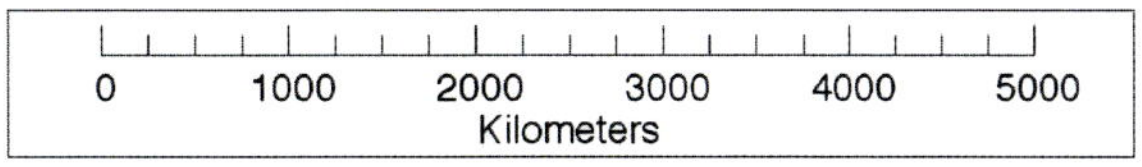

Total Solar Eclipse of 2053 Sep 12

Greatest Eclipse = 09:34:08.9 TD (= 09:32:42.6 UT1)

Eclipse Magnitude = 1.0328 Saros Series = 145

Gamma = 0.3140 Saros Member = 24 of 77

Sun at Greatest Eclipse
(Geocentric Coordinates)
R.A. = 11h23m36.1s
Dec. = +03°55'14.2"
S.D. = 00°15'53.4"
H.P. = 00°00'08.7"

Moon at Greatest Eclipse
(Geocentric Coordinates)
R.A. = 11h23m58.5s
Dec. = +04°12'57.2"
S.D. = 00°16'09.4"
H.P. = 00°59'17.8"

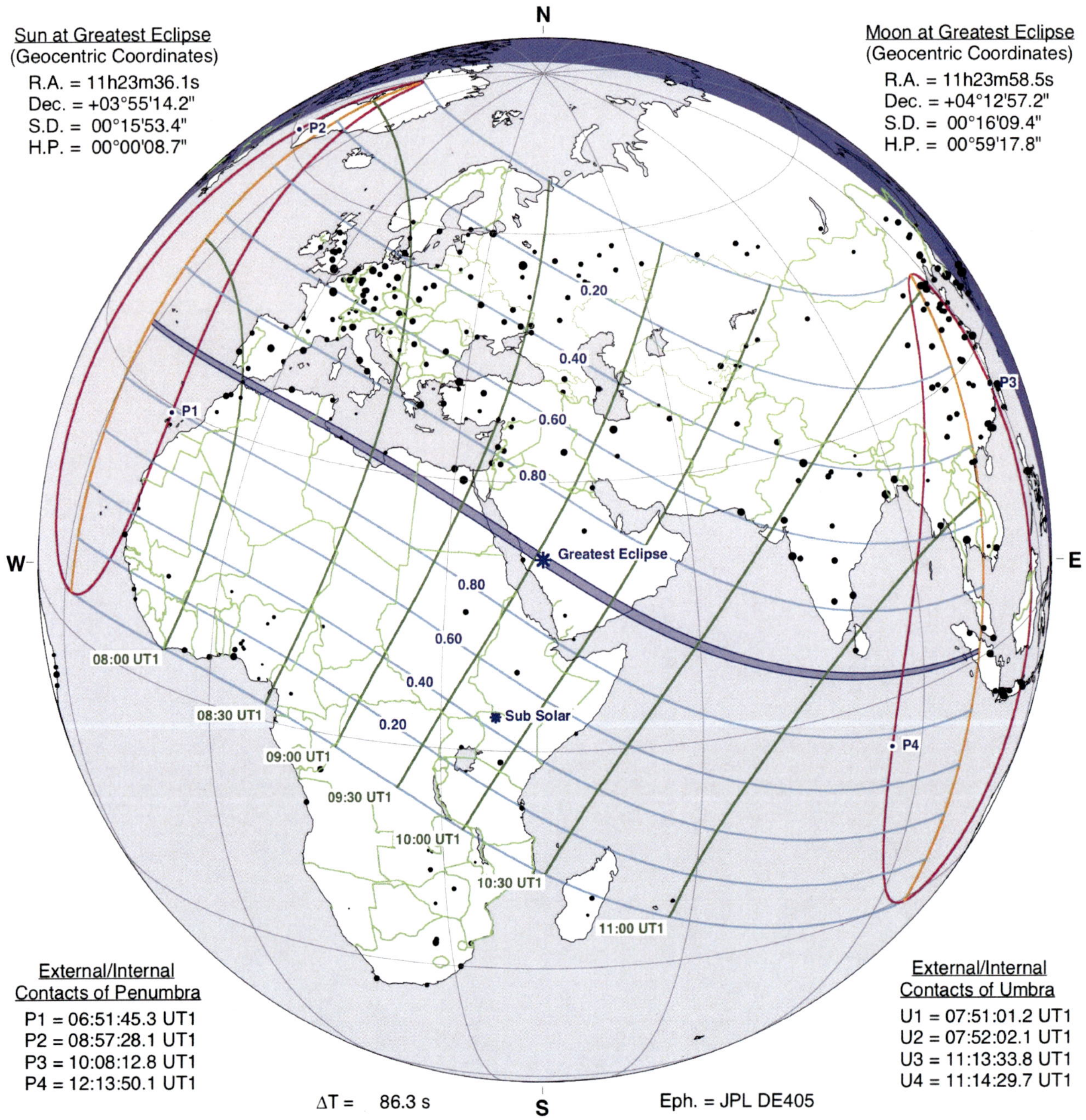

External/Internal Contacts of Penumbra
P1 = 06:51:45.3 UT1
P2 = 08:57:28.1 UT1
P3 = 10:08:12.8 UT1
P4 = 12:13:50.1 UT1

External/Internal Contacts of Umbra
U1 = 07:51:01.2 UT1
U2 = 07:52:02.1 UT1
U3 = 11:13:33.8 UT1
U4 = 11:14:29.7 UT1

ΔT = 86.3 s Eph. = JPL DE405

Circumstances at Greatest Eclipse: 09:32:42.6 UT1

Lat. = 21°28.1'N Sun Alt. = 71.6°
Long. = 041°39.8'E Sun Azm. = 198.6°
Path Width = 116.4 km Duration = 03m04.1s

Circumstances at Greatest Duration: 09:29:56.2 UT1

Lat. = 21°56.9'N Sun Alt. = 71.5°
Long. = 040°50.4'E Sun Azm. = 193.6°
Path Width = 116.3 km Duration = 03m04.2s

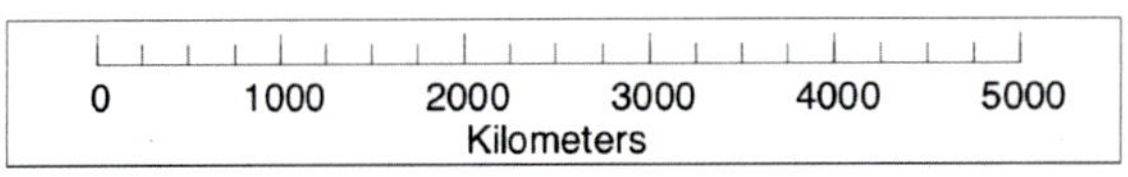

Partial Solar Eclipse of 2054 Mar 09

Greatest Eclipse = 12:33:40.5 TD (= 12:32:13.9 UT1)

Eclipse Magnitude = 0.6678 Saros Series = 150

Gamma = -1.1711 Saros Member = 19 of 71

Sun at Greatest Eclipse
(Geocentric Coordinates)

R.A. = 23h20m07.5s
Dec. = -04°17'25.4"
S.D. = 00°16'06.6"
H.P. = 00°00'08.9"

Moon at Greatest Eclipse
(Geocentric Coordinates)

R.A. = 23h21m24.6s
Dec. = -05°18'27.6"
S.D. = 00°14'55.7"
H.P. = 00°54'47.2"

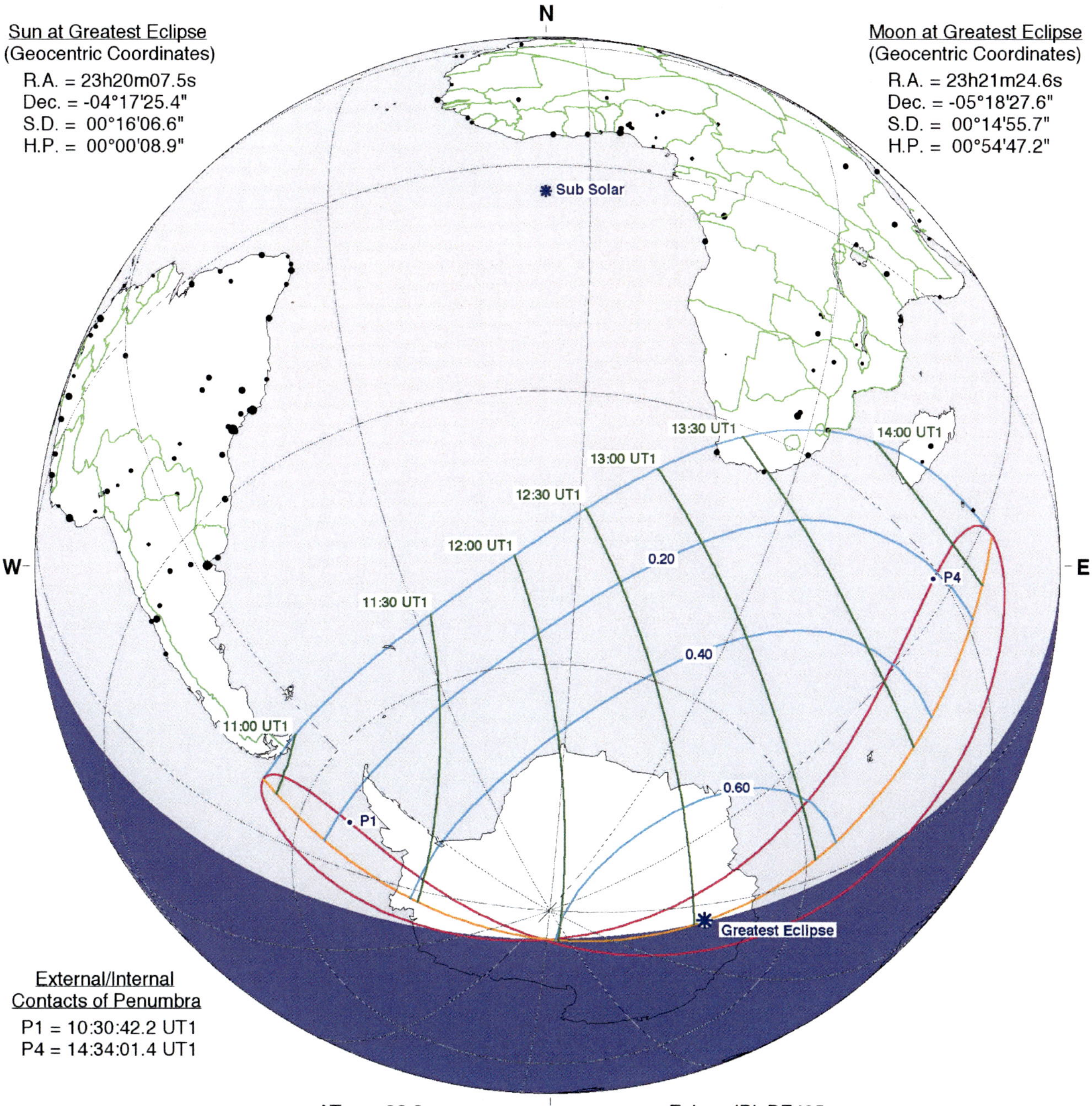

External/Internal
Contacts of Penumbra

P1 = 10:30:42.2 UT1
P4 = 14:34:01.4 UT1

ΔT = 86.6 s Eph. = JPL DE405

Circumstances at Greatest Eclipse: 12:32:13.9 UT1

Lat. = 72°02.6'S Sun Alt. = 0.0°

Long. = 097°51.8'E Sun Azm. = 256.0°

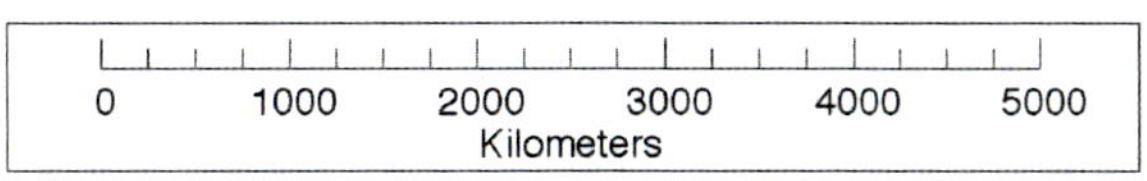

Partial Solar Eclipse of 2054 Aug 03

Greatest Eclipse = 18:04:02.1 TD (= 18:02:35.2 UT1)

Eclipse Magnitude = 0.0656 Saros Series = 117

Gamma = -1.4941 Saros Member = 71 of 71

Sun at Greatest Eclipse
(Geocentric Coordinates)
R.A. = 08h56m24.5s
Dec. = +17°17'09.2"
S.D. = 00°15'45.7"
H.P. = 00°00'08.7"

Moon at Greatest Eclipse
(Geocentric Coordinates)
R.A. = 08h55m14.2s
Dec. = +15°47'22.5"
S.D. = 00°16'41.7"
H.P. = 01°01'16.5"

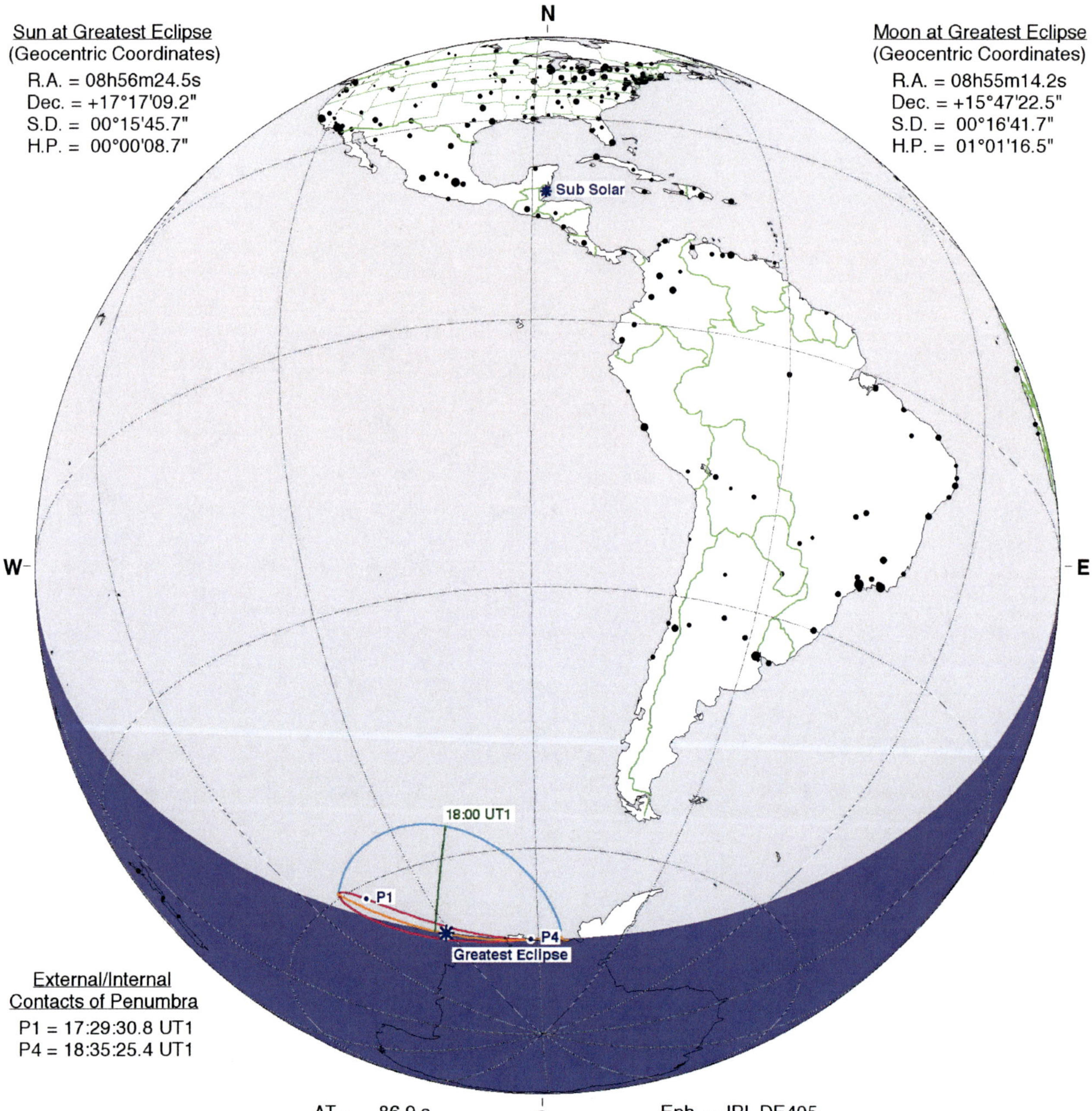

External/Internal
Contacts of Penumbra
P1 = 17:29:30.8 UT1
P4 = 18:35:25.4 UT1

ΔT = 86.9 s Eph. = JPL DE405

Circumstances at Greatest Eclipse: 18:02:35.2 UT1

Lat. = 69°46.4'S Sun Alt. = 0.0°

Long. = 121°24.7'W Sun Azm. = 30.7°

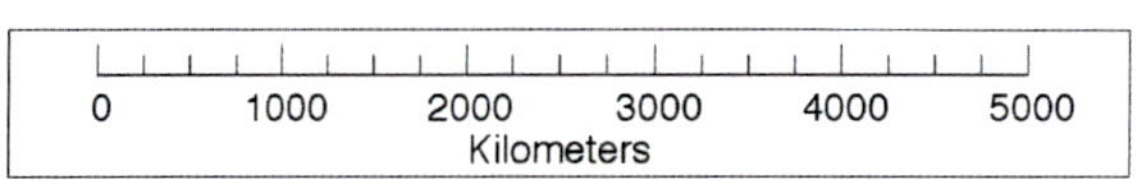

Partial Solar Eclipse of 2054 Sep 02

Greatest Eclipse = 01:09:33.7 TD (= 01:08:06.7 UT1)

Eclipse Magnitude = 0.9793
Gamma = 1.0215

Saros Series = 155
Saros Member = 8 of 71

Sun at Greatest Eclipse
(Geocentric Coordinates)
R.A. = 10h45m28.2s
Dec. = +07°52'58.6"
S.D. = 00°15'50.9"
H.P. = 00°00'08.7"

Moon at Greatest Eclipse
(Geocentric Coordinates)
R.A. = 10h46m40.4s
Dec. = +08°52'49.8"
S.D. = 00°16'42.0"
H.P. = 01°01'17.5"

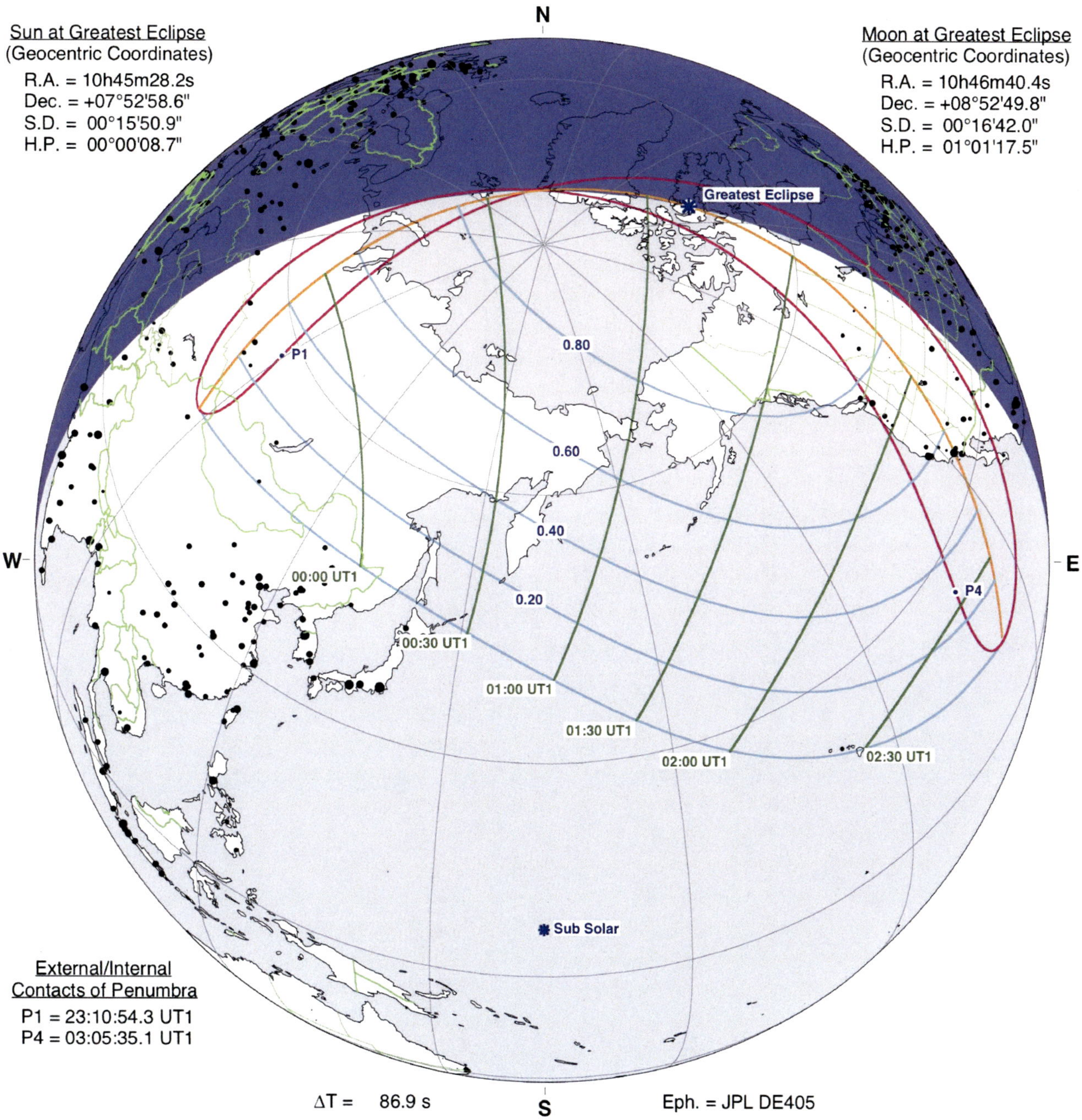

External/Internal
Contacts of Penumbra
P1 = 23:10:54.3 UT1
P4 = 03:05:35.1 UT1

ΔT = 86.9 s

Eph. = JPL DE405

Circumstances at Greatest Eclipse: 01:08:06.7 UT1

Lat. = 71°40.9'N
Long. = 082°24.8'W

Sun Alt. = 0.0°
Sun Azm. = 295.9°

0 1000 2000 3000 4000 5000
Kilometers

Partial Solar Eclipse of 2055 Jan 27

Greatest Eclipse = 17:54:05.3 TD (= 17:52:38.1 UT1)

Eclipse Magnitude = 0.6932 Saros Series = 122

Gamma = 1.1550 Saros Member = 60 of 70

Sun at Greatest Eclipse
(Geocentric Coordinates)
R.A. = 20h40m41.0s
Dec. = -18°19'18.9"
S.D. = 00°16'14.5"
H.P. = 00°00'08.9"

Moon at Greatest Eclipse
(Geocentric Coordinates)
R.A. = 20h39m58.6s
Dec. = -17°17'11.5"
S.D. = 00°14'53.3"
H.P. = 00°54'38.4"

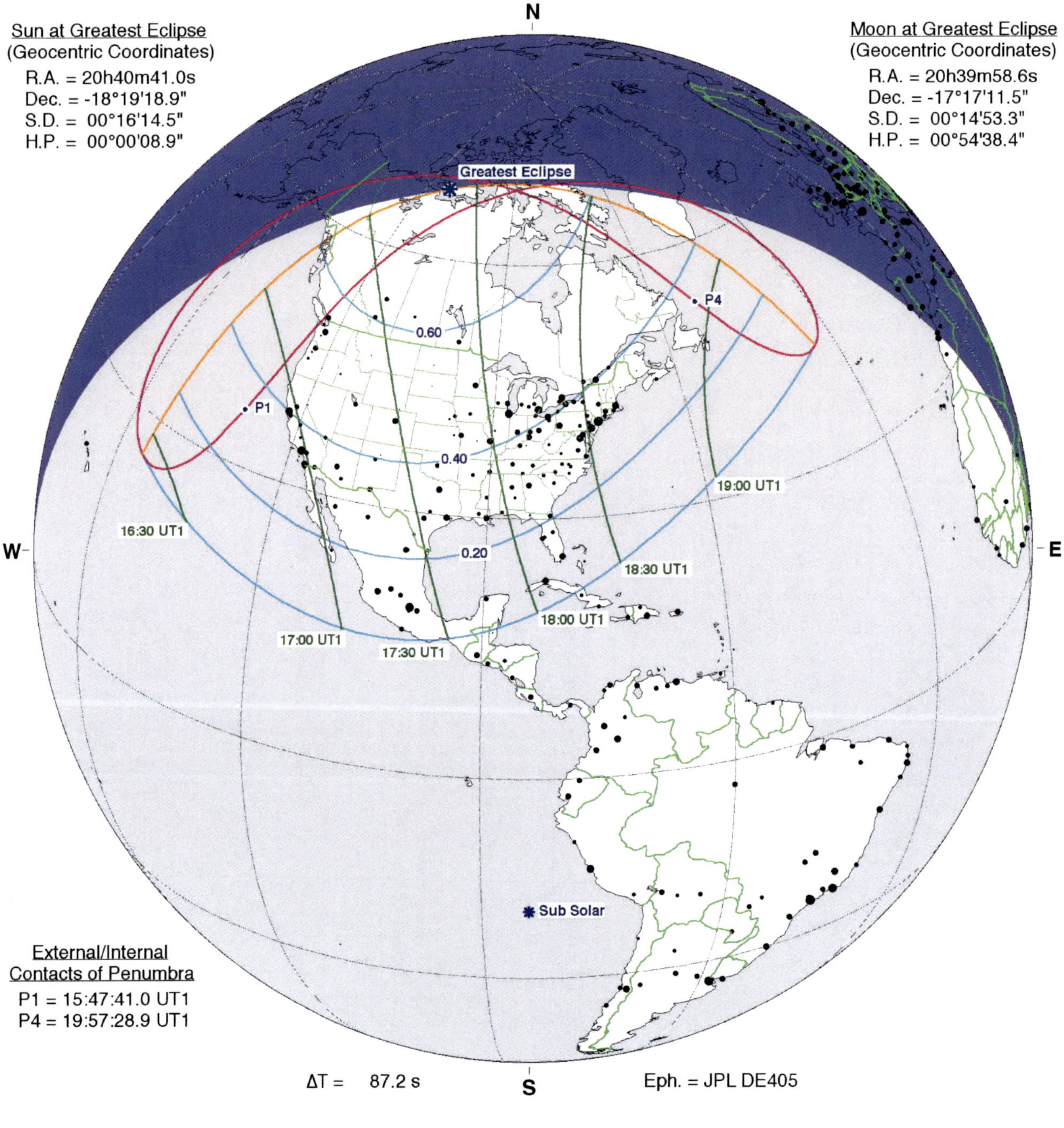

External/Internal
Contacts of Penumbra
P1 = 15:47:41.0 UT1
P4 = 19:57:28.9 UT1

ΔT = 87.2 s

Eph. = JPL DE405

Circumstances at Greatest Eclipse: 17:52:38.1 UT1

Lat. = 69°32.9'N Sun Alt. = 0.0°

Long. = 112°19.0'W Sun Azm. = 154.1°

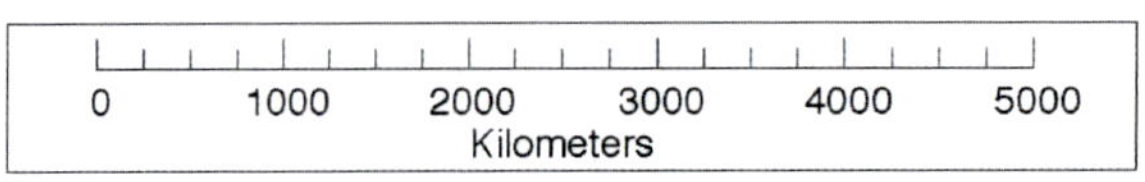

Total Solar Eclipse of 2055 Jul 24

Greatest Eclipse = 09:57:50.3 TD (= 09:56:22.7 UT1)

Eclipse Magnitude = 1.0359 Saros Series = 127

Gamma = -0.8012 Saros Member = 60 of 82

Sun at Greatest Eclipse
(Geocentric Coordinates)
R.A. = 08h15m04.2s
Dec. = +19°48'43.3"
S.D. = 00°15'44.6"
H.P. = 00°00'08.7"

Moon at Greatest Eclipse
(Geocentric Coordinates)
R.A. = 08h14m39.2s
Dec. = +19°01'42.7"
S.D. = 00°16'09.1"
H.P. = 00°59'16.7"

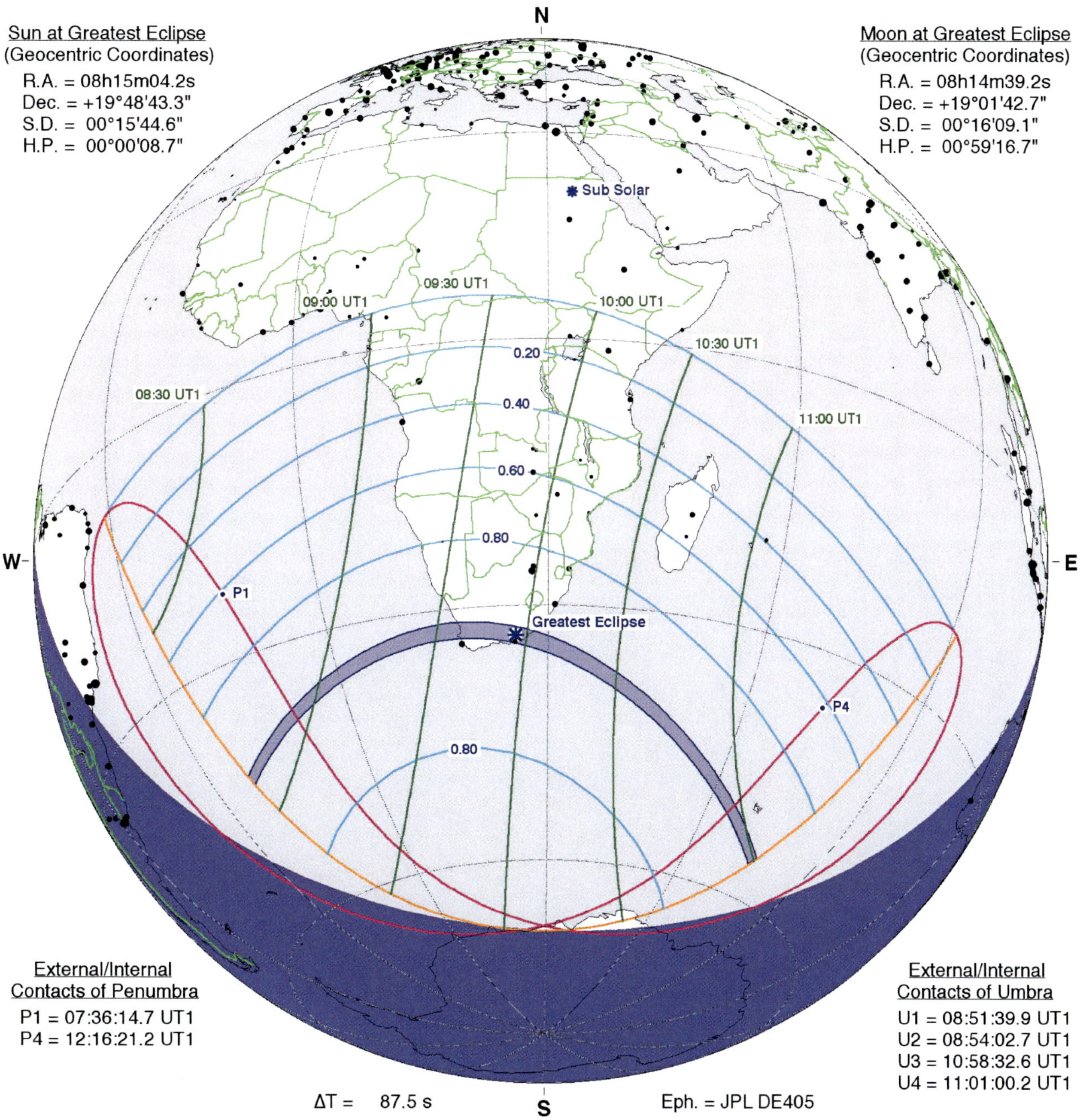

External/Internal
Contacts of Penumbra
P1 = 07:36:14.7 UT1
P4 = 12:16:21.2 UT1

External/Internal
Contacts of Umbra
U1 = 08:51:39.9 UT1
U2 = 08:54:02.7 UT1
U3 = 10:58:32.6 UT1
U4 = 11:01:00.2 UT1

ΔT = 87.5 s Eph. = JPL DE405

Circumstances at Greatest Eclipse: 09:56:22.7 UT1

Lat. = 33°16.9'S	Sun Alt. = 36.5°
Long. = 025°43.3'E	Sun Azm. = 8.0°
Path Width = 201.8 km	Duration = 03m16.8s

Circumstances at Greatest Duration: 09:56:38.1 UT1

Lat. = 33°18.1'S	Sun Alt. = 36.5°
Long. = 025°49.2'E	Sun Azm. = 7.8°
Path Width = 201.7 km	Duration = 03m16.8s

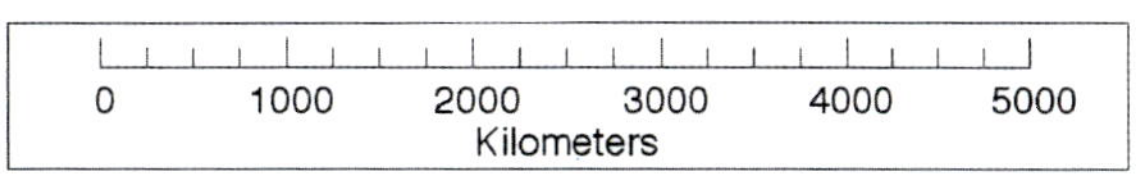

Annular Solar Eclipse of 2056 Jan 16

Greatest Eclipse = 22:16:45.2 TD (= 22:15:17.3 UT1)

Eclipse Magnitude = 0.9760 Saros Series = 132

Gamma = 0.4199 Saros Member = 48 of 71

Sun at Greatest Eclipse
(Geocentric Coordinates)
R.A. = 19h54m06.4s
Dec. = -20°50'41.3"
S.D. = 00°16'15.5"
H.P. = 00°00'08.9"

Moon at Greatest Eclipse
(Geocentric Coordinates)
R.A. = 19h53m57.0s
Dec. = -20°26'45.0"
S.D. = 00°15'38.4"
H.P. = 00°57'23.8"

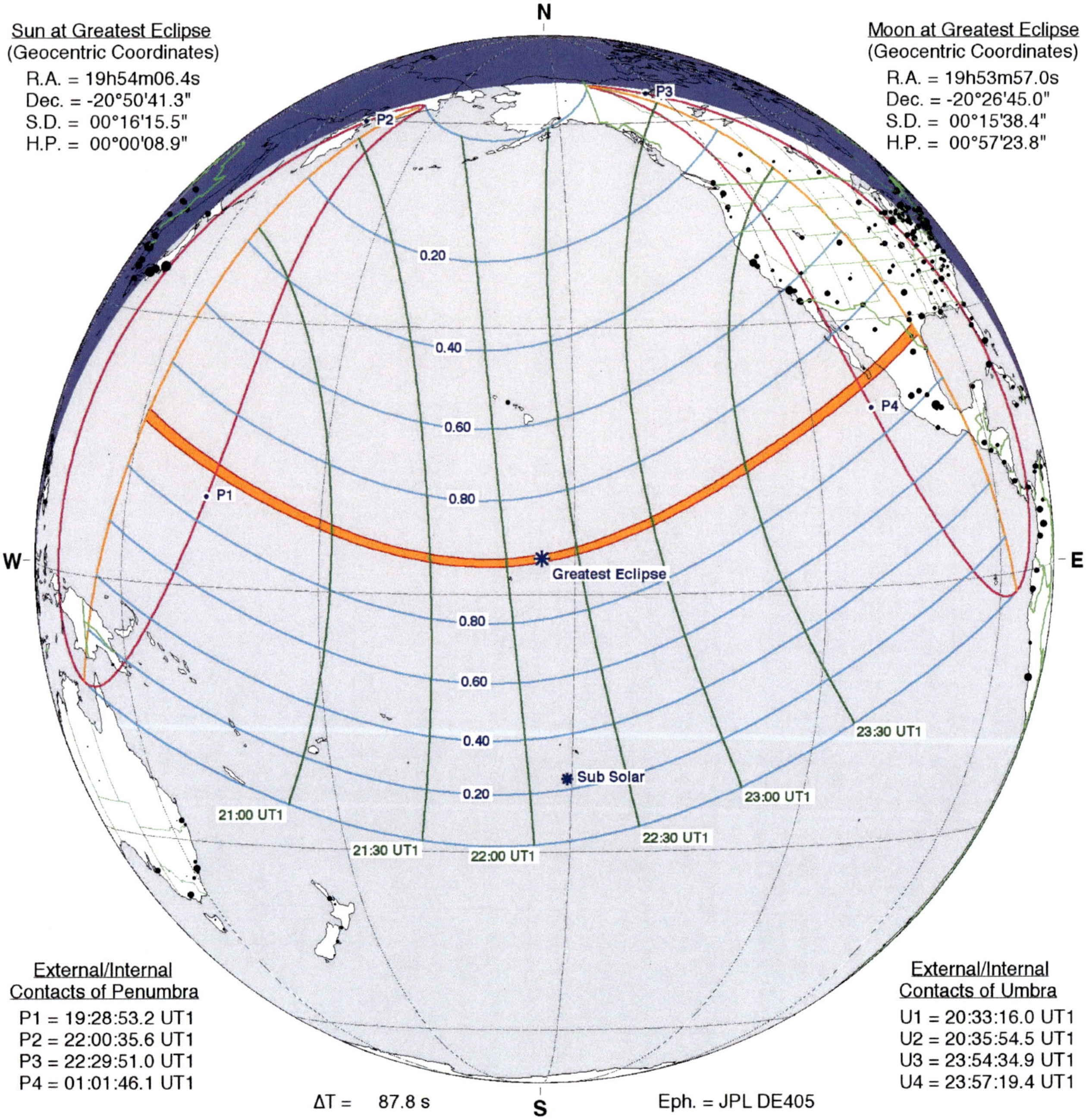

External/Internal Contacts of Penumbra
P1 = 19:28:53.2 UT1
P2 = 22:00:35.6 UT1
P3 = 22:29:51.0 UT1
P4 = 01:01:46.1 UT1

External/Internal Contacts of Umbra
U1 = 20:33:16.0 UT1
U2 = 20:35:54.5 UT1
U3 = 23:54:34.9 UT1
U4 = 23:57:19.4 UT1

ΔT = 87.8 s Eph. = JPL DE405

Circumstances at Greatest Eclipse: 22:15:17.3 UT1

Lat. = 03°54.9'N Sun Alt. = 65.1°
Long. = 153°36.2'W Sun Azm. = 175.1°
Path Width = 94.5 km Duration = 02m52.4s

Circumstances at Greatest Duration: 22:18:47.3 UT1

Lat. = 04°05.0'N Sun Alt. = 65.1°
Long. = 152°40.5'W Sun Azm. = 179.1°
Path Width = 94.5 km Duration = 02m52.5s

0 1000 2000 3000 4000 5000
Kilometers

Annular Solar Eclipse of 2056 Jul 12

Greatest Eclipse = 20:21:59.4 TD (= 20:20:31.2 UT1)

Eclipse Magnitude = 0.9878 Saros Series = 137

Gamma = -0.0426 Saros Member = 38 of 70

Sun at Greatest Eclipse
(Geocentric Coordinates)
R.A. = 07h31m42.7s
Dec. = +21°45'32.5"
S.D. = 00°15'44.0"
H.P. = 00°00'08.7"

Moon at Greatest Eclipse
(Geocentric Coordinates)
R.A. = 07h31m42.1s
Dec. = +21°43'09.5"
S.D. = 00°15'17.9"
H.P. = 00°56'08.9"

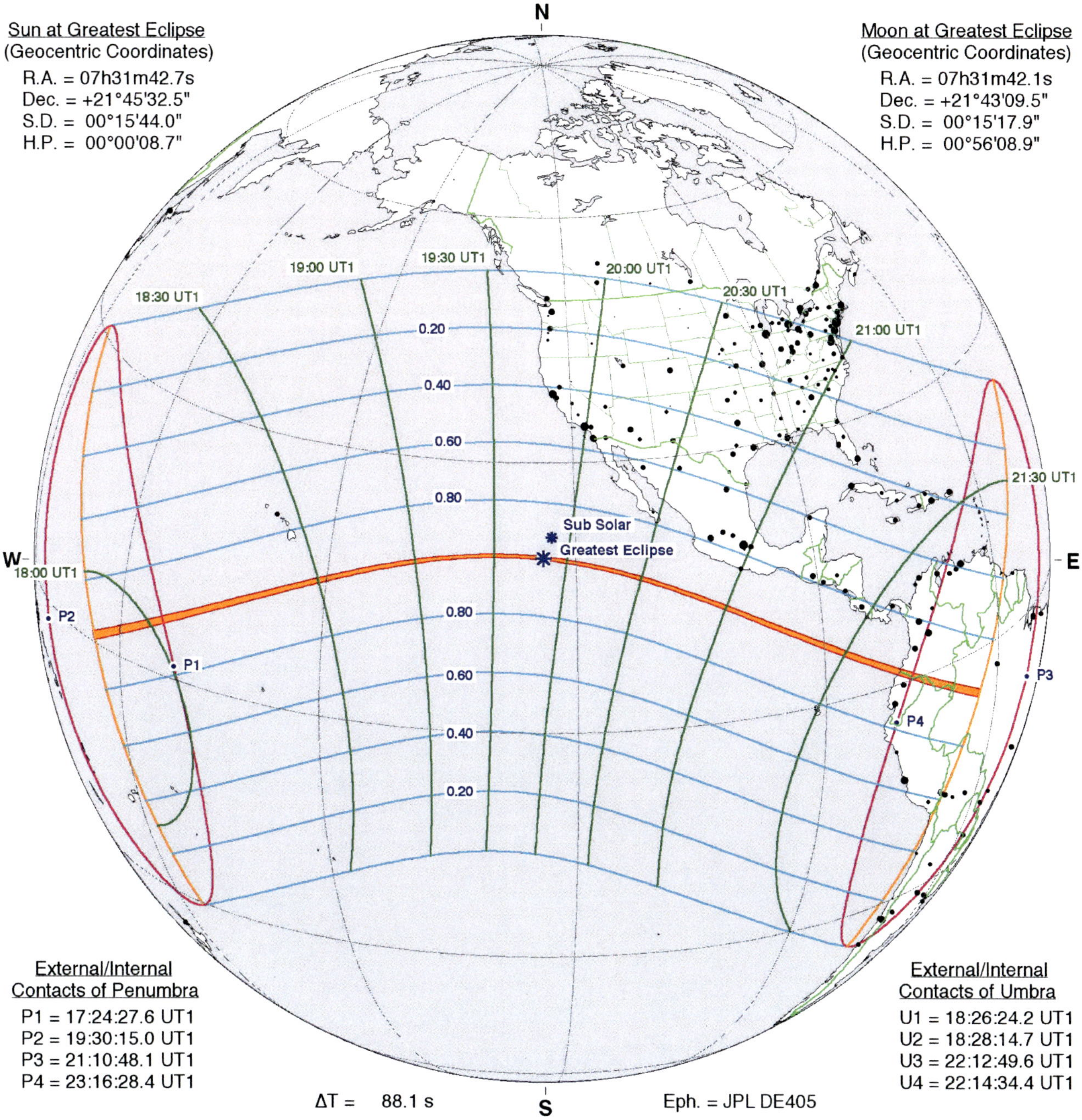

External/Internal Contacts of Penumbra
P1 = 17:24:27.6 UT1
P2 = 19:30:15.0 UT1
P3 = 21:10:48.1 UT1
P4 = 23:16:28.4 UT1

External/Internal Contacts of Umbra
U1 = 18:26:24.2 UT1
U2 = 18:28:14.7 UT1
U3 = 22:12:49.6 UT1
U4 = 22:14:34.4 UT1

ΔT = 88.1 s Eph. = JPL DE405

Circumstances at Greatest Eclipse: 20:20:31.2 UT1

Lat. = 19°26.4'N Sun Alt. = 87.7°
Long. = 123°48.5'W Sun Azm. = 3.3°
Path Width = 43.2 km Duration = 01m25.9s

Circumstances at Greatest Duration: 18:27:19.5 UT1

Lat. = 00°41.1'N Sun Alt. = 0.0°
Long. = 174°21.7'E Sun Azm. = 68.2°
Path Width = 104.6 km Duration = 01m49.4s

0 1000 2000 3000 4000 5000
Kilometers

Total Solar Eclipse of 2057 Jan 05

Greatest Eclipse = 09:47:52.2 TD (= 09:46:23.7 UT1)

Eclipse Magnitude = 1.0287
Gamma = -0.2837

Saros Series = 142
Saros Member = 25 of 72

N

Sun at Greatest Eclipse
(Geocentric Coordinates)
R.A. = 19h07m25.3s
Dec. = -22°31'37.8"
S.D. = 00°16'15.9"
H.P. = 00°00'08.9"

Moon at Greatest Eclipse
(Geocentric Coordinates)
R.A. = 19h07m26.6s
Dec. = -22°48'43.6"
S.D. = 00°16'27.8"
H.P. = 01°00'25.4"

W

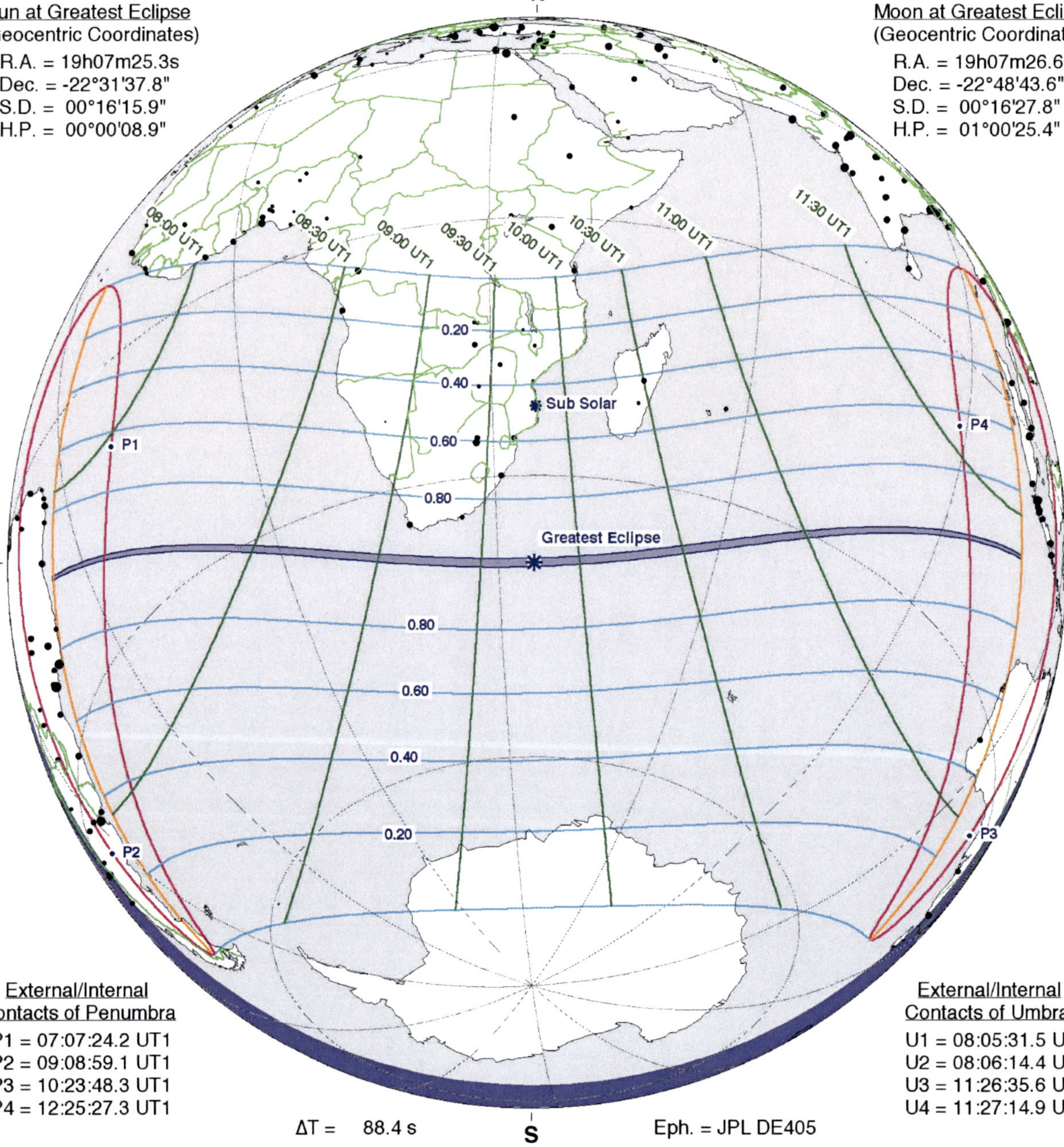

E

External/Internal Contacts of Penumbra
P1 = 07:07:24.2 UT1
P2 = 09:08:59.1 UT1
P3 = 10:23:48.3 UT1
P4 = 12:25:27.3 UT1

External/Internal Contacts of Umbra
U1 = 08:05:31.5 UT1
U2 = 08:06:14.4 UT1
U3 = 11:26:35.6 UT1
U4 = 11:27:14.9 UT1

ΔT = 88.4 s

S

Eph. = JPL DE405

Circumstances at Greatest Eclipse: 09:46:23.7 UT1

Lat. = 39°13.0'S	Sun Alt. = 73.3°
Long. = 035°06.8'E	Sun Azm. = 358.9°
Path Width = 101.6 km	Duration = 02m28.7s

Circumstances at Greatest Duration: 09:44:47.7 UT1

Lat. = 39°13.8'S	Sun Alt. = 73.3°
Long. = 034°23.1'E	Sun Azm. = 2.5°
Path Width = 101.6 km	Duration = 02m28.7s

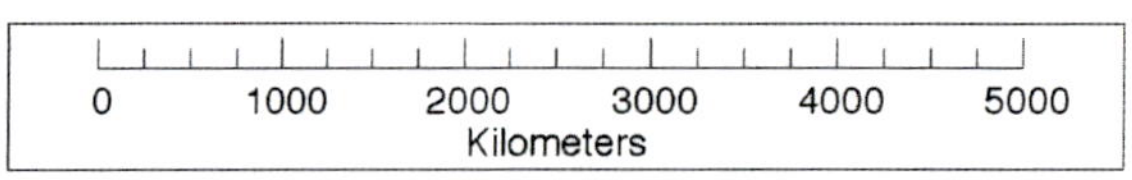

Annular Solar Eclipse of 2057 Jul 01

Greatest Eclipse = 23:40:15.3 TD (= 23:38:46.6 UT1)

Eclipse Magnitude = 0.9464 | Saros Series = 147

Gamma = 0.7455 | Saros Member = 25 of 80

Sun at Greatest Eclipse
(Geocentric Coordinates)
R.A. = 06h46m13.5s
Dec. = +23°00'23.1"
S.D. = 00°15'43.9"
H.P. = 00°00'08.6"

Moon at Greatest Eclipse
(Geocentric Coordinates)
R.A. = 06h46m10.1s
Dec. = +23°40'36.4"
S.D. = 00°14'44.6"
H.P. = 00°54'06.5"

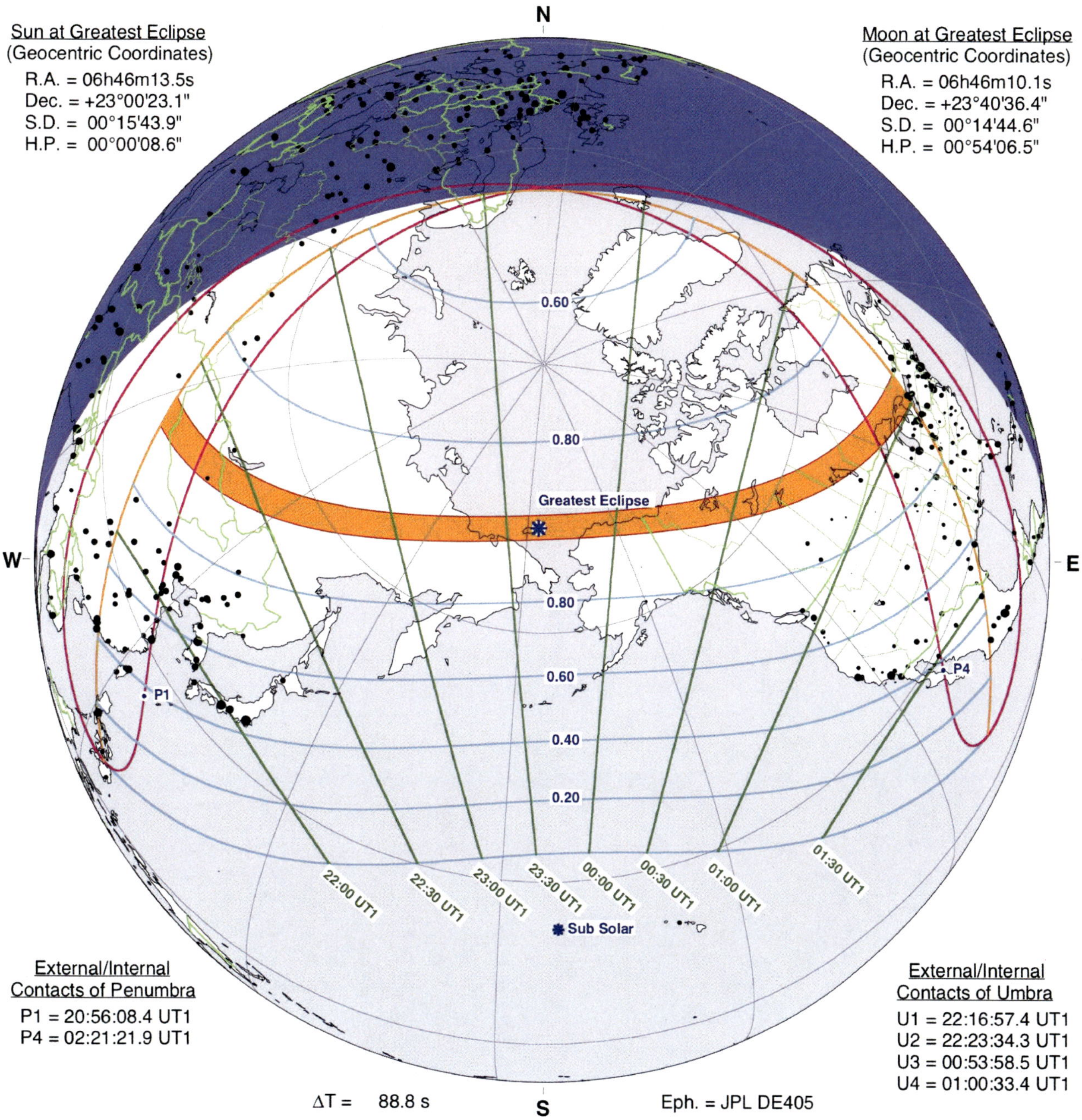

External/Internal Contacts of Penumbra
P1 = 20:56:08.4 UT1
P4 = 02:21:21.9 UT1

External/Internal Contacts of Umbra
U1 = 22:16:57.4 UT1
U2 = 22:23:34.3 UT1
U3 = 00:53:58.5 UT1
U4 = 01:00:33.4 UT1

ΔT = 88.8 s | Eph. = JPL DE405

Circumstances at Greatest Eclipse: 23:38:46.6 UT1

Lat. = 71°29.6'N | Sun Alt. = 41.5°
Long. = 176°17.0'W | Sun Azm. = 176.8°
Path Width = 298.2 km | Duration = 04m22.5s

Circumstances at Greatest Duration: 23:38:04.2 UT1

Lat. = 71°28.4'N | Sun Alt. = 41.5°
Long. = 177°10.8'W | Sun Azm. = 175.4°
Path Width = 298.2 km | Duration = 04m22.5s

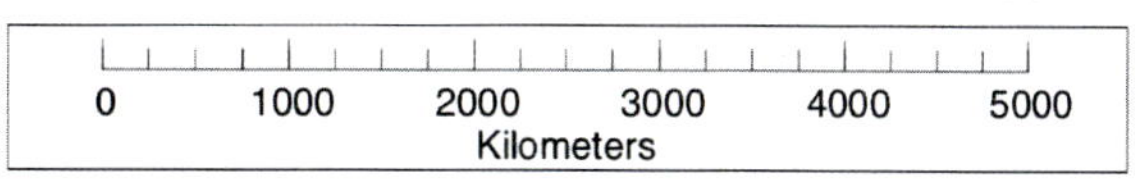

Total Solar Eclipse of 2057 Dec 26

Greatest Eclipse = 01:14:35.2 TD (= 01:13:06.1 UT1)

Eclipse Magnitude = 1.0348
Gamma = -0.9405

Saros Series = 152
Saros Member = 15 of 70

Sun at Greatest Eclipse
(Geocentric Coordinates)
R.A. = 18h20m37.6s
Dec. = -23°20'50.0"
S.D. = 00°16'15.6"
H.P. = 00°00'08.9"

Moon at Greatest Eclipse
(Geocentric Coordinates)
R.A. = 18h20m22.1s
Dec. = -24°18'21.4"
S.D. = 00°16'44.3"
H.P. = 01°01'25.8"

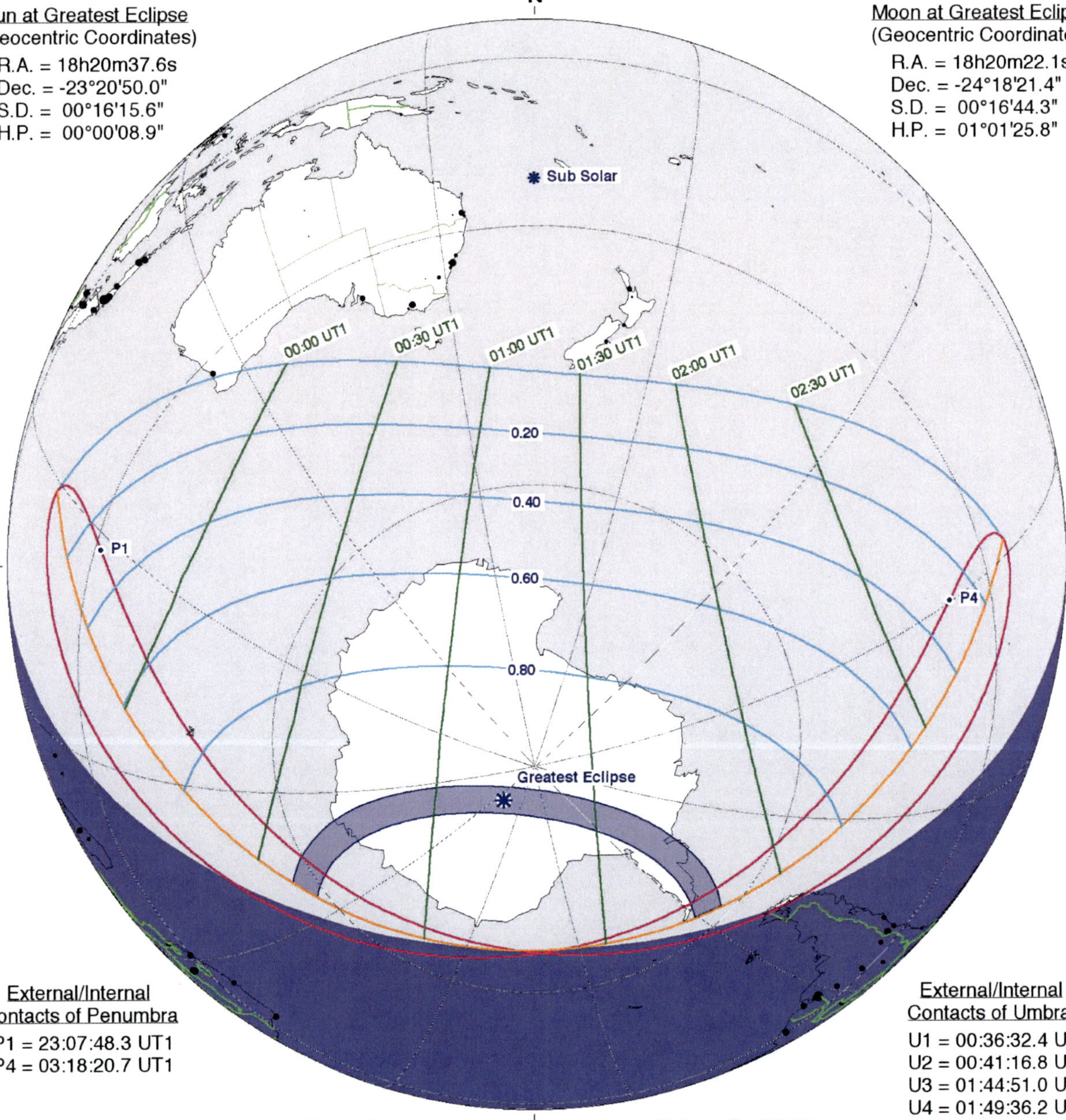

External/Internal Contacts of Penumbra
P1 = 23:07:48.3 UT1
P4 = 03:18:20.7 UT1

External/Internal Contacts of Umbra
U1 = 00:36:32.4 UT1
U2 = 00:41:16.8 UT1
U3 = 01:44:51.0 UT1
U4 = 01:49:36.2 UT1

ΔT = 89.1 s

Eph. = JPL DE405

Circumstances at Greatest Eclipse: 01:13:06.1 UT1

Lat. = 84°51.1'S
Long. = 021°42.3'E
Path Width = 354.8 km
Sun Alt. = 19.4°
Sun Azm. = 141.4°
Duration = 01m49.9s

Circumstances at Greatest Duration: 01:13:13.9 UT1

Lat. = 84°53.7'S
Long. = 021°02.3'E
Path Width = 354.8 km
Sun Alt. = 19.4°
Sun Azm. = 142.0°
Duration = 01m49.9s

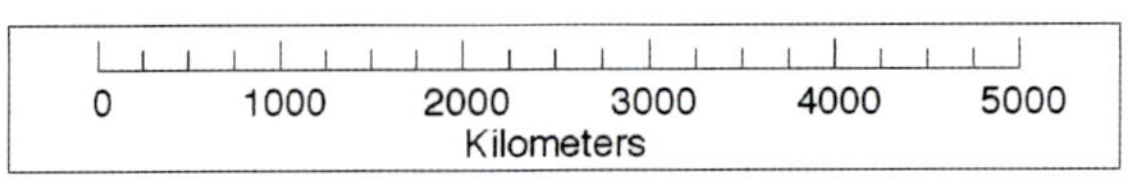

Partial Solar Eclipse of 2058 May 22

Greatest Eclipse = 10:39:25.5 TD (= 10:37:56.2 UT1)

Eclipse Magnitude = 0.4141 Saros Series = 119
Gamma = -1.3194 Saros Member = 68 of 71

Sun at Greatest Eclipse
(Geocentric Coordinates)
R.A. = 03h58m00.8s
Dec. = +20°28'40.9"
S.D. = 00°15'48.1"
H.P. = 00°00'08.7"

Moon at Greatest Eclipse
(Geocentric Coordinates)
R.A. = 03h59m32.2s
Dec. = +19°18'44.2"
S.D. = 00°15'09.0"
H.P. = 00°55'36.1"

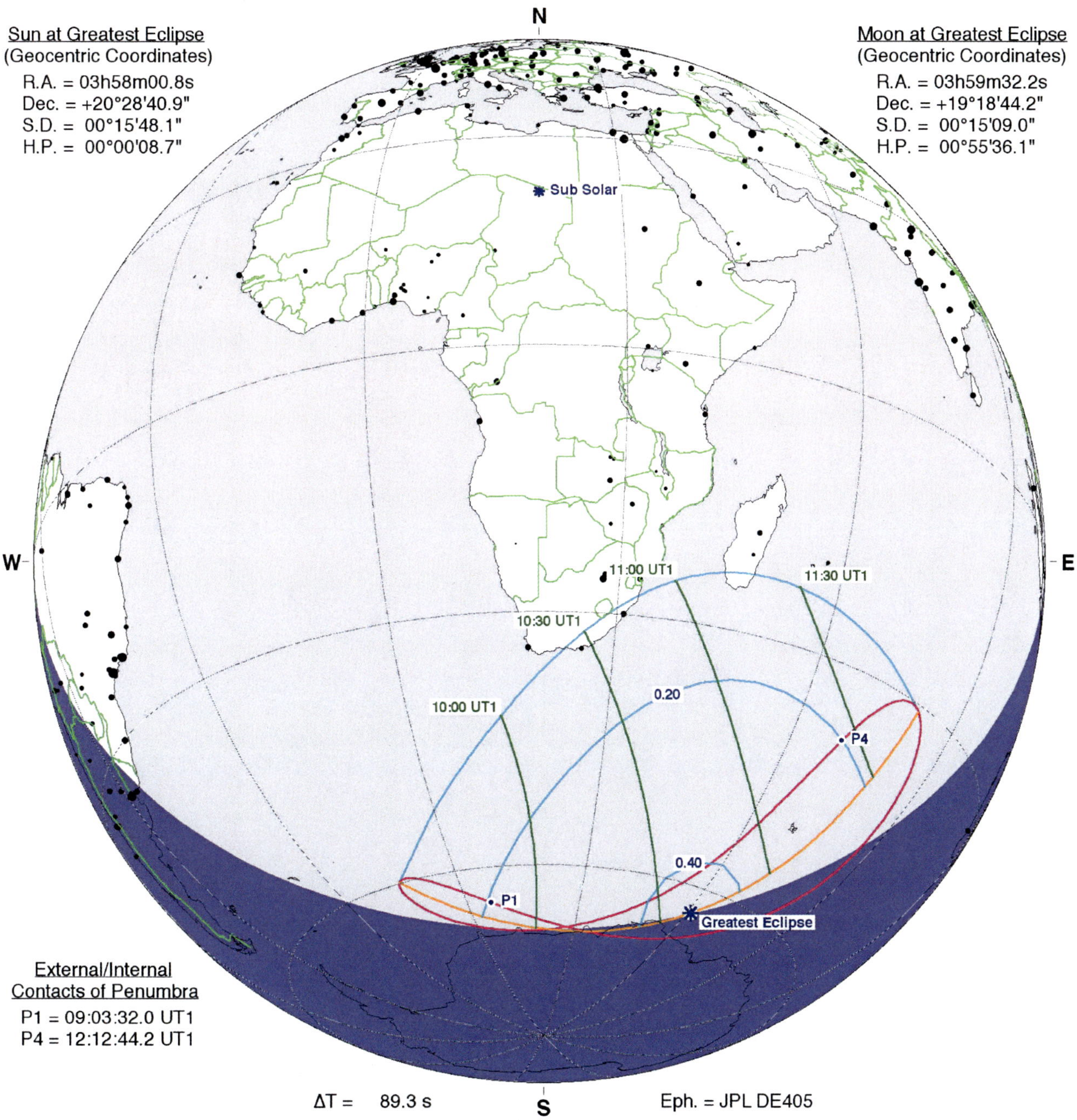

External/Internal
Contacts of Penumbra
P1 = 09:03:32.0 UT1
P4 = 12:12:44.2 UT1

ΔT = 89.3 s Eph. = JPL DE405

Circumstances at Greatest Eclipse: 10:37:56.2 UT1
Lat. = 63°31.8'S Sun Alt. = 0.0°
Long. = 061°03.2'E Sun Azm. = 321.7°

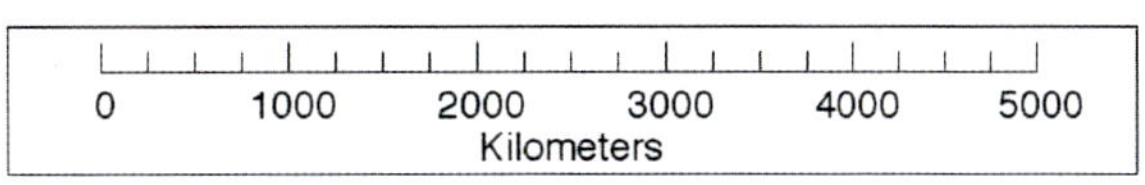

Partial Solar Eclipse of 2058 Jun 21

Greatest Eclipse = 00:19:34.6 TD (= 00:18:05.3 UT1)

Eclipse Magnitude = 0.1261 Saros Series = 157

Gamma = 1.4869 Saros Member = 1 of 70

Sun at Greatest Eclipse
(Geocentric Coordinates)
R.A. = 05h59m41.6s
Dec. = +23°25'56.0"
S.D. = 00°15'44.3"
H.P. = 00°00'08.7"

Moon at Greatest Eclipse
(Geocentric Coordinates)
R.A. = 05h59m06.9s
Dec. = +24°46'21.8"
S.D. = 00°14'50.9"
H.P. = 00°54'29.6"

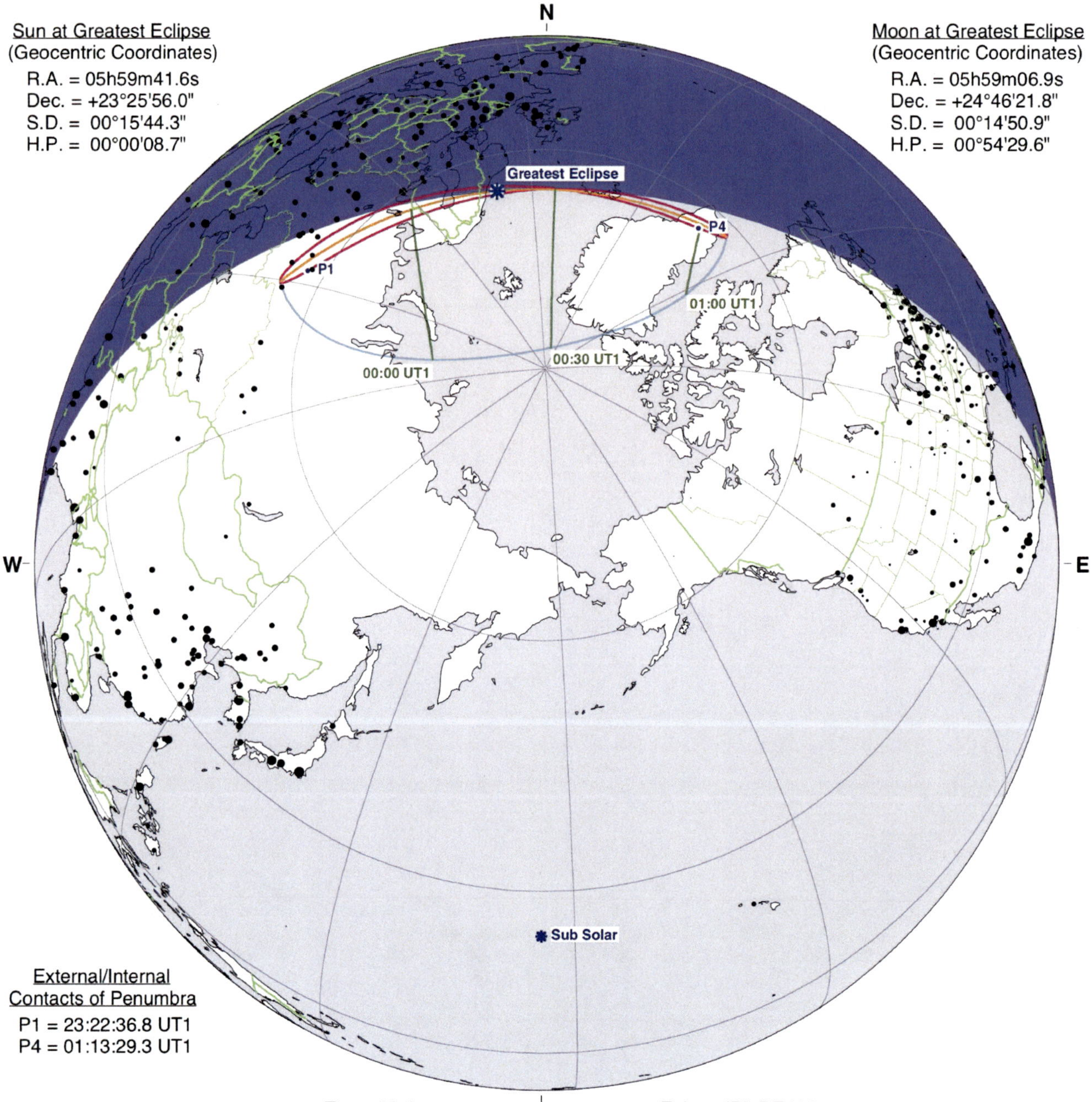

External/Internal
Contacts of Penumbra
P1 = 23:22:36.8 UT1
P4 = 01:13:29.3 UT1

ΔT = 89.4 s Eph. = JPL DE405

Circumstances at Greatest Eclipse: 00:18:05.3 UT1

Lat. = 65°57.0'N Sun Alt. = 0.0°

Long. = 009°47.8'E Sun Azm. = 12.7°

0 1000 2000 3000 4000 5000
Kilometers

Partial Solar Eclipse of 2058 Nov 16

Greatest Eclipse = 03:23:07.3 TD (= 03:21:37.7 UT1)

Eclipse Magnitude = 0.7644
Gamma = 1.1224

Saros Series = 124
Saros Member = 57 of 73

Sun at Greatest Eclipse
(Geocentric Coordinates)
R.A. = 15h26m32.8s
Dec. = -18°46'09.8"
S.D. = 00°16'10.2"
H.P. = 00°00'08.9"

Moon at Greatest Eclipse
(Geocentric Coordinates)
R.A. = 15h28m05.5s
Dec. = -17°45'10.1"
S.D. = 00°15'47.0"
H.P. = 00°57'55.4"

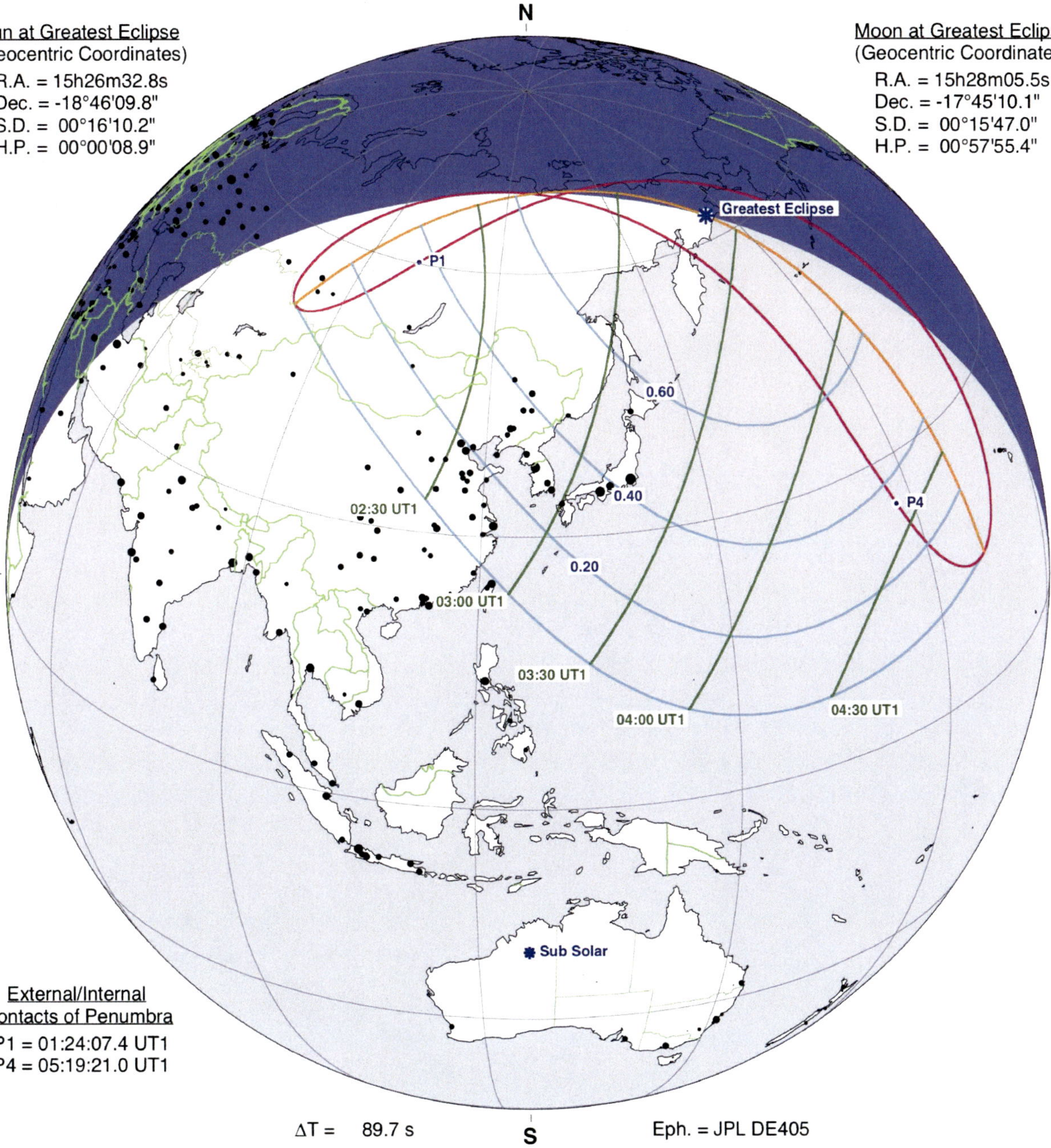

External/Internal
Contacts of Penumbra
P1 = 01:24:07.4 UT1
P4 = 05:19:21.0 UT1

ΔT = 89.7 s

Eph. = JPL DE405

Circumstances at Greatest Eclipse: 03:21:37.7 UT1

Lat. = 62°54.0'N
Long. = 174°05.5'E

Sun Alt. = 0.0°
Sun Azm. = 225.1°

0 1000 2000 3000 4000 5000
Kilometers

Total Solar Eclipse of 2059 May 11

Greatest Eclipse = 19:22:15.6 TD (= 19:20:45.6 UT1)

Eclipse Magnitude = 1.0242 | Saros Series = 129
Gamma = -0.5080 | Saros Member = 54 of 80

Sun at Greatest Eclipse
(Geocentric Coordinates)
R.A. = 03h14m47.9s
Dec. = +18°02'08.6"
S.D. = 00°15'50.2"
H.P. = 00°00'08.7"

Moon at Greatest Eclipse
(Geocentric Coordinates)
R.A. = 03h15m32.3s
Dec. = +17°34'20.5"
S.D. = 00°15'59.6"
H.P. = 00°58'41.8"

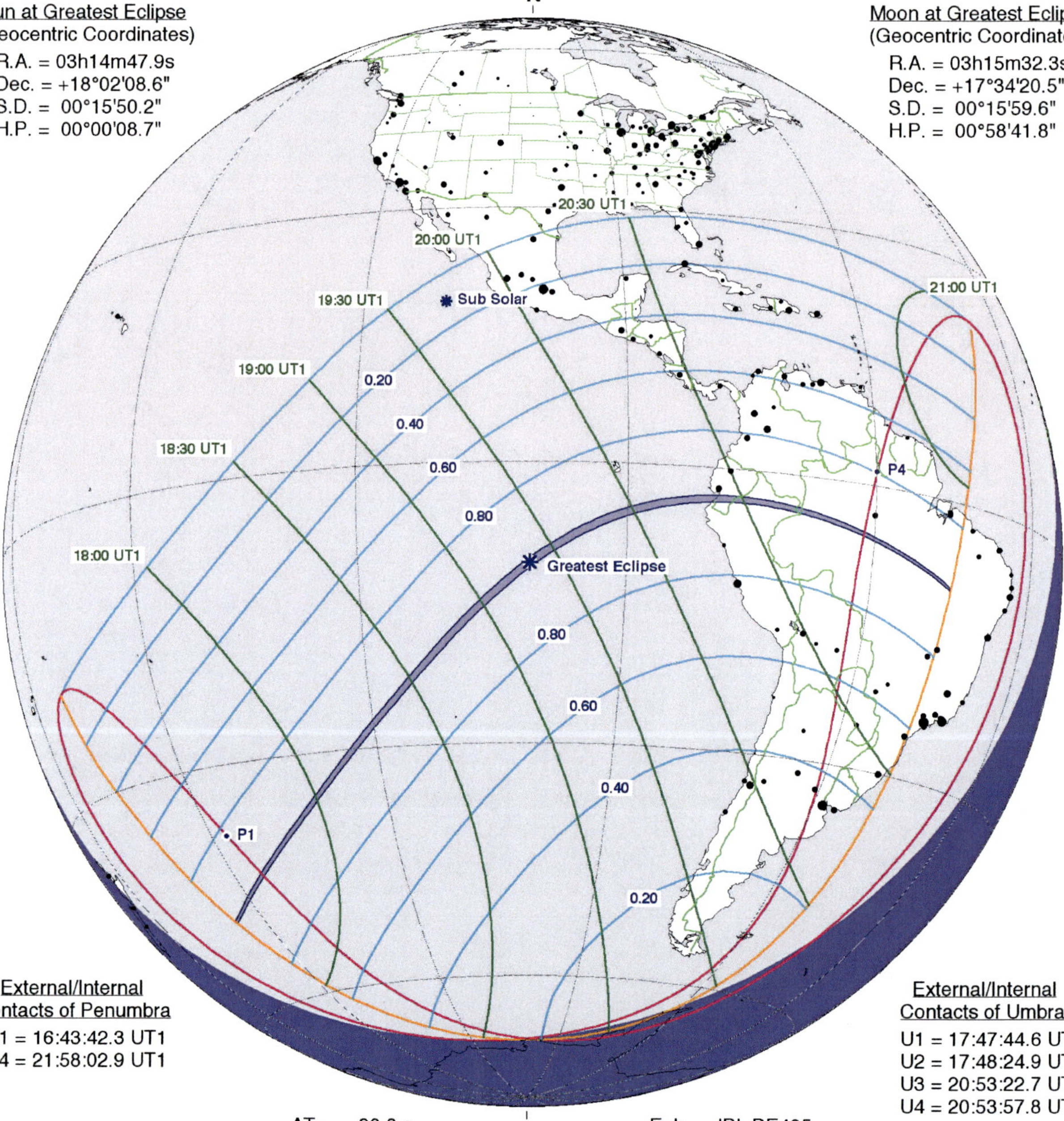

External/Internal Contacts of Penumbra
P1 = 16:43:42.3 UT1
P4 = 21:58:02.9 UT1

External/Internal Contacts of Umbra
U1 = 17:47:44.6 UT1
U2 = 17:48:24.9 UT1
U3 = 20:53:22.7 UT1
U4 = 20:53:57.8 UT1

ΔT = 90.0 s

Eph. = JPL DE405

Circumstances at Greatest Eclipse: 19:20:45.6 UT1
Lat. = 10°43.0'S | Sun Alt. = 59.4°
Long. = 100°29.3'W | Sun Azm. = 339.9°
Path Width = 94.6 km | Duration = 02m23.4s

Circumstances at Greatest Duration: 19:22:42.0 UT1
Lat. = 10°21.8'S | Sun Alt. = 59.4°
Long. = 099°59.3'W | Sun Azm. = 338.0°
Path Width = 94.8 km | Duration = 02m23.4s

0 1000 2000 3000 4000 5000
Kilometers

Annular Solar Eclipse of 2059 Nov 05

Greatest Eclipse = 09:18:14.6 TD (= 09:16:44.3 UT1)

Eclipse Magnitude = 0.9417 | Saros Series = 134

Gamma = 0.4454 | Saros Member = 46 of 71

Sun at Greatest Eclipse
(Geocentric Coordinates)
R.A. = 14h42m02.6s
Dec. = -15°43'28.3"
S.D. = 00°16'07.6"
H.P. = 00°00'08.9"

Moon at Greatest Eclipse
(Geocentric Coordinates)
R.A. = 14h42m42.7s
Dec. = -15°21'02.7"
S.D. = 00°14'58.8"
H.P. = 00°54'58.7"

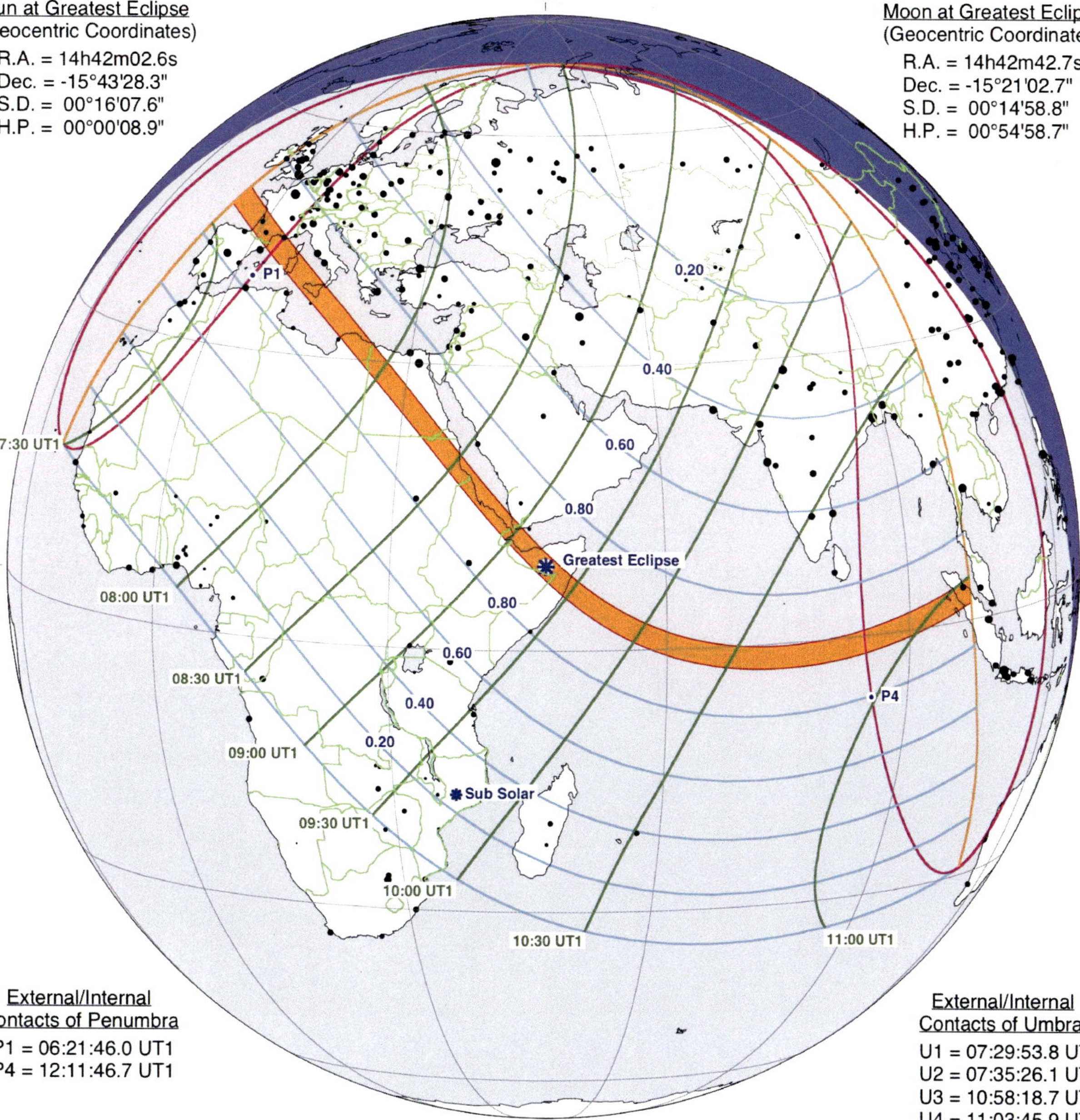

External/Internal
Contacts of Penumbra
P1 = 06:21:46.0 UT1
P4 = 12:11:46.7 UT1

External/Internal
Contacts of Umbra
U1 = 07:29:53.8 UT1
U2 = 07:35:26.1 UT1
U3 = 10:58:18.7 UT1
U4 = 11:03:45.9 UT1

ΔT = 90.3 s | Eph. = JPL DE405

Circumstances at Greatest Eclipse: 09:16:44.3 UT1

Lat. = 08°44.6'N | Sun Alt. = 63.5°
Long. = 046°59.3'E | Sun Azm. = 202.7°
Path Width = 238.3 km | Duration = 06m59.6s

Circumstances at Greatest Duration: 09:31:43.2 UT1

Lat. = 06°02.0'N | Sun Alt. = 62.3°
Long. = 050°11.1'E | Sun Azm. = 217.9°
Path Width = 243.9 km | Duration = 07m01.5s

0 1000 2000 3000 4000 5000
Kilometers

Total Solar Eclipse of 2060 Apr 30

Greatest Eclipse = 10:09:59.8 TD (= 10:08:29.2 UT1)

Eclipse Magnitude = 1.0660 Saros Series = 139

Gamma = 0.2422 Saros Member = 32 of 71

Sun at Greatest Eclipse
(Geocentric Coordinates)
R.A. = 02h33m38.4s
Dec. = +15°04'16.7"
S.D. = 00°15'52.6"
H.P. = 00°00'08.7"

Moon at Greatest Eclipse
(Geocentric Coordinates)
R.A. = 02h33m13.6s
Dec. = +15°17'46.8"
S.D. = 00°16'38.8"
H.P. = 01°01'05.8"

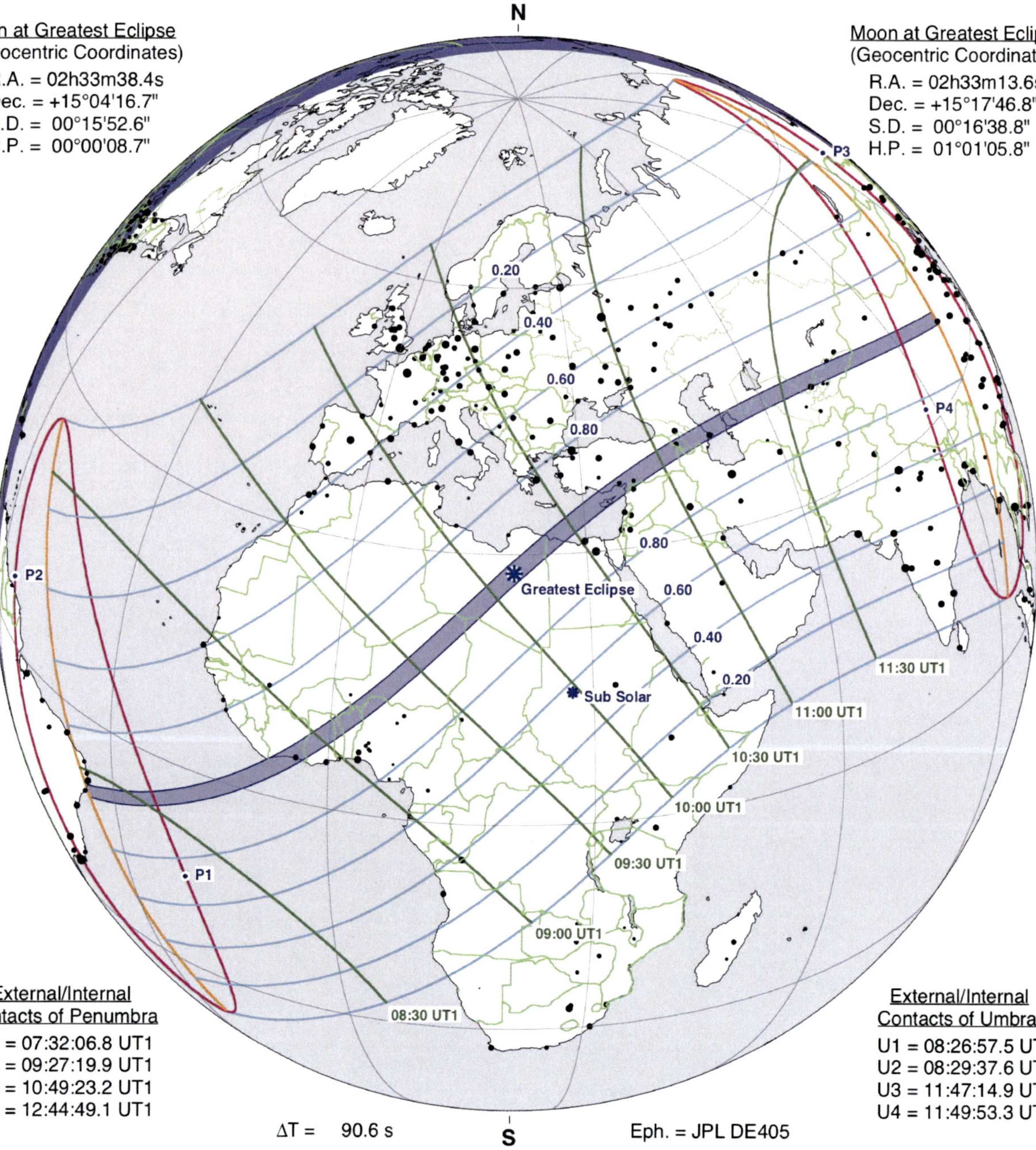

External/Internal
Contacts of Penumbra
P1 = 07:32:06.8 UT1
P2 = 09:27:19.9 UT1
P3 = 10:49:23.2 UT1
P4 = 12:44:49.1 UT1

External/Internal
Contacts of Umbra
U1 = 08:26:57.5 UT1
U2 = 08:29:37.6 UT1
U3 = 11:47:14.9 UT1
U4 = 11:49:53.3 UT1

ΔT = 90.6 s Eph. = JPL DE405

Circumstances at Greatest Eclipse: 10:08:29.2 UT1
Lat. = 27°57.4'N Sun Alt. = 75.8°
Long. = 020°49.0'E Sun Azm. = 154.1°
Path Width = 222.0 km Duration = 05m15.2s

Circumstances at Greatest Duration: 10:13:24.7 UT1
Lat. = 29°09.3'N Sun Alt. = 75.5°
Long. = 022°23.1'E Sun Azm. = 166.1°
Path Width = 221.3 km Duration = 05m15.5s

0 1000 2000 3000 4000 5000
Kilometers

Annular Solar Eclipse of 2060 Oct 24

Greatest Eclipse = 09:24:10.4 TD (= 09:22:39.4 UT1)

Eclipse Magnitude = 0.9277 Saros Series = 144

Gamma = -0.2625 Saros Member = 19 of 70

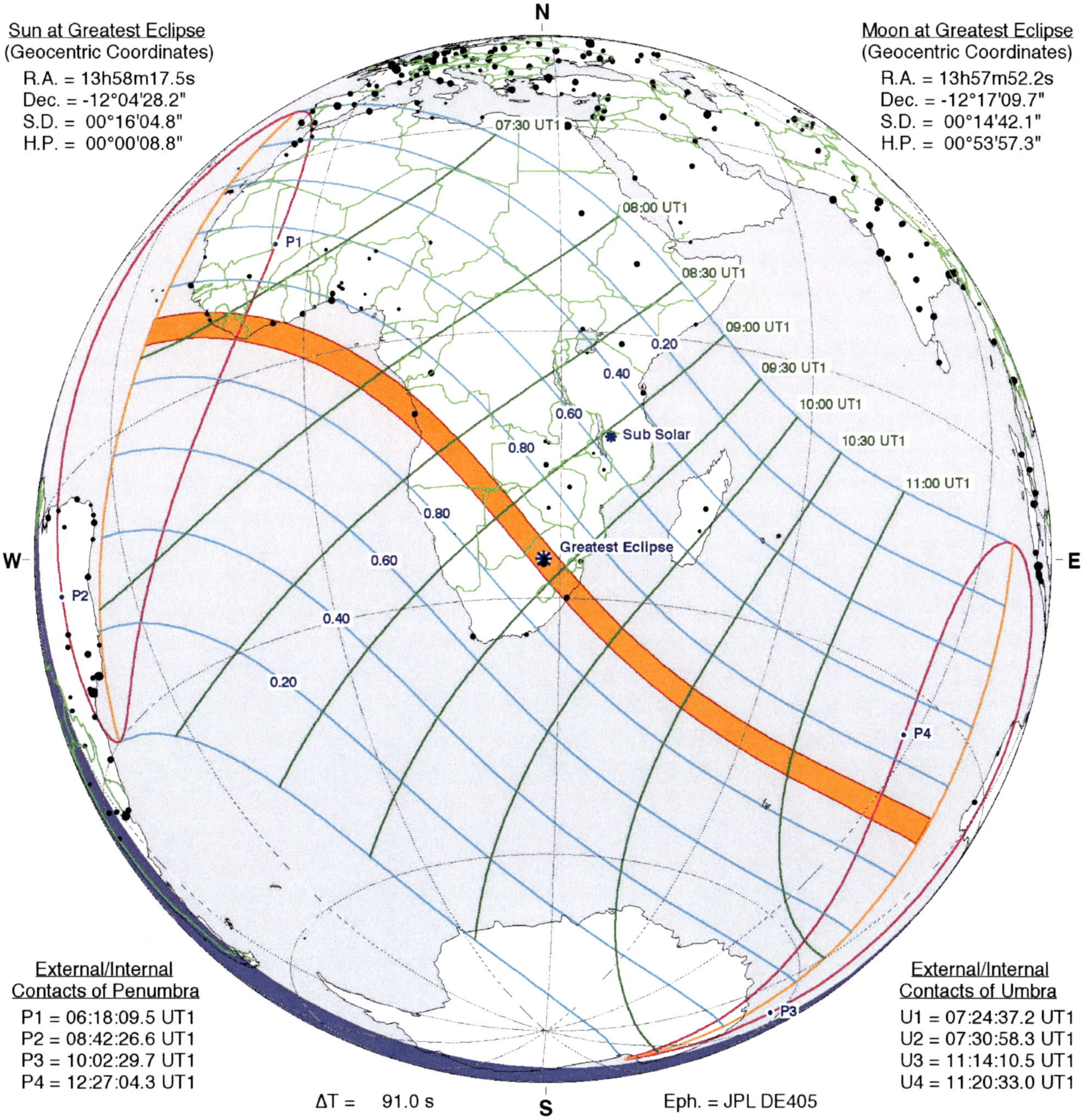

Sun at Greatest Eclipse
(Geocentric Coordinates)
R.A. = 13h58m17.5s
Dec. = -12°04'28.2"
S.D. = 00°16'04.8"
H.P. = 00°00'08.8"

Moon at Greatest Eclipse
(Geocentric Coordinates)
R.A. = 13h57m52.2s
Dec. = -12°17'09.7"
S.D. = 00°14'42.1"
H.P. = 00°53'57.3"

External/Internal Contacts of Penumbra
P1 = 06:18:09.5 UT1
P2 = 08:42:26.6 UT1
P3 = 10:02:29.7 UT1
P4 = 12:27:04.3 UT1

External/Internal Contacts of Umbra
U1 = 07:24:37.2 UT1
U2 = 07:30:58.3 UT1
U3 = 11:14:10.5 UT1
U4 = 11:20:33.0 UT1

ΔT = 91.0 s

Eph. = JPL DE405

Circumstances at Greatest Eclipse: 09:22:39.4 UT1

Lat. = 25°47.2'S Sun Alt. = 74.6°
Long. = 028°01.1'E Sun Azm. = 28.1°
Path Width = 280.5 km Duration = 08m05.8s

Circumstances at Greatest Duration: 09:39:59.5 UT1

Lat. = 29°42.0'S Sun Alt. = 72.4°
Long. = 032°03.8'E Sun Azm. = 356.6°
Path Width = 278.5 km Duration = 08m08.0s

0 1000 2000 3000 4000 5000
Kilometers

Total Solar Eclipse of 2061 Apr 20

Greatest Eclipse = 02:56:49.1 TD (= 02:55:17.8 UT1)

Eclipse Magnitude = 1.0476
Gamma = 0.9578

Saros Series = 149
Saros Member = 23 of 71

Sun at Greatest Eclipse
(Geocentric Coordinates)
R.A. = 01h53m47.8s
Dec. = +11°39'59.8"
S.D. = 00°15'55.3"
H.P. = 00°00'08.8"

Moon at Greatest Eclipse
(Geocentric Coordinates)
R.A. = 01h52m03.2s
Dec. = +12°32'19.1"
S.D. = 00°16'36.4"
H.P. = 01°00'56.9"

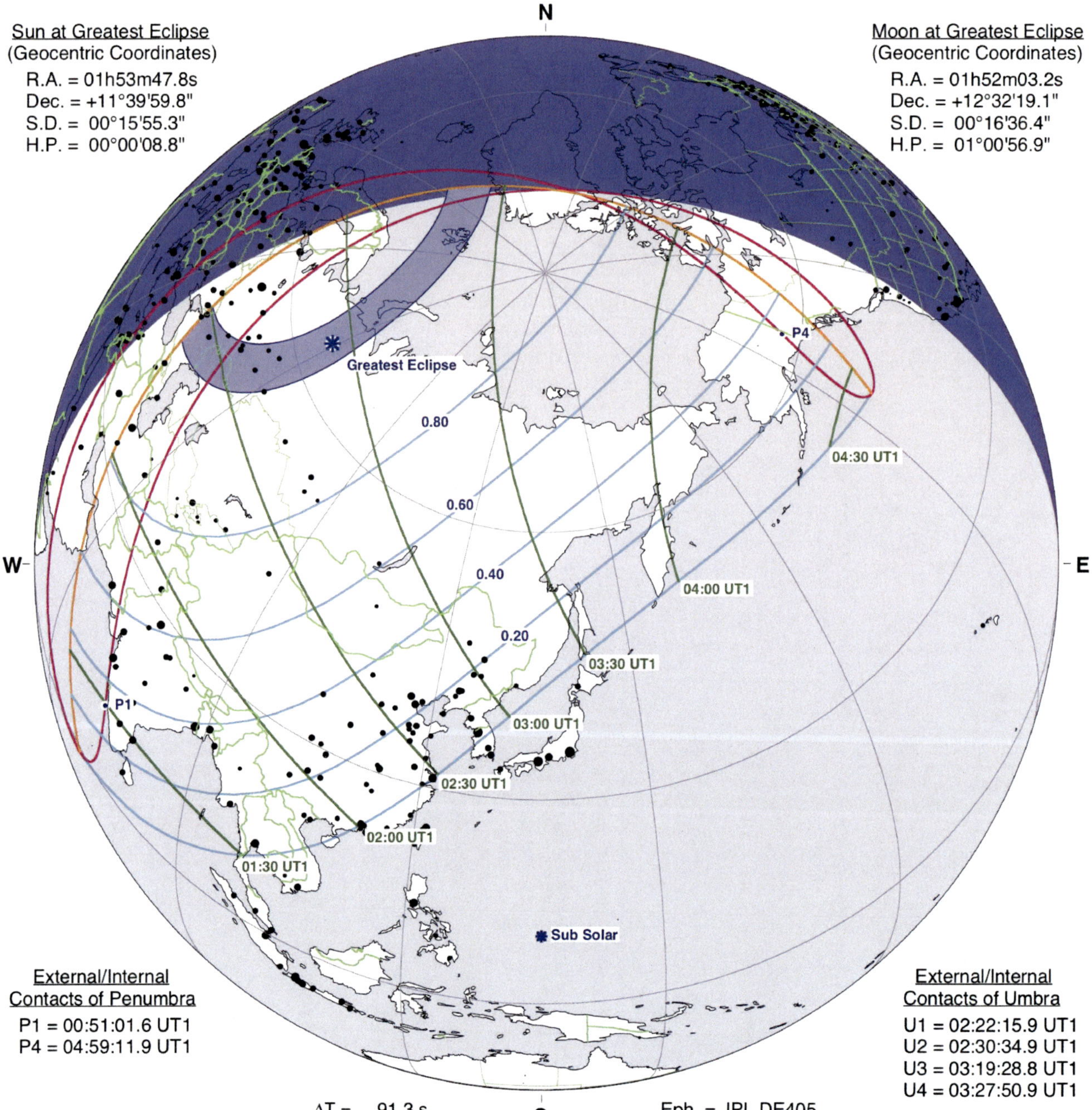

External/Internal
Contacts of Penumbra
P1 = 00:51:01.6 UT1
P4 = 04:59:11.9 UT1

External/Internal
Contacts of Umbra
U1 = 02:22:15.9 UT1
U2 = 02:30:34.9 UT1
U3 = 03:19:28.8 UT1
U4 = 03:27:50.9 UT1

ΔT = 91.3 s

Eph. = JPL DE405

Circumstances at Greatest Eclipse: 02:55:17.8 UT1
Lat. = 64°32.2'N
Sun Alt. = 16.2°
Long. = 059°03.3'E
Sun Azm. = 96.8°
Path Width = 558.9 km
Duration = 02m37.4s

Circumstances at Greatest Duration: 02:55:31.2 UT1
Lat. = 64°39.5'N
Sun Alt. = 16.2°
Long. = 059°01.8'E
Sun Azm. = 96.9°
Path Width = 558.2 km
Duration = 02m37.4s

0 1000 2000 3000 4000 5000
Kilometers

Annular Solar Eclipse of 2061 Oct 13

Greatest Eclipse = 10:32:09.7 TD (= 10:30:38.0 UT1)

Eclipse Magnitude = 0.9469 Saros Series = 154

Gamma = -0.9639 Saros Member = 9 of 71

Sun at Greatest Eclipse
(Geocentric Coordinates)
R.A. = 13h16m11.1s
Dec. = -08°03'03.6"
S.D. = 00°16'01.7"
H.P. = 00°00'08.8"

Moon at Greatest Eclipse
(Geocentric Coordinates)
R.A. = 13h14m30.5s
Dec. = -08°50'16.1"
S.D. = 00°15'07.5"
H.P. = 00°55'30.4"

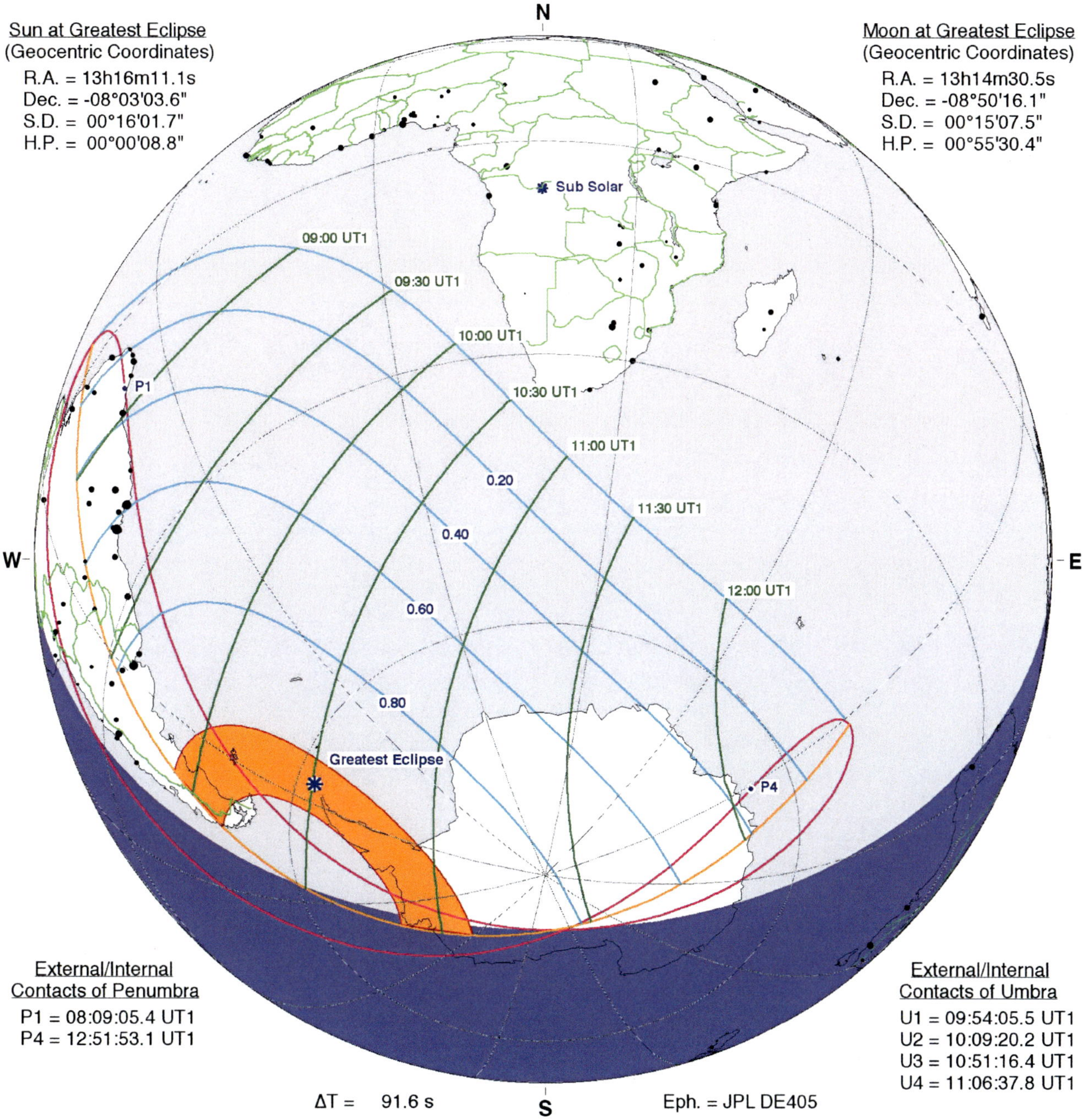

External/Internal
Contacts of Penumbra
P1 = 08:09:05.4 UT1
P4 = 12:51:53.1 UT1

External/Internal
Contacts of Umbra
U1 = 09:54:05.5 UT1
U2 = 10:09:20.2 UT1
U3 = 10:51:16.4 UT1
U4 = 11:06:37.8 UT1

ΔT = 91.6 s Eph. = JPL DE405

Circumstances at Greatest Eclipse: 10:30:38.0 UT1

Lat. = 62°08.2'S Sun Alt. = 14.9°
Long. = 054°29.2'W Sun Azm. = 79.0°
Path Width = 742.6 km Duration = 03m41.0s

Circumstances at Greatest Duration: 10:33:54.1 UT1

Lat. = 63°45.1'S Sun Alt. = 14.7°
Long. = 054°50.1'W Sun Azm. = 78.1°
Path Width = 728.7 km Duration = 03m41.0s

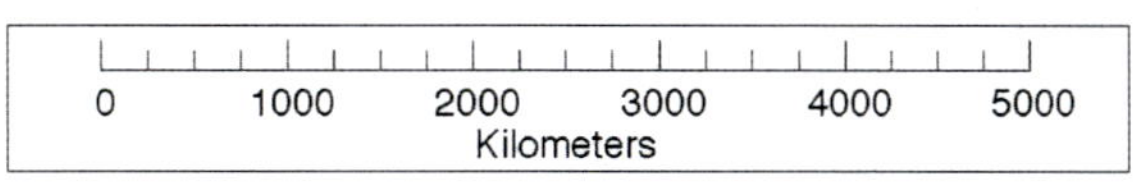

Partial Solar Eclipse of 2062 Mar 11

Greatest Eclipse = 04:26:16.2 TD (= 04:24:44.3 UT1)

Eclipse Magnitude = 0.9331 Saros Series = 121

Gamma = -1.0238 Saros Member = 63 of 71

Sun at Greatest Eclipse
(Geocentric Coordinates)
R.A. = 23h26m28.0s
Dec. = -03°36'57.3"
S.D. = 00°16'06.2"
H.P. = 00°00'08.9"

Moon at Greatest Eclipse
(Geocentric Coordinates)
R.A. = 23h28m20.0s
Dec. = -04°27'39.9"
S.D. = 00°15'26.8"
H.P. = 00°56'41.5"

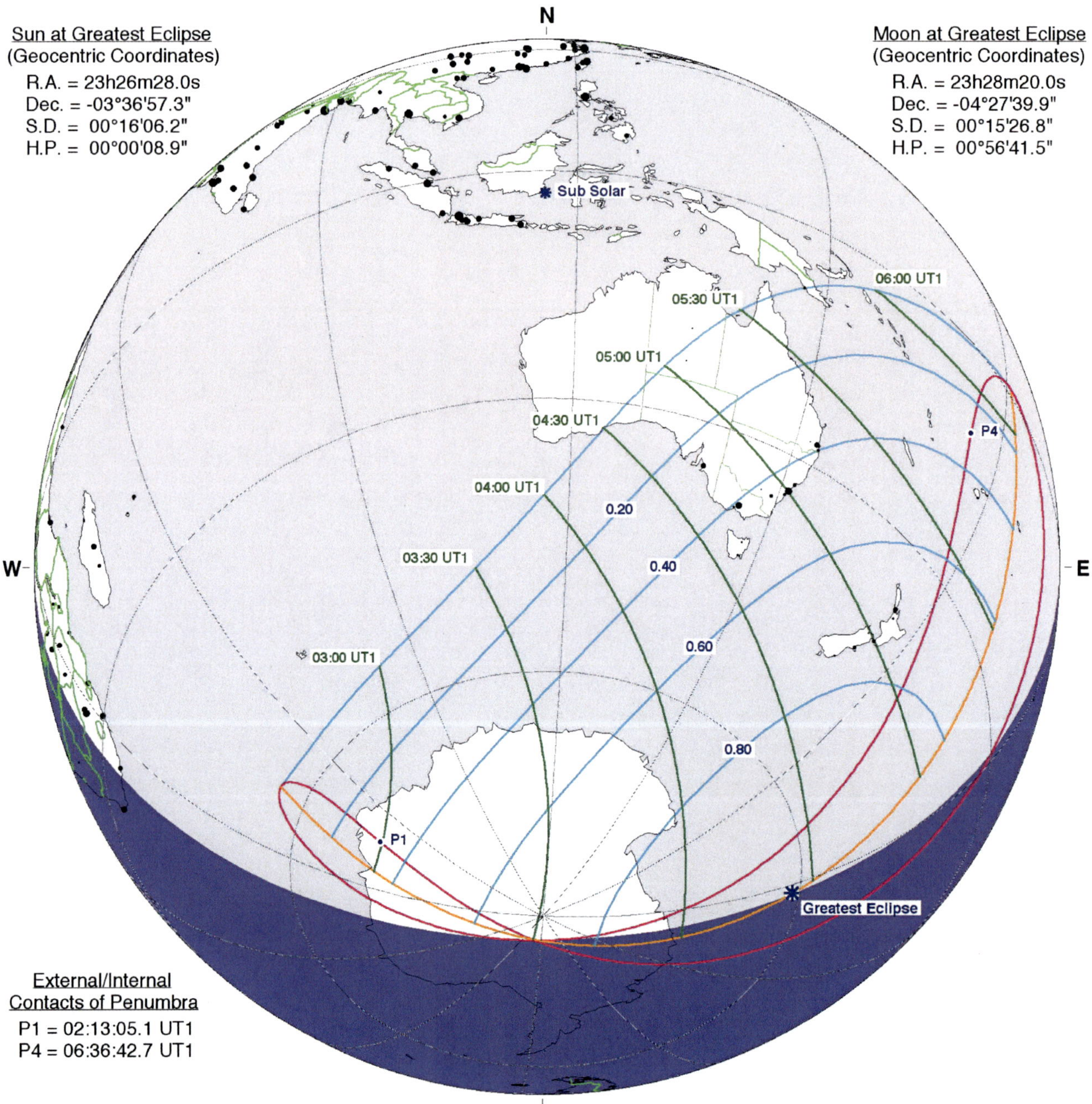

External/Internal
Contacts of Penumbra
P1 = 02:13:05.1 UT1
P4 = 06:36:42.7 UT1

ΔT = 91.9 s Eph. = JPL DE405

Circumstances at Greatest Eclipse: 04:24:44.3 UT1

Lat. = 60°58.3'S Sun Alt. = 0.0°

Long. = 147°14.6'W Sun Azm. = 262.5°

0 1000 2000 3000 4000 5000
Kilometers

Partial Solar Eclipse of 2062 Sep 03

Greatest Eclipse = 08:54:27.4 TD (= 08:52:55.1 UT1)

Eclipse Magnitude = 0.9749 Saros Series = 126

Gamma = 1.0192 Saros Member = 50 of 72

Sun at Greatest Eclipse
(Geocentric Coordinates)
R.A. = 10h50m30.3s
Dec. = +07°22'28.5"
S.D. = 00°15'51.2"
H.P. = 00°00'08.7"

Moon at Greatest Eclipse
(Geocentric Coordinates)
R.A. = 10h52m25.5s
Dec. = +08°16'29.0"
S.D. = 00°16'22.2"
H.P. = 01°00'04.6"

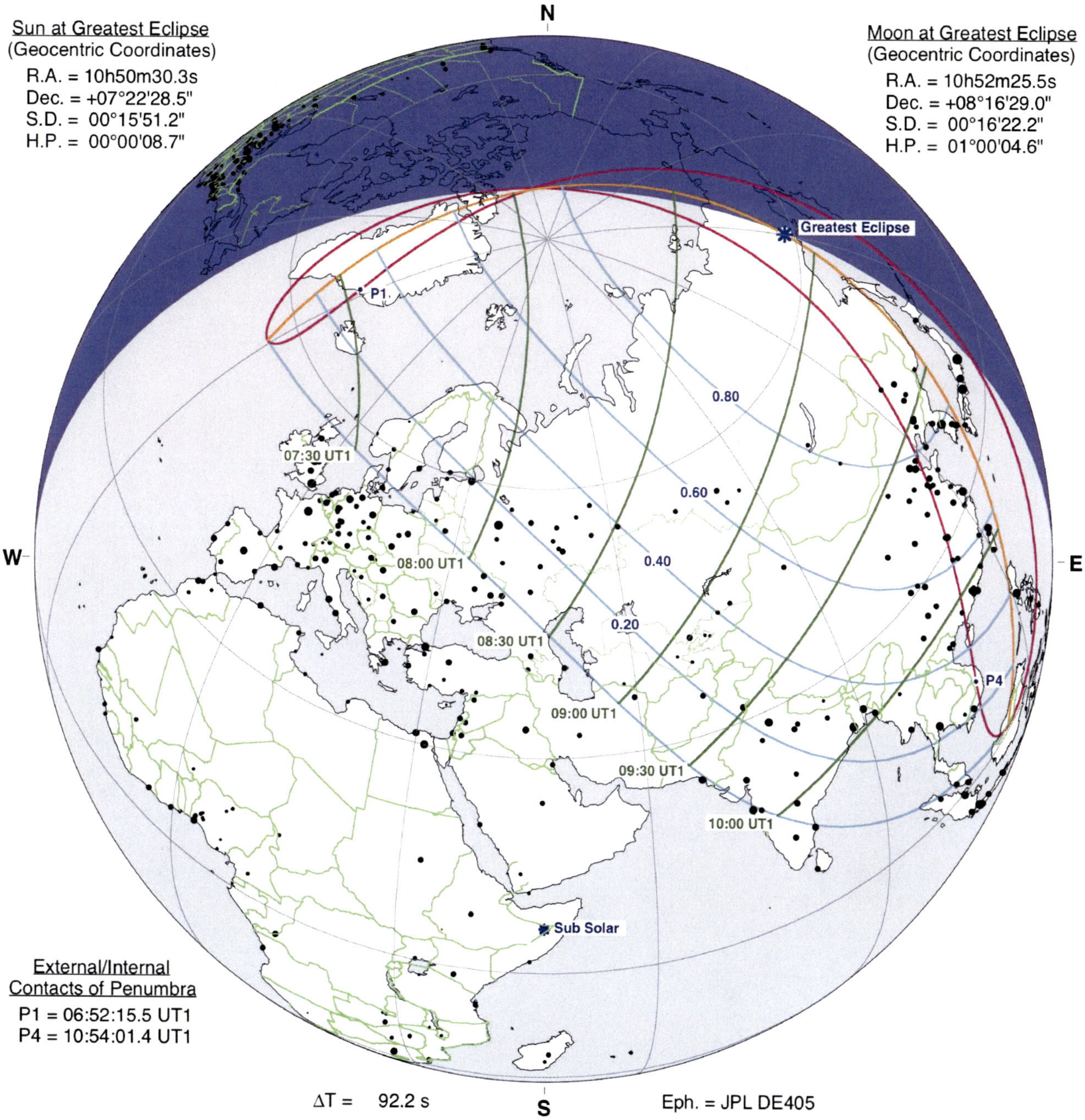

External/Internal
Contacts of Penumbra
P1 = 06:52:15.5 UT1
P4 = 10:54:01.4 UT1

ΔT = 92.2 s Eph. = JPL DE405

Circumstances at Greatest Eclipse: 08:52:55.1 UT1

Lat. = 61°18.3'N Sun Alt. = 0.0°

Long. = 150°12.6'E Sun Azm. = 285.5°

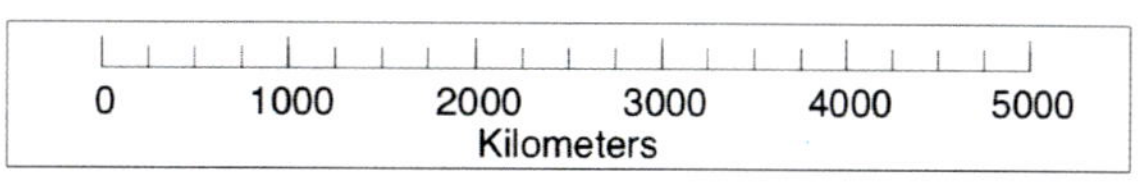

Annular Solar Eclipse of 2063 Feb 28

Greatest Eclipse = 07:43:30.0 TD (= 07:41:57.4 UT1)

Eclipse Magnitude = 0.9293 Saros Series = 131

Gamma = -0.3360 Saros Member = 53 of 70

Sun at Greatest Eclipse
(Geocentric Coordinates)
R.A. = 22h45m11.8s
Dec. = -07°54'42.4"
S.D. = 00°16'08.9"
H.P. = 00°00'08.9"

Moon at Greatest Eclipse
(Geocentric Coordinates)
R.A. = 22h45m46.2s
Dec. = -08°10'47.1"
S.D. = 00°14'47.6"
H.P. = 00°54'17.7"

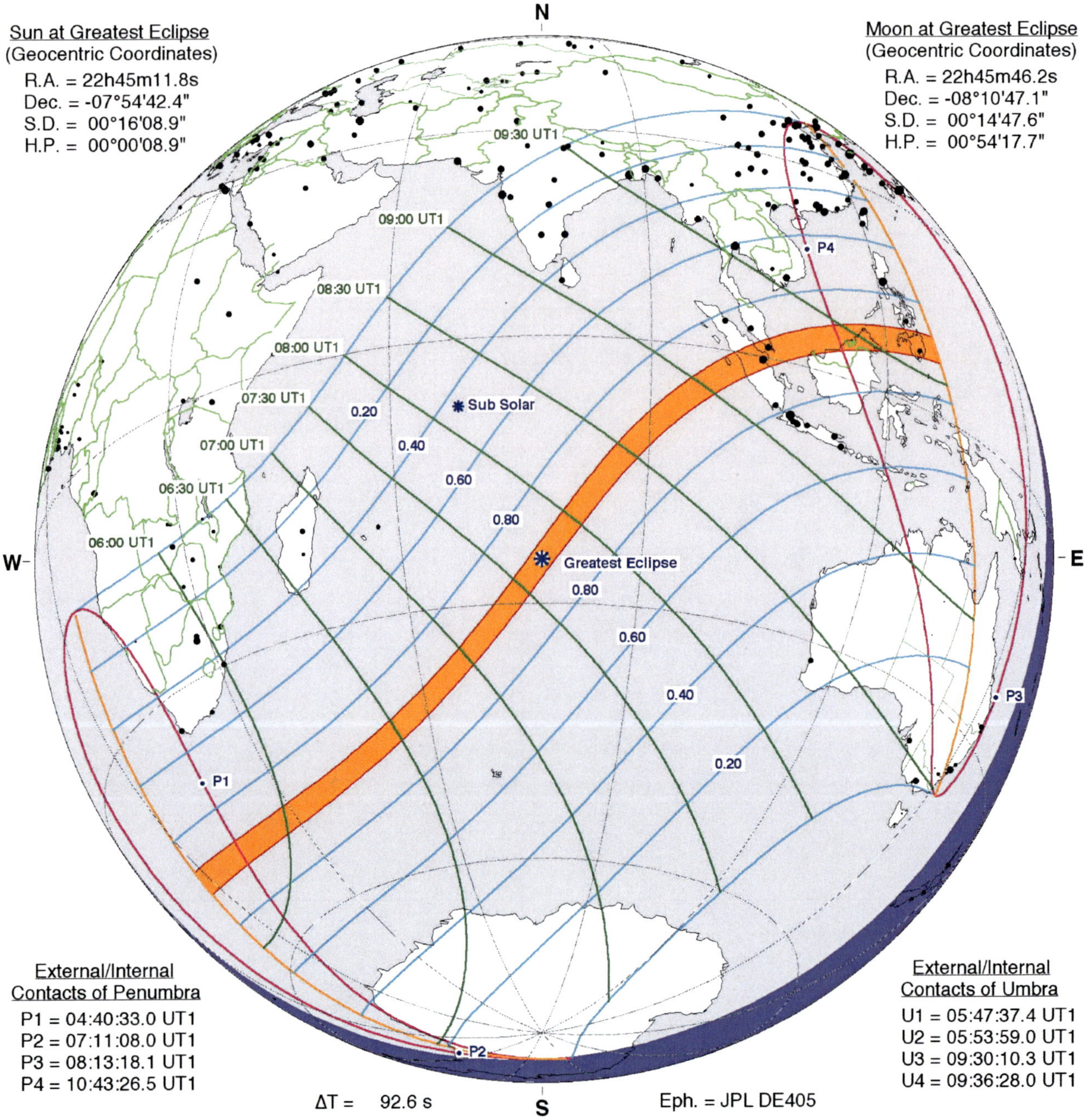

External/Internal
Contacts of Penumbra
P1 = 04:40:33.0 UT1
P2 = 07:11:08.0 UT1
P3 = 08:13:18.1 UT1
P4 = 10:43:26.5 UT1

External/Internal
Contacts of Umbra
U1 = 05:47:37.4 UT1
U2 = 05:53:59.0 UT1
U3 = 09:30:10.3 UT1
U4 = 09:36:28.0 UT1

ΔT = 92.6 s Eph. = JPL DE405

Circumstances at Greatest Eclipse: 07:41:57.4 UT1

Lat. = 25°13.9'S Sun Alt. = 70.2°
Long. = 077°37.8'E Sun Azm. = 329.4°
Path Width = 279.6 km Duration = 07m41.2s

Circumstances at Greatest Duration: 07:27:17.6 UT1

Lat. = 28°52.6'S Sun Alt. = 68.8°
Long. = 074°24.5'E Sun Azm. = 351.4°
Path Width = 276.8 km Duration = 07m42.5s

0 1000 2000 3000 4000 5000
Kilometers

Total Solar Eclipse of 2063 Aug 24

Greatest Eclipse = 01:22:10.6 TD (= 01:20:37.6 UT1)

Eclipse Magnitude = 1.0750
Gamma = 0.2771

Saros Series = 136
Saros Member = 40 of 71

Sun at Greatest Eclipse
(Geocentric Coordinates)
R.A. = 10h12m03.7s
Dec. = +11°07'34.9"
S.D. = 00°15'48.9"
H.P. = 00°00'08.7"

Moon at Greatest Eclipse
(Geocentric Coordinates)
R.A. = 10h12m34.5s
Dec. = +11°22'46.8"
S.D. = 00°16'43.4"
H.P. = 01°01'22.6"

N

W

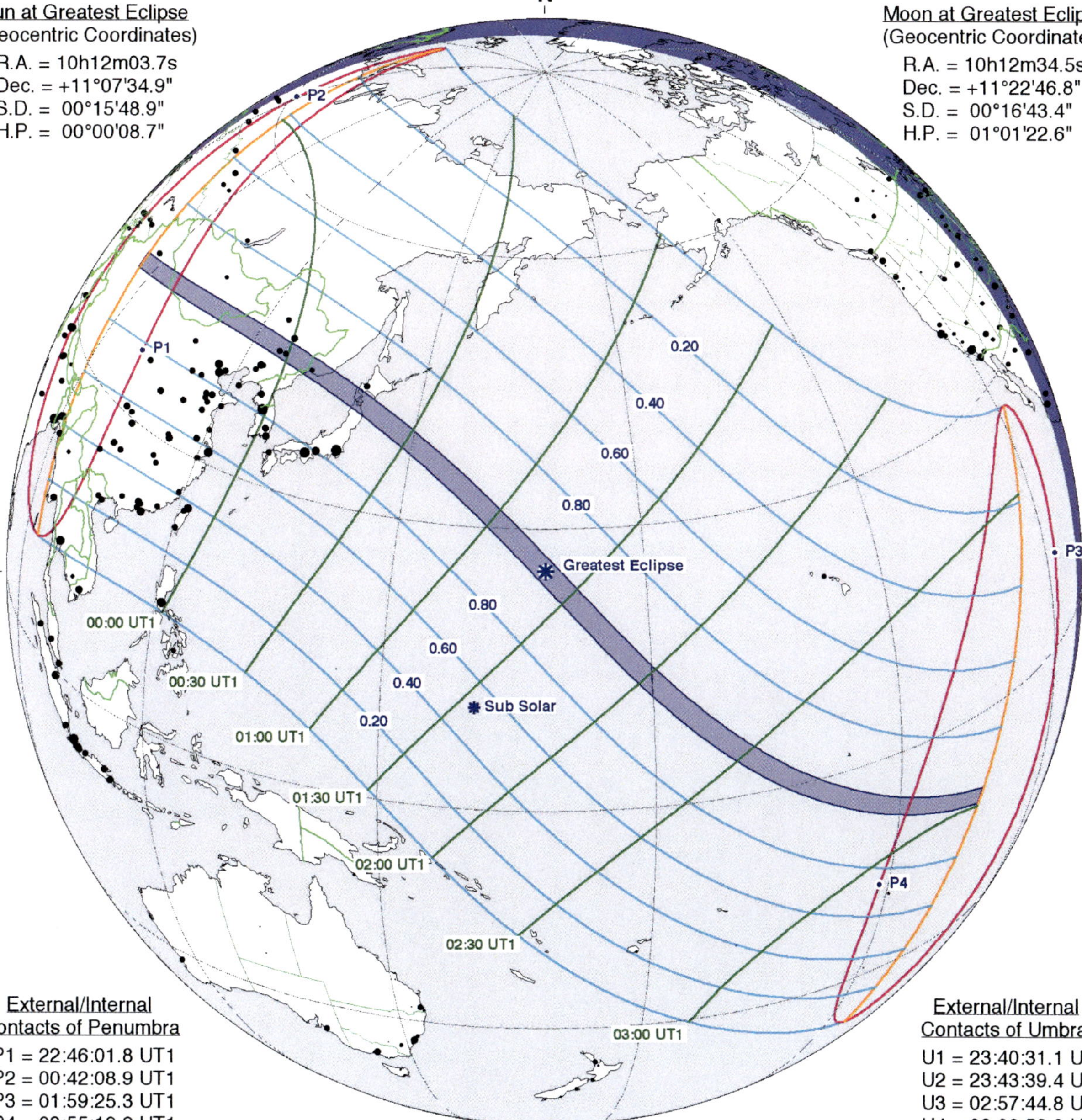

E

S

External/Internal
Contacts of Penumbra
P1 = 22:46:01.8 UT1
P2 = 00:42:08.9 UT1
P3 = 01:59:25.3 UT1
P4 = 03:55:19.2 UT1

External/Internal
Contacts of Umbra
U1 = 23:40:31.1 UT1
U2 = 23:43:39.4 UT1
U3 = 02:57:44.8 UT1
U4 = 03:00:52.8 UT1

ΔT = 92.9 s

Eph. = JPL DE405

Circumstances at Greatest Eclipse: 01:20:37.6 UT1

Lat. = 25°33.7'N	Sun Alt. = 73.8°
Long. = 168°19.2'E	Sun Azm. = 208.6°
Path Width = 252.2 km	Duration = 05m49.1s

Circumstances at Greatest Duration: 01:15:57.6 UT1

Lat. = 26°49.2'N	Sun Alt. = 73.5°
Long. = 166°56.6'E	Sun Azm. = 198.7°
Path Width = 251.3 km	Duration = 05m49.3s

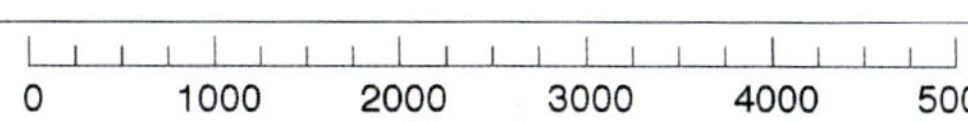

Annular Solar Eclipse of 2064 Feb 17

Greatest Eclipse = 07:00:23.3 TD (= 06:58:50.1 UT1)

Eclipse Magnitude = 0.9262
Gamma = 0.3597

Saros Series = 141
Saros Member = 26 of 70

Sun at Greatest Eclipse
(Geocentric Coordinates)
R.A. = 22h02m13.8s
Dec. = -12°01'37.5"
S.D. = 00°16'11.3"
H.P. = 00°00'08.9"

Moon at Greatest Eclipse
(Geocentric Coordinates)
R.A. = 22h01m38.9s
Dec. = -11°44'08.3"
S.D. = 00°14'47.1"
H.P. = 00°54'15.6"

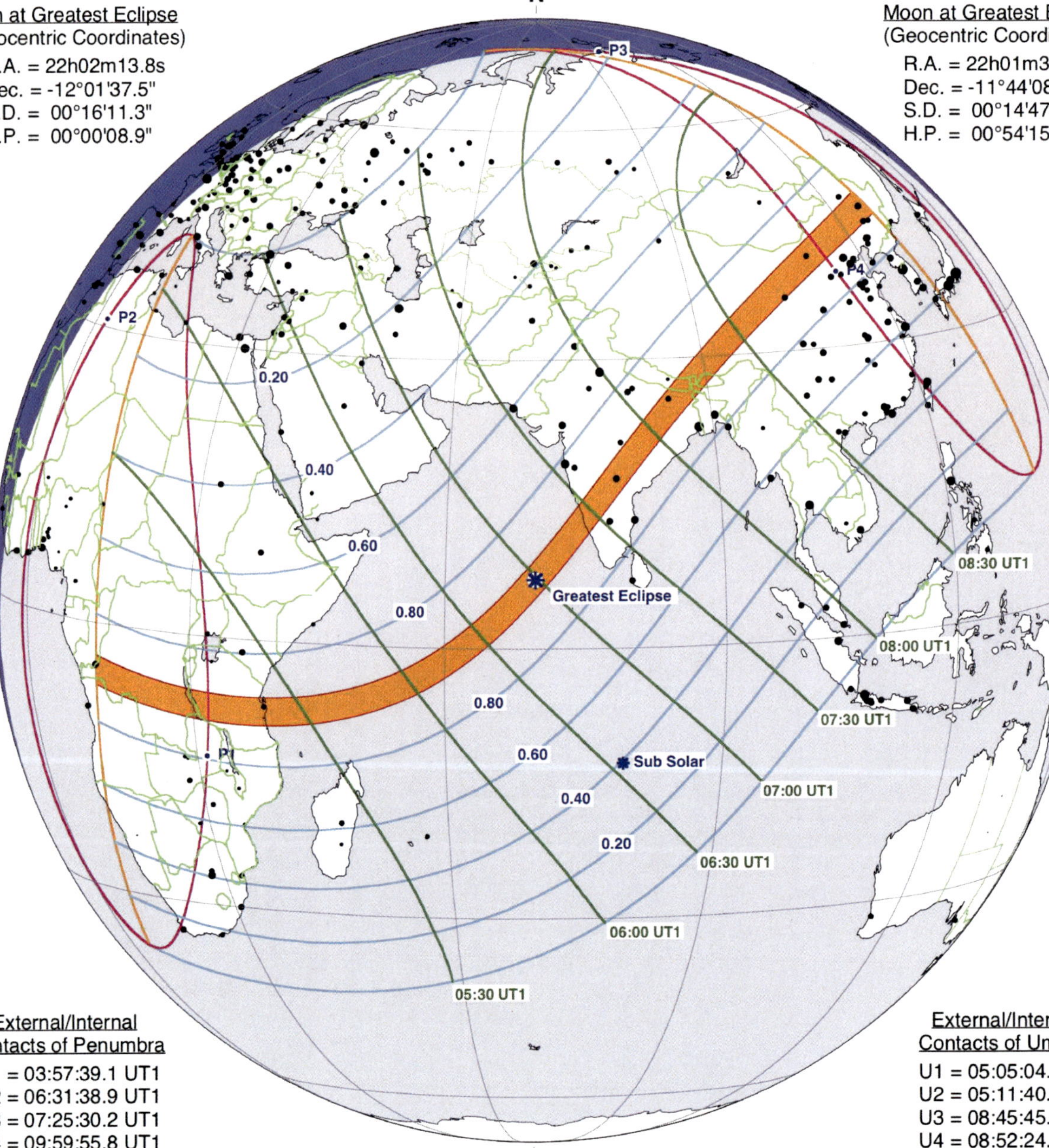

External/Internal
Contacts of Penumbra
P1 = 03:57:39.1 UT1
P2 = 06:31:38.9 UT1
P3 = 07:25:30.2 UT1
P4 = 09:59:55.8 UT1

External/Internal
Contacts of Umbra
U1 = 05:05:04.6 UT1
U2 = 05:11:40.8 UT1
U3 = 08:45:45.0 UT1
U4 = 08:52:24.6 UT1

ΔT = 93.2 s

Eph. = JPL DE405

Circumstances at Greatest Eclipse: 06:58:50.1 UT1

Lat. = 07°02.0'N
Long. = 069°37.3'E
Path Width = 294.6 km
Sun Alt. = 68.9°
Sun Azm. = 154.4°
Duration = 08m56.4s

Circumstances at Greatest Duration: 06:43:05.7 UT1

Lat. = 03°50.2'N
Long. = 066°34.3'E
Path Width = 300.9 km
Sun Alt. = 67.4°
Sun Azm. = 134.9°
Duration = 08m58.7s

0 1000 2000 3000 4000 5000
Kilometers

Total Solar Eclipse of 2064 Aug 12

Greatest Eclipse = 17:46:06.3 TD (= 17:44:32.7 UT1)

Eclipse Magnitude = 1.0495 Saros Series = 146

Gamma = -0.4652 Saros Member = 30 of 76

Sun at Greatest Eclipse
(Geocentric Coordinates)

R.A. = 09h32m49.7s
Dec. = +14°33'07.1"
S.D. = 00°15'47.0"
H.P. = 00°00'08.7"

Moon at Greatest Eclipse
(Geocentric Coordinates)

R.A. = 09h32m02.7s
Dec. = +14°07'45.3"
S.D. = 00°16'19.3"
H.P. = 00°59'54.2"

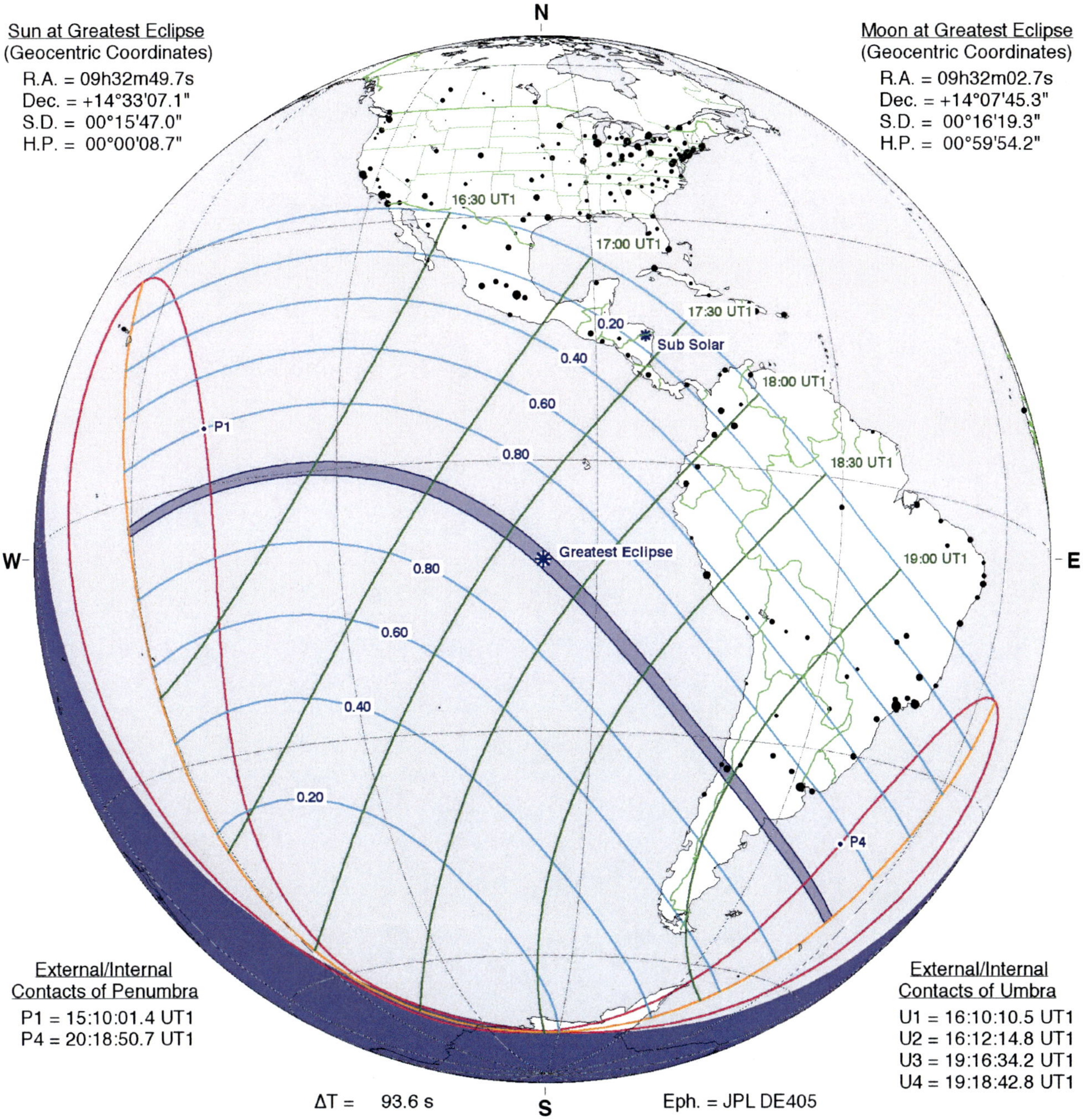

External/Internal
Contacts of Penumbra

P1 = 15:10:01.4 UT1
P4 = 20:18:50.7 UT1

External/Internal
Contacts of Umbra

U1 = 16:10:10.5 UT1
U2 = 16:12:14.8 UT1
U3 = 19:16:34.2 UT1
U4 = 19:18:42.8 UT1

ΔT = 93.6 s Eph. = JPL DE405

Circumstances at Greatest Eclipse: 17:44:32.7 UT1

Lat. = 10°55.8'S	Sun Alt. = 62.2°
Long. = 096°05.9'W	Sun Azm. = 23.8°
Path Width = 183.7 km	Duration = 04m27.7s

Circumstances at Greatest Duration: 17:40:44.1 UT1

Lat. = 10°06.0'S	Sun Alt. = 62.1°
Long. = 097°04.6'W	Sun Azm. = 28.0°
Path Width = 184.6 km	Duration = 04m27.8s

0 1000 2000 3000 4000 5000
Kilometers

Partial Solar Eclipse of 2065 Aug 02

Greatest Eclipse = 05:34:16.6 TD (= 05:32:42.3 UT1)

Eclipse Magnitude = 0.4903
Gamma = -1.2758

Saros Series = 156
Saros Member = 4 of 69

Sun at Greatest Eclipse
(Geocentric Coordinates)
R.A. = 08h51m52.4s
Dec. = +17°35'43.5"
S.D. = 00°15'45.5"
H.P. = 00°00'08.7"

Moon at Greatest Eclipse
(Geocentric Coordinates)
R.A. = 08h50m03.4s
Dec. = +16°28'16.4"
S.D. = 00°15'28.9"
H.P. = 00°56'49.3"

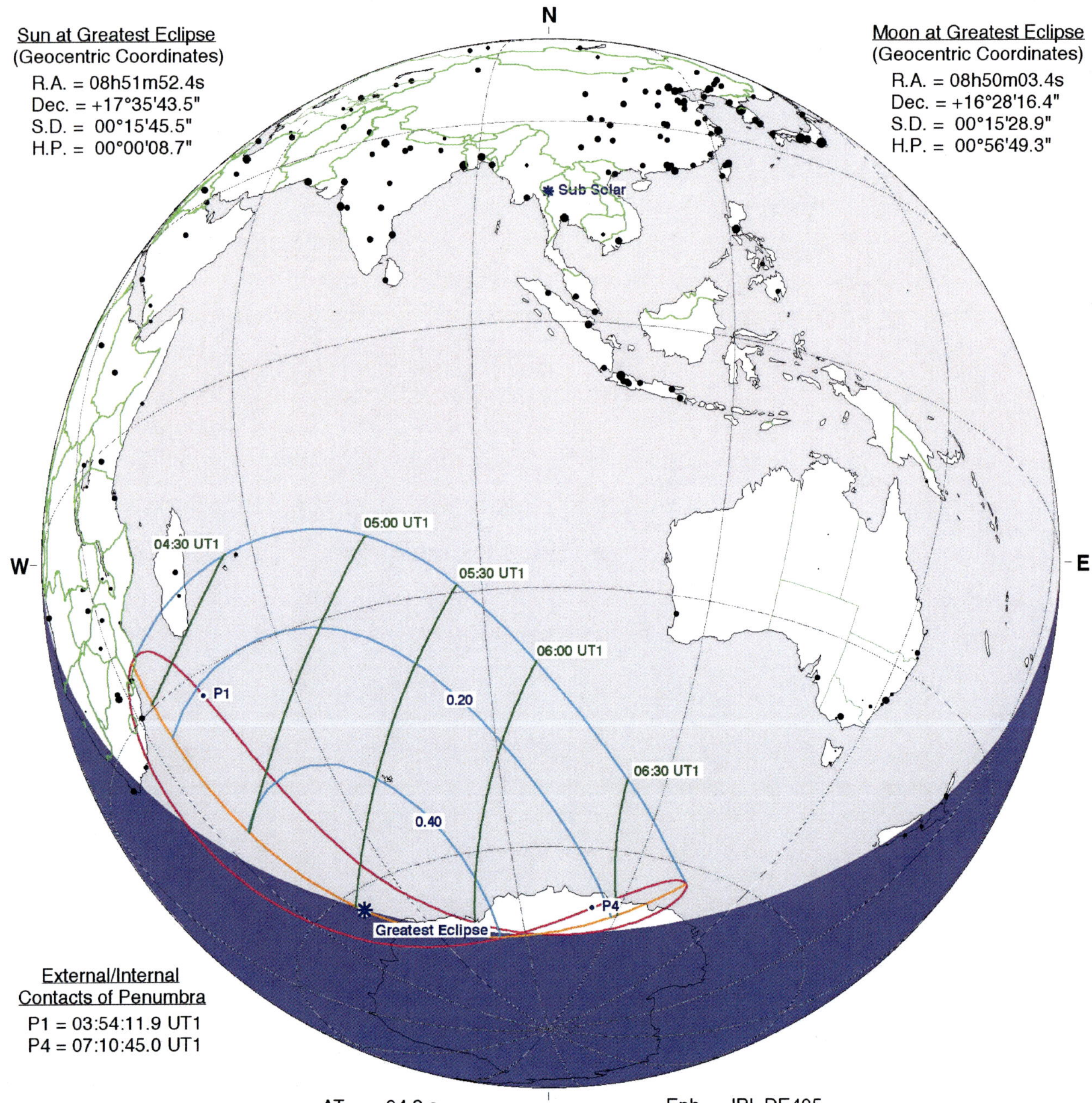

External/Internal
Contacts of Penumbra
P1 = 03:54:11.9 UT1
P4 = 07:10:45.0 UT1

ΔT = 94.3 s

Eph. = JPL DE405

Circumstances at Greatest Eclipse: 05:32:42.3 UT1

Lat. = 62°42.7'S
Long. = 046°24.5'E

Sun Alt. = 0.0°
Sun Azm. = 48.7°

0 1000 2000 3000 4000 5000
Kilometers

Partial Solar Eclipse of 2065 Dec 27

Greatest Eclipse = 08:39:55.7 TD (= 08:38:21.2 UT1)

Eclipse Magnitude = 0.8769 | Saros Series = 123
Gamma = -1.0688 | Saros Member = 56 of 70

Sun at Greatest Eclipse
(Geocentric Coordinates)
R.A. = 18h26m44.9s
Dec. = -23°17'20.3"
S.D. = 00°16'15.7"
H.P. = 00°00'08.9"

Moon at Greatest Eclipse
(Geocentric Coordinates)
R.A. = 18h27m25.5s
Dec. = -24°21'42.8"
S.D. = 00°16'37.3"
H.P. = 01°01'00.2"

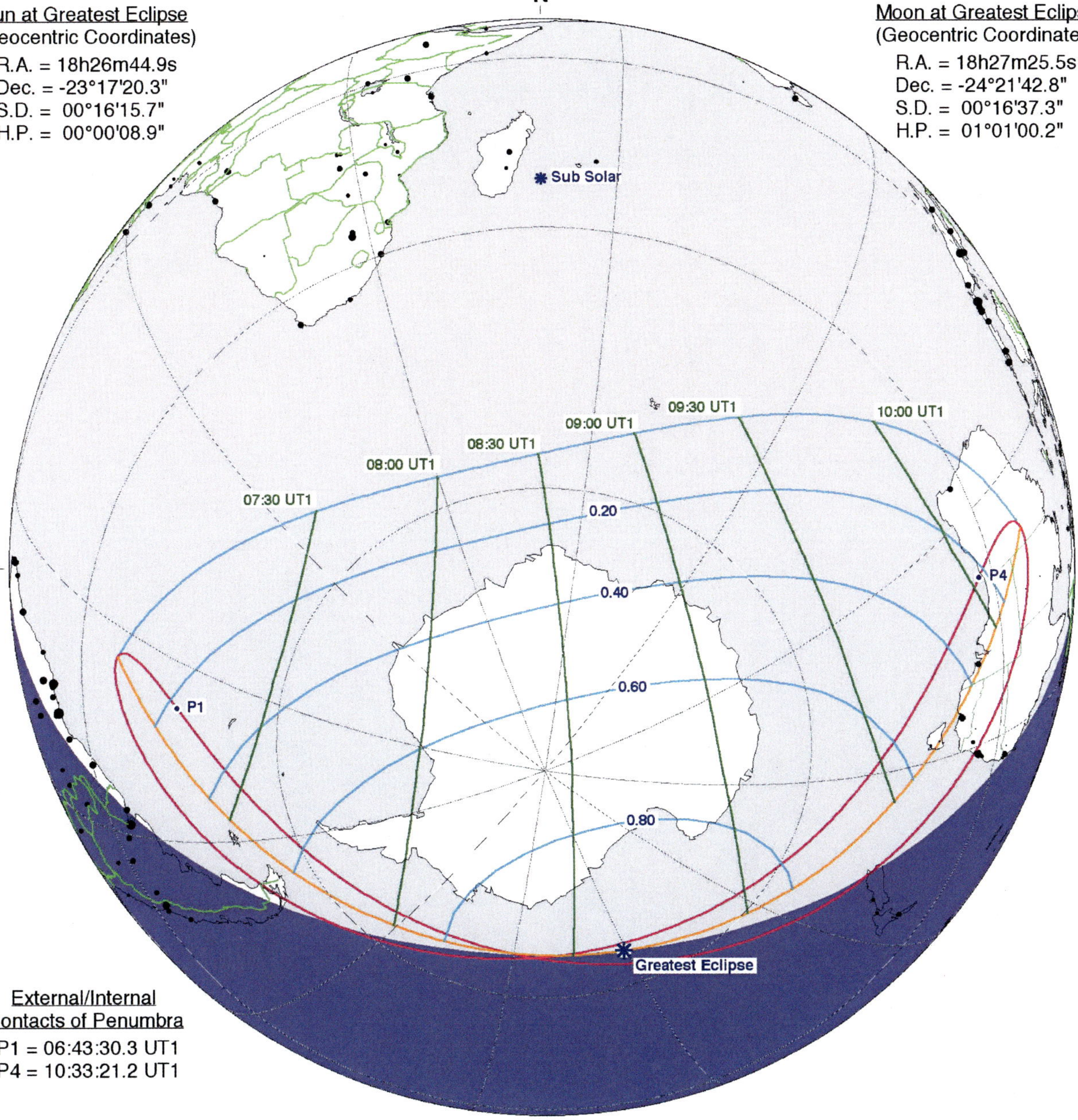

External/Internal
Contacts of Penumbra
P1 = 06:43:30.3 UT1
P4 = 10:33:21.2 UT1

ΔT = 94.6 s

Eph. = JPL DE405

Circumstances at Greatest Eclipse: 08:38:21.2 UT1
Lat. = 65°23.8'S | Sun Alt. = 0.0°
Long. = 149°17.5'W | Sun Azm. = 198.3°

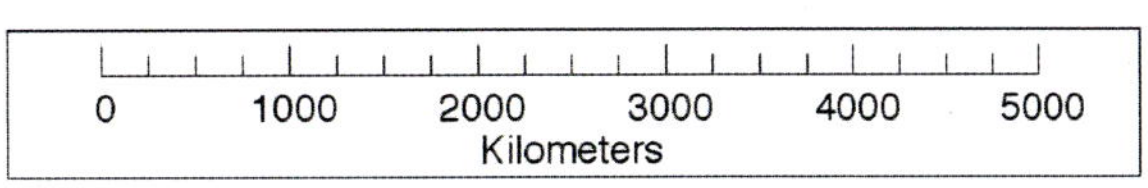

Partial Solar Eclipse of 2065 Feb 05

Greatest Eclipse = 09:52:25.5 TD (= 09:50:51.6 UT1)

Eclipse Magnitude = 0.9123 Saros Series = 151

Gamma = 1.0336 Saros Member = 17 of 72

Sun at Greatest Eclipse
(Geocentric Coordinates)
R.A. = 21h18m22.7s
Dec. = -15°41'30.6"
S.D. = 00°16'13.3"
H.P. = 00°00'08.9"

Moon at Greatest Eclipse
(Geocentric Coordinates)
R.A. = 21h16m47.2s
Dec. = -14°47'52.5"
S.D. = 00°15'25.7"
H.P. = 00°56'37.5"

N

W

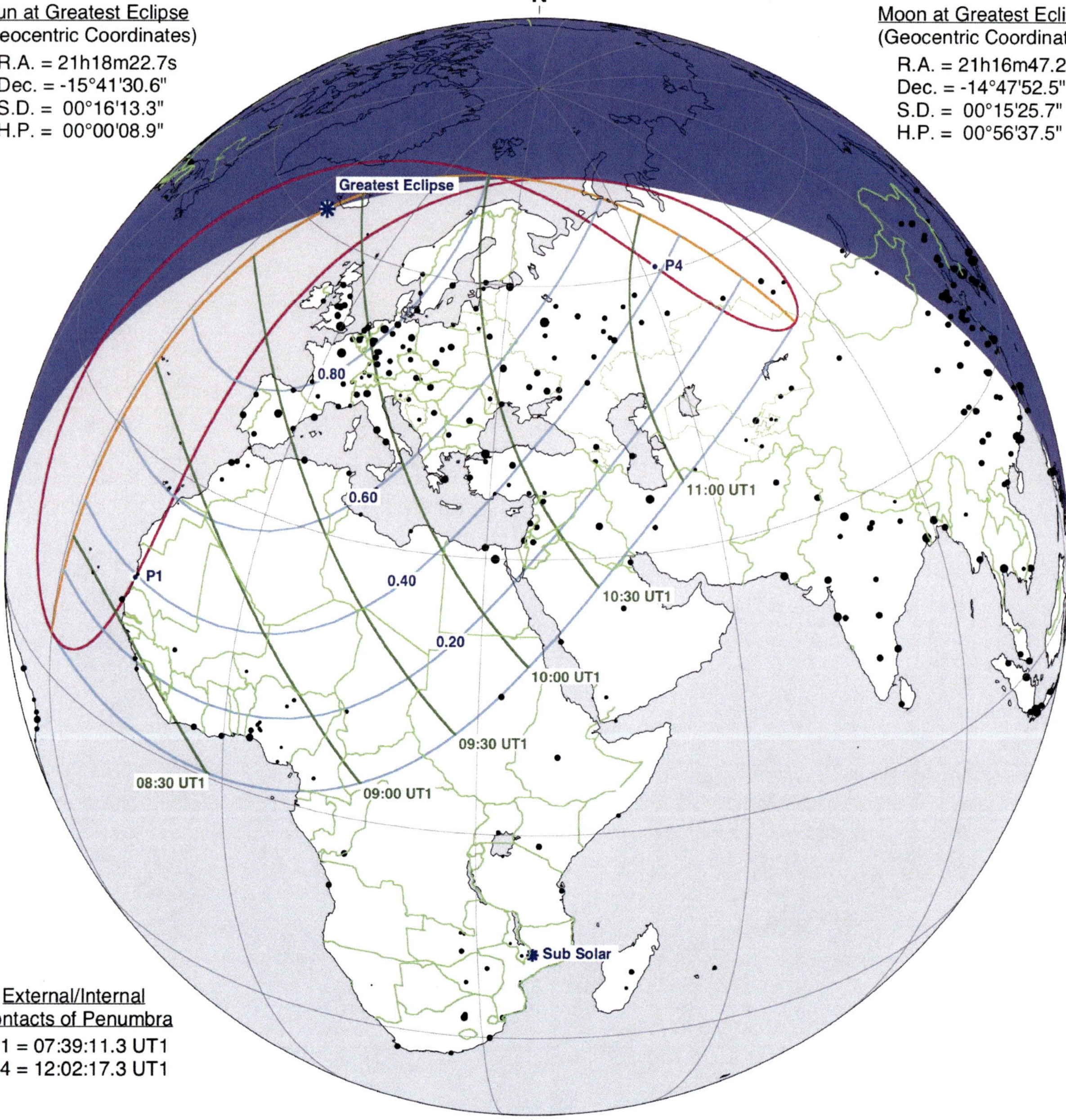

E

External/Internal
Contacts of Penumbra
P1 = 07:39:11.3 UT1
P4 = 12:02:17.3 UT1

ΔT = 93.9 s

S

Eph. = JPL DE405

Circumstances at Greatest Eclipse: 09:50:51.6 UT1

Lat. = 62°09.7'N Sun Alt. = 0.0°

Long. = 022°01.5'W Sun Azm. = 125.4°

0 1000 2000 3000 4000 5000
Kilometers

Partial Solar Eclipse of 2065 Jul 03

Greatest Eclipse = 17:33:52.5 TD (= 17:32:18.3 UT1)

Eclipse Magnitude = 0.1639 Saros Series = 118

Gamma = 1.4619 Saros Member = 71 of 72

Sun at Greatest Eclipse
(Geocentric Coordinates)
R.A. = 06h53m43.9s
Dec. = +22°51'26.7"
S.D. = 00°15'43.9"
H.P. = 00°00'08.6"

Moon at Greatest Eclipse
(Geocentric Coordinates)
R.A. = 06h54m50.6s
Dec. = +24°10'43.8"
S.D. = 00°15'05.3"
H.P. = 00°55'22.6"

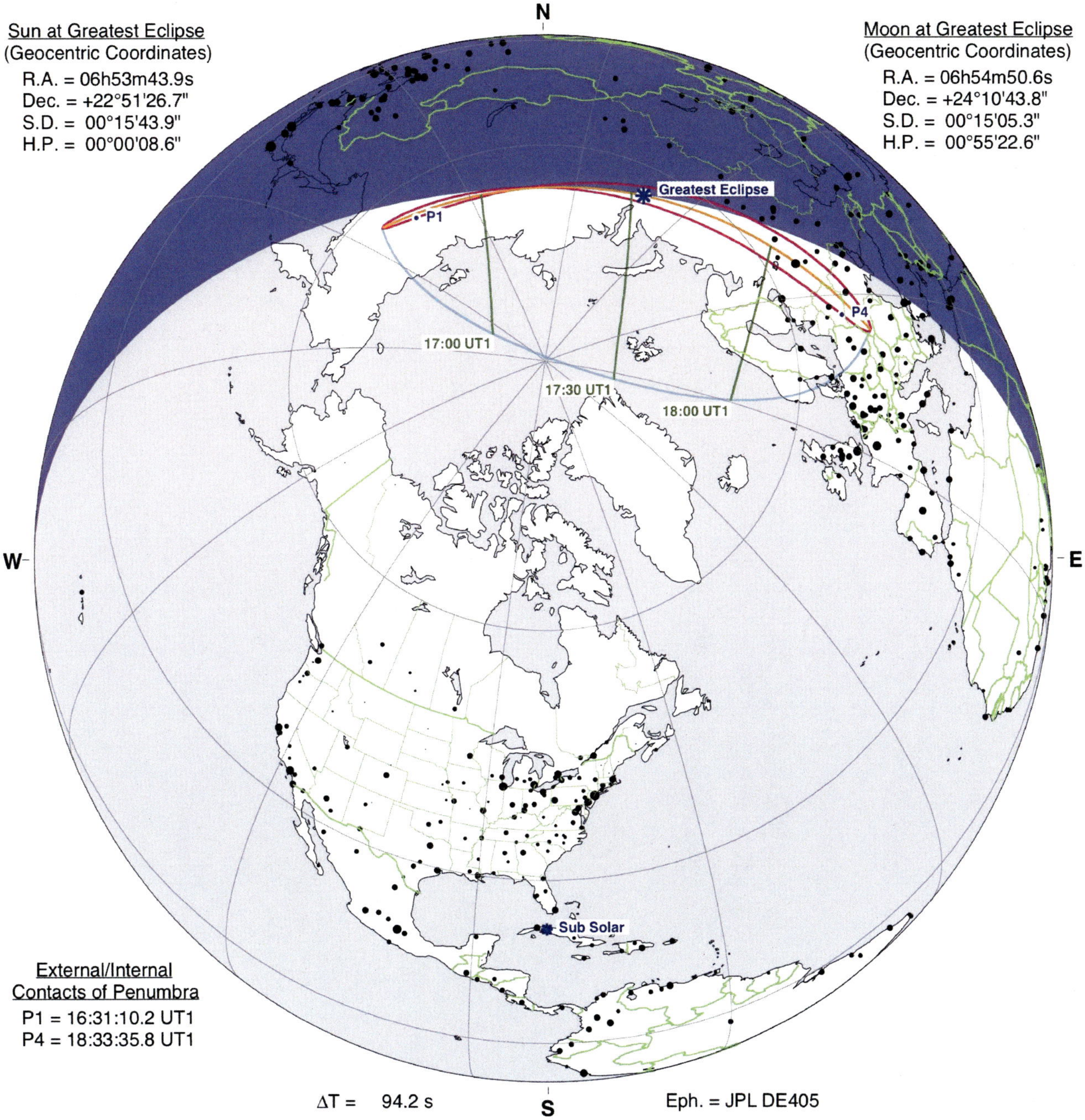

External/Internal
Contacts of Penumbra
P1 = 16:31:10.2 UT1
P4 = 18:33:35.8 UT1

ΔT = 94.2 s Eph. = JPL DE405

Circumstances at Greatest Eclipse: 17:32:18.3 UT1

Lat. = 64°49.3'N Sun Alt. = 0.0°

Long. = 071°43.9'E Sun Azm. = 335.9°

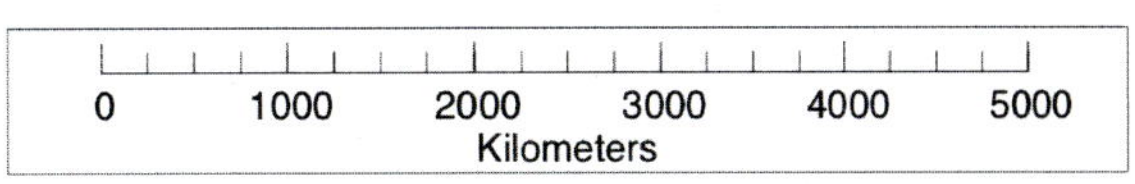

Annular Solar Eclipse of 2066 Jun 22

Greatest Eclipse = 19:25:47.7 TD (= 19:24:12.8 UT1)

Eclipse Magnitude = 0.9435 Saros Series = 128

Gamma = 0.7330 Saros Member = 61 of 73

Sun at Greatest Eclipse
(Geocentric Coordinates)
R.A. = 06h07m28.7s
Dec. = +23°25'11.2"
S.D. = 00°15'44.2"
H.P. = 00°00'08.7"

Moon at Greatest Eclipse
(Geocentric Coordinates)
R.A. = 06h07m48.1s
Dec. = +24°04'22.4"
S.D. = 00°14'42.0"
H.P. = 00°53'57.0"

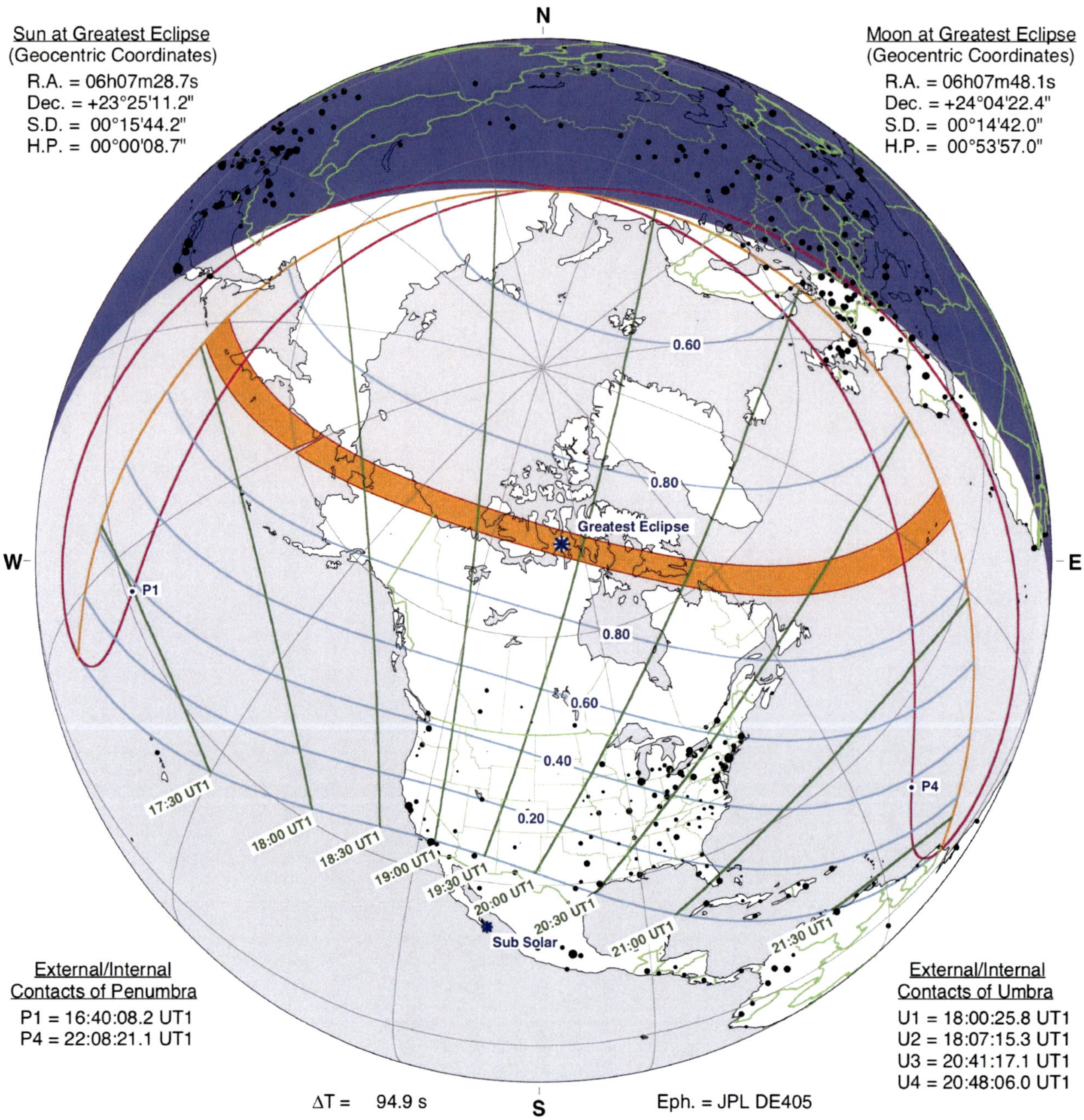

External/Internal
Contacts of Penumbra
P1 = 16:40:08.2 UT1
P4 = 22:08:21.1 UT1

External/Internal
Contacts of Umbra
U1 = 18:00:25.8 UT1
U2 = 18:07:15.3 UT1
U3 = 20:41:17.1 UT1
U4 = 20:48:06.0 UT1

ΔT = 94.9 s Eph. = JPL DE405

Circumstances at Greatest Eclipse: 19:24:12.8 UT1

Lat. = 70°07.9'N Sun Alt. = 42.6°
Long. = 096°29.9'W Sun Azm. = 197.5°
Path Width = 308.5 km Duration = 04m39.8s

Circumstances at Greatest Duration: 19:21:23.9 UT1

Lat. = 70°31.1'N Sun Alt. = 42.5°
Long. = 099°39.3'W Sun Azm. = 192.6°
Path Width = 308.3 km Duration = 04m39.8s

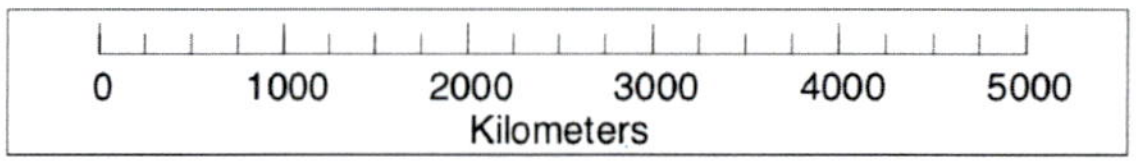

Total Solar Eclipse of 2066 Dec 17

Greatest Eclipse = 00:23:39.9 TD (= 00:22:04.7 UT1)

Eclipse Magnitude = 1.0416 Saros Series = 133

Gamma = -0.4043 Saros Member = 48 of 72

Sun at Greatest Eclipse
(Geocentric Coordinates)
R.A. = 17h39m46.4s
Dec. = -23°20'56.0"
S.D. = 00°16'15.1"
H.P. = 00°00'08.9"

Moon at Greatest Eclipse
(Geocentric Coordinates)
R.A. = 17h39m53.3s
Dec. = -23°45'32.9"
S.D. = 00°16'39.9"
H.P. = 01°01'09.6"

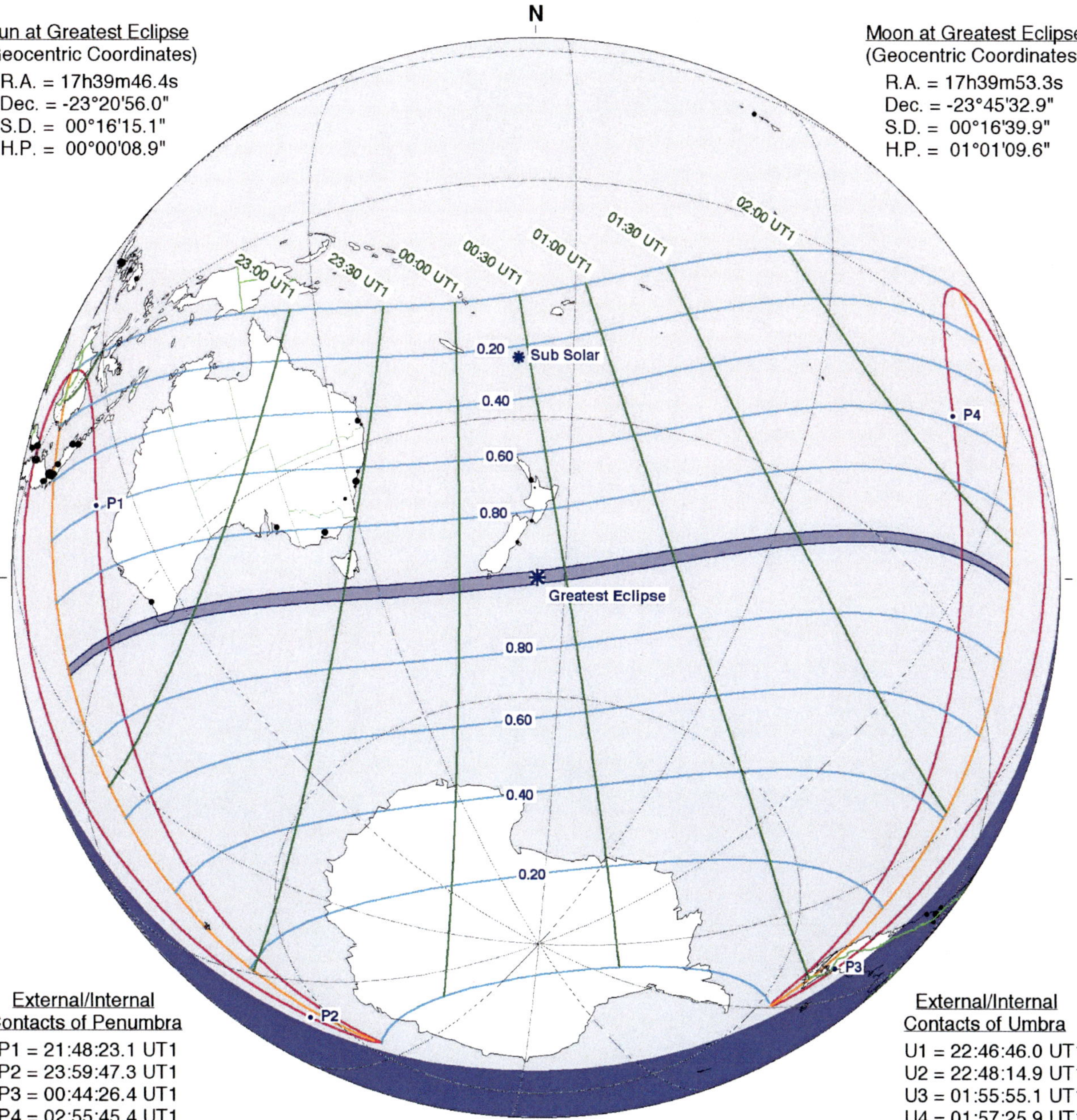

External/Internal
Contacts of Penumbra
P1 = 21:48:23.1 UT1
P2 = 23:59:47.3 UT1
P3 = 00:44:26.4 UT1
P4 = 02:55:45.4 UT1

External/Internal
Contacts of Umbra
U1 = 22:46:46.0 UT1
U2 = 22:48:14.9 UT1
U3 = 01:55:55.1 UT1
U4 = 01:57:25.9 UT1

ΔT = 95.3 s Eph. = JPL DE405

Circumstances at Greatest Eclipse: 00:22:04.7 UT1

Lat. = 47°22.1'S Sun Alt. = 65.9°
Long. = 175°38.4'E Sun Azm. = 355.1°
Path Width = 152.2 km Duration = 03m14.5s

Circumstances at Greatest Duration: 00:21:56.4 UT1

Lat. = 47°22.5'S Sun Alt. = 65.9°
Long. = 175°33.7'E Sun Azm. = 355.3°
Path Width = 152.2 km Duration = 03m14.5s

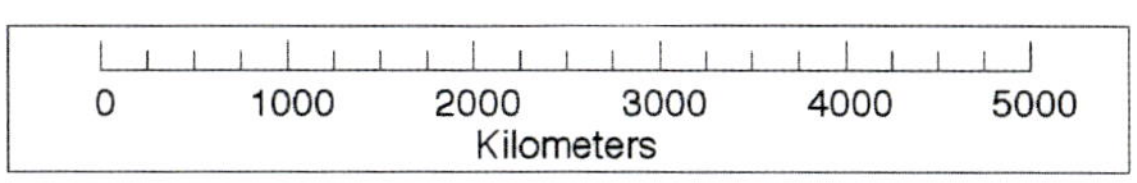

Annular Solar Eclipse of 2067 Jun 11

Greatest Eclipse = 20:42:26.4 TD (= 20:40:50.8 UT1)

Eclipse Magnitude = 0.9670 Saros Series = 138

Gamma = -0.0387 Saros Member = 34 of 70

Sun at Greatest Eclipse
(Geocentric Coordinates)
R.A. = 05h20m58.3s
Dec. = +23°07'36.6"
S.D. = 00°15'45.1"
H.P. = 00°00'08.7"

Moon at Greatest Eclipse
(Geocentric Coordinates)
R.A. = 05h20m58.0s
Dec. = +23°05'29.3"
S.D. = 00°15'00.0"
H.P. = 00°55'03.2"

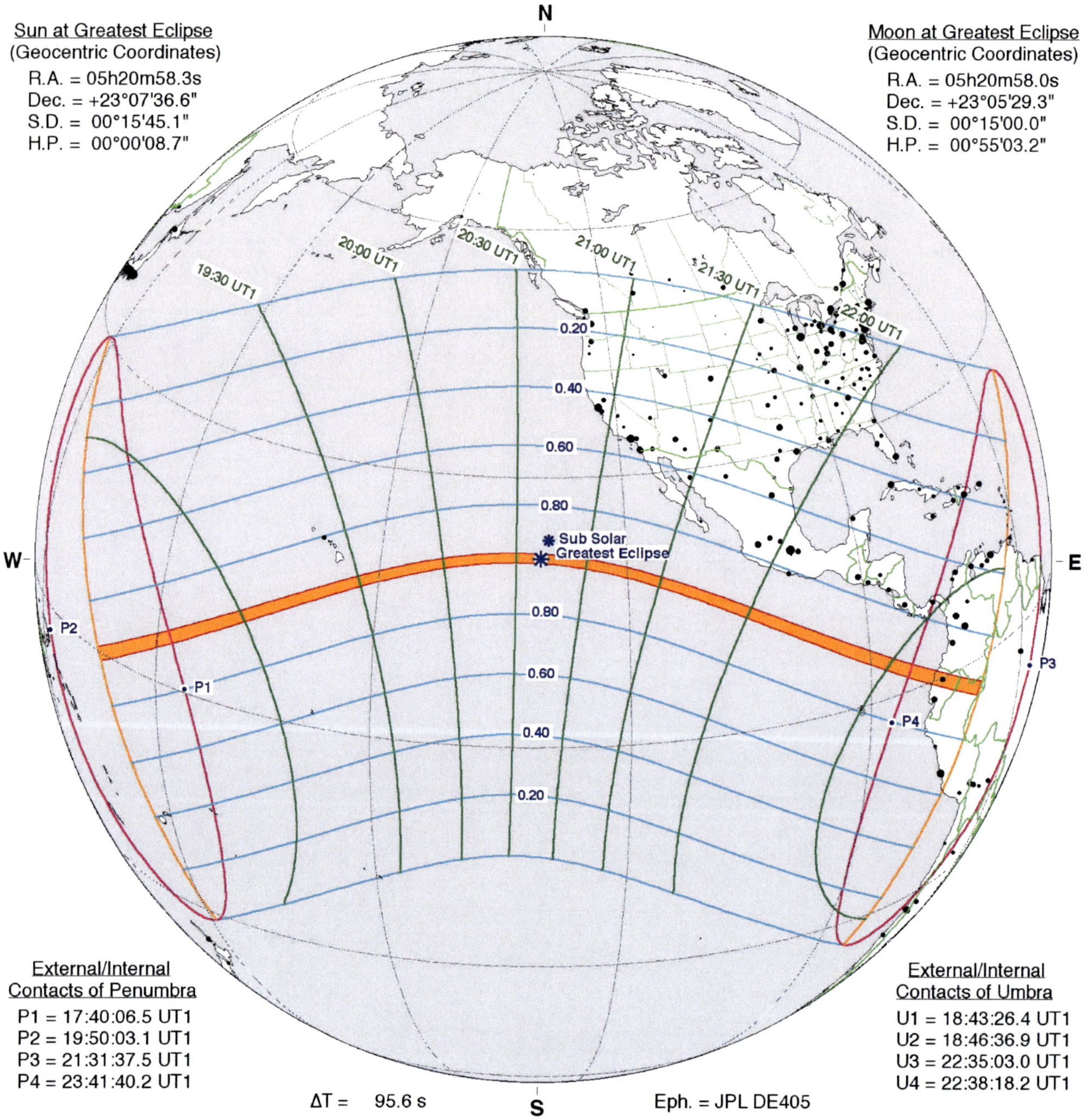

External/Internal
Contacts of Penumbra
P1 = 17:40:06.5 UT1
P2 = 19:50:03.1 UT1
P3 = 21:31:37.5 UT1
P4 = 23:41:40.2 UT1

External/Internal
Contacts of Umbra
U1 = 18:43:26.4 UT1
U2 = 18:46:36.9 UT1
U3 = 22:35:03.0 UT1
U4 = 22:38:18.2 UT1

ΔT = 95.6 s Eph. = JPL DE405

Circumstances at Greatest Eclipse: 20:40:50.8 UT1

Lat. = 21°02.4'N Sun Alt. = 87.9°
Long. = 130°19.2'W Sun Azm. = 2.0°
Path Width = 118.9 km Duration = 04m05.2s

Circumstances at Greatest Duration: 20:42:21.3 UT1

Lat. = 21°00.9'N Sun Alt. = 87.8°
Long. = 129°53.9'W Sun Azm. = 342.6°
Path Width = 119.0 km Duration = 04m05.2s

0 1000 2000 3000 4000 5000
Kilometers

Hybrid Solar Eclipse of 2067 Dec 06

Greatest Eclipse = 14:03:43.3 TD (= 14:02:07.3 UT1)

Eclipse Magnitude = 1.0011 Saros Series = 143

Gamma = 0.2845 Saros Member = 26 of 72

Sun at Greatest Eclipse
(Geocentric Coordinates)
R.A. = 16h52m45.8s
Dec. = -22°31'49.1"
S.D. = 00°16'13.8"
H.P. = 00°00'08.9"

Moon at Greatest Eclipse
(Geocentric Coordinates)
R.A. = 16h52m46.9s
Dec. = -22°15'09.9"
S.D. = 00°15'59.7"
H.P. = 00°58'42.2"

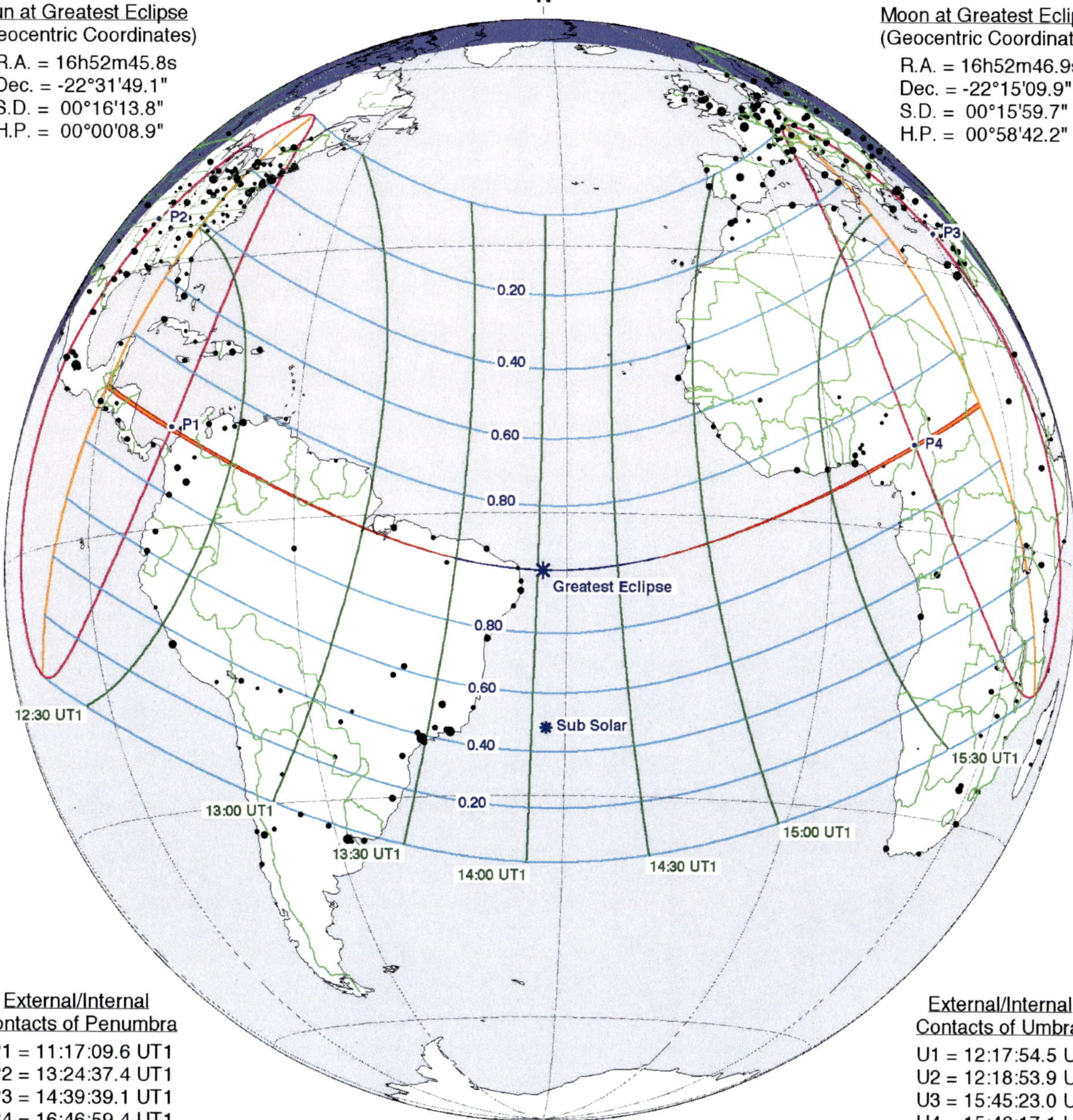

External/Internal
Contacts of Penumbra
P1 = 11:17:09.6 UT1
P2 = 13:24:37.4 UT1
P3 = 14:39:39.1 UT1
P4 = 16:46:59.4 UT1

External/Internal
Contacts of Umbra
U1 = 12:17:54.5 UT1
U2 = 12:18:53.9 UT1
U3 = 15:45:23.0 UT1
U4 = 15:46:17.1 UT1

ΔT = 96.0 s Eph. = JPL DE405

Circumstances at Greatest Eclipse: 14:02:07.3 UT1

Lat. = 06°02.8'S Sun Alt. = 73.5°
Long. = 032°31.2'W Sun Azm. = 180.9°
Path Width = 4.1 km Duration = 00m07.5s

Circumstances at Greatest Duration: 12:18:24.2 UT1

Lat. = 16°11.9'N Sun Alt. = 0.0°
Long. = 089°56.5'W Sun Azm. = 113.5°
Path Width = 56.2 km Duration = 00m59.0s

0 1000 2000 3000 4000 5000
Kilometers

Total Solar Eclipse of 2068 May 31

Greatest Eclipse = 03:56:39.1 TD (= 03:55:02.7 UT1)

Eclipse Magnitude = 1.0110 Saros Series = 148

Gamma = -0.7970 Saros Member = 24 of 75

Sun at Greatest Eclipse
(Geocentric Coordinates)
R.A. = 04h35m49.8s
Dec. = +22°01'13.9"
S.D. = 00°15'46.5"
H.P. = 00°00'08.7"

Moon at Greatest Eclipse
(Geocentric Coordinates)
R.A. = 04h35m58.7s
Dec. = +21°15'11.0"
S.D. = 00°15'47.8"
H.P. = 00°57'58.6"

N

W

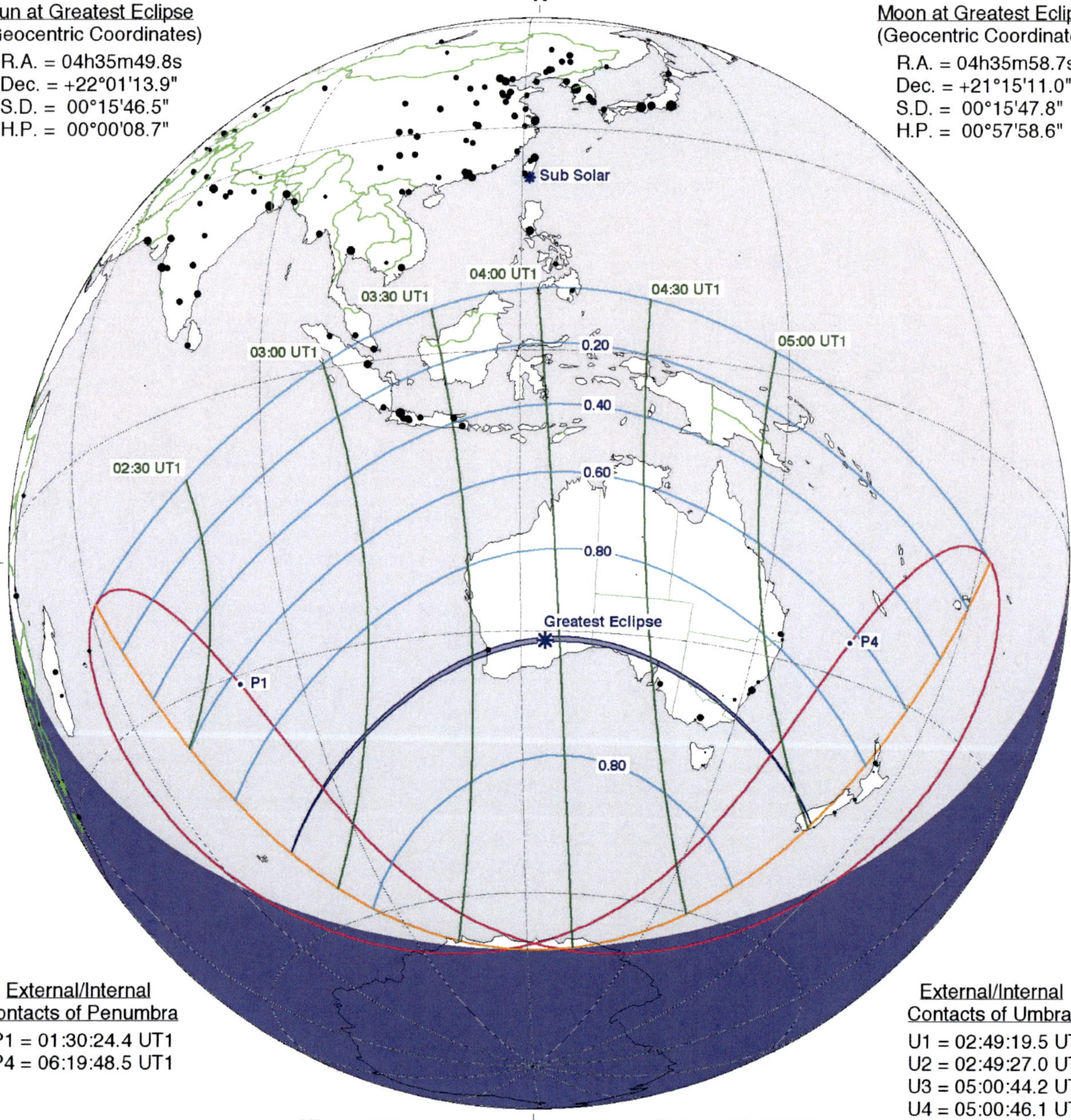

E

External/Internal Contacts of Penumbra
P1 = 01:30:24.4 UT1
P4 = 06:19:48.5 UT1

External/Internal Contacts of Umbra
U1 = 02:49:19.5 UT1
U2 = 02:49:27.0 UT1
U3 = 05:00:44.2 UT1
U4 = 05:00:46.1 UT1

ΔT = 96.3 s

S

Eph. = JPL DE405

Circumstances at Greatest Eclipse: 03:55:02.7 UT1

Lat. = 31°00.8'S Sun Alt. = 36.9°
Long. = 123°05.1'E Sun Azm. = 357.2°
Path Width = 62.8 km Duration = 01m05.8s

Circumstances at Greatest Duration: 03:53:13.6 UT1

Lat. = 31°03.9'S Sun Alt. = 36.9°
Long. = 122°25.8'E Sun Azm. = 358.5°
Path Width = 62.8 km Duration = 01m05.9s

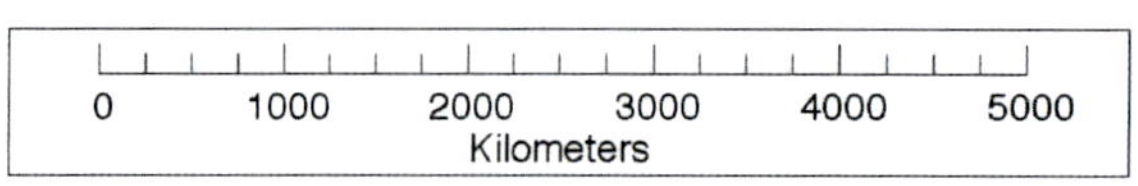

Partial Solar Eclipse of 2068 Nov 24

Greatest Eclipse = 21:32:29.6 TD (= 21:30:52.9 UT1)

Eclipse Magnitude = 0.9109   Saros Series = 153

Gamma = 1.0299   Saros Member = 12 of 70

Sun at Greatest Eclipse
(Geocentric Coordinates)

R.A. = 16h05m39.1s
Dec. = -20°49'55.6"
S.D. = 00°16'12.0"
H.P. = 00°00'08.9"

Moon at Greatest Eclipse
(Geocentric Coordinates)

R.A. = 16h06m01.8s
Dec. = -19°53'06.5"
S.D. = 00°15'08.3"
H.P. = 00°55'33.5"

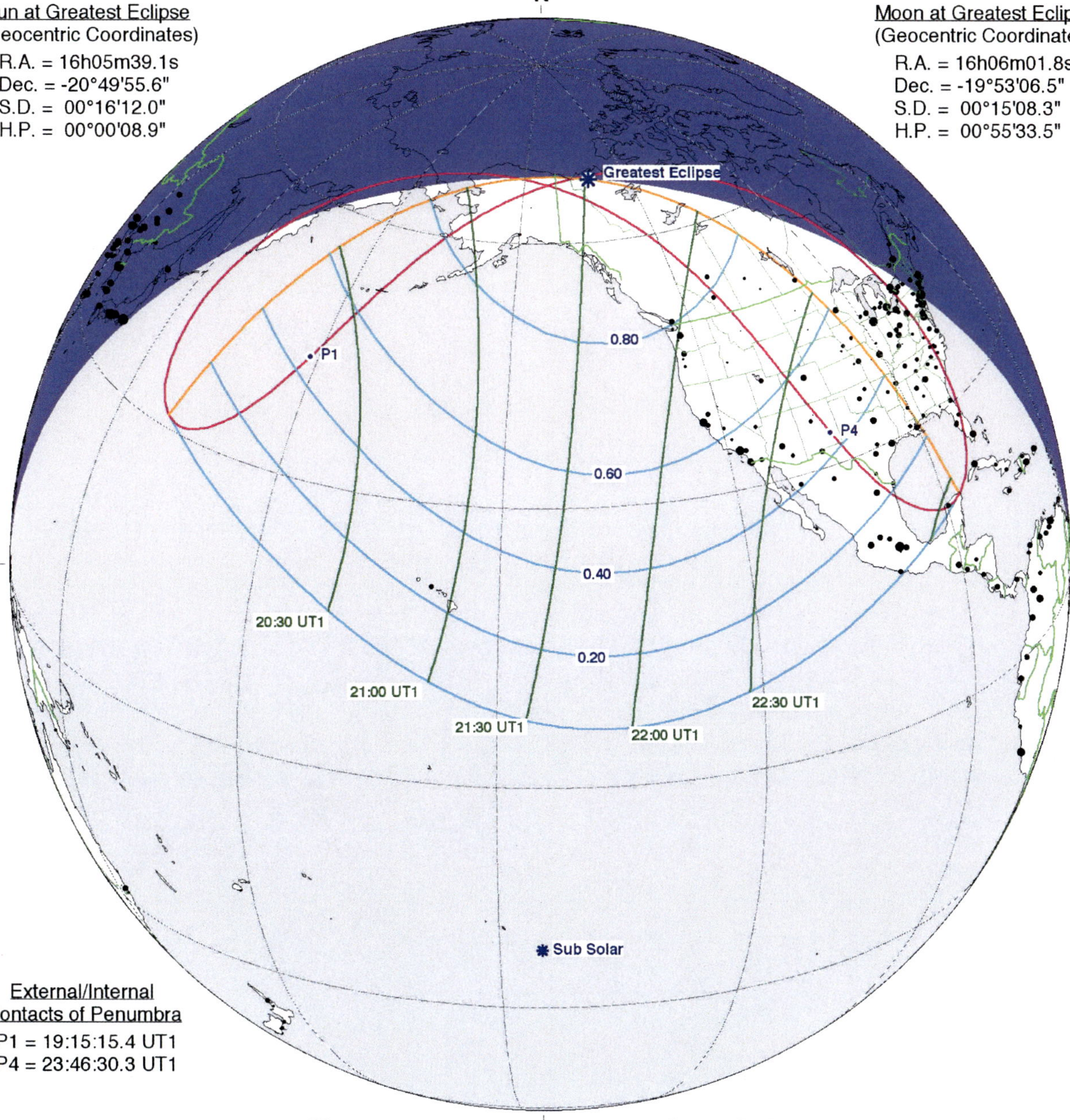

External/Internal
Contacts of Penumbra

P1 = 19:15:15.4 UT1
P4 = 23:46:30.3 UT1

ΔT = 96.7 s   Eph. = JPL DE405

Circumstances at Greatest Eclipse: 21:30:52.9 UT1

Lat. = 68°31.0'N   Sun Alt. = 0.0°
Long. = 131°13.5'W   Sun Azm. = 193.8°

0	1000	2000	3000	4000	5000

Kilometers

Partial Solar Eclipse of 2069 Apr 21

Greatest Eclipse = 10:11:08.9 TD (= 10:09:31.9 UT1)

Eclipse Magnitude = 0.8992
Gamma = 1.0624

Saros Series = 120
Saros Member = 64 of 71

Sun at Greatest Eclipse
(Geocentric Coordinates)
R.A. = 01h58m57.2s
Dec. = +12°07'52.1"
S.D. = 00°15'55.0"
H.P. = 00°00'08.8"

Moon at Greatest Eclipse
(Geocentric Coordinates)
R.A. = 01h57m49.5s
Dec. = +13°10'46.5"
S.D. = 00°16'43.2"
H.P. = 01°01'21.7"

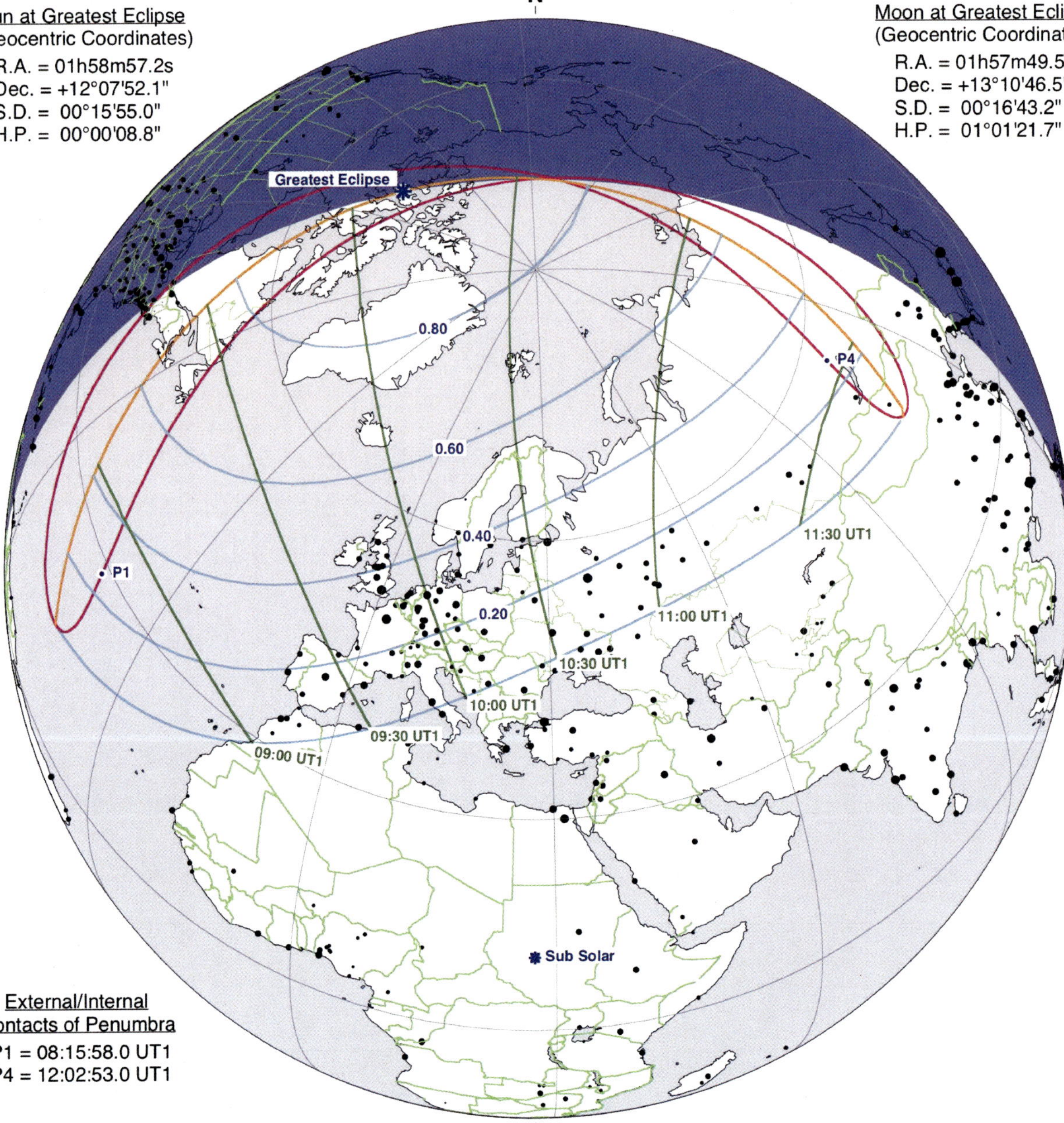

External/Internal
Contacts of Penumbra
P1 = 08:15:58.0 UT1
P4 = 12:02:53.0 UT1

ΔT = 97.0 s

Eph. = JPL DE405

Circumstances at Greatest Eclipse: 10:09:31.9 UT1

Lat. = 71°02.7'N
Long. = 101°25.2'W

Sun Alt. = 0.0°
Sun Azm. = 49.7°

0 1000 2000 3000 4000 5000
Kilometers

Partial Solar Eclipse of 2069 May 20

Greatest Eclipse = 17:53:17.8 TD (= 17:51:40.7 UT1)

Eclipse Magnitude = 0.0879
Gamma = -1.4852

Saros Series = 158
Saros Member = 1 of 70

Sun at Greatest Eclipse
(Geocentric Coordinates)
R.A. = 03h52m35.6s
Dec. = +20°12'26.5"
S.D. = 00°15'48.3"
H.P. = 00°00'08.7"

Moon at Greatest Eclipse
(Geocentric Coordinates)
R.A. = 03h53m19.8s
Dec. = +18°43'03.9"
S.D. = 00°16'32.8"
H.P. = 01°00'43.6"

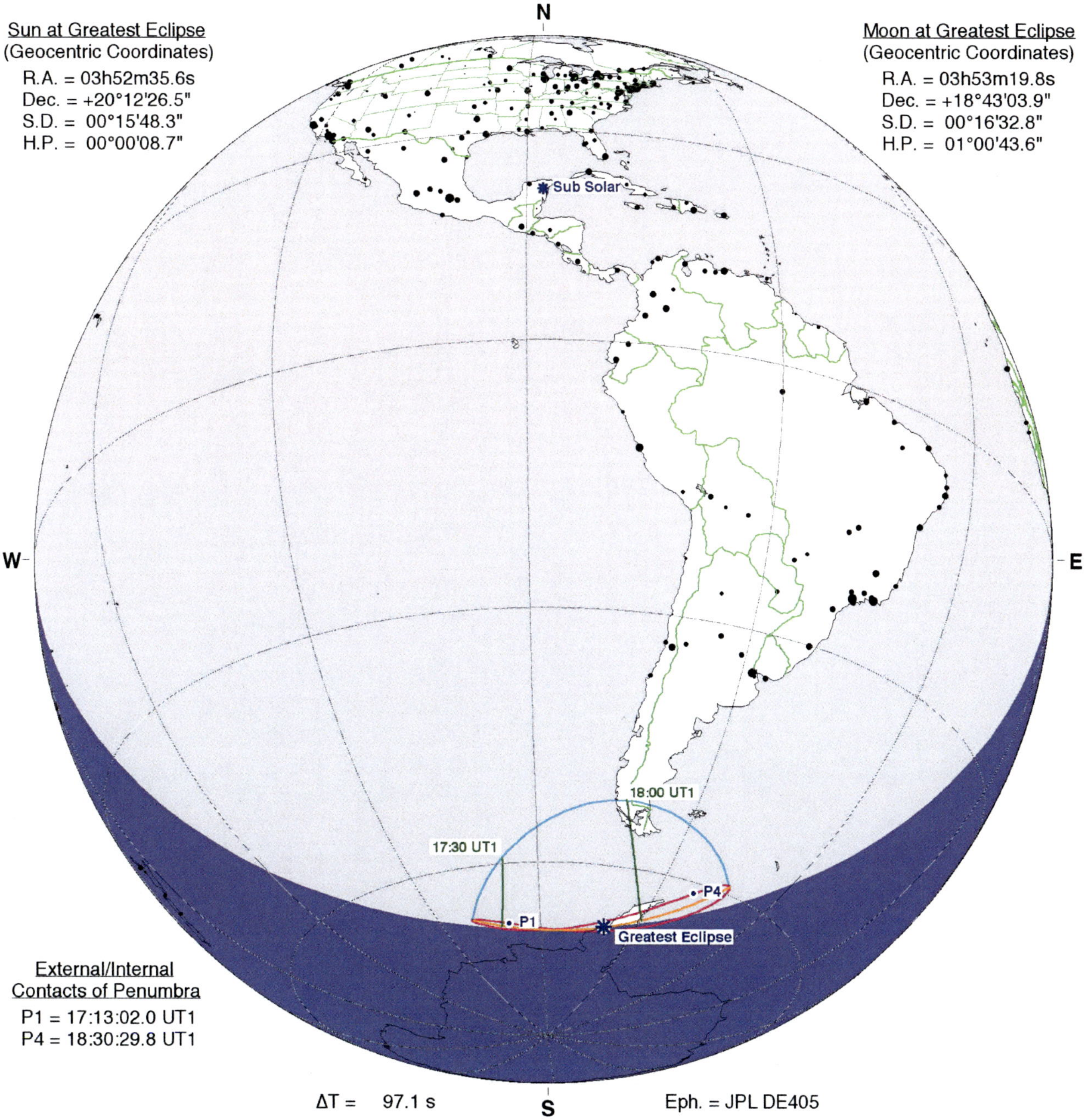

External/Internal
Contacts of Penumbra
P1 = 17:13:02.0 UT1
P4 = 18:30:29.8 UT1

ΔT = 97.1 s

Eph. = JPL DE405

Circumstances at Greatest Eclipse: 17:51:40.7 UT1

Lat. = 68°45.6'S
Long. = 070°03.9'W

Sun Alt. = 0.0°
Sun Azm. = 342.5°

0 1000 2000 3000 4000 5000
Kilometers

Partial Solar Eclipse of 2069 Oct 15

Greatest Eclipse = 04:19:56.3 TD (= 04:18:19.0 UT1)

Eclipse Magnitude = 0.5298
Gamma = -1.2524

Saros Series = 125
Saros Member = 57 of 73

Sun at Greatest Eclipse
(Geocentric Coordinates)
R.A. = 13h22m54.2s
Dec. = -08°43'06.9"
S.D. = 00°16'02.2"
H.P. = 00°00'08.8"

Moon at Greatest Eclipse
(Geocentric Coordinates)
R.A. = 13h21m37.5s
Dec. = -09°48'03.1"
S.D. = 00°14'45.3"
H.P. = 00°54'09.1"

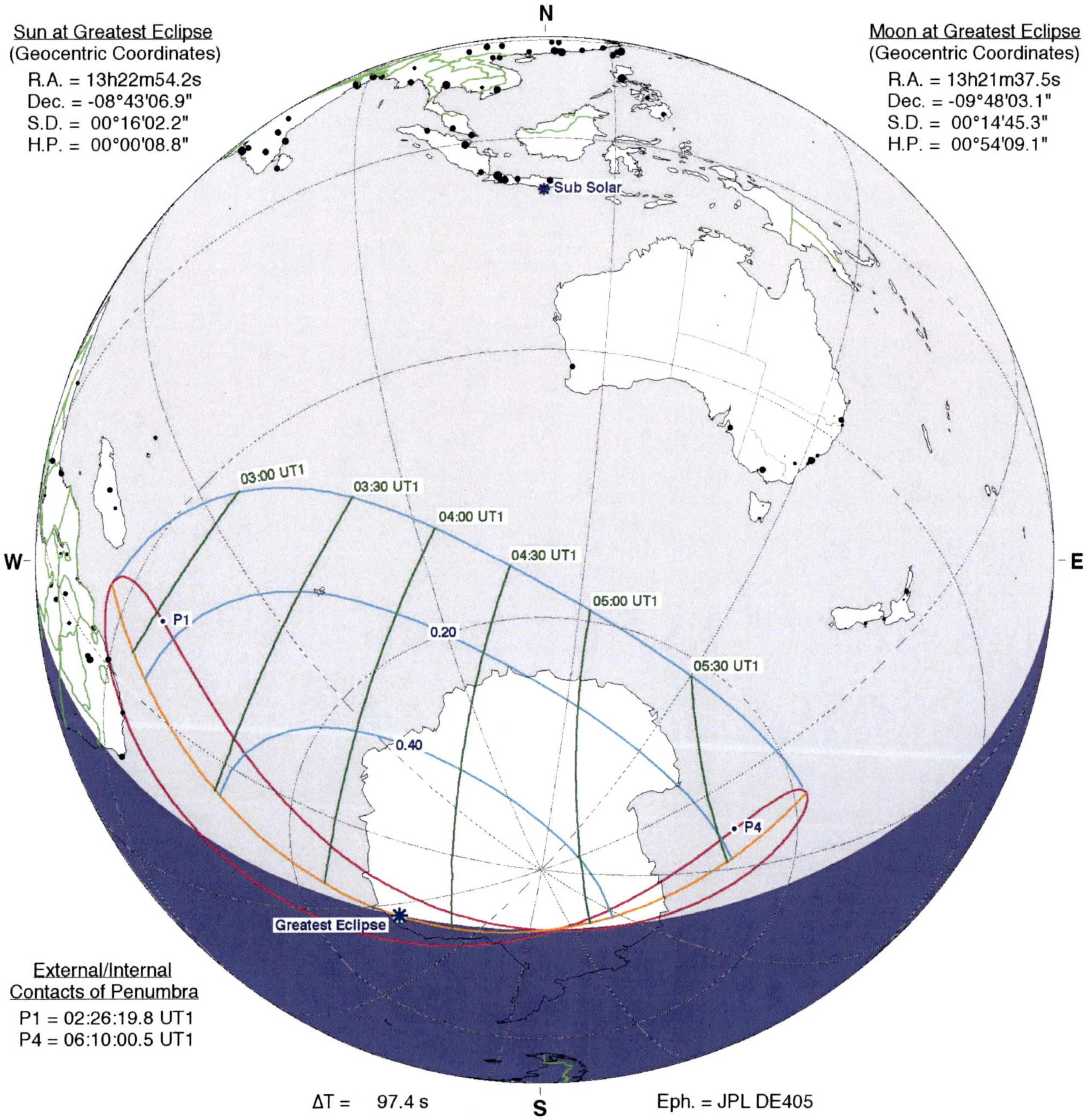

External/Internal
Contacts of Penumbra
P1 = 02:26:19.8 UT1
P4 = 06:10:00.5 UT1

ΔT = 97.4 s

Eph. = JPL DE405

Circumstances at Greatest Eclipse: 04:18:19.0 UT1

Lat. = 71°38.6'S
Long. = 005°36.9'W

Sun Alt. = 0.0°
Sun Azm. = 118.8°

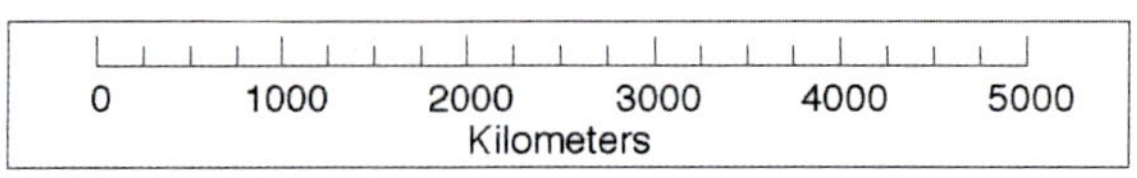

Total Solar Eclipse of 2070 Apr 11

Greatest Eclipse = 02:36:09.4 TD (= 02:34:31.7 UT1)

Eclipse Magnitude = 1.0472 Saros Series = 130

Gamma = 0.3652 Saros Member = 55 of 73

Sun at Greatest Eclipse
(Geocentric Coordinates)
R.A. = 01h19m45.0s
Dec. = +08°24'18.3"
S.D. = 00°15'57.8"
H.P. = 00°00'08.8"

Moon at Greatest Eclipse
(Geocentric Coordinates)
R.A. = 01h19m20.0s
Dec. = +08°45'25.6"
S.D. = 00°16'27.4"
H.P. = 01°00'23.9"

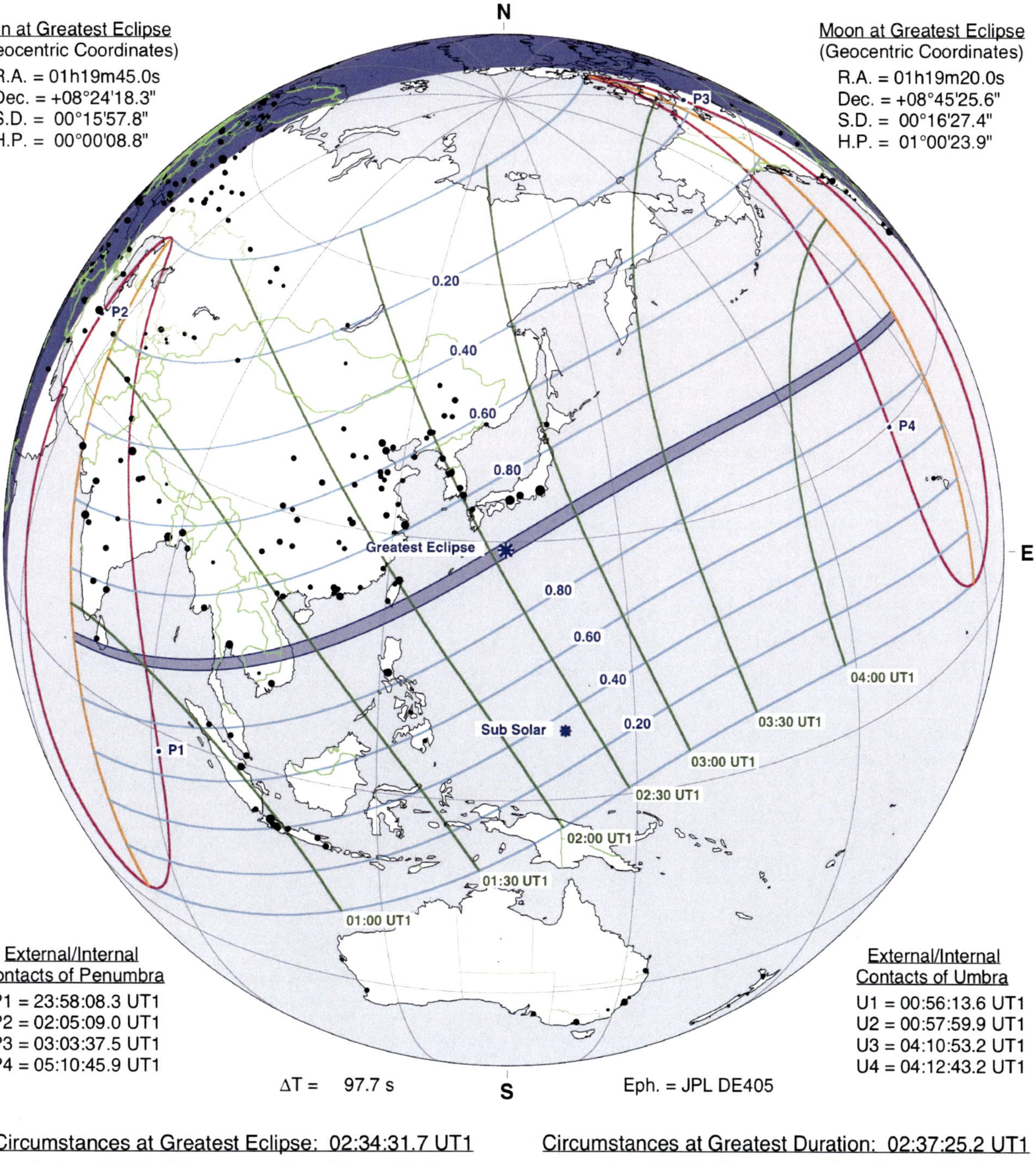

External/Internal
Contacts of Penumbra
P1 = 23:58:08.3 UT1
P2 = 02:05:09.0 UT1
P3 = 03:03:37.5 UT1
P4 = 05:10:45.9 UT1

External/Internal
Contacts of Umbra
U1 = 00:56:13.6 UT1
U2 = 00:57:59.9 UT1
U3 = 04:10:53.2 UT1
U4 = 04:12:43.2 UT1

ΔT = 97.7 s Eph. = JPL DE405

Circumstances at Greatest Eclipse: 02:34:31.7 UT1

Lat. = 29°03.2'N Sun Alt. = 68.4°
Long. = 134°54.7'E Sun Azm. = 161.7°
Path Width = 168.2 km Duration = 04m04.2s

Circumstances at Greatest Duration: 02:37:25.2 UT1

Lat. = 29°33.4'N Sun Alt. = 68.3°
Long. = 135°54.6'E Sun Azm. = 166.5°
Path Width = 167.9 km Duration = 04m04.3s

0 1000 2000 3000 4000 5000
Kilometers

Annular Solar Eclipse of 2070 Oct 04

Greatest Eclipse = 07:08:56.8 TD (= 07:07:18.8 UT1)

Eclipse Magnitude = 0.9731
Gamma = -0.4950

Saros Series = 135
Saros Member = 42 of 71

<u>Sun at Greatest Eclipse</u>
(Geocentric Coordinates)
R.A. = 12h42m00.6s
Dec. = -04°30'57.6"
S.D. = 00°15'59.1"
H.P. = 00°00'08.8"

<u>Moon at Greatest Eclipse</u>
(Geocentric Coordinates)
R.A. = 12h41m27.3s
Dec. = -04°57'29.9"
S.D. = 00°15'20.7"
H.P. = 00°56'19.0"

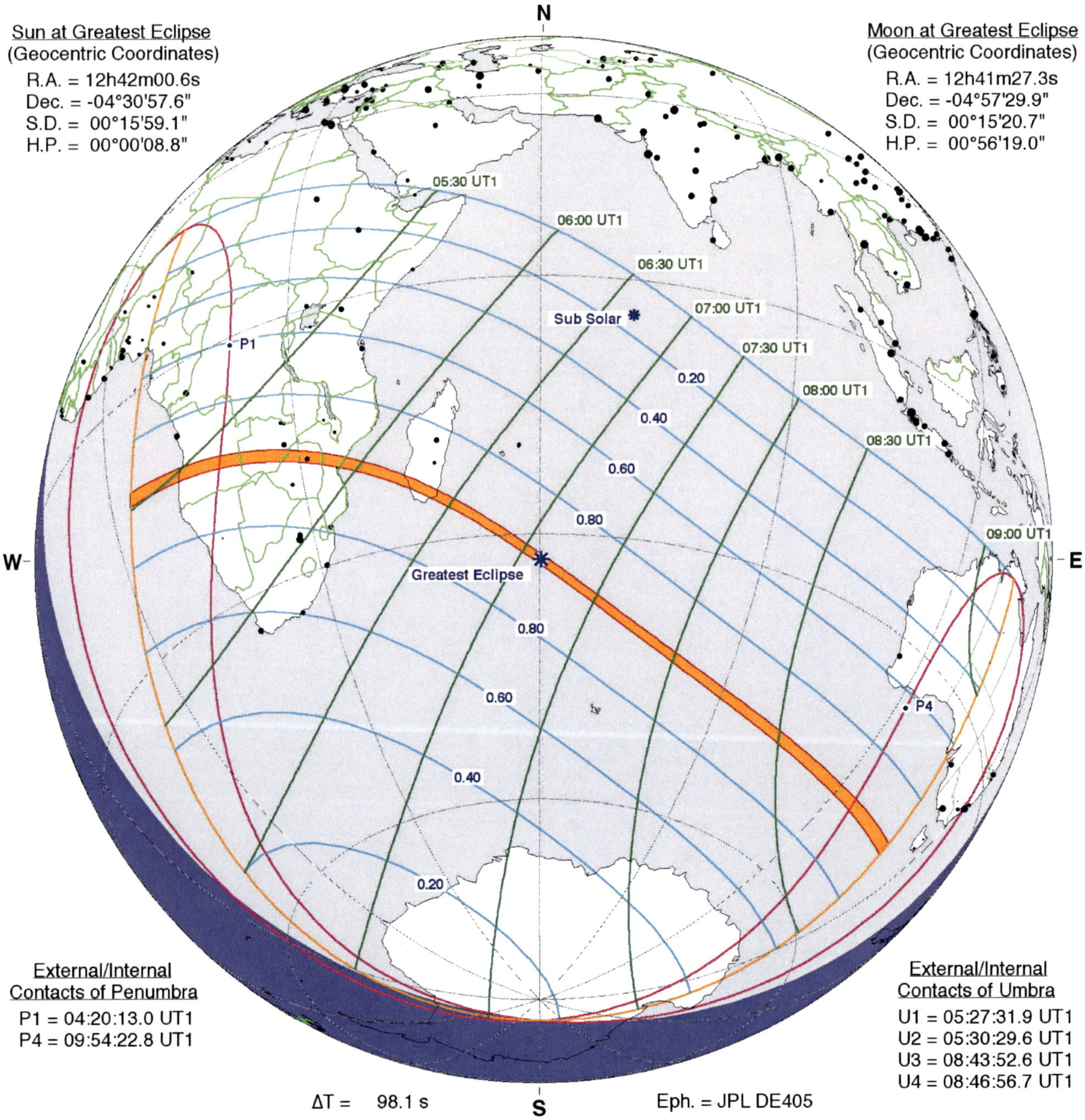

<u>External/Internal Contacts of Penumbra</u>
P1 = 04:20:13.0 UT1
P4 = 09:54:22.8 UT1

<u>External/Internal Contacts of Umbra</u>
U1 = 05:27:31.9 UT1
U2 = 05:30:29.6 UT1
U3 = 08:43:52.6 UT1
U4 = 08:46:56.7 UT1

ΔT = 98.1 s

Eph. = JPL DE405

<u>Circumstances at Greatest Eclipse: 07:07:18.8 UT1</u>

Lat. = 32°50.9'S	Sun Alt. = 60.1°		
Long. = 060°14.5'E	Sun Azm. = 20.6°		
Path Width = 110.2 km	Duration = 02m44.1s		

<u>Circumstances at Greatest Duration: 07:43:06.0 UT1</u>

Lat. = 39°12.0'S	Sun Alt. = 53.8°
Long. = 072°52.3'E	Sun Azm. = 340.4°
Path Width = 112.1 km	Duration = 02m45.3s

0 1000 2000 3000 4000 5000
Kilometers

Annular Solar Eclipse of 2071 Mar 31

Greatest Eclipse = 15:01:06.4 TD (= 14:59:27.9 UT1)

Eclipse Magnitude = 0.9919 | Saros Series = 140
Gamma = -0.3739 | Saros Member = 32 of 71

Sun at Greatest Eclipse
(Geocentric Coordinates)
R.A. = 00h40m27.6s
Dec. = +04°21'06.6"
S.D. = 00°16'00.8"
H.P. = 00°00'08.8"

Moon at Greatest Eclipse
(Geocentric Coordinates)
R.A. = 00h40m53.3s
Dec. = +04°00'40.0"
S.D. = 00°15'38.9"
H.P. = 00°57'25.9"

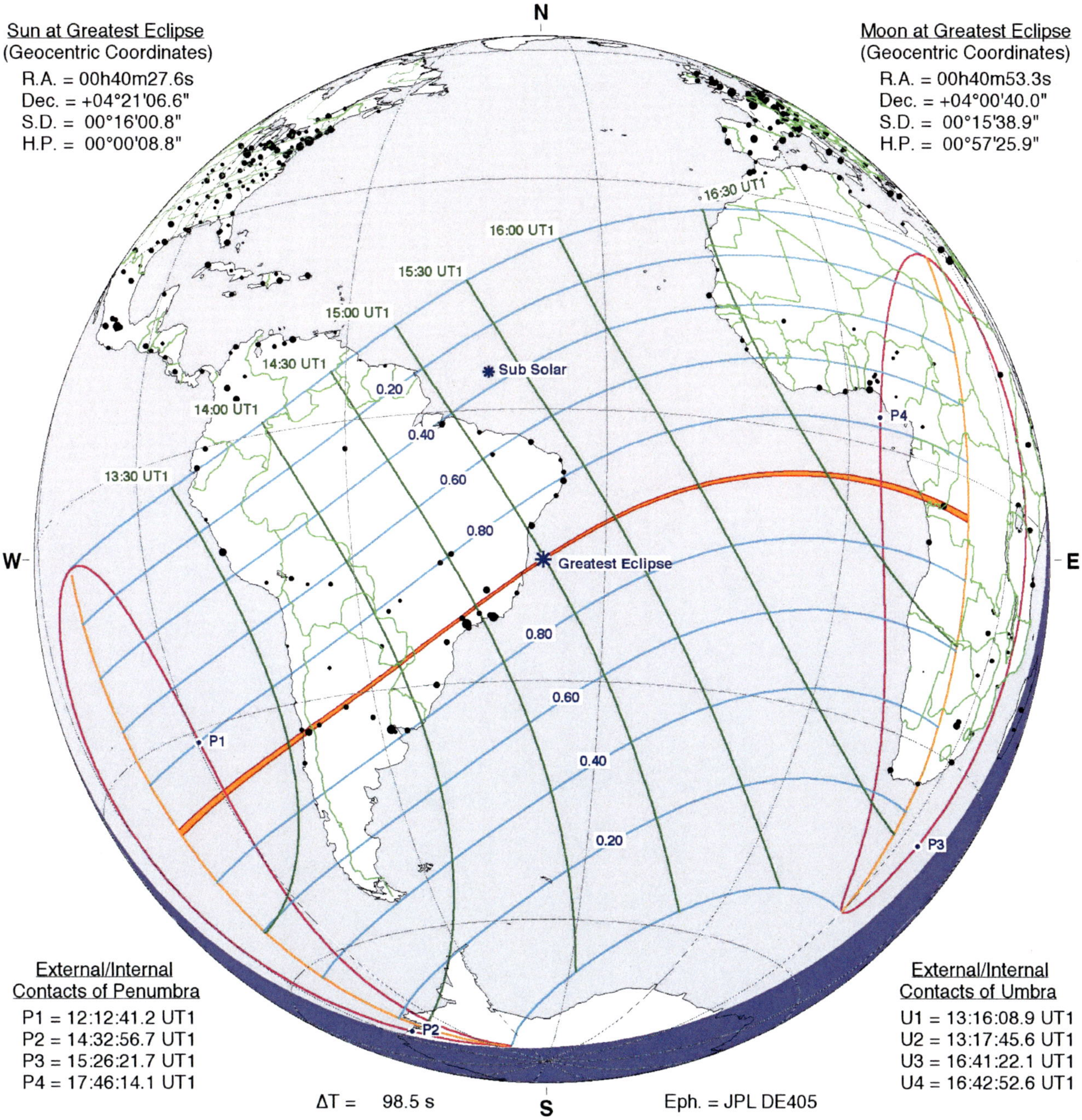

External/Internal Contacts of Penumbra
P1 = 12:12:41.2 UT1
P2 = 14:32:56.7 UT1
P3 = 15:26:21.7 UT1
P4 = 17:46:14.1 UT1

External/Internal Contacts of Umbra
U1 = 13:16:08.9 UT1
U2 = 13:17:45.6 UT1
U3 = 16:41:22.1 UT1
U4 = 16:42:52.6 UT1

ΔT = 98.5 s

Eph. = JPL DE405

Circumstances at Greatest Eclipse: 14:59:27.9 UT1

Lat. = 16°41.9'S | Sun Alt. = 67.9°
Long. = 037°09.6'W | Sun Azm. = 342.0°
Path Width = 30.7 km | Duration = 00m51.7s

Circumstances at Greatest Duration: 13:16:57.2 UT1

Lat. = 39°24.8'S | Sun Alt. = 0.0°
Long. = 104°39.1'W | Sun Azm. = 84.4°
Path Width = 91.6 km | Duration = 01m30.7s

0 1000 2000 3000 4000 5000
Kilometers

Total Solar Eclipse of 2071 Sep 23

Greatest Eclipse = 17:20:28.0 TD (= 17:18:49.2 UT1)

Eclipse Magnitude = 1.0333 Saros Series = 145

Gamma = 0.2620 Saros Member = 25 of 77

Sun at Greatest Eclipse
(Geocentric Coordinates)
R.A. = 12h02m56.7s
Dec. = -00°19'08.2"
S.D. = 00°15'56.2"
H.P. = 00°00'08.8"

Moon at Greatest Eclipse
(Geocentric Coordinates)
R.A. = 12h03m15.7s
Dec. = -00°04'20.0"
S.D. = 00°16'12.3"
H.P. = 00°59'28.3"

N

W 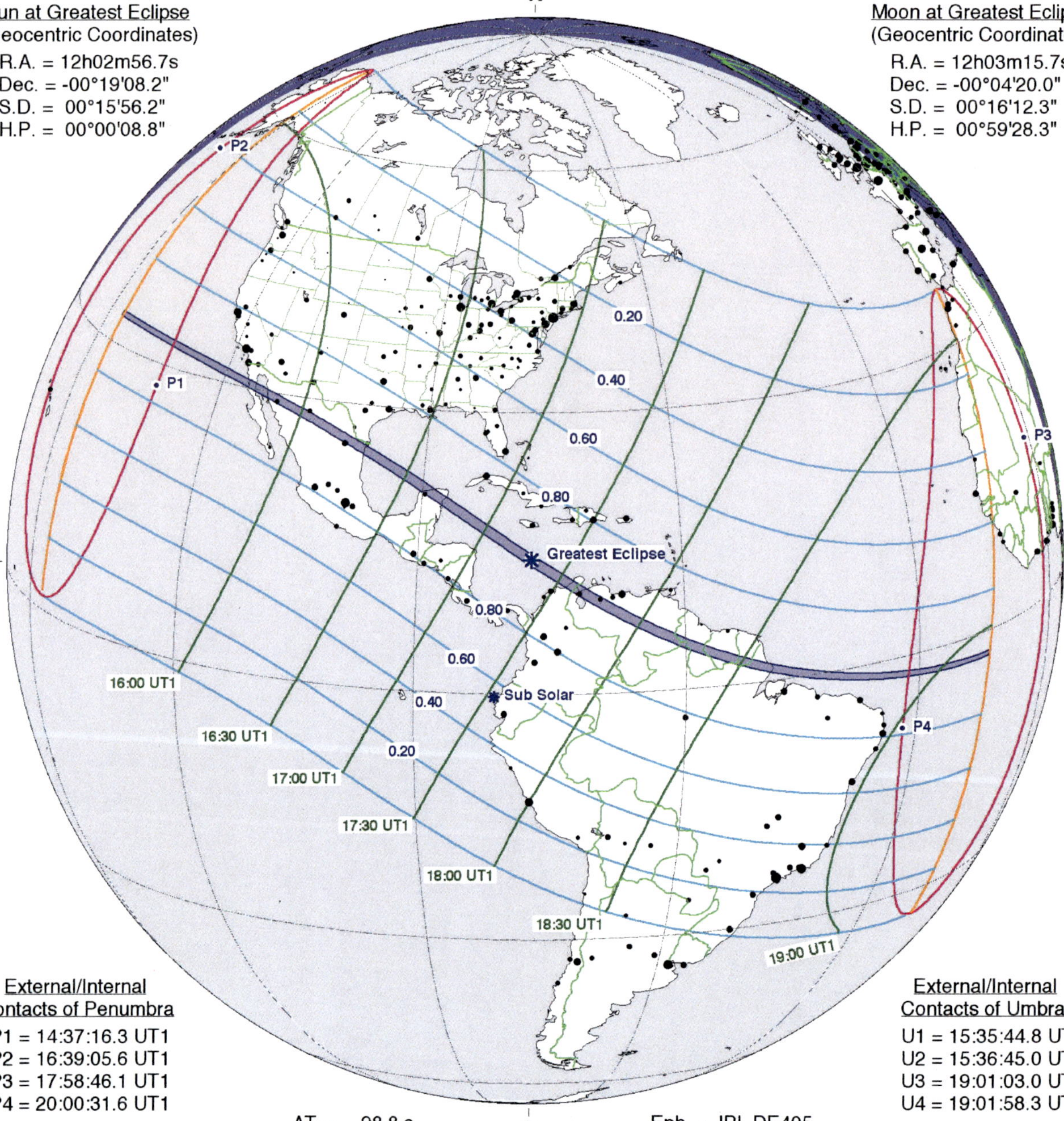 E

S

External/Internal Contacts of Penumbra
P1 = 14:37:16.3 UT1
P2 = 16:39:05.6 UT1
P3 = 17:58:46.1 UT1
P4 = 20:00:31.6 UT1

External/Internal Contacts of Umbra
U1 = 15:35:44.8 UT1
U2 = 15:36:45.0 UT1
U3 = 19:01:03.0 UT1
U4 = 19:01:58.3 UT1

ΔT = 98.8 s Eph. = JPL DE405

Circumstances at Greatest Eclipse: 17:18:49.2 UT1

Lat. = 14°13.5'N	Sun Alt. = 74.7°
Long. = 076°54.7'W	Sun Azm. = 198.2°
Path Width = 116.1 km	Duration = 03m10.8s

Circumstances at Greatest Duration: 17:17:08.4 UT1

Lat. = 14°30.8'N	Sun Alt. = 74.7°
Long. = 077°22.7'W	Sun Azm. = 194.7°
Path Width = 116.0 km	Duration = 03m10.9s

0 1000 2000 3000 4000 5000
Kilometers

Partial Solar Eclipse of 2072 Mar 19

Greatest Eclipse = 20:10:31.1 TD (= 20:08:51.9 UT1)

Eclipse Magnitude = 0.7199 Saros Series = 150

Gamma = -1.1405 Saros Member = 20 of 71

Sun at Greatest Eclipse
(Geocentric Coordinates)
R.A. = 00h00m16.3s
Dec. = +00°01'46.1"
S.D. = 00°16'03.7"
H.P. = 00°00'08.8"

Moon at Greatest Eclipse
(Geocentric Coordinates)
R.A. = 00h01m32.5s
Dec. = -00°57'26.5"
S.D. = 00°14'53.9"
H.P. = 00°54'40.7"

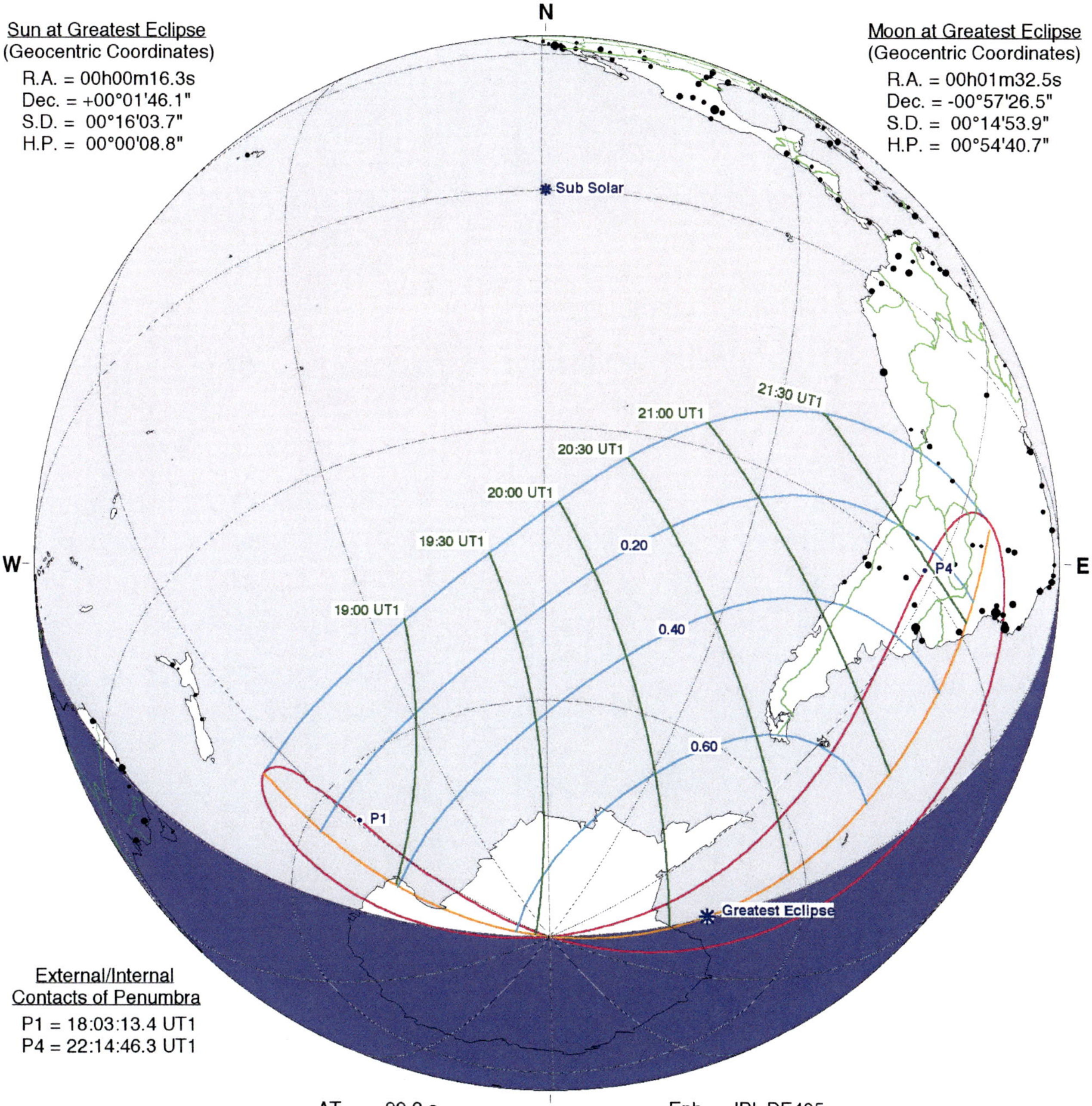

External/Internal
Contacts of Penumbra
P1 = 18:03:13.4 UT1
P4 = 22:14:46.3 UT1

ΔT = 99.2 s Eph. = JPL DE405

Circumstances at Greatest Eclipse: 20:08:51.9 UT1
Lat. = 72°11.0'S Sun Alt. = 0.0°
Long. = 030°32.2'W Sun Azm. = 270.1°

0 1000 2000 3000 4000 5000
Kilometers

Total Solar Eclipse of 2072 Sep 12

Greatest Eclipse = 08:59:20.2 TD (= 08:57:40.7 UT1)

Eclipse Magnitude = 1.0558 Saros Series = 155
Gamma = 0.9655 Saros Member = 9 of 71

Sun at Greatest Eclipse
(Geocentric Coordinates)
R.A. = 11h24m57.8s
Dec. = +03°46'31.0"
S.D. = 00°15'53.4"
H.P. = 00°00'08.7"

Moon at Greatest Eclipse
(Geocentric Coordinates)
R.A. = 11h26m09.4s
Dec. = +04°42'50.2"
S.D. = 00°16'42.8"
H.P. = 01°01'20.3"

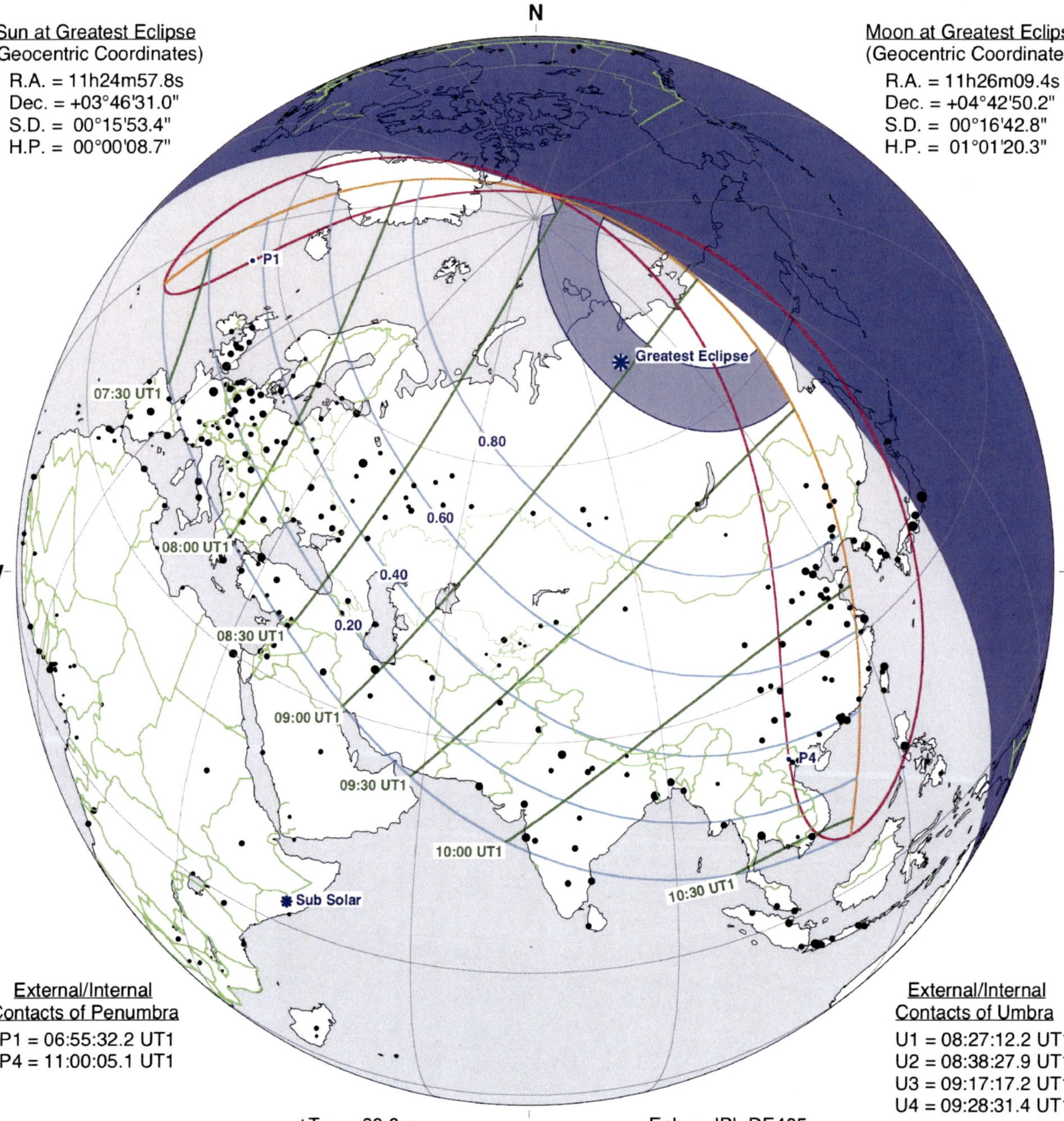

External/Internal
Contacts of Penumbra
P1 = 06:55:32.2 UT1
P4 = 11:00:05.1 UT1

External/Internal
Contacts of Umbra
U1 = 08:27:12.2 UT1
U2 = 08:38:27.9 UT1
U3 = 09:17:17.2 UT1
U4 = 09:28:31.4 UT1

ΔT = 99.6 s Eph. = JPL DE405

Circumstances at Greatest Eclipse: 08:57:40.7 UT1
Lat. = 69°46.3'N Sun Alt. = 14.4°
Long. = 101°49.9'E Sun Azm. = 240.0°
Path Width = 732.4 km Duration = 03m12.9s

Circumstances at Greatest Duration: 08:57:45.4 UT1
Lat. = 69°44.0'N Sun Alt. = 14.4°
Long. = 101°52.5'E Sun Azm. = 240.1°
Path Width = 732.7 km Duration = 03m12.9s

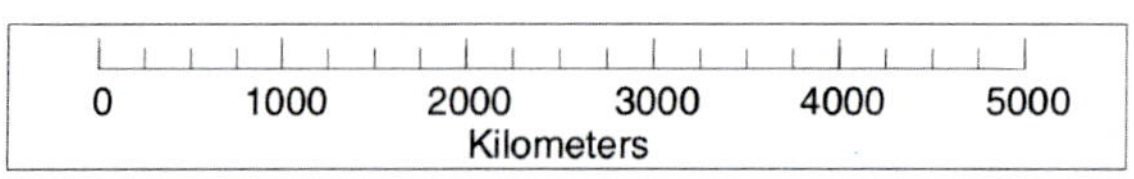

Partial Solar Eclipse of 2073 Feb 07

Greatest Eclipse = 01:55:59.0 TD (= 01:54:19.1 UT1)

Eclipse Magnitude = 0.6768 Saros Series = 122

Gamma = 1.1651 Saros Member = 61 of 70

Sun at Greatest Eclipse
(Geocentric Coordinates)
R.A. = 21h25m15.3s
Dec. = -15°09'16.6"
S.D. = 00°16'13.0"
H.P. = 00°00'08.9"

Moon at Greatest Eclipse
(Geocentric Coordinates)
R.A. = 21h24m18.6s
Dec. = -14°07'10.1"
S.D. = 00°14'54.9"
H.P. = 00°54'44.3"

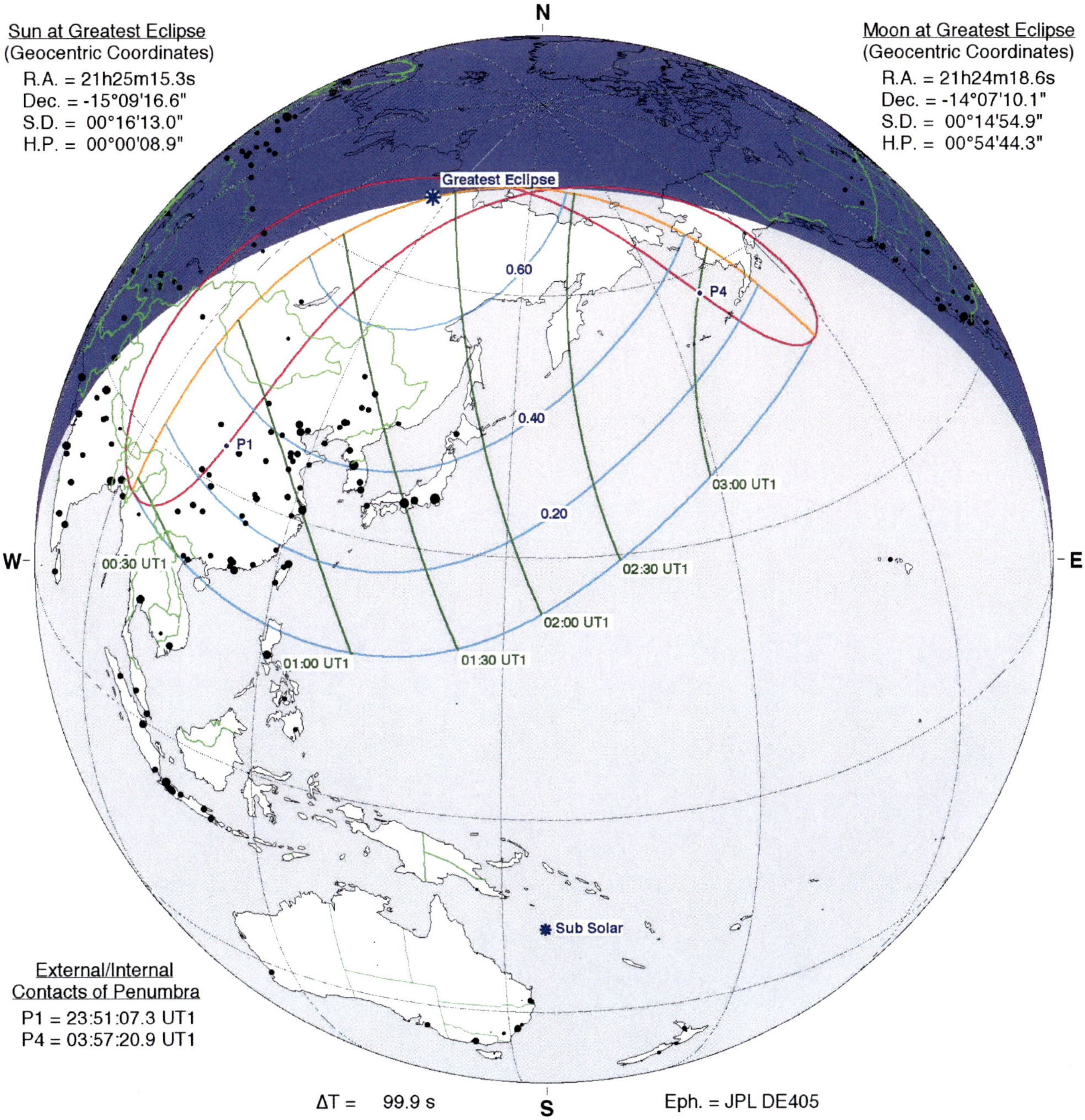

External/Internal
Contacts of Penumbra
P1 = 23:51:07.3 UT1
P4 = 03:57:20.9 UT1

ΔT = 99.9 s Eph. = JPL DE405

Circumstances at Greatest Eclipse: 01:54:19.1 UT1

Lat. = 70°28.0'N Sun Alt. = 0.0°

Long. = 114°44.4'E Sun Azm. = 141.4°

0 1000 2000 3000 4000 5000
Kilometers

Total Solar Eclipse of 2073 Aug 03

Greatest Eclipse = 17:15:22.9 TD (= 17:13:42.6 UT1)

Eclipse Magnitude = 1.0294 Saros Series = 127

Gamma = -0.8763 Saros Member = 61 of 82

Sun at Greatest Eclipse
(Geocentric Coordinates)
R.A. = 08h57m50.6s
Dec. = +17°11'06.4"
S.D. = 00°15'45.7"
H.P. = 00°00'08.7"

Moon at Greatest Eclipse
(Geocentric Coordinates)
R.A. = 08h57m11.2s
Dec. = +16°20'19.0"
S.D. = 00°16'06.2"
H.P. = 00°59'05.8"

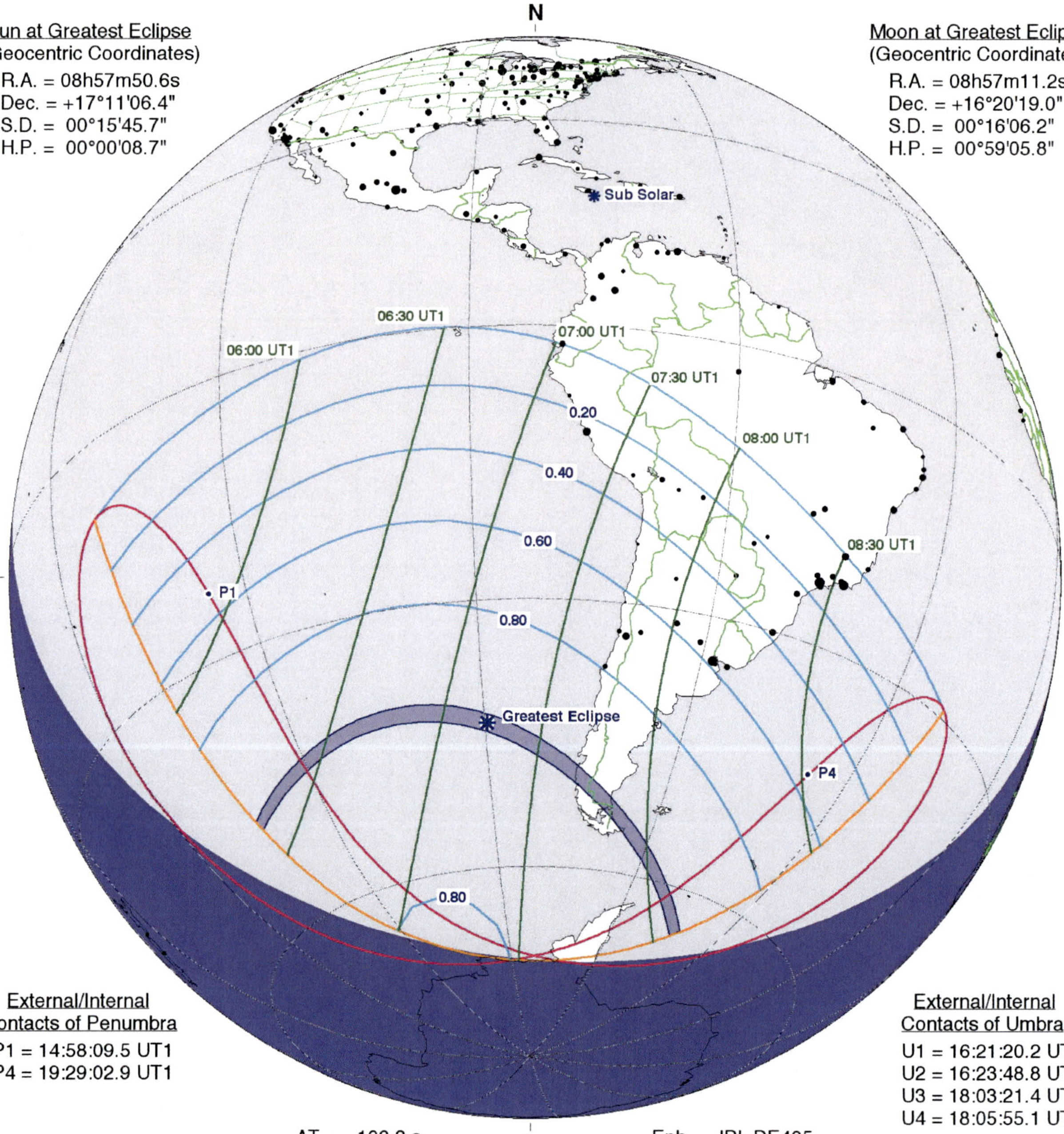

External/Internal Contacts of Penumbra
P1 = 14:58:09.5 UT1
P4 = 19:29:02.9 UT1

External/Internal Contacts of Umbra
U1 = 16:21:20.2 UT1
U2 = 16:23:48.8 UT1
U3 = 18:03:21.4 UT1
U4 = 18:05:55.1 UT1

ΔT = 100.3 s Eph. = JPL DE405

Circumstances at Greatest Eclipse: 17:13:42.6 UT1

Lat. = 43°14.5'S	Sun Alt. = 28.5°
Long. = 089°34.3'W	Sun Azm. = 13.8°
Path Width = 206.4 km	Duration = 02m29.2s

Circumstances at Greatest Duration: 17:14:07.2 UT1

Lat. = 43°17.7'S	Sun Alt. = 28.5°
Long. = 089°22.9'W	Sun Azm. = 13.5°
Path Width = 206.2 km	Duration = 02m29.2s

0 1000 2000 3000 4000 5000
Kilometers

Annular Solar Eclipse of 2074 Jan 27

Greatest Eclipse = 06:44:15.3 TD (= 06:42:34.7 UT1)

Eclipse Magnitude = 0.9798 Saros Series = 132

Gamma = 0.4251 Saros Member = 49 of 71

Sun at Greatest Eclipse
(Geocentric Coordinates)

R.A. = 20h40m20.9s
Dec. = -18°20'28.4"
S.D. = 00°16'14.5"
H.P. = 00°00'08.9"

Moon at Greatest Eclipse
(Geocentric Coordinates)

R.A. = 20h40m04.5s
Dec. = -17°56'22.6"
S.D. = 00°15'41.1"
H.P. = 00°57'33.8"

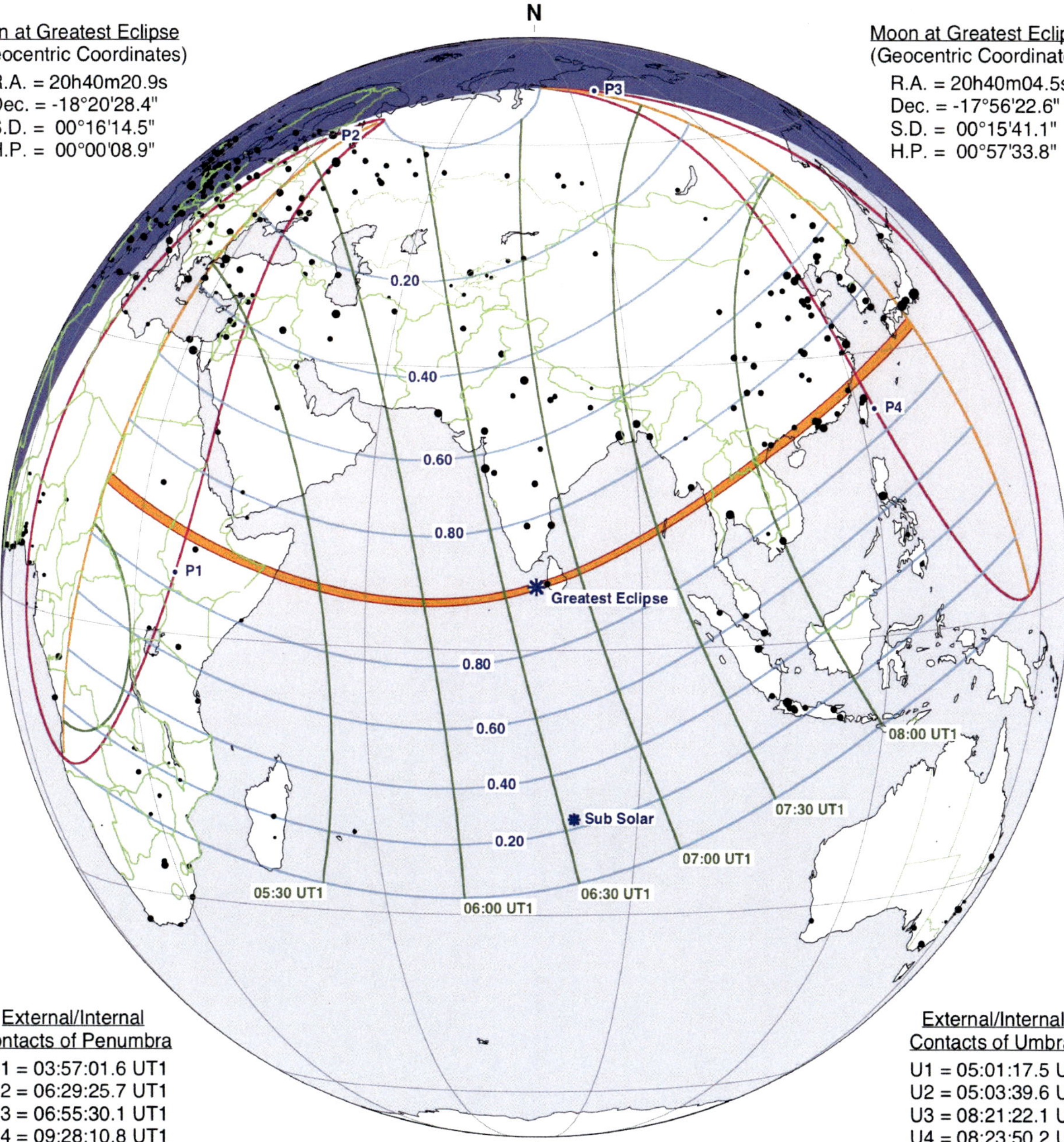

External/Internal
Contacts of Penumbra

P1 = 03:57:01.6 UT1
P2 = 06:29:25.7 UT1
P3 = 06:55:30.1 UT1
P4 = 09:28:10.8 UT1

External/Internal
Contacts of Umbra

U1 = 05:01:17.5 UT1
U2 = 05:03:39.6 UT1
U3 = 08:21:22.1 UT1
U4 = 08:23:50.2 UT1

ΔT = 100.6 s Eph. = JPL DE405

Circumstances at Greatest Eclipse: 06:42:34.7 UT1

Lat. = 06°33.8'N Sun Alt. = 64.8°
Long. = 078°36.5'E Sun Azm. = 171.3°
Path Width = 79.3 km Duration = 02m21.1s

Circumstances at Greatest Duration: 08:22:36.3 UT1

Lat. = 32°32.7'N Sun Alt. = 0.0°
Long. = 135°18.3'E Sun Azm. = 248.1°
Path Width = 134.4 km Duration = 02m21.6s

0 1000 2000 3000 4000 5000
Kilometers

Annular Solar Eclipse of 2074 Jul 24

Greatest Eclipse = 03:10:32.0 TD (= 03:08:51.0 UT1)

Eclipse Magnitude = 0.9838
Gamma = -0.1242

Saros Series = 137
Saros Member = 39 of 70

Sun at Greatest Eclipse
(Geocentric Coordinates)
R.A. = 08h15m33.5s
Dec. = +19°47'02.4"
S.D. = 00°15'44.6"
H.P. = 00°00'08.7"

Moon at Greatest Eclipse
(Geocentric Coordinates)
R.A. = 08h15m29.9s
Dec. = +19°40'09.4"
S.D. = 00°15'15.0"
H.P. = 00°55'58.2"

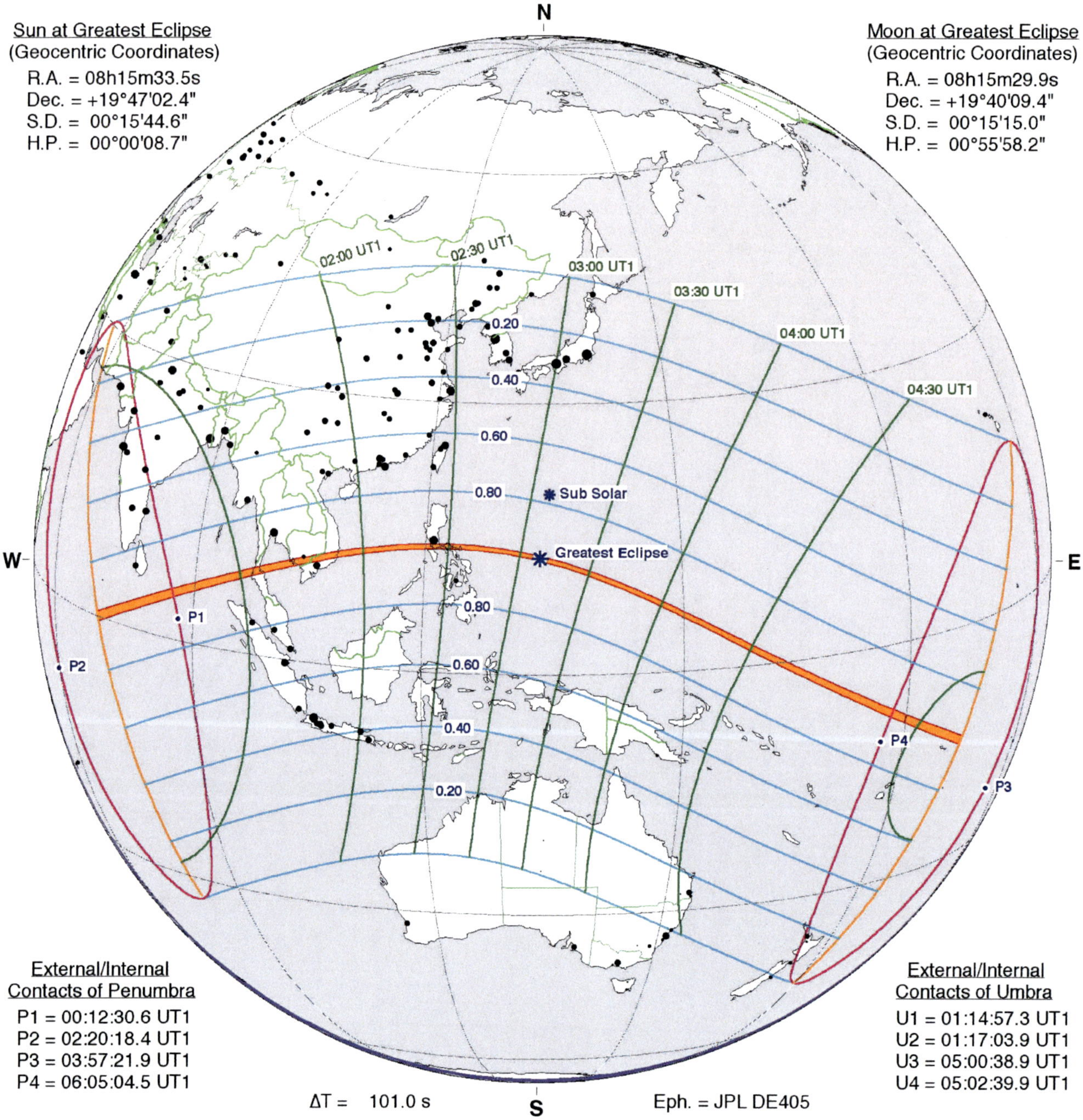

External/Internal Contacts of Penumbra
P1 = 00:12:30.6 UT1
P2 = 02:20:18.4 UT1
P3 = 03:57:21.9 UT1
P4 = 06:05:04.5 UT1

External/Internal Contacts of Umbra
U1 = 01:14:57.3 UT1
U2 = 01:17:03.9 UT1
U3 = 05:00:38.9 UT1
U4 = 05:02:39.9 UT1

ΔT = 101.0 s

Eph. = JPL DE405

Circumstances at Greatest Eclipse: 03:08:51.0 UT1

Lat. = 12°46.9'N	Sun Alt. = 82.9°		
Long. = 133°32.8'E	Sun Azm. = 6.9°		
Path Width = 57.8 km	Duration = 01m57.4s		

Circumstances at Greatest Duration: 01:16:00.6 UT1

Lat. = 00°03.8'S	Sun Alt. = 0.0°
Long. = 072°40.8'E	Sun Azm. = 70.2°
Path Width = 117.5 km	Duration = 02m06.4s

0 1000 2000 3000 4000 5000
Kilometers

Total Solar Eclipse of 2075 Jan 16

Greatest Eclipse = 18:36:04.3 TD (= 18:34:22.9 UT1)

Eclipse Magnitude = 1.0311 Saros Series = 142

Gamma = -0.2799 Saros Member = 26 of 72

Sun at Greatest Eclipse
(Geocentric Coordinates)

R.A. = 19h55m06.1s
Dec. = -20°47'51.3"
S.D. = 00°16'15.5"
H.P. = 00°00'08.9"

Moon at Greatest Eclipse
(Geocentric Coordinates)

R.A. = 19h55m12.8s
Dec. = -21°04'40.9"
S.D. = 00°16'29.7"
H.P. = 01°00'32.2"

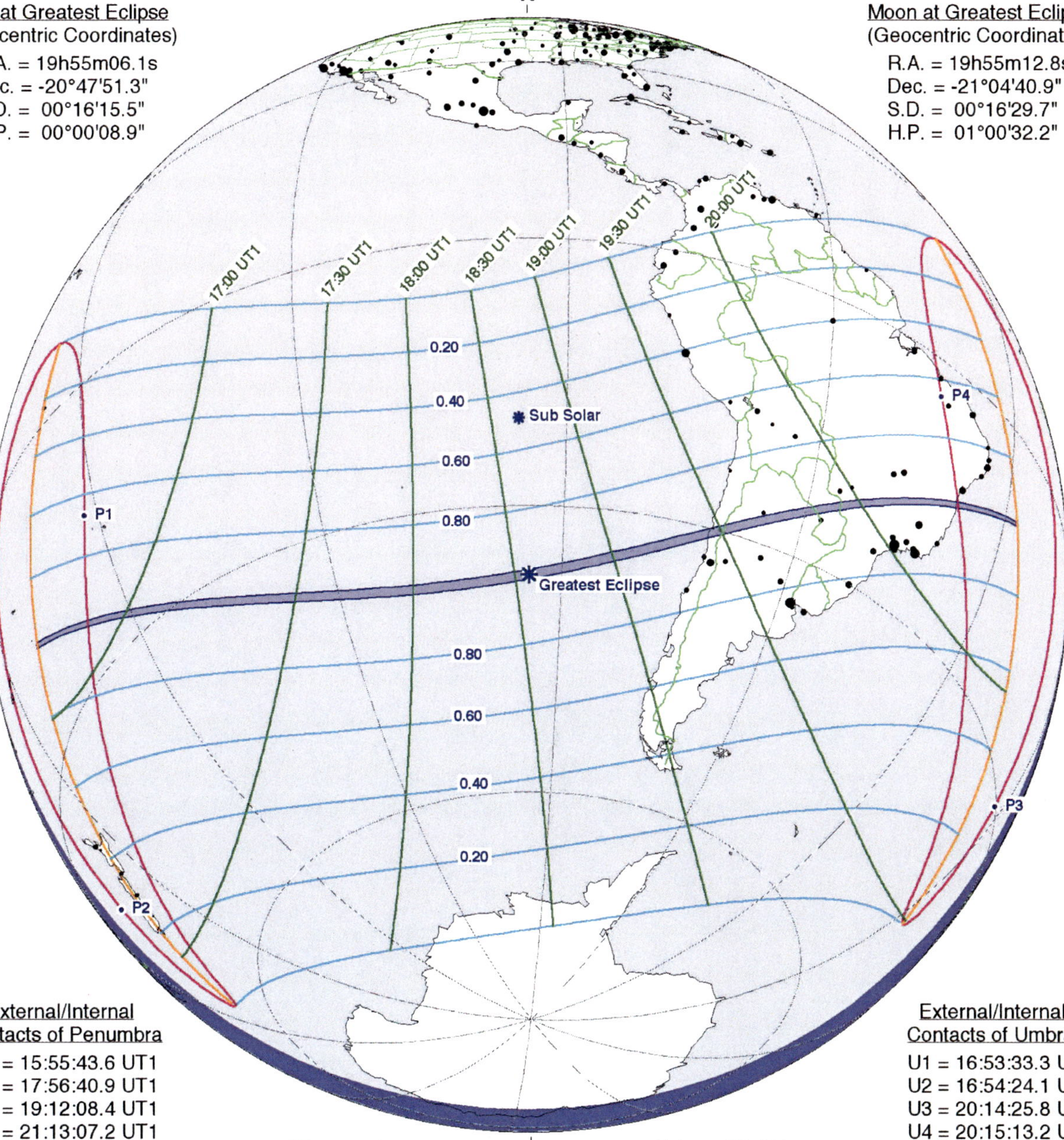

External/Internal Contacts of Penumbra

P1 = 15:55:43.6 UT1
P2 = 17:56:40.9 UT1
P3 = 19:12:08.4 UT1
P4 = 21:13:07.2 UT1

External/Internal Contacts of Umbra

U1 = 16:53:33.3 UT1
U2 = 16:54:24.1 UT1
U3 = 20:14:25.8 UT1
U4 = 20:15:13.2 UT1

ΔT = 101.4 s Eph. = JPL DE405

Circumstances at Greatest Eclipse: 18:34:22.9 UT1

Lat. = 37°10.4'S Sun Alt. = 73.5°
Long. = 094°16.9'W Sun Azm. = 353.8°
Path Width = 109.7 km Duration = 02m42.0s

Circumstances at Greatest Duration: 18:31:58.4 UT1

Lat. = 37°18.7'S Sun Alt. = 73.5°
Long. = 095°19.6'W Sun Azm. = 359.3°
Path Width = 109.7 km Duration = 02m42.1s

0 1000 2000 3000 4000 5000
Kilometers

Annular Solar Eclipse of 2075 Jul 13

Greatest Eclipse = 06:05:44.0 TD (= 06:04:02.2 UT1)

Eclipse Magnitude = 0.9467	Saros Series = 147
Gamma = 0.6583	Saros Member = 26 of 80

Sun at Greatest Eclipse
(Geocentric Coordinates)
R.A. = 07h30m57.3s
Dec. = +21°47'03.5"
S.D. = 00°15'44.0"
H.P. = 00°00'08.7"

Moon at Greatest Eclipse
(Geocentric Coordinates)
R.A. = 07h31m05.6s
Dec. = +22°22'29.6"
S.D. = 00°14'43.7"
H.P. = 00°54'03.3"

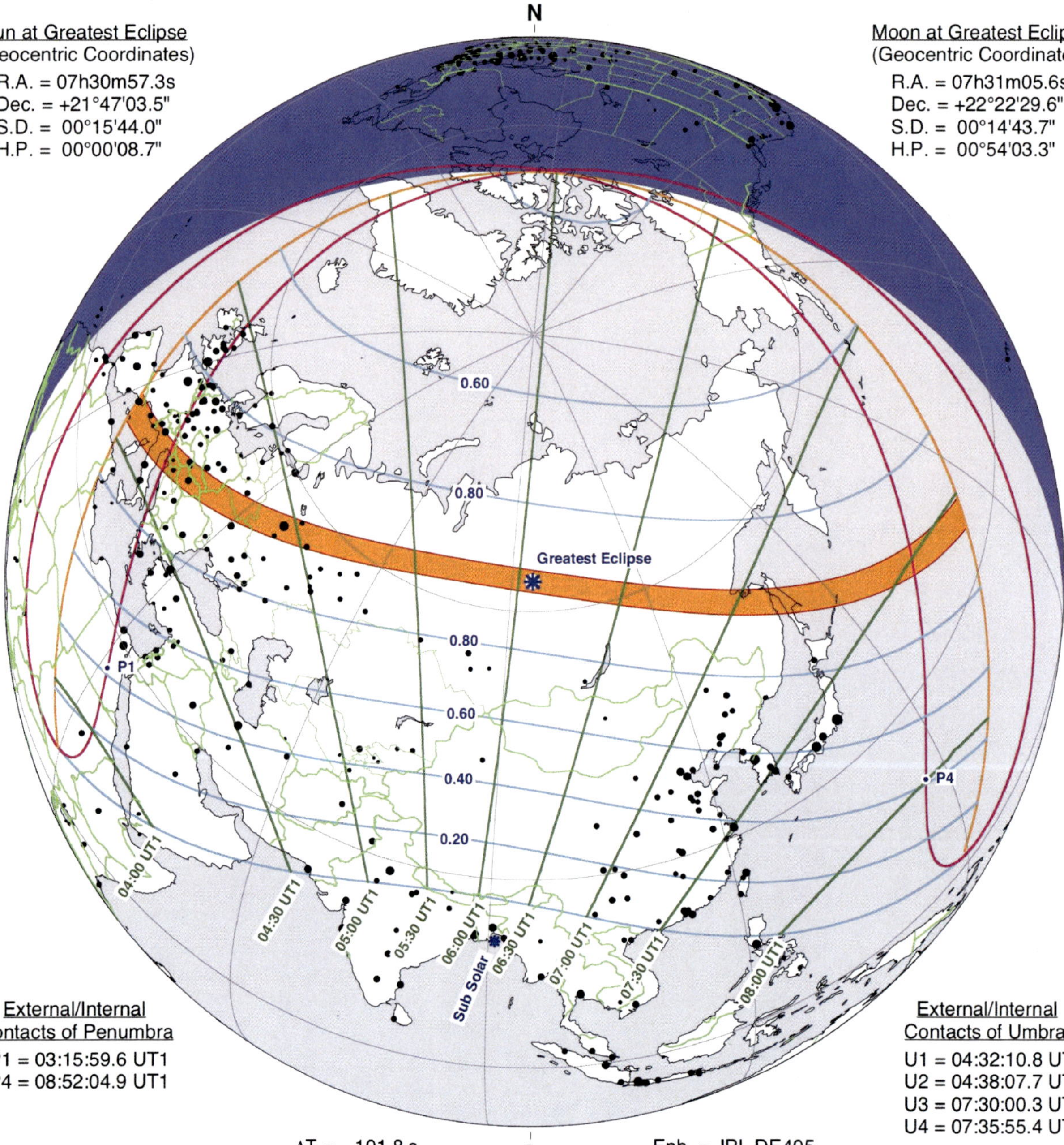

External/Internal Contacts of Penumbra
P1 = 03:15:59.6 UT1
P4 = 08:52:04.9 UT1

External/Internal Contacts of Umbra
U1 = 04:32:10.8 UT1
U2 = 04:38:07.7 UT1
U3 = 07:30:00.3 UT1
U4 = 07:35:55.4 UT1

ΔT = 101.8 s

Eph. = JPL DE405

Circumstances at Greatest Eclipse: 06:04:02.2 UT1

Lat. = 63°07.6'N	Sun Alt. = 48.5°
Long. = 095°00.8'E	Sun Azm. = 186.4°
Path Width = 261.9 km	Duration = 04m44.6s

Circumstances at Greatest Duration: 06:01:41.2 UT1

Lat. = 63°14.5'N	Sun Alt. = 48.5°
Long. = 093°06.0'E	Sun Azm. = 182.9°
Path Width = 261.8 km	Duration = 04m44.6s

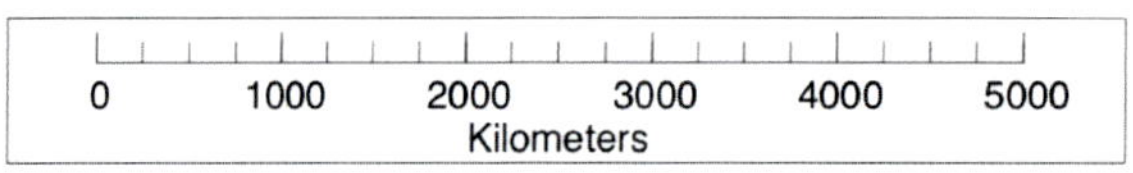

Total Solar Eclipse of 2076 Jan 06

Greatest Eclipse = 10:07:27.5 TD (= 10:05:45.3 UT1)

Eclipse Magnitude = 1.0342
Gamma = -0.9373

Saros Series = 152
Saros Member = 16 of 70

Sun at Greatest Eclipse
(Geocentric Coordinates)

R.A. = 19h09m11.6s
Dec. = -22°28'36.7"
S.D. = 00°16'15.9"
H.P. = 00°00'08.9"

Moon at Greatest Eclipse
(Geocentric Coordinates)

R.A. = 19h09m16.9s
Dec. = -23°26'00.6"
S.D. = 00°16'43.8"
H.P. = 01°01'24.1"

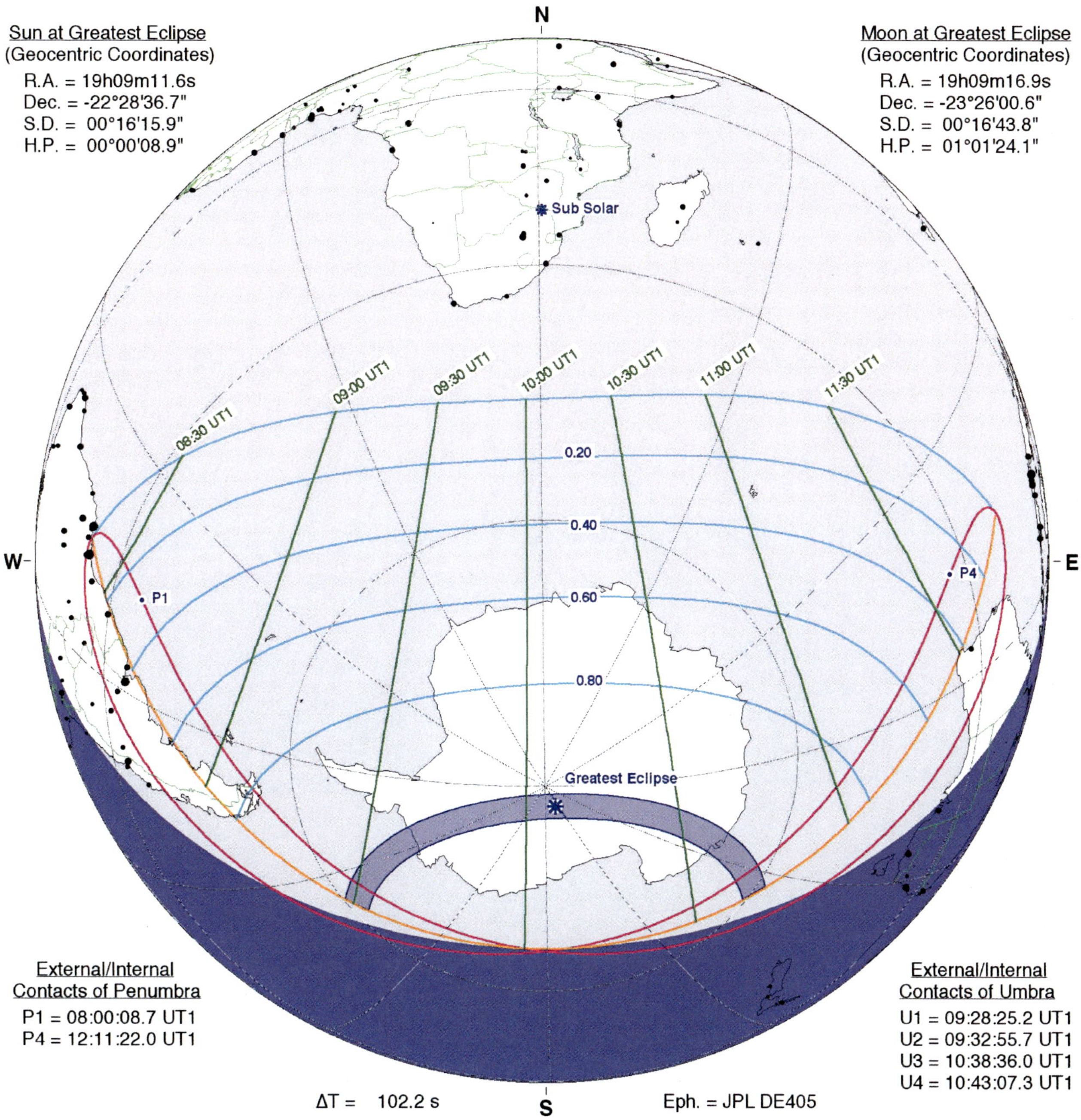

External/Internal Contacts of Penumbra

P1 = 08:00:08.7 UT1
P4 = 12:11:22.0 UT1

External/Internal Contacts of Umbra

U1 = 09:28:25.2 UT1
U2 = 09:32:55.7 UT1
U3 = 10:38:36.0 UT1
U4 = 10:43:07.3 UT1

ΔT = 102.2 s

Eph. = JPL DE405

Circumstances at Greatest Eclipse: 10:05:45.3 UT1

Lat. = 87°10.6'S
Long. = 173°55.4'W
Path Width = 340.4 km
Sun Alt. = 19.9°
Sun Azm. = 203.5°
Duration = 01m49.5s

Circumstances at Greatest Duration: 10:05:54.8 UT1

Lat. = 87°08.4'S
Long. = 175°33.7'W
Path Width = 340.4 km
Sun Alt. = 19.9°
Sun Azm. = 205.0°
Duration = 01m49.5s

0 1000 2000 3000 4000 5000
Kilometers

Partial Solar Eclipse of 2076 Jun 01

Greatest Eclipse = 17:31:21.9 TD (= 17:29:39.4 UT1)

Eclipse Magnitude = 0.2897 Saros Series = 119

Gamma = -1.3897 Saros Member = 69 of 71

Sun at Greatest Eclipse
(Geocentric Coordinates)
R.A. = 04h42m27.8s
Dec. = +22°14'01.6"
S.D. = 00°15'46.3"
H.P. = 00°00'08.7"

Moon at Greatest Eclipse
(Geocentric Coordinates)
R.A. = 04h43m42.6s
Dec. = +20°58'42.6"
S.D. = 00°15'11.7"
H.P. = 00°55'45.9"

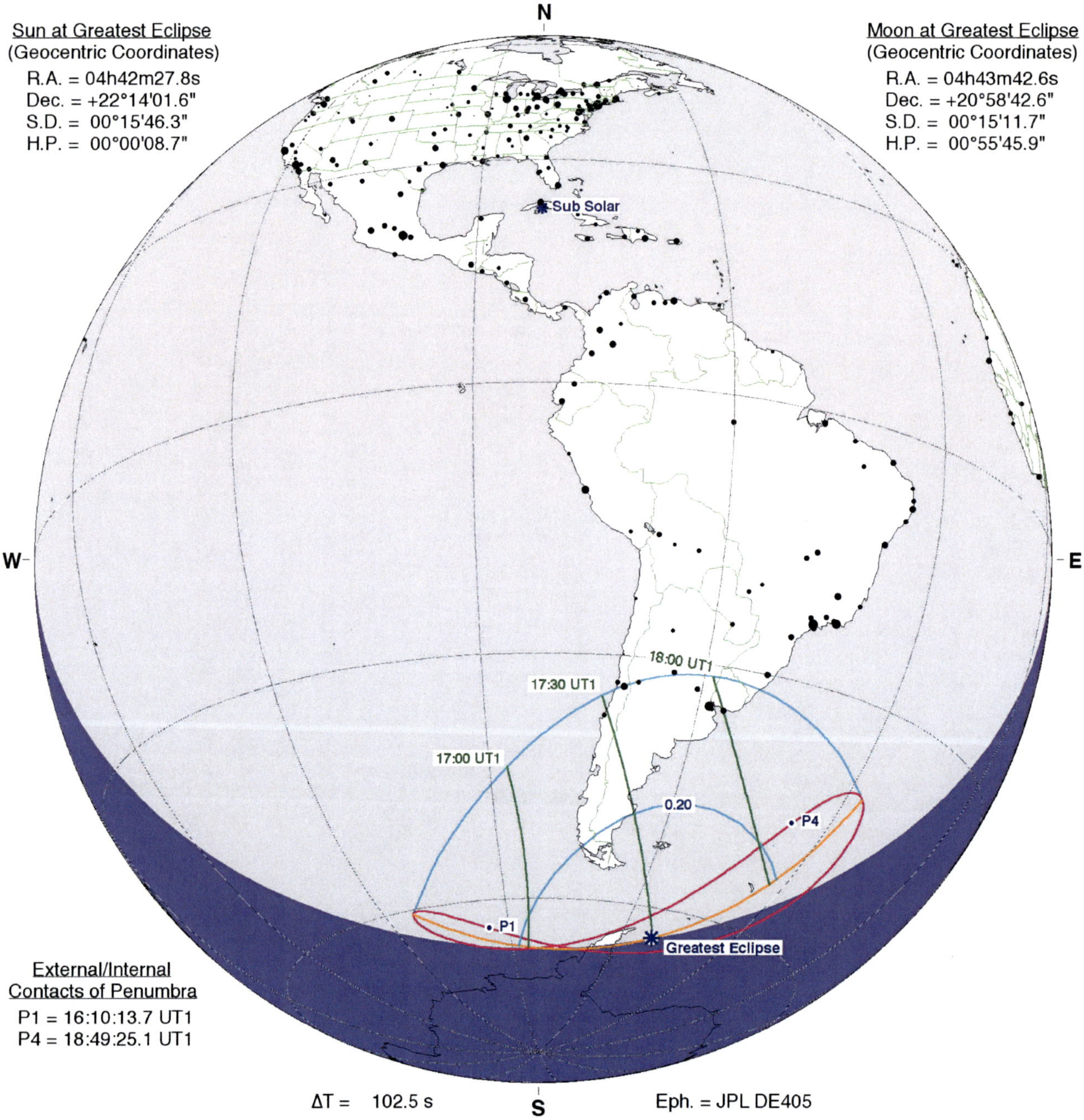

External/Internal
Contacts of Penumbra
P1 = 16:10:13.7 UT1
P4 = 18:49:25.1 UT1

ΔT = 102.5 s Eph. = JPL DE405

Circumstances at Greatest Eclipse: 17:29:39.4 UT1

Lat. = 64°23.0'S Sun Alt. = 0.0°

Long. = 051°25.6'W Sun Azm. = 331.1°

0 1000 2000 3000 4000 5000
Kilometers

Partial Solar Eclipse of 2076 Jul 01

Greatest Eclipse = 06:50:43.3 TD (= 06:49:00.8 UT1)

Eclipse Magnitude = 0.2746
Gamma = 1.4005

Saros Series = 157
Saros Member = 2 of 70

Sun at Greatest Eclipse
(Geocentric Coordinates)
R.A. = 06h44m59.8s
Dec. = +23°01'35.5"
S.D. = 00°15'43.9"
H.P. = 00°00'08.6"

Moon at Greatest Eclipse
(Geocentric Coordinates)
R.A. = 06h44m53.2s
Dec. = +24°17'50.8"
S.D. = 00°14'52.7"
H.P. = 00°54'36.1"

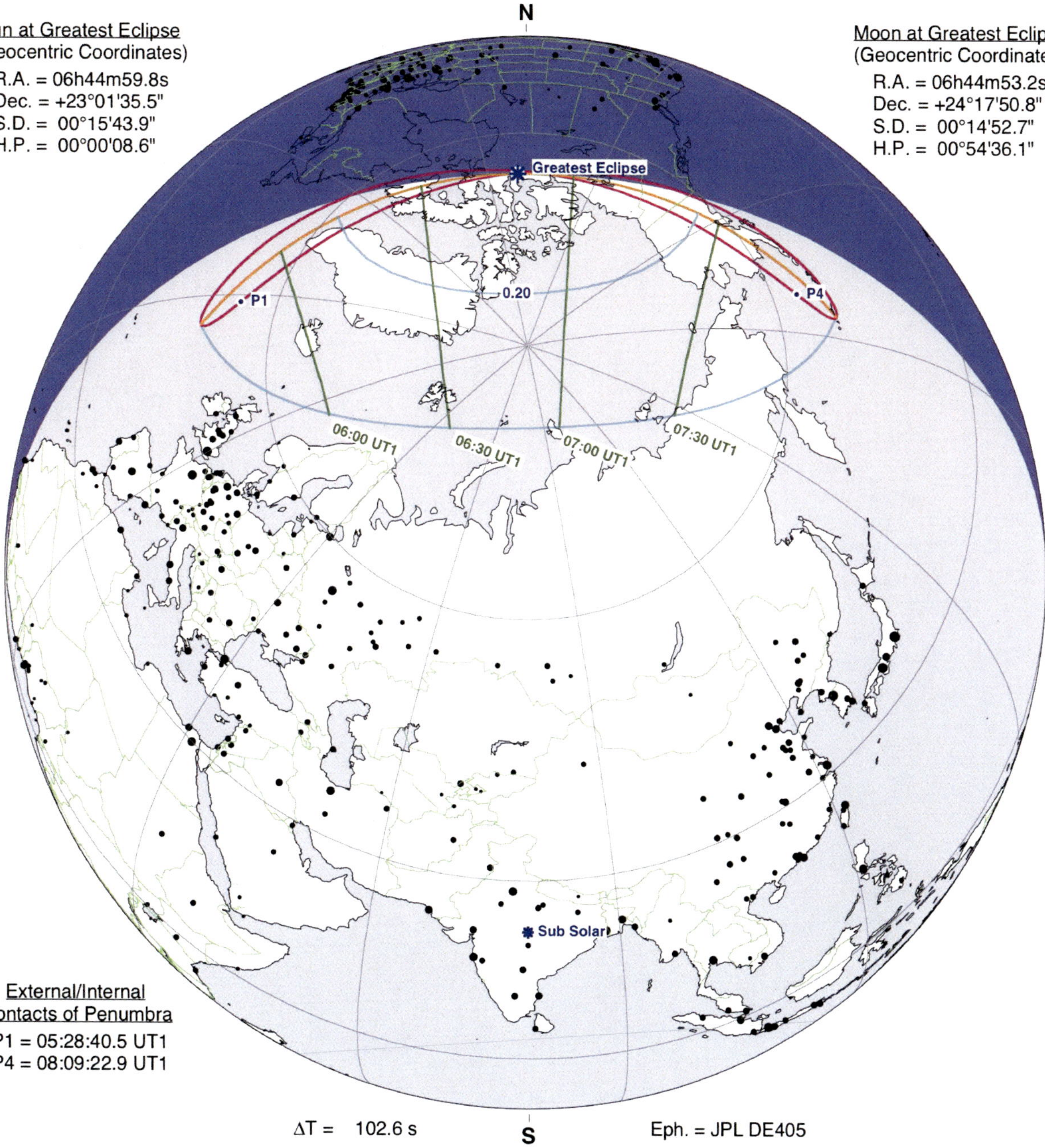

External/Internal
Contacts of Penumbra
P1 = 05:28:40.5 UT1
P4 = 08:09:22.9 UT1

ΔT = 102.6 s

Eph. = JPL DE405

Circumstances at Greatest Eclipse: 06:49:00.8 UT1
Lat. = 66°57.0'N
Long. = 098°18.8'W
Sun Alt. = 0.0°
Sun Azm. = 2.7°

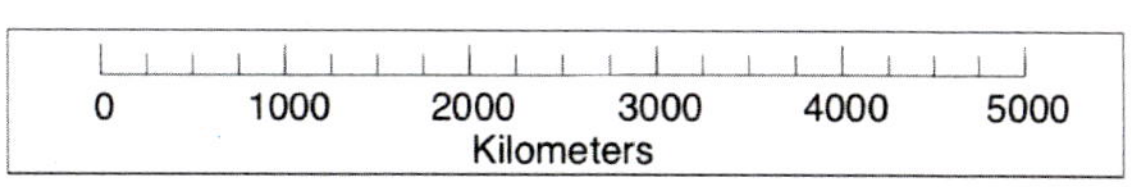

Partial Solar Eclipse of 2076 Nov 26

Greatest Eclipse = 11:43:00.9 TD (= 11:41:18.0 UT1)

Eclipse Magnitude = 0.7315
Gamma = 1.1401

Saros Series = 124
Saros Member = 58 of 73

Sun at Greatest Eclipse
(Geocentric Coordinates)
R.A. = 16h12m39.7s
Dec. = -21°08'26.9"
S.D. = 00°16'12.3"
H.P. = 00°00'08.9"

Moon at Greatest Eclipse
(Geocentric Coordinates)
R.A. = 16h13m56.6s
Dec. = -20°05'16.9"
S.D. = 00°15'44.2"
H.P. = 00°57'45.3"

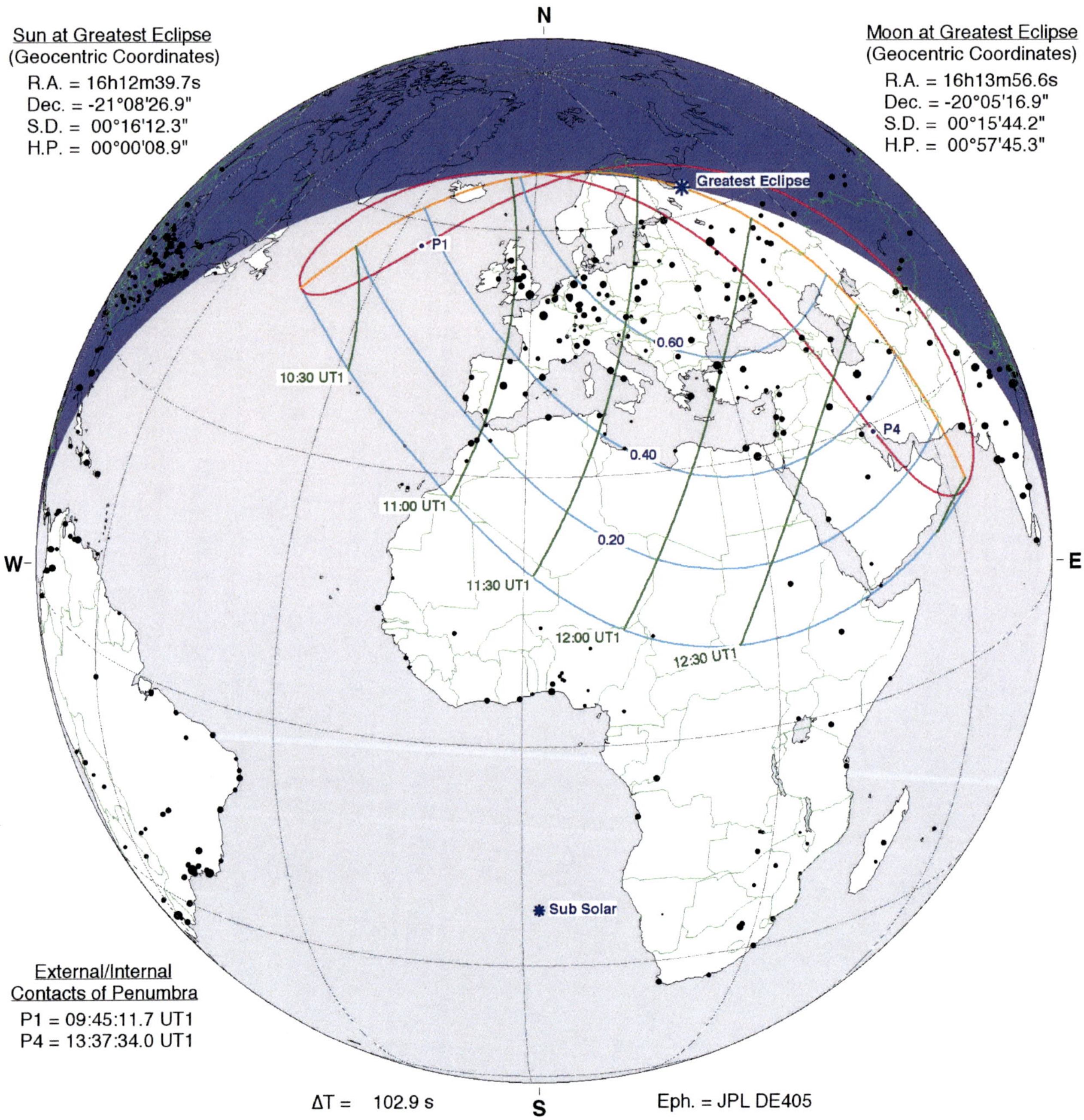

External/Internal
Contacts of Penumbra
P1 = 09:45:11.7 UT1
P4 = 13:37:34.0 UT1

ΔT = 102.9 s

Eph. = JPL DE405

Circumstances at Greatest Eclipse: 11:41:18.0 UT1

Lat. = 63°43.8'N
Long. = 039°55.2'E

Sun Alt. = 0.0°
Sun Azm. = 215.4°

0 1000 2000 3000 4000 5000
Kilometers

Total Solar Eclipse of 2077 May 22

Greatest Eclipse = 02:46:05.3 TD (= 02:44:22.0 UT1)

Eclipse Magnitude = 1.0290 Saros Series = 129

Gamma = -0.5725 Saros Member = 55 of 80

Sun at Greatest Eclipse
(Geocentric Coordinates)
R.A. = 03h58m18.6s
Dec. = +20°29'25.4"
S.D. = 00°15'48.1"
H.P. = 00°00'08.7"

Moon at Greatest Eclipse
(Geocentric Coordinates)
R.A. = 03h59m01.0s
Dec. = +19°57'18.2"
S.D. = 00°16'02.6"
H.P. = 00°58'52.8"

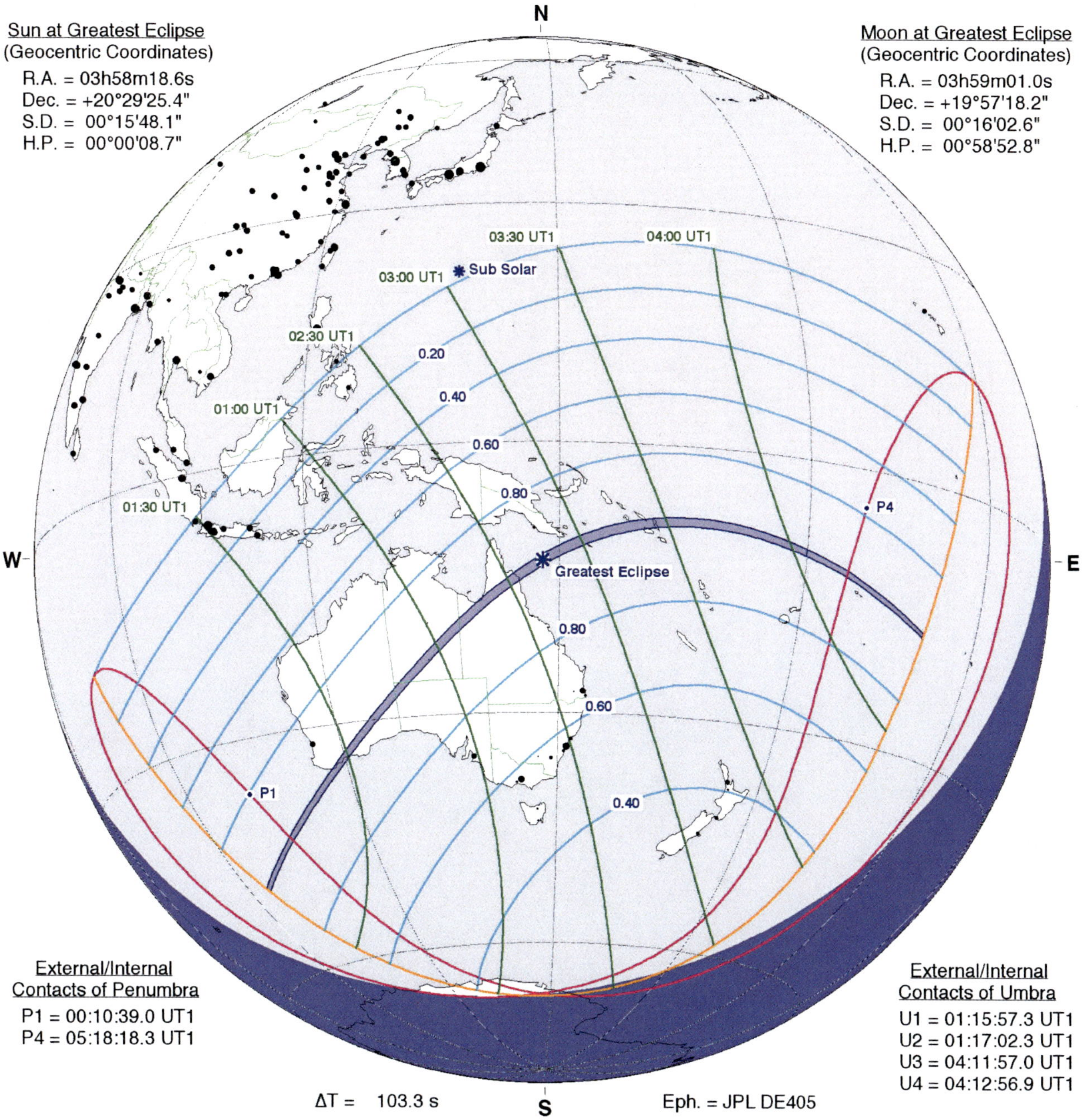

External/Internal
Contacts of Penumbra
P1 = 00:10:39.0 UT1
P4 = 05:18:18.3 UT1

External/Internal
Contacts of Umbra
U1 = 01:15:57.3 UT1
U2 = 01:17:02.3 UT1
U3 = 04:11:57.0 UT1
U4 = 04:12:56.9 UT1

ΔT = 103.3 s Eph. = JPL DE405

Circumstances at Greatest Eclipse: 02:44:22.0 UT1
Lat. = 13°06.2'S Sun Alt. = 55.0°
Long. = 148°07.3'E Sun Azm. = 343.5°
Path Width = 118.9 km Duration = 02m53.7s

Circumstances at Greatest Duration: 02:46:17.2 UT1
Lat. = 12°48.9'S Sun Alt. = 55.0°
Long. = 148°38.9'E Sun Azm. = 341.8°
Path Width = 119.2 km Duration = 02m53.7s

0 1000 2000 3000 4000 5000
Kilometers

Annular Solar Eclipse of 2077 Nov 15

Greatest Eclipse = 17:07:56.2 TD (= 17:06:12.6 UT1)

Eclipse Magnitude = 0.9371 | Saros Series = 134

Gamma = 0.4705 | Saros Member = 47 of 71

<u>Sun at Greatest Eclipse</u>
(Geocentric Coordinates)
R.A. = 15h26m24.3s
Dec. = -18°45'33.3"
S.D. = 00°16'10.1"
H.P. = 00°00'08.9"

<u>Moon at Greatest Eclipse</u>
(Geocentric Coordinates)
R.A. = 15h27m01.5s
Dec. = -18°21'22.3"
S.D. = 00°14'56.9"
H.P. = 00°54'51.8"

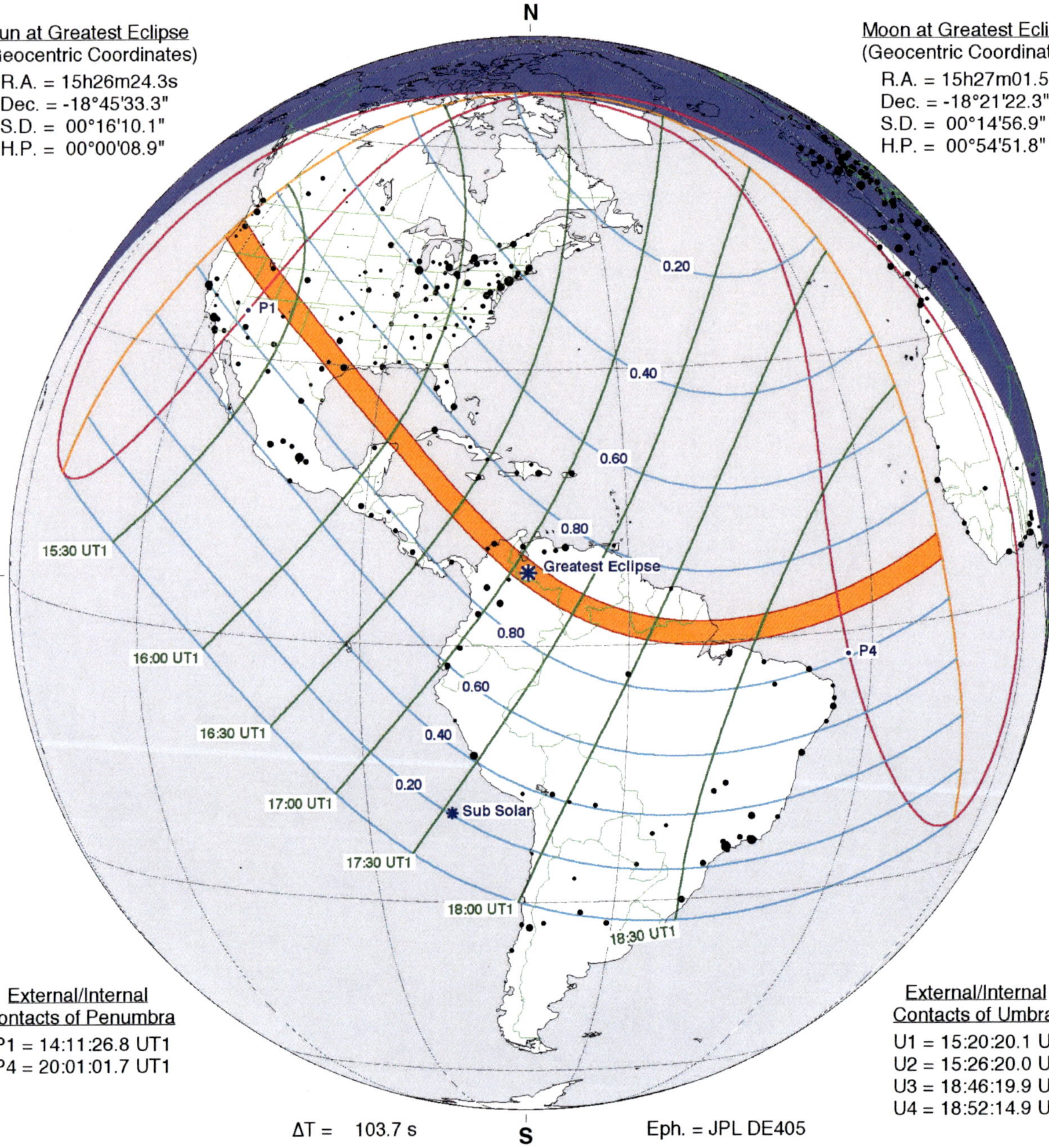

<u>External/Internal Contacts of Penumbra</u>
P1 = 14:11:26.8 UT1
P4 = 20:01:01.7 UT1

<u>External/Internal Contacts of Umbra</u>
U1 = 15:20:20.1 UT1
U2 = 15:26:20.0 UT1
U3 = 18:46:19.9 UT1
U4 = 18:52:14.9 UT1

ΔT = 103.7 s

Eph. = JPL DE405

<u>Circumstances at Greatest Eclipse: 17:06:12.6 UT1</u>
Lat. = 07°47.5'N | Sun Alt. = 61.9°
Long. = 071°00.9'W | Sun Azm. = 199.1°
Path Width = 262.5 km | Duration = 07m53.9s

<u>Circumstances at Greatest Duration: 17:19:19.0 UT1</u>
Lat. = 05°48.5'N | Sun Alt. = 61.0°
Long. = 068°06.8'W | Sun Azm. = 211.6°
Path Width = 267.6 km | Duration = 07m55.9s

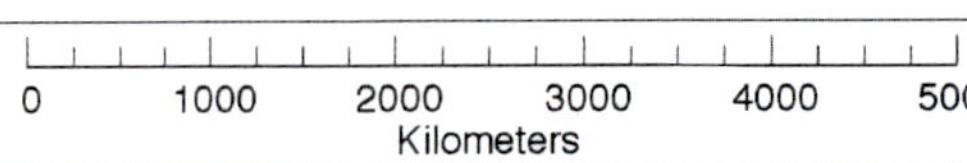

Total Solar Eclipse of 2078 May 11

Greatest Eclipse = 17:56:54.8 TD (= 17:55:10.7 UT1)

Eclipse Magnitude = 1.0701 | Saros Series = 139

Gamma = 0.1838 | Saros Member = 33 of 71

Sun at Greatest Eclipse
(Geocentric Coordinates)

R.A. = 03h16m09.4s
Dec. = +18°07'17.6"
S.D. = 00°15'50.2"
H.P. = 00°00'08.7"

Moon at Greatest Eclipse
(Geocentric Coordinates)

R.A. = 03h15m52.6s
Dec. = +18°17'46.7"
S.D. = 00°16'39.9"
H.P. = 01°01'09.6"

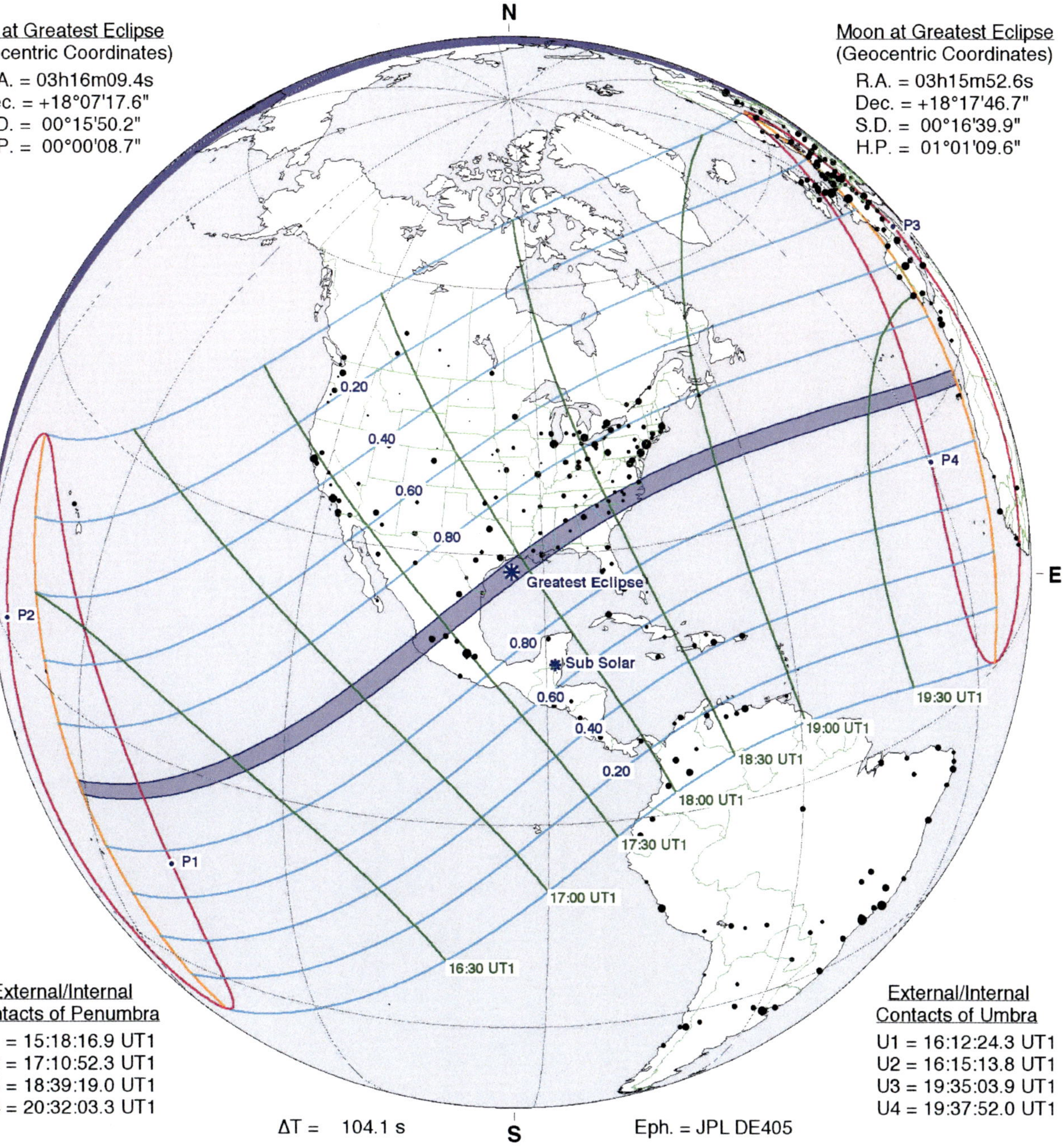

External/Internal Contacts of Penumbra

P1 = 15:18:16.9 UT1
P2 = 17:10:52.3 UT1
P3 = 18:39:19.0 UT1
P4 = 20:32:03.3 UT1

External/Internal Contacts of Umbra

U1 = 16:12:24.3 UT1
U2 = 16:15:13.8 UT1
U3 = 19:35:03.9 UT1
U4 = 19:37:52.0 UT1

ΔT = 104.1 s

Eph. = JPL DE405

Circumstances at Greatest Eclipse: 17:55:10.7 UT1

Lat. = 28°08.5'N | Sun Alt. = 79.3°
Long. = 093°55.1'W | Sun Azm. = 157.9°
Path Width = 232.2 km | Duration = 05m40.5s

Circumstances at Greatest Duration: 18:00:33.5 UT1

Lat. = 29°15.9'N | Sun Alt. = 78.8°
Long. = 092°06.2'W | Sun Azm. = 174.7°
Path Width = 231.6 km | Duration = 05m40.9s

0 1000 2000 3000 4000 5000
Kilometers

Annular Solar Eclipse of 2078 Nov 04

Greatest Eclipse = 16:55:44.4 TD (= 16:53:59.9 UT1)

Eclipse Magnitude = 0.9255
Gamma = -0.2285

Saros Series = 144
Saros Member = 20 of 70

Sun at Greatest Eclipse
(Geocentric Coordinates)
R.A. = 14h40m53.9s
Dec. = -15°38'07.6"
S.D. = 00°16'07.5"
H.P. = 00°00'08.9"

Moon at Greatest Eclipse
(Geocentric Coordinates)
R.A. = 14h40m33.5s
Dec. = -15°49'24.5"
S.D. = 00°14'42.4"
H.P. = 00°53'58.5"

N

W

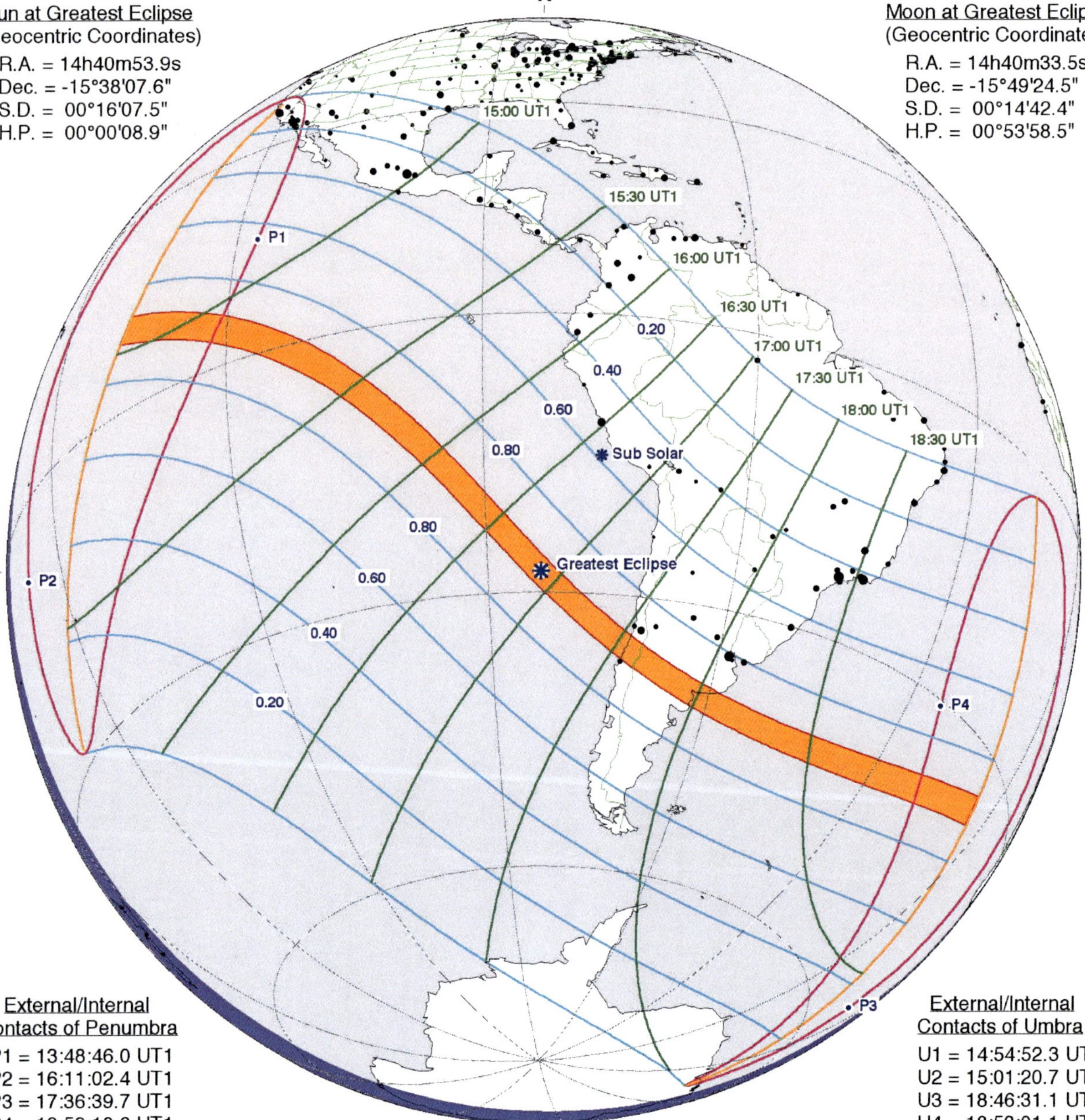

E

S

External/Internal Contacts of Penumbra
P1 = 13:48:46.0 UT1
P2 = 16:11:02.4 UT1
P3 = 17:36:39.7 UT1
P4 = 19:59:10.6 UT1

External/Internal Contacts of Umbra
U1 = 14:54:52.3 UT1
U2 = 15:01:20.7 UT1
U3 = 18:46:31.1 UT1
U4 = 18:53:01.1 UT1

ΔT = 104.5 s

Eph. = JPL DE405

Circumstances at Greatest Eclipse: 16:53:59.9 UT1

Lat. = 27°49.8'S	Sun Alt. = 76.6°
Long. = 083°31.2'W	Sun Azm. = 25.4°
Path Width = 287.5 km	Duration = 08m29.1s

Circumstances at Greatest Duration: 17:11:45.0 UT1

Lat. = 31°24.5'S	Sun Alt. = 74.0°
Long. = 078°59.0'W	Sun Azm. = 349.2°
Path Width = 286.0 km	Duration = 08m31.9s

0 1000 2000 3000 4000 5000
Kilometers

Total Solar Eclipse of 2079 May 01

Greatest Eclipse = 10:50:12.8 TD (= 10:48:28.0 UT1)

Eclipse Magnitude = 1.0512 Saros Series = 149
Gamma = 0.9081 Saros Member = 24 of 71

Sun at Greatest Eclipse
(Geocentric Coordinates)
R.A. = 02h35m18.8s
Dec. = +15°12'06.8"
S.D. = 00°15'52.6"
H.P. = 00°00'08.7"

Moon at Greatest Eclipse
(Geocentric Coordinates)
R.A. = 02h33m47.0s
Dec. = +16°02'36.5"
S.D. = 00°16'34.7"
H.P. = 01°00'50.6"

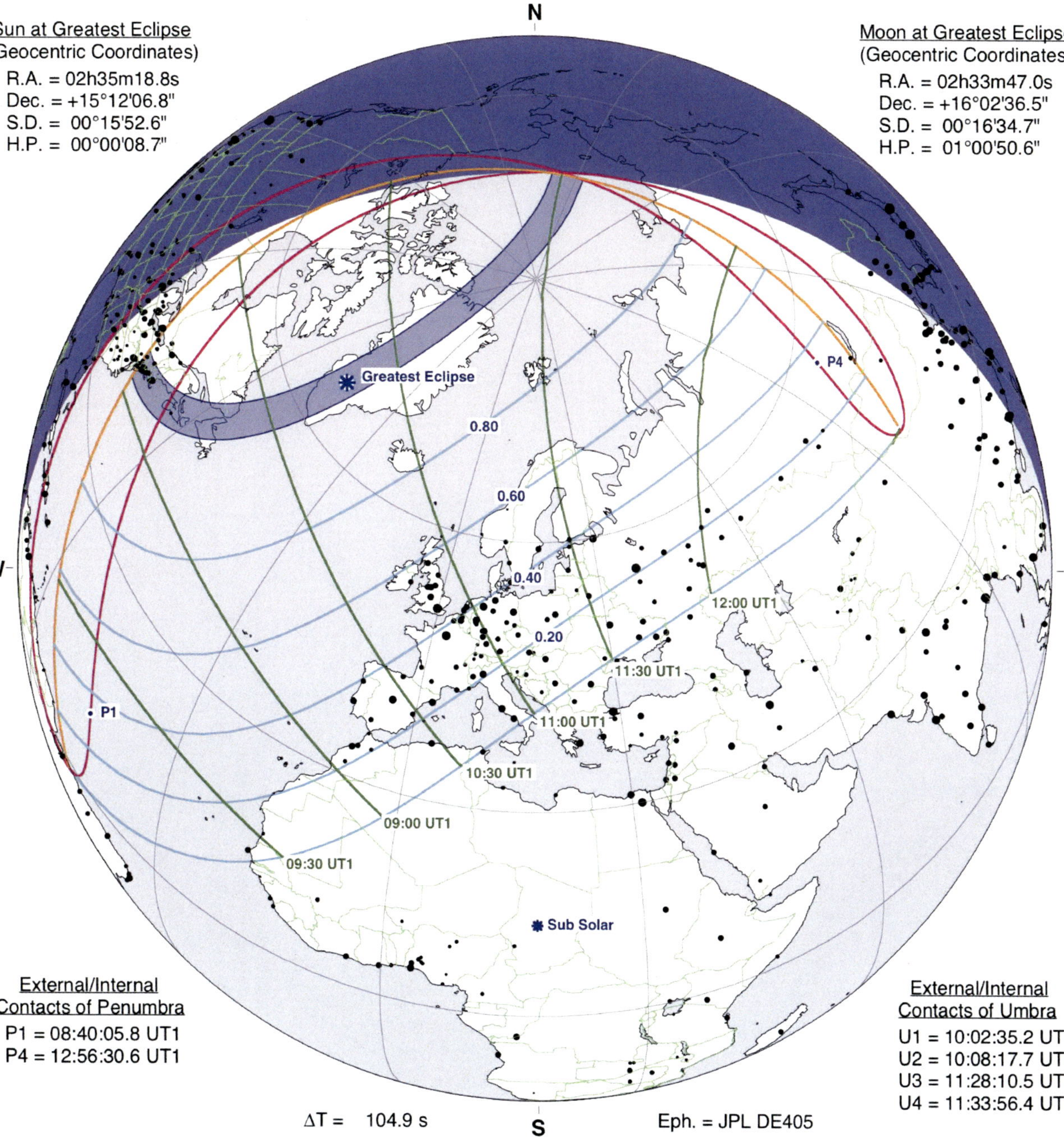

External/Internal
Contacts of Penumbra
P1 = 08:40:05.8 UT1
P4 = 12:56:30.6 UT1

External/Internal
Contacts of Umbra
U1 = 10:02:35.2 UT1
U2 = 10:08:17.7 UT1
U3 = 11:28:10.5 UT1
U4 = 11:33:56.4 UT1

ΔT = 104.9 s Eph. = JPL DE405

Circumstances at Greatest Eclipse: 10:48:28.0 UT1

Lat. = 66°10.0'N Sun Alt. = 24.4°
Long. = 046°32.7'W Sun Azm. = 108.2°
Path Width = 405.7 km Duration = 02m54.9s

Circumstances at Greatest Duration: 10:49:13.2 UT1

Lat. = 66°33.8'N Sun Alt. = 24.4°
Long. = 046°25.3'W Sun Azm. = 108.7°
Path Width = 404.8 km Duration = 02m54.9s

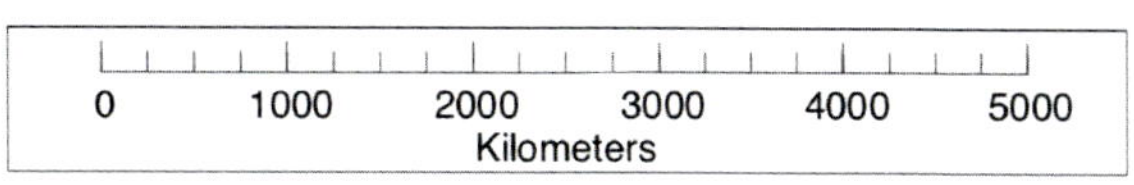

Annular Solar Eclipse of 2079 Oct 24

Greatest Eclipse = 18:11:21.4 TD (= 18:09:36.1 UT1)

Eclipse Magnitude = 0.9484 Saros Series = 154

Gamma = -0.9243 Saros Member = 10 of 71

Sun at Greatest Eclipse
(Geocentric Coordinates)
R.A. = 13h57m22.1s
Dec. = -11°59'23.6"
S.D. = 00°16'04.6"
H.P. = 00°00'08.8"

Moon at Greatest Eclipse
(Geocentric Coordinates)
R.A. = 13h55m50.0s
Dec. = -12°45'30.1"
S.D. = 00°15'09.9"
H.P. = 00°55'39.3"

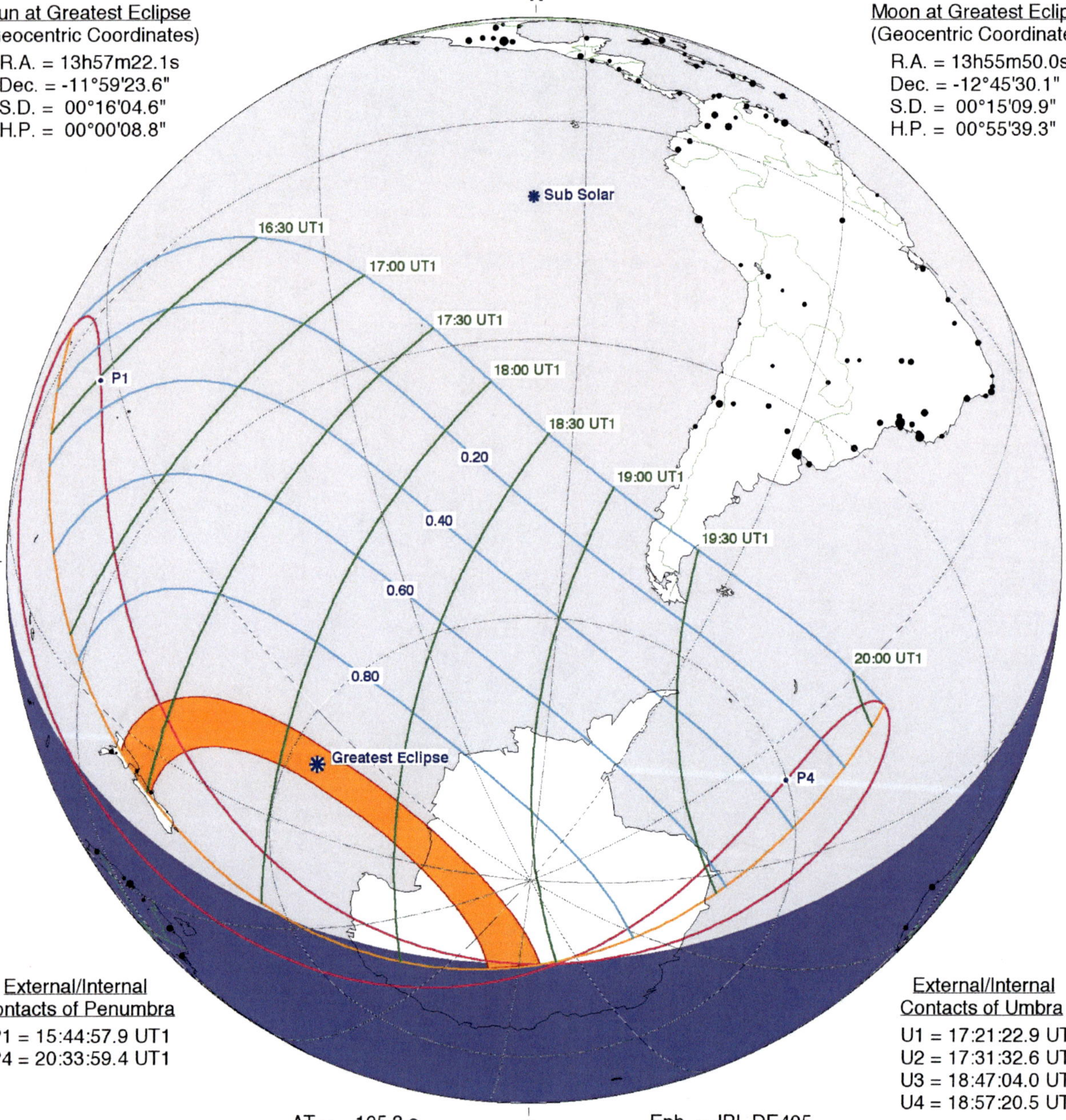

External/Internal
Contacts of Penumbra
P1 = 15:44:57.9 UT1
P4 = 20:33:59.4 UT1

External/Internal
Contacts of Umbra
U1 = 17:21:22.9 UT1
U2 = 17:31:32.6 UT1
U3 = 18:47:04.0 UT1
U4 = 18:57:20.5 UT1

ΔT = 105.3 s Eph. = JPL DE405

Circumstances at Greatest Eclipse: 18:09:36.1 UT1

Lat. = 63°24.8'S Sun Alt. = 22.0°
Long. = 160°46.3'W Sun Azm. = 72.1°
Path Width = 494.6 km Duration = 03m38.9s

Circumstances at Greatest Duration: 18:16:11.2 UT1

Lat. = 66°36.3'S Sun Alt. = 21.7°
Long. = 160°30.0'W Sun Azm. = 69.0°
Path Width = 484.5 km Duration = 03m39.0s

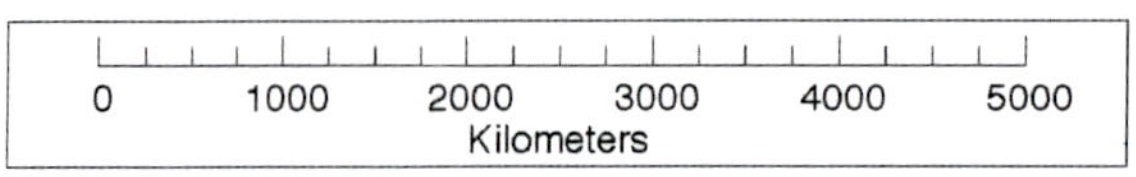

Partial Solar Eclipse of 2080 Mar 21

Greatest Eclipse = 12:20:15.4 TD (= 12:18:29.8 UT1)

Eclipse Magnitude = 0.8734 Saros Series = 121

Gamma = -1.0578 Saros Member = 64 of 71

<u>Sun at Greatest Eclipse</u>
(Geocentric Coordinates)
R.A. = 00h06m37.3s
Dec. = +00°43'02.5"
S.D. = 00°16'03.3"
H.P. = 00°00'08.8"

<u>Moon at Greatest Eclipse</u>
(Geocentric Coordinates)
R.A. = 00h08m33.2s
Dec. = -00°09'04.7"
S.D. = 00°15'24.0"
H.P. = 00°56'31.2"

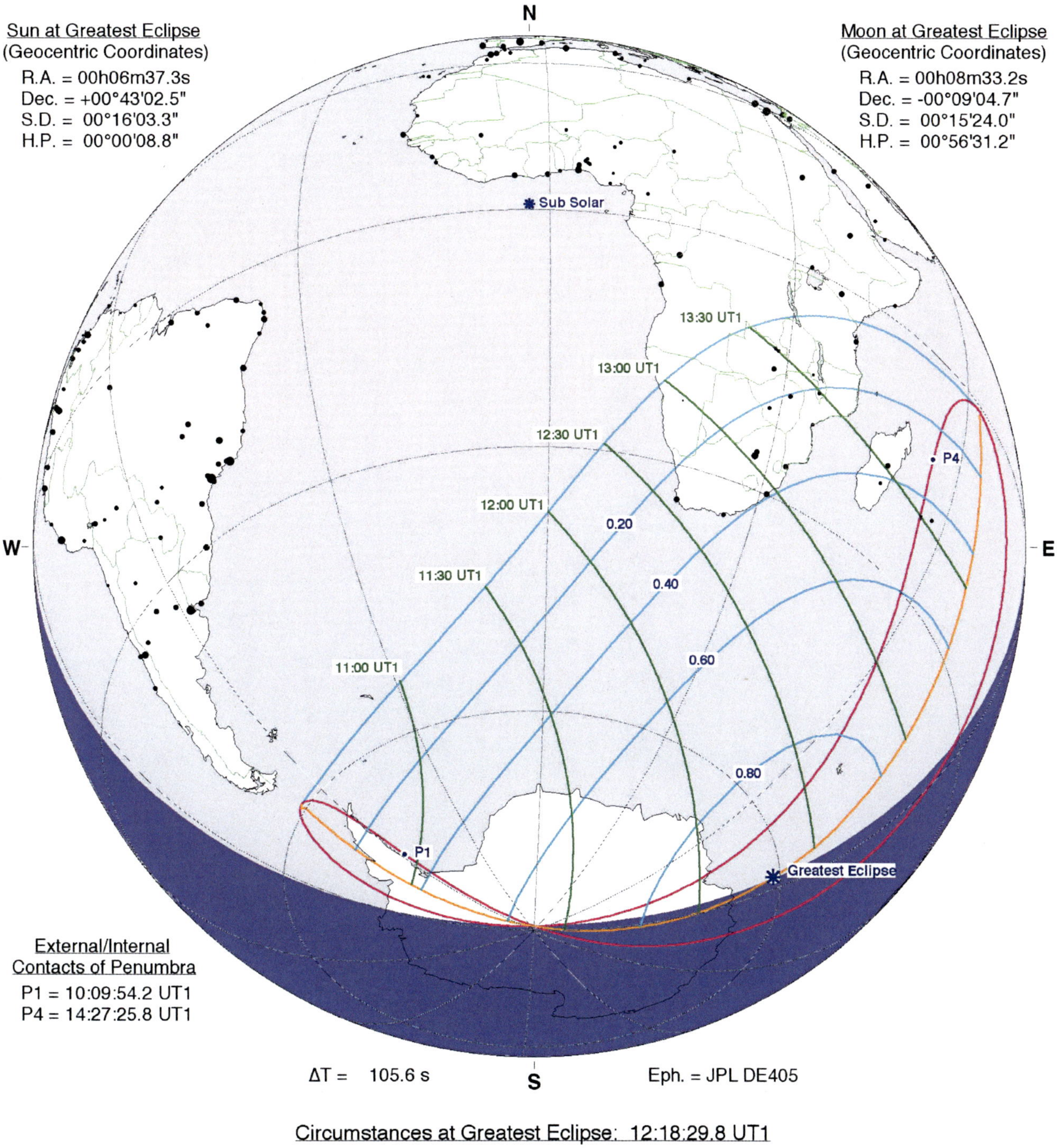

<u>External/Internal Contacts of Penumbra</u>
P1 = 10:09:54.2 UT1
P4 = 14:27:25.8 UT1

ΔT = 105.6 s Eph. = JPL DE405

<u>Circumstances at Greatest Eclipse: 12:18:29.8 UT1</u>

Lat. = 60°55.4'S Sun Alt. = 0.0°
Long. = 085°42.9'E Sun Azm. = 271.5°

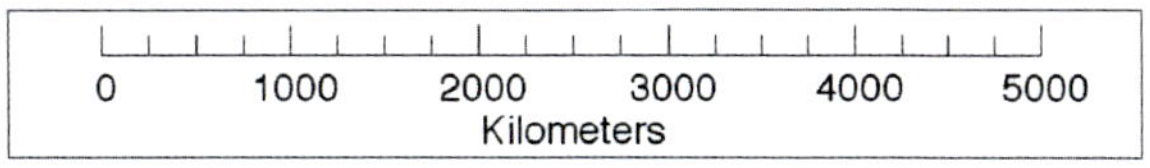

Partial Solar Eclipse of 2080 Sep 13

Greatest Eclipse = 16:38:09.2 TD (= 16:36:23.2 UT1)

Eclipse Magnitude = 0.8743 Saros Series = 126

Gamma = 1.0724 Saros Member = 51 of 72

Sun at Greatest Eclipse
(Geocentric Coordinates)
R.A. = 11h29m55.2s
Dec. = +03°14'46.9"
S.D. = 00°15'53.7"
H.P. = 00°00'08.7"

Moon at Greatest Eclipse
(Geocentric Coordinates)
R.A. = 11h31m59.3s
Dec. = +04°11'17.0"
S.D. = 00°16'24.7"
H.P. = 01°00'13.7"

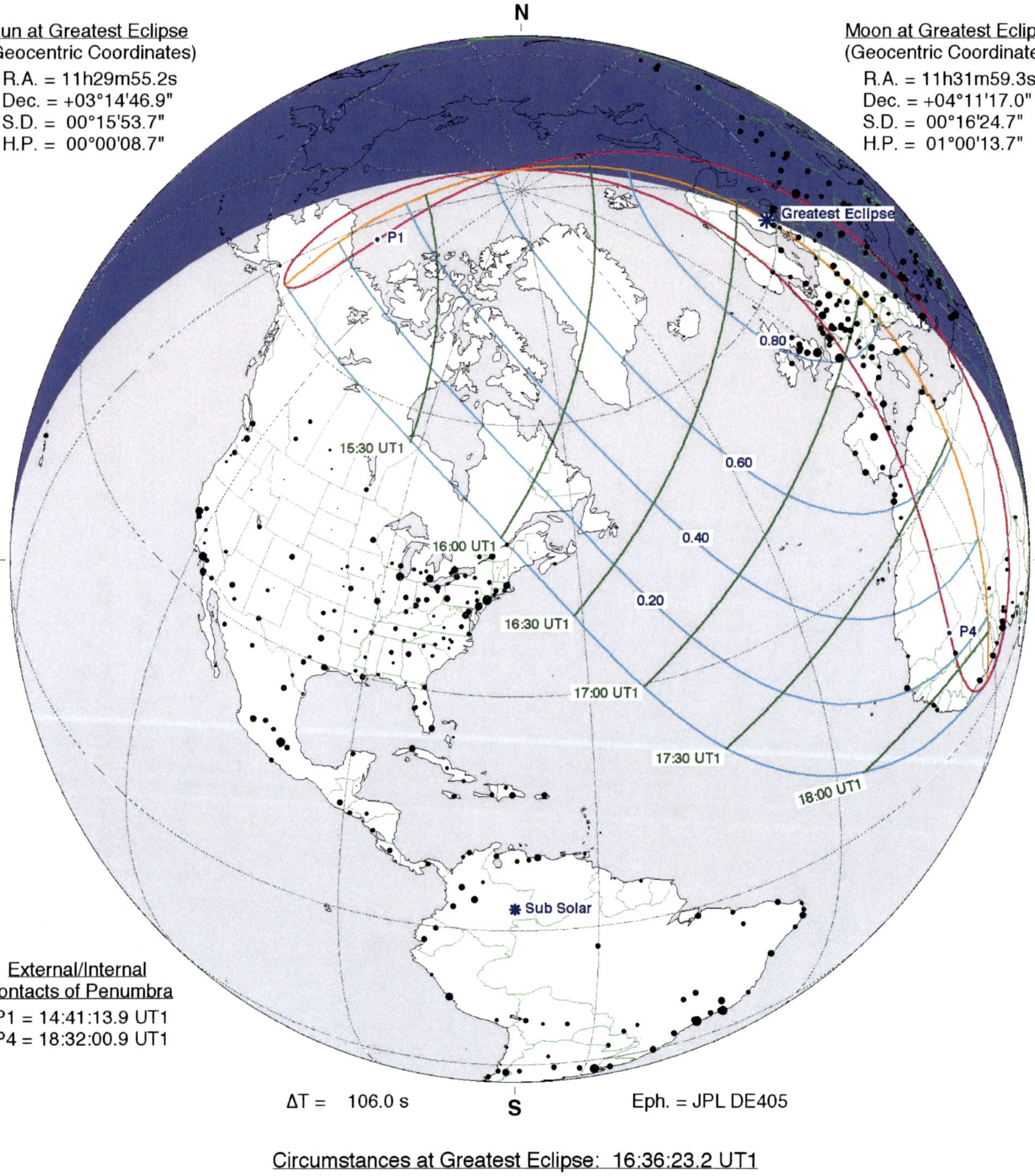

External/Internal
Contacts of Penumbra
P1 = 14:41:13.9 UT1
P4 = 18:32:00.9 UT1

ΔT = 106.0 s

Eph. = JPL DE405

Circumstances at Greatest Eclipse: 16:36:23.2 UT1

Lat. = 61°07.4'N Sun Alt. = 0.0°

Long. = 025°36.9'E Sun Azm. = 276.7°

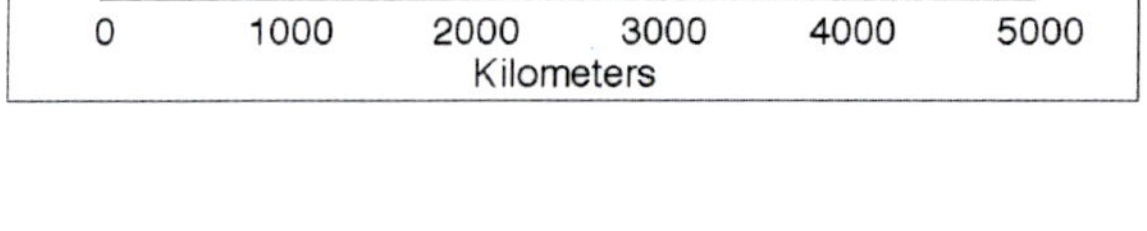

Annular Solar Eclipse of 2081 Mar 10

Greatest Eclipse = 15:23:30.7 TD (= 15:21:44.3 UT1)

Eclipse Magnitude = 0.9304 Saros Series = 131

Gamma = -0.3653 Saros Member = 54 of 70

Sun at Greatest Eclipse
(Geocentric Coordinates)
R.A. = 23h25m55.3s
Dec. = -03°40'25.8"
S.D. = 00°16'06.3"
H.P. = 00°00'08.9"

Moon at Greatest Eclipse
(Geocentric Coordinates)
R.A. = 23h26m33.6s
Dec. = -03°57'43.0"
S.D. = 00°14'46.5"
H.P. = 00°54'13.5"

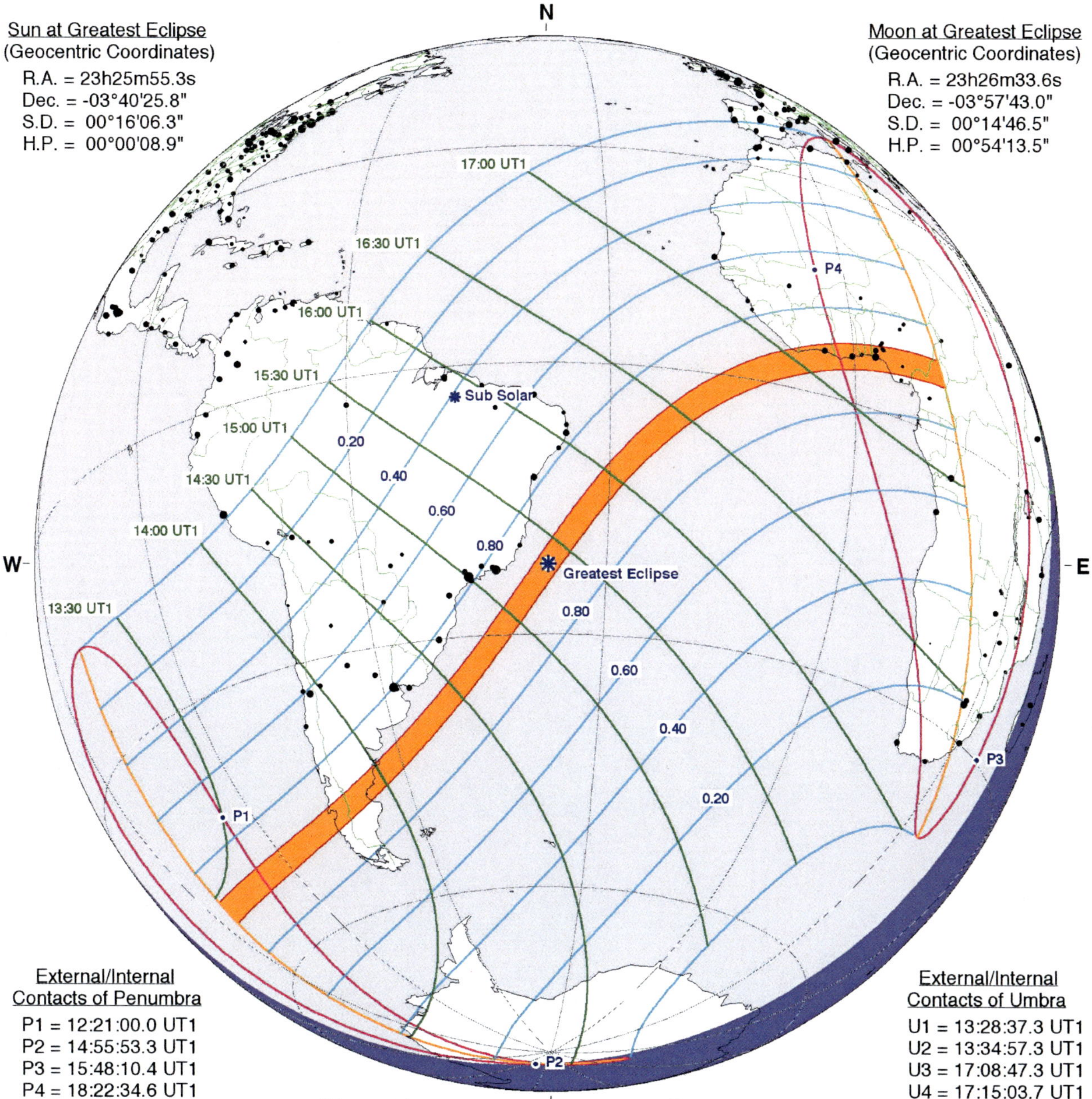

External/Internal
Contacts of Penumbra
P1 = 12:21:00.0 UT1
P2 = 14:55:53.3 UT1
P3 = 15:48:10.4 UT1
P4 = 18:22:34.6 UT1

External/Internal
Contacts of Umbra
U1 = 13:28:37.3 UT1
U2 = 13:34:57.3 UT1
U3 = 17:08:47.3 UT1
U4 = 17:15:03.7 UT1

ΔT = 106.4 s Eph. = JPL DE405

Circumstances at Greatest Eclipse: 15:21:44.3 UT1
Lat. = 22°23.8'S Sun Alt. = 68.4°
Long. = 036°55.8'W Sun Azm. = 328.7°
Path Width = 277.4 km Duration = 07m36.3s

Circumstances at Greatest Duration: 15:13:02.4 UT1
Lat. = 24°36.9'S Sun Alt. = 68.0°
Long. = 038°40.2'W Sun Azm. = 340.8°
Path Width = 275.1 km Duration = 07m36.7s

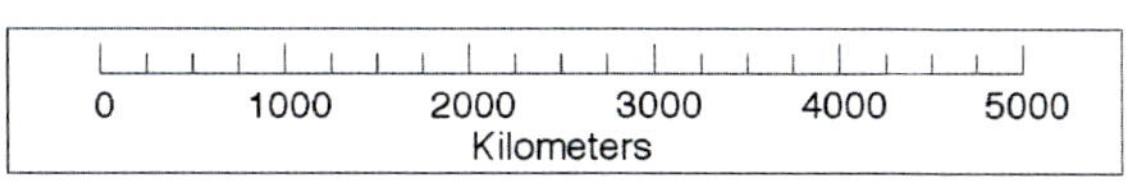

Total Solar Eclipse of 2081 Sep 03

Greatest Eclipse = 09:07:30.5 TD (= 09:05:43.7 UT1)

Eclipse Magnitude = 1.0720 Saros Series = 136

Gamma = 0.3379 Saros Member = 41 of 71

Sun at Greatest Eclipse (Geocentric Coordinates)

R.A. = 10h52m00.4s
Dec. = +07°13'15.0"
S.D. = 00°15'51.2"
H.P. = 00°00'08.7"

Moon at Greatest Eclipse (Geocentric Coordinates)

R.A. = 10h52m39.6s
Dec. = +07°31'30.8"
S.D. = 00°16'43.4"
H.P. = 01°01'22.4"

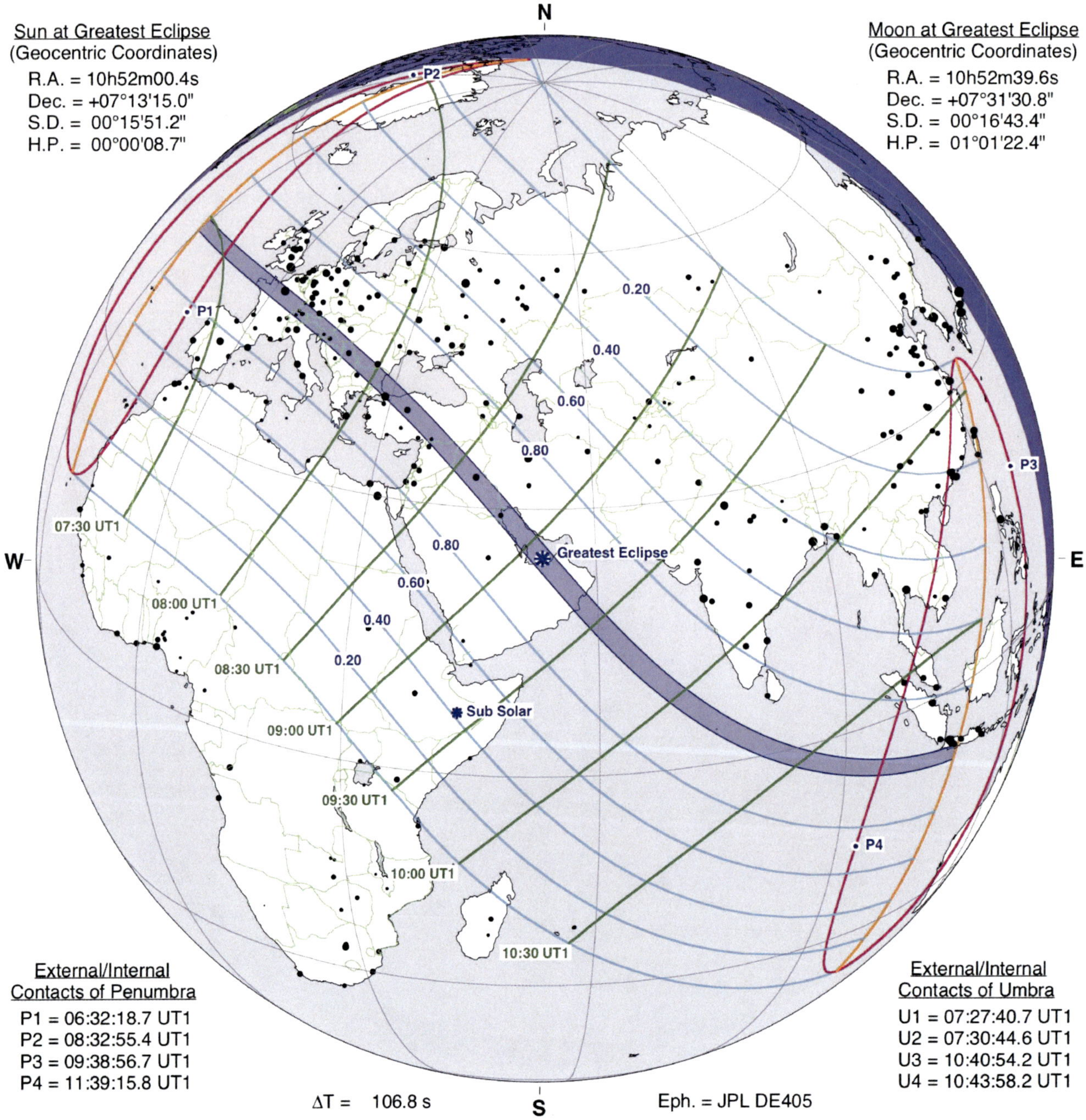

External/Internal Contacts of Penumbra

P1 = 06:32:18.7 UT1
P2 = 08:32:55.4 UT1
P3 = 09:38:56.7 UT1
P4 = 11:39:15.8 UT1

External/Internal Contacts of Umbra

U1 = 07:27:40.7 UT1
U2 = 07:30:44.6 UT1
U3 = 10:40:54.2 UT1
U4 = 10:43:58.2 UT1

ΔT = 106.8 s Eph. = JPL DE405

Circumstances at Greatest Eclipse: 09:05:43.7 UT1

Lat. = 24°37.7'N Sun Alt. = 70.1°
Long. = 053°25.2'E Sun Azm. = 210.6°
Path Width = 247.3 km Duration = 05m33.1s

Circumstances at Greatest Duration: 09:02:43.3 UT1

Lat. = 25°29.5'N Sun Alt. = 70.0°
Long. = 052°34.9'E Sun Azm. = 205.3°
Path Width = 246.6 km Duration = 05m33.2s

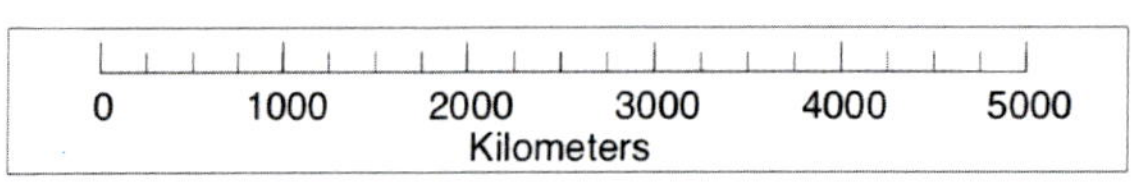

Annular Solar Eclipse of 2082 Feb 27

Greatest Eclipse = 14:46:59.8 TD (= 14:45:12.6 UT1)

Eclipse Magnitude = 0.9298   Saros Series = 141

Gamma = 0.3361   Saros Member = 27 of 70

Sun at Greatest Eclipse
(Geocentric Coordinates)

R.A. = 22h44m00.6s
Dec. = -08°01'49.1"
S.D. = 00°16'09.0"
H.P. = 00°00'08.9"

Moon at Greatest Eclipse
(Geocentric Coordinates)

R.A. = 22h43m26.2s
Dec. = -07°45'43.0"
S.D. = 00°14'48.3"
H.P. = 00°54'19.9"

N

W

E

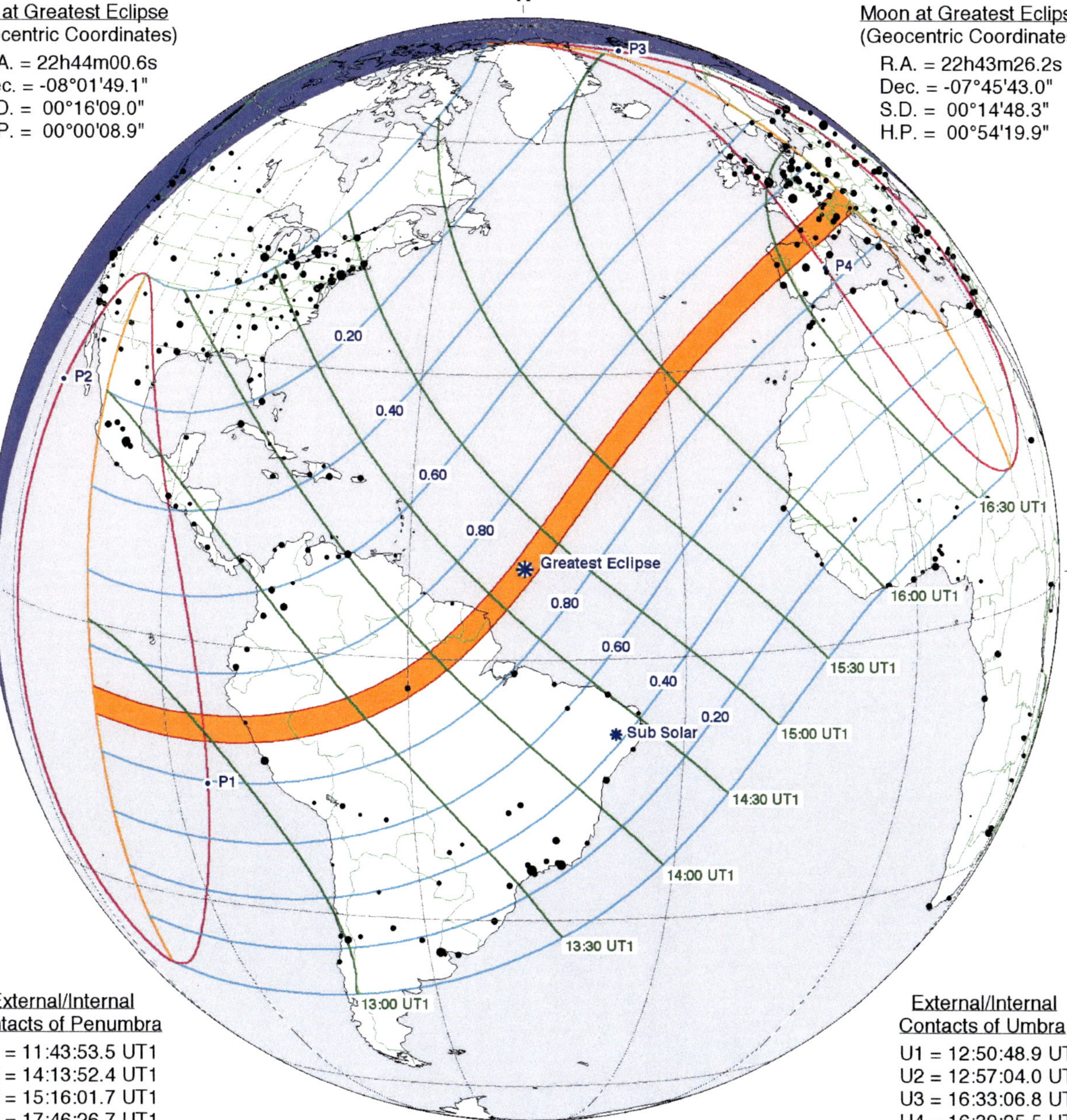

S

External/Internal
Contacts of Penumbra

P1 = 11:43:53.5 UT1
P2 = 14:13:52.4 UT1
P3 = 15:16:01.7 UT1
P4 = 17:46:26.7 UT1

External/Internal
Contacts of Umbra

U1 = 12:50:48.9 UT1
U2 = 12:57:04.0 UT1
U3 = 16:33:06.8 UT1
U4 = 16:39:25.5 UT1

ΔT = 107.2 s

Eph. = JPL DE405

Circumstances at Greatest Eclipse: 14:45:12.6 UT1

Lat. = 09°25.5'N   Sun Alt. = 70.3°
Long. = 047°20.5'W   Sun Azm. = 152.1°
Path Width = 276.9 km   Duration = 08m11.8s

Circumstances at Greatest Duration: 14:32:21.4 UT1

Lat. = 06°32.5'N   Sun Alt. = 69.3°
Long. = 049°44.1'W   Sun Azm. = 134.5°
Path Width = 281.5 km   Duration = 08m12.9s

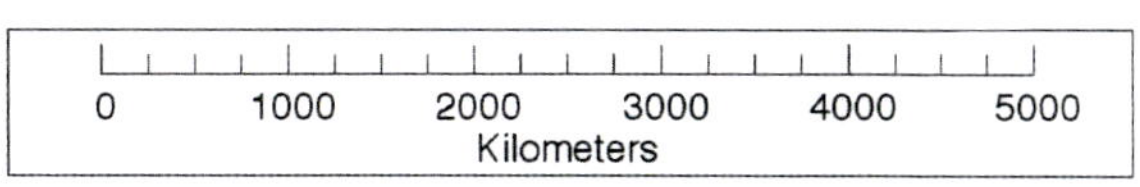

Total Solar Eclipse of 2082 Aug 24

Greatest Eclipse = 01:16:20.6 TD (= 01:14:33.0 UT1)

Eclipse Magnitude = 1.0452
Gamma = -0.4004

Saros Series = 146
Saros Member = 31 of 76

Sun at Greatest Eclipse
(Geocentric Coordinates)
R.A. = 10h13m33.7s
Dec. = +10°59'07.1"
S.D. = 00°15'48.9"
H.P. = 00°00'08.7"

Moon at Greatest Eclipse
(Geocentric Coordinates)
R.A. = 10h12m50.5s
Dec. = +10°37'45.2"
S.D. = 00°16'16.8"
H.P. = 00°59'44.8"

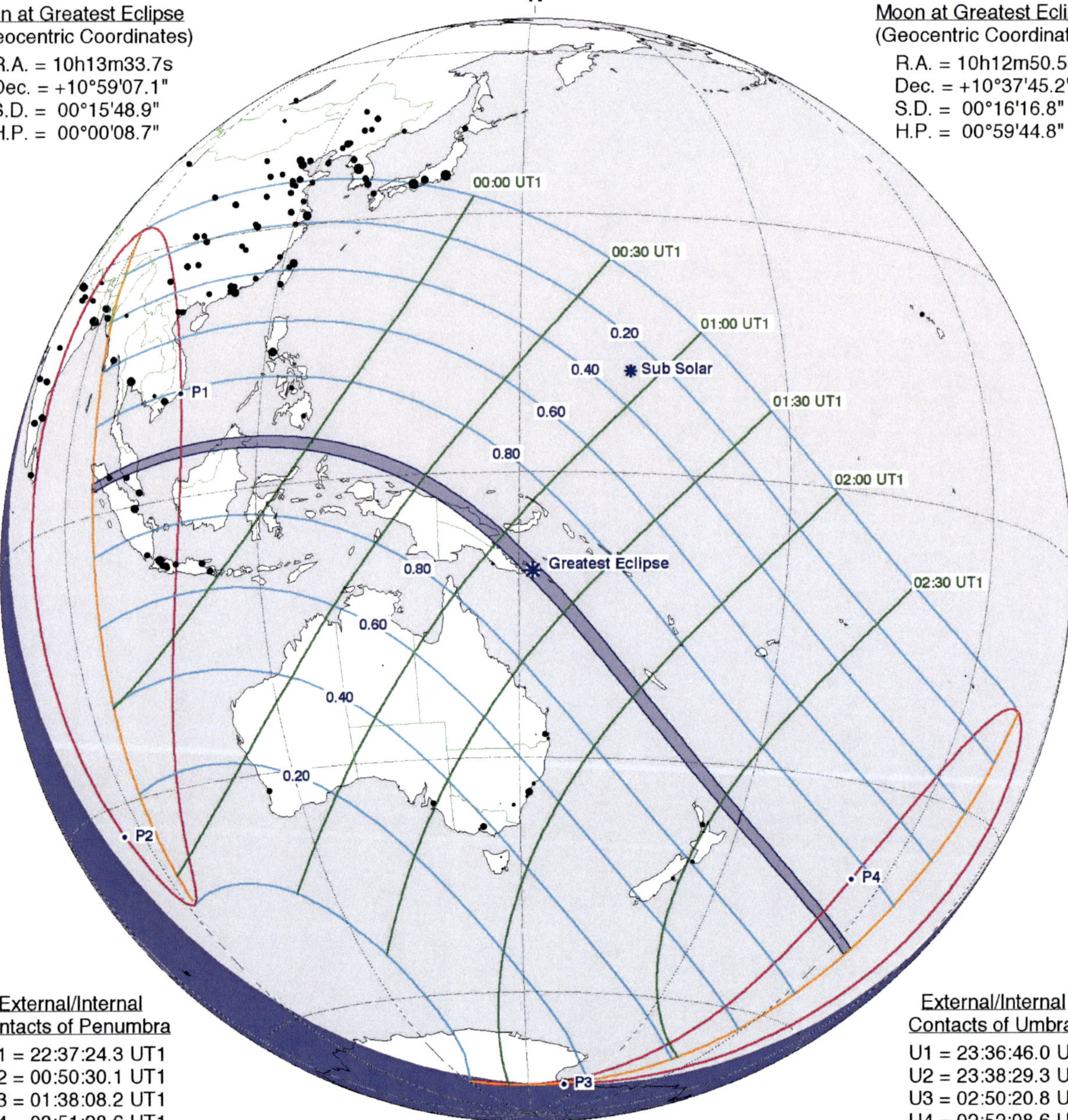

External/Internal
Contacts of Penumbra
P1 = 22:37:24.3 UT1
P2 = 00:50:30.1 UT1
P3 = 01:38:08.2 UT1
P4 = 03:51:28.6 UT1

External/Internal
Contacts of Umbra
U1 = 23:36:46.0 UT1
U2 = 23:38:29.3 UT1
U3 = 02:50:20.8 UT1
U4 = 02:52:08.6 UT1

ΔT = 107.6 s

Eph. = JPL DE405

Circumstances at Greatest Eclipse: 01:14:33.0 UT1

Lat. = 10°16.7'S Sun Alt. = 66.3°
Long. = 151°32.4'E Sun Azm. = 26.3°
Path Width = 162.9 km Duration = 04m01.0s

Circumstances at Greatest Duration: 01:11:33.1 UT1

Lat. = 09°33.4'S Sun Alt. = 66.3°
Long. = 150°48.6'E Sun Azm. = 30.3°
Path Width = 163.4 km Duration = 04m01.1s

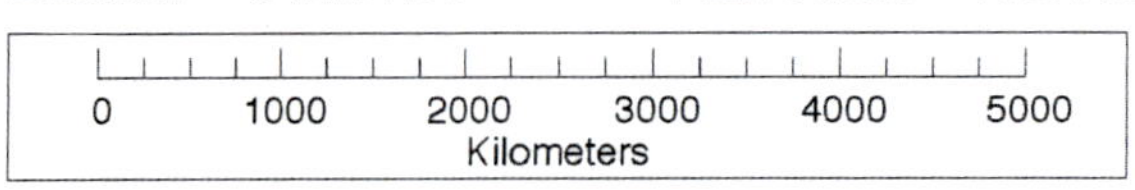

Partial Solar Eclipse of 2083 Feb 16

Greatest Eclipse = 18:06:36.2 TD (= 18:04:48.2 UT1)

Eclipse Magnitude = 0.9433   Saros Series = 151
Gamma = 1.0170   Saros Member = 18 of 72

Sun at Greatest Eclipse
(Geocentric Coordinates)
R.A. = 22h01m38.8s
Dec. = -12°04'40.8"
S.D. = 00°16'11.4"
H.P. = 00°00'08.9"

Moon at Greatest Eclipse
(Geocentric Coordinates)
R.A. = 21h59m56.3s
Dec. = -11°12'50.5"
S.D. = 00°15'28.4"
H.P. = 00°56'47.2"

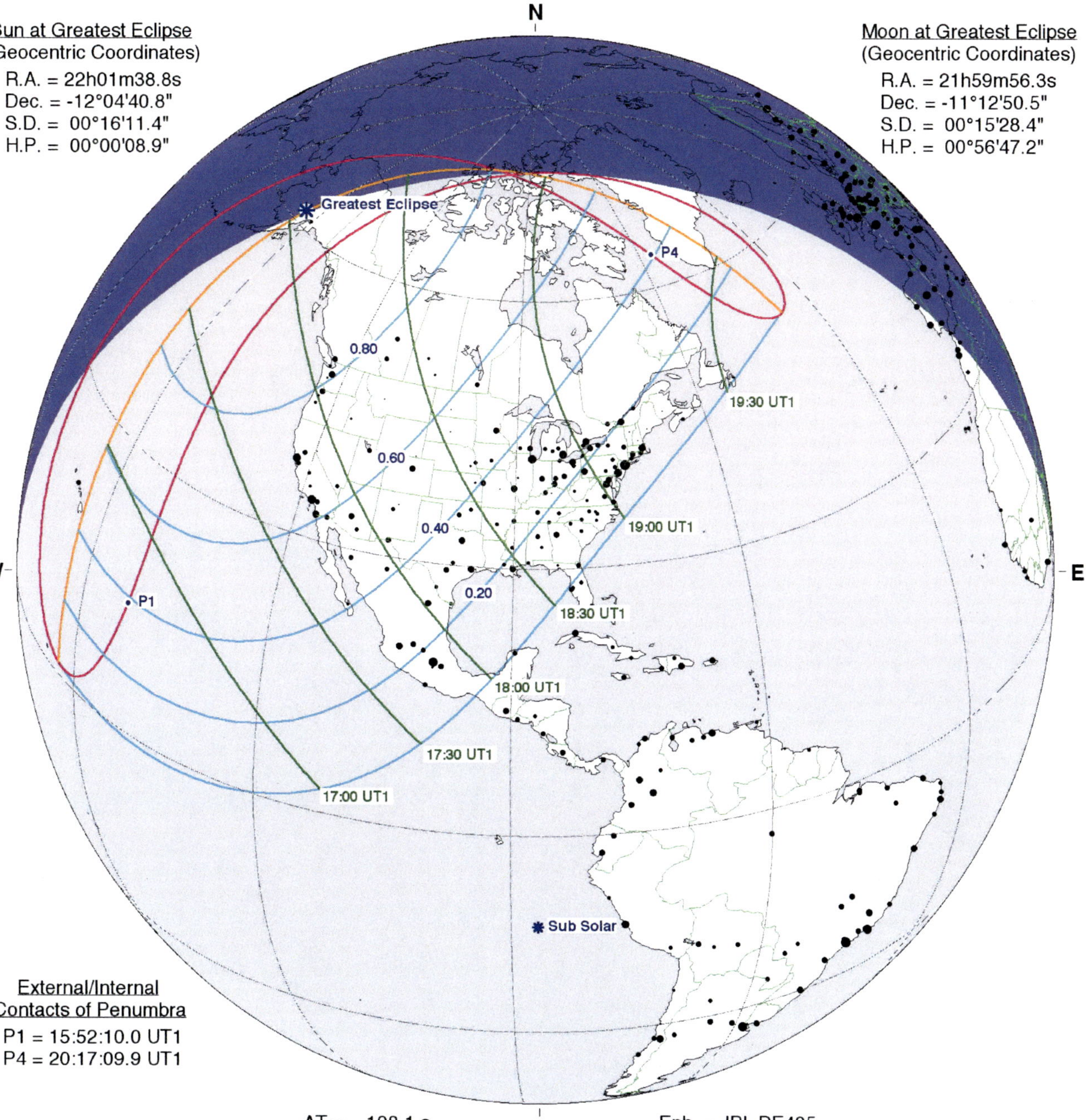

External/Internal
Contacts of Penumbra
P1 = 15:52:10.0 UT1
P4 = 20:17:09.9 UT1

ΔT = 108.1 s   Eph. = JPL DE405

Circumstances at Greatest Eclipse: 18:04:48.2 UT1

Lat. = 61°36.4'N   Sun Alt. = 0.0°
Long. = 154°19.9'W   Sun Azm. = 116.1°

0	1000	2000	3000	4000	5000

Kilometers

Partial Solar Eclipse of 2083 Jul 15

Greatest Eclipse = 00:14:22.9 TD (= 00:12:34.5 UT1)

Eclipse Magnitude = 0.0169
Gamma = 1.5464

Saros Series = 118
Saros Member = 72 of 72

Sun at Greatest Eclipse
(Geocentric Coordinates)
R.A. = 07h38m24.0s
Dec. = +21°30'20.2"
S.D. = 00°15'44.0"
H.P. = 00°00'08.7"

Moon at Greatest Eclipse
(Geocentric Coordinates)
R.A. = 07h40m00.2s
Dec. = +22°52'33.7"
S.D. = 00°15'02.8"
H.P. = 00°55'13.4"

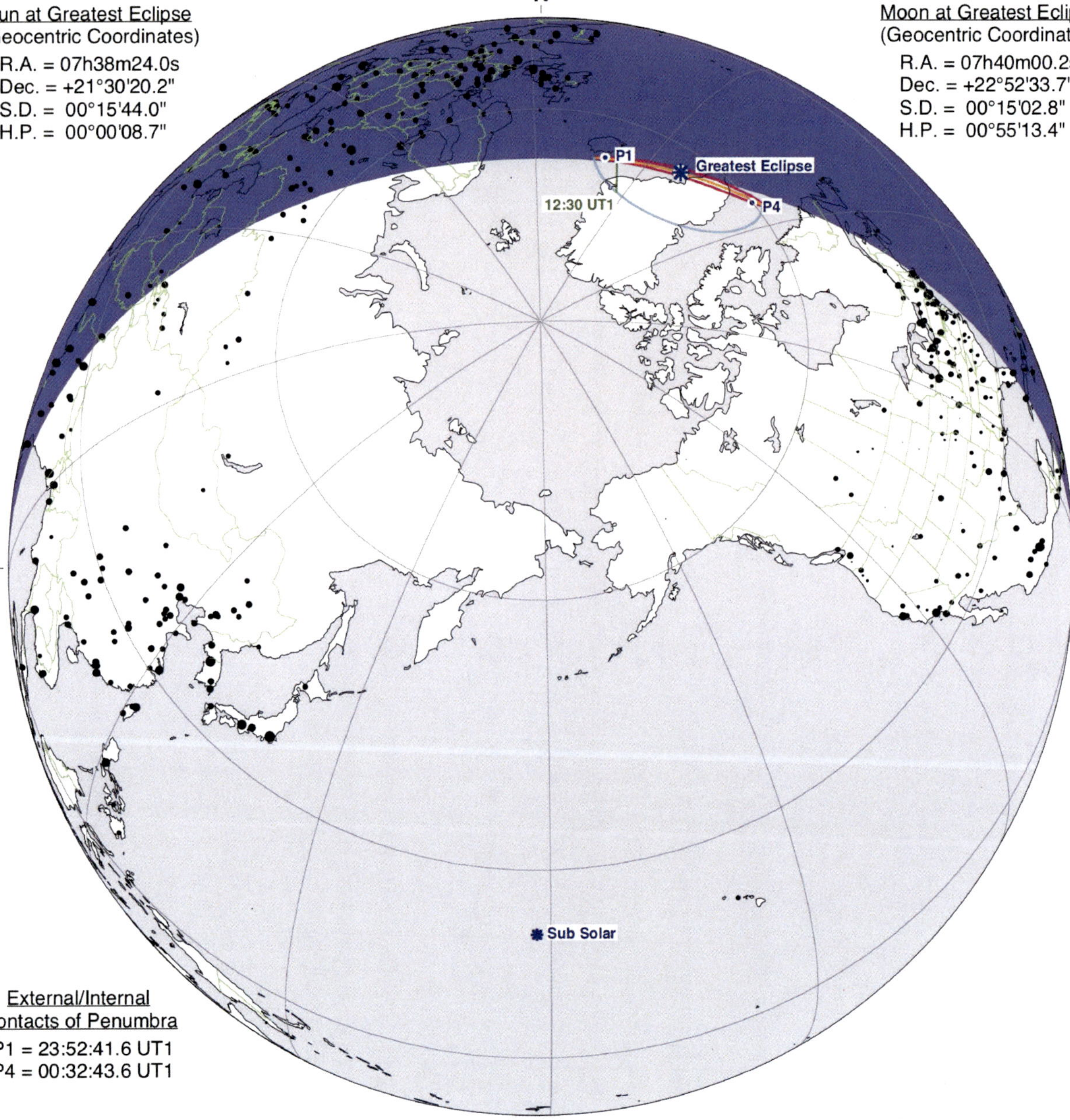

External/Internal
Contacts of Penumbra
P1 = 23:52:41.6 UT1
P4 = 00:32:43.6 UT1

ΔT = 108.4 s

Eph. = JPL DE405

Circumstances at Greatest Eclipse: 00:12:34.5 UT1

Lat. = 63°57.1'N
Long. = 037°57.0'W

Sun Alt. = 0.0°
Sun Azm. = 326.6°

0 1000 2000 3000 4000 5000
Kilometers

©2016 F. Espenak
www.EclipseWise.com

Partial Solar Eclipse of 2083 Aug 13

Greatest Eclipse = 12:34:41.2 TD (= 12:32:52.7 UT1)

Eclipse Magnitude = 0.6146 Saros Series = 156

Gamma = -1.2064 Saros Member = 5 of 69

Sun at Greatest Eclipse
(Geocentric Coordinates)
R.A. = 09h33m34.6s
Dec. = +14°29'20.4"
S.D. = 00°15'47.0"
H.P. = 00°00'08.7"

Moon at Greatest Eclipse
(Geocentric Coordinates)
R.A. = 09h31m40.1s
Dec. = +13°27'06.6"
S.D. = 00°15'25.9"
H.P. = 00°56'37.9"

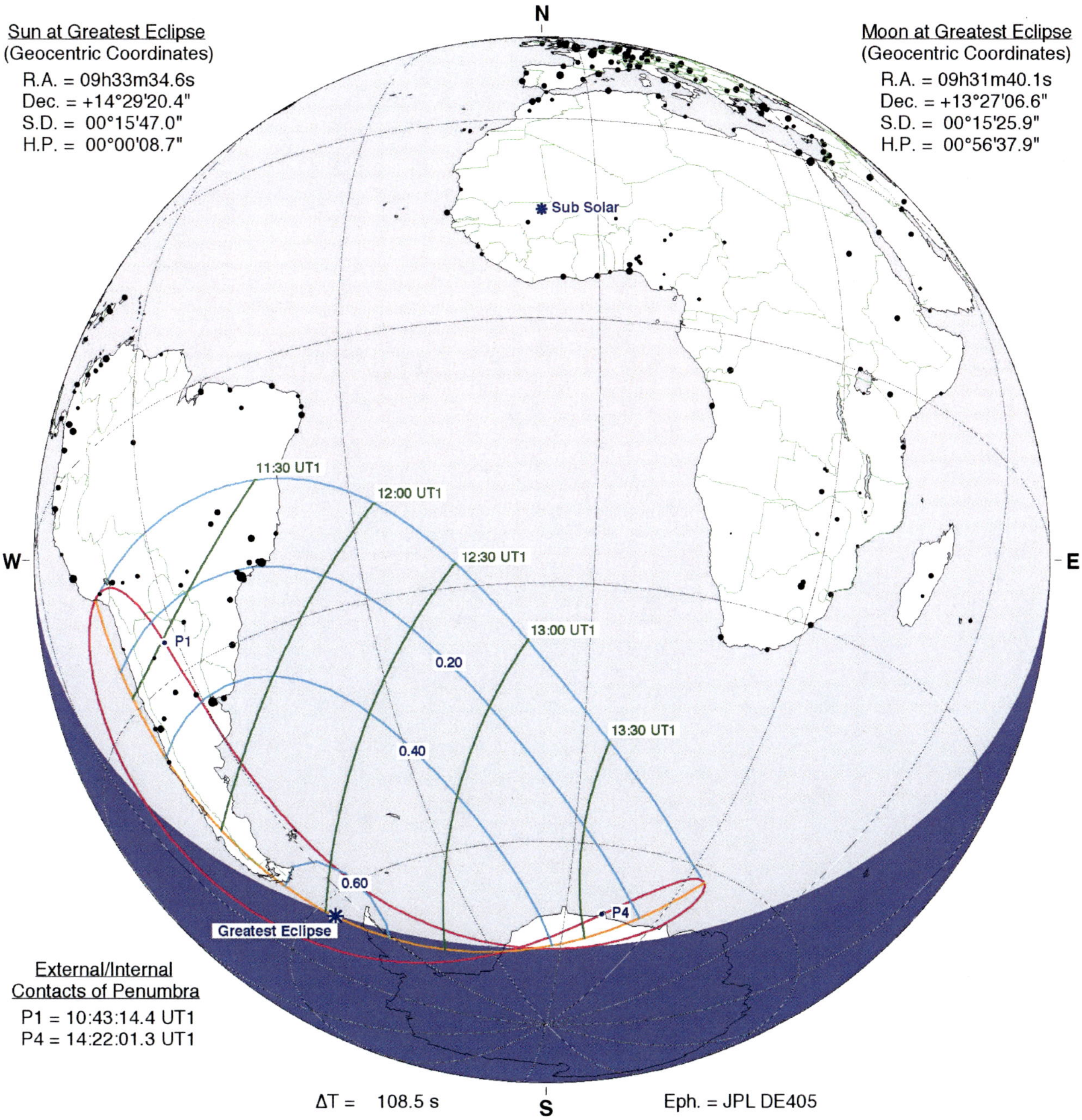

External/Internal
Contacts of Penumbra
P1 = 10:43:14.4 UT1
P4 = 14:22:01.3 UT1

ΔT = 108.5 s Eph. = JPL DE405

Circumstances at Greatest Eclipse: 12:32:52.7 UT1

Lat. = 62°05.6'S Sun Alt. = 0.0°
Long. = 067°42.4'W Sun Azm. = 57.7°

0 1000 2000 3000 4000 5000
Kilometers

Partial Solar Eclipse of 2084 Jan 07

Greatest Eclipse = 17:30:23.5 TD (= 17:28:34.7 UT1)

Eclipse Magnitude = 0.8723
Gamma = -1.0715

Saros Series = 123
Saros Member = 57 of 70

Sun at Greatest Eclipse
(Geocentric Coordinates)
R.A. = 19h15m11.8s
Dec. = -22°18'17.9"
S.D. = 00°16'15.9"
H.P. = 00°00'08.9"

Moon at Greatest Eclipse
(Geocentric Coordinates)
R.A. = 19h16m15.2s
Dec. = -23°21'56.0"
S.D. = 00°16'38.6"
H.P. = 01°01'05.0"

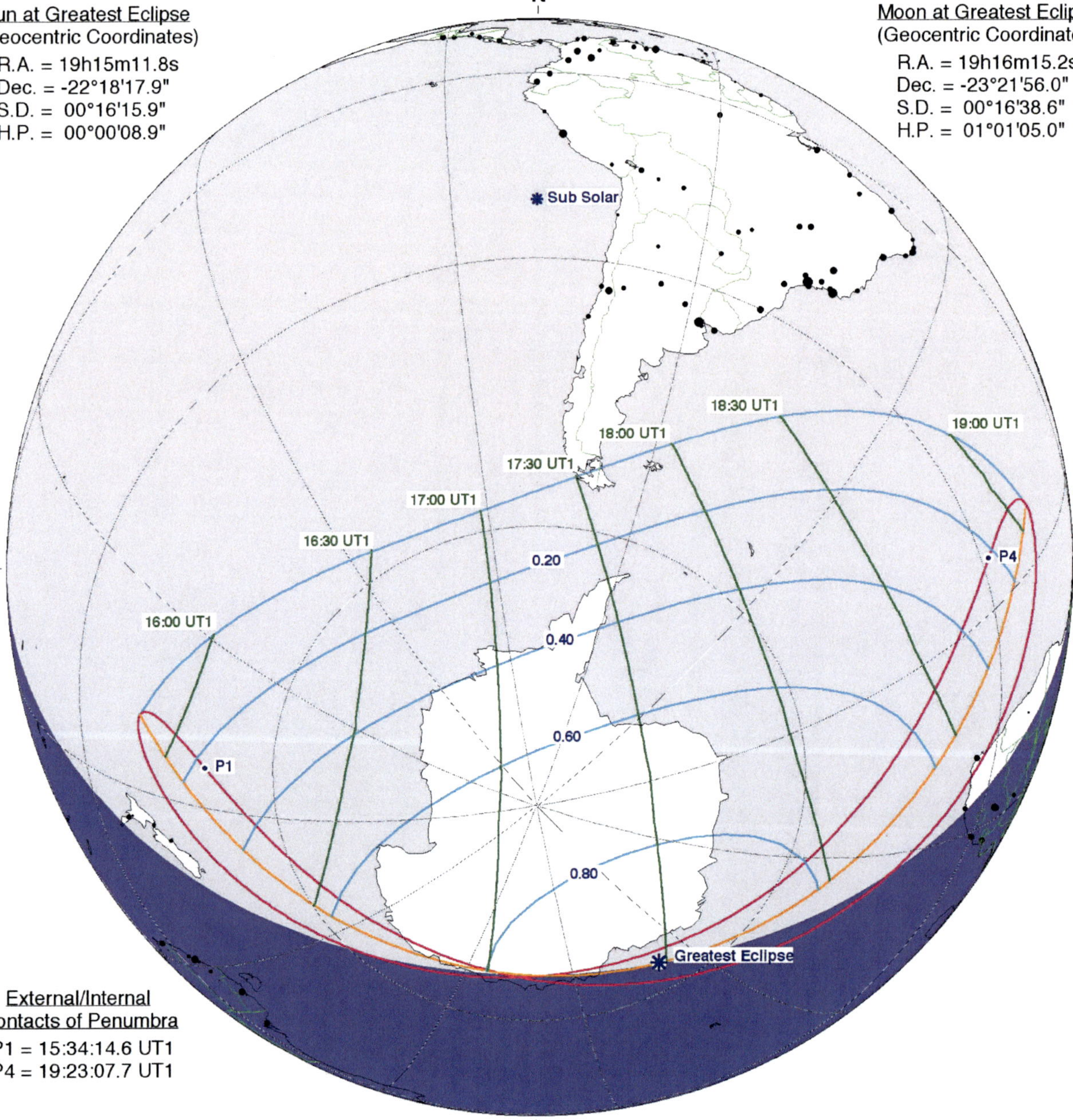

External/Internal
Contacts of Penumbra
P1 = 15:34:14.6 UT1
P4 = 19:23:07.7 UT1

ΔT = 108.8 s

Eph. = JPL DE405

Circumstances at Greatest Eclipse: 17:28:34.7 UT1

Lat. = 64°24.6'S
Long. = 068°17.7'E

Sun Alt. = 0.0°
Sun Azm. = 208.5°

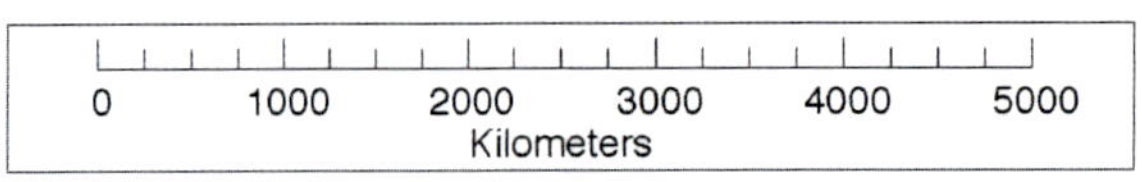

Annular Solar Eclipse of 2084 Jul 03

Greatest Eclipse = 01:50:25.9 TD (= 01:48:36.6 UT1)

Eclipse Magnitude = 0.9421 Saros Series = 128

Gamma = 0.8208 Saros Member = 62 of 73

Sun at Greatest Eclipse
(Geocentric Coordinates)
R.A. = 06h52m43.5s
Dec. = +22°52'33.4"
S.D. = 00°15'43.9"
H.P. = 00°00'08.6"

Moon at Greatest Eclipse
(Geocentric Coordinates)
R.A. = 06h53m20.0s
Dec. = +23°35'54.8"
S.D. = 00°14'41.9"
H.P. = 00°53'56.6"

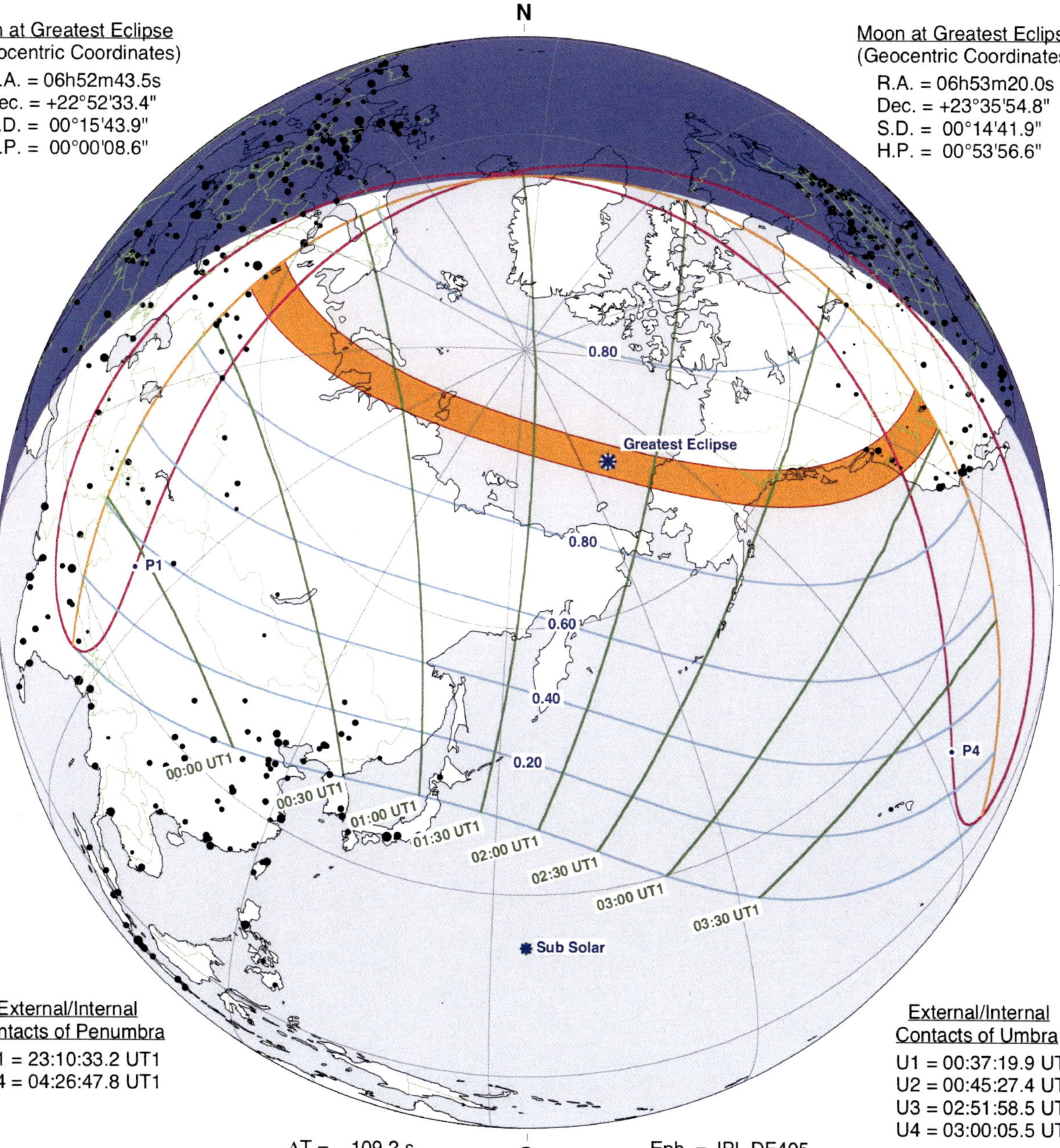

External/Internal Contacts of Penumbra
P1 = 23:10:33.2 UT1
P4 = 04:26:47.8 UT1

External/Internal Contacts of Umbra
U1 = 00:37:19.9 UT1
U2 = 00:45:27.4 UT1
U3 = 02:51:58.5 UT1
U4 = 03:00:05.5 UT1

ΔT = 109.2 s Eph. = JPL DE405

Circumstances at Greatest Eclipse: 01:48:36.6 UT1

Lat. = 75°00.6'N Sun Alt. = 34.5°
Long. = 169°17.6'W Sun Azm. = 222.0°
Path Width = 377.4 km Duration = 04m25.1s

Circumstances at Greatest Duration: 01:45:34.2 UT1

Lat. = 75°58.3'N Sun Alt. = 34.5°
Long. = 172°56.6'W Sun Azm. = 216.7°
Path Width = 376.5 km Duration = 04m25.2s

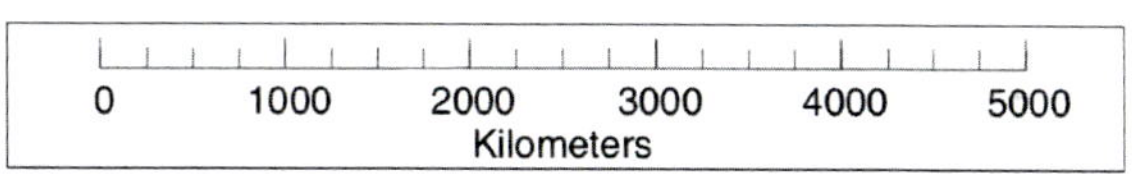

Total Solar Eclipse of 2084 Dec 27

Greatest Eclipse = 09:13:48.0 TD (= 09:11:58.3 UT1)

Eclipse Magnitude = 1.0396
Gamma = -0.4094

Saros Series = 133
Saros Member = 49 of 72

Sun at Greatest Eclipse
(Geocentric Coordinates)
R.A. = 18h28m34.2s
Dec. = -23°15'58.0"
S.D. = 00°16'15.7"
H.P. = 00°00'08.9"

Moon at Greatest Eclipse
(Geocentric Coordinates)
R.A. = 18h28m50.3s
Dec. = -23°40'38.6"
S.D. = 00°16'38.7"
H.P. = 01°01'05.5"

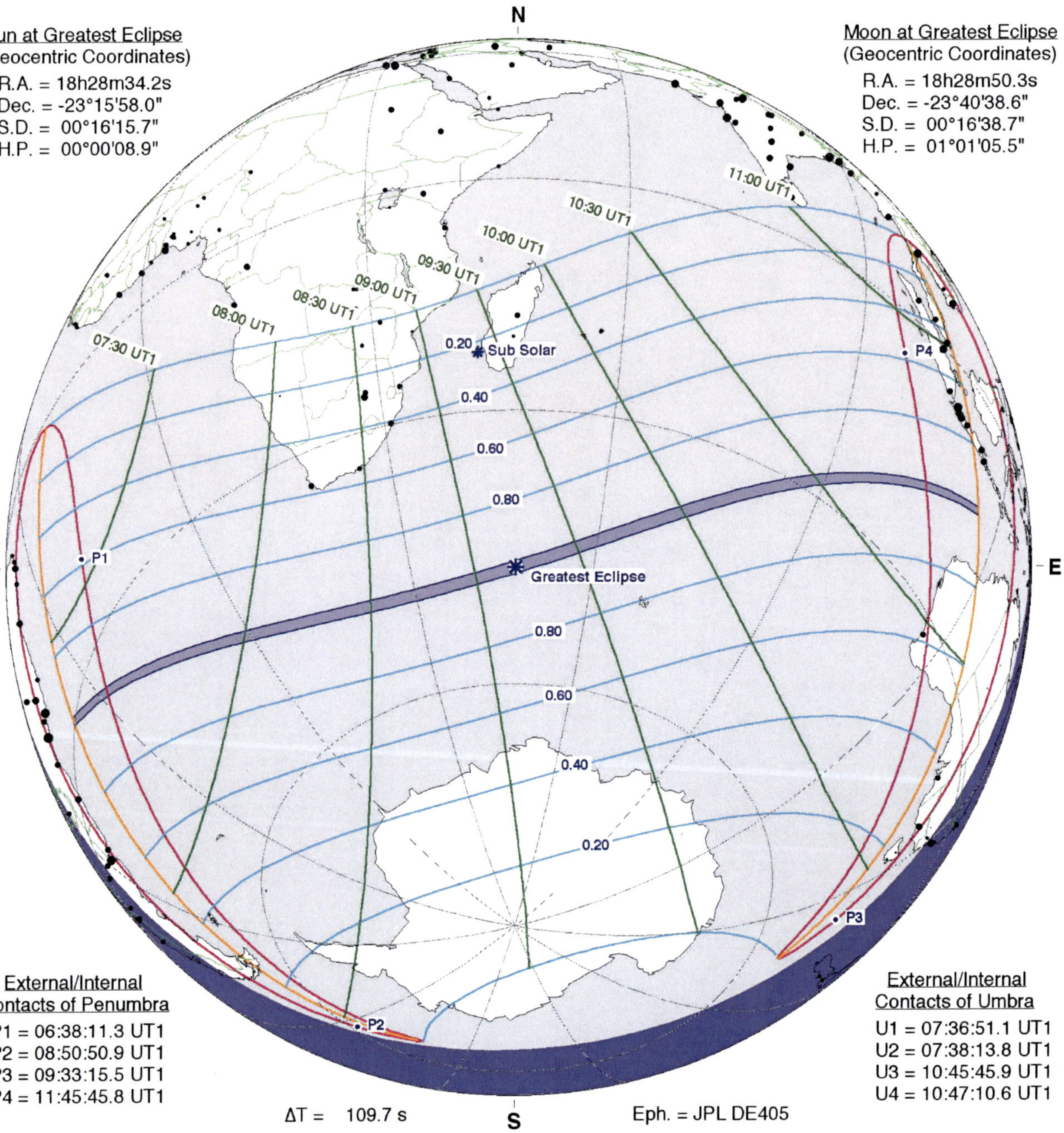

External/Internal
Contacts of Penumbra
P1 = 06:38:11.3 UT1
P2 = 08:50:50.9 UT1
P3 = 09:33:15.5 UT1
P4 = 11:45:45.8 UT1

External/Internal
Contacts of Umbra
U1 = 07:36:51.1 UT1
U2 = 07:38:13.8 UT1
U3 = 10:45:45.9 UT1
U4 = 10:47:10.6 UT1

ΔT = 109.7 s

Eph. = JPL DE405

Circumstances at Greatest Eclipse: 09:11:58.3 UT1

Lat. = 47°19.4'S	Sun Alt. = 65.6°
Long. = 047°26.9'E	Sun Azm. = 348.6°
Path Width = 145.7 km	Duration = 03m04.4s

Circumstances at Greatest Duration: 09:10:59.4 UT1

Lat. = 47°25.8'S	Sun Alt. = 65.6°
Long. = 046°54.5'E	Sun Azm. = 350.3°
Path Width = 145.7 km	Duration = 03m04.4s

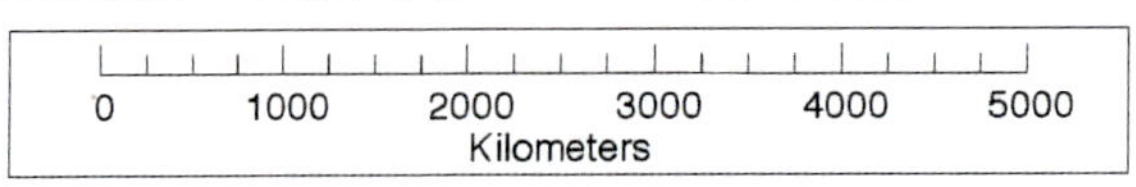

Annular Solar Eclipse of 2085 Jun 22

Greatest Eclipse = 03:21:15.9 TD (= 03:19:25.8 UT1)

Eclipse Magnitude = 0.9704 — Saros Series = 138

Gamma = 0.0453 — Saros Member = 35 of 70

Sun at Greatest Eclipse
(Geocentric Coordinates)
R.A. = 06h06m22.6s
Dec. = +23°25'12.0"
S.D. = 00°15'44.3"
H.P. = 00°00'08.7"

Moon at Greatest Eclipse
(Geocentric Coordinates)
R.A. = 06h06m23.8s
Dec. = +23°27'40.5"
S.D. = 00°15'02.4"
H.P. = 00°55'11.8"

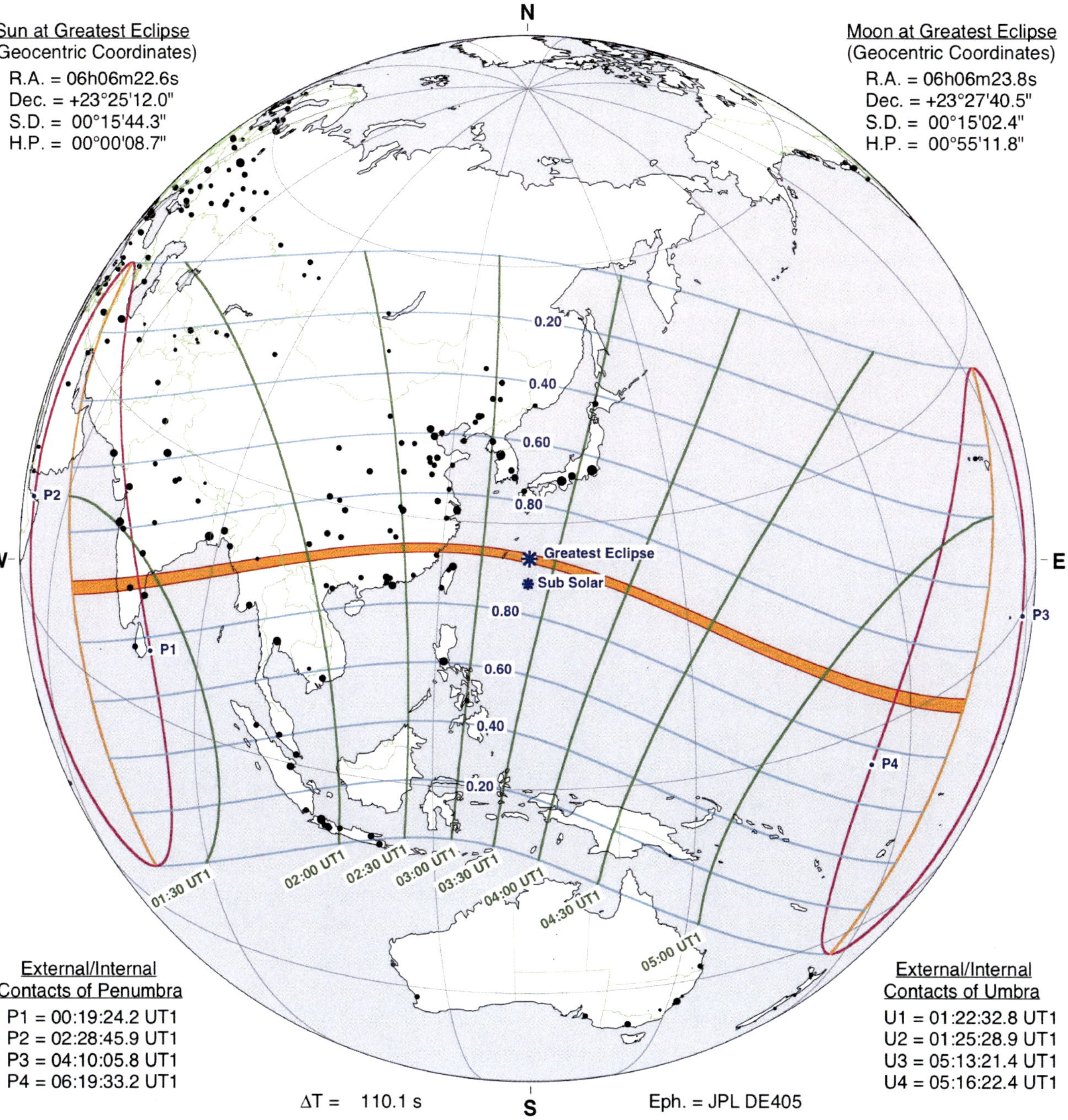

External/Internal Contacts of Penumbra
P1 = 00:19:24.2 UT1
P2 = 02:28:45.9 UT1
P3 = 04:10:05.8 UT1
P4 = 06:19:33.2 UT1

External/Internal Contacts of Umbra
U1 = 01:22:32.8 UT1
U2 = 01:25:28.9 UT1
U3 = 05:13:21.4 UT1
U4 = 05:16:22.4 UT1

ΔT = 110.1 s — Eph. = JPL DE405

Circumstances at Greatest Eclipse: 03:19:25.8 UT1

Lat. = 26°09.1'N	Sun Alt. = 87.3°
Long. = 131°01.3'E	Sun Azm. = 186.1°
Path Width = 106.4 km	Duration = 03m29.0s

Circumstances at Greatest Duration: 03:15:53.1 UT1

Lat. = 26°20.3'N	Sun Alt. = 86.7°
Long. = 129°57.6'E	Sun Azm. = 152.8°
Path Width = 106.3 km	Duration = 03m29.0s

0 1000 2000 3000 4000 5000
Kilometers

Annular Solar Eclipse of 2085 Dec 16

Greatest Eclipse = 22:37:47.8 TD (= 22:35:57.3 UT1)

Eclipse Magnitude = 0.9971 Saros Series = 143

Gamma = 0.2786 Saros Member = 27 of 72

Sun at Greatest Eclipse
(Geocentric Coordinates)

R.A. = 17h41m09.8s
Dec. = -23°21'25.3"
S.D. = 00°16'15.0"
H.P. = 00°00'08.9"

Moon at Greatest Eclipse
(Geocentric Coordinates)

R.A. = 17h41m05.1s
Dec. = -23°05'11.3"
S.D. = 00°15'57.1"
H.P. = 00°58'32.5"

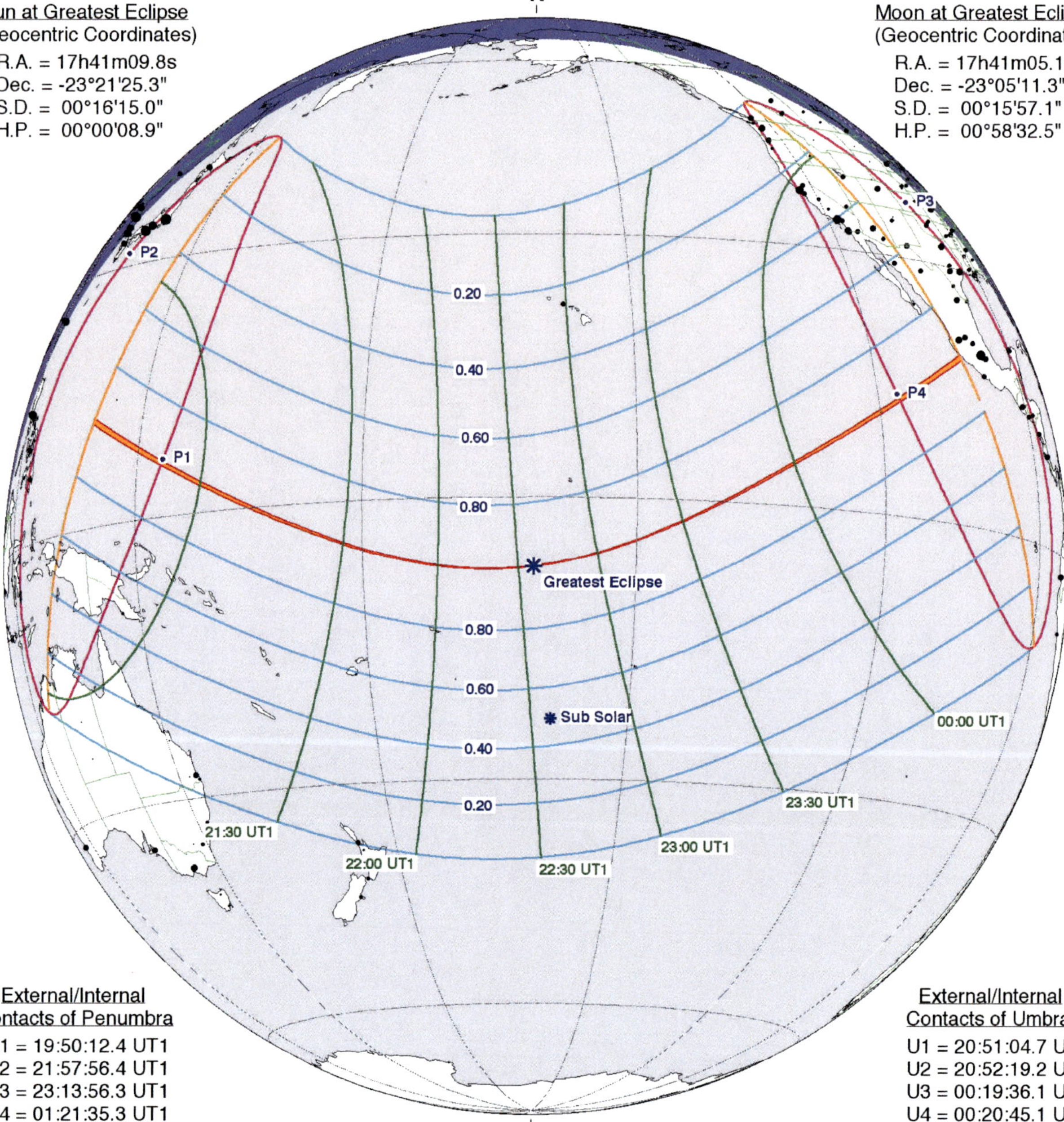

External/Internal Contacts of Penumbra

P1 = 19:50:12.4 UT1
P2 = 21:57:56.4 UT1
P3 = 23:13:56.3 UT1
P4 = 01:21:35.3 UT1

External/Internal Contacts of Umbra

U1 = 20:51:04.7 UT1
U2 = 20:52:19.2 UT1
U3 = 00:19:36.1 UT1
U4 = 00:20:45.1 UT1

ΔT = 110.5 s Eph. = JPL DE405

Circumstances at Greatest Eclipse: 22:35:57.3 UT1

Lat. = 07°15.5'S Sun Alt. = 73.9°
Long. = 161°03.2'W Sun Azm. = 176.5°
Path Width = 10.4 km Duration = 00m19.2s

Circumstances at Greatest Duration: 20:51:41.9 UT1

Lat. = 11°25.9'N Sun Alt. = 0.0°
Long. = 141°05.0'E Sun Azm. = 113.9°
Path Width = 70.3 km Duration = 01m14.1s

0 1000 2000 3000 4000 5000
Kilometers

Total Solar Eclipse of 2086 Jun 11

Greatest Eclipse = 11:07:13.9 TD (= 11:05:22.9 UT1)

Eclipse Magnitude = 1.0174 Saros Series = 148

Gamma = -0.7215 Saros Member = 25 of 75

Sun at Greatest Eclipse
(Geocentric Coordinates)
R.A. = 05h20m59.8s
Dec. = +23°07'28.2"
S.D. = 00°15'45.2"
H.P. = 00°00'08.7"

Moon at Greatest Eclipse
(Geocentric Coordinates)
R.A. = 05h20m54.1s
Dec. = +22°25'37.5"
S.D. = 00°15'51.0"
H.P. = 00°58'10.3"

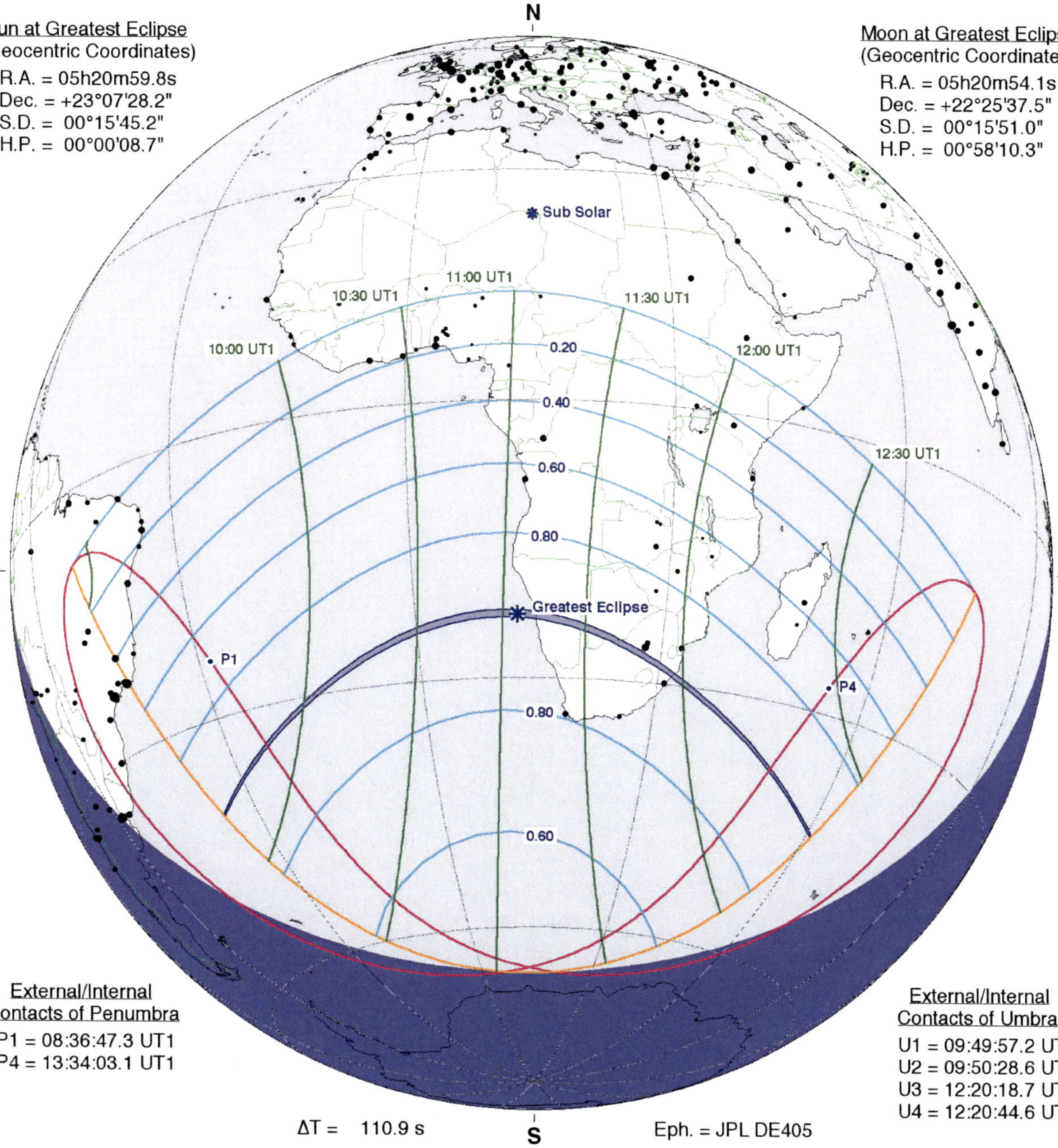

External/Internal
Contacts of Penumbra
P1 = 08:36:47.3 UT1
P4 = 13:34:03.1 UT1

External/Internal
Contacts of Umbra
U1 = 09:49:57.2 UT1
U2 = 09:50:28.6 UT1
U3 = 12:20:18.7 UT1
U4 = 12:20:44.6 UT1

ΔT = 110.9 s Eph. = JPL DE405

Circumstances at Greatest Eclipse: 11:05:22.9 UT1

Lat. = 23°12.1'S	Sun Alt. = 43.7°
Long. = 012°13.5'E	Sun Azm. = 1.8°
Path Width = 85.9 km	Duration = 01m48.1s

Circumstances at Greatest Duration: 11:03:17.5 UT1

Lat. = 23°10.4'S	Sun Alt. = 43.6°
Long. = 011°33.2'E	Sun Azm. = 3.3°
Path Width = 86.0 km	Duration = 01m48.1s

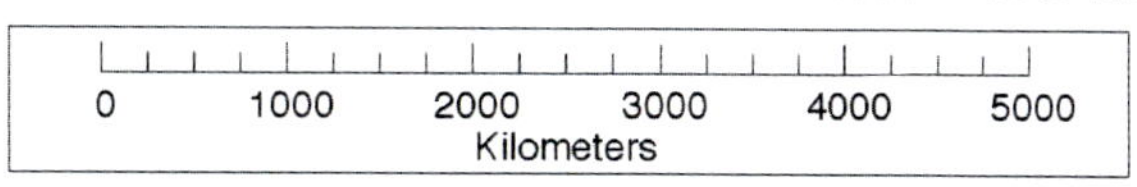

Partial Solar Eclipse of 2086 Dec 06

Greatest Eclipse = 05:38:55.4 TD (= 05:37:04.0 UT1)

Eclipse Magnitude = 0.9271 Saros Series = 153

Gamma = 1.0194 Saros Member = 13 of 70

Sun at Greatest Eclipse
(Geocentric Coordinates)
R.A. = 16h52m56.6s
Dec. = -22°31'57.0"
S.D. = 00°16'13.7"
H.P. = 00°00'08.9"

Moon at Greatest Eclipse
(Geocentric Coordinates)
R.A. = 16h53m00.7s
Dec. = -21°35'36.6"
S.D. = 00°15'06.1"
H.P. = 00°55'25.4"

N

W

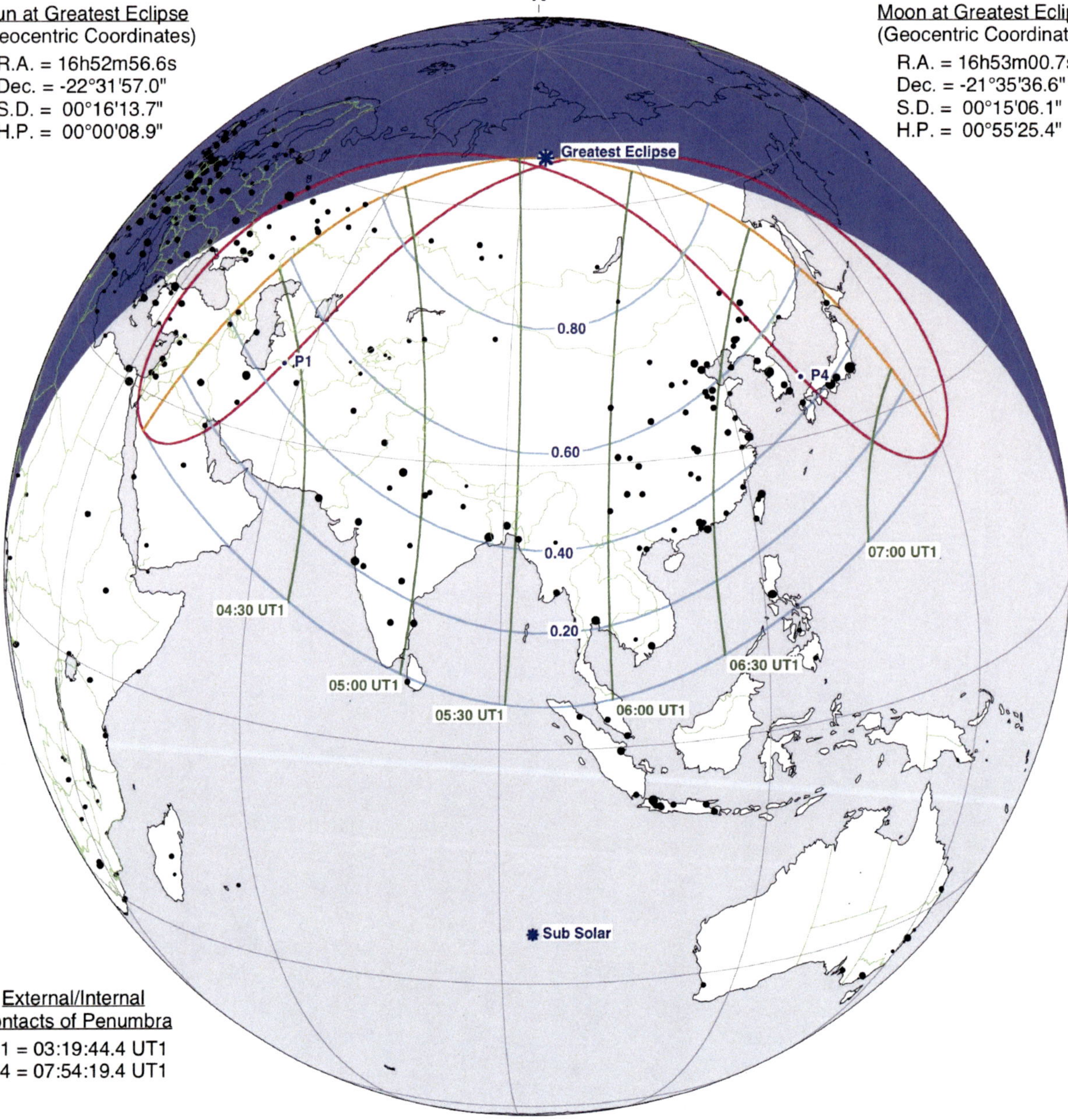

E

External/Internal
Contacts of Penumbra
P1 = 03:19:44.4 UT1
P4 = 07:54:19.4 UT1

ΔT = 111.4 s

S

Eph. = JPL DE405

Circumstances at Greatest Eclipse: 05:37:04.0 UT1

Lat. = 67°26.7'N Sun Alt. = 0.0°

Long. = 095°59.5'E Sun Azm. = 182.3°

0	1000	2000	3000	4000	5000

Kilometers

Partial Solar Eclipse of 2087 May 02

Greatest Eclipse = 18:04:42.0 TD (= 18:02:50.3 UT1)

Eclipse Magnitude = 0.8011 Saros Series = 120

Gamma = 1.1139 Saros Member = 65 of 71

Sun at Greatest Eclipse
(Geocentric Coordinates)
R.A. = 02h40m34.0s
Dec. = +15°36'24.7"
S.D. = 00°15'52.3"
H.P. = 00°00'08.7"

Moon at Greatest Eclipse
(Geocentric Coordinates)
R.A. = 02h39m33.9s
Dec. = +16°43'04.4"
S.D. = 00°16'43.4"
H.P. = 01°01'22.6"

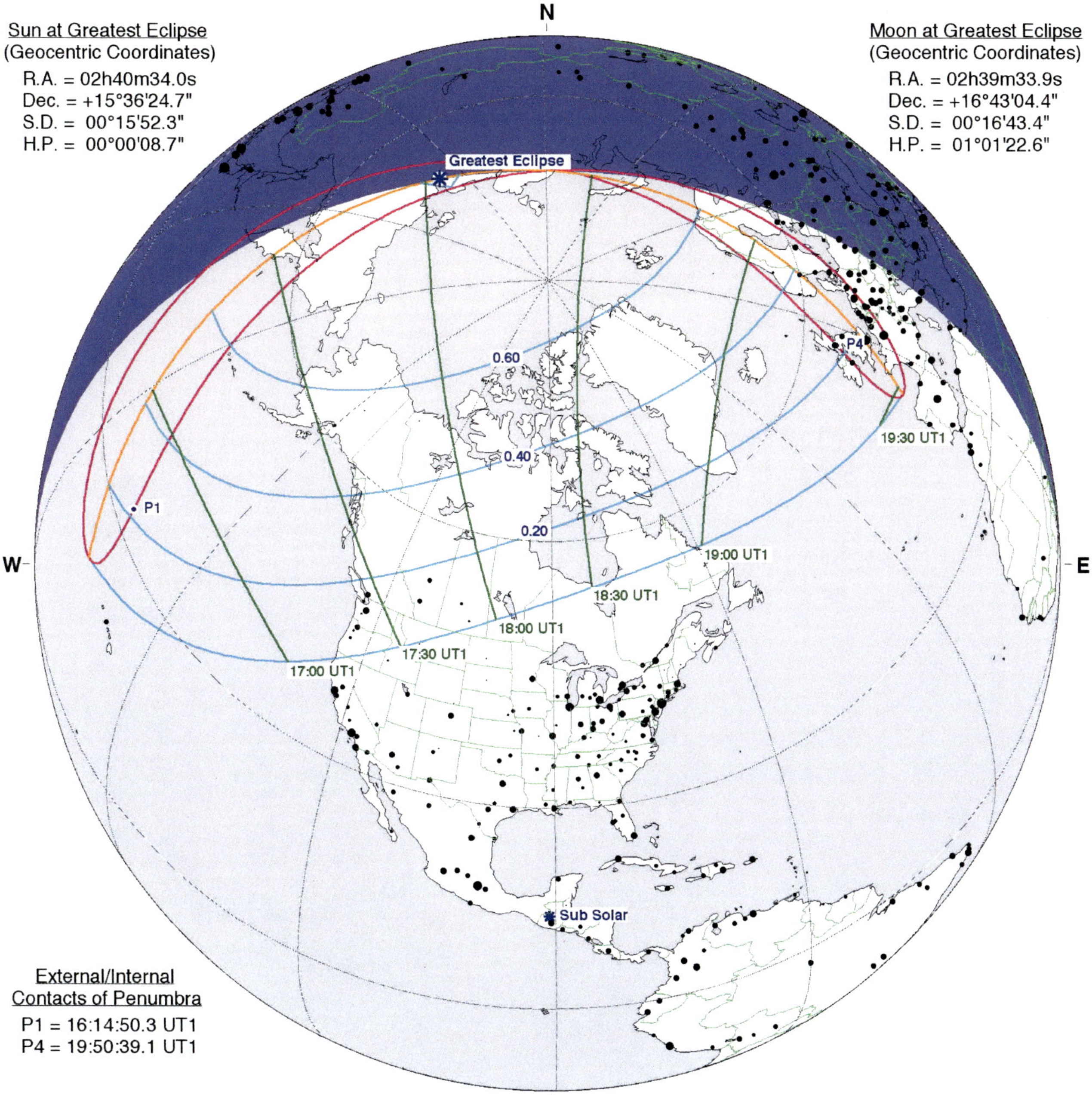

External/Internal
Contacts of Penumbra
P1 = 16:14:50.3 UT1
P4 = 19:50:39.1 UT1

ΔT = 111.7 s Eph. = JPL DE405

Circumstances at Greatest Eclipse: 18:02:50.3 UT1

Lat. = 70°17.6'N Sun Alt. = 0.0°

Long. = 127°20.2'E Sun Azm. = 37.1°

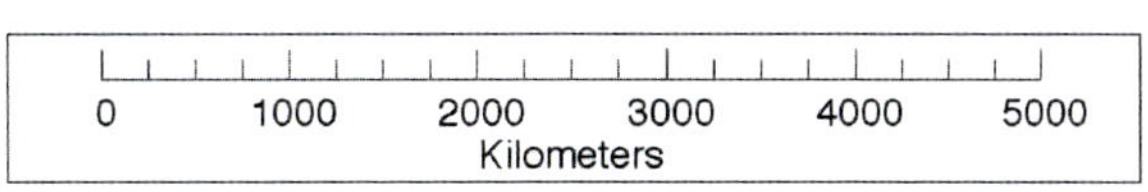

Partial Solar Eclipse of 2087 Jun 01

Greatest Eclipse = 01:27:14.3 TD (= 01:25:22.5 UT1)

Eclipse Magnitude = 0.2146
Gamma = -1.4186

Saros Series = 158
Saros Member = 2 of 70

Sun at Greatest Eclipse
(Geocentric Coordinates)
R.A. = 04h37m04.0s
Dec. = +22°03'32.2"
S.D. = 00°15'46.5"
H.P. = 00°00'08.7"

Moon at Greatest Eclipse
(Geocentric Coordinates)
R.A. = 04h37m20.9s
Dec. = +20°37'32.1"
S.D. = 00°16'34.5"
H.P. = 01°00'49.8"

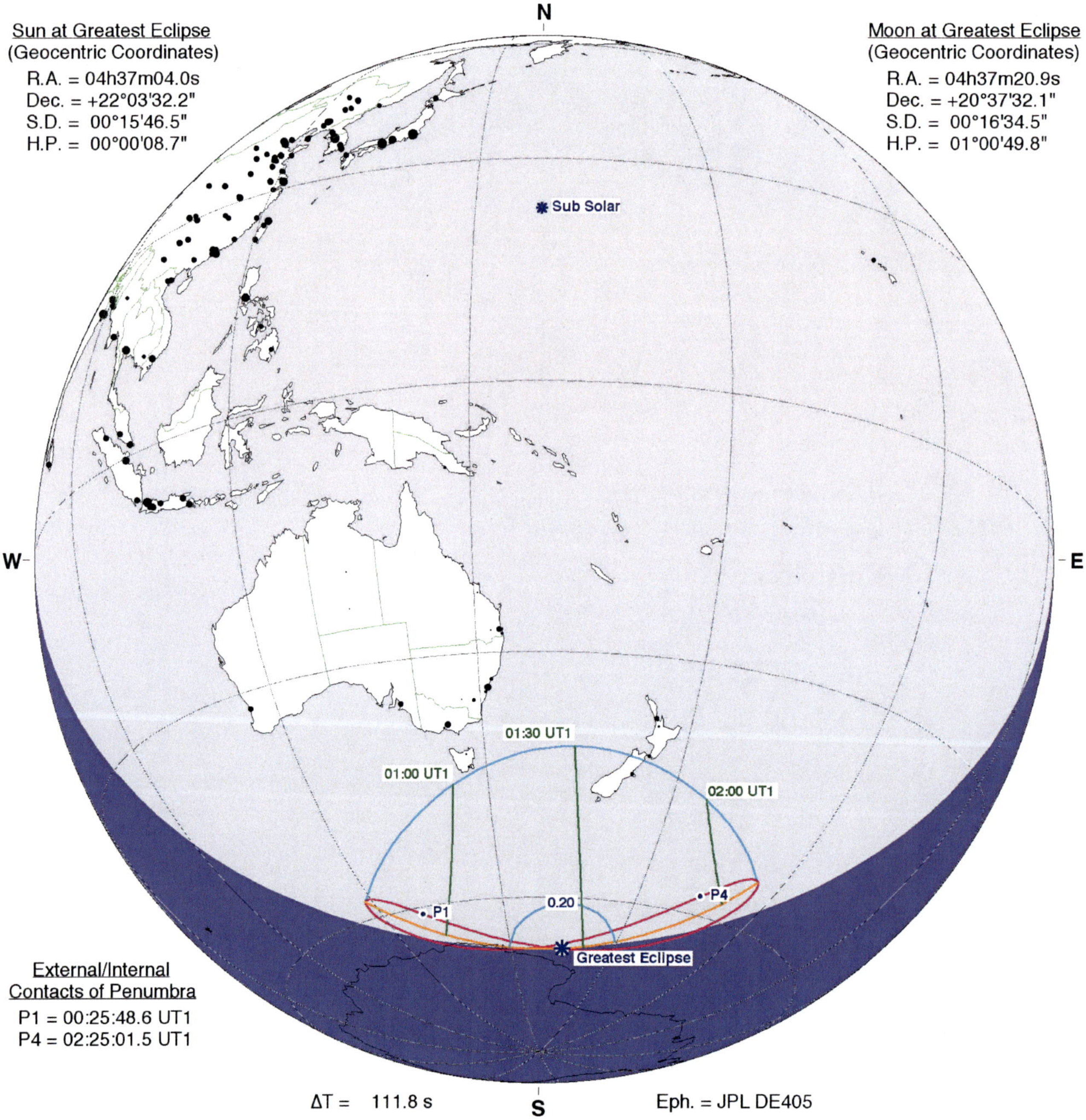

External/Internal
Contacts of Penumbra
P1 = 00:25:48.6 UT1
P4 = 02:25:01.5 UT1

ΔT = 111.8 s

Eph. = JPL DE405

Circumstances at Greatest Eclipse: 01:25:22.5 UT1

Lat. = 67°47.3'S
Long. = 165°07.4'E

Sun Alt. = 0.0°
Sun Azm. = 353.5°

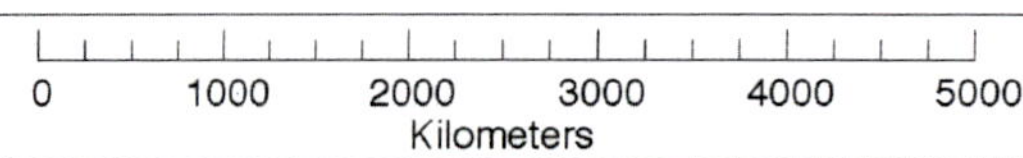

Partial Solar Eclipse of 2087 Oct 26

Greatest Eclipse = 11:46:56.7 TD (= 11:45:04.6 UT1)

Eclipse Magnitude = 0.4696

Gamma = -1.2882

Saros Series = 125

Saros Member = 58 of 73

Sun at Greatest Eclipse
(Geocentric Coordinates)

R.A. = 14h04m17.3s
Dec. = -12°36'18.7"
S.D. = 00°16'05.1"
H.P. = 00°00'08.8"

Moon at Greatest Eclipse
(Geocentric Coordinates)

R.A. = 14h03m06.4s
Dec. = -13°43'47.4"
S.D. = 00°14'46.3"
H.P. = 00°54'12.8"

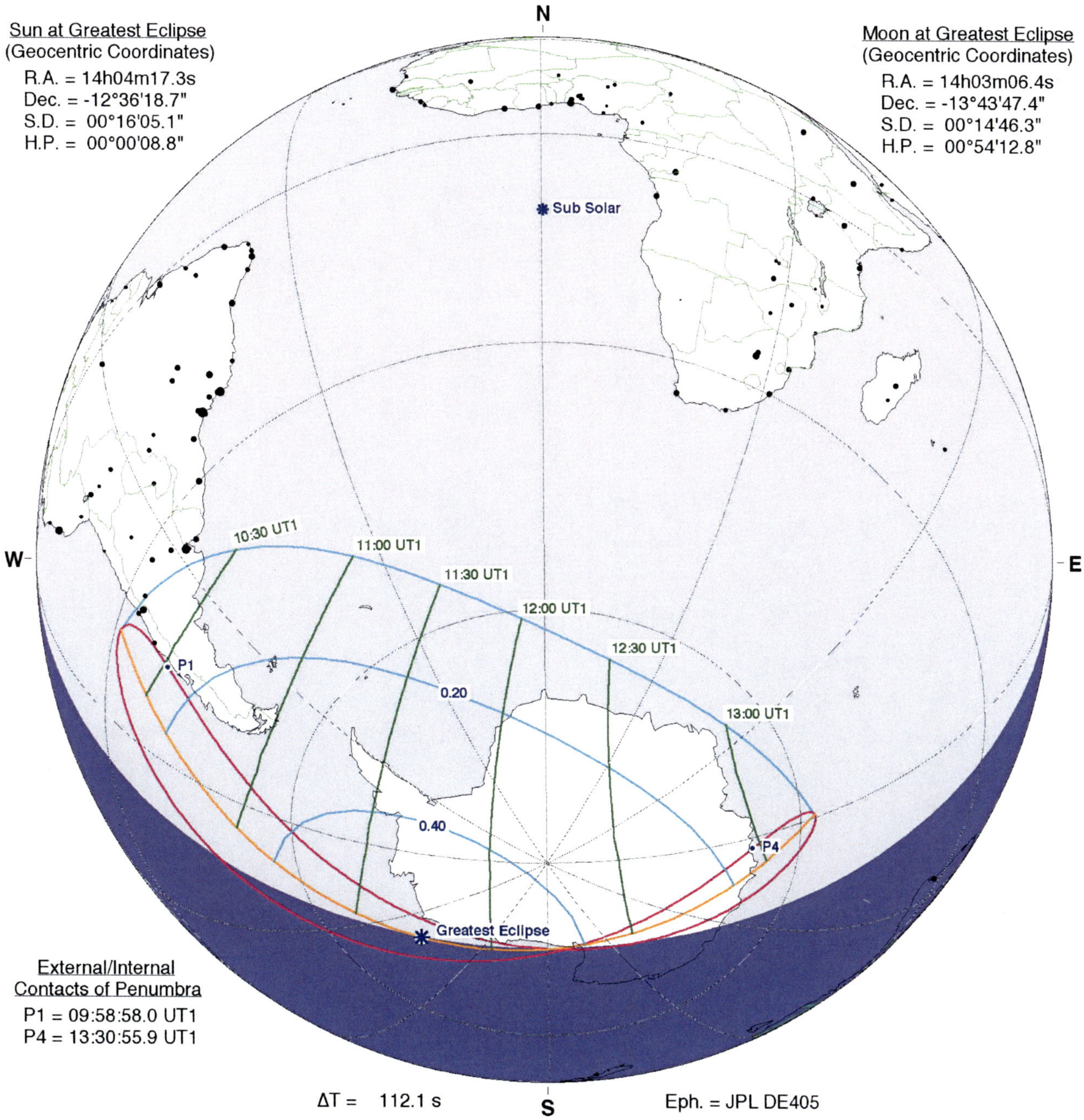

External/Internal
Contacts of Penumbra

P1 = 09:58:58.0 UT1
P4 = 13:30:55.9 UT1

ΔT = 112.1 s

Eph. = JPL DE405

Circumstances at Greatest Eclipse: 11:45:04.6 UT1

Lat. = 71°01.3'S

Long. = 130°47.4'W

Sun Alt. = 0.0°

Sun Azm. = 132.1°

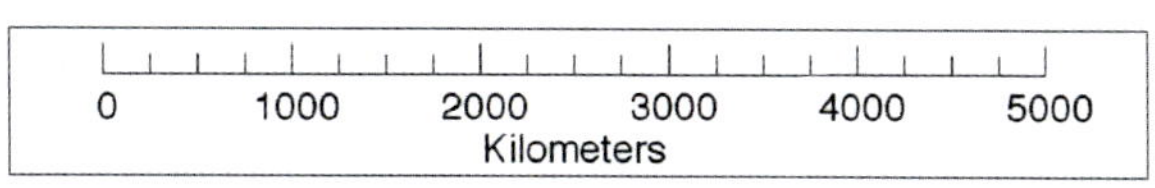

www.EclipseWise.com

Total Solar Eclipse of 2088 Apr 21

Greatest Eclipse = 10:31:49.5 TD (= 10:29:56.9 UT1)

Eclipse Magnitude = 1.0474
Gamma = 0.4135

Saros Series = 130
Saros Member = 56 of 73

Sun at Greatest Eclipse
(Geocentric Coordinates)
R.A. = 02h00m30.3s
Dec. = +12°16'07.2"
S.D. = 00°15'54.9"
H.P. = 00°00'08.8"

Moon at Greatest Eclipse
(Geocentric Coordinates)
R.A. = 02h00m04.9s
Dec. = +12°40'11.3"
S.D. = 00°16'25.1"
H.P. = 01°00'15.3"

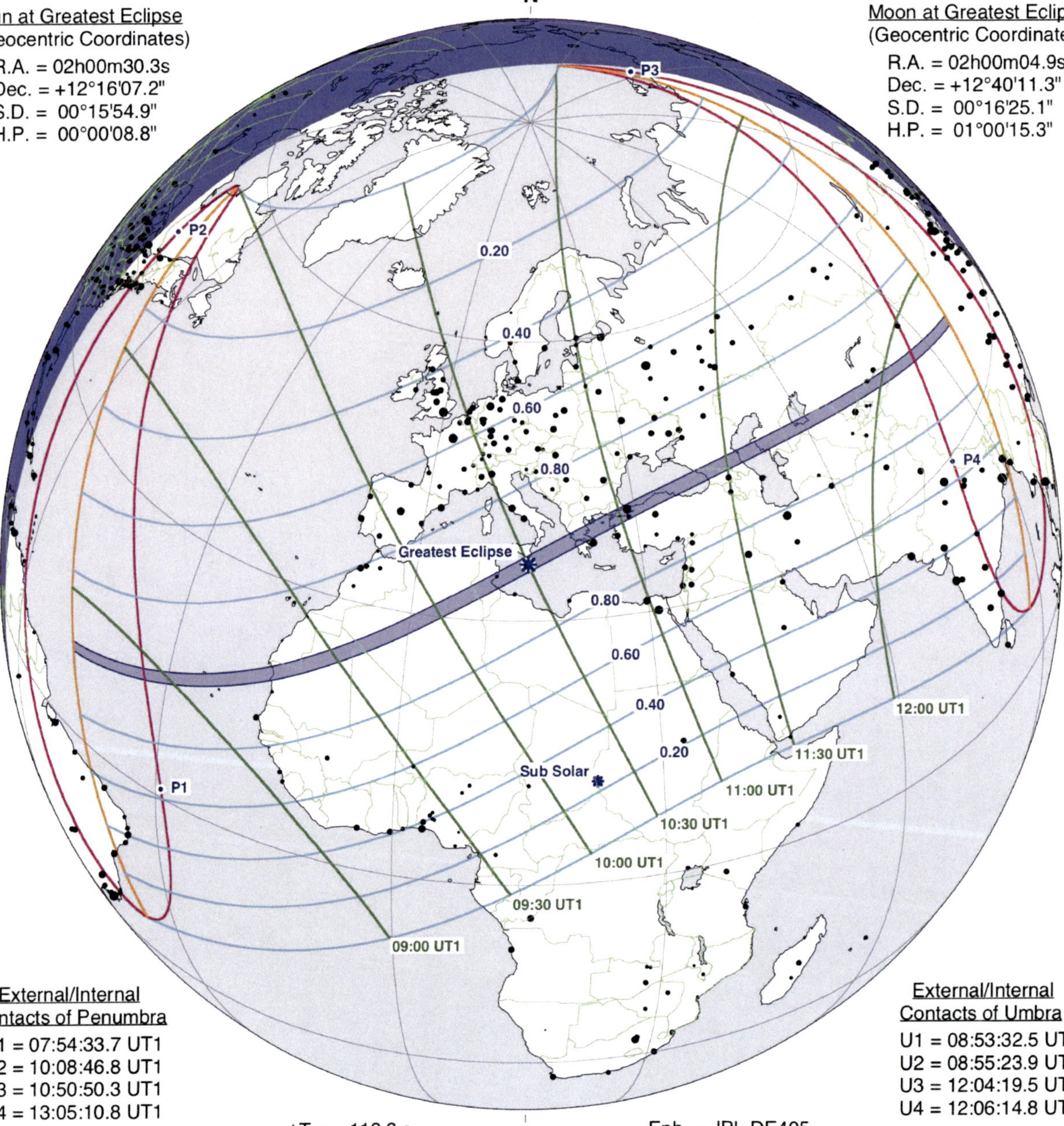

External/Internal
Contacts of Penumbra
P1 = 07:54:33.7 UT1
P2 = 10:08:46.8 UT1
P3 = 10:50:50.3 UT1
P4 = 13:05:10.8 UT1

External/Internal
Contacts of Umbra
U1 = 08:53:32.5 UT1
U2 = 08:55:23.9 UT1
U3 = 12:04:19.5 UT1
U4 = 12:06:14.8 UT1

ΔT = 112.6 s

Eph. = JPL DE405

Circumstances at Greatest Eclipse: 10:29:56.9 UT1

Lat. = 35°59.8'N
Long. = 014°50.1'E
Path Width = 173.1 km
Sun Alt. = 65.4°
Sun Azm. = 162.6°
Duration = 03m57.8s

Circumstances at Greatest Duration: 10:33:08.9 UT1

Lat. = 36°31.2'N
Long. = 016°06.1'E
Path Width = 172.8 km
Sun Alt. = 65.3°
Sun Azm. = 167.6°
Duration = 03m58.0s

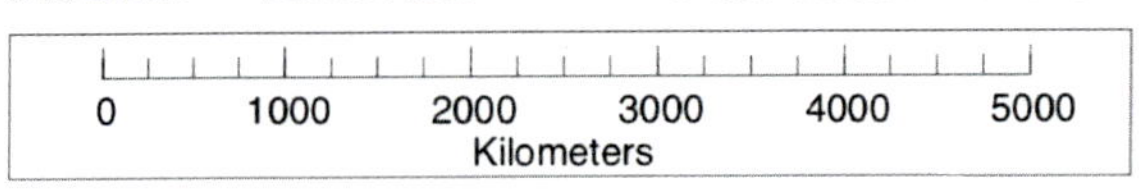

Annular Solar Eclipse of 2088 Oct 14

Greatest Eclipse = 14:48:05.1 TD (= 14:46:12.1 UT1)

Eclipse Magnitude = 0.9727 | Saros Series = 135

Gamma = -0.5349 | Saros Member = 43 of 71

Sun at Greatest Eclipse
(Geocentric Coordinates)
R.A. = 13h22m16.6s
Dec. = -08°39'19.3"
S.D. = 00°16'02.1"
H.P. = 00°00'08.8"

Moon at Greatest Eclipse
(Geocentric Coordinates)
R.A. = 13h21m42.6s
Dec. = -09°08'15.8"
S.D. = 00°15'23.5"
H.P. = 00°56'29.4"

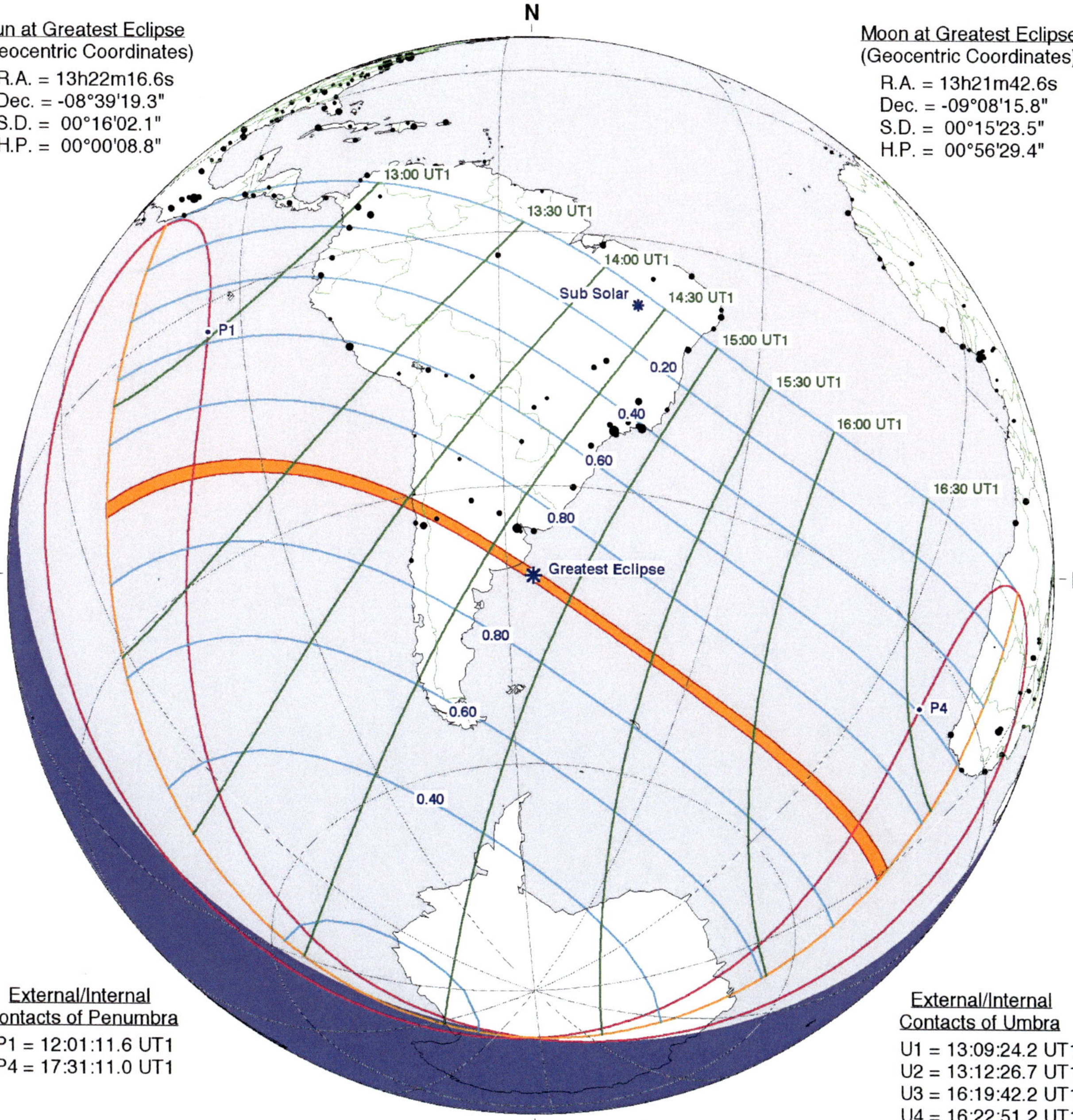

External/Internal Contacts of Penumbra
P1 = 12:01:11.6 UT1
P4 = 17:31:11.0 UT1

External/Internal Contacts of Umbra
U1 = 13:09:24.2 UT1
U2 = 13:12:26.7 UT1
U3 = 16:19:42.2 UT1
U4 = 16:22:51.2 UT1

ΔT = 113.0 s | Eph. = JPL DE405

Circumstances at Greatest Eclipse: 14:46:12.1 UT1

Lat. = 39°39.3'S | Sun Alt. = 57.4°
Long. = 056°15.9'W | Sun Azm. = 20.8°
Path Width = 115.2 km | Duration = 02m38.2s

Circumstances at Greatest Duration: 15:26:14.1 UT1

Lat. = 46°24.9'S | Sun Alt. = 49.8°
Long. = 039°00.3'W | Sun Azm. = 334.8°
Path Width = 119.0 km | Duration = 02m39.8s

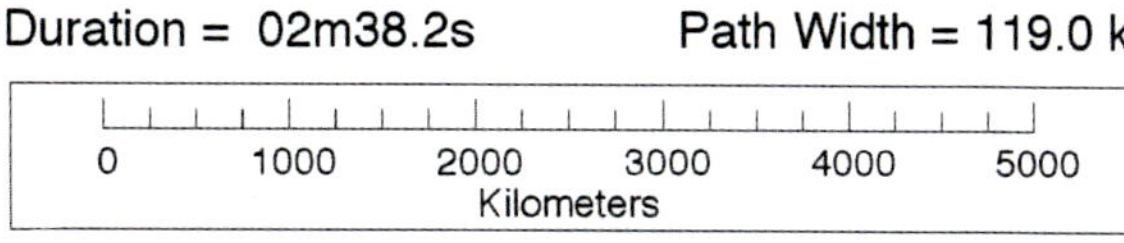

Annular Solar Eclipse of 2089 Apr 10

Greatest Eclipse = 22:44:41.5 TD (= 22:42:48.1 UT1)

Eclipse Magnitude = 0.9919
Gamma = -0.3319

Saros Series = 140
Saros Member = 33 of 71

Sun at Greatest Eclipse
(Geocentric Coordinates)
R.A. = 01h20m36.9s
Dec. = +08°29'24.5"
S.D. = 00°15'57.8"
H.P. = 00°00'08.8"

Moon at Greatest Eclipse
(Geocentric Coordinates)
R.A. = 01h20m58.3s
Dec. = +08°11'12.9"
S.D. = 00°15'35.9"
H.P. = 00°57'14.7"

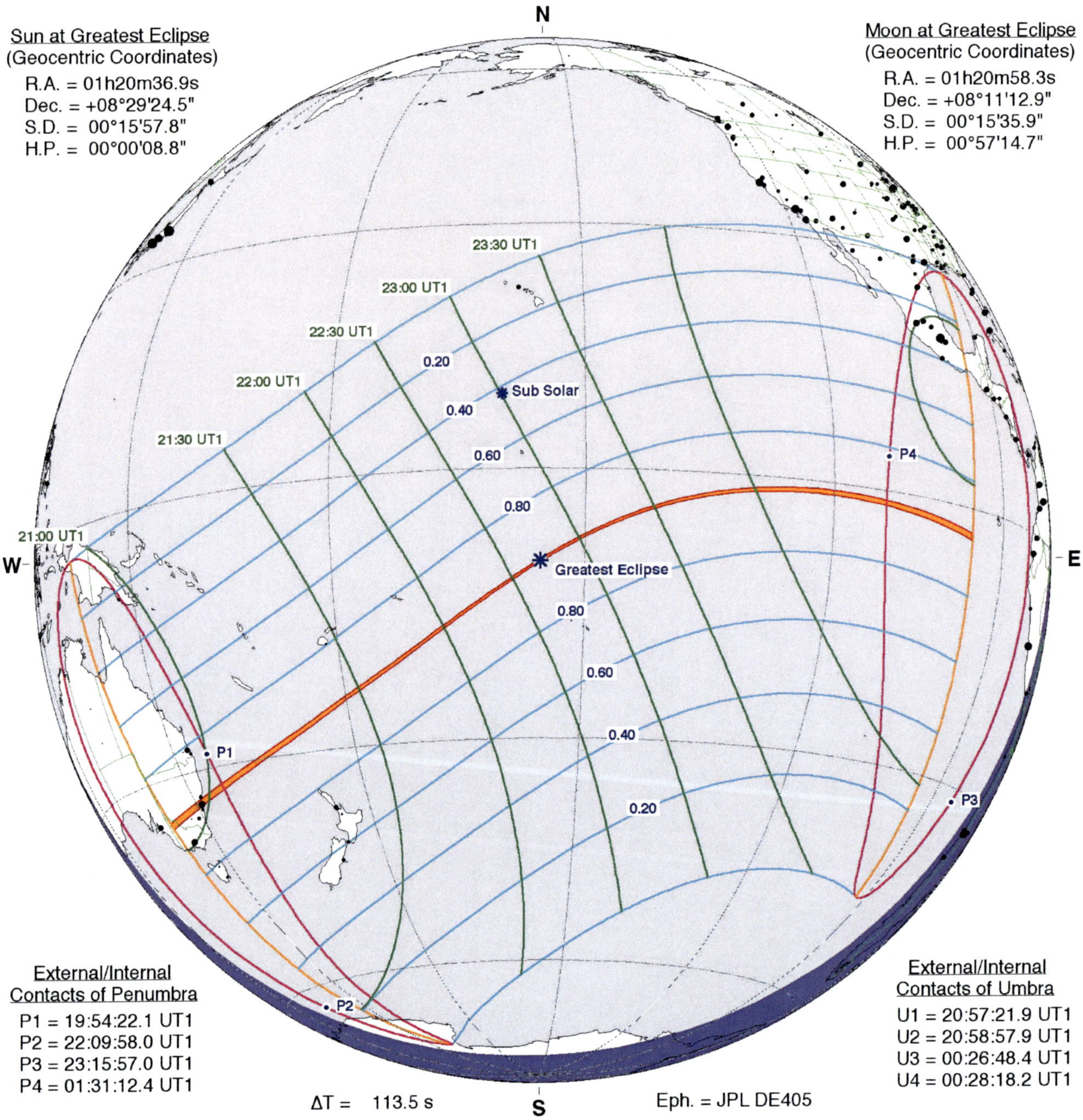

External/Internal Contacts of Penumbra
P1 = 19:54:22.1 UT1
P2 = 22:09:58.0 UT1
P3 = 23:15:57.0 UT1
P4 = 01:31:12.4 UT1

External/Internal Contacts of Umbra
U1 = 20:57:21.9 UT1
U2 = 20:58:57.9 UT1
U3 = 00:26:48.4 UT1
U4 = 00:28:18.2 UT1

ΔT = 113.5 s

Eph. = JPL DE405

Circumstances at Greatest Eclipse: 22:42:48.1 UT1

Lat. = 10°11.8'S Sun Alt. = 70.6°
Long. = 155°02.6'W Sun Azm. = 343.7°
Path Width = 30.0 km Duration = 00m53.2s

Circumstances at Greatest Duration: 20:58:09.9 UT1

Lat. = 35°22.6'S Sun Alt. = 0.0°
Long. = 141°46.8'E Sun Azm. = 79.6°
Path Width = 89.8 km Duration = 01m32.4s

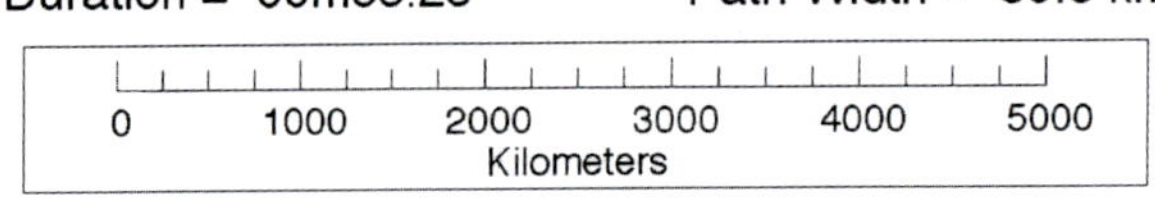

Total Solar Eclipse of 2089 Oct 04

Greatest Eclipse = 01:15:23.2 TD (= 01:13:29.3 UT1)

Eclipse Magnitude = 1.0333 Saros Series = 145

Gamma = 0.2167 Saros Member = 26 of 77

Sun at Greatest Eclipse
(Geocentric Coordinates)
R.A. = 12h42m34.2s
Dec. = -04°34'29.0"
S.D. = 00°15'59.1"
H.P. = 00°00'08.8"

Moon at Greatest Eclipse
(Geocentric Coordinates)
R.A. = 12h42m49.6s
Dec. = -04°22'10.5"
S.D. = 00°16'15.0"
H.P. = 00°59'38.5"

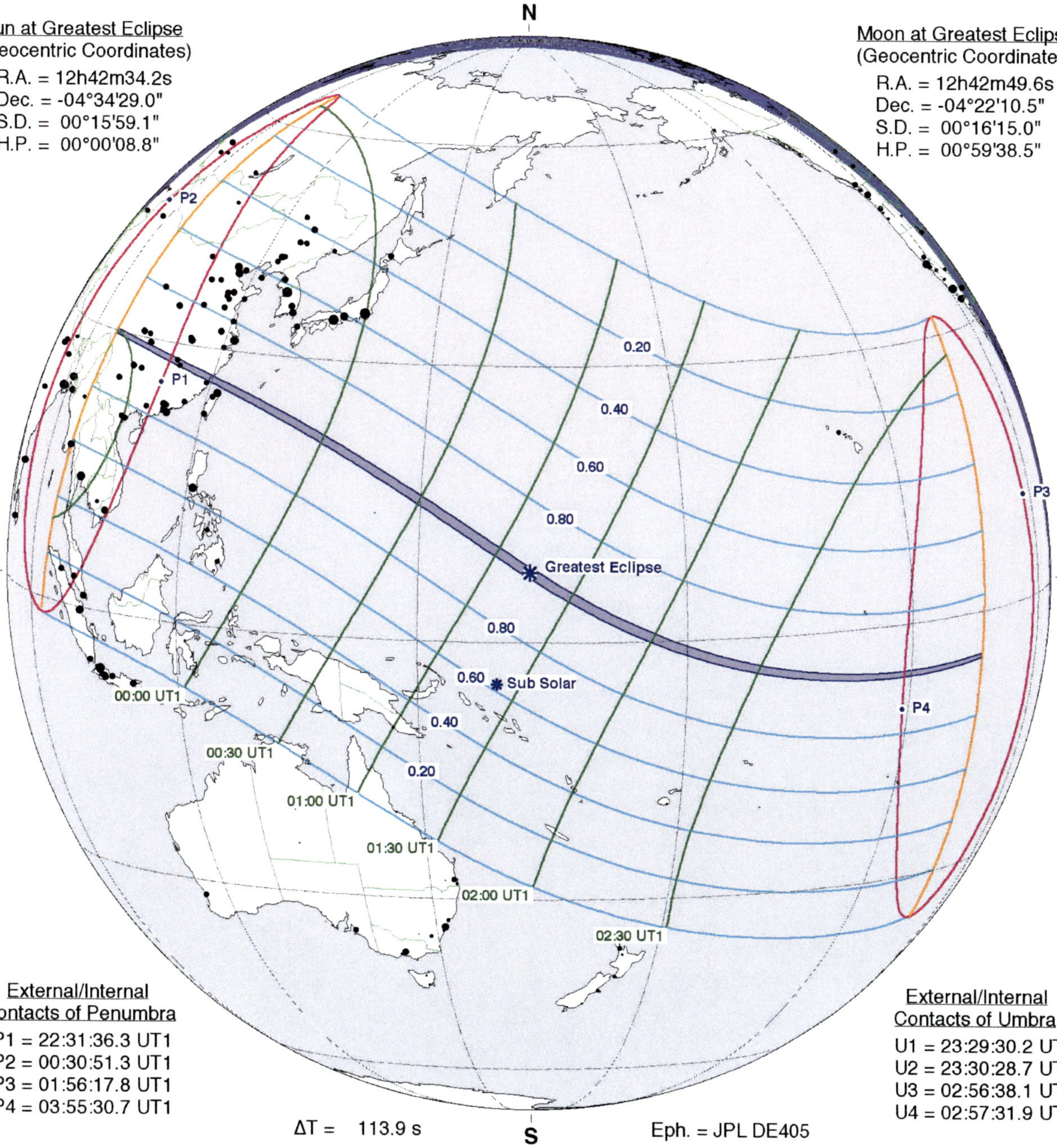

External/Internal
Contacts of Penumbra
P1 = 22:31:36.3 UT1
P2 = 00:30:51.3 UT1
P3 = 01:56:17.8 UT1
P4 = 03:55:30.7 UT1

External/Internal
Contacts of Umbra
U1 = 23:29:30.2 UT1
U2 = 23:30:28.7 UT1
U3 = 02:56:38.1 UT1
U4 = 02:57:31.9 UT1

ΔT = 113.9 s Eph. = JPL DE405

Circumstances at Greatest Eclipse: 01:13:29.3 UT1
Lat. = 07°25.5'N Sun Alt. = 77.4°
Long. = 162°31.1'E Sun Azm. = 197.3°
Path Width = 114.9 km Duration = 03m14.2s

Circumstances at Greatest Duration: 01:13:02.0 UT1
Lat. = 07°29.9'N Sun Alt. = 77.4°
Long. = 162°23.7'E Sun Azm. = 196.2°
Path Width = 114.9 km Duration = 03m14.2s

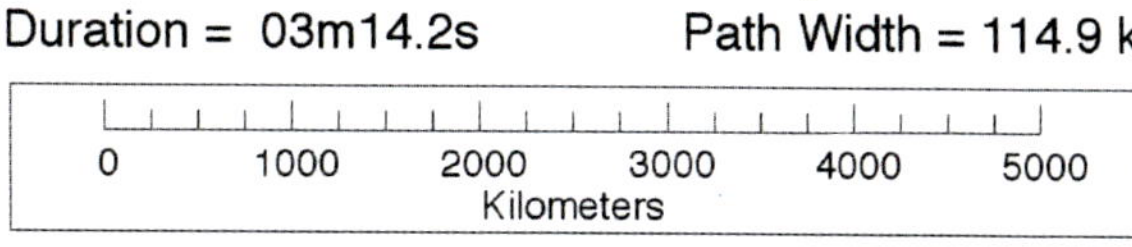

Partial Solar Eclipse of 2090 Mar 31

Greatest Eclipse = 03:38:07.9 TD (= 03:36:13.5 UT1)

Eclipse Magnitude = 0.7843
Gamma = -1.1028

Saros Series = 150
Saros Member = 21 of 71

Sun at Greatest Eclipse
(Geocentric Coordinates)
R.A. = 00h40m11.0s
Dec. = +04°19'18.8"
S.D. = 00°16'00.8"
H.P. = 00°00'08.8"

Moon at Greatest Eclipse
(Geocentric Coordinates)
R.A. = 00h41m23.0s
Dec. = +03°22'02.4"
S.D. = 00°14'52.2"
H.P. = 00°54'34.6"

N

W

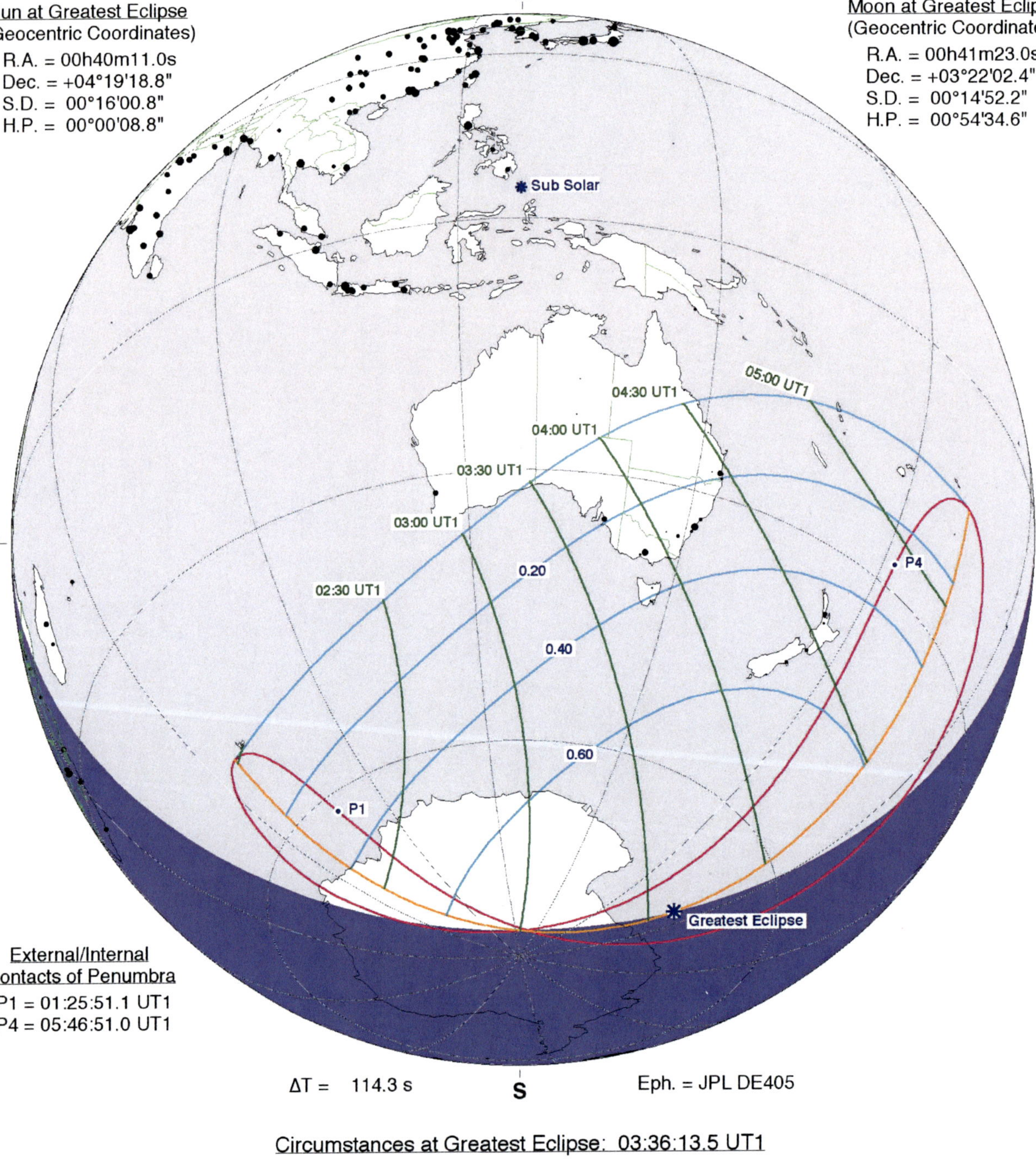

E

S

External/Internal
Contacts of Penumbra
P1 = 01:25:51.1 UT1
P4 = 05:46:51.0 UT1

ΔT = 114.3 s

Eph. = JPL DE405

Circumstances at Greatest Eclipse: 03:36:13.5 UT1

Lat. = 72°03.9'S
Long. = 156°36.2'W

Sun Alt. = 0.0°
Sun Azm. = 284.2°

0 1000 2000 3000 4000 5000
Kilometers

Total Solar Eclipse of 2090 Sep 23

Greatest Eclipse = 16:56:36.3 TD (= 16:54:41.5 UT1)

Eclipse Magnitude = 1.0562 Saros Series = 155

Gamma = 0.9157 Saros Member = 10 of 71

Sun at Greatest Eclipse
(Geocentric Coordinates)
R.A. = 12h04m19.6s
Dec. = -00°28'06.5"
S.D. = 00°15'56.2"
H.P. = 00°00'08.8"

Moon at Greatest Eclipse
(Geocentric Coordinates)
R.A. = 12h05m28.3s
Dec. = +00°25'15.5"
S.D. = 00°16'43.4"
H.P. = 01°01'22.6"

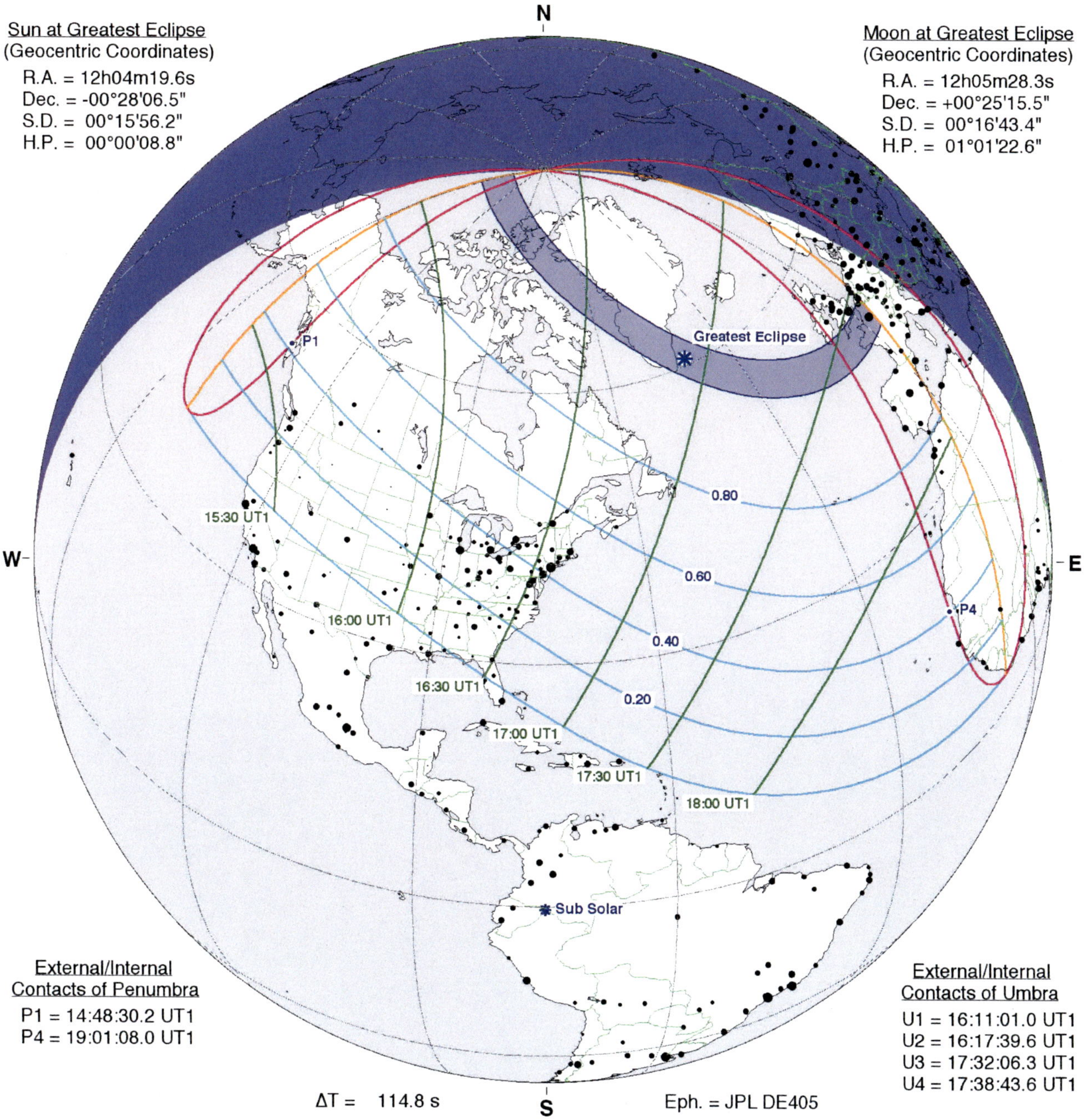

External/Internal Contacts of Penumbra
P1 = 14:48:30.2 UT1
P4 = 19:01:08.0 UT1

External/Internal Contacts of Umbra
U1 = 16:11:01.0 UT1
U2 = 16:17:39.6 UT1
U3 = 17:32:06.3 UT1
U4 = 17:38:43.6 UT1

ΔT = 114.8 s Eph. = JPL DE405

Circumstances at Greatest Eclipse: 16:54:41.5 UT1

Lat. = 60°41.2'N Sun Alt. = 23.2°
Long. = 040°45.3'W Sun Azm. = 218.5°
Path Width = 462.9 km Duration = 03m35.9s

Circumstances at Greatest Duration: 16:54:48.5 UT1

Lat. = 60°38.7'N Sun Alt. = 23.2°
Long. = 040°40.8'W Sun Azm. = 218.6°
Path Width = 463.1 km Duration = 03m35.9s

0 1000 2000 3000 4000 5000
Kilometers

Partial Solar Eclipse of 2091 Feb 18

Greatest Eclipse = 09:54:39.8 TD (= 09:52:44.6 UT1)

Eclipse Magnitude = 0.6558
Gamma = 1.1779

Saros Series = 122
Saros Member = 62 of 70

Sun at Greatest Eclipse
(Geocentric Coordinates)
R.A. = 22h08m17.5s
Dec. = -11°28'13.5"
S.D. = 00°16'11.1"
H.P. = 00°00'08.9"

Moon at Greatest Eclipse
(Geocentric Coordinates)
R.A. = 22h07m09.8s
Dec. = -10°25'58.4"
S.D. = 00°14'56.6"
H.P. = 00°54'50.7"

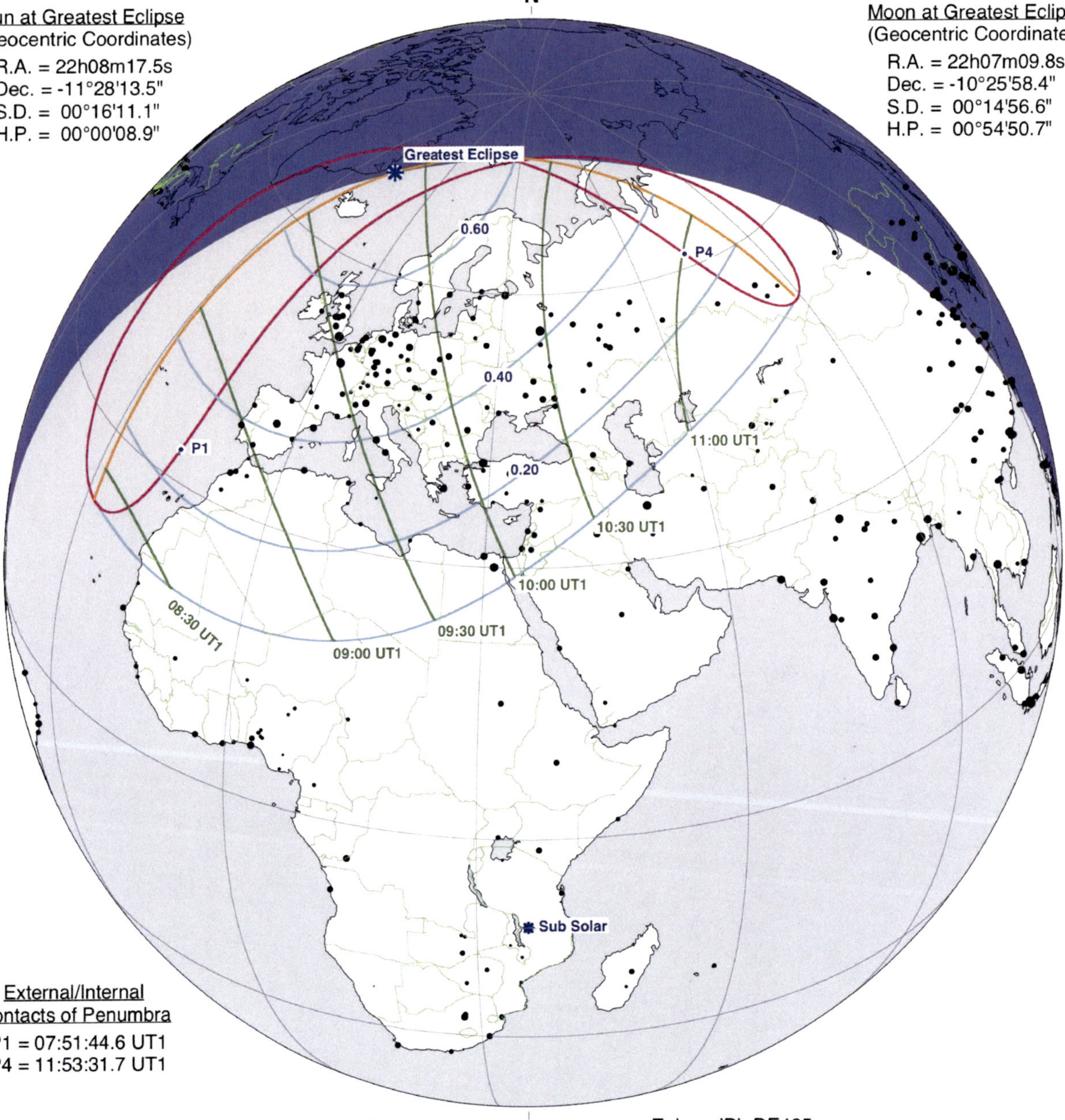

External/Internal
Contacts of Penumbra
P1 = 07:51:44.6 UT1
P4 = 11:53:31.7 UT1

ΔT = 115.1 s

Eph. = JPL DE405

Circumstances at Greatest Eclipse: 09:52:44.6 UT1

Lat. = 71°13.5'N
Long. = 018°02.1'W

Sun Alt. = 0.0°
Sun Azm. = 128.2°

0 1000 2000 3000 4000 5000
Kilometers

Total Solar Eclipse of 2091 Aug 15

Greatest Eclipse = 00:34:42.9 TD (= 00:32:47.3 UT1)

Eclipse Magnitude = 1.0216 Saros Series = 127

Gamma = -0.9490 Saros Member = 62 of 82

<u>Sun at Greatest Eclipse</u>
(Geocentric Coordinates)
R.A. = 09h39m24.9s
Dec. = +14°00'16.0"
S.D. = 00°15'47.2"
H.P. = 00°00'08.7"

<u>Moon at Greatest Eclipse</u>
(Geocentric Coordinates)
R.A. = 09h38m32.0s
Dec. = +13°05'59.9"
S.D. = 00°16'03.2"
H.P. = 00°58'54.9"

N

W

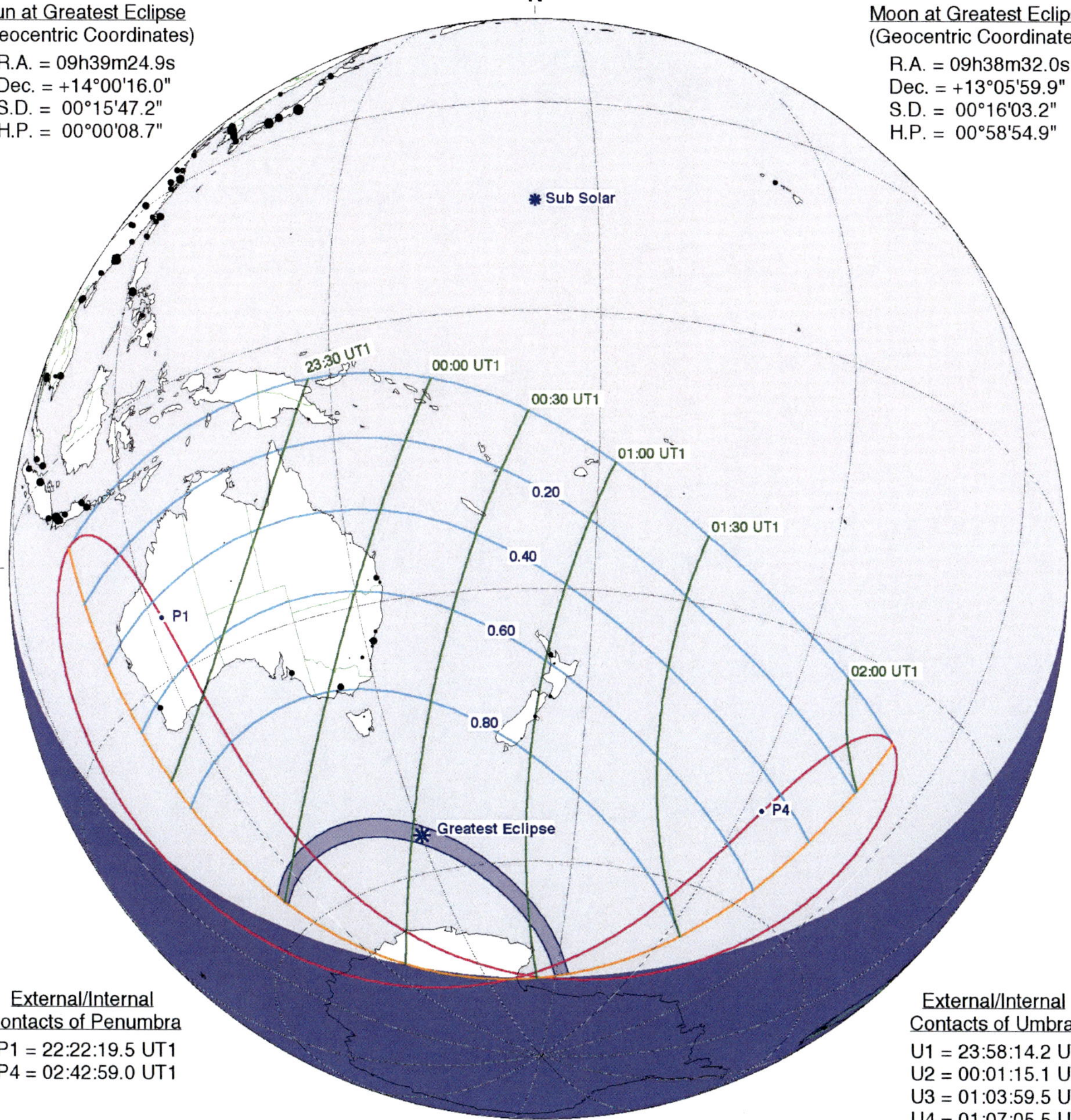

E

S

<u>External/Internal Contacts of Penumbra</u>
P1 = 22:22:19.5 UT1
P4 = 02:42:59.0 UT1

<u>External/Internal Contacts of Umbra</u>
U1 = 23:58:14.2 UT1
U2 = 00:01:15.1 UT1
U3 = 01:03:59.5 UT1
U4 = 01:07:05.5 UT1

ΔT = 115.6 s Eph. = JPL DE405

<u>Circumstances at Greatest Eclipse: 00:32:47.3 UT1</u>

Lat. = 55°34.9'S Sun Alt. = 17.8°
Long. = 150°12.1'E Sun Azm. = 23.2°
Path Width = 236.5 km Duration = 01m38.1s

<u>Circumstances at Greatest Duration: 00:33:18.2 UT1</u>

Lat. = 55°41.5'S Sun Alt. = 17.8°
Long. = 150°31.0'E Sun Azm. = 22.8°
Path Width = 236.0 km Duration = 01m38.1s

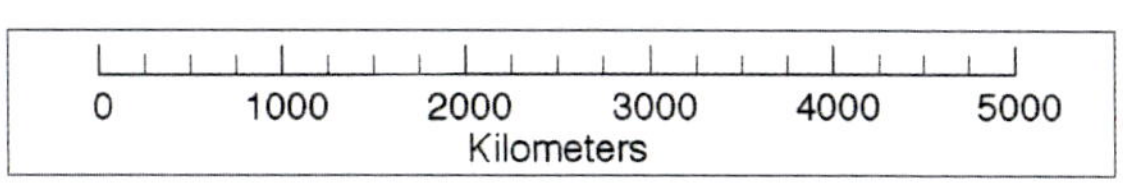

Annular Solar Eclipse of 2092 Feb 07

Greatest Eclipse = 15:10:20.2 TD (= 15:08:24.1 UT1)

Eclipse Magnitude = 0.9840
Gamma = 0.4322

Saros Series = 132
Saros Member = 50 of 71

Sun at Greatest Eclipse
(Geocentric Coordinates)
R.A. = 21h25m01.6s
Dec. = -15°10'15.3"
S.D. = 00°16'13.1"
H.P. = 00°00'08.9"

Moon at Greatest Eclipse
(Geocentric Coordinates)
R.A. = 21h24m39.5s
Dec. = -14°45'56.9"
S.D. = 00°15'43.8"
H.P. = 00°57'43.8"

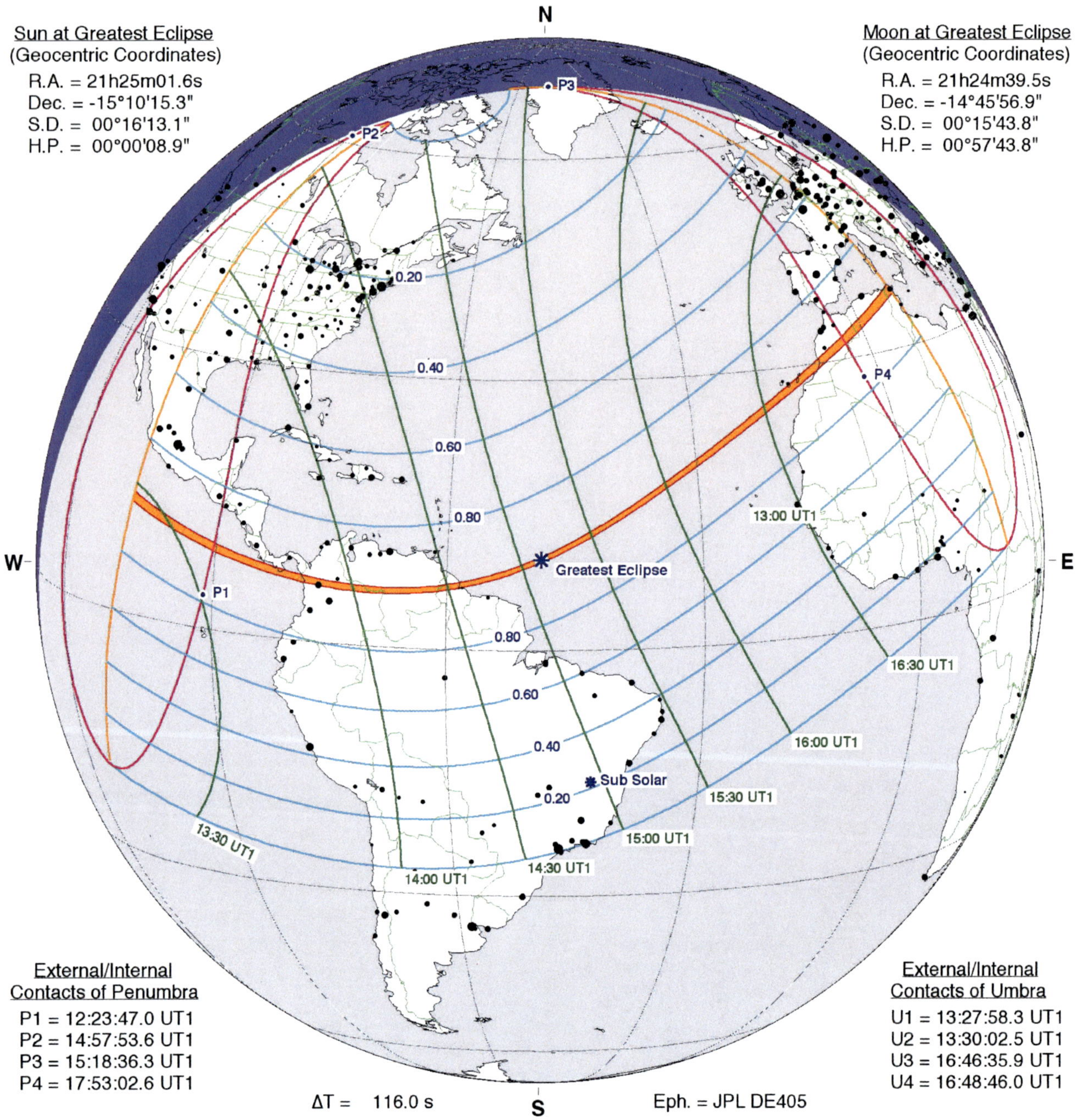

External/Internal Contacts of Penumbra
P1 = 12:23:47.0 UT1
P2 = 14:57:53.6 UT1
P3 = 15:18:36.3 UT1
P4 = 17:53:02.6 UT1

External/Internal Contacts of Umbra
U1 = 13:27:58.3 UT1
U2 = 13:30:02.5 UT1
U3 = 16:46:35.9 UT1
U4 = 16:48:46.0 UT1

ΔT = 116.0 s

Eph. = JPL DE405

Circumstances at Greatest Eclipse: 15:08:24.1 UT1

Lat. = 09°56.2'N	Sun Alt. = 64.3°
Long. = 049°00.6'W	Sun Azm. = 167.9°
Path Width = 62.5 km	Duration = 01m47.7s

Circumstances at Greatest Duration: 13:29:00.3 UT1

Lat. = 12°50.4'N	Sun Alt. = 0.0°
Long. = 105°12.6'W	Sun Azm. = 105.6°
Path Width = 113.9 km	Duration = 01m57.5s

0 1000 2000 3000 4000 5000
Kilometers

Annular Solar Eclipse of 2092 Aug 03

Greatest Eclipse = 09:59:32.8 TD (= 09:57:36.3 UT1)

Eclipse Magnitude = 0.9794 | Saros Series = 137

Gamma = -0.2044 | Saros Member = 40 of 70

Sun at Greatest Eclipse
(Geocentric Coordinates)
R.A. = 08h58m14.3s
Dec. = +17°09'21.7"
S.D. = 00°15'45.7"
H.P. = 00°00'08.7"

Moon at Greatest Eclipse
(Geocentric Coordinates)
R.A. = 08h58m05.6s
Dec. = +16°58'10.4"
S.D. = 00°15'12.2"
H.P. = 00°55'47.9"

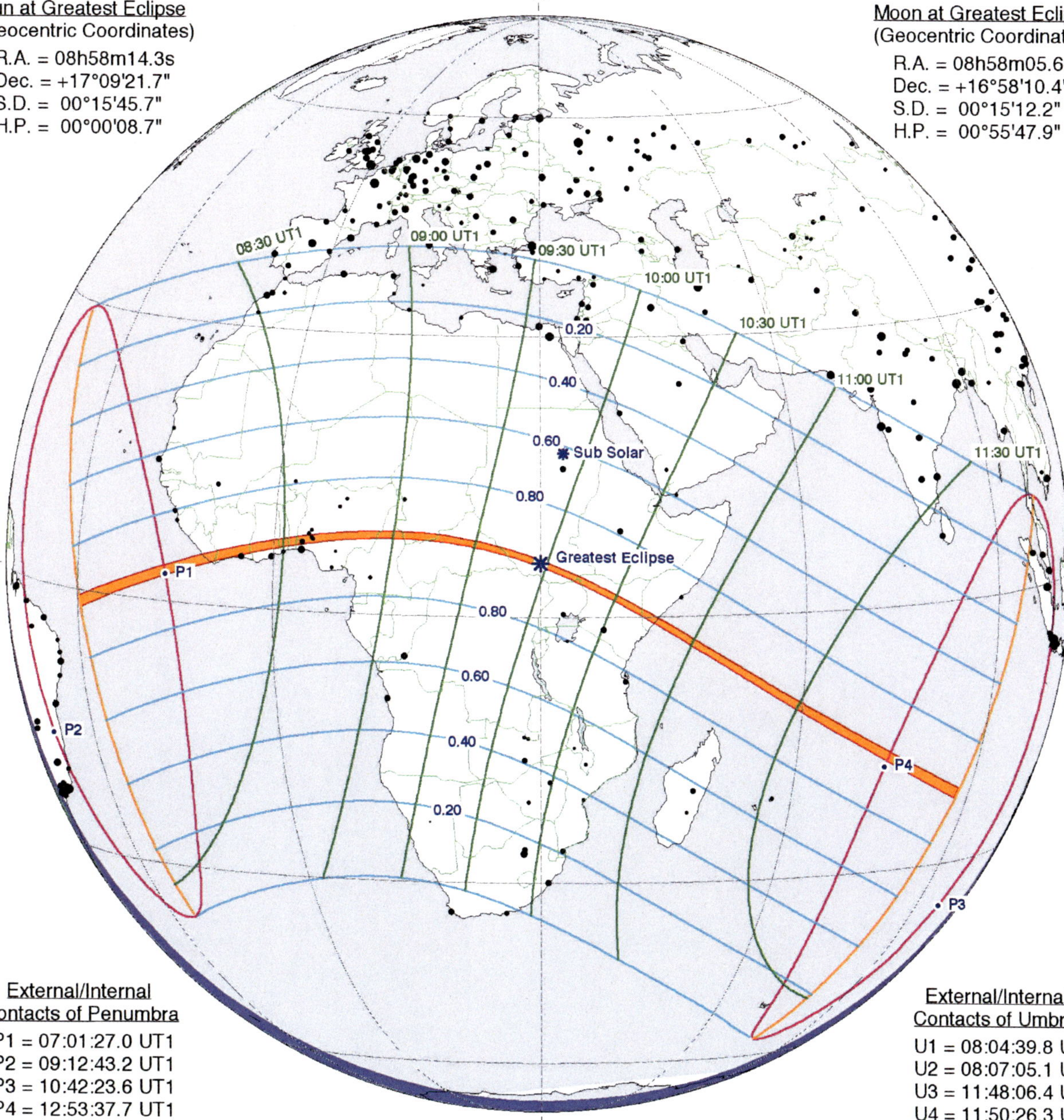

External/Internal Contacts of Penumbra
P1 = 07:01:27.0 UT1
P2 = 09:12:43.2 UT1
P3 = 10:42:23.6 UT1
P4 = 12:53:37.7 UT1

External/Internal Contacts of Umbra
U1 = 08:04:39.8 UT1
U2 = 08:07:05.1 UT1
U3 = 11:48:06.4 UT1
U4 = 11:50:26.3 UT1

ΔT = 116.5 s

Eph. = JPL DE405

Circumstances at Greatest Eclipse: 09:57:36.3 UT1

Lat. = 05°35.1'N | Sun Alt. = 78.2°
Long. = 030°01.2'E | Sun Azm. = 10.1°
Path Width = 74.7 km | Duration = 02m31.4s

Circumstances at Greatest Duration: 09:16:14.1 UT1

Lat. = 08°03.3'N | Sun Alt. = 65.4°
Long. = 019°05.1'E | Sun Azm. = 66.0°
Path Width = 80.1 km | Duration = 02m33.9s

0 1000 2000 3000 4000 5000
Kilometers

Total Solar Eclipse of 2093 Jan 27

Greatest Eclipse = 03:22:16.1 TD (= 03:20:19.1 UT1)

Eclipse Magnitude = 1.0340
Gamma = -0.2737

Saros Series = 142
Saros Member = 27 of 72

Sun at Greatest Eclipse
(Geocentric Coordinates)
R.A. = 20h41m22.6s
Dec. = -18°16'28.3"
S.D. = 00°16'14.6"
H.P. = 00°00'08.9"

Moon at Greatest Eclipse
(Geocentric Coordinates)
R.A. = 20h41m33.8s
Dec. = -18°32'48.9"
S.D. = 00°16'31.4"
H.P. = 01°00'38.7"

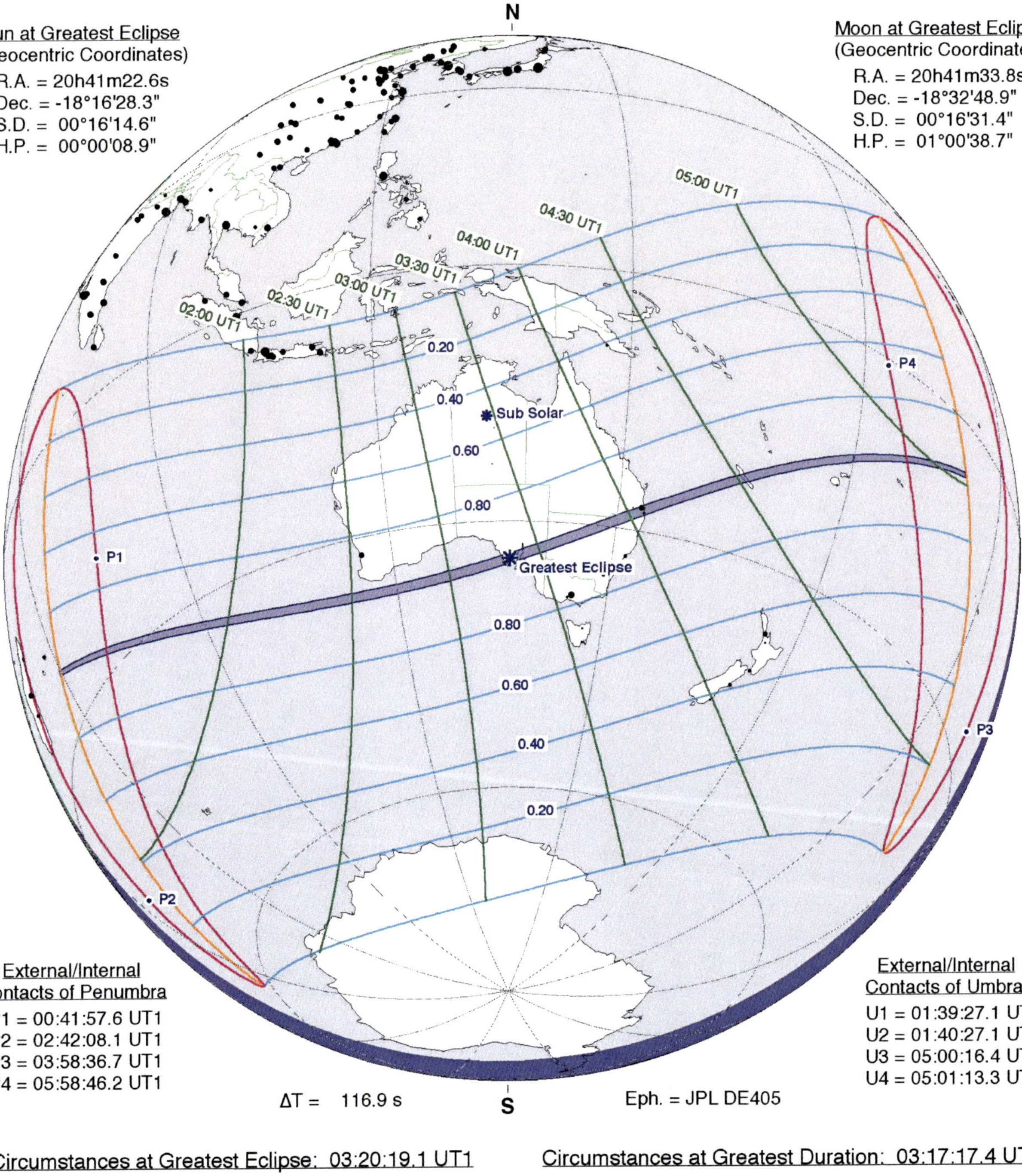

External/Internal
Contacts of Penumbra
P1 = 00:41:57.6 UT1
P2 = 02:42:08.1 UT1
P3 = 03:58:36.7 UT1
P4 = 05:58:46.2 UT1

External/Internal
Contacts of Umbra
U1 = 01:39:27.1 UT1
U2 = 01:40:27.1 UT1
U3 = 05:00:16.4 UT1
U4 = 05:01:13.3 UT1

ΔT = 116.9 s

Eph. = JPL DE405

Circumstances at Greatest Eclipse: 03:20:19.1 UT1

Lat. = 34°07.3'S
Long. = 136°07.0'E
Path Width = 119.3 km
Sun Alt. = 73.9°
Sun Azm. = 349.5°
Duration = 02m58.0s

Circumstances at Greatest Duration: 03:17:17.4 UT1

Lat. = 34°25.2'S
Long. = 134°53.8'E
Path Width = 119.2 km
Sun Alt. = 73.8°
Sun Azm. = 356.4°
Duration = 02m58.1s

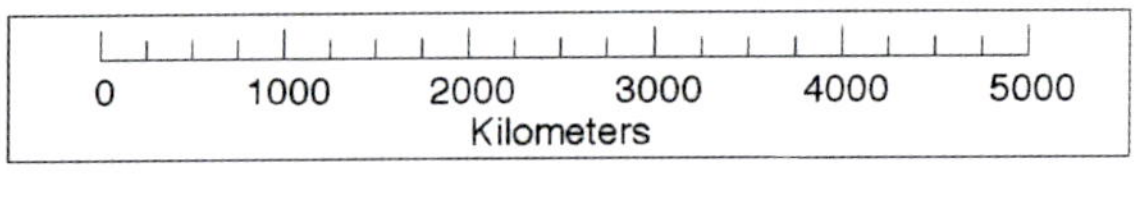

Annular Solar Eclipse of 2093 Jul 23

Greatest Eclipse = 12:32:03.8 TD (= 12:30:06.4 UT1)

Eclipse Magnitude = 0.9463 | Saros Series = 147

Gamma = 0.5717 | Saros Member = 27 of 80

Sun at Greatest Eclipse
(Geocentric Coordinates)

R.A. = 08h14m45.4s
Dec. = +19°49'29.6"
S.D. = 00°15'44.6"
H.P. = 00°00'08.7"

Moon at Greatest Eclipse
(Geocentric Coordinates)

R.A. = 08h15m01.4s
Dec. = +20°20'03.3"
S.D. = 00°14'43.0"
H.P. = 00°54'00.8"

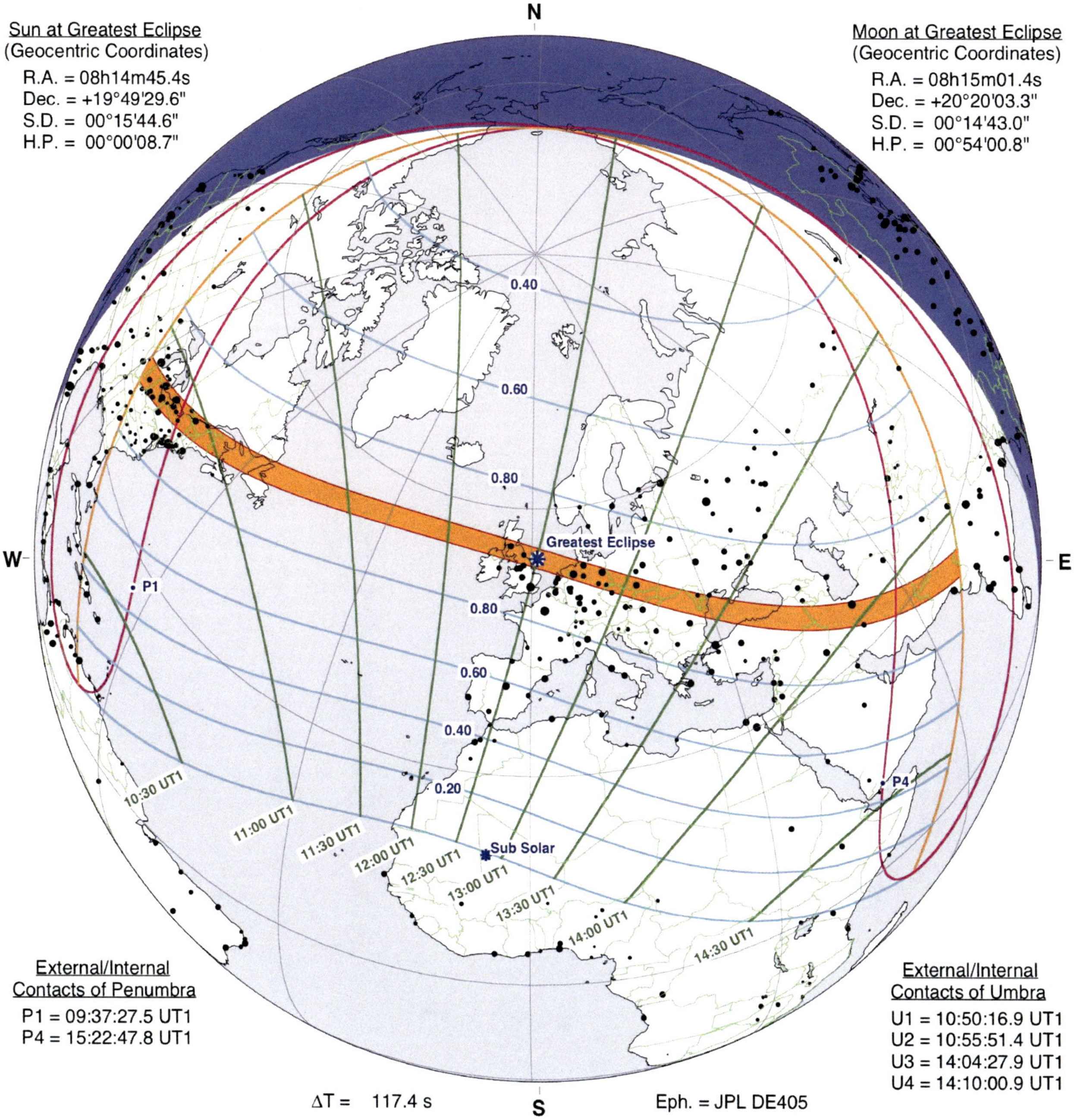

External/Internal
Contacts of Penumbra

P1 = 09:37:27.5 UT1
P4 = 15:22:47.8 UT1

External/Internal
Contacts of Umbra

U1 = 10:50:16.9 UT1
U2 = 10:55:51.4 UT1
U3 = 14:04:27.9 UT1
U4 = 14:10:00.9 UT1

ΔT = 117.4 s

Eph. = JPL DE405

Circumstances at Greatest Eclipse: 12:30:06.4 UT1

Lat. = 54°33.7'N | Sun Alt. = 54.9°
Long. = 001°02.6'E | Sun Azm. = 191.3°
Path Width = 241.3 km | Duration = 05m11.3s

Circumstances at Greatest Duration: 12:25:45.4 UT1

Lat. = 54°57.1'N | Sun Alt. = 54.8°
Long. = 001°26.7'W | Sun Azm. = 185.4°
Path Width = 241.0 km | Duration = 05m11.4s

0 1000 2000 3000 4000 5000
Kilometers

Total Solar Eclipse of 2094 Jan 16

Greatest Eclipse = 18:59:03.4 TD (= 18:57:05.6 UT1)

Eclipse Magnitude = 1.0342 | Saros Series = 152
Gamma = -0.9333 | Saros Member = 17 of 70

Sun at Greatest Eclipse
(Geocentric Coordinates)
R.A. = 19h56m48.4s
Dec. = -20°43'02.8"
S.D. = 00°16'15.5"
H.P. = 00°00'08.9"

Moon at Greatest Eclipse
(Geocentric Coordinates)
R.A. = 19h57m12.4s
Dec. = -21°39'54.3"
S.D. = 00°16'43.2"
H.P. = 01°01'21.9"

N

W

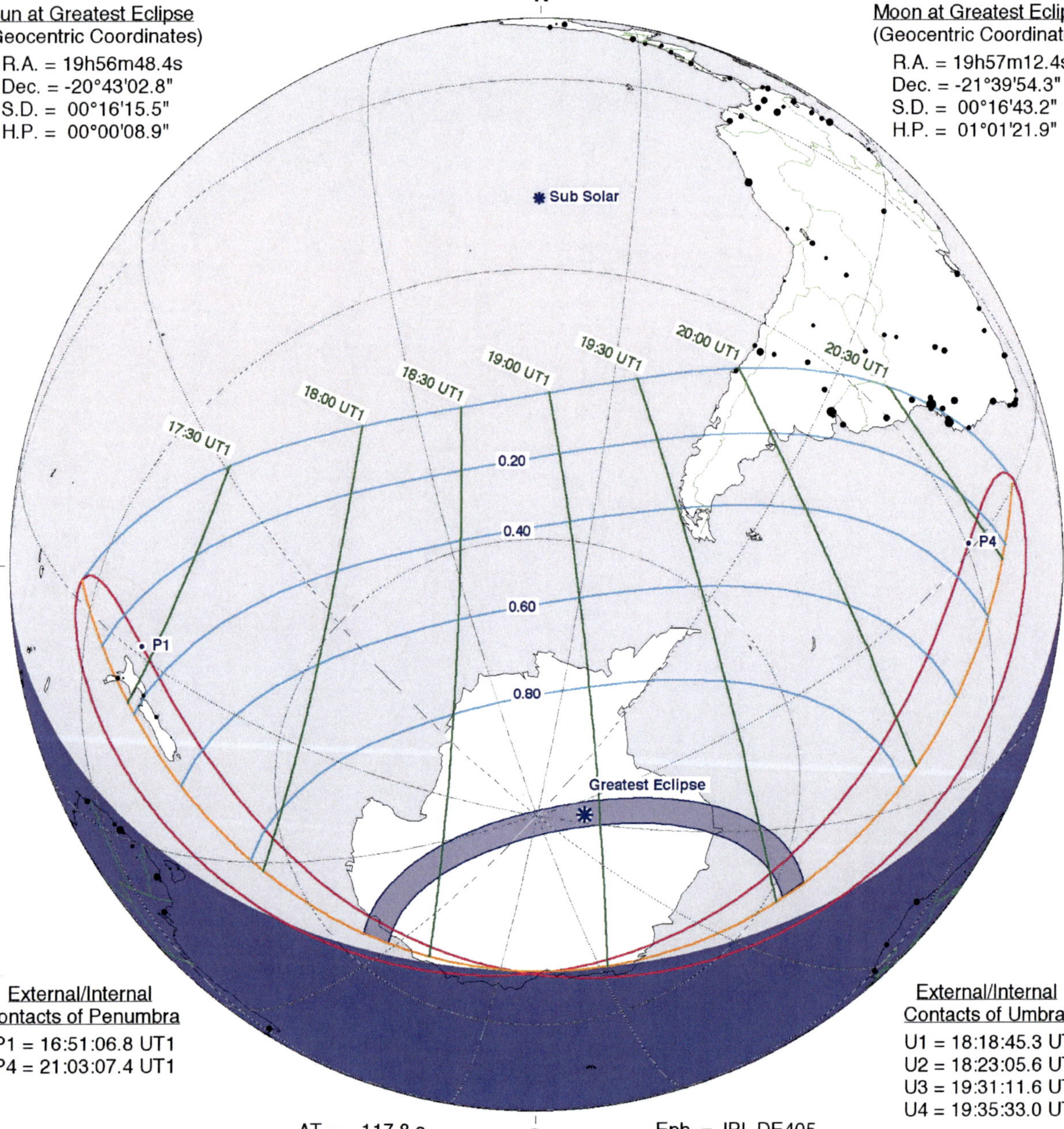

E

External/Internal
Contacts of Penumbra
P1 = 16:51:06.8 UT1
P4 = 21:03:07.4 UT1

External/Internal
Contacts of Umbra
U1 = 18:18:45.3 UT1
U2 = 18:23:05.6 UT1
U3 = 19:31:11.6 UT1
U4 = 19:35:33.0 UT1

ΔT = 117.8 s

S

Eph. = JPL DE405

Circumstances at Greatest Eclipse: 18:57:05.6 UT1
Lat. = 84°46.4'S | Sun Alt. = 20.5°
Long. = 010°51.9'W | Sun Azm. = 267.1°
Path Width = 329.4 km | Duration = 01m51.5s

Circumstances at Greatest Duration: 18:57:16.6 UT1
Lat. = 84°40.3'S | Sun Alt. = 20.5°
Long. = 010°57.2'W | Sun Azm. = 267.1°
Path Width = 329.5 km | Duration = 01m51.5s

0 1000 2000 3000 4000 5000
Kilometers

Partial Solar Eclipse of 2094 Jun 13

Greatest Eclipse = 00:22:11.3 TD (= 00:20:13.1 UT1)

Eclipse Magnitude = 0.1618 | Saros Series = 119

Gamma = -1.4613 | Saros Member = 70 of 71

Sun at Greatest Eclipse
(Geocentric Coordinates)
R.A. = 05h27m41.4s
Dec. = +23°13'07.5"
S.D. = 00°15'45.0"
H.P. = 00°00'08.7"

Moon at Greatest Eclipse
(Geocentric Coordinates)
R.A. = 05h28m34.2s
Dec. = +21°52'30.5"
S.D. = 00°15'14.5"
H.P. = 00°55'56.1"

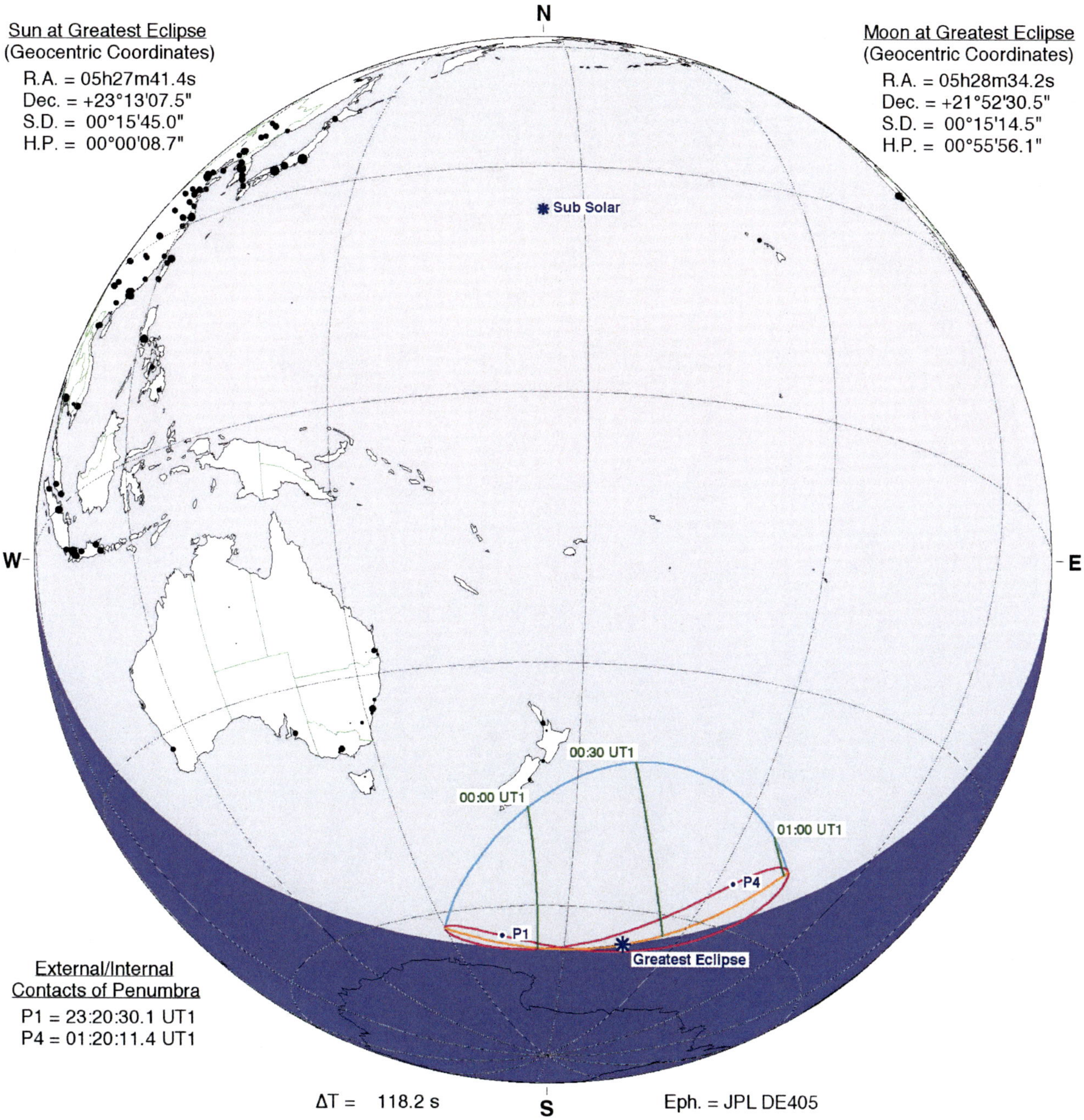

External/Internal
Contacts of Penumbra
P1 = 23:20:30.1 UT1
P4 = 01:20:11.4 UT1

ΔT = 118.2 s

Eph. = JPL DE405

Circumstances at Greatest Eclipse: 00:20:13.1 UT1

Lat. = 65°18.3'S | Sun Alt. = 0.0°

Long. = 163°56.0'W | Sun Azm. = 340.7°

0 1000 2000 3000 4000 5000
Kilometers

Partial Solar Eclipse of 2094 Jul 12

Greatest Eclipse = 13:24:34.9 TD (= 13:22:36.6 UT1)

Eclipse Magnitude = 0.4225
Gamma = 1.3149

Saros Series = 157
Saros Member = 3 of 70

Sun at Greatest Eclipse
(Geocentric Coordinates)
R.A. = 07h29m49.1s
Dec. = +21°49'23.2"
S.D. = 00°15'43.9"
H.P. = 00°00'08.7"

Moon at Greatest Eclipse
(Geocentric Coordinates)
R.A. = 07h30m06.1s
Dec. = +23°01'02.4"
S.D. = 00°14'54.5"
H.P. = 00°54'43.0"

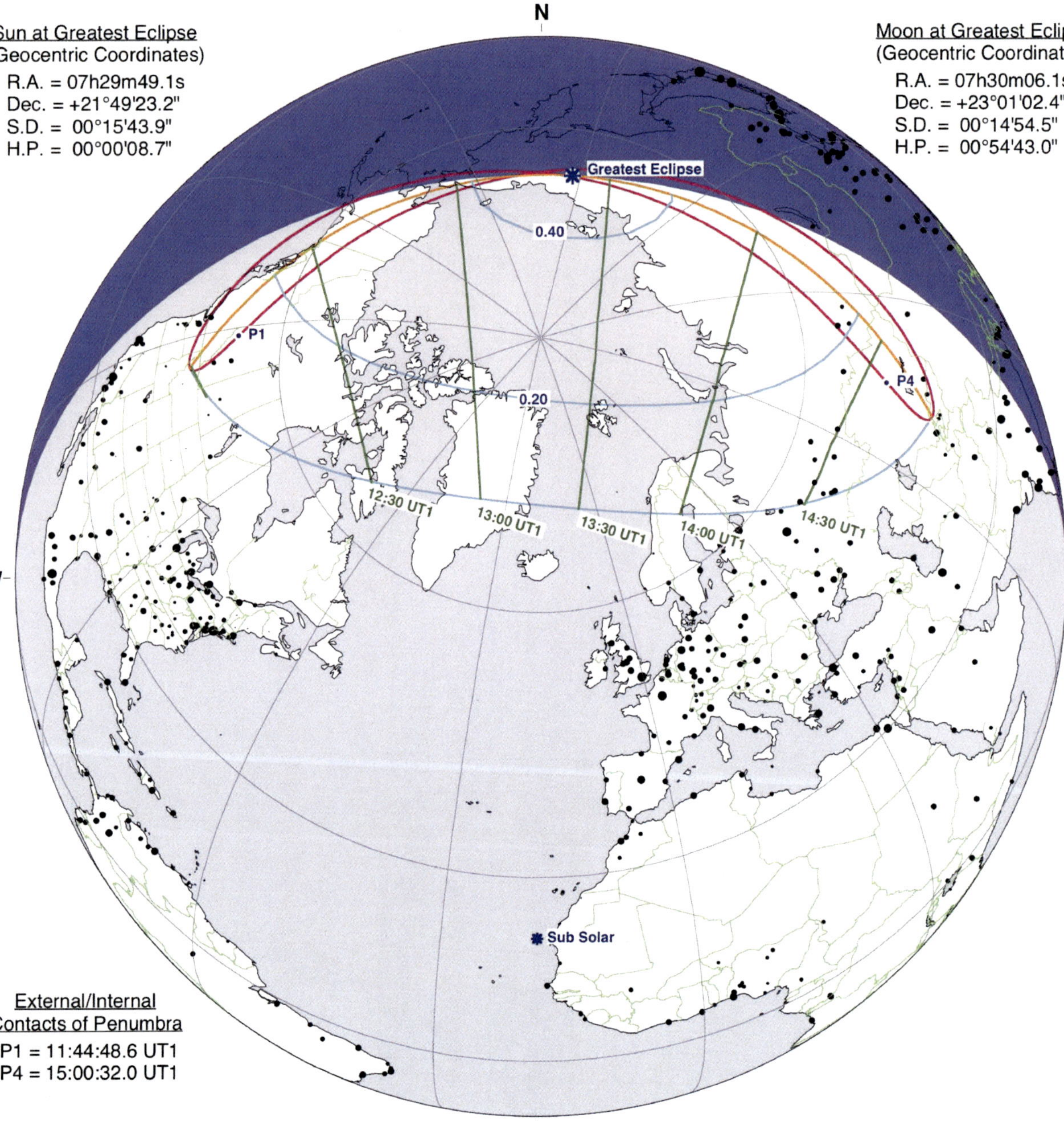

External/Internal
Contacts of Penumbra
P1 = 11:44:48.6 UT1
P4 = 15:00:32.0 UT1

ΔT = 118.3 s

Eph. = JPL DE405

Circumstances at Greatest Eclipse: 13:22:36.6 UT1

Lat. = 67°58.1'N
Long. = 152°27.1'E

Sun Alt. = 0.0°
Sun Azm. = 352.2°

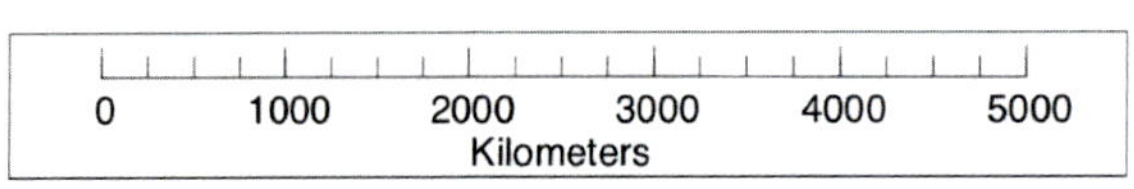

Partial Solar Eclipse of 2094 Dec 07

Greatest Eclipse = 20:05:55.6 TD (= 20:03:56.9 UT1)

Eclipse Magnitude = 0.7046
Gamma = 1.1547

Saros Series = 124
Saros Member = 59 of 73

Sun at Greatest Eclipse
(Geocentric Coordinates)
R.A. = 17h00m09.4s
Dec. = -22°42'52.2"
S.D. = 00°16'13.9"
H.P. = 00°00'08.9"

Moon at Greatest Eclipse
(Geocentric Coordinates)
R.A. = 17h01m06.4s
Dec. = -21°37'52.2"
S.D. = 00°15'41.5"
H.P. = 00°57'35.2"

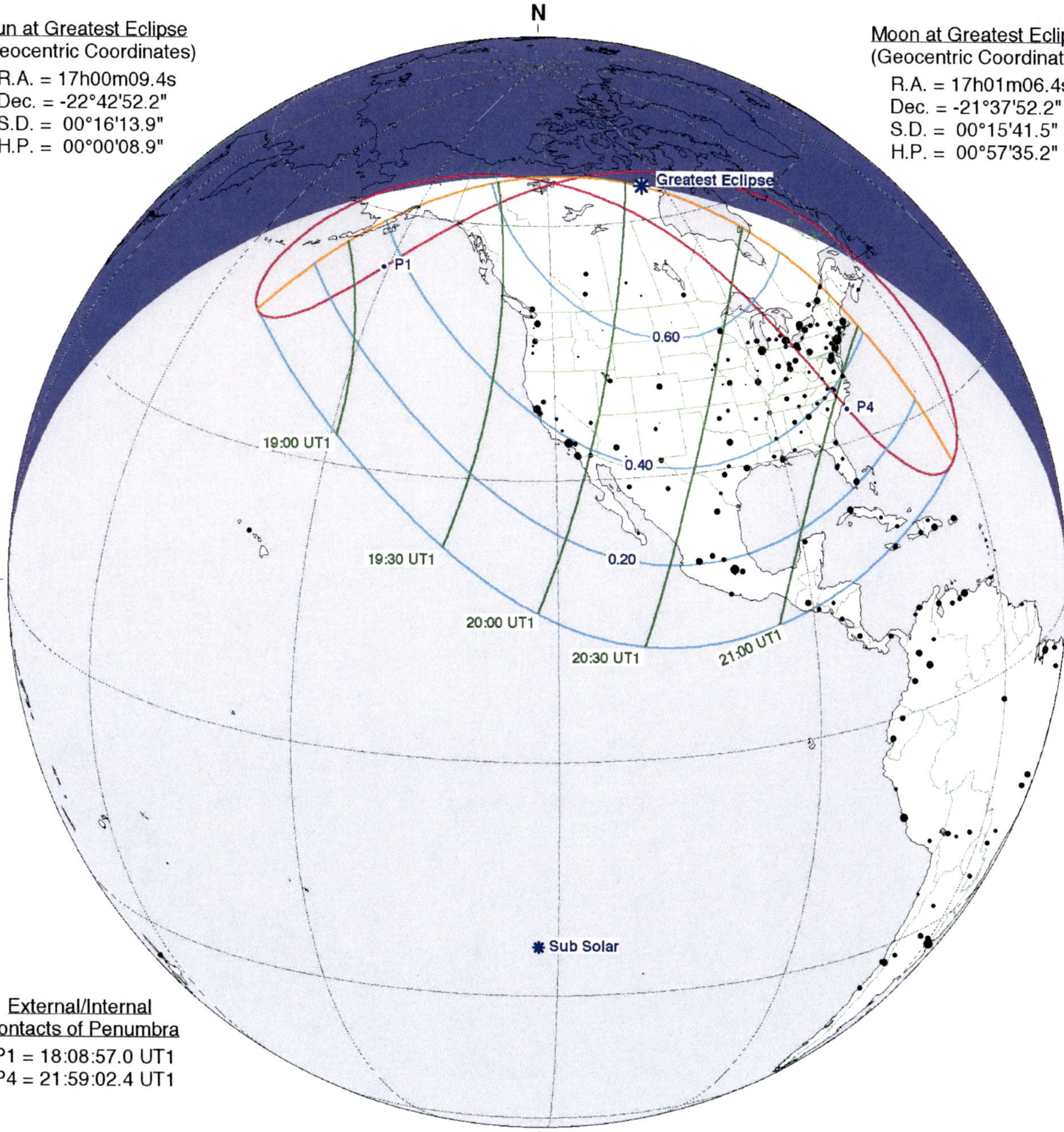

External/Internal
Contacts of Penumbra
P1 = 18:08:57.0 UT1
P4 = 21:59:02.4 UT1

ΔT = 118.7 s

Eph. = JPL DE405

Circumstances at Greatest Eclipse: 20:03:56.9 UT1

Lat. = 64°40.1'N
Long. = 095°17.3'W

Sun Alt. = 0.0°
Sun Azm. = 205.5°

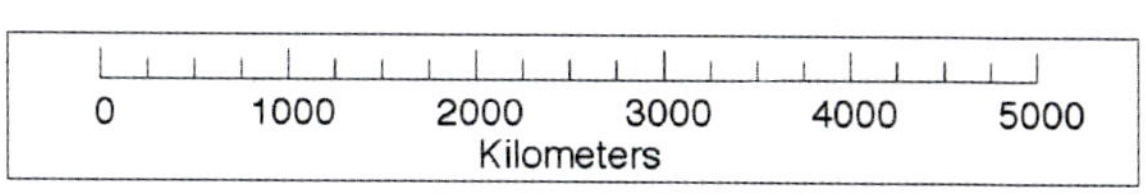

Total Solar Eclipse of 2095 Jun 02

Greatest Eclipse = 10:07:39.9 TD (= 10:05:40.7 UT1)

Eclipse Magnitude = 1.0332 Saros Series = 129
Gamma = -0.6396 Saros Member = 56 of 80

Sun at Greatest Eclipse
(Geocentric Coordinates)
R.A. = 04h42m53.4s
Dec. = +22°14'41.8"
S.D. = 00°15'46.4"
H.P. = 00°00'08.7"

Moon at Greatest Eclipse
(Geocentric Coordinates)
R.A. = 04h43m30.2s
Dec. = +21°37'59.7"
S.D. = 00°16'05.6"
H.P. = 00°59'03.8"

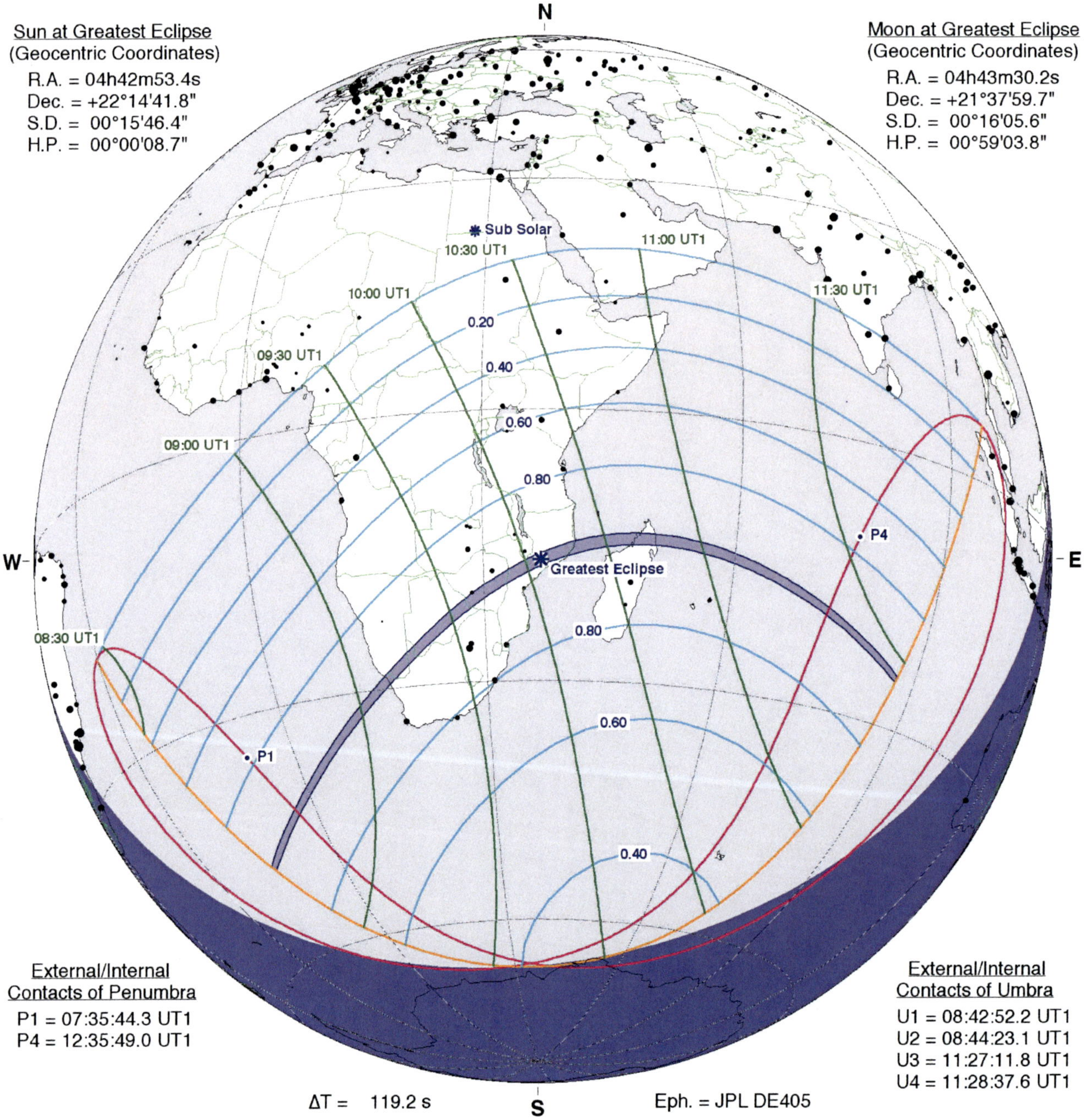

External/Internal
Contacts of Penumbra
P1 = 07:35:44.3 UT1
P4 = 12:35:49.0 UT1

External/Internal
Contacts of Umbra
U1 = 08:42:52.2 UT1
U2 = 08:44:23.1 UT1
U3 = 11:27:11.8 UT1
U4 = 11:28:37.6 UT1

ΔT = 119.2 s

Eph. = JPL DE405

Circumstances at Greatest Eclipse: 10:05:40.7 UT1

Lat. = 16°42.3'S Sun Alt. = 50.1°
Long. = 036°51.2'E Sun Azm. = 347.4°
Path Width = 144.8 km Duration = 03m18.4s

Circumstances at Greatest Duration: 10:06:58.5 UT1

Lat. = 16°33.3'S Sun Alt. = 50.1°
Long. = 037°14.1'E Sun Azm. = 346.3°
Path Width = 145.0 km Duration = 03m18.5s

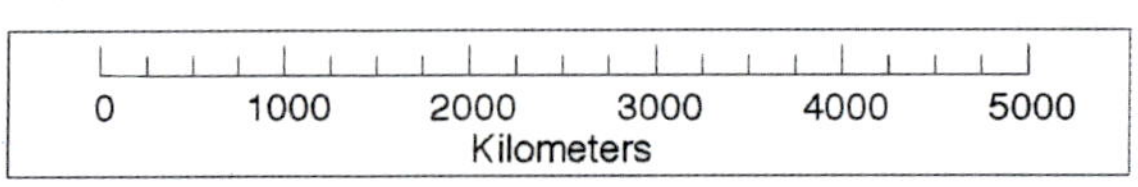

Annular Solar Eclipse of 2095 Nov 27

Greatest Eclipse = 01:02:57.4 TD (= 01:00:57.8 UT1)

Eclipse Magnitude = 0.9330 Saros Series = 134

Gamma = 0.4903 Saros Member = 48 of 71

Sun at Greatest Eclipse
(Geocentric Coordinates)

R.A. = 16h12m24.6s
Dec. = -21°07'41.4"
S.D. = 00°16'12.2"
H.P. = 00°00'08.9"

Moon at Greatest Eclipse
(Geocentric Coordinates)

R.A. = 16h12m56.4s
Dec. = -20°41'58.0"
S.D. = 00°14'55.2"
H.P. = 00°54'45.3"

N

W

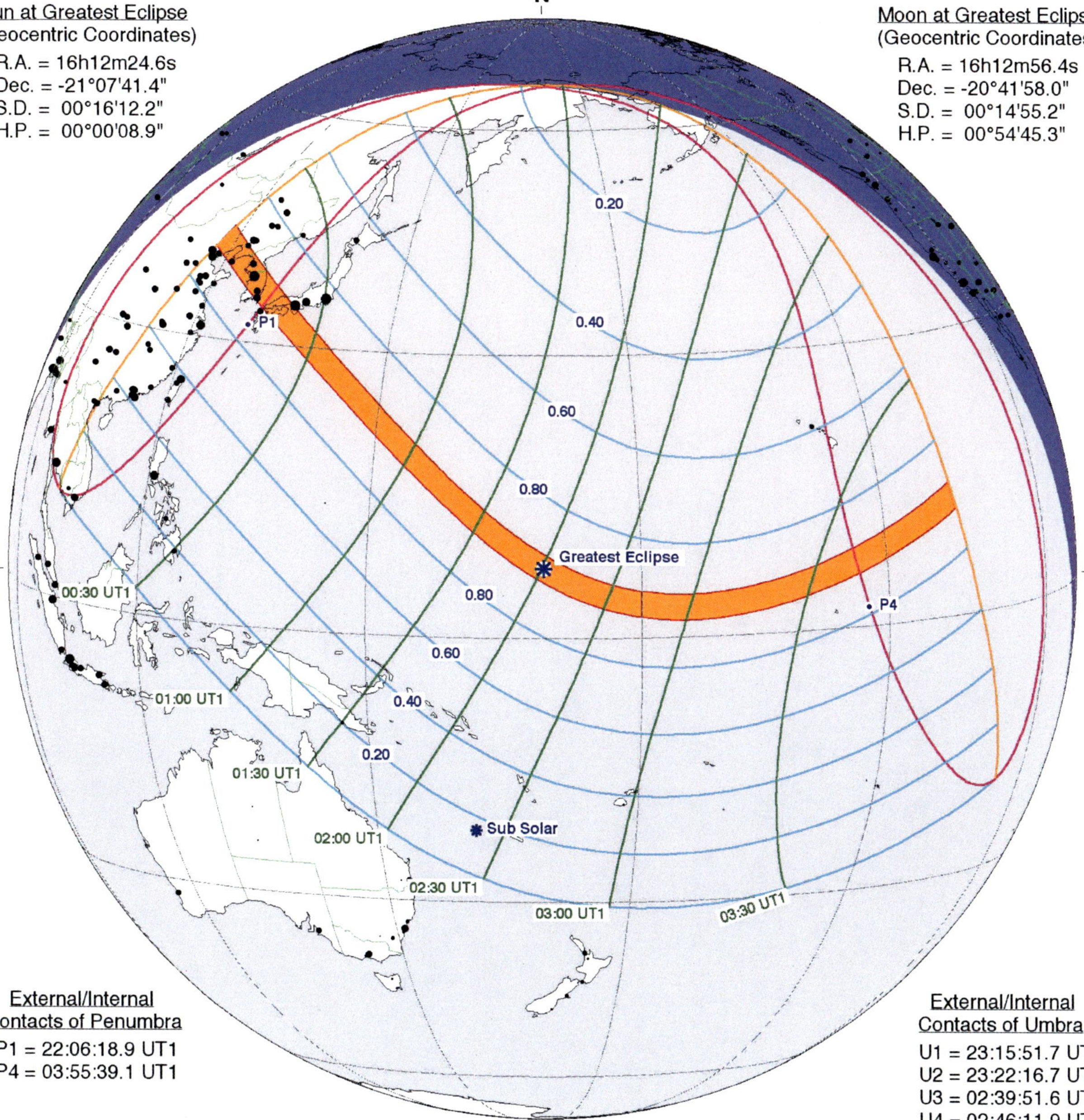

E

S

External/Internal
Contacts of Penumbra

P1 = 22:06:18.9 UT1
P4 = 03:55:39.1 UT1

External/Internal
Contacts of Umbra

U1 = 23:15:51.7 UT1
U2 = 23:22:16.7 UT1
U3 = 02:39:51.6 UT1
U4 = 02:46:11.9 UT1

ΔT = 119.6 s Eph. = JPL DE405

Circumstances at Greatest Eclipse: 01:00:57.8 UT1

Lat. = 07°14.4'N Sun Alt. = 60.6°
Long. = 169°30.1'E Sun Azm. = 195.1°
Path Width = 284.8 km Duration = 08m46.7s

Circumstances at Greatest Duration: 01:11:24.9 UT1

Lat. = 05°59.6'N Sun Alt. = 60.1°
Long. = 171°53.4'E Sun Azm. = 204.6°
Path Width = 288.6 km Duration = 08m48.4s

0 1000 2000 3000 4000 5000
Kilometers

Total Solar Eclipse of 2096 May 22

Greatest Eclipse = 01:37:14.1 TD (= 01:35:14.1 UT1)

Eclipse Magnitude = 1.0737

Gamma = 0.1196

Saros Series = 139

Saros Member = 34 of 71

Sun at Greatest Eclipse
(Geocentric Coordinates)

R.A. = 03h59m45.5s
Dec. = +20°33'28.2"
S.D. = 00°15'48.1"
H.P. = 00°00'08.7"

Moon at Greatest Eclipse
(Geocentric Coordinates)

R.A. = 03h59m36.3s
Dec. = +20°40'26.9"
S.D. = 00°16'40.8"
H.P. = 01°01'13.0"

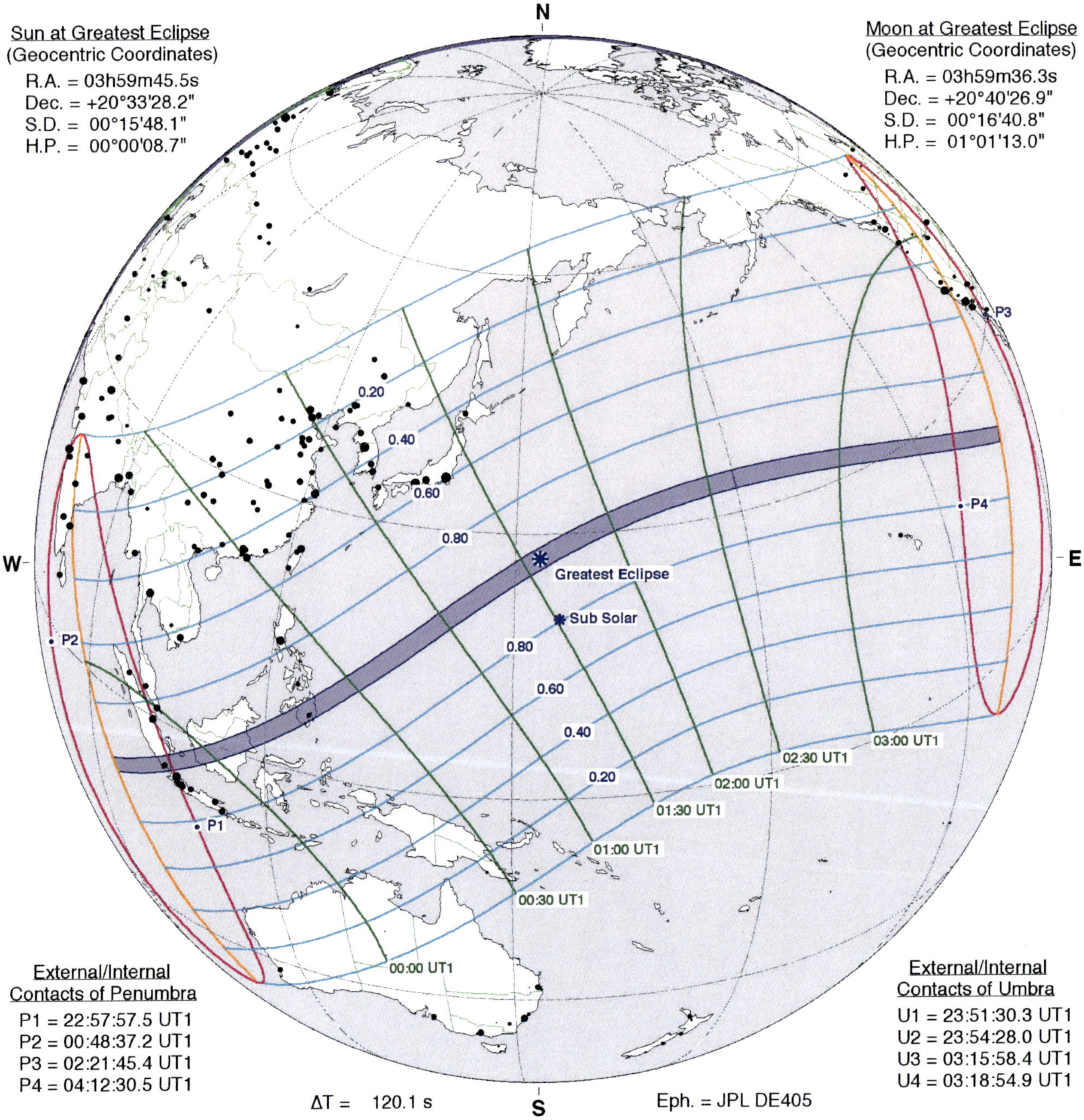

External/Internal Contacts of Penumbra

P1 = 22:57:57.5 UT1
P2 = 00:48:37.2 UT1
P3 = 02:21:45.4 UT1
P4 = 04:12:30.5 UT1

External/Internal Contacts of Umbra

U1 = 23:51:30.3 UT1
U2 = 23:54:28.0 UT1
U3 = 03:15:58.4 UT1
U4 = 03:18:54.9 UT1

ΔT = 120.1 s

Eph. = JPL DE405

Circumstances at Greatest Eclipse: 01:35:14.1 UT1

Lat. = 27°16.0'N
Long. = 153°07.8'E
Path Width = 240.9 km
Sun Alt. = 83.0°
Sun Azm. = 162.3°
Duration = 06m06.5s

Circumstances at Greatest Duration: 01:40:24.6 UT1

Lat. = 28°08.3'N
Long. = 154°56.2'E
Path Width = 240.6 km
Sun Alt. = 82.4°
Sun Azm. = 185.9°
Duration = 06m06.9s

0 1000 2000 3000 4000 5000
Kilometers

Annular Solar Eclipse of 2096 Nov 15

Greatest Eclipse = 00:36:14.8 TD (= 00:34:14.3 UT1)

Eclipse Magnitude = 0.9237 Saros Series = 144

Gamma = -0.2018 Saros Member = 21 of 70

Sun at Greatest Eclipse
(Geocentric Coordinates)
R.A. = 15h25m10.4s
Dec. = -18°40'58.6"
S.D. = 00°16'10.0"
H.P. = 00°00'08.9"

Moon at Greatest Eclipse
(Geocentric Coordinates)
R.A. = 15h24m54.6s
Dec. = -18°51'10.6"
S.D. = 00°14'42.9"
H.P. = 00°54'00.1"

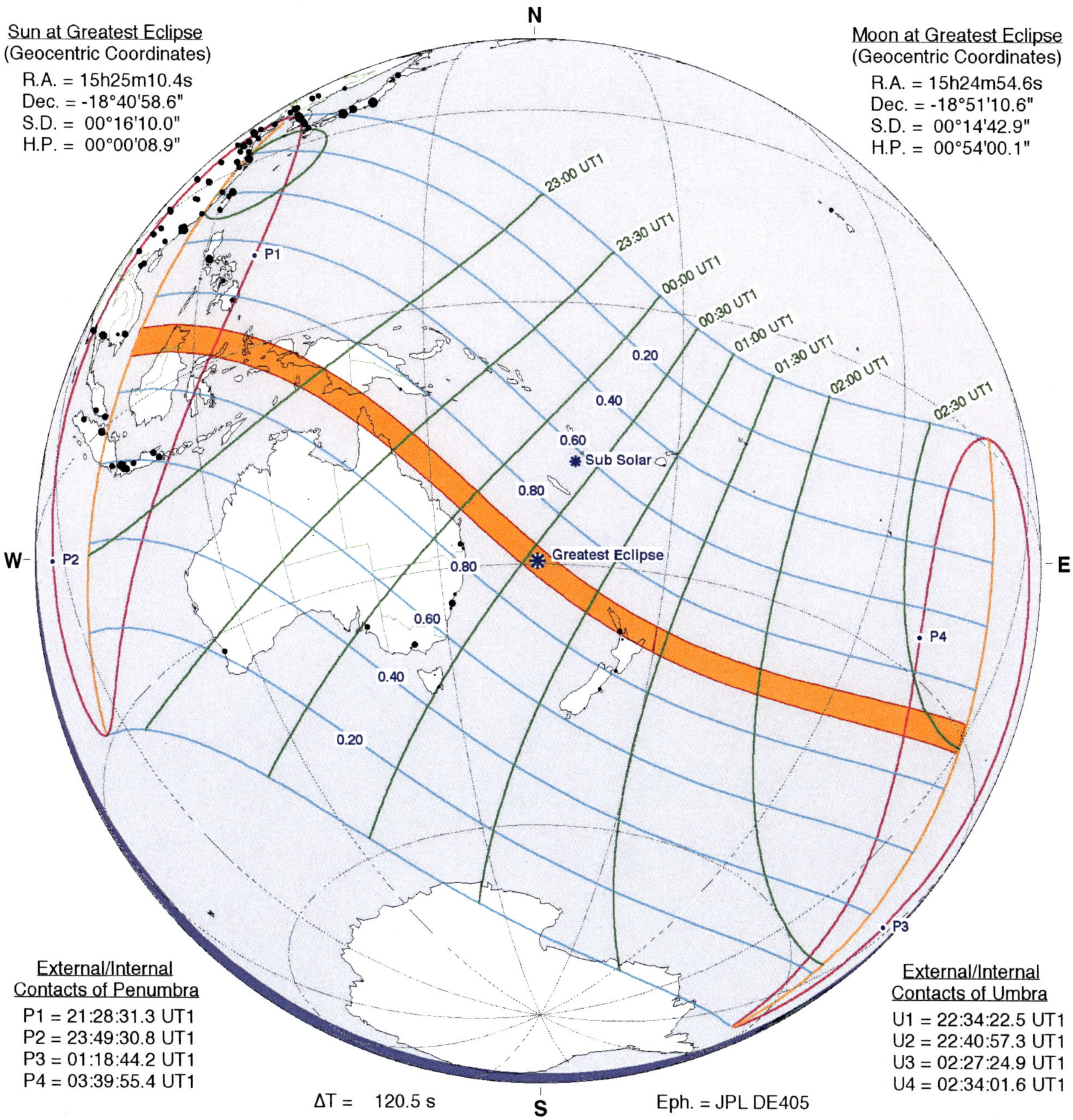

External/Internal
Contacts of Penumbra
P1 = 21:28:31.3 UT1
P2 = 23:49:30.8 UT1
P3 = 01:18:44.2 UT1
P4 = 03:39:55.4 UT1

External/Internal
Contacts of Umbra
U1 = 22:34:22.5 UT1
U2 = 22:40:57.3 UT1
U3 = 02:27:24.9 UT1
U4 = 02:34:01.6 UT1

ΔT = 120.5 s Eph. = JPL DE405

Circumstances at Greatest Eclipse: 00:34:14.3 UT1

Lat. = 29°43.4'S Sun Alt. = 78.2°
Long. = 162°59.4'E Sun Azm. = 21.8°
Path Width = 293.6 km Duration = 08m52.5s

Circumstances at Greatest Duration: 00:50:32.4 UT1

Lat. = 32°31.4'S Sun Alt. = 75.7°
Long. = 167°30.9'E Sun Azm. = 344.5°
Path Width = 292.7 km Duration = 08m55.4s

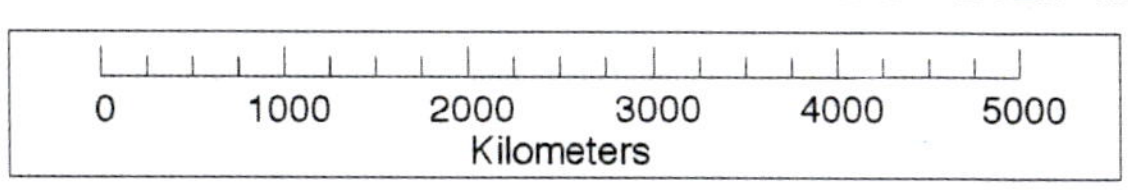

Total Solar Eclipse of 2097 May 11

Greatest Eclipse = 18:34:31.4 TD (= 18:32:30.4 UT1)

Eclipse Magnitude = 1.0538
Gamma = 0.8516

Saros Series = 149
Saros Member = 25 of 71

Sun at Greatest Eclipse
(Geocentric Coordinates)
R.A. = 03h17m49.7s
Dec. = +18°13'35.1"
S.D. = 00°15'50.2"
H.P. = 00°00'08.7"

Moon at Greatest Eclipse
(Geocentric Coordinates)
R.A. = 03h16m33.2s
Dec. = +19°01'53.1"
S.D. = 00°16'32.8"
H.P. = 01°00'43.7"

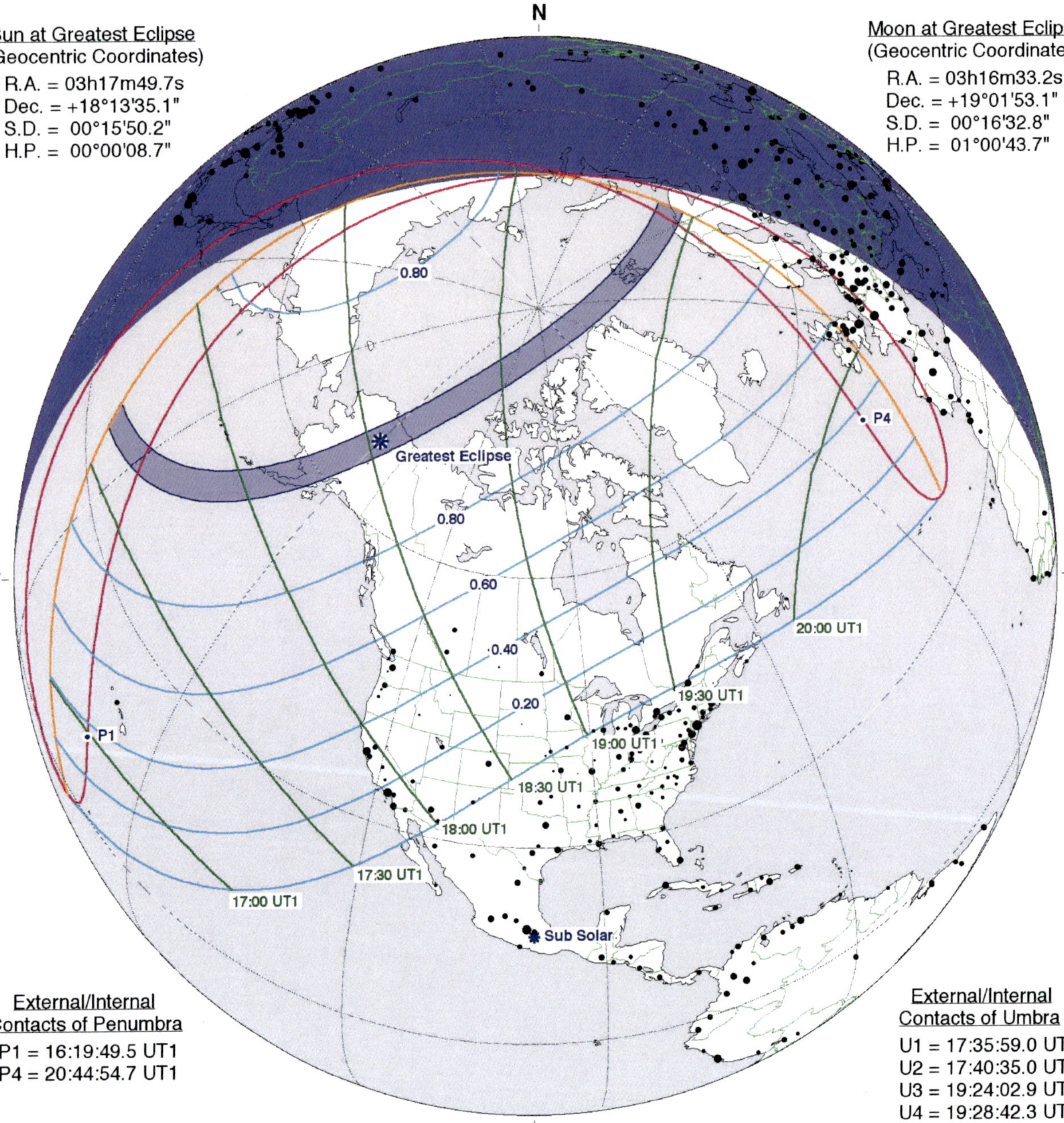

External/Internal Contacts of Penumbra
P1 = 16:19:49.5 UT1
P4 = 20:44:54.7 UT1

External/Internal Contacts of Umbra
U1 = 17:35:59.0 UT1
U2 = 17:40:35.0 UT1
U3 = 19:24:02.9 UT1
U4 = 19:28:42.3 UT1

ΔT = 121.0 s

Eph. = JPL DE405

Circumstances at Greatest Eclipse: 18:32:30.4 UT1

Lat. = 67°24.7'N
Long. = 149°50.0'W
Path Width = 339.5 km

Sun Alt. = 31.3°
Sun Azm. = 120.5°
Duration = 03m10.0s

Circumstances at Greatest Duration: 18:33:49.5 UT1

Lat. = 68°02.4'N
Long. = 149°12.2'W
Path Width = 338.6 km

Sun Alt. = 31.3°
Sun Azm. = 121.8°
Duration = 03m10.0s

0 1000 2000 3000 4000 5000
Kilometers

Annular Solar Eclipse of 2097 Nov 04

Greatest Eclipse = 02:01:25.2 TD (= 01:59:23.7 UT1)

Eclipse Magnitude = 0.9494 Saros Series = 154

Gamma = -0.8926 Saros Member = 11 of 71

Sun at Greatest Eclipse
(Geocentric Coordinates)
R.A. = 14h40m01.3s
Dec. = -15°33'59.2"
S.D. = 00°16'07.3"
H.P. = 00°00'08.9"

Moon at Greatest Eclipse
(Geocentric Coordinates)
R.A. = 14h38m39.0s
Dec. = -16°19'33.5"
S.D. = 00°15'12.3"
H.P. = 00°55'48.3"

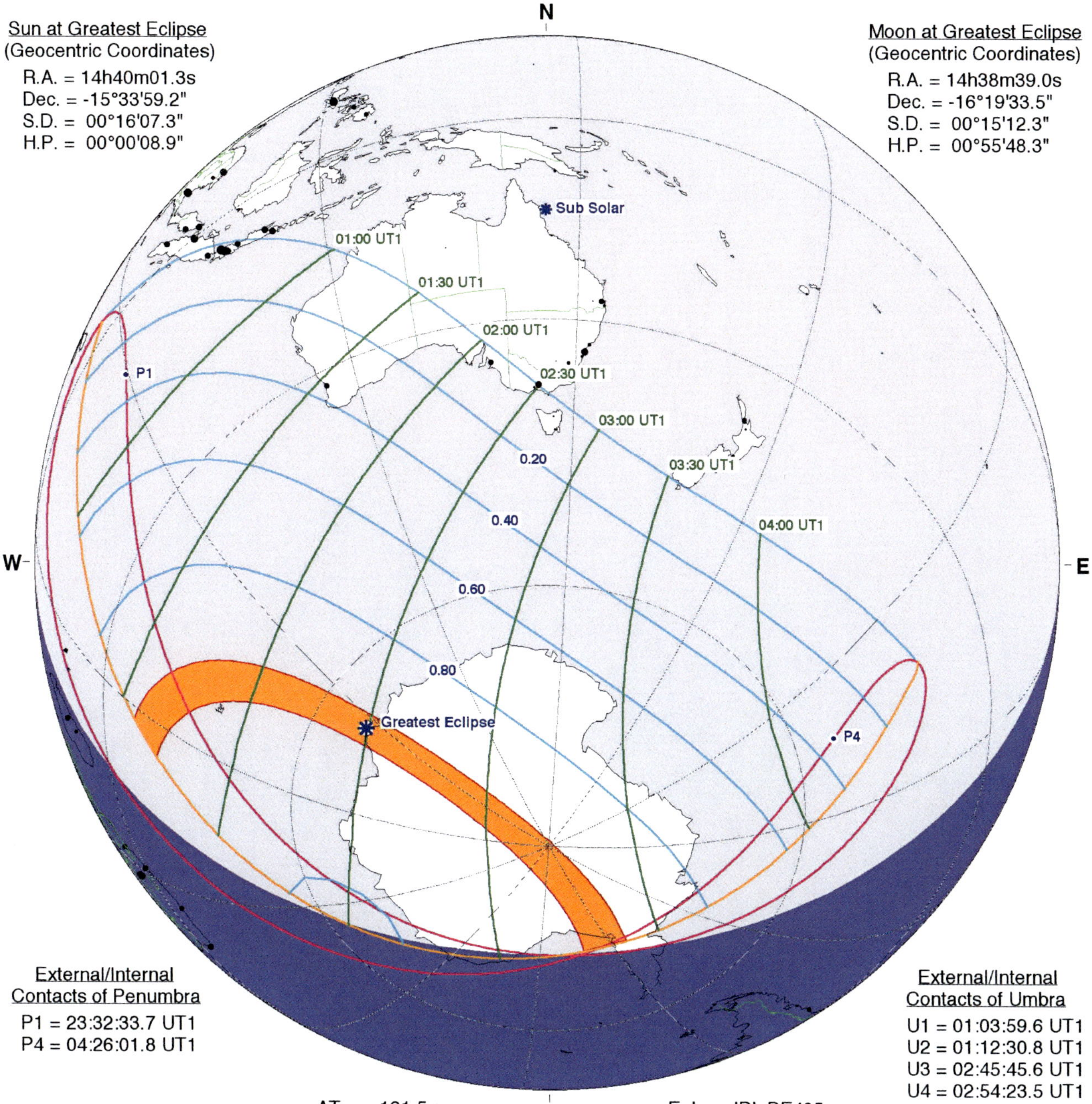

External/Internal
Contacts of Penumbra
P1 = 23:32:33.7 UT1
P4 = 04:26:01.8 UT1

External/Internal
Contacts of Umbra
U1 = 01:03:59.6 UT1
U2 = 01:12:30.8 UT1
U3 = 02:45:45.6 UT1
U4 = 02:54:23.5 UT1

ΔT = 121.5 s

Eph. = JPL DE405

Circumstances at Greatest Eclipse: 01:59:23.7 UT1

Lat. = 65°46.2'S Sun Alt. = 26.4°
Long. = 086°28.0'E Sun Azm. = 68.0°
Path Width = 410.7 km Duration = 03m35.6s

Circumstances at Greatest Duration: 02:08:26.3 UT1

Lat. = 70°04.1'S Sun Alt. = 25.9°
Long. = 088°06.1'E Sun Azm. = 62.2°
Path Width = 403.5 km Duration = 03m35.8s

0 1000 2000 3000 4000 5000
Kilometers

Partial Solar Eclipse of 2098 Apr 01

Greatest Eclipse = 20:02:30.8 TD (= 20:00:29.0 UT1)

Eclipse Magnitude = 0.7984
Gamma = -1.1005

Saros Series = 121
Saros Member = 65 of 71

Sun at Greatest Eclipse
(Geocentric Coordinates)
R.A. = 00h46m32.1s
Dec. = +04°59'38.4"
S.D. = 00°16'00.4"
H.P. = 00°00'08.8"

Moon at Greatest Eclipse
(Geocentric Coordinates)
R.A. = 00h48m30.7s
Dec. = +04°05'18.4"
S.D. = 00°15'21.2"
H.P. = 00°56'20.9"

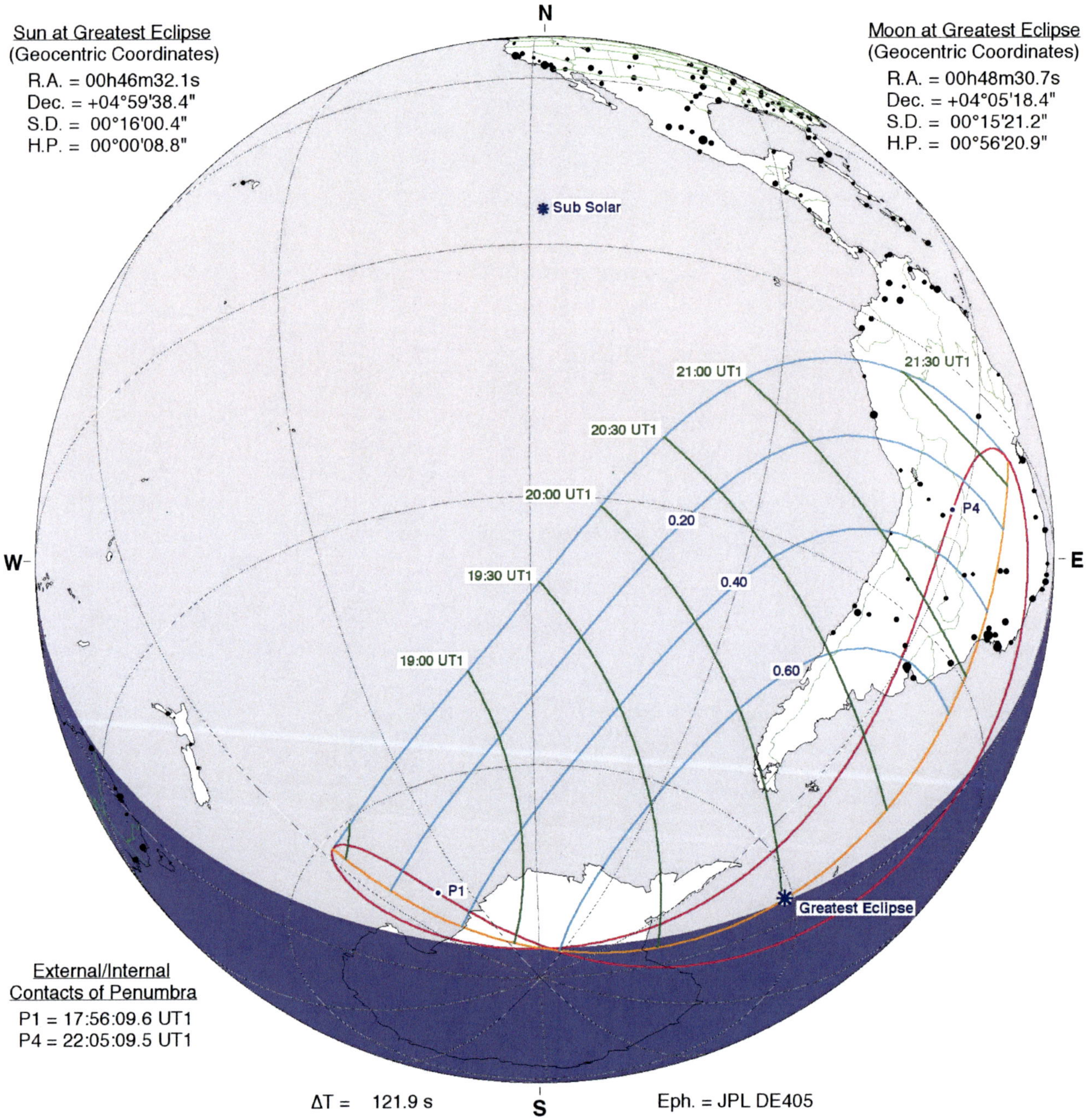

External/Internal
Contacts of Penumbra
P1 = 17:56:09.6 UT1
P4 = 22:05:09.5 UT1

ΔT = 121.9 s

Eph. = JPL DE405

Circumstances at Greatest Eclipse: 20:00:29.0 UT1

Lat. = 61°02.5'S
Long. = 038°24.1'W

Sun Alt. = 0.0°
Sun Azm. = 280.4°

0 1000 2000 3000 4000 5000
Kilometers

Partial Solar Eclipse of 2098 Sep 25

Greatest Eclipse = 00:31:16.2 TD (= 00:29:13.8 UT1)

Eclipse Magnitude = 0.7871
Gamma = 1.1184

Saros Series = 126
Saros Member = 52 of 72

Sun at Greatest Eclipse
(Geocentric Coordinates)
R.A. = 12h09m17.5s
Dec. = -01°00'22.0"
S.D. = 00°15'56.5"
H.P. = 00°00'08.8"

Moon at Greatest Eclipse
(Geocentric Coordinates)
R.A. = 12h11m27.7s
Dec. = -00°01'23.2"
S.D. = 00°16'27.0"
H.P. = 01°00'22.3"

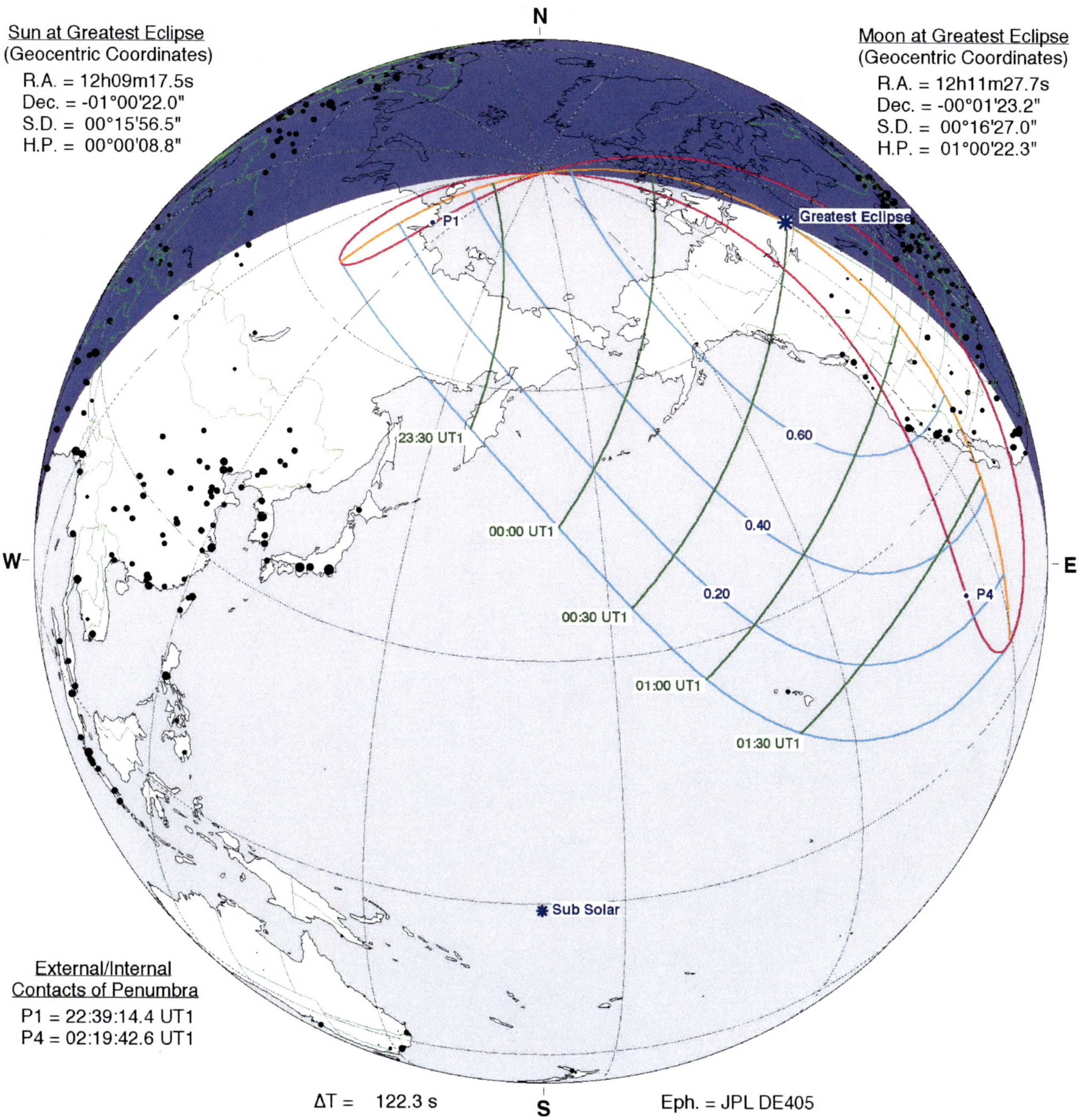

External/Internal
Contacts of Penumbra
P1 = 22:39:14.4 UT1
P4 = 02:19:42.6 UT1

ΔT = 122.3 s

Eph. = JPL DE405

Circumstances at Greatest Eclipse: 00:29:13.8 UT1

Lat. = 61°05.6'N
Long. = 101°17.4'W

Sun Alt. = 0.0°
Sun Azm. = 267.9°

0 1000 2000 3000 4000 5000
Kilometers

Partial Solar Eclipse of 2098 Oct 24

Greatest Eclipse = 10:36:10.8 TD (= 10:34:08.4 UT1)

Eclipse Magnitude = 0.0057
Gamma = -1.5407

Saros Series = 164
Saros Member = 1 of 80

Sun at Greatest Eclipse
(Geocentric Coordinates)
R.A. = 13h57m42.1s
Dec. = -12°01'06.6"
S.D. = 00°16'04.5"
H.P. = 00°00'08.8"

Moon at Greatest Eclipse
(Geocentric Coordinates)
R.A. = 13h55m00.2s
Dec. = -13°22'41.3"
S.D. = 00°16'04.0"
H.P. = 00°58'57.8"

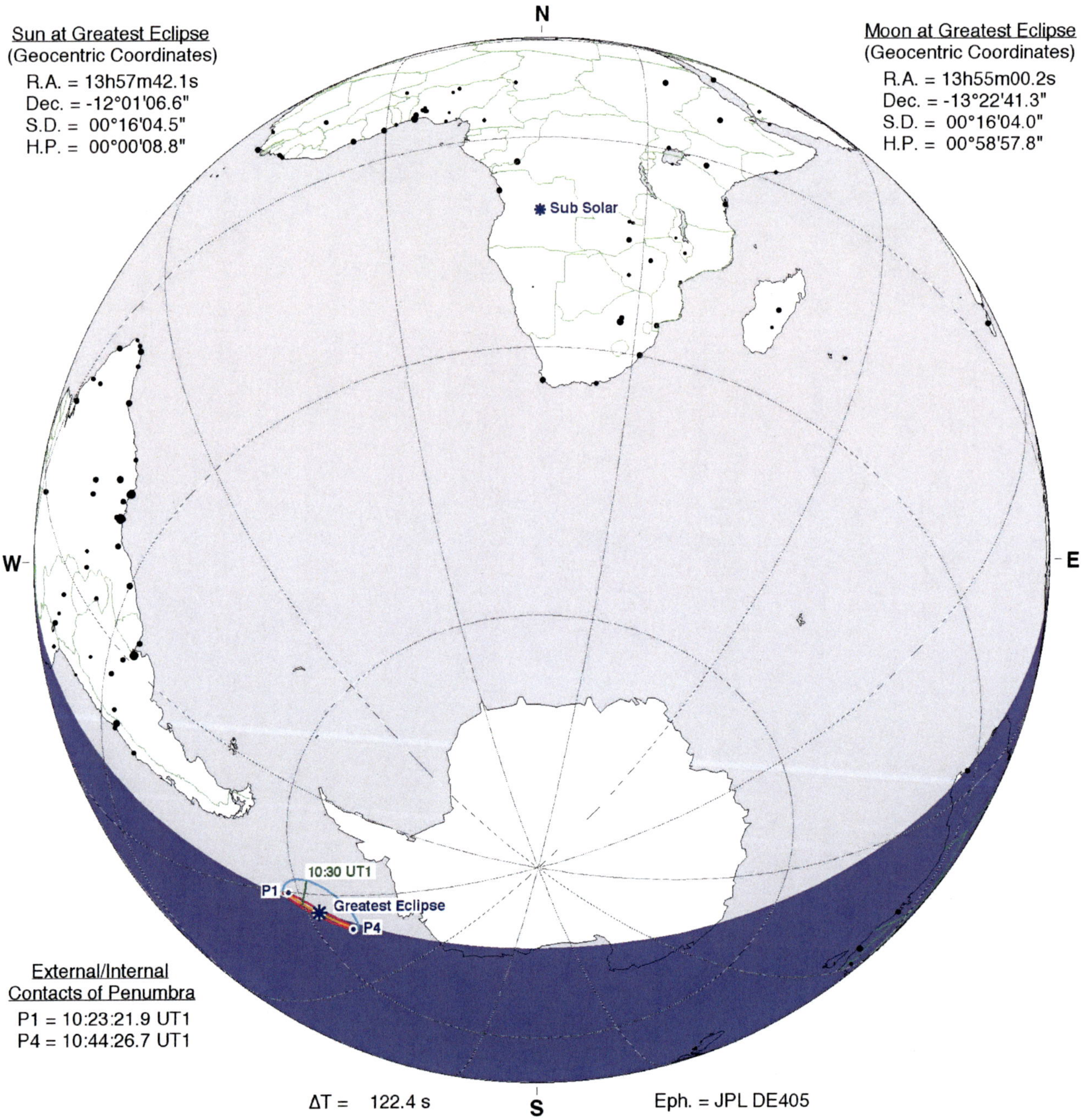

External/Internal
Contacts of Penumbra
P1 = 10:23:21.9 UT1
P4 = 10:44:26.7 UT1

ΔT = 122.4 s

Eph. = JPL DE405

Circumstances at Greatest Eclipse: 10:34:08.4 UT1

Lat. = 61°45.7'S
Long. = 095°47.2'W

Sun Alt. = 0.0°
Sun Azm. = 116.1°

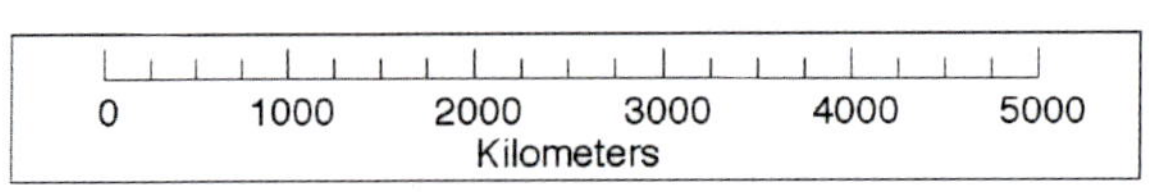

Annular Solar Eclipse of 2099 Mar 21

Greatest Eclipse = 22:54:32.0 TD (= 22:52:29.2 UT1)

Eclipse Magnitude = 0.9318 Saros Series = 131

Gamma = -0.4016 Saros Member = 55 of 70

Sun at Greatest Eclipse
(Geocentric Coordinates)
R.A. = 00h06m00.9s
Dec. = +00°39'05.5"
S.D. = 00°16'03.5"
H.P. = 00°00'08.8"

Moon at Greatest Eclipse
(Geocentric Coordinates)
R.A. = 00h06m43.3s
Dec. = +00°20'09.1"
S.D. = 00°14'45.5"
H.P. = 00°54'09.7"

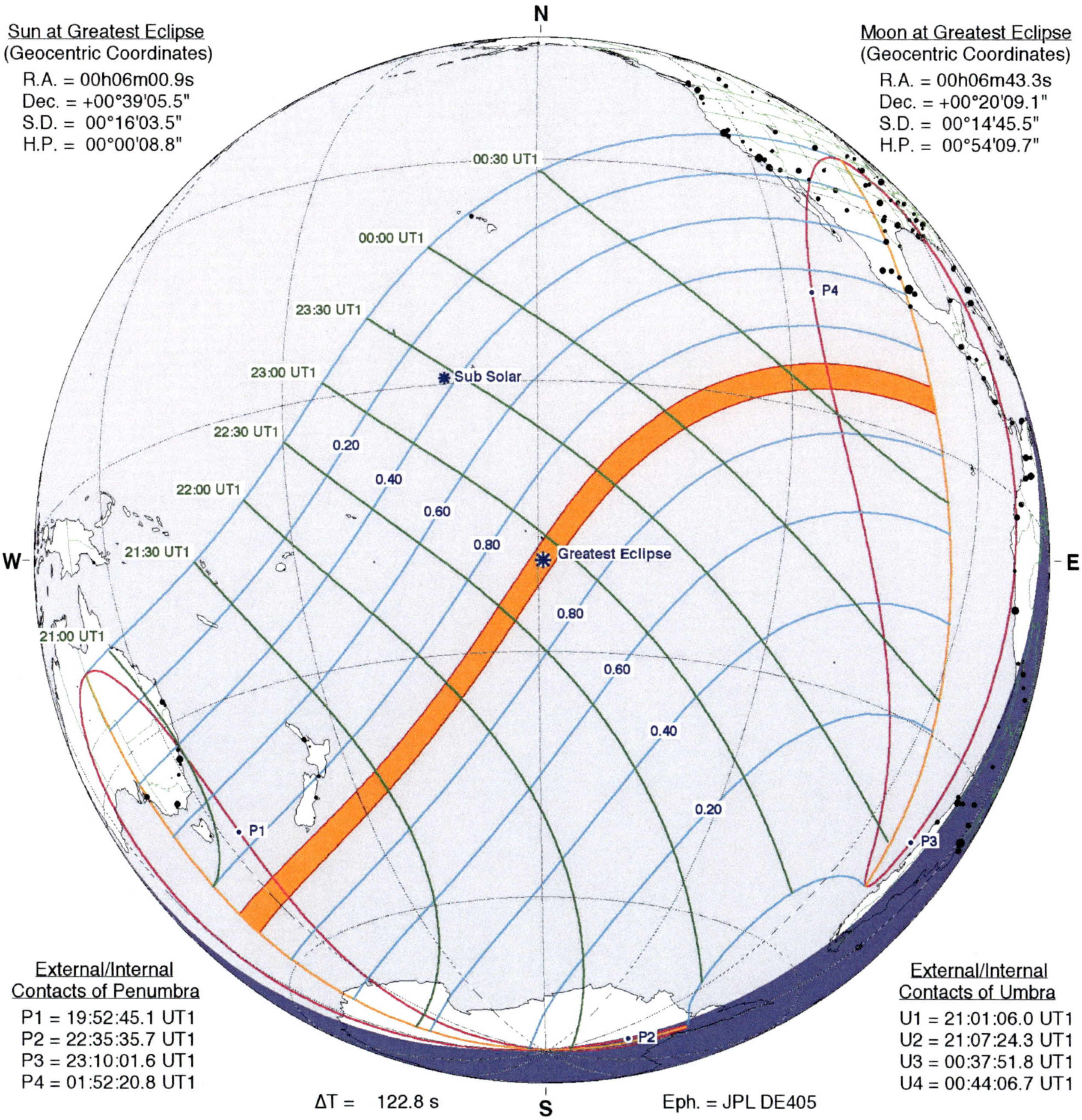

External/Internal Contacts of Penumbra
P1 = 19:52:45.1 UT1
P2 = 22:35:35.7 UT1
P3 = 23:10:01.6 UT1
P4 = 01:52:20.8 UT1

External/Internal Contacts of Umbra
U1 = 21:01:06.0 UT1
U2 = 21:07:24.3 UT1
U3 = 00:37:51.8 UT1
U4 = 00:44:06.7 UT1

ΔT = 122.8 s Eph. = JPL DE405

Circumstances at Greatest Eclipse: 22:52:29.2 UT1

Lat. = 20°01.2'S Sun Alt. = 66.2°
Long. = 149°22.4'W Sun Azm. = 328.9°
Path Width = 275.0 km Duration = 07m32.3s

Circumstances at Greatest Duration: 22:52:12.1 UT1

Lat. = 20°05.6'S Sun Alt. = 66.2°
Long. = 149°25.7'W Sun Azm. = 329.3°
Path Width = 274.9 km Duration = 07m32.3s

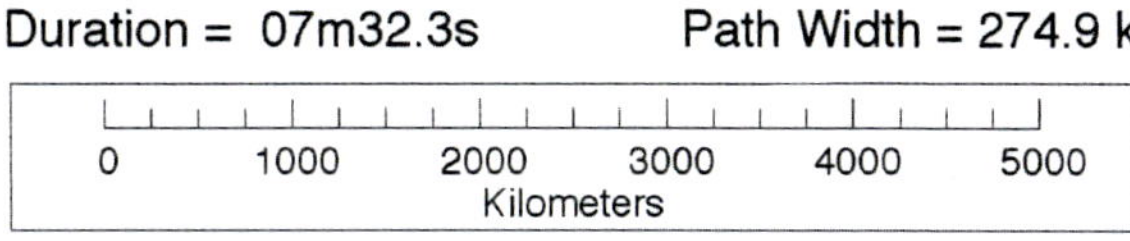

Total Solar Eclipse of 2099 Sep 14

Greatest Eclipse = 16:57:53.0 TD (= 16:55:49.7 UT1)

Eclipse Magnitude = 1.0684   Saros Series = 136
Gamma = 0.3942   Saros Member = 42 of 71

Sun at Greatest Eclipse
(Geocentric Coordinates)
R.A. = 11h31m25.7s
Dec. = +03°05'04.1"
S.D. = 00°15'53.8"
H.P. = 00°00'08.7"

Moon at Greatest Eclipse
(Geocentric Coordinates)
R.A. = 11h32m12.4s
Dec. = +03°26'11.8"
S.D. = 00°16'43.1"
H.P. = 01°01'21.6"

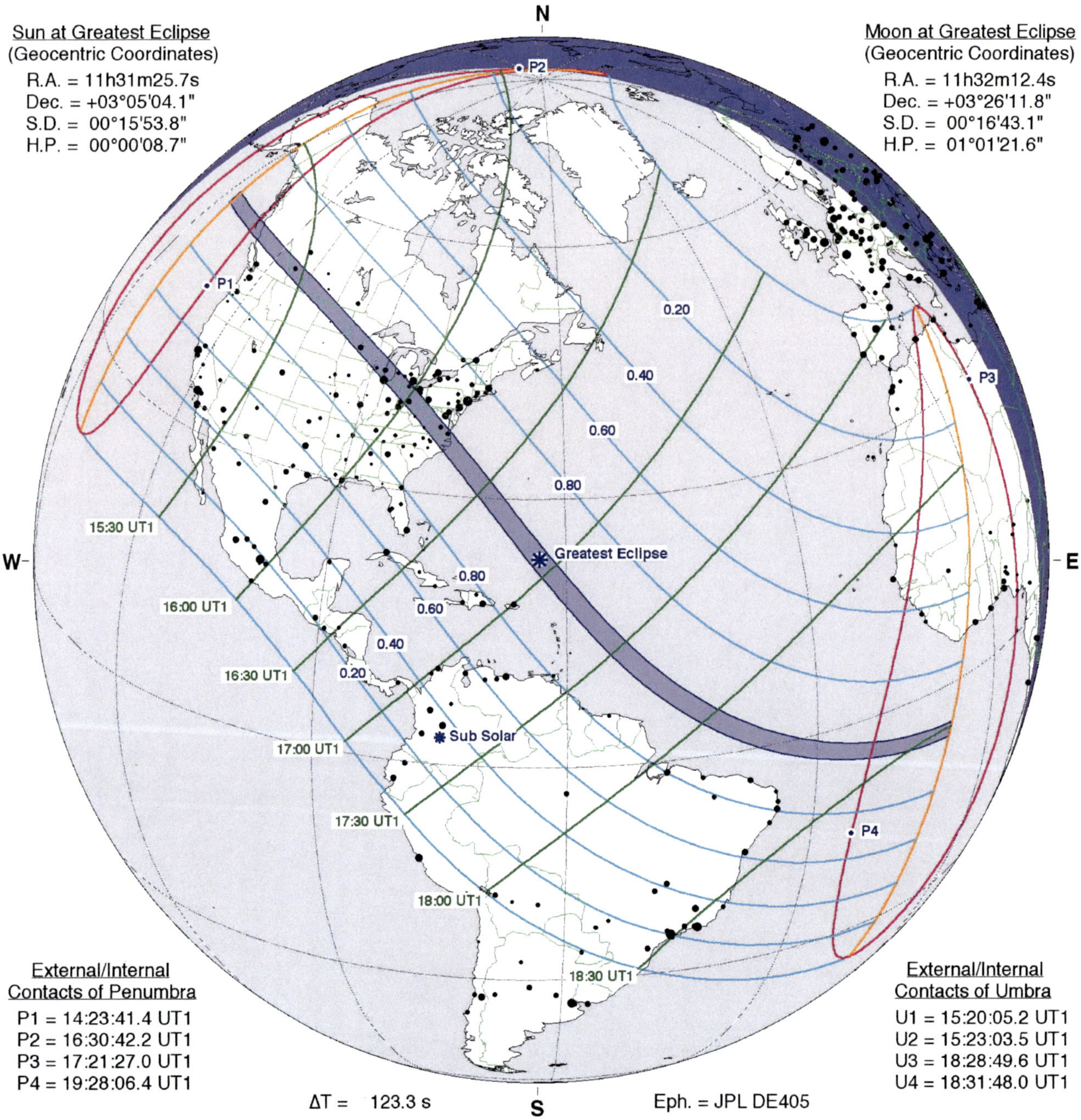

External/Internal Contacts of Penumbra
P1 = 14:23:41.4 UT1
P2 = 16:30:42.2 UT1
P3 = 17:21:27.0 UT1
P4 = 19:28:06.4 UT1

External/Internal Contacts of Umbra
U1 = 15:20:05.2 UT1
U2 = 15:23:03.5 UT1
U3 = 18:28:49.6 UT1
U4 = 18:31:48.0 UT1

ΔT = 123.3 s   Eph. = JPL DE405

Circumstances at Greatest Eclipse: 16:55:49.7 UT1
Lat. = 23°21.8'N   Sun Alt. = 66.7°
Long. = 063°08.2'W   Sun Azm. = 211.5°
Path Width = 241.0 km   Duration = 05m17.8s

Circumstances at Greatest Duration: 16:54:46.0 UT1
Lat. = 23°40.5'N   Sun Alt. = 66.6°
Long. = 063°25.2'W   Sun Azm. = 209.9°
Path Width = 240.7 km   Duration = 05m17.8s

0 1000 2000 3000 4000 5000
Kilometers

Annular Solar Eclipse of 2100 Mar 10

Greatest Eclipse = 22:28:11.0 TD (= 22:26:07.3 UT1)

Eclipse Magnitude = 0.9338 Saros Series = 141

Gamma = 0.3077 Saros Member = 28 of 70

Sun at Greatest Eclipse
(Geocentric Coordinates)
R.A. = 23h24m46.6s
Dec. = -03°47'43.4"
S.D. = 00°16'06.4"
H.P. = 00°00'08.9"

Moon at Greatest Eclipse
(Geocentric Coordinates)
R.A. = 23h24m14.3s
Dec. = -03°33'06.4"
S.D. = 00°14'49.6"
H.P. = 00°54'24.7"

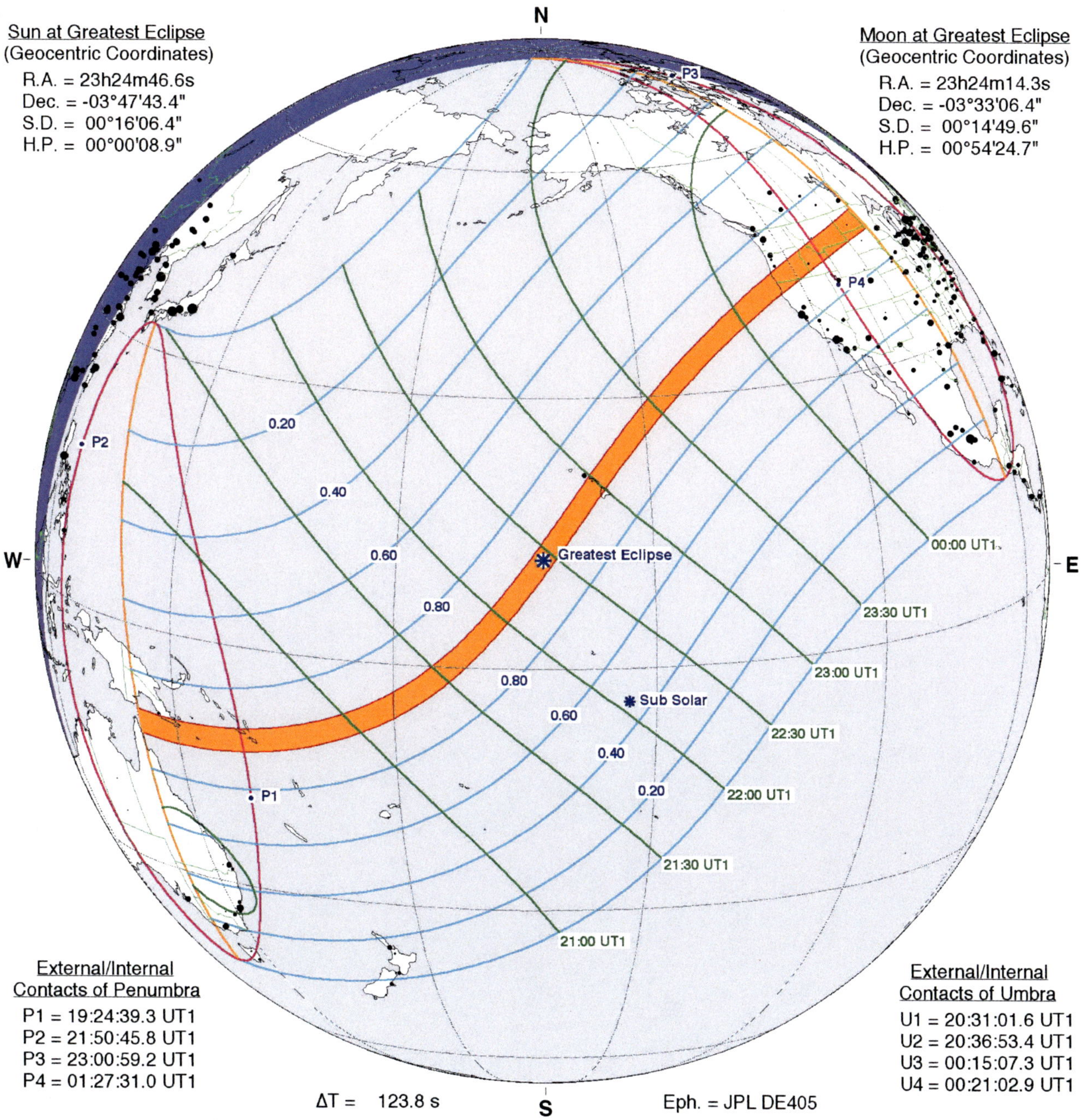

External/Internal
Contacts of Penumbra
P1 = 19:24:39.3 UT1
P2 = 21:50:45.8 UT1
P3 = 23:00:59.2 UT1
P4 = 01:27:31.0 UT1

External/Internal
Contacts of Umbra
U1 = 20:31:01.6 UT1
U2 = 20:36:53.4 UT1
U3 = 00:15:07.3 UT1
U4 = 00:21:02.9 UT1

ΔT = 123.8 s

Eph. = JPL DE405

Circumstances at Greatest Eclipse: 22:26:07.3 UT1

Lat. = 11°57.3'N Sun Alt. = 72.0°
Long. = 162°45.8'W Sun Azm. = 150.6°
Path Width = 257.5 km Duration = 07m29.2s

Circumstances at Greatest Duration: 22:19:52.0 UT1

Lat. = 10°27.3'N Sun Alt. = 71.8°
Long. = 163°54.1'W Sun Azm. = 140.8°
Path Width = 259.2 km Duration = 07m29.4s

0 1000 2000 3000 4000 5000
Kilometers

Total Solar Eclipse of 2100 Sep 04

Greatest Eclipse = 08:49:20.3 TD (= 08:47:16.0 UT1)

Eclipse Magnitude = 1.0402
Gamma = -0.3384

Saros Series = 146
Saros Member = 32 of 76

Sun at Greatest Eclipse
(Geocentric Coordinates)
R.A. = 10h53m24.7s
Dec. = +07°04'34.9"
S.D. = 00°15'51.3"
H.P. = 00°00'08.7"

Moon at Greatest Eclipse
(Geocentric Coordinates)
R.A. = 10h52m46.7s
Dec. = +06°46'49.6"
S.D. = 00°16'14.1"
H.P. = 00°59'35.2"

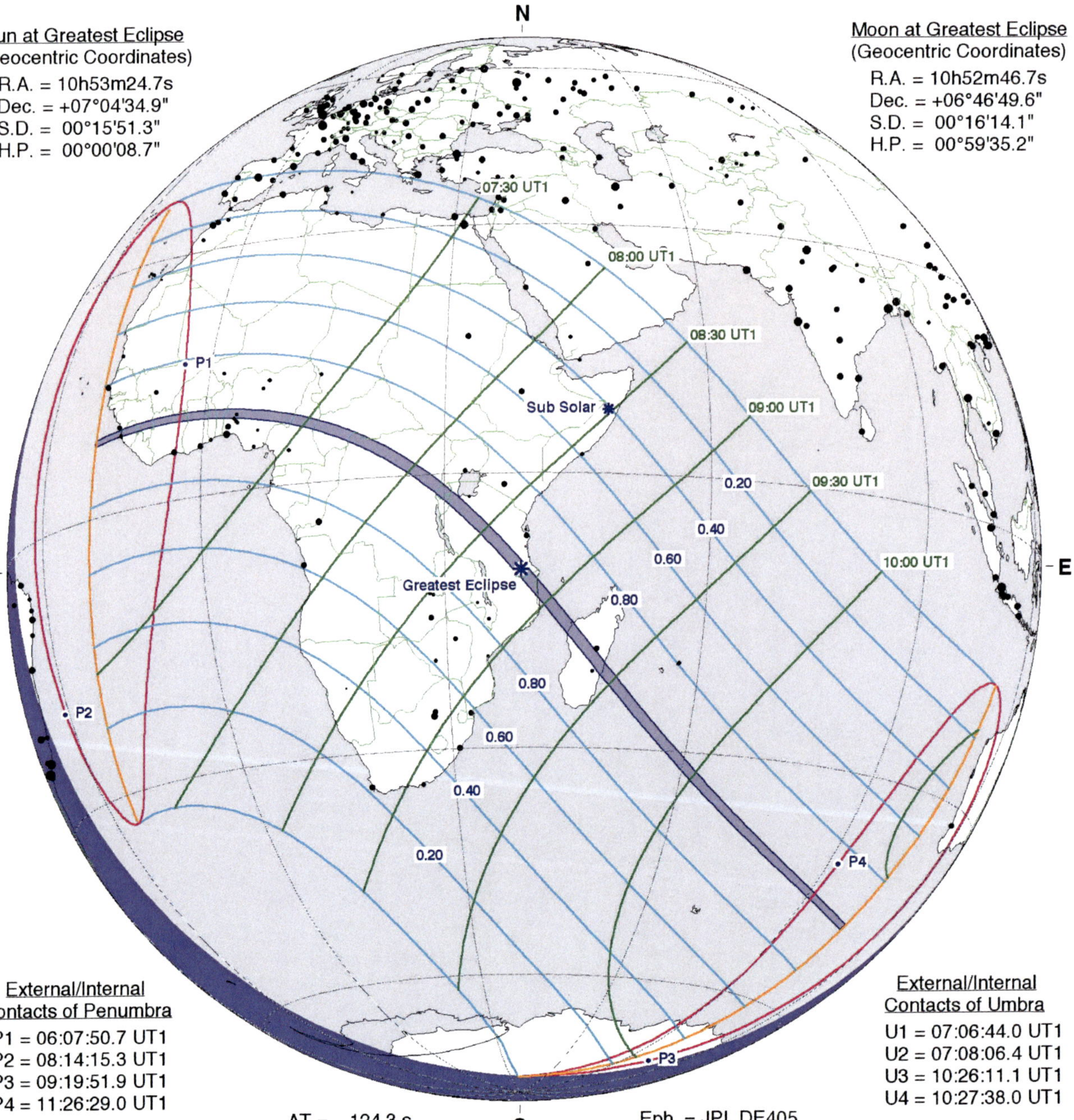

External/Internal
Contacts of Penumbra
P1 = 06:07:50.7 UT1
P2 = 08:14:15.3 UT1
P3 = 09:19:51.9 UT1
P4 = 11:26:29.0 UT1

External/Internal
Contacts of Umbra
U1 = 07:06:44.0 UT1
U2 = 07:08:06.4 UT1
U3 = 10:26:11.1 UT1
U4 = 10:27:38.0 UT1

ΔT = 124.3 s

Eph. = JPL DE405

Circumstances at Greatest Eclipse: 08:47:16.0 UT1

Lat. = 10°28.1'S
Long. = 038°39.7'E
Path Width = 142.2 km
Sun Alt. = 70.2°
Sun Azm. = 28.2°
Duration = 03m32.5s

Circumstances at Greatest Duration: 08:45:50.6 UT1

Lat. = 10°06.1'S
Long. = 038°19.7'E
Path Width = 142.3 km
Sun Alt. = 70.1°
Sun Azm. = 30.5°
Duration = 03m32.5s

0 1000 2000 3000 4000 5000
Kilometers

Appendix D

Central Solar Eclipse Maps: 2001 to 2100

Key to Central Solar Eclipse Maps

Key to Continental Solar Eclipse Maps (Appendix D)

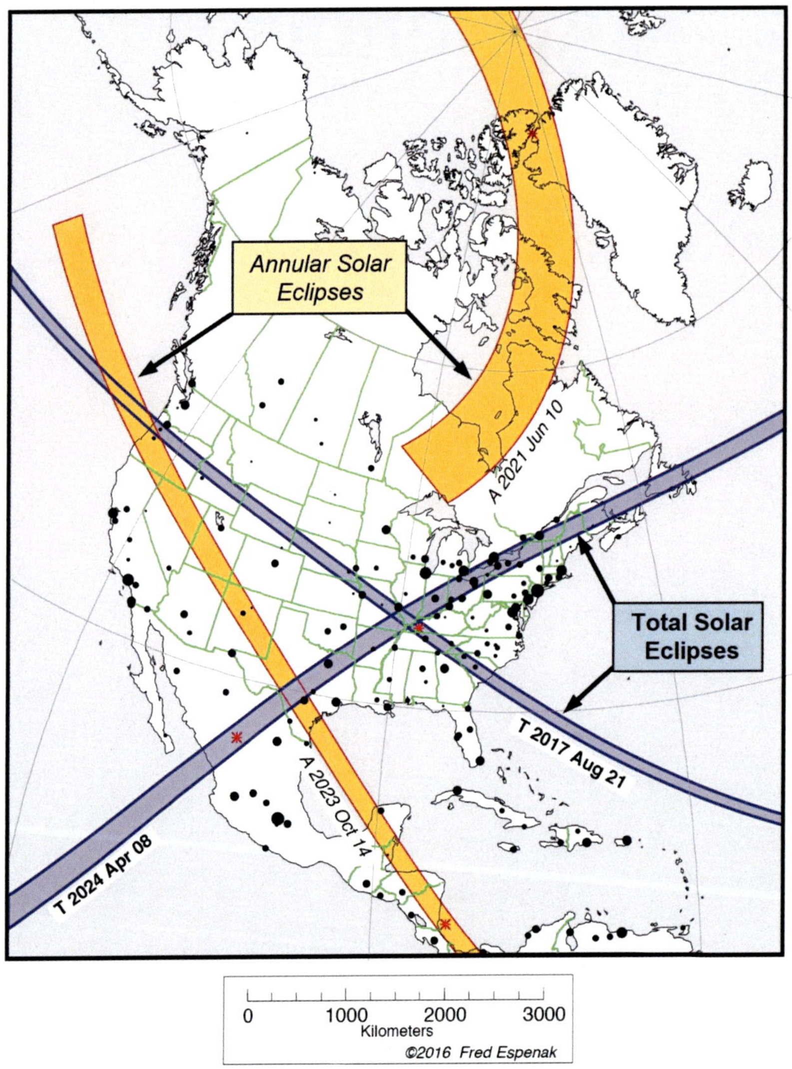

Total Eclipses –	the path is darkly shaded (blue in Color & Deluxe Editions); the date appears in bold and is preceded by the letter **T**
Annular Eclipses –	the path is lightly shaded (orange in Color & Deluxe Editions); the date appears in italics and is preceded by the letter **A**
Hybrid Eclipses –	the path is darkly shaded (or blue) along the total sections and lightly shaded (or orange) along the annular sections; the date appears in bold-italics and is preceded by the letter **H**

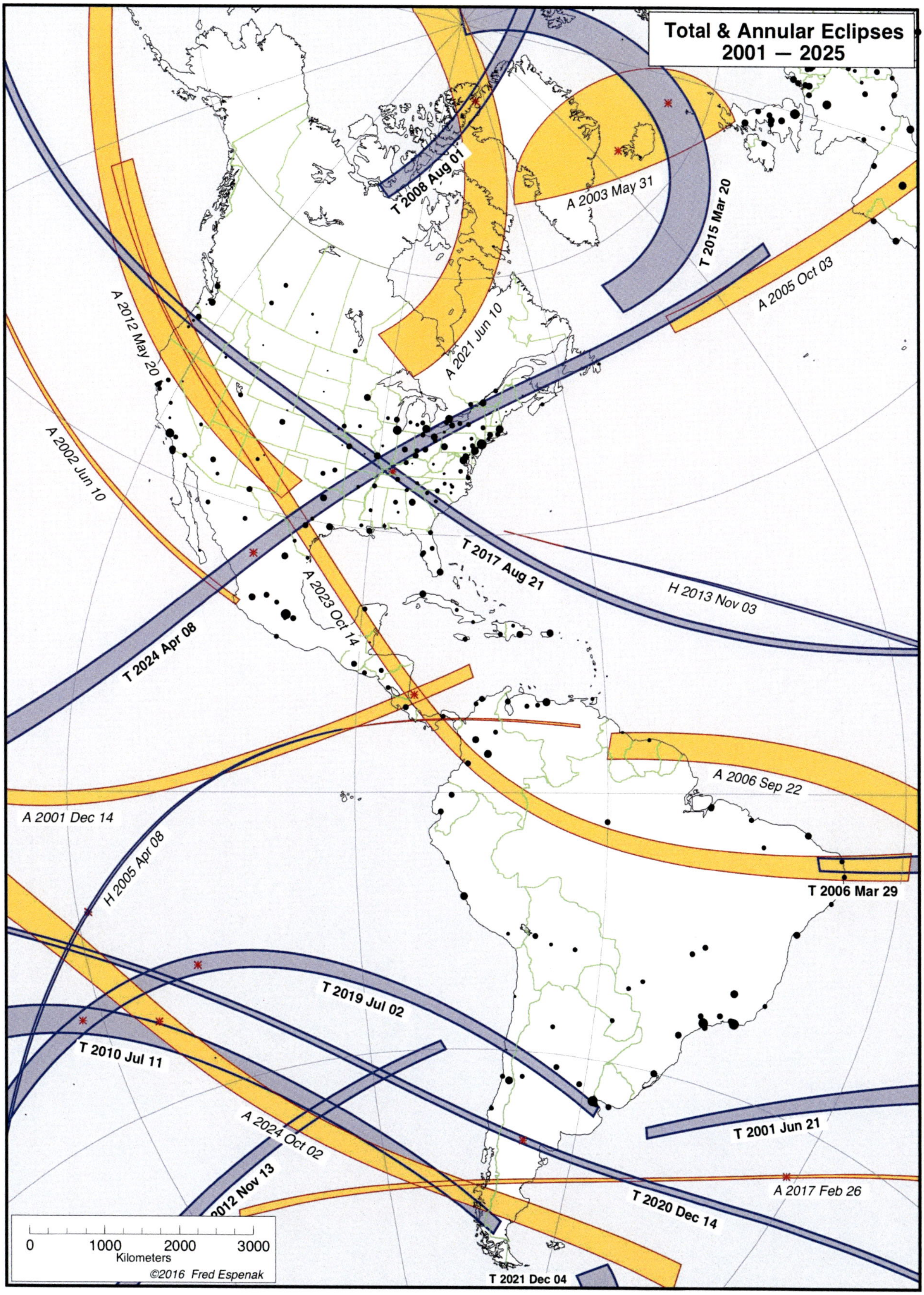

Total & Annular Eclipses
2001 — 2025
T 2008 Aug 01
A 2003 May 31
T 2015 Mar 20
A 2005 Oct 03
A 2021 Jun 10
A 2012 May 20
A 2002 Jun 10
T 2017 Aug 21
H 2013 Nov 03
A 2023 Oct 14
T 2024 Apr 08
A 2006 Sep 22
A 2001 Dec 14
H 2005 Apr 08
T 2006 Mar 29
T 2019 Jul 02
T 2010 Jul 11
A 2024 Oct 02
T 2001 Jun 21
2012 Nov 13
A 2017 Feb 26
T 2020 Dec 14
T 2021 Dec 04
0 1000 2000 3000
Kilometers
©2016 Fred Espenak

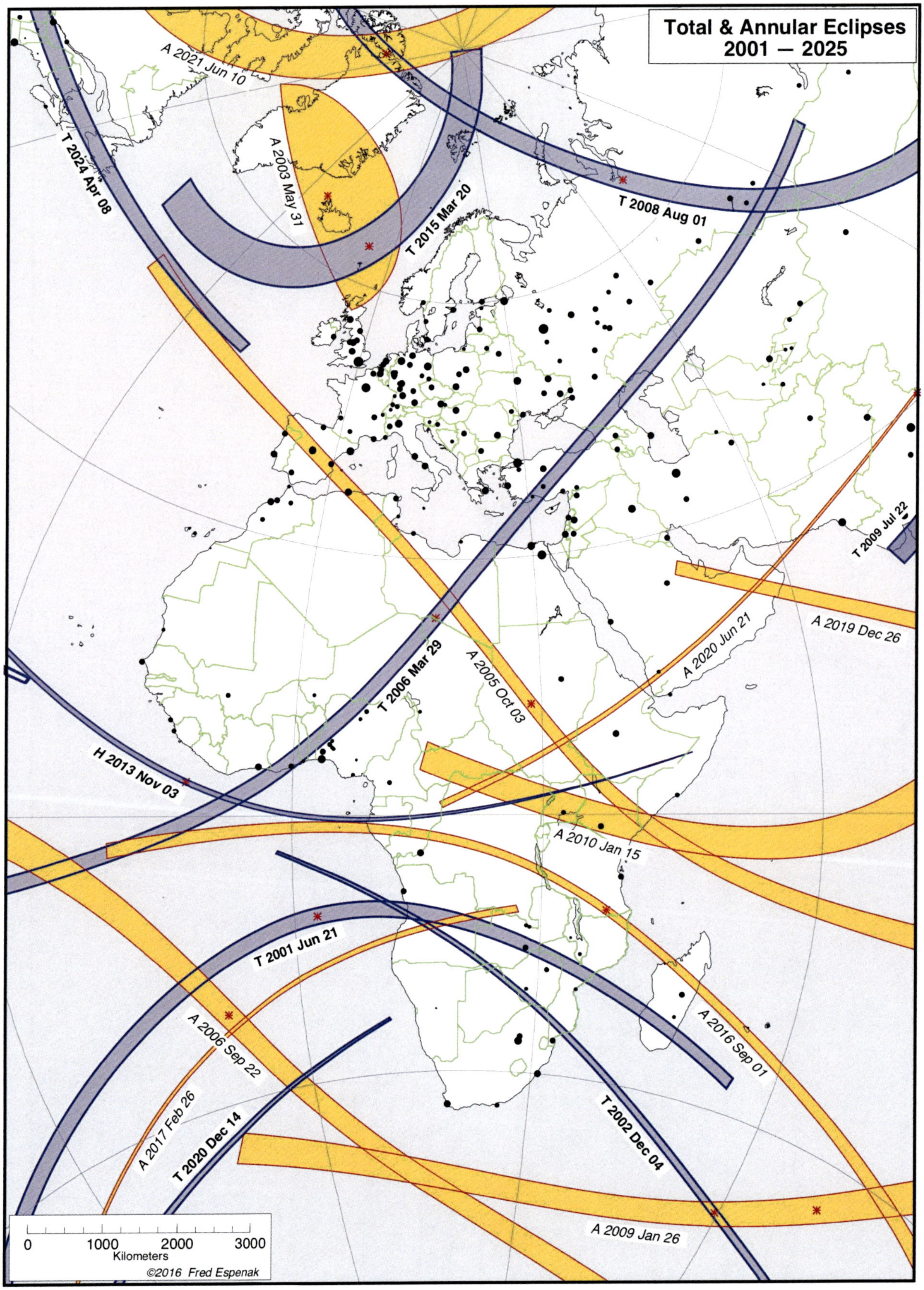

Total & Annular Eclipses
2001 — 2025
A 2021 Jun 10
A 2003 May 31
T 2024 Apr 08
T 2015 Mar 20
T 2008 Aug 01
T 2009 Jul 22
A 2019 Dec 26
A 2020 Jun 21
T 2006 Mar 29
A 2005 Oct 03
H 2013 Nov 03
A 2010 Jan 15
T 2001 Jun 21
A 2006 Sep 22
A 2016 Sep 01
A 2017 Feb 26
T 2020 Dec 14
T 2002 Dec 04
A 2009 Jan 26
0
1000
2000
3000
Kilometers
©2016 Fred Espenak

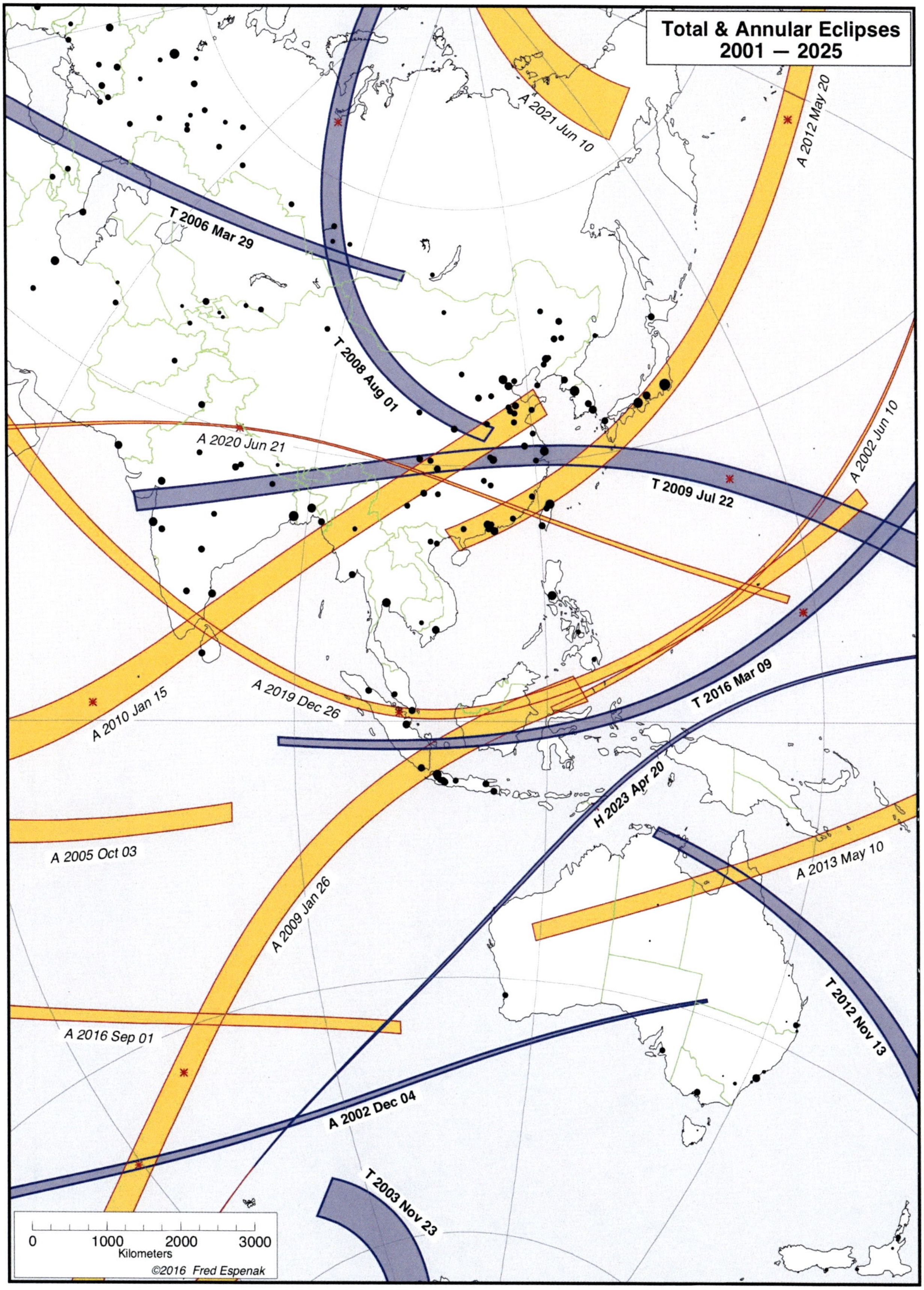

Total & Annular Eclipses
2001 — 2025
A 2021 Jun 10
A 2012 May 20
T 2006 Mar 29
T 2008 Aug 01
A 2020 Jun 21
T 2009 Jul 22
A 2002 Jun 10
A 2010 Jan 15
A 2019 Dec 26
T 2016 Mar 09
H 2023 Apr 20
A 2005 Oct 03
A 2013 May 10
A 2009 Jan 26
A 2016 Sep 01
T 2012 Nov 13
A 2002 Dec 04
T 2003 Nov 23
0
1000
2000
3000
Kilometers
©2016 Fred Espenak

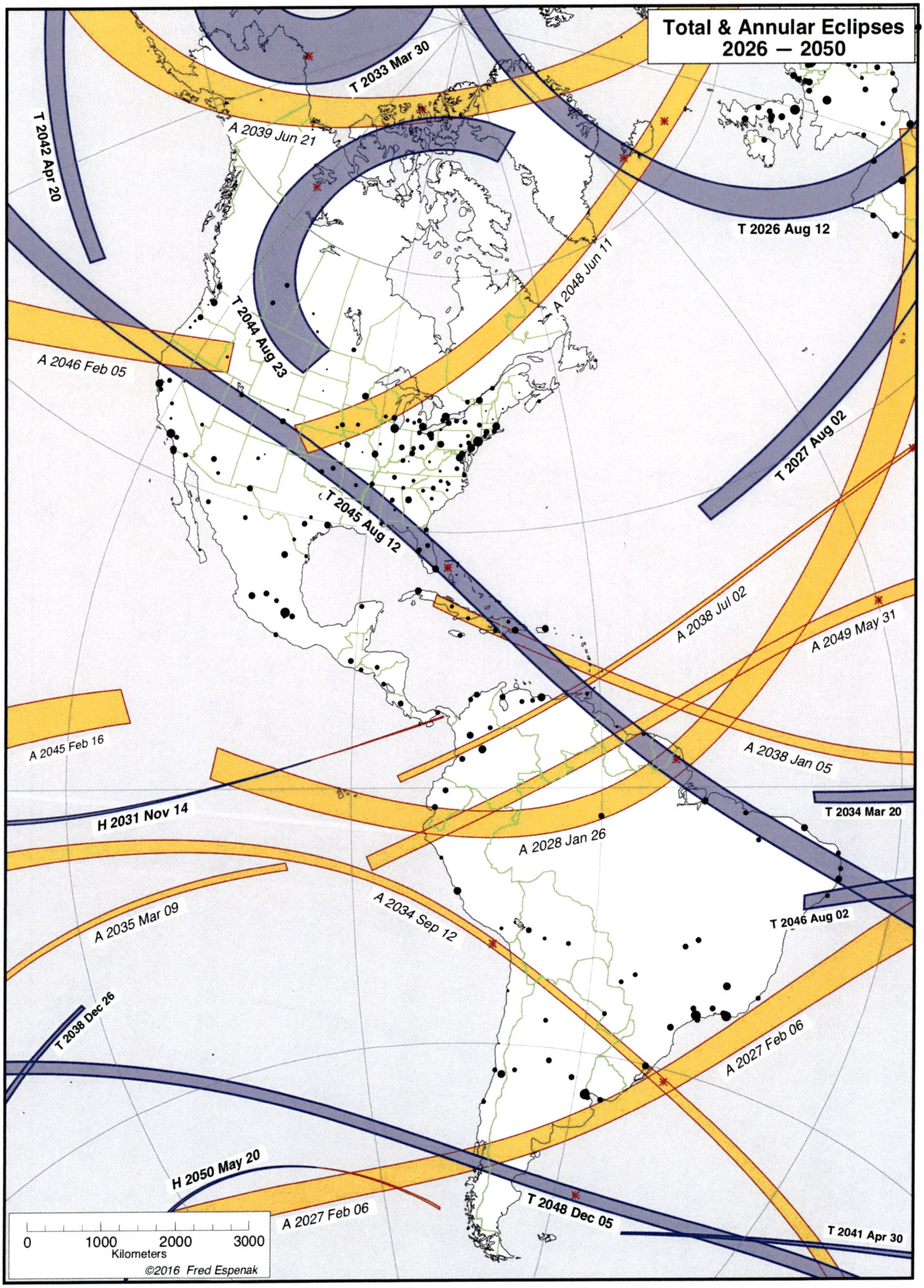
Total & Annular Eclipses
2026 — 2050
T 2033 Mar 30
A 2039 Jun 21
T 2042 Apr 20
T 2026 Aug 12
A 2048 Jun 11
T 2044 Aug 23
A 2046 Feb 05
T 2027 Aug 02
T 2045 Aug 12
A 2038 Jul 02
A 2049 May 31
A 2045 Feb 16
A 2038 Jan 05
H 2031 Nov 14
T 2034 Mar 20
A 2028 Jan 26
A 2035 Mar 09
A 2034 Sep 12
T 2046 Aug 02
T 2038 Dec 26
A 2027 Feb 06
H 2050 May 20
T 2048 Dec 05
A 2027 Feb 06
T 2041 Apr 30
0
1000
2000
3000
Kilometers
©2016 Fred Espenak

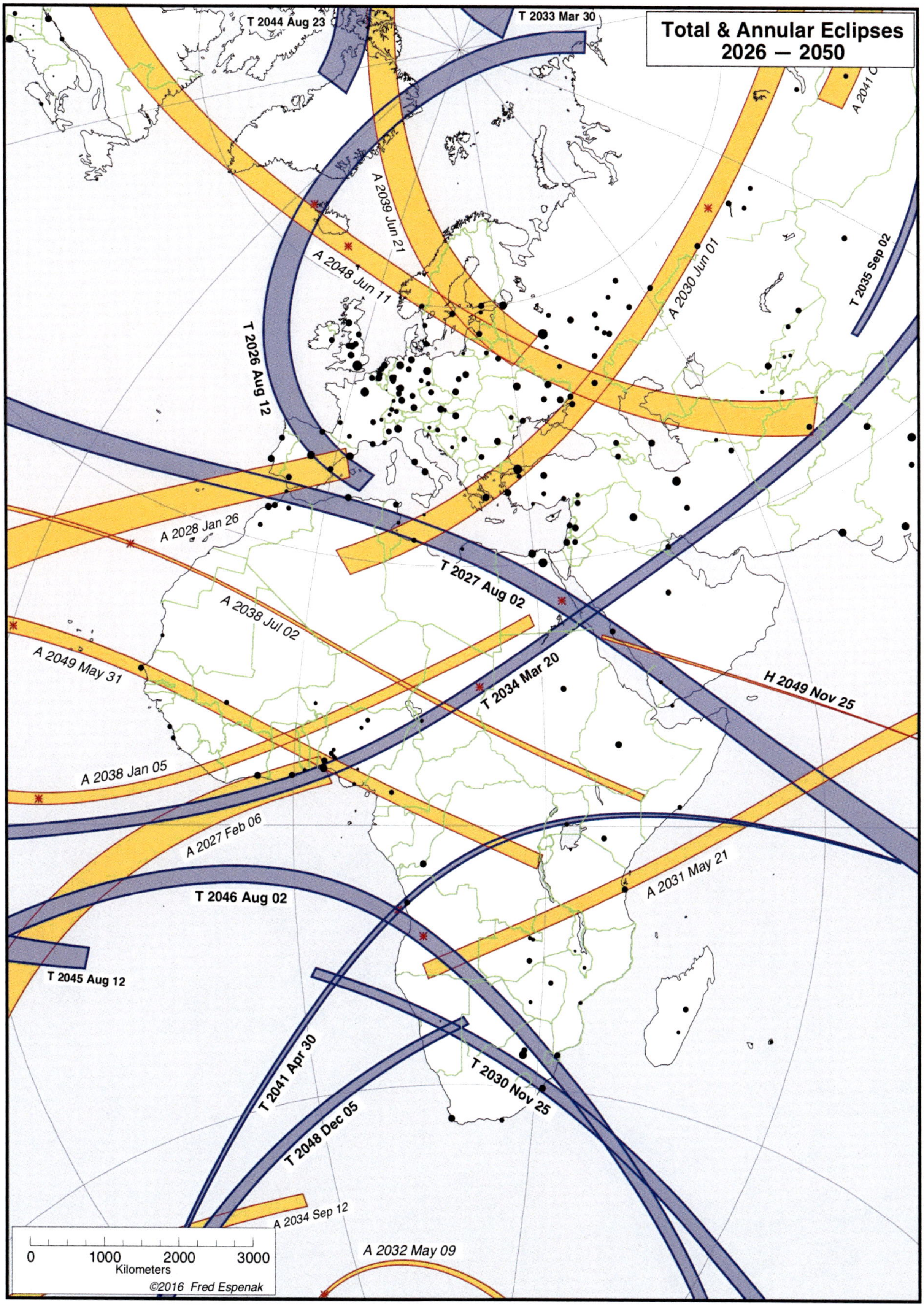
Total & Annular Eclipses
2026 — 2050
T 2044 Aug 23
T 2033 Mar 30
A 2039 Jun 21
A 2048 Jun 11
A 2030 Jun 01
T 2035 Sep 02
T 2026 Aug 12
A 2028 Jan 26
T 2027 Aug 02
A 2038 Jul 02
A 2049 May 31
T 2034 Mar 20
H 2049 Nov 25
A 2038 Jan 05
A 2027 Feb 06
A 2031 May 21
T 2046 Aug 02
T 2045 Aug 12
T 2041 Apr 30
T 2030 Nov 25
T 2048 Dec 05
A 2034 Sep 12
A 2032 May 09
0 1000 2000 3000
Kilometers
©2016 Fred Espenak

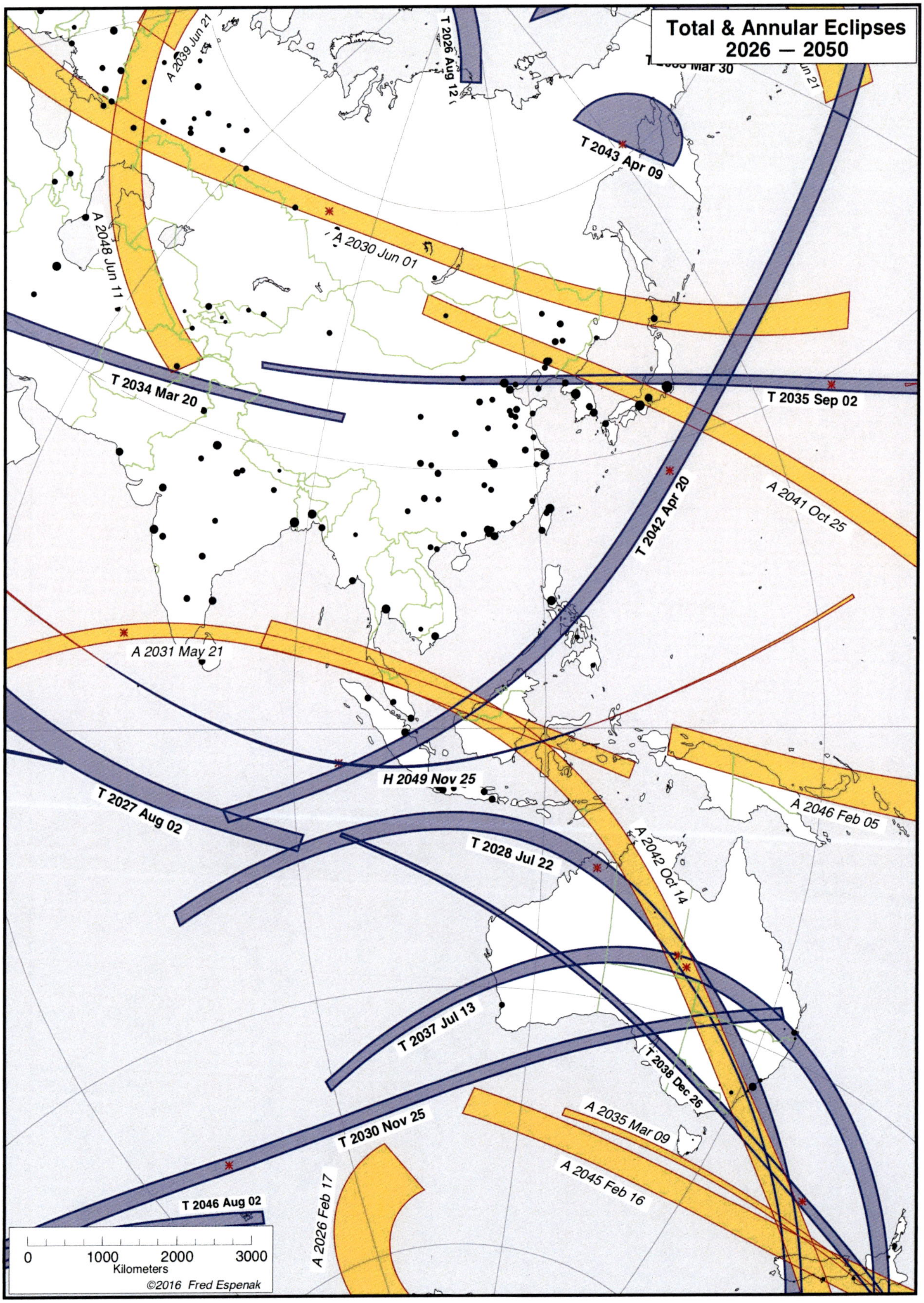
Total & Annular Eclipses
2026 — 2050
A 2039 Jun 21
T 2026 Aug 12
T 2043 Apr 09
A 2030 Jun 01
A 2048 Jun 11
T 2034 Mar 20
T 2035 Sep 02
T 2042 Apr 20
A 2041 Oct 25
A 2031 May 21
H 2049 Nov 25
T 2027 Aug 02
A 2046 Feb 05
T 2028 Jul 22
A 2042 Oct 14
T 2037 Jul 13
T 2038 Dec 26
A 2035 Mar 09
T 2030 Nov 25
A 2045 Feb 16
A 2026 Feb 17
T 2046 Aug 02
0 1000 2000 3000
Kilometers
©2016 Fred Espenak

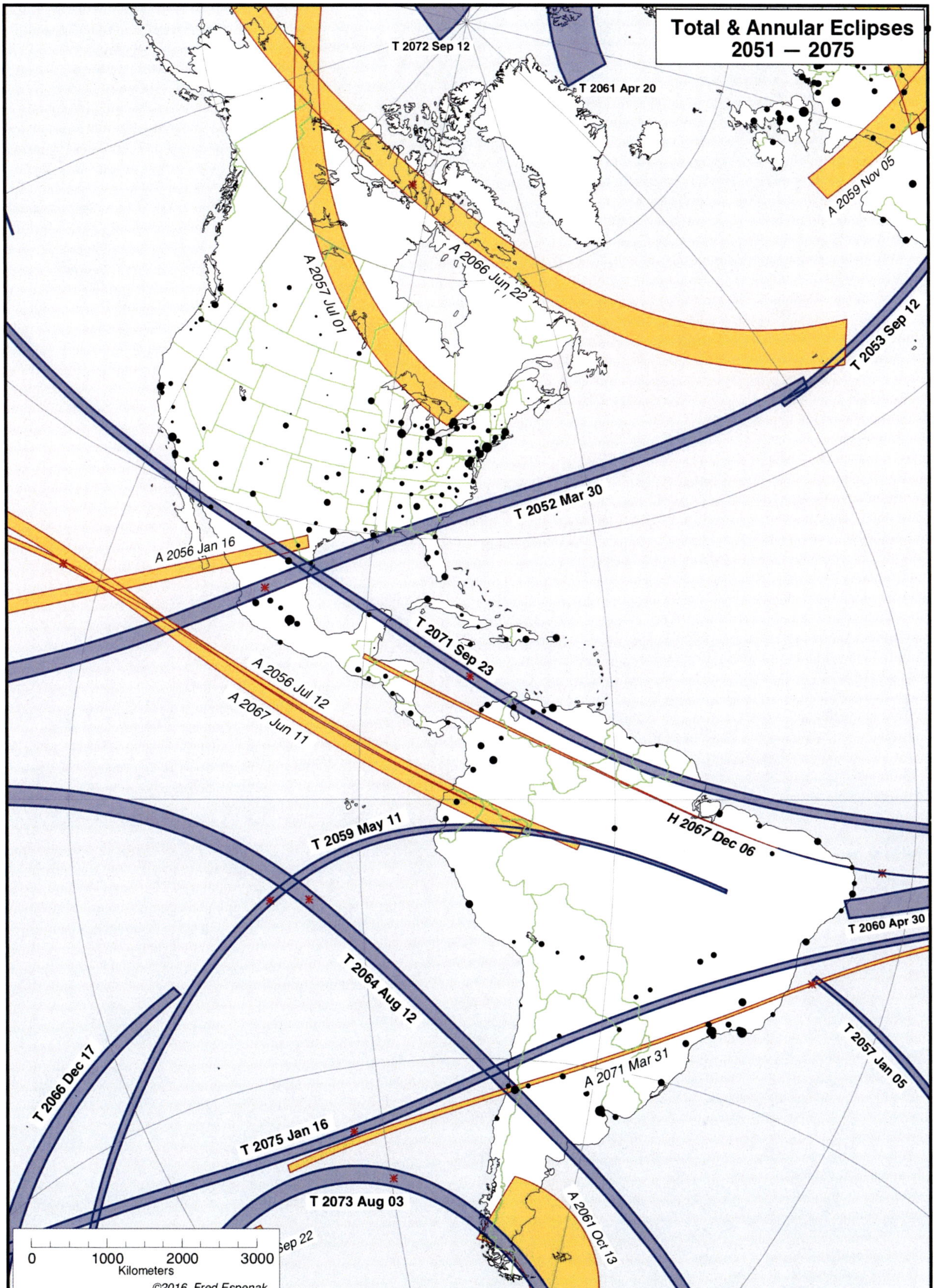
Total & Annular Eclipses
2051 — 2075
T 2072 Sep 12
T 2061 Apr 20
A 2059 Nov 05
A 2066 Jun 22
A 2057 Jul 01
T 2053 Sep 12
T 2052 Mar 30
A 2056 Jan 16
T 2071 Sep 23
A 2056 Jul 12
A 2067 Jun 11
T 2059 May 11
H 2067 Dec 06
T 2060 Apr 30
T 2064 Aug 12
T 2057 Jan 05
T 2066 Dec 17
A 2071 Mar 31
T 2075 Jan 16
T 2073 Aug 03
A 2061 Oct 13
Sep 22
0 1000 2000 3000
Kilometers
©2016 Fred Espenak

Total & Annular Eclipses
2051 — 2075
T 2072 Sep 12
A 2057 Jul 01
A 2075 Jul 13
T 2063 Aug 24
T 2061 Apr 20
A 2066 Jun 22
T 2052 Mar 30
T 2060 Apr 30
T 2053 Sep 12
A 2059 Nov 05
T 2053 Sep 12
T 2060 Apr 30
A 2074 Jan 27
T 2071 Sep 23
H 2067 Dec 06
A 2071 Mar 31
A 2064 Feb 17
A 2060 Oct 24
A 2070 Oct 04
T 2055 Jul 24
T 2057 Jan 05
A 2053 Mar 20
A 2063 Feb 28
0 1000 2000 3000
Kilometers
©2016 Fred Espenak

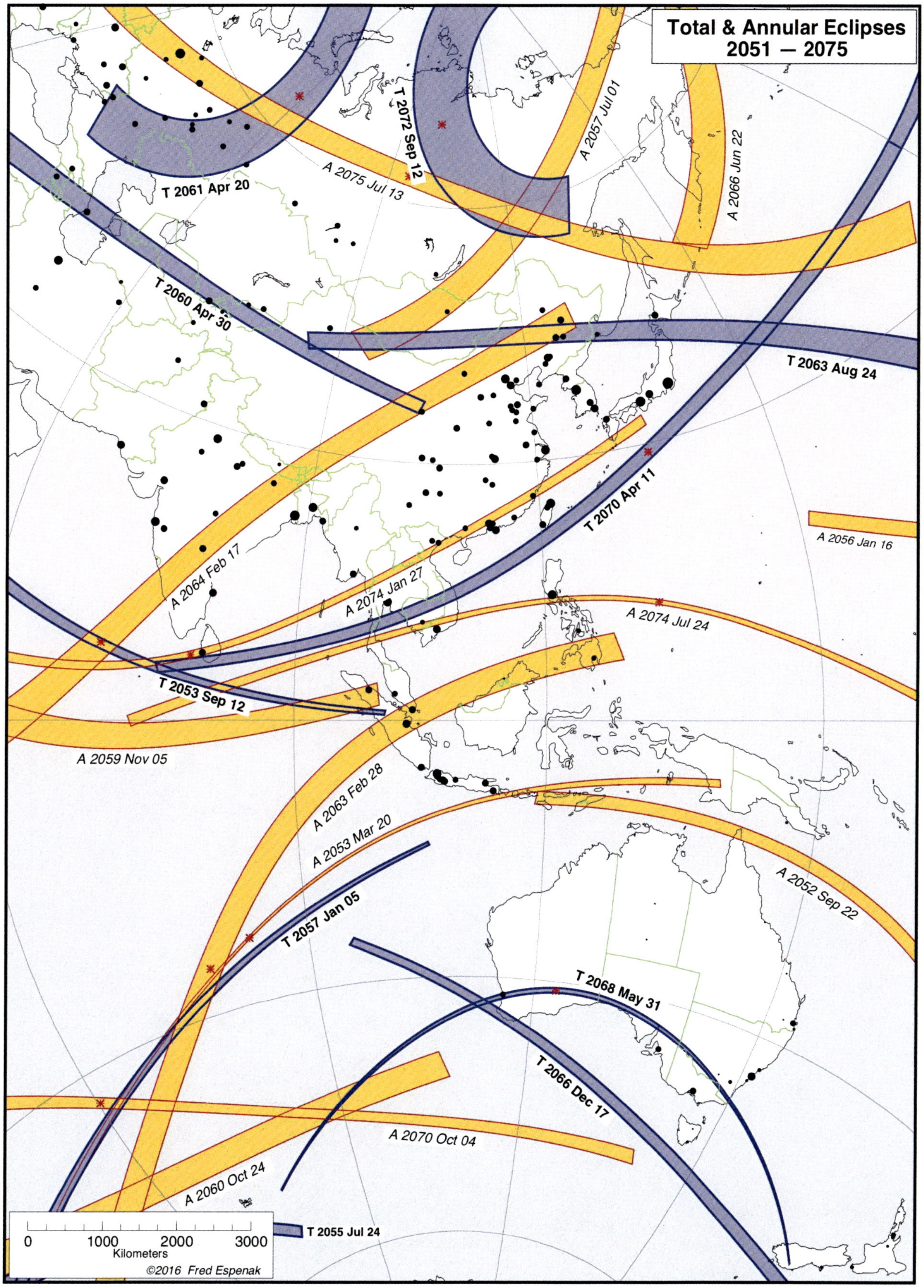
Total & Annular Eclipses
2051 — 2075
T 2061 Apr 20
T 2072 Sep 12
A 2075 Jul 13
A 2057 Jul 01
A 2066 Jun 22
T 2060 Apr 30
T 2063 Aug 24
T 2070 Apr 11
A 2056 Jan 16
A 2064 Feb 17
A 2074 Jan 27
A 2074 Jul 24
T 2053 Sep 12
A 2059 Nov 05
A 2063 Feb 28
A 2053 Mar 20
A 2052 Sep 22
T 2057 Jan 05
T 2068 May 31
T 2066 Dec 17
A 2070 Oct 04
A 2060 Oct 24
T 2055 Jul 24
0 1000 2000 3000
Kilometers
©2016 Fred Espenak

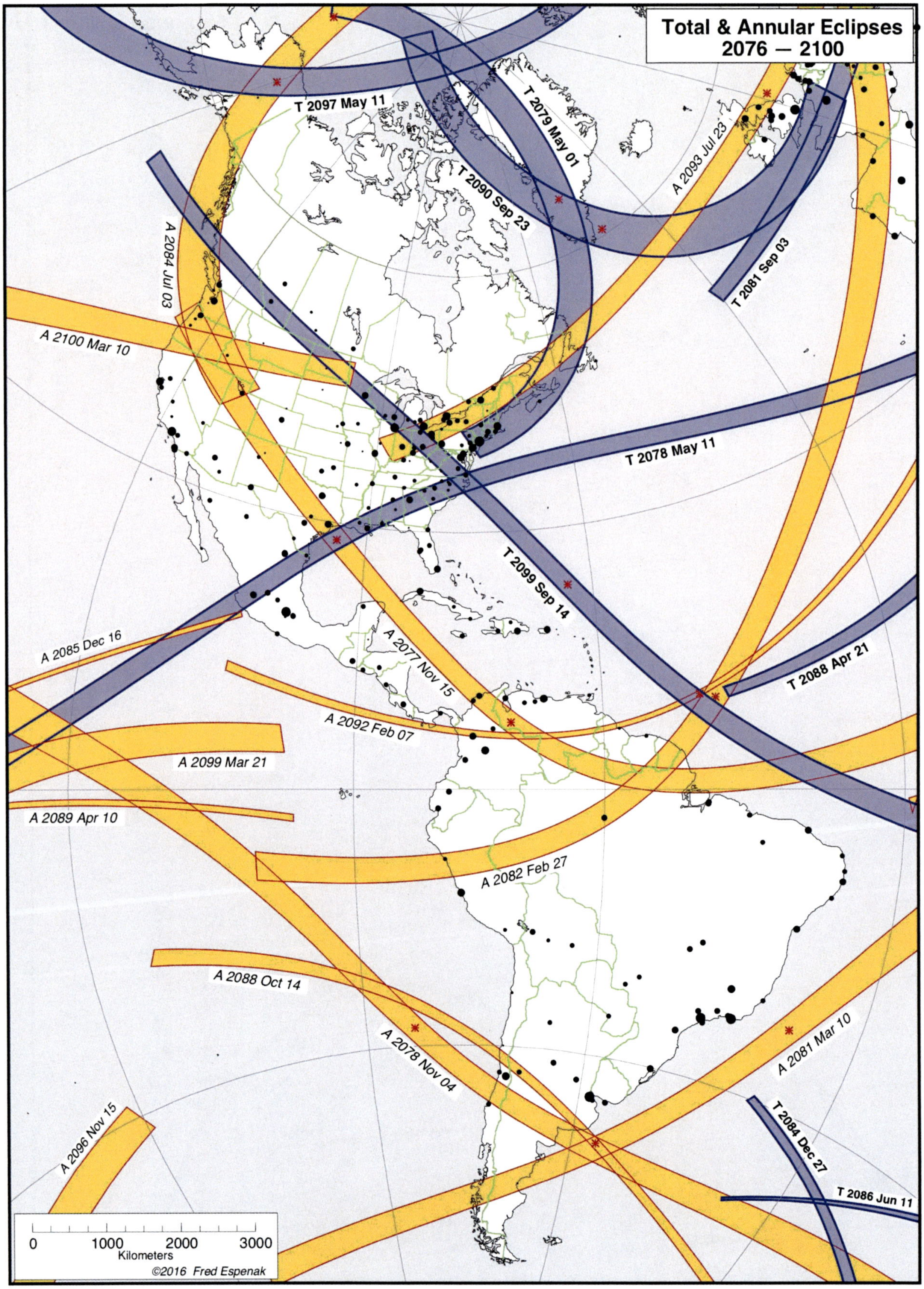
Total & Annular Eclipses
2076 — 2100
T 2097 May 11
T 2079 May 01
A 2093 Jul 23
T 2090 Sep 23
A 2084 Jul 03
T 2081 Sep 03
A 2100 Mar 10
T 2078 May 11
T 2099 Sep 14
A 2085 Dec 16
A 2077 Nov 15
T 2088 Apr 21
A 2092 Feb 07
A 2099 Mar 21
A 2089 Apr 10
A 2082 Feb 27
A 2088 Oct 14
A 2078 Nov 04
A 2081 Mar 10
T 2084 Dec 27
A 2096 Nov 15
T 2086 Jun 11
0 1000 2000 3000
Kilometers
©2016 Fred Espenak

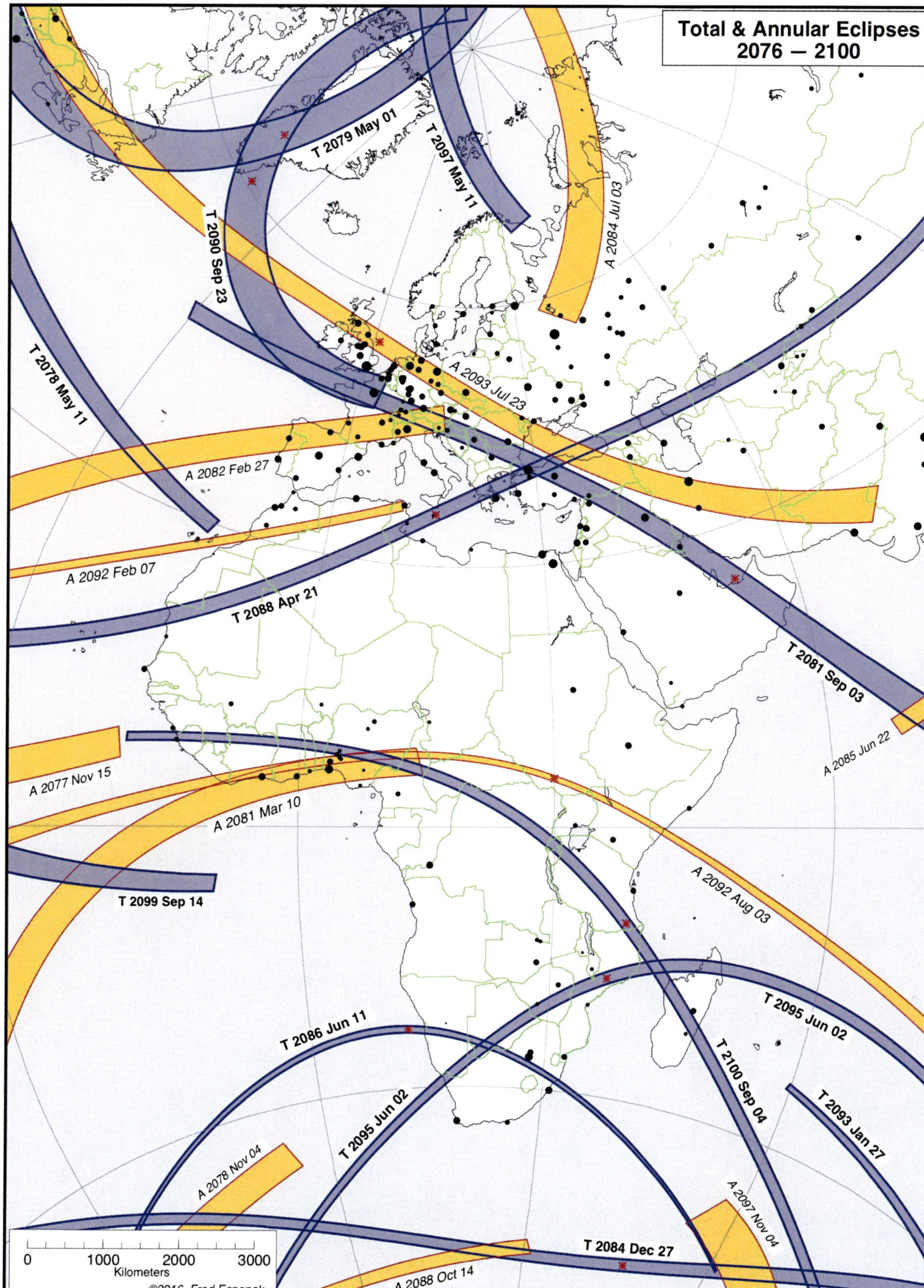
Total & Annular Eclipses
2076 — 2100
T 2079 May 01
T 2097 May 11
A 2084 Jul 03
T 2090 Sep 23
T 2078 May 11
A 2093 Jul 23
A 2082 Feb 27
A 2092 Feb 07
T 2088 Apr 21
T 2081 Sep 03
A 2085 Jun 22
A 2077 Nov 15
A 2081 Mar 10
T 2099 Sep 14
A 2092 Aug 03
T 2095 Jun 02
T 2086 Jun 11
T 2100 Sep 04
T 2093 Jan 27
T 2095 Jun 02
A 2078 Nov 04
A 2097 Nov 04
T 2084 Dec 27
A 2088 Oct 14
0
1000
2000
3000
Kilometers
©2016 Fred Espenak

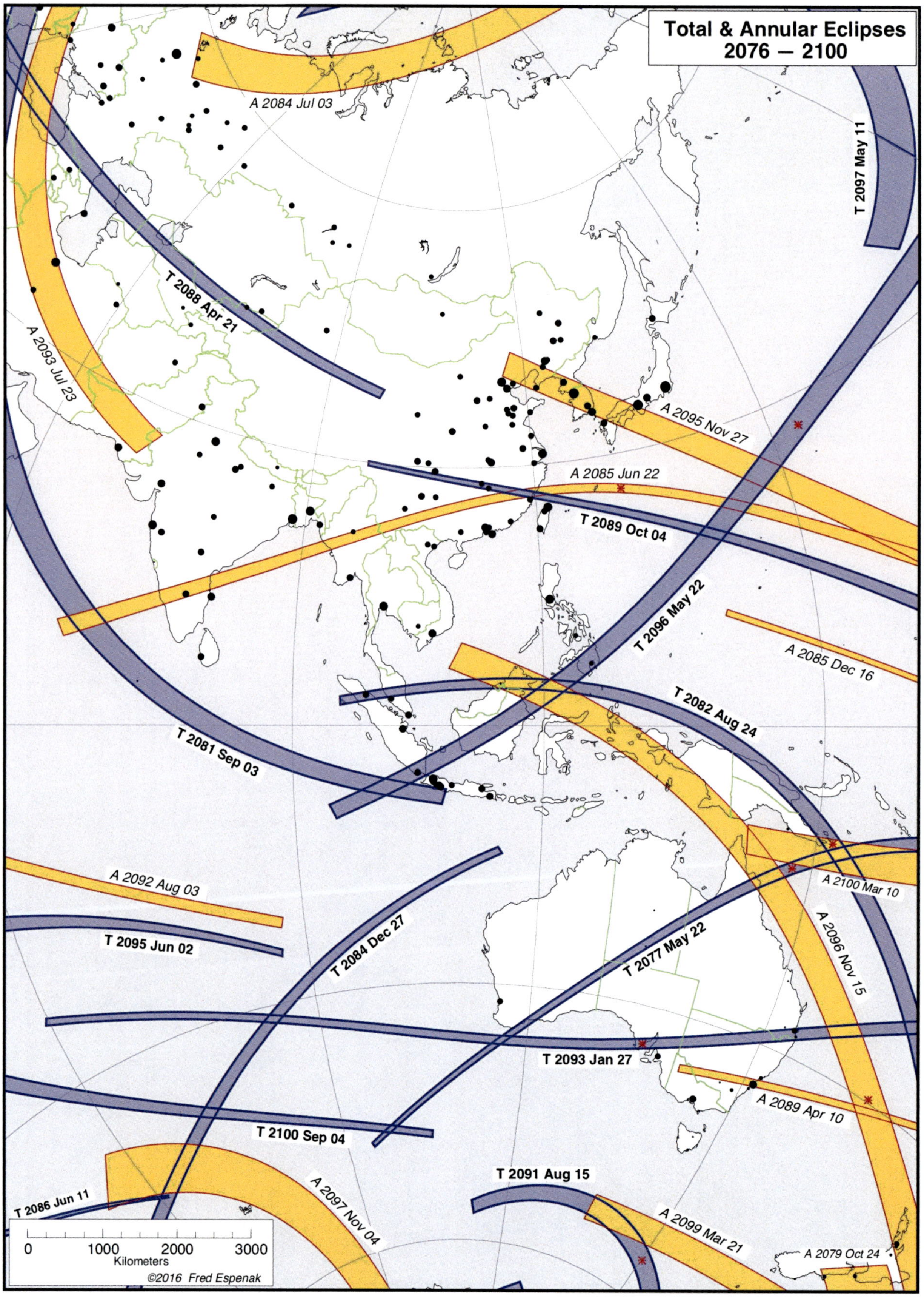
Total & Annular Eclipses
2076 — 2100
A 2084 Jul 03
T 2097 May 11
T 2088 Apr 21
A 2093 Jul 23
A 2095 Nov 27
A 2085 Jun 22
T 2089 Oct 04
T 2096 May 22
A 2085 Dec 16
T 2082 Aug 24
T 2081 Sep 03
A 2100 Mar 10
A 2092 Aug 03
T 2095 Jun 02
T 2084 Dec 27
T 2077 May 22
A 2096 Nov 15
T 2093 Jan 27
A 2089 Apr 10
T 2100 Sep 04
T 2091 Aug 15
T 2086 Jun 11
A 2097 Nov 04
A 2099 Mar 21
A 2079 Oct 24
0 1000 2000 3000
Kilometers
©2016 Fred Espenak

Astropixels Publications

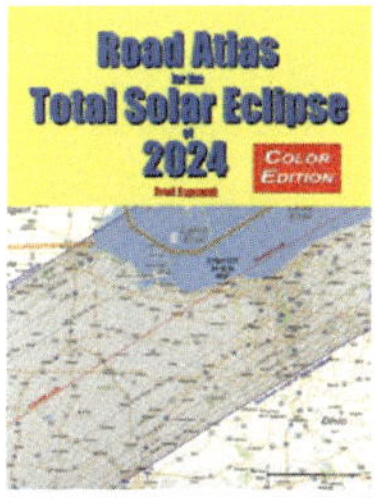

Road Atlas for the Total Solar Eclipse of 2024 (Fred Espenak) contains a comprehensive series of 26 high-resolution maps of the path of totality across the USA, Mexico and Canada. The large scale (1 inch = 22 miles) shows both major and minor roads, towns and cities, rivers, lakes, parks, national forests, wilderness areas and mountain ranges. The duration of totality is plotted in 30-second steps, making it easy to estimate the length of the total eclipse from any location in the eclipse path.

21st Century Canon of Solar Eclipses (Fred Espenak) contains maps and data for all 224 solar eclipses occurring during the 100-year period from 2001 through 2100. Appendix A is comprehensive catalog listing the essential characteristics of each eclipse. Appendix B contains maps depicting the geographic regions of visibility of each eclipse with 12 maps per page. Appendix C has detailed full-page maps of every eclipse from 2017 through 2066. Appendix D plots the track of every central eclipse (total, annular and hybrid) on large-scale maps with countries borders and major cities.

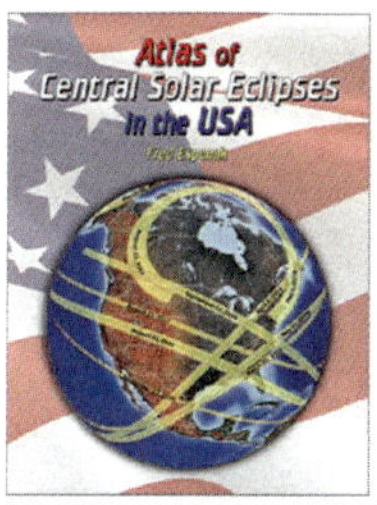

Atlas of Central Solar Eclipses in the USA (Fred Espenak) contains of a series of 499 global maps showing the track of every total and annular solar eclipse across the USA from 1001 through 3000. It is accompanied by a catalog that lists the major characteristics of each eclipse. A set of 20 detailed maps, each covering a 50-year period and centered on the lower 48 states, shows the path of every total and annular eclipse. The maps include state boundaries and major cities. These maps also cover southern Canada and northern Mexico.

Thousand Year Canon of Solar Eclipses 1501 to 2500 (Fred Espenak) contains maps and data for each of the 2,389 solar eclipses occurring over the ten-century period centered on the present era. A comprehensive catalog lists the essential characteristics of each eclipse while a series of global maps show the exact geographic extent of each eclipse.

Thousand Year Canon of Lunar Eclipses 1501 to 2500 (Fred Espenak) contains diagrams, maps and data for each of the 2,424 lunar eclipses occurring over the ten-century period centered on the present era. A comprehensive catalog lists the essential characteristics of each eclipse while a series of diagrams and maps illustrate the Moon-shadow geometry and geographic regions of visibility of each eclipse.

Totality: The Great American Eclipses of 2017 and 2024 (Mark Littmann & Fred Espenak) is the ultimate guide to the most stunning of celestial sights, total eclipses of the Sun The book provides information, photographs, and illustrations to help the public understand and safely enjoy all aspects of these eclipses including how to observe a total eclipse of the Sun, how to photograph and video record an eclipse, why solar eclipses happen, and more. Several chapters focus exclusively on the total eclipses of 2017 and 2024 though the USA.

For complete details and ordering information on the above ***Astropixels Publications***, visit:

astropixels.com/pubs/

Made in the USA
Las Vegas, NV
25 May 2024

90348088R00178